Gene Expression

Volume 1

Gene Expression

Volume 1

Bacterial Genomes

Benjamin Lewin

Editor, *CELL*

A Wiley–Interscience Publication

JOHN WILEY & SONS

LONDON NEW YORK SYDNEY TORONTO

Reprinted November 1975
Reprinted May 1976

Library of Congress Cataloging in Publication Data:

Lewin, Benjamin M.
Gene expression.

CONTENTS: v. 1. Bacterial genomes.
1. Gene expression. 2. Molecular genetics. I. Title.

QH450.L48 575.2′1 73–14382
ISBN 0 471 53167 7 Cloth bound
ISBN 0 471 53168 5 Paper bound

Printed in Great Britain by William Clowes & Sons Limited
London, Colchester and Beccles

For my father

PREFACE

When I began to write this book, I intended it to be a revision of "The Molecular Basis of Gene Expression". But it soon became obvious that a revision alone could not serve the purpose of describing our present ideas of how genes are expressed in the bacterial cell. So much has been discovered since 1969 that little now remains of the manuscript written then; this is, therefore, essentially a new book.

In the preface to my earlier book, I asserted that "although important problems remain to be solved, it seems likely that our present conceptual framework will retain its validity, and future research result, by and large, in additions and in changes of detail rather than in basic revision." But even though this point of view has been vindicated in the sense that research in the past few years has indeed added to rather than replaced our previous ideas, so much has been discovered that the additions have overwhelmed our previous framework.

The view of molecular biology which this book presents therefore owes as much to research performed during this decade as before. Indeed, I can make no claim to provide any enduring account of molecular biology. No doubt our present ideas, seemingly sophisticated, will themselves appear outdated only too rapidly. The aim of this book is therefore to give a report of our progress in explaining how genes are expressed and controlled in bacteria. The second volume of Gene Expression attempts to provide an analogous account of our more primitive knowledge of eucaryotic cells. An exception to this general division of material is that since the mechanisms of protein and probably nucleic acid synthesis seem to be similar in all organisms, these processes are discussed largely in this volume so that the second volume can concentrate upon the features peculiar to the control of gene activity in eucaryotic cells. Discussion in this book is for the most part devoted to the systems of bacteria although some of the more general systems of bacteriophages are also included.

The starting point of this book is the fusion of genetic and biochemical concepts into the definition of the gene as a length of DNA which codes for a single protein chain. The level of exposition is thus moderately advanced and assumes some familiarity with the techniques of microbial genetics and nucleic acid biochemistry. Because a solely conceptual account of molecular biology would tend to present too absolute a picture, I have endeavoured to point out the limitations of the experiments upon which our ideas rely; it is easy to state conclusions but important to realise the intricate, sometimes fragile, chain of deductions upon which they rest.

I have tried to draw a broad picture of the processes involved in gene expression which will be useful both to students and to those engaged in research in these and related topics. I have in general presented this material in a logical rather than historical sequence; this means that experiments are discussed in an order which may sometimes differ from that in which they were performed. I have also attempted to convey the sense of intellectual excitement which provided the impetus for this research; to this end I have tried to indicate briefly our state of knowledge when critical experiments were performed and have on occasion digressed into a historical exposition. Molecular biology continues to generate enthusiasm, of course, and I have not confined this account to a description of what seems to have been established; I think it is equally important to communicate the problems with which we are now concerned and to include some of the speculations presently occupying molecular biologists.

This book starts with a discussion of the genetic code, which must surely be central to any account of gene expression. The details of how protein chains are elongated, initiated and terminated occupy the next two chapters. Our ideas on these topics are similar to, although naturally somewhat more developed than, the account which I presented four years ago. A notable advance, however, can be seen in the research described in the next chapter, which discusses the structure, function and biosynthesis of the ribosome. We can see here the promise that in time the position and role of each component of the bacterial ribosome will be fully defined. The last chapter of this section of the book describes the general conclusions which we can formulate about the functions of transfer RNA; although many more tRNAs have now been sequenced, no rules relating structure to function have yet been discovered and our best hopes for further progress must lie with the characterization of more mutant tRNAs.

A considerable advance in our knowledge is reviewed in the next four chapters, which discuss the transcription of RNA and its control. An inevitable source of disappointment in any such account must be that we still know so little of the enzymic processes implicated in the synthesis of RNA. However, we have a much better idea of the extent to which the activity of RNA polymerase itself may be controlled in bacterial cells and during phage infection; and our knowledge of the control systems which act on bacterial operons is greatly extended. Even the simplest model, the lactose operon, is subject to at least two control systems, one positive and one negative; and a yet more complex network of interactions takes place in other bacterial operons and in phage lambda. Accompanying this tracing of formal control networks, there have been advances in understanding the molecular interactions between regulator proteins and DNA, and the structures of various recognition elements of DNA are now under study.

The enzymes of DNA replication have at last come into the limelight and the first chapter of the last section of the book is devoted to a discussion of the characteristics of DNA replication as we now see them and of the different enzyme activities which appear to be implicated. The subsequent chapter discussing the modification and repair of DNA is able to consider the palindromic sequences which are recognised in DNA; and we can now delineate more clearly the actions of the different repair systems. Recombination remains better understood in formal than in enzymic terms. The concluding chapter expands my earlier account of how the cell cycle relates to replication into a tentative discussion of molecular mechanisms.

I have followed a more stringent policy in citing references than is usual. The references cited are intended, of course, to attribute results to researchers and to indicate the important original papers upon which this account of molecular biology is based. I have made no attempt, however, to provide an exhaustive list of the literature; so many confirmatory papers have been published that such a survey would tend only to obscure the principal lines of research and to turn the text into little more than a list of references. In trying to point the reader toward those papers which may perhaps form the mainstream of molecular biology, I have endeavoured wherever possible to avoid citing contributions to obscure symposia and to ensure that as many as possible of the references cited are contained in journals which are readily available. Even within these restrictions, the number of papers which I have felt it imperative to cite has nearly doubled since 1969, from about 800 then to some 1500 today, a tribute to the speed at which research in molecular biology is advancing.

I gratefully acknowledge permission from many authors to reproduce their results; these are cited individually. I am indebted to many of my colleagues for their assistance and in particular to Dr. Max Gottesman for his helpful criticisms of several chapters.

It is once again a pleasure to acknowledge the immeasurable debt I owe to my father, Dr Sherry Lewin, to whom this book is dedicated, for his unfailing enthusiasm which first made me aware of the excitement of molecular biology; I am deeply indebted to him for innumerable stimulating discussions, not least his criticisms of the manuscripts of this book and its counterpart. It is a pleasure also to thank my wife, Ann, for her unflagging encouragement during the long gestation period of the two volumes of Gene Expression.

BENJAMIN LEWIN

CONTENTS

SYNTHESIS OF PROTEIN

1 The Genetic Code

2 Synthesis of the Polypeptide Chain

3 Punctuation in the Message

4 The Ribosome

5 Transfer RNA

CONTROL OF TRANSCRIPTION

6 Transcription of Phage Genomes

11 Modification and Repair of DNA

12 Recombination Between DNA Duplexes

13 The Cell Division Cycle

Synthesis of Protein

CHAPTER 1

The Genetic Code

Equivalance of Gene and Protein

Definition of the Gene

Our present concept of the gene unifies the genetical and biochemical studies of the past century, initiated by Mendel's discovery that inheritance is carried by particulate factors and by Miescher's discovery of nucleic acids. Once the demonstration that the units of heredity pass from one generation to the next in a predictable manner had been equated with the visible behaviour of chromosomes, the correspondence between genes and chromosomes soon evolved into the idea that each chromosome consists of a linear array of genes (reviewed by Stern, 1970). The genetic definition of the gene by the complementation test implies that it is a unit all of whose parts must be present in the same chromosome. The demonstration that DNA carries inheritance in bacteria and bacteriophages suggested that the DNA of chromosomes provides their genetic material. The biochemical definition of the gene as a sequence of DNA which is responsible for the synthesis of a single protein chain thus provides a molecular basis for the formal genetic test.

Any particular mutation can be identified by the genetic locus at which it maps. Mutations which map at separate loci thus identify different genes. Different mutations which map at the same locus identify the alleles of a gene, any one of which may occupy its genetic locus. However, when two mutations influencing the same characteristic lie close together on the genetic map, recombination frequencies may be too low to distinguish whether the mutants are alleles or whether they lie in adjacent genes influencing the same character of the phenotype.

Allelic mutants can be distinguished from those in different genes by a complementation test. When two homozygous mutants are crossed to give the heterozygote which contains one copy of each mutation, they generate a mutant phenotype if they are allelic but complement to give the wild type character if they are located in different genes. Crossing allelic mutants gives a heterozygote:

$$\frac{a}{a} \times \frac{b}{b} \rightarrow \frac{a}{b}$$

in which both copies of the gene are mutant. But crossing mutants located in different genes generates an organism which contains one wild type (+) copy and one mutant copy of each gene:

$$\frac{a+}{a+} \times \frac{+b}{+b} \rightarrow \frac{a+}{+b}$$

The gene can therefore be defined as a genetic unit by the results of a complementation test; alleles do not show complementation when crossed with each other, whereas mutants in different genes do so.

Metabolic processes in living organisms are mediated and controlled by enzymes and the idea that heredity might be exercised by a connection between genes and enzymes is long established. (At one time, indeed, fashionable opinion was that genes *were* enzymes). The research of Beadle and Tatum in 1948 on nutritional mutants of the fungus Neurospora crassa placed the idea that genes specify enzymes on a firm basis. When the fungus is grown on a substrate which is metabolized through an ordered pathway of reactions, mutation can block the process at any step of the pathway. Each metabolic step is catalysed by a particular enzyme; since each of the inactivating mutations behaves as a single gene in inheritance, Beadle and Tatum proposed the one gene—one enzyme hypothesis: each gene is responsible for the synthesis of one particular enzyme.

Direct evidence that genes code for proteins was first provided by Ingram in 1957 when he showed that the trait of sickle cell anaemia, caused by mutation in a single gene, can be accounted for at the molecular level by an alteration in the amino acid composition of the protein haemoglobin. Because some proteins comprise more than one type of polypeptide chain, the one gene—one enzyme theory has been rephrased since its original formulation. The functional unit of haemoglobin, for example, consists of two α-chains and two β-chains. Genetic studies of mutant haemoglobins have shown that synthesis of the α-chains and β-chains is governed by different genetic loci. This has led to the more precise definition that one gene is a segment of genetic material which specifies one polypeptide chain.

One early concept visualised the gene as the unit of genetic recombination, but in 1945 Lewis demonstrated that mutants defined by the complementation test as alleles affecting the size of the eye of Drosophila can recombine with each other. Recombination between alleles generates the reciprocal products of a gene which carries both mutations and a wild type gene which carries neither mutation. A heterozygote in which one gene carries both mutations and its allele is wild type differs in phenotype from a heterozygote in which each allele carries one of the two mutations.

The only difference between these two genotypes lies in the molecular arrangement of the mutant sites, for in both the cell has one wild type and one mutant sequence at each site. When the mutants are in *cis* array on the same chromosome the cell is wild type; but when they are in *trans* array on the homologous chromosomes the cell is mutant. This comparison provides the *cis/trans* complementation test, whose molecular basis is illustrated in figure 1.1. The right half of the figure shows the *trans* complementation test discussed above which distinguishes allelic from non-allelic mutations. The left side shows the control, the *cis* arrangement. The term *cistron* is often used to describe the unit defined by the comparison between the phenotypes of the *cis* and *trans* arrangements of heterozygotes; and is equivalent to the older term, gene.

The implication of the *cis/trans* test is that a gene is a unit which can function only when all its parts are together in a single segment of the genome, its function being impaired by mutation at any point along its length. When two mutant sites lie in different genes, they therefore show complementation to give a wild phenotype in either *cis* or *trans* arrangement; as the lower part of the figure shows, in each case the cell has a wild type copy of each gene and can therefore synthesise a functional molecule of each protein. But as the upper part of the figure shows, when the two sites are in the same gene they display wild phenotype only in the *cis* arrangement, when the other chromosome carries a wild type gene (left). In the *trans* arrangement (right), both of the alleles are mutant and the cell produces two different types of mutant protein and no wild type protein.

The *cis/trans* test therefore distinguishes allelic from non-allelic mutants; allelic mutants are wild type in *cis* arrangement but mutant in *trans* array, whereas non-allelic mutants are wild type in both instances. Any unidentified mutation can therefore be tested for location in a gene by a *cis/trans* complementation test with a mutant in the gene. The concept that a gene species a single polypeptide chain explains the implication of the *cis/trans* test that all its parts must be continuous on the same chromosome; if each allele independently directs synthesis of its corresponding polypeptide chain, mutation at any point may result in the production of a non-functional protein, irrespective of the wild type sequence elsewhere.

Nucleic Acids as the Genetic Material

That chromosomes of higher organisms are composed of both nucleic acid and protein has been known almost since their discovery. That the nucleic acid is DNA, rather than RNA, was not realised until 1924 when Feulgen applied the stain fuchsin to cells; their chromosomes showed the purple colour which is produced by the interaction of DNA with the dye. Many early theories of heredity took the view that genetic specificity must reside in the protein component—nucleic acids were thought to have too little variety to be able to specify the enormous numbers of different proteins. Direct

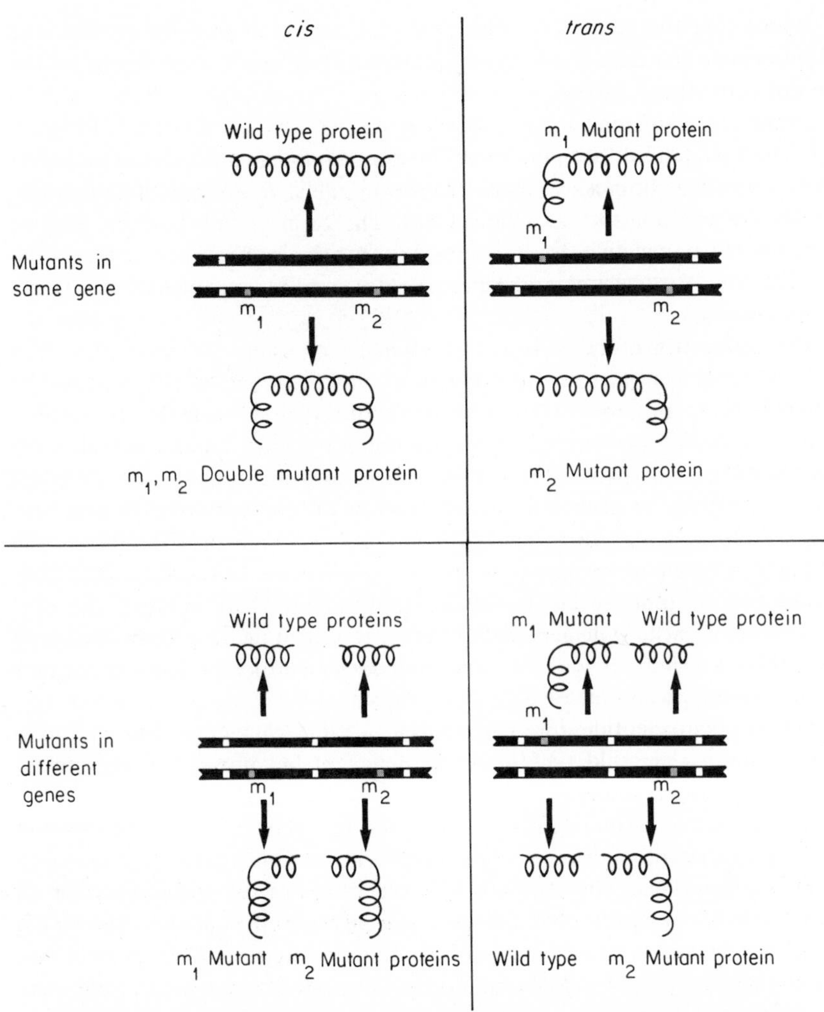

Figure 1.1: The cis/trans test for the gene in heterozygous diploid cells. Upper: If two mutant sites, m1 and m2, are located in the same gene the phenotype of the cell depends upon their arrangement. If they are in the cis array—both are located on the same segment of DNA as shown on the left—one gene carries no mutations and synthesises wild type protein; the other synthesises a protein with two mutations. If the mutations are in trans array—so that each is located on a different molecule of DNA as shown on the right—only mutant proteins are made, one corresponding to each mutation. Lower: If the two mutant sites are located in different genes, it makes no difference whether they are in cis arrangement (left) or trans arrangement (right). In each case, one active copy of each gene is present together with one inactive copy of each; it does not matter whether the two active copies are located on the same segment of DNA or on different molecules. When two

evidence that the genetic material is nucleic acid derives from studies with microorganisms; data from higher organisms, although strongly suggestive, do not constitute a formal proof.

Transformation between bacteria was discovered when Griffith found in 1928 that when mice are injected with dead virulent Pneumococcus together with a live but harmless mutant, they die as the result of production of live bacteria of the virulent form. The inference is that although dead, the virulent strain contains genetic information which can be transferred in some way to use the live organization of the other bacterial cell type. Studies of cell free preparations by Avery, Macleod and McCarty in 1944 showed that the transforming factor is DNA.

Confirmation that DNA is the genetic material was provided by the classic experiment in which Hershey and Chase (1951a, b) radioactively labelled either the protein or the nucleic acid components of T2 bacteriophages. Only the nucleic acid of the virus enters the cells of the host bacterium, Escherichia coli, to provide the genetic information which must be required for the subsequent reproduction of the phages. Most viruses, bacterial cells and the cells of higher organisms utilise DNA as their genetic material; RNA constitutes the genetic information of a few small viruses.

It is a common view that molecular biology began in 1953 with the elucidation of the structure of DNA. Diffraction patterns obtained by Wilkins et al. from X-ray crystallography proved to be consistent with a structure in which two very long polynucleotide chains are wound around each other to form a double helix. The model building of Watson and Crick (1953a, 1953b) suggested that this might be achieved if the two complementary strands are oriented in opposite directions, one running 5′ to 3′ and the other 3′ to 5′, held together only by hydrogen bonding between complementary base pairs.

This suggests the structure for DNA represented by the illustration of figure 1.2. A covalently linked chain of deoxyribose-phosphate provides the backbone of the molecule, with the nucleotide bases located inside. DNA usually contains four types of base: the two purines adenine and guanine; and the two pyrimidines thymine and cytosine. Watson and Crick suggested that the double helix can remain the constant width demanded by its diffraction pattern only if a purine in one strand always partners a pyrimidine in the other; purines never complement purines and pyrimidines never complement

---

mutations lie in different genes, a cell containing one copy of each is therefore wild type no matter whether their arrangement is cis or trans. But when two mutations are located in the same gene, the cell can make active protein only when their arrangement is cis; in the trans arrangement two mutant proteins are made. When two mutations show wild type behaviour in cis array but are inactive in trans they must therefore be located in the same gene

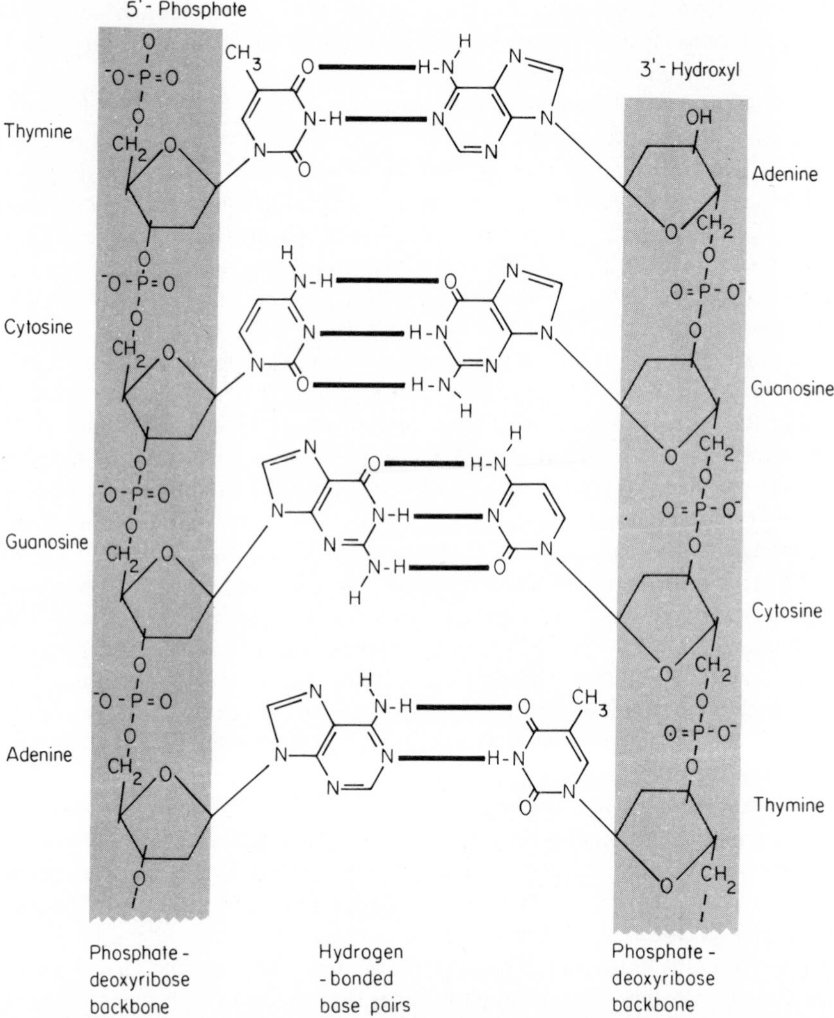

Figure 1.2: Two dimentional projection of the hydrogen bonded anti-parallel chains of a double helix. Negative charges would usually be neutralised by counter-ions. Projection not to scale

pyrimidines. As figure 1.2 shows, these interactions are specific, for guanine can form three hydrogen bonds with cytosine and adenine can form two hydrogen bonds with thymine; these are the complementary base pairs. The specificity of their formation explains the equality found in the base composition of DNA; the number of adenine residues is always the same as the number of thymines and the number of guanines equals the number of cytosines (reviewed by Chargaff, 1971).

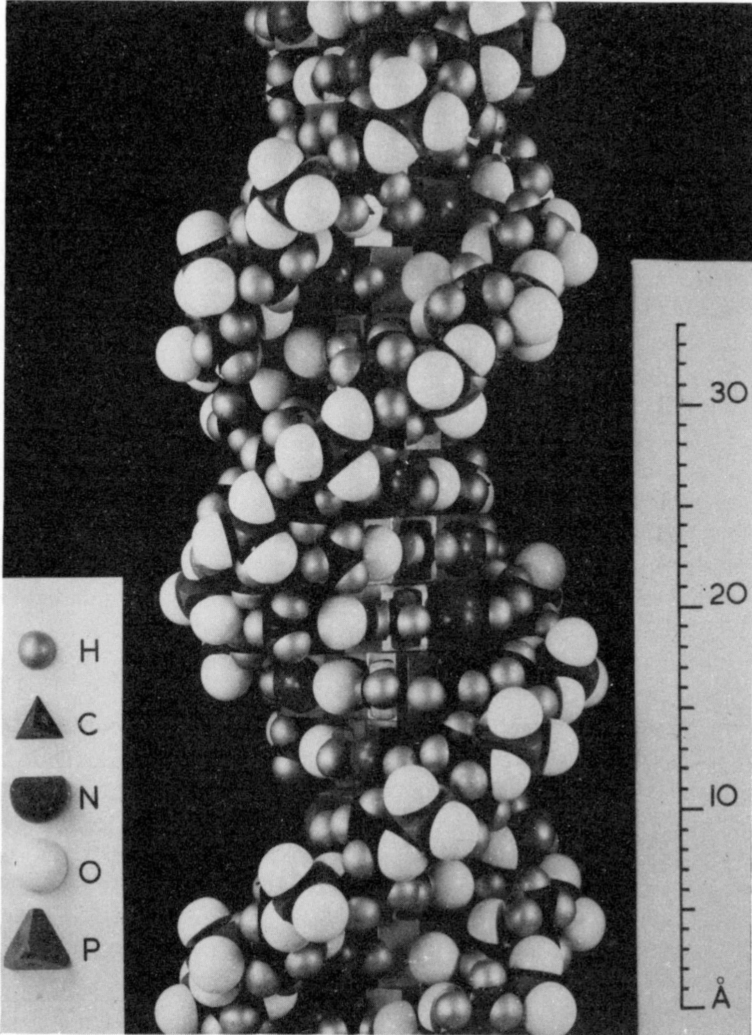

Figure 1.3: A scale space-filling model of the DNA double helix. By kind permission of Professor M. F. H. Wilkins

The planar purine and pyrimidine bases in the interior of the duplex are aligned parallel to each other and the planes of the hetercyclic rings are roughly perpendicular to the axis of the helix. The bases are therefore said to be "stacked" above each other in this parallel array. As the model of figure 1.3 shows, there is a distance of 3·4Å between each base pair so that ten base pairs make up each complete turn of the helix in 34Å. The duplex has a narrow groove of about 12Å across and a wide groove of about 22Å. Although hydrogen bonding to form the complementary base pairs explains the specificity of strand association, forces other than hydrogen bonding—such as hydrophobic interactions from base stacking—contribute to the stability of the duplex structure (reviewed by Marmur, Rownd and Schildkraut, 1963; Lewin, 1967, 1973).

The phosphodiester backbone of the helix is highly negatively charged—one charge per phosphate group—and in physiological conditions this charge is neutralized by interactions with positive groups. In bacteria, the positive charges may be provided by cations. In the chromosomes of higher organisms, basic proteins are associated with DNA; these histone proteins possess sequences of amino acids whose positive charges interact electrostatically with the negatively charged phosphate groups of the double helix (see chapter three of volume two).

Because the two strands of DNA are joined only by hydrogen bonds and hydrophobic interactions, they can unwind without breaking any covalent bonds; the single strands which result can direct the formation of replicas by their ability to hydrogen bond specifically to their complementary bases, adenine to thymine and guanine with cytosine. The specificity of complementary base pairing accounts not only for the reproduction of DNA but also for its expression. RNA also possesses adenine, guanine and cytosine, but has uracil, which lacks the C3-methyl group of thymine, as its fourth base; the sugar groups of the nucleotides of RNA possess a hydroxyl group at the 2′ position of the ribose ring where DNA has hydrogen.

Although RNA is generally encountered as a single strand polynucleotide chain, its bases have the same specificity of hydrogen bonding as DNA and recognition between complementary sequences to give duplex regions can occur between RNA and DNA and within RNA molecules, where it is important in the structure of bacteriophage RNAs (see chapter 3), ribosomal RNAs (see chapter 4) and transfer RNAs (see chapter 5). Short nucleotide sequences of transfer RNA molecules also hydrogen bond to complementary sequences in messenger RNAs (see chapter 2).

Colinearity of Gene and Protein

The concept that genes act by specifying proteins implies that the segment of DNA comprising a gene must in some way be responsible for the sequence of amino acids which constitute the protein for which it codes. (It seems likely that the secondary and tertiary structures of a protein are determined by its primary structure.) The simplest and most obvious way for a length of nucleic acid to specify a protein is for them to be colinear; the sequence of nucleotides along the gene should correspond to the sequence of amino acids making up the protein. This is equivalent to saying that the sequential structure of the chromosome extends to within genes as well as between them and leads to the picture of chromosomes as structures which contain long molecules of DNA, successive regions coding for different proteins.

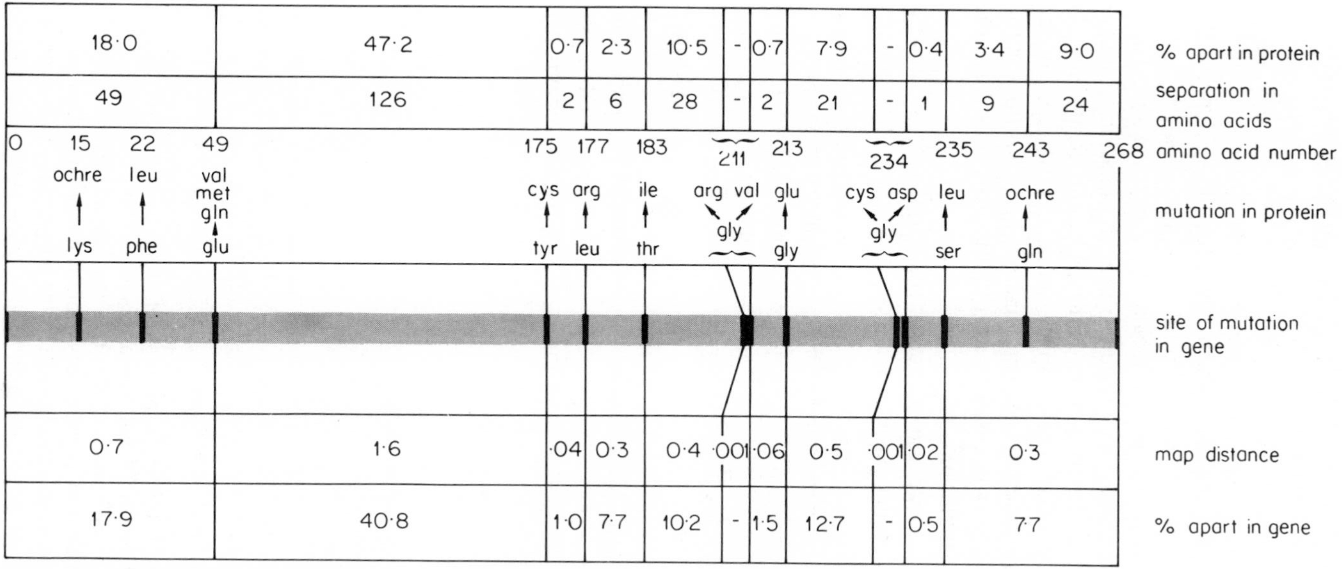

% apart in protein: 18·0 | 47·2 | 0·7 | 2·3 | 10·5 | - | 0·7 | 7·9 | - | 0·4 | 3·4 | 9·0

separation in amino acids: 49 | 126 | 2 | 6 | 28 | - | 2 | 21 | - | 1 | 9 | 24

amino acid number: 0 15 22 49 175 177 183 211 213 234 235 243 268

mutation in protein:
lys → ochre (15); phe → leu (22); glu → val, met, gln (49); tyr → cys (175); leu → arg (177); thr → ile (183); gly → arg, val (211); gly → glu (213); gly → cys, asp (234); ser → leu (235); gln → ochre (243)

site of mutation in gene

map distance: 0·7 | 1·6 | ·04 | 0·3 | 0·4 | ·001 | ·06 | 0·5 | ·001 | ·02 | 0·3

% apart in gene: 17·9 | 40·8 | 1·0 | 7·7 | 10·2 | - | 1·5 | 12·7 | - | 0·5 | 7·7

Figure 1.4: Colinearity of the tryptophan synthetase A gene of E. coli and its protein. The positions of mutants in the gene are shown to scale; the two outside mutants do not cause amino acid changes and are probably very close to the ends of the gene, where they are shown. Map distances are given in recombination units. Three mutants which do not recombine cause changes in amino acid 49; they presumably occur at the same nucleotide. Two mutants which map extremely close together but do recombine cause changes in amino acid 211; they must lie in different nucleotides of this codon. The same is true of the two mutants which change the amino acid at position 234. There is a good correlation between the relative distances apart of mutant sites in the gene (very lowest line) and the separation of mutated acids in the protein (very upper line). Agreement is not exact because the recombination frequencies vary along the genetic material and do not reflect an accurate measure of distance along **DNA**; but the comparison between the per cent distances separating amino acids and mutations respectively is close enough to suggest that gene and protein are colinear. On average, 0·018 map units represent one amino acid. Data of Yanofsky et al. (1967) and Yanofsky and Horn (1972)

Two approaches have been successfully employed to demonstrate by experiment the colinearity between nucleic acid and protein. One has been to map the position of point mutations—which change only a single base—within a gene and to correlate them with alterations in the amino acids of the corresponding protein. Using the gene which codes for the A chain of the tryptophan synthetase enzyme of E.coli, Yanofsky et al. (1964, 1967) have isolated and mapped mutants by genetic means. By degrading the altered protein produced by each mutant into peptides through digestion with proteolytic enzymes, the peptide differing from wild type can be isolated and its amino acid sequence determined. The complete amino acid sequence of the protein is now known and, as can be seen from figure 1.4, the positions of the altered amino acids in the protein closely match the positions in the genetic map of the corresponding mutations.

Another approach has made use of *amber* mutations, which cause a premature termination of protein synthesis; if gene and protein are colinear,

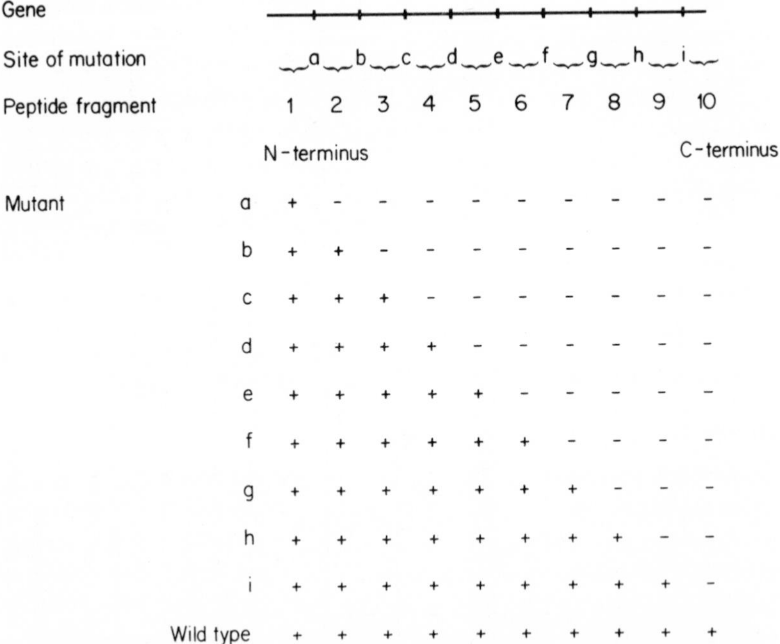

Gene
Site of mutation
a b c d e f g h i
Peptide fragment
1 2 3 4 5 6 7 8 9 10
N-terminus
C-terminus
Mutant
a + - - - - - - - - -
b + + - - - - - - - -
c + + + - - - - - - -
d + + + + - - - - - -
e + + + + + - - - - -
f + + + + + + - - - -
g + + + + + + + - - -
h + + + + + + + + - -
i + + + + + + + + + -
Wild type + + + + + + + + + +

Figure 1.5: Correlation between the position of amber mutations in the head protein gene of phage T4 and the extent of protein synthesised; (+) signifies that a peptide is present, (–) that it is absent. Each amber mutant directs the synthesis of a certain length of the protein starting at its $-NH_2$ terminal end; this length corresponds to the position of the mutant on the genetic map of the gene. This demonstrates that a gene is colinear with the protein which it specifies

an amber mutation located, say, halfway along a gene should allow only the first half of the protein to be made. Sarabhai, Stretton and Brenner (1964) used chemical mutagens to induce ten different mutations in a gene specifying the head protein of phage T4; the proteins synthesised in phage-infected bacteria were then labelled with radioactive amino acids, the cells lysed, nucleic acid removed and the mixture degraded with trypsin or chymotrypsin. The resulting peptides were characterized by high voltage electrophoresis on paper and located by autoradiography. (An advantage of this system is that the peptides derived from the head protein can readily be recognised in these digests so that they need not be purified from the other proteins present.)

Different amber mutants proved to terminate protein synthesis at different points along the head protein. Proteins are synthesized from the $-NH_2$ to the –COOH terminus and in each mutant, the N-terminal peptide of the protein is found but the number of peptides successive to it varies with the particular mutant. Different mutants have different peptides present, but ordering these peptides into a "hierarchy" from the $-NH_2$ to the –COOH terminal end of the protein demonstrates that each mutant allows synthesis of the different lengths of protein shown in figure 1.5.

By determining the molecular weight of each amber fragment, Celis et al. (1973) constructed a translational map of the gene; comparison of this map with the genetic map determined by recombination frequencies shows that the length of protein produced by each mutant correlates well with the distance of the mutation from the beginning of the gene. Gene and protein are therefore colinear. For much of the gene, a constant recombination frequency of 0·012 units corresponds to each amino acid (compared with the average of 0·018 units per amino acid in the tryptophan synthetase gene).

Nature of the Code for Amino Acids

Overlapping and Non-Overlapping Codes

Because DNA is different in nature from the proteins which it specifies, the content of the genome is often described as genetic "information". It is information in the sense that the sequence of nucleotides in DNA is not important because it is a polynucleotide chain as such, but because it specifies—codes for—certain sequences of amino acids which constitute active proteins. These proteins in turn play the catalytic and structural roles which determine the structure of cell. (If the genetic material consisted of enzymes which were simply reproduced to give more enzymes, it would not constitute such "information".) Although the DNA of an organism contains the information for all the proteins which it will need to make, different genes may be active under different conditions in bacteria and in the different cells types of eucaryotes. The ability of the cell to control its genes in this manner is also part of its genetic information. DNA therefore constitutes a "store" of information, bits of which are used by the cell as and when it is appropriate.

Given a sequence of DNA, therefore, we can in principle deduce the proteins to which it corresponds if we know how the amino acid sequence of a protein is derived from the nucleotide sequence of a gene; their relationship constitutes the genetic code. Early ideas about the code envisaged some sort of stereochemical relationship between nucleic acid and the protein which it specifies —the nucleic acid might act as a physical template upon which the amino acids would first be assembled and subsequently linked together by covalent bonds. But although some theories hold that this type of relationship must have applied during the evolution of the code, the transfer of genetic information in species living today is mediated by a complex apparatus which does not use such direct mechanisms. The first step in gene expression is the formation of a single stranded RNA which has the same sequence of bases as that of one strand of the DNA duplex sequence from which it is derived; this is achieved by complementary base pairing with the other strand. It is for this reason that the genetic code is depicted in terms of U, C, A and G, the four bases of RNA.

As twenty amino acids are commonly found in proteins, but there are only four types of base in a nucleic acid, each amino acid must be specified by more than one base (reviewed by Ycas, 1969). The minimum number of bases able to provide sufficient information is three for each amino acid:

coding ratio (nucleotides/amino acid):	1	2	3	4
amino acids specifiable:	$4^1 = 4$	$4^2 = 16$	$4^3 = 64$	$4^4 = 256$

Since there are sixty-four possible combinations of three nucleotides, although there are only twenty amino acids, if the coding ratio is three, either some triplet sequences must be without meaning or different sequences must in some instances code for the same amino acid.

There are in principle two ways in which a number of nucleotides, say three, could specify one amino acid. An idea which was popular when it was thought that the nucleic acid templates and protein products might be physically related was that the code might be *overlapping*. Thus the sequence

AGUCA

could specify the three amino acids *a*, *b*, *c* thus:

$$a = \text{AGU}$$
$$b = \ \ \text{GUC}$$
$$c = \ \ \ \ \text{UCA}$$

In a *non-overlapping* code, successive triplets specify successive amino acids in the protein. The two types of code can be distinguished directly by the effect of a single base change in the nucleic acid. Figure 1.6 shows that in an overlapping code such a mutation changes all the amino acids whose overlapping triplets contain the altered base; whether three or only two amino acids are affected depends upon whether the code is fully-overlapping (as shown in the figure) or partially overlapping, in which case the last base of one triplet would be the first base of the next triplet. But in a non-overlapping code only one amino acid would be replaced. Comparisons of normal and mutated proteins have shown that such mutational changes—known as point mutations—invariably affect only one amino acid and that its neighbours on either side remain unaltered.

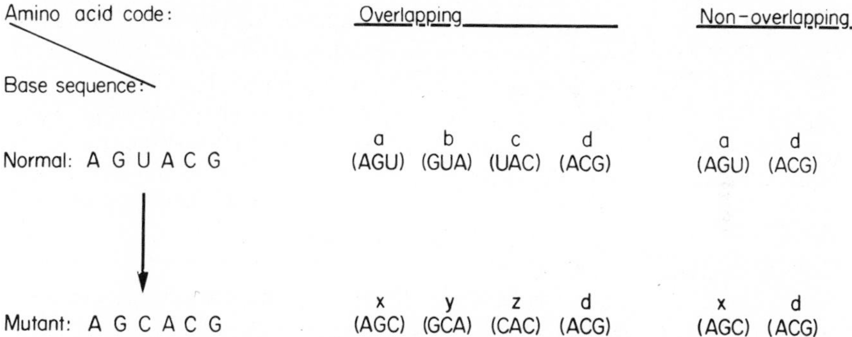

Figure 1.6: The effect of mutation in a single base upon overlapping (left) or non-overlapping codes (right). In a fully overlapping code, the alteration of one base alters three of the amino acids specified. In a non-overlapping code, only one amino acid is altered

If the code is non-overlapping, there must be right and wrong ways of reading it since otherwise any sequence, say:

A C U A C U A C U A C U A C U

can be read in any one of three phases as

ACU ACU ACU ACU ACU ACU
CUA CUA CUA CUA CUA CUA
UAC UAC UAC UAC UAC UAC

depending on the starting point. This difficulty might be overcome either by reading the sequence from some fixed starting point, or because the nature of the code itself precludes any such ambiguity.

Perhaps the most elegant solution to this problem was the formulation by Crick et al. in 1957 of a code "without commas", so called because the phase of reading is inherent in the sequence. In this code, certain sequences of three bases represent an amino acid, but others have no such meaning, so that only some of the sixty-four possible triplets would actually specify amino acids.

At every point in a gene, adjacent triplets make sense in only one phase; the overlapping triplets which they also form have no meaning:

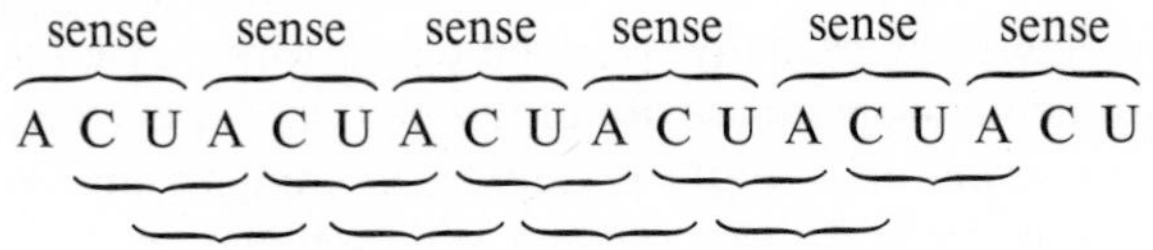

these combinations lack meaning

The four possible triplets AAA, CCC etc. cannot represent amino acids, since if they were to code for successive identical amino acids, sequences of the form XXXXXX would make sense in the wrong phase. The sixty remaining triplets can be grouped into twenty sets of three, each comprising cyclic permutations of the same three nucleotide bases; only one of these three can represent an amino acid. For example, if

A C U A C U codes for *a-a*
then the overlaps C U A
U A C

are nonsense and have no meaning. Crick et al. demonstrated that there are eight possible basic solutions which allow construction of such a code.

Non-overlap requires each amino acid to be related to a different set of three nucleotides and Crick et al. suggested that this may be achieved by a mechanism using "adaptor" molecules which exist to recognise only the sense sequences which code for amino acids. He visualised the adaptor as a trinucleotide attached to the amino acid; its bases would be complementary to the correct sequence and could hydrogen bond with it, but would unable to form stable attachments to the overlapping nonsense triplets (The adaptor molecule—now known as transfer RNA—has proved to be rather more complex than this but does work by specific hydrogen bonding between complementary triplets.)

Fixed Reading Frame of the Code

But despite the simplicity with which the commaless code accounts for the excess number of triplets (sixty-four) compared with amino acids (twenty), Crick et al. showed in 1961 that the code is read in triplets from a fixed starting point which defines the phase of reading. This conclusion was derived from a genetic analysis of the *rII* system of phage T4, using mutants induced by acridines. (The use of this system has been extensively reviewed by Barnett et al., 1967.) Wild type phages grow on both the B and K12 strains of E.coli, but a phage which has lost the function of either of the two cistrons of the *rII* region will not grow at all on K12 and produces a different type of plaque —the *r* or rapid lysis phenotype—on E.coli B:

	E.coli B	E.coli K12
wild type phage	normal (small fuzzy plaque)	normal (small fuzzy plaque)
mutant *rII* phage	large plaque (*r* type)	no growth

Some types of mutant—termed leaky—show partial function in growing on strain K12, but when tested on E.coli B do not show the true wild type plaque.

Acridines act as mutagens by causing addition or deletion of a base pair in the nucleotide sequence of DNA. Whereas mutants produced by mutagens which cause the substitution of one base for another are often leaky—typically about half are—those produced by acridine mutation usually lack the gene function completely. The difference between the two types of mutagen can be accounted for if the genetic code is read in non-overlapping triplets from a fixed point. In this case, misreading occurs after an addition or deletion because the phase of the message has been shifted by one nucleotide residue; the amino acid content of the corresponding protein is completely altered from this point on; as figure 1.7 shows, this results in complete lack of function. (This would not happen if the phase were defined by the nucleotide sequence in a commaless code.) In a base substitution mutant, however, where only one or at most a few bases have been altered, only a few amino acids are changed in the protein. This may leave some residual protein activity, generating the leaky phenotype.

Acridine mutants cannot be induced to revert by mutagens which cause base substitution, but can do so upon further mutation with acridine. If an acridine mutant is produced by, say, addition of a base, it should revert to wild type by deletion of that base. But in FC_0, a mutant induced by acridine, this reversion can be achieved by producing a second mutation at another site in the same gene, instead of by reverting the original site of mutation. This second mutation is termed a *suppressor* (see later in this chapter). If FC_0 were formed by base addition, the original phase of reading could be restored by deleting another base somewhere nearby. Misreading then takes place only between the two sites of mutation; figure 1.7 shows that the parts of the gene outside them are read in the original phase.

By forming double mutants through recombination between FC_0 and a phage carrying a second mutation, Crick et al. identified eight suppressors of the first mutation, all located within a short region of the same *rII* cistron. All these suppressors are themselves non-leaky mutants, so that their reversion to wild type can in turn be studied by the same procedure used to follow reversion of FC_0. They also revert to wild type when a double mutant is formed, the suppressors of the suppressors again comprising non-leaky *rII* mutants. If,

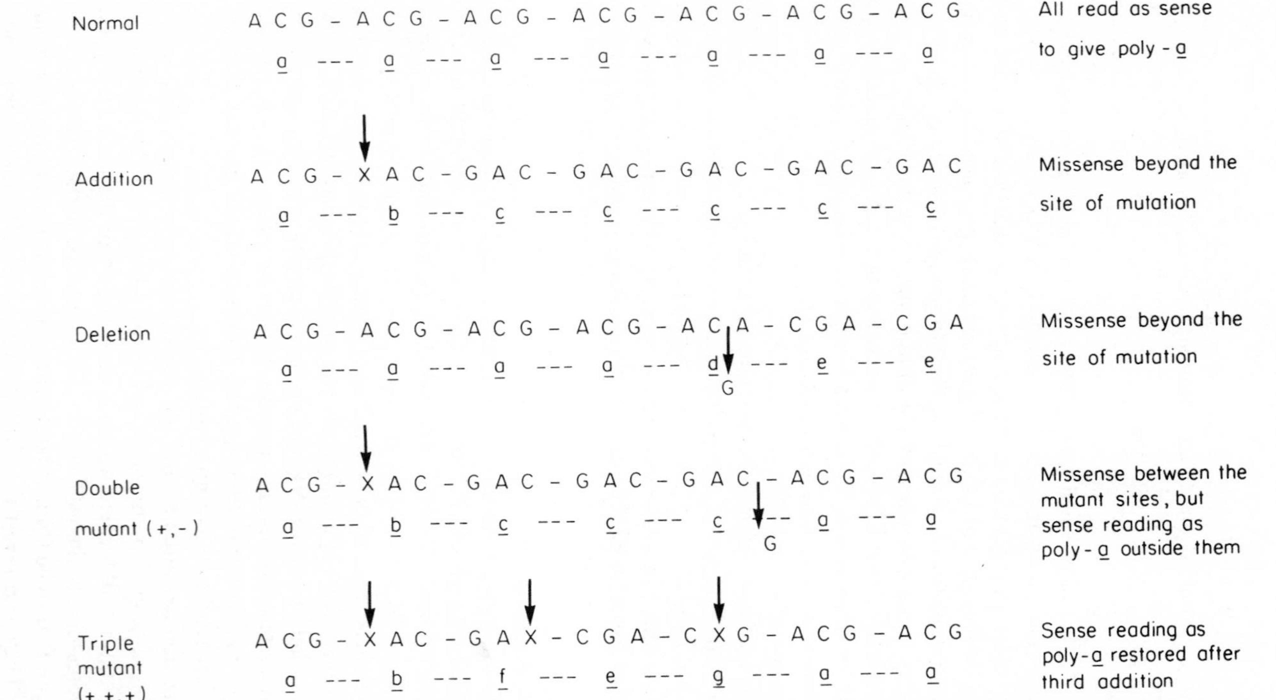

Figure 1.7: The effect of additions and deletions on the genetic code. Additions are shown as X, deletions as ↓. Either an addition or a deletion alters the reading frame so that the sequence to the right of the alteration is read in triplets but in the wrong phase; this produces a missense protein. When an addition is followed by a deletion, the correct phase is restored beyond the second site of mutation; the protein contains a missense region only between the two mutant sites. When there are three successive additions the correct reading frame is restored beyond the last site and the protein contains a missense region only between the first and last mutations

for example, the addition mutant shown in figure 1.7 is one non-leaky mutant, its combination with the deletion phage—which by itself is also a non-leaky mutant—generates a double mutant in which the misreading is suppressed.

Defining FC_0 as (+)—that is base addition—and its suppressors as (–)—that is base deletion—enables various double mutants of different (+) and (–) combinations to be constructed. No (++) or (– –) sequences show any activity, but (+–) and (–+) combinations are active. This accords with the idea that the genetic code is read in groups of nucleotides from a fixed point, although it does not indicate how many nucleotides represent each amino acid. One consequence of this conclusion is that the genetic code must include "punctuation" signals which indicate where the reading frame starts.

Triple mutants of the type (+++) or (– – –) also show suppression of the mutant characteristic and are wild type; this suggests that the code is read in triplets, since three additions or deletions then correspond in net effect to the insertion or removal of one amino acid, returning the reading phase to normal after the third site of mutation; see figure 1.7. (To be more precise, in 1961 these results implied that the code must be read in triplets or some higher multiple of three, since at that time it was not known how many base pairs an acridine adds or deletes in DNA, although we now know it is only one.)

This theory predicts that the active double or triple mutants should differ from wild type in the sequence of amino acids corresponding to the part of the gene between the two outside mutants. In fact, the multiple mutants which are active on E.coli K12 show a variety of phenotypes on strain B, ranging from the apparently wild type to those producing *r* type mutant plaques. This can be explained by the presence of a sequence of altered amino acids in the protein—its length depending on the distance apart of the sites of mutation—which produces a pseudo-wild type activity, its precise extent depending upon the amino acids involved.

The proteins synthesised by the *rII* cistrons have proved difficult to isolate, although McClain and Champe (1970) have more recently identified some fragments of the *rIIB* protein and Weintraub and Frankel (1972) have shown that this is a membrane protein. Direct confirmation of this theory has therefore been provided by experiments with another system. Comparisons of the lysozyme proteins synthesised by different phage T4 DNAs have confirmed that the protein in a double mutant differs from wild type only in the amino acid sequence between two sites of mutation. In one double mutant, a deletion followed by an addition results in a protein which differs from wild type in only five amino acids, shown in figure 1.11.

Assignment of Codons to Amino Acids

Three types of approach have been used to assign the actual nucleotide sequences within the triplet code words—*codons*—to their respective amino acids (reviewed by Woese, 1967 and Ycas, 1969). The first is to use cell free extracts to achieve protein synthesis under direction of specific oligonucleotide templates in vitro. The earliest successful experiment was reported by Nirenberg and Matthaei in 1961; their system consisted of ribosomes and aminoacyl-tRNAs from E.coli, with ATP as energy source.

Addition of synthetic polynucleotides to such systems stimulates the incorporation into polypeptides of specific amino acids; the effect of the template on the incorporation of any particular amino acid can be followed by using an incubation mixture in which this amino acid carries a C^{14} radioactive label but the other nineteen amino acids are unlabelled. Incorporation into protein is usually measured by precipitating the polypeptide synthesized and counting its radioactivity. Nirenberg and Mattaei found that the homopolymer poly-U promotes the uptake of phenylalanine; this suggests that the codon for phenylalanine is UUU. Subsequent experiments have shown that poly-A encourages uptake of lysine and poly-C stimulates incorporation of proline (see Speyer, 1963). Poly-G is not easy to test because it is difficult to prepare a suitable physical form to use as template.

The next advantage with such systems was the use of oligonucleotides containing more than one type of base; such copolymers may direct the incorporation of several amino acids into protein. Polynucleotides of known overall composition but of random sequence are easily prepared by providing for the enzyme polynucleotide phosphorylase a mixture of the appropriate nucleoside diphosphates; correlation between the frequencies of occurence of the various nucleotide triplets and the proportions of different amino acids in the polypeptide synthesized gives the nucleotide compositions of the codons. But the use of random copolymers suffers from the limitation that it shows only the nucleotide composition and not the sequence within the codon.

A more powerful use of cell free systems has been derived from Khorana's synthesis of polynucleotides which consist of a known sequence of polynucleotides. A copolymer containing two bases in alternating sequence causes the interdependent incorporation of only two amino acids when used as template. Jones, Nishimura and Khorana (1966) found that the kinetics of incorporation of the two amino acids are similar and approximately equimolar amounts of each are incorporated; the incorporation of either amino acid is stimulated in the presence of the other and very much reduced by its absence. (The relatively low incorporation of one amino acid which takes place in the presence of the other is probably caused by some endogenous content in the extract system.) One such experiment was the stimulation of valine and cysteine incorporation by poly-UG shown in figure 1.8.

If the nucleotide sequence

U G U G U G U G U G U G U G U G U G

directs synthesis of the polypeptide

cys – val – cys – val – cys – val

one of the codons UGU and GUG must represent cysteine and the other valine. Other data, such as those gained from the responses of amino acids to random copolymers, can distinguish between these alternatives. These results also demonstrate that the coding ratio must almost certainly be three and in all events must be an odd multiple of it.

Further research has taken advantage of polynucleotides consisting of repeating tri- and tetranucleotide sequences and analysis of the amino acids incorporated has defined the meaning of about half of the sixty four codons (Khorana et al., 1966; Kossel, Morgan and Khorana, 1967). For example,

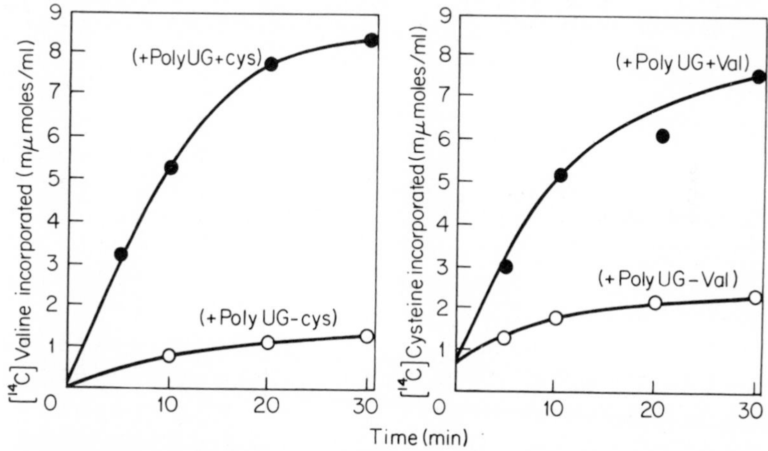

Figure 1.8: Stimulation of amino acid incorporation by poly-UG. Equimolar amounts of cysteine and valine are incorporated and the utilization of the two amino acids is interdependent; each can only be incorporated into polypeptide if the other is present. No third amino acid responds to the poly-UG template to any significant extent. This indicates that one of the codons UGU and GUG represents cysteine and the other species valine. Data of Jones, Nishimura and Khorana (1966).

poly-AAG directs the synthesis of the three homo-polypeptides polylysine, polyarginine and polyglutamate. This can be explained if reading starts at a random triplet at the beginning of the message:

A A G A A G A A G A A G A A G A A G A A G
lys – lys – lys – lys – lys – lys – lys
arg – arg – arg – arg – arg – arg – arg
glu – glu – glu – glu – glu – glu – glu

The codons AAG, AGA and GAA therefore represent lysine, arginine and glutamic acid respectively. (The nomenclature used for polynucleotides is: poly-A,B,C (1:2:3) has a random sequence with the bases in the order stated;

Trinucleotide	ala	arg	asp	asn	cys	glu	gln	gly	his	ile	leu	lys	met	phe	pro	ser	thr	trp	tyr	val
	Radioactive amino acid used to charge tRNA (C^{14} or H^3)																			
CpGpU	---	1·73	---	---	---	---	---	0·11	---	---	---	---	---	---	---	---	---	---	---	---
CpGpG	---	1·49	0·01	---	---	0·02	---	0·02	---	---	---	---	---	---	0·08	---	---	---	0·01	---
GpUpC	---	0·01	0·04	0·05	---	---	---	---	0·02	---	---	0·08	---	---	0·01	---	0·03	---	---	0·75
GpUpA	---	---	---	0·02	---	---	---	---	---	---	---	0·08	---	---	---	---	---	---	---	1·33
GpCpG	---	---	0·01	0·02	---	---	---	---	---	---	---	---	0·10	---	---	---	---	---	---	1·08
GpCpC	0·18	---	0·02	---	---	---	---	---	---	---	---	0·03	---	---	---	---	0·15*	---	---	---
GpCpA	2·42	0·07	0·01	---	---	0·06	0·05	---	0·01	---	---	---	---	---	---	---	0·14*	---	---	---
GpCpG	3·51	---	---	---	---	---	---	---	0·05	---	---	---	---	---	---	---	0·47*	---	0·01	---
ApGpU	---	0·01	---	0·04		---	---	---	0·03	---	---	---	---	---	---	0·27	---	---	---	0·08
ApGpC	0·43	---	0·03	0·10		---	---	---	0·03	---	---	---	---	---	---	0·17	---	---	0·01	0·02
ApCpA	---	0·10	0·01	---		0·19	---	---	---	---	---	---	---	---	0·01	0·03	---	---	0·02	0·02
ApUpG	---	---	---	0·02		---	---	---	---	---	---	---	1·00	---	---	---	0·01	---	---	---
UpApA	---	---	---	---		0·02	---	---	---	---	---	0·10	---	---	0·02	0·02	---	---	0·01	0·03
UpApG	---	0·06	0·01	0·12		0·04	---	---	---	---	---	---	---	---	---	---	---	0·03	0·03	---
UpGpA	---	---	0·07	0·14		0·10	---	---	---	0·02	---	---	---	---	---	---	---	0·03	0·03	---

* Indicates preparation contaminated with radioactive alanine

--- Indicates a zero or negative binding after subtraction of background binding in absence of added triplet.

Numbers show the $\mu\mu$ moles of amino acid bound in a standard reaction mixture

Figure 1.9: The Nirenberg binding assay for codon meanings. A nucleotide triplet is added to ribosomes in the presence of a preparation of aminoacyl-tRNAs which have been charged with one radioactive amino acid; the number of counts retained on a nitrocellulose filter is measured for each of the amino acids in turn; unbound aminoacyl-tRNAs are washed through the filter. The triplets GCU and CGG specify arginine, which is the only amino acid to respond significantly to them. GUC, GUA and GCG all specify valine. Although GCA and GCG specify arginine, it is difficult to decide from these data whether GCC also codes for this amino acid. No firm assignments can be made for the codons AGU, AGC and AGA, although the data do not disagree with the conclusion of other studies that the first two code for serine and the last for glutamic acid. Codon AUG binds only methionine. None of the three nonsense triplets stimulates the binding of any amino acid to an appreciable extent. These data are examples taken from the work of Brimacombe et al. (1965) and Nirenberg et al. (1966); the other triplets have also been assayed in this way. It is not always possible to arrive at a firm assignment of a codon meaning from the Nirenberg ribosome binding assay alone, but when viewed in the light of assignments made by other methods, the data gained in this way support the code shown in figure 1.10

if no figures are given, the bases are present in equimolar amounts. Poly-ABC, with no commas between the bases, has the sequence ABC repeated.)

Since errors occur in cell free incorporation systems, a check is provided by the assignment of codons through an entirely different technique developed by Nirenberg and Leder in 1964. A trinucleotide of known sequence-derived either by degradation of RNA or prepared synthetically—causes specific molecules of aminoacyl-tRNA (transfer RNA carrying amino acids) to bind to ribosomes. Ribosomes are retained on filters of cellulose nitrate, but aminoacyl-tRNAs are not; the aminoacyl-tRNA species attached to ribosomes can therefore be separated from the unbound transfer molecules, which are removed by washing the filter with a salt solution. Each trinucleotide can be assayed by adding it to twenty different incubations, each containing ribosomes and one radioactive aminoacyl-tRNA species. This experiment has been performed for most of the triplets—an example is shown in figure 1.9—and the results are in fairly good agreement with the data obtained from polynucleotide directed amino acid incorporation experiments (summarized by Brimacombe et al., 1965; Nirenberg et al., 1966). The code which has been worked out from both sets of studies is shown in figure 1.10.

But do these triplets code for the same amino acids in vivo? Three types of experiment have suggested that most, and probably all, of these codons are used in E.coli with the meanings shown in the figure. Codon meanings can be deduced indirectly by analysing the amino acid changes produced in proteins by mutations. The tryptophan synthetase A protein of E.coli, for example, has been subjected to a detailed analysis of the amino acid substitutions which result from the mutation of a single nucleotide induced by a mutagen which causes a specific type of base change (see Berger, Brammar and Yanofsky, 1968). For example, if we know that tryptophan is coded by UGG and is mutated to arginine by inducing a U → C base change, then CGG must be a codon used in vivo for arginine.

A more productive way to use mutated proteins is to compare the amino acid sequences of wild type and frameshift mutants. Analysis of acridine induced double mutants in the lysozyme of phage T4 has proved very successful. When the sites of addition and deletion are fairly close to each other, the tryptic fingerprints of double mutant and wild type proteins differ only in those peptides derived from the frameshift region; these can be isolated and sequenced with comparative ease. Using the codon assignments derived from studies in vitro, Streisinger et al. (1966) found that for every such double mutant a unique series of bases can be assigned to code for the wild type amino acid sequence and, with appropriate addition and deletion, for the double mutant amino acid sequence. Figure 1.11 shows an example; a summary of the codons which have been shown to be used in vivo by phage T4 has been given by Ocada, Amagase and Tsugita (1970).

The only direct way to assign codon meanings, of course, is to compare

First	Second: U	C	A	G	Third
U	phe	ser	tyr	cys	U
	phe	ser	tyr	cys	C
	leu	ser	CT (ochre)	CT	A
	leu	ser	CT (amber)	try	G
C	leu	pro	his	arg	U
	leu	pro	his	arg	C
	leu	pro	gln	arg	A
	leu	pro	gln	arg	G
A	ile	thr	asn	ser	U
	ile	thr	asn	ser	C
	ile	thr	lys	arg	A
	met,fmet	thr	lys	arg	G
G	val	ala	asp	gly	U
	val	ala	asp	gly	C
	val	ala	glu	gly	A
	val	ala	glu	gly	G

Figure 1.10: The genetic code. Codon meanings have been derived largely from studies with E.coli but are probably universal (see also figure 5.20). CT indicates codons which cause termination of protein synthesis; fmet is a species which is used to initiate protein synthesis in bacteria

	thr	lys	ser	pro	ser	leu	asn	ala
Wild	AC-	AA$^{A}_{G}$ ↓	AGU	CCA	UCA	CUU	AAU ↓	GC-
Mutant	AC-	AAA	GUC	CAU	CAC	UUA	AU$^{A}_{G}$	GC-
	thr	lys	val	his	his	leu	met	ala

Figure 1.11: Amino acid sequences in a wild type and double acridine mutant of T4 lysozyme. The two amino acid sequences detected experimentally can be related only by the nucleotide sequences shown, with the deletion and addition sixteen nucleotides apart. This confirms that these codon meanings are used in vivo. (At the sites of addition and deletion it is not known whether A or G is present; these are therefore shown as $^{A}_{G}$)

ala---try---arg---ser---tyr---leu---asn---met---glu---leu---thr---ile---pro---ile---phe---ala---thr---asn---ser---asp

U-G-G•C-G-U•U-C-G•U-A-C•U-U-A•A-A-U•A-U-G•G-A-A•U-U-A•A-C-U•A-U-U•C-C-A•A-U-U•U-U-C•G-C-U•A-C-G•A-A-C•U-C-C•G

Figure 1.12: Nucleotide sequence of a fragment of the RNA of phage R17 and the corresponding sequence of amino acids in positions 81–100 of the coat protein. The bonds linking nucleotide within a codon are marked by a dash (—) and those joining codons by a dot (.). Data of Adams et al. (1969)

the nucleotide sequence of a messenger RNA (or DNA gene) with the amino acid sequence of the protein for which it codes. It is difficult to isolate specific messengers in amounts sufficient for such analysis, but RNA phages have proved more amenable to this treatment because the RNA contained in the phage particle—which is easy to isolate—itself directs protein synthesis. The f2 group of phages—which includes phages f2, MS2 and R17—has RNA genomes which are of the order of 3300 nucleotides long; this is too large to sequence directly but the specific fragments which are produced by enzymic degradation of the phage RNA can be sequenced and identified.

The first sequence was reported by Adams et al. (1969), a length of 57 nucleotides of the gene which codes for the coat protein of phage R17; as figure 1.12 shows this sequence appears to represent the nineteen amino acids in positions 81–99. Subsequent reports have correlated the nucleotide sequences of other regions of the phage with the protein coded and the complete coat protein gene of phage MS2 has since been sequenced by Jou et al. (1972). As more sequences are defined, it appears increasingly likely that the codon meanings assigned by in vitro studies are all used in the living bacterial cell.

Nonsense Codons for Termination

Suppression of Nonsense and Missense Mutations

Three codons do not specify amino acids but instead act as signals to terminate synthesis of the polypeptide chain. The *nonsense* triplets UAA, UAG, UGA do not stimulate the binding of any aminoacyl-tRNAs in the Nirenberg binding assay; and when these trinucleotide sequences are present in synthetic RNA templates used to direct protein synthesis in vitro, they cause release of the polypeptide chain synthesized as far as the codon (reviewed by Beaudet and Caskey, 1972). Experiments to define the role of these three codons have made use of mutants of E.coli or phage T4 in which a codon specifying an amino acid in some protein is changed to a nonsense triplet; more recently, sequencing studies have directly identified termination codons at the ends of the genes of RNA phages.

When a base is mutated in the DNA of the genome, a codon in RNA is changed and this may result in the substitution of a different amino acid from wild type in the corresponding polypeptide chain. If this substitution is acceptable to the protein there is no phenotypic effect, but if it is unacceptable a missense mutation can be detected; the activity of the protein produced by the mutant gene depends upon the physico-chemical relationship between the original amino acid and its replacement. When the base substitution creates a nonsense codon, however, synthesis of the polypeptide chain is terminated prematurely at the mutant site.

Mutations of either type can be restored to the original phenotype either by reversion at the original mutant site or by mutation at a different locus.

In the former case, the organism regains its original genotype; the latter situation is termed *suppression*. Suppressor loci may be situated within the same cistron as the original mutation (although at a different site) or in a different cistron (either on the same or a different chromosome). Suppression falls into three main classes.

Indirect suppression is trivial; the primary mutation is circumvented rather than repaired. This can happen by opening another metabolic pathway, substituting the function of the product of the mutant gene by some other cell component, or changing cytoplasmic conditions to stabilise a previously unstable product.

Intragenic suppression falls into two classes. When a mutation has caused an amino acid substitution which inactivates the protein, a second substitution at a different point in the same polypeptide chain may restore the protein function. An example of this situation occurs in the tryptophan synthetase A protein of E.coli; substitution of a glutamic acid for a glycine at one site inactivates the protein, but a second substitution some thirty-six amino acids distant of cysteine by tyrosine—which also by itself causes inactivity—restores activity to the protein.

When a base has been added or deleted, the correct code meaning can be restored by introduction of a second, compensating frameshift in the same gene. Combinations of the (+−) form between T4 *rII* acridine mutants which are themselves inactive can therefore result in the synthesis of a protein which is active, although it differs from the wild type by the amino acids coded by the nucleotide sequence between the two mutant sites.

Information—or intergenic—suppression relies on a second mutation which concerns one of the factors controlling the mechanism of transfer of genetic information from DNA to protein. The meaning of a codon is altered during gene expression and this may be caused by mutation in the genome—in which case it is genetic—or by the external effect of some environmental agent, in which case the suppression is phenotypic. Such suppression can in theory take place at any one of several levels of action. Transcription might be altered so that mistakes are made in the production of a messenger, or the messenger itself might be modified after transcription by, for example, a methylation causing recognition by the wrong anticodons. Alteration in components of the translation machinery could comprise changes in the specificity of an amino-acyl-tRNA activating enzyme, alteration by mutation of the anticodon of a rRNA (see chapter 5) ,or some modification of the ribosome which influences the accuracy of translation (see chapter 4).

Nonsense mutations were first distinguished from missense mutations by a genetic test constructed by Benzer and Champe (1961, 1962). A deletion in the middle of the *rII* region of phage T4—the *r*1589 deletion—joins the A and B cistrons together so that the two proteins usually synthesised (one from each cistron) are replaced by a single protein produced by the joint cistron shown in figure 1.13, This protein has B activity, so that its B segment is presumably unaffected by the attachment of the A region.

The nature of mutants in the *rII* cistron can therefore be tested by placing them in series with *r*1589 by genetic recombination. Testing the double mutant for B protein activity shows whether the mutant in A is of the missense or nonsense type; a nonsense mutant allows no B activity, because synthesis of the polypeptide chain is terminated in the A region, but a missense mutation in the A region does not influence B activity.

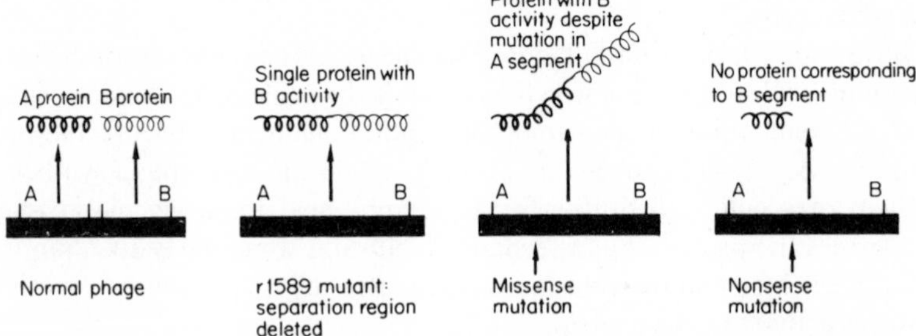

Figure 1.13: The r1589 test for nonsense mutations. In normal phage, the A and B cistrons each produce a separate protein. The r1589 deletion synthesises a single protein representing both cistrons but which has the activity characteristic of the B protein. A missense mutation in the A region does not affect the activity of the B region of the combined protein. But the presence of a nonsense mutation terminates protein synthesis in the A region so that protein for the B region is no longer made

The behaviour of the phage nonsense mutants isolated by this test depends upon which strain of E.coli is used as the bacterial host; in *non-permissive* strains the phage cannot grow because the protein chain is prematurely terminated, but growth can occur in other, *permissive*, (or *suppressor*) strains. Benzer and Champe suggested that this dependence on the host bacterial strain can be accounted for if the suppressor strain contains an unusual (mutant) aminoacyl-tRNA which responds to the nonsense codon by inserting an amino acid into the growing polypeptide chain, whereas the non-permissive strain lacks this species and terminates protein synthesis at the nonsense codon.

Nucleotide Sequences of Nonsense Codons

Nonsense mutants can be classified according the suppressor loci to which they respond. They fall into three classes. The mutants found by Benzer and Champe all respond to the same suppressor and have been termed *amber*. Brenner and Beckwith (1965) found another class, termed *ochre*, which are not suppressed by amber suppressors. However, suppressors of the ochre mutants also suppress amber mutants; this implies that, although the two types of nonsense mutant are distinct, they must be related. A third class of nonsense mutant (which lacks any agreed exotic name) was isolated by Sambrook, Fan and Brenner (1967) and by Zipser (1967); its suppressors do not act on either amber or ochre codons.

The nucleotide sequences of the amber and ochre codons were first deduced by Brenner, Stretton and Kaplan (1965) by experiments in inducing and reverting nonsense codons in the *rII* cistron with the chemical mutagens hydroxylamine and 2-aminopurine. Hydroxylamine causes the replacement of G/C base pairs by A/T base pairs and 2-aminopurine causes both this reaction and the reverse replacement of A/T by G/C. The patterns of production of the mutants show that each nonsense codon must contain at least one A/T base pair in DNA and that the ochre must contain another which is matched by a G/C pair at the same site in amber.

The orientation of the base pairs on the DNA double helix can be distinguished by making use of the need of the phage to express the *rII* function before replication. This means that the phage cannot grow on K12 bacteria, on which *rII* function is essential, if the coding strand of DNA specifies a nonsense codon; but mutations to the anticoding strand do not inhibit growth until a later generation of phage, when replication has transferred the mutation to the daughter coding strand. As a control, phage can be grown on B type bacteria, where *rII* nonsense mutants do grow to show an altered phenotype. The characteristics of the phages induced by hydroxylamine imply that amber triplets must contain two bases derived from A/T base pairs in opposite orientation in the DNA, that is one A and one U on the messenger. This led to the conclusion that the nucleotide compositions of these codons must be:

amber	U, A, G/C
ochre	U, A, A/U

A second type of experiment was used to resolve the sequences of the nucleotides within the codons. Studies on mutants in which a purine is replaced only by the other purine, or a pyrimidine only by the other pyrimidine, showed that in T4 head protein, ambers are derived only from glutamine codons (CAA or CAG) or tryptophan (UGG). This suggests that the amber codon is UAG; the ochre must therefore be UAA. (These experiments have been discussed in more detail by Stretton, 1965 and Lewin, 1970).

Studies on the suppression of nonsense triplets have confirmed that a transfer RNA may be mutated so that it recognises certain nonsense codon(s) instead of—or in addition to—the codons to which it usually responds. There are many different suppressor strains and the various amino acids which they insert are always related to the appropriate nonsense codons by a single base change, just as nonsense mutants are always derived by mutations of amino acids which are specified by codons differing in only one base from UAA, UAG or UGA.

The suppressor tRNA which inserts tyrosine in response to UAG has been sequenced and comparison with the wild type molecule shows that the mutation is located in the anticodon and enables the suppressor to recognise UAG instead of the usual UAU or UAC. A further mutation in the anticodon of this rRNA changes its coding properties again so that it recognises both UAA and UAG. Such experiments confirm the sequences assigned to the amber and ochre nonsense codons; and this form of suppression is discussed on page 211.

Biochemical experiments with synthetic templates and with both nonsense mutant and wild type RNA phages have demonstrated that the effect of nonsense codons is mediated by protein factors which recognise their sequences as the signals for releasing completed proteins from transfer RNA (see page 111). That all three nonsense codons may be used as termination signals at the ends of genes has been shown by determination of the nucleotide sequences of RNA phages (see page 86); we do not know for certain the extent to which each of the three nonsense triplets is used within the bacterial cell, but this problem is discussed on page 114.

The chain terminating role of the codons UAA, UAG and UGA has been defined conclusively only in bacteria. However, it seems very likely that at least the UAA and UAG codons and probably also the UGA triplet have the same meaning in eucaryotic cells. Biochemical experiments have failed to identify any transfer RNAs which respond to the ochre and amber codons (see for example figure 5.20); and mammalian cells contain a protein terminator factor which responds to all three nonsense codons. Genetic experiments have been possible only with yeast, in which Stewart et al. (1972) and Stewart and Sherman (1972, 1973) identified two classes of mutant in cytochrome c which cause complete inactivation; these appear to be nonsense mutants and their codons may be deduced from the amino acids which can be inserted when a second mutation generates a reversion to give an active protein. One class appears to comprise UAA and the other UAG. Hatfield (1972) has found that the UGA triplet appears to stimulate binding of mammalian arg-tRNA and ser-tRNA in the Nirenberg assay; but this alone cannot be taken as evidence that this codon represents an amino acid, since binding (to ribosomes of E.coli) may reflect insufficiently stringent conditions of incubation. Ochre and amber codons therefore appear likely to have a universal meaning of chain termination; the role of the UGA codon cannot at present be defined, although genetic experiments with yeast show that it does not code for tryptophan.

Evolution of the Code

Universal Nature of the Code

The genetic code appears to be universal, so that the codon meanings which have been defined in E.coli are the same in all living organisms. The coding responses of transfer RNAs of different species—see page 204—support the conclusion that codon assignments are invariable. The only known difference in meanings is a nuance of expression of the AUG initiation codon, which directs incorporation of formyl-methionine in bacteria and methionine in eucaryotes; but since the formyl moiety and sometimes also the methionine group are later removed, the sequence of the mature protein does not depend upon the interpretation of the initiation codon. And it has not yet been demonstrated that UGA represents nonsense in eucaryotes.

That messenger RNA of one species can be correctly translated in vitro and in vivo by the synthetic apparatus of another species provides strong evidence that all codon meanings are universal (see chapter 5 of volume two). Of course, a different set of components forms the translation apparatus in each species; and miscoding may be displayed by systems comprising heterologous components—for example, ribosomes of one species and transfer RNA of another—because incorrect interactions are then possible (see Hunter and Jackson, 1970). But the overall set of interactions comprising the translation apparatus of any one species achieves the same interpretation of each codon, appropriate irrespective of the source of the messenger. Any model to account for the evolution of the code must therefore explain why it is the same in all species.

The code is highly degenerate since sixty-one of the possible sixty-four triplets code for only twenty amino acids. Indeed, it is only because most amino acids are represented by more than one codon that it is possible for all species to use the same code, despite the drastic variations between the base compositions of their DNAs—the two extremes seem to be Tetrahymena pyriformis which has only 25% G + C and Micrococcus lysodeikticus which has as much as 72%. But the amino acid composition of proteins does not reflect this range of variation, for it tends to remain roughly similar for all species. This implies that although the code itself remains the same, different organisms will use to different extents the various codons which are available to represent any particular amino acid.

Not all amino acids are represented by the same number of codons; some have only one codon whilst others have as many as six:

1 codon—met, try
2 codons—glu, gln, asp, asn, cys, lys, his, tyr, phe
3 codons—ile, nonsense
4 codons—val, pro, thr, ala, gly
6 codons—leu, ser, arg

From what is known about the amino acid composition of proteins in general—that is taking no account of the different organisms from which they derive—the frequency with which an amino acid occurs seems to be in reasonably good agreement with its number of codons. Indeeed, King and Jukes (1969) have shown that, given the triplet base mechanism, the natural genetic code comes close to the optimal coding arrangement to provide for these frequencies, as illustrated in figure 1.14. King and Jukes have suggested that it may be misleading to think of codon assignments as having attained their present meanings in order best to match the amino acid composition of living material; the relationship rather may be the reverse, the composition of proteins reflecting the availability of amino acids, given their codon assignments.

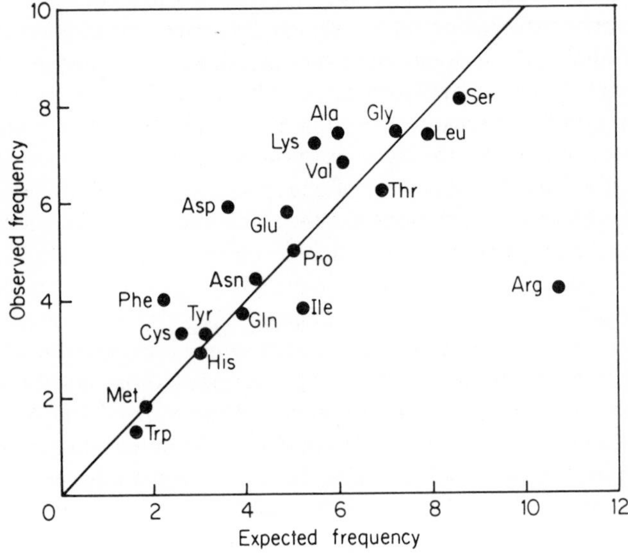

Figure 1.14: Amino acid frequencies in protein compared with the frequencies expected from their relative number of codons. There is good agreement with the sole exception of arginine. Data of King and Jukes (1969)

Any explanation for the evolution of the code must account for the pattern in its degeneracy; the arrangement of codons representing the same or similar amino acids is not random. The third nucleotide of a codon is the most degenerate; the base in this position is either irrelevant to the meaning of the code or a distinction is made only between purine and pyrimidine. This relationship follows from the nature of the base pairing between the codon of messenger RNA and the anticodon of transfer RNA (and the wobble hypothesis is discussed on page 202). The result is that alteration of the final nucleotide in a triplet very often fails to produce any change in the amino acid for which it codes. As figure 1.10 shows, the codons are thus clustered in sets representing a single amino acid.

In addition to the clustering of codons representing a particular amino acid, the codons for functionally related amino acids appear themselves to be related to some extent. For example, codons with U as their second base (first column of figure 1.10) represent the most hydrophobic amino acids; this arrangement means that a mutation in the second base of these codons results either in no change of amino acid or in the replacement of one amino acid by another related to it functionally.

Stereochemical Interactions in the Primitive Code

Two general classes of theory have been proposed to explain the evolution of the code (reviewed by Woese, 1969; and Ycas, 1969). *Mechanistic* models suggest that the original code, from which the present code must have evolved, was based upon some physico-chemical relationship between an amino acid and its codon. This implies that the code could be constructed in only one, or at most a very few, ways. *Selective* models do not depend upon a fixed relationship between codon and amino acid, but argue that the initial set of codon assignments was free to vary. Selective forces would act to modify this starting set during the course of evolution and the final set of codons must therefore be optimal with respect to whatever criteria were used by selection. As such, selective theories do not account for the universal nature of the code; to explain this they must assume that the code developed at a very early stage during evolution and that all living organisms are descended from a single cell line, or, more precisely, from a single inbreeding population. Of course, solely mechanistic and solely selective models represent extremes and there are intermediate models with features of both. The different selective models which have been proposed differ in the nature of the selective forces to which they ascribe the evolution of the code.

One dilemma in accounting for the evolution of the code is that the apparatus which is used to translate the nucleotide sequences of mRNA into protein is itself dependent upon the code. As Monod (1972) has observed, the code is meaningless unless it is translated. The information which specifies the proteins which synthesise and translate RNA is coded by triplet sequences in the genome. This means that the translation of the code depends upon the products of such translation. How did this circular situation evolve?

Several proposals have been made that the original code might have been determined by a stereochemical fit between an amino acid and its codon, the nucleic acid acting as a physical template for assembly of the protein (see for example, Woese, 1967, 1968). Initially, there might have been an autocatalytic cycle in which a polynucleotide and polypeptide assisted each other to replicate, so that their relationship was at first reciprocal and only subsequently became unidirectional. Presumably, some of the polypeptides produced by stereochemical fit might evolve catalytic activities to improve the process of their own production and that of other proteins. The complex system now used for protein synthesis might have evolved gradually to improve the efficiency of translating the code, keeping the relationship between triplets and amino acids that first defined the original polypeptides.

Simulations of the type of conditions which may have composed the prebiotic environment—including simple compounds such as water, methane and ammonia—have suggested that polymers of amino acids are likely to have preceded the formation of polynucleotides. One difficulty in conducting any experimental analysis is that the nucleotides and amino acids involved under primitive conditions may not have been the same as those employed today. But many of the amino acids and nucleotides have been formed in such simulation experiments (see, for example, Fuller, Sanchez and Orgel, 1972).

There is some evidence also that complexes of the general type which may have been involved can be formed between the present components of the code. Lacey and Pruitt (1969) have observed the formation of a complex between poly-L-lysine and mononucleotides; they suggested that the nucleotides form an array of two chains which resemble duplex DNA, except that the structure of each nucleotide chain is stabilised by the interaction with a polylysine chain instead of by covalent linkage between the nucleotides. Woese (1969) has suggested that the physico-chemical properties of trinucleotides may be correlated with the physico-chemical relationships between the amino acids for which they code. But it is at present controversial whether an amino acid could interact with its codon with sufficient stereochemical specificity of fit to account for the development of the code.

In spite of objections that amino acids do not interact specifically with triplets, it is certainly attractive to visualise some sort of physical interaction as the basis for evolution from the prebiotic environment. It is difficult to see any other form of interaction which would allow the products of translation to evolve gradually into the translation apparatus found in cells today. Unless we invoke some sort of stereochemical fit as the basis for the primitive code, we are left with no satisfactory way to overcome the dilemma that we cannot select for a more accurate code—by whatever criterion—until we have a code on which selection can act. Where else might this first code come from if not direct interaction between nucleotides and amino acids?

Forces of Selection on the Code

Most simple mutations—that is excluding gross chromosomal rearrangements—are caused by the substitution of a single nucleotide by another. Mutations tend to exert a deleterious effect on the cell and Sonneborn (1965) has suggested that cell lines which can reduce their burden of harmful mutations are likely to be at a selective advantage. Indeed, the genetic code shows the features expected of any code "buffered" against mutational change. Such a code must be highly degenerate, since unassigned codons would probably interfere with protein synthesis. And there should be maximum connection between the various codons assigned to any particular amino acid, that is to say, they should differ by as little as possible (one base of the triplet). This achieves the maximum probability that a mutation is "silent"—that it does not change the amino acid specified because the original and mutated triplets code for the same amino acid.

The burden of mutation can be further reduced by increasing the likelihood that if there is to be a change in the amino acid specified, the replacement should have similar properties to the original species; this is achieved if related codons specify functionally related amino acids. Epstein (1966) has observed that it is particularly important that the code should minimise replacements which involve changes between hydrophobic amino acids, which tend to lie in the interior of protein molecules, and hydrophilic, which tend to be positioned on the surface. In fact, the actual dictionary of codon meanings comes close to the maximum connection possible. The mutational-buffer theory is purely selective since it no way depends upon the absolute relationship between a codon and an amino acid, but demands only that the codons for similar amino acids should be related to each other.

But it is precisely this facet of the model which comprises its main drawback; it does not account for the universal nature of the code. It seems unlikely that only one set of codon assignments should be optimal; even under similar conditions, independently evolving species might well evolve different optimal sets. Allowing for the wide variety of environmental conditions, it seems even less likely that evolution would result in only one code. Another disadvantage is that the model requires codons to change their meanings during evolution; whilst any particular amino substitution may benefit some proteins, it is likely to create a series of deleterious mutations in the other proteins coded by the genome.

An alternative model is to suppose that the selective forces of evolution act to minimize errors made in the translation of codons into amino acids. This model is not entirely selective since it depends upon the mechanistic features involved in translation. Woese (1965, 1968) has argued that translation is at present accomplished by a very specific and thus very highly evolved system, so that the processes involved in translation in primitive cells must have been very different. He suggested that the translation apparatus of a primitive cell is likely to have been relatively inaccurate, with a high degree of ambiguity in the assignment of amino acids to codons. Errors would be so sommon that the protein molecules produced by any particular gene would have no unique structure, but rather would comprise a group of "statistical proteins "related to some theoretical primary structure, and therefore to each other, by varying degrees of closeness. It would therefore be comparatively easy to alter codon assignments—this would have little harmful effect upon such an error-prone system. At this stage, the system would probably involve groups of related amino acids, for example hydrophobic and hydrophilic, rather than specific species, which would be introduced later.

The chance that an error will occur when a codon is translated seems to vary with the codon and Woese suggested that evolution would favour a dictionary in which the more error-prone codons represent amino acids which are functionally less important to proteins. Amino acids which tend to be involved in the mechanisms of enzyme catalysis—those with polar side chains—would tend to be assigned to the least error-prone codons. Also, codons which are likely to be mistaken for each other during translation would tend to be assigned to the same, or at least to functionally related, amino acids. Woese observed that the probabilities that misreading will occur at each of the three nucleotides in a codon are 100:10:1 from position three:one:two. Position three is the most degenerate in the genetic code, so that codons differing in this base position tend to be assigned to related amino acids. The second base is the least error-prone and probably the least involved in such restraints.

As evolution progressed, the cell would make more precise distinctions amongst amino acids and begin to recognise individual species rather than just functionally similar groups. Selection would favour the development of a less error-prone and more complex translation apparatus; and only when the codon catalogue has become (virtually) completely ordered in its amino acids would it be possible to evolve a really efficient translation apparatus. (The ways in which this might happen have been discussed by Woese, 1967 and Orgel, 1968). Any cell succeeding in this development would have such an enormous selective advantage that all living cells would be descendents of this single cell line. As translation evolves into a more sophisticated form, the effect of changing any particular codon assignment increases greatly; eventually, such change becomes lethal and is no longer possible. Evolution of the code itself must therefore have been confined to the earlier and more fluid stage of codon assignments.

Crick (1968) has suggested that the code might start in a primitive form in which a small number of triplets coded for comparatively few amino acids. The assignment of codons to amino acids would be random—so this is a purely selective model. Crick argued that the code is most likely to have started as a triplet one, since if it commenced with some other coding ratio, all previous messages would become nonsense, and presumably lethal to the cell, with the switch to triplet reading. (This argument disposes of all models which suppose that the code might have started as, say, sixteen dinucleotides coding for few amino acids, later expanding to a triplet system coding more amino acids.) It is not clear just which amino acids would be present in this initial code, although some of the more complex species present today are more unlikely to have existed under prebiotic conditions.

In an intermediate phase, these primitive amino acids would take over most of the triplets of the code so as to reduce to a minimum the number of codons without meaning; the extension of codons for any particular amino acid would probably be to those related to the currently assigned codons. The final code would form as new amino acids replaced some of the primitive ones. This would happen most easily if the amino acid was related to that previously coded by the triplet and if the organism coded only a few, rather crude proteins. As the number and sophistication of proteins increased, a time would be reached when the substitution of new amino acids would disrupt too many pre-existing proteins and would therefore be lethal. At this point, the code would be "frozen"; this theory has accordingly been termed the "frozen accident" model. To explain the universality of the code, it is again necessary to postulate that this event must have happened at an early stage of evolution, with all cells descended from the single population involved.

There is therefore no entirely satisfactory explanation for the universal nature of the code; and, indeed, no way to elucidate its evolution since this must have occured at a very early stage and even the simplest organisms living today possess a highly sophisticated system for translating messengers into proteins. Even the mechanistic theories of stereochemical fit do not provide entirely satisfactory explanation for universality; we may suppose that, even if the first stage of evolution of the code was a direct interaction between polypeptide and polynucleotide, this direct relationship must have been lost as the polypeptides gained the catalytic activities which were eventually to develop into the translation apparatus. It is reasonable to suppose that there would still be opportunity for some evolution of codon meanings during this process, which again drives us back to the idea that the first cell to develop an "efficient" code must be the progenitor of all organisms.

Stereochemical fit alone is therefore inadequate as an explanation of universality, because it seems probable that some, at least, of the twenty amino acids in present use may have been added to the repertoire of the cell during the stage of expansion after the original stereochemical fit had begun to be superceded by a translation apparatus. Whilst this idea overcomes some of the difficulties in allowing for specificity of stereochemical fit—for it means that only a few amino acids need be distinguished, not necessarily by the sixty-four codons in use today but perhaps by a much smaller selection of triplets— at the same time it reopens the question of the universal origin of the code by alllowing selection to operate before the code is finalised. We do not know, of course, which, if any, of the selective criteria so far considered may have been used during this evolution, but we may be sure that the organised degeneracy of the code is not fortuitous.

In summary, then, we may imagine the initial stages of evolution of the code to have arisen from some interaction between polypeptide and polynucleotide; it is otherwise difficult to see how a code which depends on its own translation products could have evolved. An early priority in evolution must have been the development of some system for holding together the successful replicating system—that is, an early cell membrane. Selective processes must have acted to develop a more accurate translation apparatus from this starting point; we can only speculate on the nature of these selective forces and on to what extent they may have enlarged and altered the codon dictionary.

CHAPTER 2

Synthesis of the Polypeptide Chain

Components of the Apparatus for Protein Synthesis

Sequential Synthesis of Proteins

The simplest way to interpret the genetic code is for amino acids to be added sequentially to the growing polypeptide chain as non-overlapping triplets are read from the starting point. (The alternative that all the amino acids are positioned on the template, after which peptide bonds are synthesised to connect them would reopen the question of how triplets are recognised only in the correct phase.) Dintzis (1961) and Naughton and Dintzis (1962) confirmed experimentally that the synthesis of haemoglobin in reticulocytes is sequential; their technique was to replace one of the amino acid precursors by an isotopic radioactive form during incubation.

If amino acids are positioned before peptide bond synthesis, the label should be taken up equally into all parts of the protein. But if synthesis is sequential, the amino acids in those parts of the protein chains completed before the switch to radioactive conditions should be normal and only those parts of the chain synthesised after the substitution can contain the isotope. The point where radioactive replace normal amino acid is different for each chain synthesised. However, if the radioactive content of completed haemoglobin molecules is analysed, only the far end should be labelled after a short time of incubation. Figure 2.1 shows that as incubation proceeds and more polypeptide chains are completed, the radioactive label moves towards the starting end.

The procedure used in these experiments is to replace C^{14}-leucine by H^3-leucine in the incubation mixture and then to allow protein synthesis to continue for varying lengths of time before completed haemoglobin proteins are isolated and degraded to peptides by treatment with trypsin. When the peptides are fingerprinted by two dimensional chromatography, the levels of each radioactive isotope can be counted and the H^3-label expressed relative to the C^{14}-label, which acts as an internal control of the absolute level of leucine radioactivity—some peptides might have one or two leucine residues, others three or four.

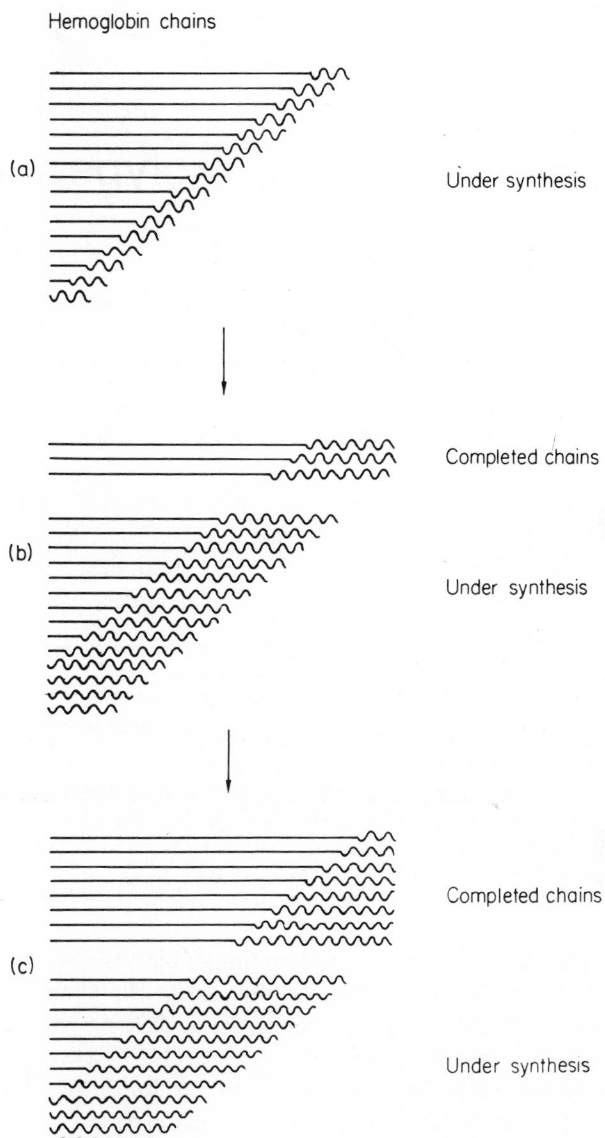

Figure 2.1: Labelling of haemoglobin chains after increasing periods of incubation with radioactive amino acids. Radioactively labelled portions of the protein chain are indicated by a jagged line. (a) the label simultaneously enters protein chains at all stages of synthesis. (b) the first chains to be completed in radioactive conditions have labelled amino acids at their termini. (c) as the period of incubation increases more chains are completed in radioactive conditions; a gradient of radioactivity thus runs from the terminus to the start of the protein chain

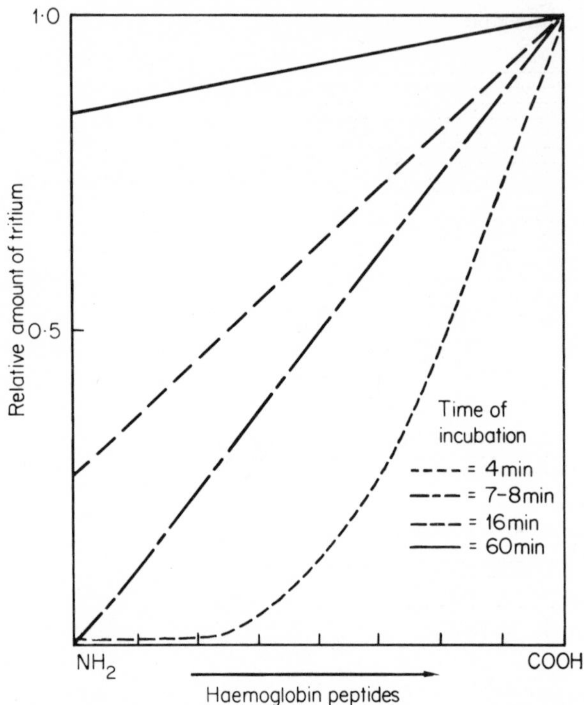

Figure 2.2: Incorporation of of radioactive leucine into haemoglobin after different periods of incubation. There is a gradient from the C-terminus to the N-terminus of the protein, which implies that synthesis takes place sequentially from N-terminus to C-terminus. As the time of incubation is increased, the total amount of radioactivity is increased but the gradient remains. Results are expressed as the ratio of an H^3 label in leucine to an internal control of C^{14} (see text). Data of Dintzis (1961)

Examining the radioactive ratios as a function of the position of the peptide in the protein chain shows that peptides located at the C-terminus of the protein possess the highest level of tritium; figure 2.2 shows that this falls steadily towards the other end of the chain. This implies that protein synthesis proceeds sequentially from the N- to the C-terminus, for the termini of chains are first labelled by the radioactive amino acids. As the period of incubation is increased, the level of radioactivity rises in all peptides, but the gradient from N- to C-terminus is always maintained.

Messenger RNA as the Template

The equivalance between gene and protein implies that synthesis of a protein must utilise the information contained in the sequence of DNA which specifies it. The genetic code provides the rules which govern the conversion of the sequence of nucleotides of the gene into the sequence of amino acids in its protein. However, this transfer does not take place directly on DNA itself, but utilises as an intermediate in information transfer an RNA copy of the gene.

One of the concepts which arose early in ideas about protein synthesis was that amino acids might be assembled into protein on a nucleic acid "template" carrying the appropriate sequences of trinucleotides (see Crick, 1958). This template is now known as messenger RNA (mRNA) and is a molecule of (single stranded) RNA which bears an exact copy of the nucleotide sequence of the gene(s) on DNA specifying the protein(s) for which it codes (reviewed by Lipmann, 1963). The *transcription* of messenger RNA takes place from one strand of the duplex comprising the genes which it represents, so that its sequence is complementary to one strand of the DNA and identical with the other. The processes by which mRNA directs the formation of proteins are known as *translation* and require the participation of the other RNA containing components of the protein synthesis apparatus, ribosomes and transfer RNA.

Messenger RNA represents only a small proportion of cellular RNA, so that it has proved difficult to characterize. It was first identified in bacterial cells infected with bacteriophages as an RNA component specified by the invading phage. In phage infected cells, protein synthesis not accompanied by net synthesis of RNA, but as Volkin and Astrachan (1957) observed, a minor fraction—about 3%—of the RNA is both rapidly synthesised and rapidly degraded, as judged by its rate of incorporation of a P^{32} label. The base composition of this RNA in E.coli cells infected with either phage T2 or phage T7 reflects that of the phage rather than the cellular DNA. This suggests that the infecting phage genome directs the production of phage specific proteins through the mediation of some unstable species of RNA molecule.

More direct evidence for the production of unstable messenger RNAs by phage DNA was provided by the work of Brenner, Jacob and Meselson (1961), who found that a new RNA with a relatively rapid turnover is synthesised in E.coli cells after infection with phage T2. This RNA has a nucleotide composition corresponding to that of the phage DNA and sediments in the range 14S-16S. Although sedimentation coefficients provide only imprecise estimates of RNA length, this should be the right order of magnitude to code for a long polypeptide chain. The RNA becomes associated with the pre-existing ribosomes of the bacterial cell and is continually replaced. At the same time, Gros et al. (1961) detected a comparable species of RNA in uninfected E.coli which becomes pulse labelled and associates with the ribosomes engaged in protein synthesis.

Experiments involving the hybridization of messenger RNA to denatured DNA have shown that, as expected, it constitutes a copy of the nucleotide sequences of DNA. Experiments with the rather crude techniques then available were used by Hall and Spiegelman (1961) to show that T2 mRNA anneals with T2 DNA but not with other DNAs. When the RNA of bacterial cells is extracted shortly after infection with T2, newly synthesised molecules can be identified by their incorporation of a P^{32} label. After mixing with heat denatured phage DNA, centrifugation to equilibrium on CsCl density gradient separates the resulting RNA-DNA double stranded hybrid. Experiments using more stringent conditions for hybridization have since been used to characterize different classes of RNA coded by phage DNA (see page 262).

The isolation of DNA corresponding to particular bacterial genes—by its incorporation into an episome or phage DNA—has made possible the development of more precise hybridization assays for the corresponding RNA molecules (see page 282). The critical test for any putative messenger is to characterize the protein(s) whose synthesis it directs; using the DNAs of phages carrying bacterial genes, it is possible first to transcribe the genes into RNA and then to translate the RNA into proteins. All stages of gene expression can be followed in these systems.

Messenger RNAs of eucaryotic cells have been less well characterized and individual messengers have been isolated only more recently. Their synthesis involves the transcription of a large precursor molecule from which the messenger sequence is cleaved in the nucleus; it is then transported to the cytoplasm where it is translated by the ribosomes in the same way as bacterial mRNA (see chapter 5 of volume 2).

Transfer of genetic information from gene to protein through messenger RNA was until recently thought to be unidirectional. This was the implication usually ascribed to the "central dogma" proposed by Crick (1958); genetic information can be perpetuated in the form of nucleic acid but cannot be retrieved from the amino acid sequences of proteins. The dogma has thus often been expressed in the form:

$$\circlearrowleft \text{DNA} \rightarrow \text{RNA} \rightarrow \text{protein}$$

implying that genetic information can be maintained by the replication of DNA or expressed by the synthesis of RNA and then protein.

Both the steps of expression, transcription and translation, have generally been regarded as irreversible; this implies that it is impossible to utilise the information of RNA sequences to synthesize DNA. But more recent research has suggested that the original form of the dogma is more accurate, for the nucleotide sequences of messenger RNA can be converted to duplex DNA by the reverse transcriptase activities (and other associated enzymes) discovered in RNA tumour viruses (for review of the dogma, see Crick, 1970).

We do not know whether such mechanisms are used only by RNA tumour viruses during infection or may also operate in uninfected eucaryotic cells; and reverse transcription does not seem to take place in bacteria. The usual flow of genetic information may therefore be unidirectional in most cells. But since at least some genetic information (that of the tumour viruses) may be perpetuated by transcription of RNA from DNA followed by reverse transcription of DNA from RNA, it is necessary to reformulate the dogma. Because the translation of nucleotide sequences in RNA to amino acid sequences in proteins is irreversible, the dogma can be stated as:

$$\circlearrowright \text{DNA} \rightleftharpoons \text{RNA} \rightarrow \text{protein}$$

Activating Amino Acids with Transfer RNA

Transfer RNA is the smallest species of RNA in the cell and sediments at about 4S; this corresponds to a molecule some 75–85 nucleotides long. (Its earlier name of soluble RNA [sRNA] has been replaced by the functional description, tRNA.) This is the "adaptor" species, first proposed by Crick et al. (1957), which is responsible for fitting an amino acid to its correct nucleotide triplet on the messenger. Each transfer RNA thus has the two properties of specifically recognising both its particular amino acid and the codon representing it. The first function is achieved by its interaction with the amino acid activating enzymes; the second is mediated by the ribosome, which enables a triplet in messenger RNA to be recognised by a complementary trinucleotide sequence of bases in tRNA, the *anticodon*.

All transfer RNAs have the same nucleotide sequence at their 3′ hydroxyl end and this common sequence of pCpCpA-OH is required for their interaction with aminoacids to form aminoacyl-tRNAs. Treatment with periodate shows that loss of the hydroxyl groups on the ribose of the 3′ terminal adenosine abolishes the ability of tRNA to accept amino acids. Preiss et al. (1959) showed that aminoacyl-tRNA is resistant to such treatment; and Zachau, Acs and Lipmann (1958) demonstrated directly that the amino acid is accepted into an ester link by charging tRNA with radioactively labelled leucine and then degrading the molecule with ribonuclease. A fragment of 3′ adenyl-leucine can be recovered; this implies that the terminus of aminoacyl-tRNA has the structure shown in figure 2.3.

Formation of aminoacyl-tRNA takes place in two stages; first the amino acid is activated by reaction with ATP and then the aminoacyl-adenylate is transferred to transfer RNA (reviewed by Novelli, 1967). Activation by reaction with ATP is a common step in biosynthetic pathways; and Hoagland, Keller and Zamecnik (1956) first showed that a soluble extract of rat liver catalyses an exchange of P^{32}-pyrophosphate with ATP in the presence of amino acids, although no AMP is released. This suggests that the amino acids are activated by formation of an enzyme bound complex with AMP, with release of pyrophosphate. Hoagland et al. (1957) confirmed this mechanism

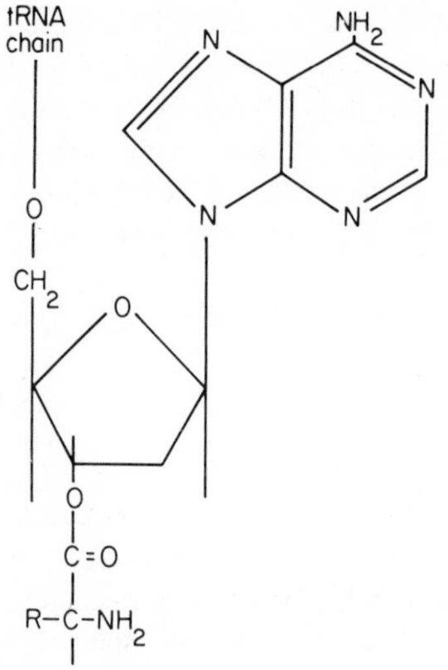

Figure 2.3: The 3′ terminus of aminoacyl-tRNA. An amino acid is linked through an ester bond to the 3′ hydroxyl position of the ribose of the terminal adenylate residue of the polynucleotide chain

by cleaving the activation complex with hydroxylamine; when the tryptophan present is labelled with O^{18} in its carboxyl group, the label is found in the AMP isolated after the hydrolysis. Direct proof was provided when Kingdon et al. (1958) isolated a complex of tryptophan–AMP. The reaction is quite specific for ATP and no other nucleoside triphosphate can be substituted.

Cytoplasmic tRNA becomes labelled with C^{14}-amino acids upon incubation in the presence of ATP and the amino acid activating system and Hoagland et al. (1958) suggested that the formation of ATP can be represented by the two reactions:

(1) ATP + amino acid + enzyme $\longrightarrow$ enzyme–amino acid–AMP + PP_i

(2) enzyme–amino acid–AMP + tRNA $\rightarrow$ aminoacyl-tRNA + enzyme + AMP

This two step mechanism was confirmed when Lagerkvist, Rymo and Waldenstrom (1966) isolated the complex formed in reaction (1) by valine and showed that addition of the appropriate tRNA results in the reaction with amino acid shown in (2). The charging of tRNA with amino acid can therefore be represented as illustrated in figure 2.4 and is discussed in more detail in Chapter 5. A historical review of the discovery of the activation of amino acids has been given by Zamecnik (1969).

According to the adaptor hypothesis, it is the nucleotide sequence of the anticodon alone in an aminoacyl-tRNA which recognises its particular codon. Chapeville et al. (1962) confirmed that the only stage in protein synthesis at

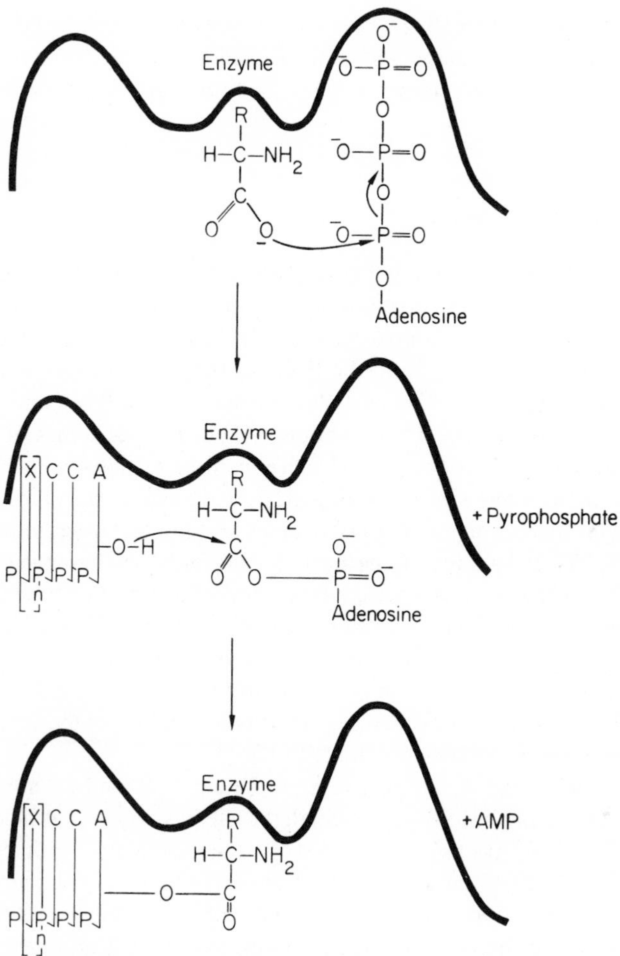

Figure 2.4: Formation of aminoacyl-tRNA by aminoacyl-tRNA synthetase. The first reaction between amino acid and ATP yields an enzyme bound aminoacyl-adenylate moiety. The activated amino acid is then transferred to tRNA and AMP is released

which the identity of the amino acid plays a role is the charging of aminoacyl-tRNA; after this it is the transfer molecule which is recognised. It is possible to perform a reductive desulfuration of cysteine with Raney nickel to give alanine whilst the amino acid is attached to its transfer molecule ($tRNA_{cys}$). This produces the hybrid species alanyl-$tRNA_{cys}$. Using poly-UG (which directs the incorporation of cysteine and valine into a polypeptide chain) as template, the hybrid responds as would cys-$tRNA_{cys}$ but incorporates alanine in place of cysteine. Fahnestock, Weissbach and Rich (1972) have shown that recognition of poly-U by $tRNA_{phe}$ is unaltered even after the α-NH_2 group of the amino acid is deaminated with nitrous acid to give phenyllactyl-$tRNA_{phe}$. It is therefore the tRNA alone and not its amino acid which recognises the messenger.

The Polyribosome Complex

Formation of peptide bonds between the amino acids assembled through their transfer RNAs at the appropriate codons on mRNA is catalysed by the ribosome. Ribosomes are compact ribonucleoprotein particles which account for an appreciable proportion of the RNA of the cell (up to 80–90% in exponentially growing E.coli). The particles consist solely of RNA and protein; in bacteria the RNA/protein ratio is about 6:4, but the ribosomes of eucaryotic cells have about equal proportions of each constituent. Ribosomes are generally characterized by their size on sedimentation; bacterial ribosomes sediment at about 70S whereas mammalian ribosomes are slightly larger, sedimenting at about 80S. The behaviour of the particles depends on the concentration of magnesium ions; when this is lowered they dissociate into two subunits (depending on the precise ionic milieu but in general starting at about 1·5 mM Mg^{2+} for bacterial ribosomes). The two subunits of bacterial ribosomes sediment at about 50S and 30S; the mammalian subunits sediment at about 60S and 40S. The larger is roughly spherical and is about twice the size of the smaller, which is more asymmetrical (Tissieres et al., 1959).

When a short dose of radioactive amino acids is chased by normal amino acids, its passage can be followed through the various stages of its incorporation into protein. After injecting pulse doses of C^{14}-leucine intravenously into rats, Littlefield et al. (1955) removed each of the three lobes of the liver at various times after the injection and homogenised the preparation to yield fractions of ribosomes, membranes and soluble proteins. Within 2–3 minutes the label is incorporated into the ribosome fraction; it then leaves this fraction and can be recovered from the membranes. Over a longer period of time, a pulse label is found only in the soluble proteins of the cell sap. These results indicate that proteins are synthesised by ribosomes, after which they are passed by the endoplasmic reticulum with which the ribosomes are associated to the soluble proteins of the cytoplasm. Fewer stages are of course demanded in bacteria, where ribosomes do not seem to be attached to membranes and proteins are released into the cell upon synthesis.

One of the first ideas about the function of the ribosome was that it might contain the template specifying proteins. But the isolation of messenger RNA and the demonstration that ribosomes of one species may faithfully translate messengers from another have seen the development of the concept that ribosomes are part of the apparatus of the cell responsible for assembling amino acids into proteins under the direction of messenger templates. The ribosome and the proteins which associate with it constitutes a complex organelle which performs all of the many catalytic activities required for protein synthesis.

Although it is the 70S ribosome particle which is active in protein synthesis, ribosomes obtained from cells by isolation of the fraction associated with radioactive amino acids sediment rather more rapidly, generally in the range of 140–200S. The sedimenting unit is a complex of messenger RNA with ribosomes, known as the polyribosome (often abbreviated to polysome). The number of ribosomes attached to any particular copy of each messenger

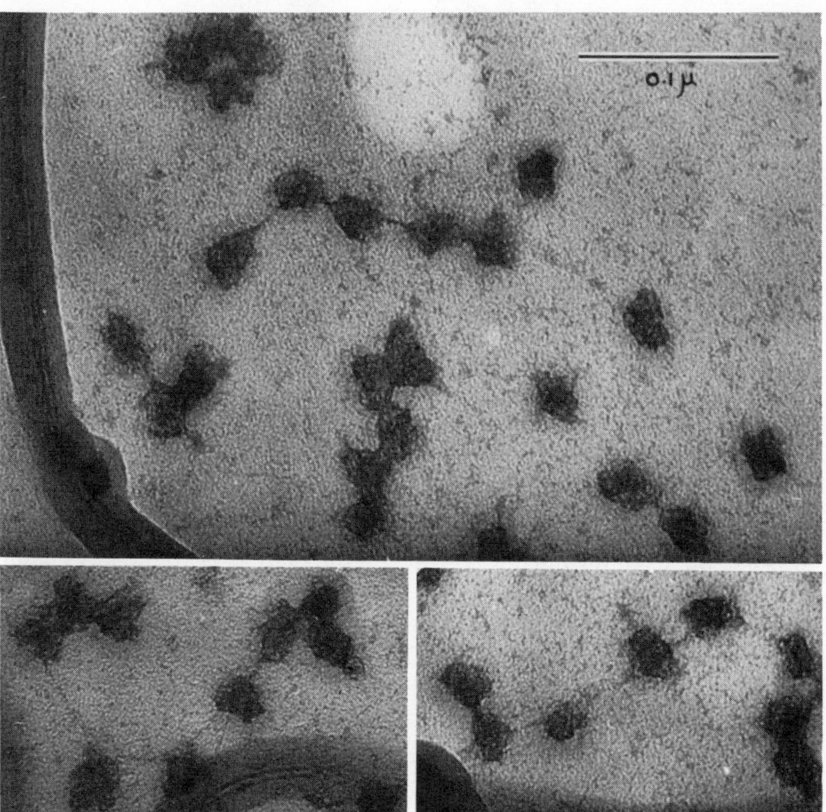

Figure 2.5: Electron micrograph of pentasomes synthesising haemoglobin in reticulocytes. Polyribosomes were stained with uranyl acetate and are shown at a magnification of ×268,000. The upper photograph shows three pentasomes in the form of a rosette, an extended linear array, and a compact linear array. The two lower photographs both show polysomes in an extended configuration. Photographs kindly provided by Dr. H. S. Slayter from the data of Slayter et al. (1963)

appears to be a matter of statistical probability rather than an absolute invariable, the average being characteristic for each messenger. The major polysomes of reticulocytes synthesizing haemoglobin possess five ribosomes, although some messengers have four and some have six (Warner et al., 1963). Figure 2.5 shows an electron micrograph of pentasomes. Bacterial messengers, however, may carry much larger numbers of ribosomes, to be measured in tens rather than single units (see page 384).

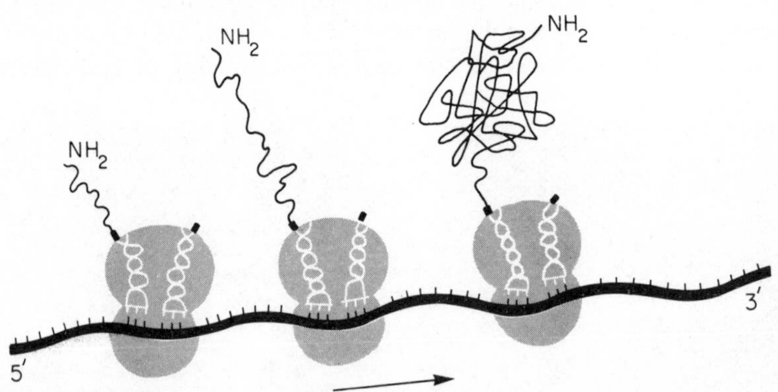

Figure 2.6: Ribosomes translate the messenger from 5′ to 3′ terminus. One messenger may be under translation by several ribosomes at a time; the average spacing between ribosomes in the haemoglobin system is 150 nucleotides but is probably much less in bacterial systems. As each ribosome progresses, the length of protein synthesized increases and it probably begins to take up its tertiary conformation before synthesis is complete.

Messenger RNA is always read in triplets from its 5′ end to the 3′ terminus, the same direction in which it is synthesized. Analysis of frameshift mutants has shown that the amino acid sequences in double mutants and wild type cells can be related only if the triplets are orientated so that their 5′ to 3′ polarity parallels the $-NH_2$ to $-COOH$ protein polarity (see page 23). The correspondence between sequences of RNA phages and the proteins for which they code also are consistent with this direction of translation.

As a ribosome moves towards the end of its messenger, the polypeptide chain progressively lengthens with each amino acid added. When the ribosome has moved away from the initiation sequence, another may attach and start synthesis of the next polypeptide chain. A messenger therefore bears a series of ribosomes, each carrying a successively greater length of the polypeptide chain under synthesis as illustrated in figure 2.6.

The conformation of proteins appears to be dictated solely by their sequence of amino acids; and a protein chain under synthesis on a ribosome probably begins to take up its active conformation as it is synthesized. By following the resistance of globin protein under synthesis on reticulocyte polysomes to degradation with proteolytic enzymes, Malkin and Rich (1967) showed that the most recently added 30–34 residues are protected by the structure of the ribosome but that the remainder of the protein is susceptible to attack. The free part of the protein may be able to start folding into its conformation. By following the reactivity of polysomes synthesising β-galactosidase with antiserum against the protein, Hamlin and Zabin (1972) showed that nascent proteins on the ribosomes possess sufficient conformation to generate immunological reaction.

Addition of Amino Acids to the Polypeptide Chain

Ribosomal Sites for Transfer RNA

Most experiments on the processes involved in protein synthesis as the ribosome translates a messenger have been performed with in vitro systems (for review see Lengyel and Soll, 1969). Crude cell-free extracts which can synthesize protein can be prepared by breaking cells, and removing unbroken cells and debris by low speed centrifugation and small molecules by dialysis. These systems can be directed either by their endogenous messenger species or by added natural or synthetic mRNA. Translation is assayed by following the incorporation of a radioactively labelled amino acid into protein. More purified systems can be prepared by a high speed centrifugation, after which tRNA molecules and the aminoacyl-tRNA synthetases are located in the supernatant and ribosomes are found in the pellet.

In addition to activities which are intrinsic functions of the ribosome, other necessary enzyme activities are present as supernatant protein factors which may become associated with ribosomes during protein synthesis. The factors required for chain elongation and termination are found in the supernatant of a high speed centrifugation and those involved in initiation are associated with the ribosomes in the pellet. Certain other components must also be present. GTP is essential and no other nucleotide can replace it; although it has been known for some time that there is a close relation between the hydrolysis of GTP and the incorporation of amino acids into protein, it is only recently that the details of how GTP is used have been understood. It is also necessary to add ATP and an ATP generating system, the divalent cation Mg^{2+} and the univalent cations K^+ and NH_4^+.

When the concentration of magnesium ions is reduced, the ribosomes of polysomes separate into subunits as do free ribosomes. Gilbert (1963) demonstrated that when this dissociation takes place in a poly-U directed system synthesising phenylalanine, the 30S subunit remains associated with the poly-U messenger whilst the 50S subunit retains the polypeptide. Oligopeptidyl-tRNAs of different chain lengths are found in such systems, which indicates that the growing polypeptide chain is attached to tRNA. This gives the picture shown in figure 2.6 of a ribosome engaged in protein synthesis as a 30S subunit attached to the messenger with transfer RNAs bearing incoming amino acids or the polypeptides under synthesis attached through the 50S subunit.

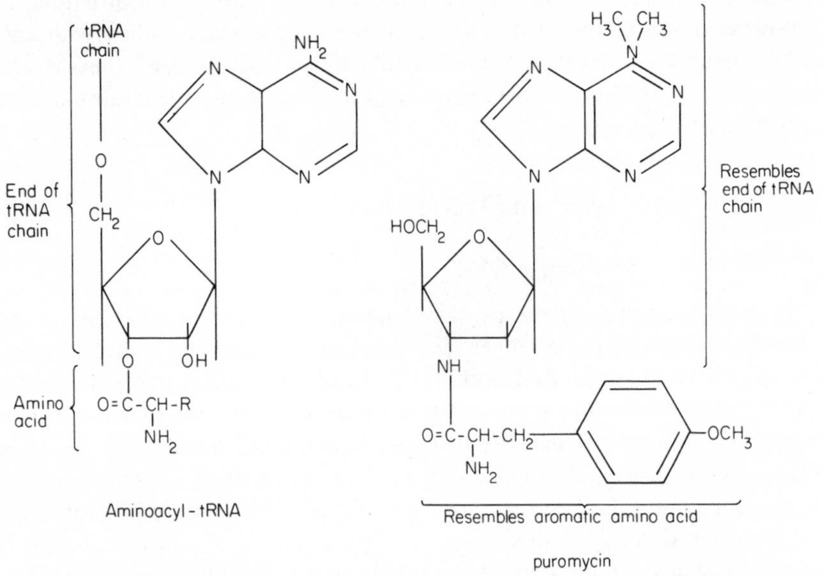

Figure 2.7: Aminoacyl-tRNA and its analogue puromycin

The number of molecules of transfer RNA attached to a ribosome at one time has been a matter of controversy, although it now seems that only two are usually involved. Using conditions in which ribosomes can only initiate protein synthesis and therefore cannot move along the RNA of phage Qβ, Roufa, Skogerson and Leder (1970) tested radioactively labelled aminoacyl-tRNAs for activity in binding to the ribosome–mRNA complex; only molecules representing the first two amino acids of the protein can bind. Tai and Davis (1972) showed that free ribosomes have a general affinity for tRNA, but that the transfer molecules are readily washed off; in contrast, ribosomes engaged in protein synthesis have two firmly bound tRNAs which cannot be washed off.

The two sites at which tRNAs can bind to the ribosome can be defined by the availability of bound aminoacyl-tRNAs to react with puromycin, an antibiotic inhibitor of protein synthesis whose reaction with polysomes synthesising protein has been used as a model system for polypeptide synthesis. Puromycin bears a chemical resemblance to an amino acid attached to the terminal adenosine of transfer RNA, shown in figure 2.7, and reacts with the ribosome as would an incoming aminoacyl-tRNA. The effect of the reaction is to release a nascent polypeptide as polypeptidyl-puromycin, thus terminating protein synthesis prematurely; this accounts for the inhibition which the antibiotic exerts. The puromycin reaction has similar requirements and is inhibited in the same manner as protein synthesis proper, justifying its use as a model system.

With aminoacyl-tRNA as substrate, the puromycin reaction depends partly on the presence of GTP and supernatant factors, the extent of dependence varying amongst different preparations of ribosomes. Traut and Monro (1964) accordingly suggested that the ribosome might exist in either of two states, depending on the stage of protein synthesis; conversion between these states demands GTP and supernatant factors. In one state, the amino acids attached to tRNA can react with puromycin; in the other they cannot.

Figure 2.8 depicts the two sites of the ribosome; the site which is entered by an incoming aminoacyl-tRNA—or its puromycin analogue—is known as the A site (sometimes called the entry site). When the A site contains a molecule of tRNA, puromycin cannot bind and the amino acids attached to the tRNA do not react with it. The ribosomal site from which the polypeptide part of peptidyl-tRNA (tRNA containing the protein chain so far synthesised) is donated is referred to as the P site (sometimes called the donor site). When the P site is occupied by aminoacyl- or peptidyl-tRNA, the A site is free to accept an incoming molecule of aminoacyl-tRNA or puromycin, so that the bound transfer RNA is now reactive.

Protein synthesis starts by a reaction between the aminoacyl-tRNAs bearing the first two amino acids of the protein; the carboxyl group of the first amino acid—whose tRNA is bound in the P site—forms a bond to the amino group of the second—whose tRNA is bound in the A site. This synthesises a dipeptide attached to the second tRNA. This dipeptide is in turn transferred to the third aminoacyl-tRNA, to generate a tripeptide. This sets the general pattern of protein synthesis; the polypeptide chain synthesised so far is always transferred from the tRNA to which it is attached (that of the last amino acid added to the chain) to the amino acid attached to the next tRNA. Donation of the polypeptide can only take place from a polypeptidyl-tRNA bound in the P site to an aminoacyl-tRNA bound in the A site.

But condensation between a polypeptide chain attached to tRNA in the P site and aminoacyl-tRNA in the A site generates a ribosome which has the peptide chain attached to the tRNA in the A site. Such a ribosome cannot accept another aminoacyl-tRNA until after a *translocation*, when it advances one triplet to move the peptidyl-tRNA into the P site and to move the next codon to be read into the A site. When an aminoacyl- or peptidyl-tRNA is bound in the A site, therefore, GTP and a protein factor must be supplied to assist its translocation into the reactive P site.

The ribosome therefore follows a cycle in peptide bond formation and the present version of the cycle originally suggested by Nishizuka and Lipmann

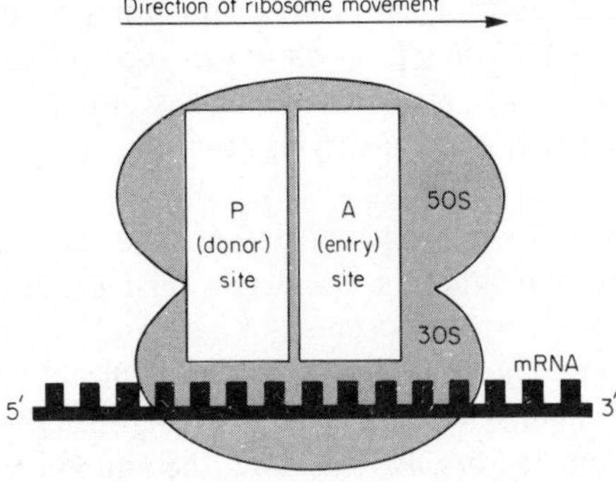

Aminoacyl-tRNA can only enter the A site on the right; peptides for bond formation can only be donated from the P site on the left.

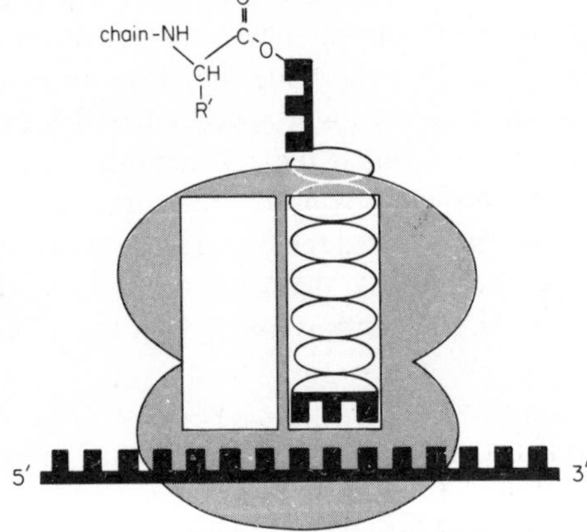

When charged tRNA is bound at the A site, it cannot react with puromycin

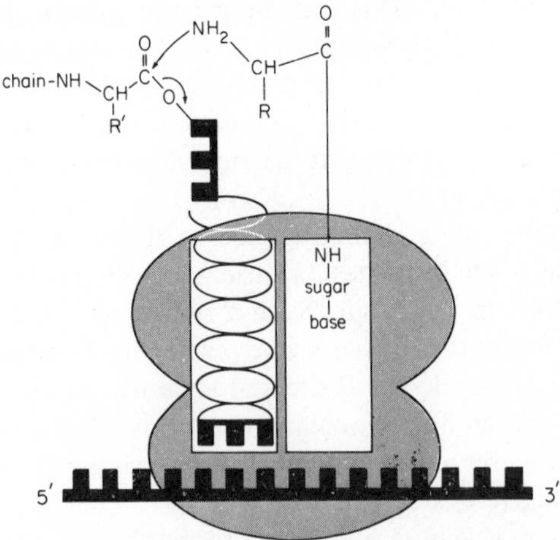

When charged tRNA is bound at the P site, puromycin can enter the A site and react to form peptidyl-puromycin

Figure 2.8: The two tRNA binding sites of the ribosome

in 1966 is shown in figure 2.9. When a ribosome has a peptidyl-tRNA in its P site, it is ready to accept an aminoacyl-tRNA into its A site, as shown in parts (a) and (d). After peptide bond formation, the now uncharged tRNA is expelled from the P site to give the ribosome in state (c), which has peptidyl-tRNA in its A site. Translocation, which is catalysed by GTP and a protein factor, restores the state of the ribosome at the beginning of the cycle; comparison of states (d) and (a) shows that they are identical except that the polypeptide is one amino acid longer and the ribosome is one codon further along the message. The progress of this cycle is accompanied by changes in the reactivity of peptidyl-tRNA to puromycin; ribosomes in states (a) or (d) are reactive, but ribosomes in states (b) or (c) must be incubated to allow a translocation first.

A model for ribosome action proposed by Bretscher (1968a) suggests that both the A and P tRNA binding sites can be envisaged as shown in figure 2.10 to consist of two subsites which are separated when the ribosome subunits dissociate. In a 70S ribosome, the A site therefore consists of the subsites 30a and 50a; and the P site comprises the association of 30p and 50p. In this case, the ribosome subunits might also be able to associate to create the hybrid structures 30a:50p or 30p:50a. This would allow translocation to take place in a two step movement in which one subunit of the ribosome "swings round" relative to the other in order to create an intermediate hybrid structure, after which the other subunit rotates to reform the normal A and P sites. Path 1 seems to be more likely because it would allow an earlier recognition of the approaching aminoacyl-tRNA and would also account more readily for chain initiation (see next chapter).

Formation of the Transfer Complex with Aminoacyl-tRNA

Although the ribosome is the cell organelle which synthesises proteins, it does not possess all the catalytic activities demanded for polypeptide synthesis and "supernatant factors" must associate with it during the cycle of protein synthesis to provide these additional functions. These factors have been isolated from a variety of systems, ranging from bacteria to mammalian cells. The factors needed for chain elongation in E.coli have been separated into three proteins by Lucas-Lenard and Lipmann (1967) and are known as the G factor, Ts factor and Tu factor. (The names for the latter two factors are related because in early preparations they were obtained as a single fraction, the T factor, which has since been split into a stable component, Ts, and an unstable component, Tu.)

Because Tu is heat labile, elongation factors have also been prepared from the bacterium Bacillus stearothermophilus, which lives at much higher temperatures than E.coli and can conduct protein synthesis at 70°C. Ono et al. (1969a) have found that three factors are required in this system also; and the use of heterologous reaction mixtures in which these factors replace the E.coli

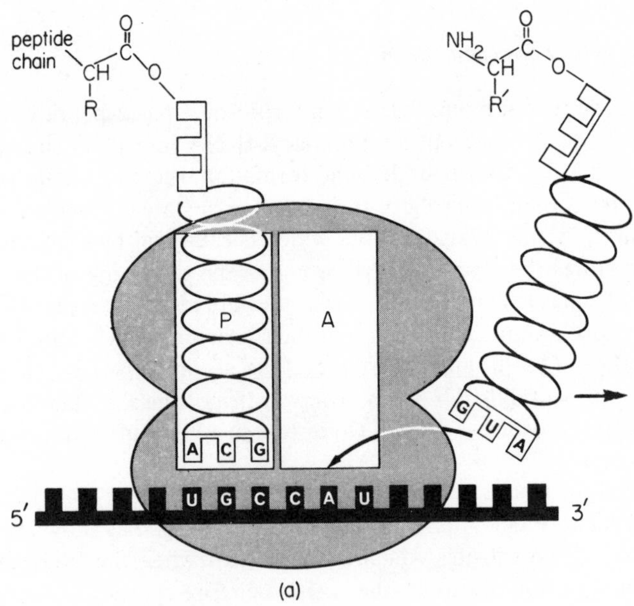

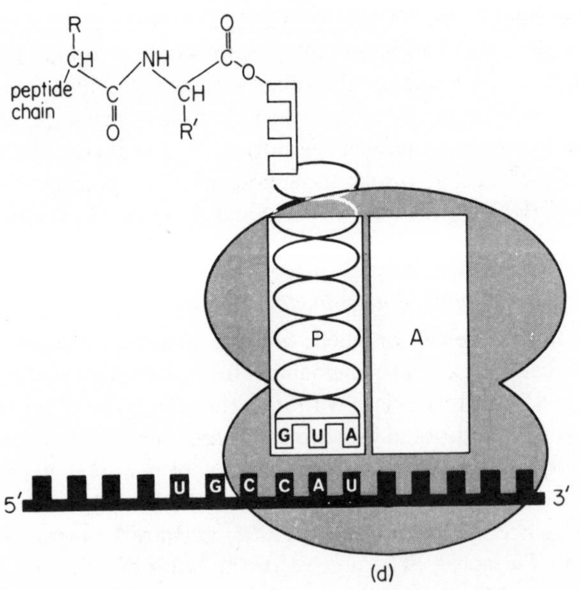

Figure 2.9: The ribosome cycle of peptide bond synthesis (a) The ribosome P site is filled by a tRNA carrying the polypeptide chain synthesised so far. The aminoacyl-tRNA corresponding to the next codon is entering the A site; this binding requires the T factors and GTP. (b) The ribosome carries peptidyl-tRNA in its P site and aminoacyl-tRNA in its A site. Pepetide bond synthesis is accomplished by transfer of the polypeptide chain attached to the tRNA in the P site to the aminoacyl-tRNA in the A site. This reaction is catalysed by an enzyme—peptidyl transferase—

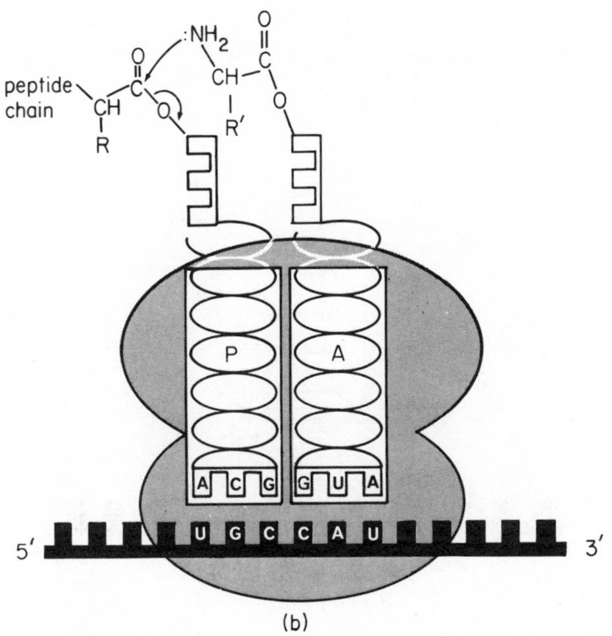

(b)

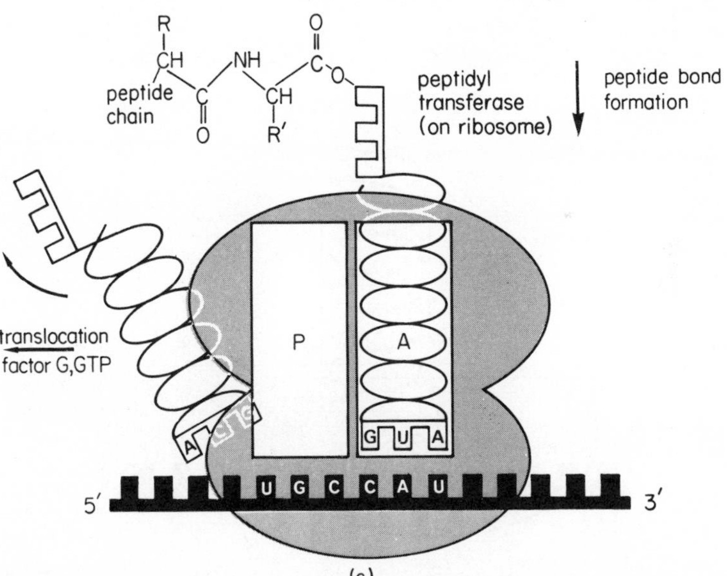

(c)

which is part of the 50S subunit. (c) Peptide bond synthesis generates peptidyl-tRNA in the A site. Translocation to move the peptidyl-tRNA into the P site requires G factor and GTP. (d) After translocation, the ribosome is restored to the state prevailing at the beginning of the cycle—it has peptidyl-tRNA in its P site and can accept the next aminoacyl-tRNA into its A site. But the polypeptide chain is one amino acid longer and the ribosomes is one triplet further along the message

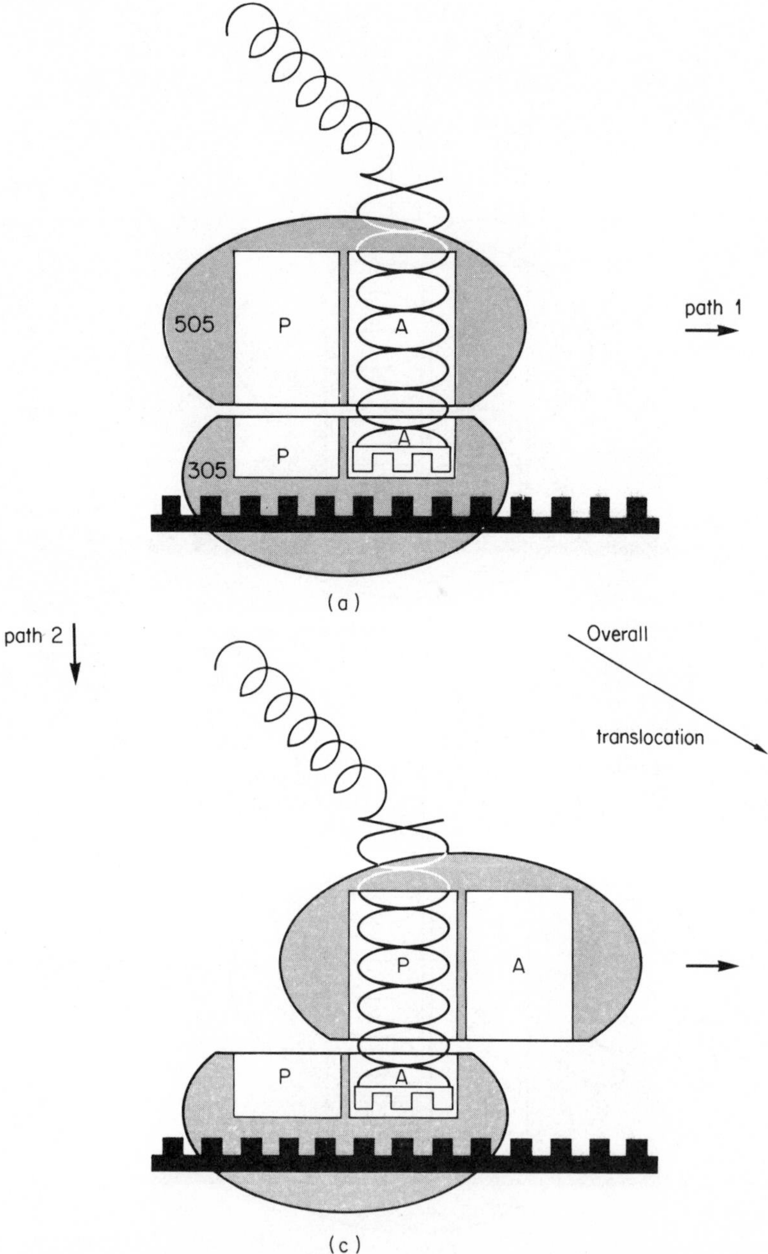

Figure 2.10: The hybrid site model for translocation of the ribosome in protein synthesis. The A and P sites each consist of subsites located on the 50S and 30S ribosome subunits. Translocation of a polypeptidyl-tRNA from the A site to the P site takes place by a two stage reaction in which one of the subunits first moves relative to the other. Path 2: the 50S subunit moves first, giving a structure in which the polypeptidyl tRNA is located in the hybrid 50p:30a site shown in part (c): movement of the 30S subunit

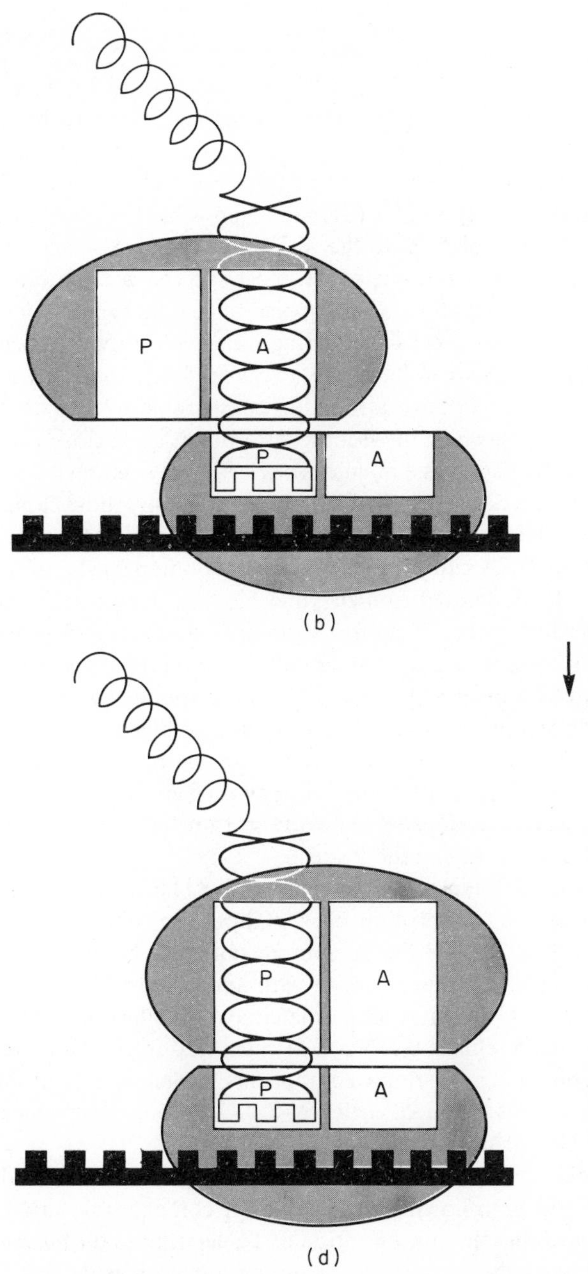

follows. Path 1: movement of the 30S subunit one triplet along the messenger creates hybrid structure (b) where the polypeptidyl-tRNA is located a site comprising 50a:30p. Translocation is completed when the larger subunit moves to join the 30S subunit. This path has the advantage that the next incoming aminoacyl-tRNA could be recognised by hybrid (b) before completion of translocation. After Bretscher (1968a)

factors has shown that S1, S3 and S2 are equivalent to Ts, Tu and G respectively. The synthesis of these factors in the cell seems to be linked to the production of ribosomes (see also page 137), probably on the basis of one molecule of factor for every ribosome (Gordon and Weissbach, 1970).

The factors Tu (S3) and Ts (S1) are responsible for transferring incoming aminoacyl-tRNA molecules to the ribosome. The G (S2) factor promotes the translocation of the ribosome which must succeed peptide bond formation. The functions of elongation factors appear to be rather similar in most systems, although mammalian cells—including rat liver, rabbit reticulocytes and human tonsils—appear to have only two transfer factors, EF1 and EF2. The first of these seems to have a function comparable to that of the combined T factor (Tu + Ts), whilst the second corresponds quite closely to the bacterial G factor. Protein synthesis in eucaryotic cells has been reviewed by Moldave (1972). However, despite similarities of function, only those elongation factors derived from closely related organisms can function in a heterologous system.

Aminoacyl-tRNA can bind to ribosomes spontaneously, in the absence of T factors—this is termed non-enzymic binding. However, in conditions of enzymic binding—when T factor is present—aminoacyl-tRNA enters the A site of the ribosome and can do so only when the P site is already occupied. Springer and Grunberg-Manago (1972) have found that binding conditions are less stringent in the absence of T factors, as judged by the reactivity of the bound aminoacyl-tRNA to puromycin. The control exerted by the T factors on entry to the A site within the cell may thus ensure that aminoacyl-tRNA binds only when the ribosome is ready to transfer a polypeptide from polypeptidyl-tRNA occupying the P site.

The combined T factor preparation binds GTP to form a complex which can be detected by adsorption on a millipore membrane filter; subsequent work has shown that it is the Tu component which binds the nucleotide. The Tu–GTP complex can in turn react with aminoacyl-tRNA to form a further complex which Gordon (1968) isolated by gel filtration. The addition of aminoacyl-tRNA elutes the Tu–GTP complex from millipore filters as a complex containing all three components which is not absorbed by the filters. It is this aminoacyl-tRNA–Tu–GTP complex which places the aminoacyl-tRNA on the ribosome.

The amount of binding of GTP by Tu is considerably enhanced by the addition of Ts and by following the kinetics of complex formation, Ertel et al. (1968) showed that the concentration of Tu determines the *amount* of complex formed whereas the quantity of Ts present determines its *rate* of formation. This implies that the action of Ts is to catalyse formation of the complex between Tu and GTP.

Although this interpretation is correct in outline, the Tu–GTP complex is not in fact formed by direct reaction between Tu and GTP. The commercial preparations of GTP used to form complexes with Tu contained appreciable quantities of GDP as a contaminant; and Lockwood, Hattman and Maitra (1969) found that H^3-GTP freed of GDP by passage through a Dowex column does not bind to Tu very efficiently; H^3-GDP binds far more effectively. The GTP which appears to bind in purified preparations seems to do so only as the result of its hydrolysis to GDP.

The complex of Tu-GDP is inactive in protein synthesis, however, and must first be converted to Tu-GTP by the action of the factor Ts, whose role in protein synthesis is to promote this interchange. The effect of Ts therefore depends upon the form in which Tu is present. The complex of Tu–GDP cannot bind aminoacyl-tRNA, but can do so when Ts and GTP are added. But Ts is no longer needed if the Tu–GDP complex is converted before reaction to Tu-GTP by treatment with phosphoenolpyruvate and pyruvate kinase, for the Tu–GTP complex can interact rapidly with aminoacyl-tRNA. Waterson,

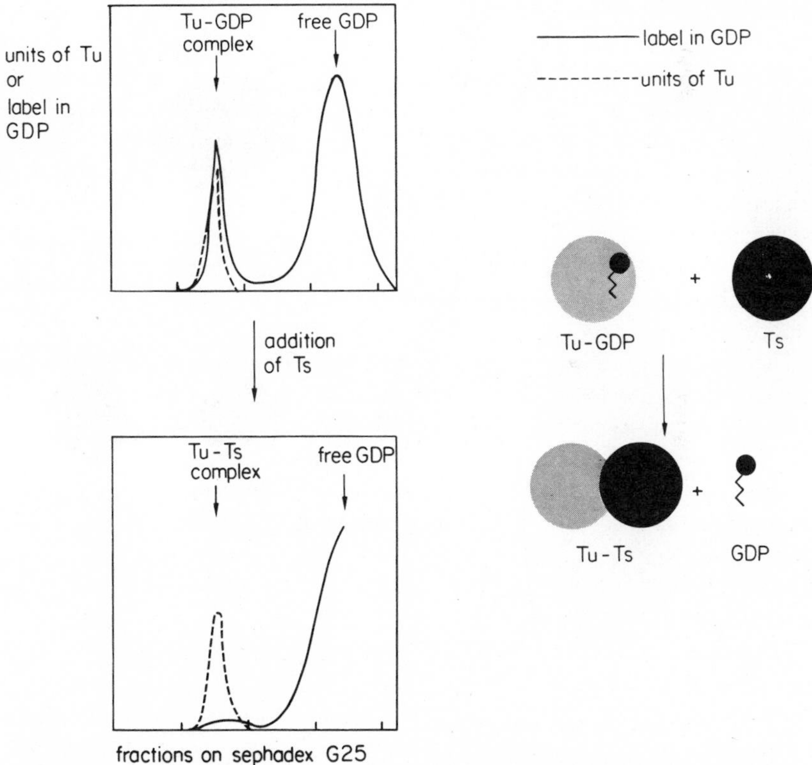

Figure 2.11: Addition of TS to Tu–GDP complex displaces GDP. Upper: in the absence of TS some GDP is present in free form and some elutes from column as complex with Tu. Lower: addition of Ts displaces all the GDP from its complex with Tu so that it elutes at the free position. Data of Miller and Weissbach (1970)

Beaud and Lengyel (1970) reported that S1 (Ts) does not increase the rate of protein synthesis when S3 (Tu) is complexed with GTP, which implies that this conversion is the only role of Ts in protein synthesis.

The overall reaction of Ts is therefore to promote exchange of Tu–GDP with free GTP to generate Tu–GTP. Miller and Weissbach (1970a) found that when Ts is added to Tu–H^3-GDP, the radioactivity no longer elutes from a

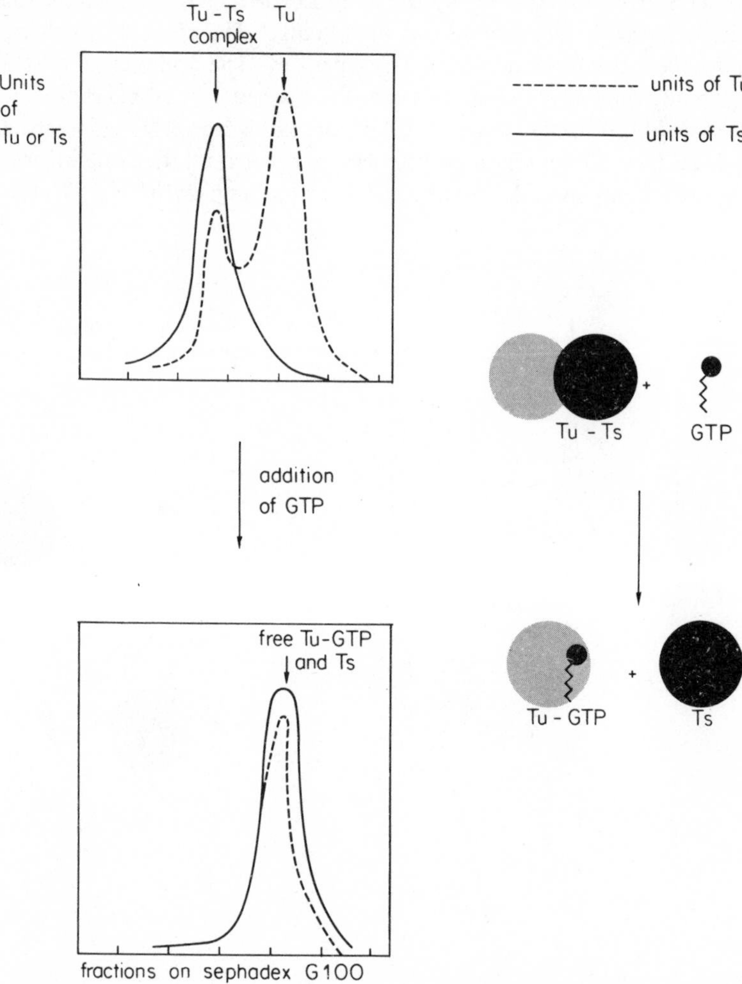

Figure 2.12: GTP displaces Ts from Ts–Tu complex. Upper: all the Ts is in the form of a complex which elutes rapidly. There is some free Tu. Lower: addition of GTP releases Ts from Tu so that rapidly eluting complex disappears; the two proteins are no longer bound to each other and elute at a lighter position from the column (both happen to elute at about the same position when free). Data of Weissbach et al. (1969)

sephadex column together with the protein but is released as shown in figure 2.11. The reaction is stoichiometric, the amount of GDP displaced being (about) equal to the amount of Ts added. Lucas-Lenard, Tao and Haenni (1969) observed that Tu and Ts cannot remain associated as a Tu–Ts complex when GTP is present; as figure 2.12 shows, the addition of GTP displaces Ts from its complex with Tu to form the Tu–GTP complex. This complex can then react with aminoacyl-tRNA to form the ternary complex which transfers its aminoacyl-tRNA moiety to the ribosome. This complex also can be detected by gel filtration as shown in figure 2.13. The reaction sequence which

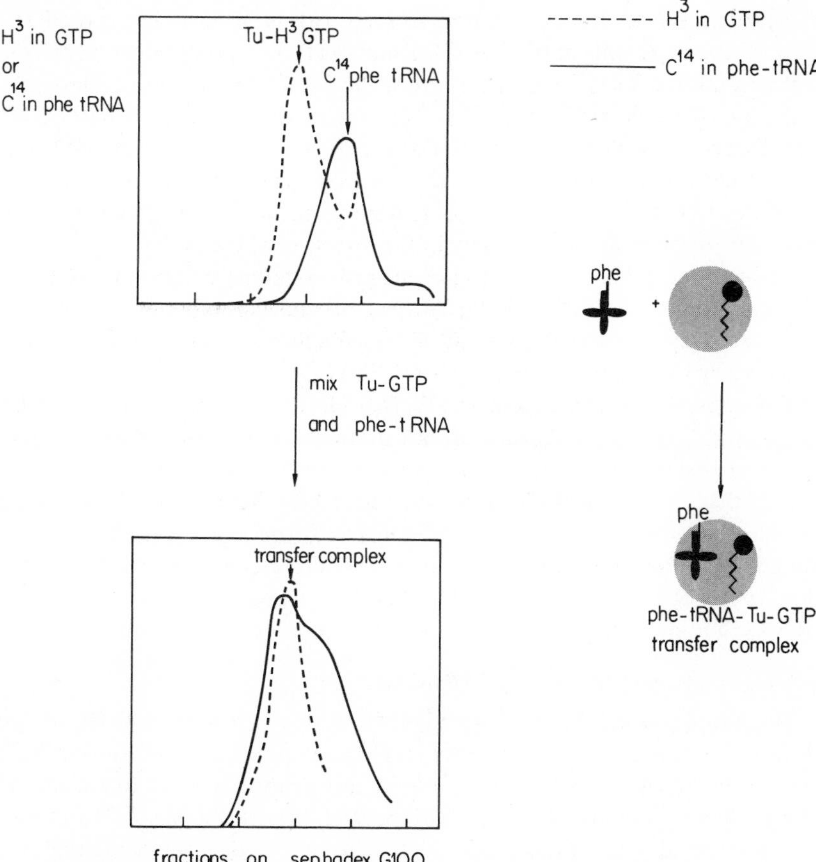

Figure 2.13: Tu–GTP binds aminoacyl-tRNA. Upper: individual elution patterns of Tu–GTP (measured by H^3 label in GTP) and phe-tRNA (measured by C^{14} in amino acid moiety) superimposed on same axes. Lower: elution pattern after mixing Tu–GTP with phe-tRNA. The aminoacyl-tRNA forms a transfer complex and elutes together with the Tu-GTP instead of at its previous position. Data of Weissbach et al. (1969)

prepares aminoacyl-tRNA for binding to the ribosome can therefore be represented as:

$$\text{Ts} + \text{Tu–GDP} \rightleftharpoons \text{Ts–Tu} + \text{GDP} \qquad \text{(figure 2.11)}$$

$$\text{Ts–Tu} + \text{GTP} \rightleftharpoons \text{Tu–GTP} + \text{Ts} \qquad \text{(figure 2.12)}$$

$$\text{aminoacyl-tRNA} + \text{Tu–GTP} \rightarrow \text{aminoacyl-tRNA–Tu–GTP} \qquad \text{(figure 2.13)}$$

When the GTP moiety of the aminoacyl-tRNA–Tu–GTP complex is cleaved at a later stage in protein synthesis (see below) the factor is released in the form of a Tu–GDP complex. Tu does not exist, therefore, in the form of a protein species free to combine as such with GDP or GTP; it oscillates between the two forms of binding GDP and GTP. The catalytic action of Ts in regenerating the Tu–GTP complex from the Tu–GDP complex achieves the cyclic use of Tu shown in figure 2.16. The interactions between Tu and GDP and Ts are reversible in vitro, but the reaction with aminoacyl-tRNA is irreversible and it is at the point of the third reaction above that the molecule is omitted to entering the cycle of protein synthesis. In the cell, of course, these reactions presumably proceed only in the direction of the cycle.

Tu has been prepared in a homogeneous crystalline condition and has a molecular weight of 42,000–44,000 daltons, this molecule binding one molecule of GDP. Ts has a molecular weight of 28,500 daltons and one cysteine which is necessary for catalytic activity. Using NEM (N-ethyl-maleimide) to inhibit –SH groups, Miller, Hachmann and Weissbach (1971) found that the ability of Tu to bind GDP can be destroyed, but that the complex of Tu–GDP is protected against this reaction. Ts also confers protection when added to Tu. This implies that Ts reacts with Tu at the same site where GDP is bound and utilises the same sulfhydryl group. The (different) sulfhydryl group needed to bind aminoacyl-tRNA is not protected in the Tu–GDP complex, which suggests that Tu has one binding site for GDP, GTP or Ts and another for aminoacyl-tRNA. Binding at the first site controls the activity of the second site.

Binding Aminoacyl-tRNA to the Ribosome

The complex which Tu–GTP (or S3–GTP) forms with aminoacyl-tRNA was shown by Lucas-Lenard and Haenni (1968) and Ono et al. (1969a) to transfer aminoacyl-tRNA to the ribosome; and by testing the reactivity to puromycin of phe-tRNA bound to ribosomes with poly-U, Weissbach, Redfield and Brot (1971a) were able to confirm that most, if not all, of the aminoacyl-tRNA is bound to the proper A site for entry. When the aminoacyl-tRNA–Tu-GTP moiety binds to ribosomes, the GTP is cleaved to yield phosphate and a Tu-GDP complex.

The role of GTP in protein synthesis has been studied by substituting for it an analogue which cannot undergo hydrolysis; GMP-PCP (5′ guanylyl-methylene-diphosphonate) has the structure shown in figure 2.14 in which a methylene bridge replaces the oxygen which links the β and γ phosphates in GTP. GMP-PCP can replace GTP in promoting the T factor dependent binding of aminoacyl-tRNA to ribosomes, but this substitution does not allow peptide bond formation to take place. Although the *presence* of GTP is needed for binding to take place, therefore, its *hydrolysis* is not required until some later point, either prior to or concomitant with peptide bond synthesis (see Ono et al., 1969b).

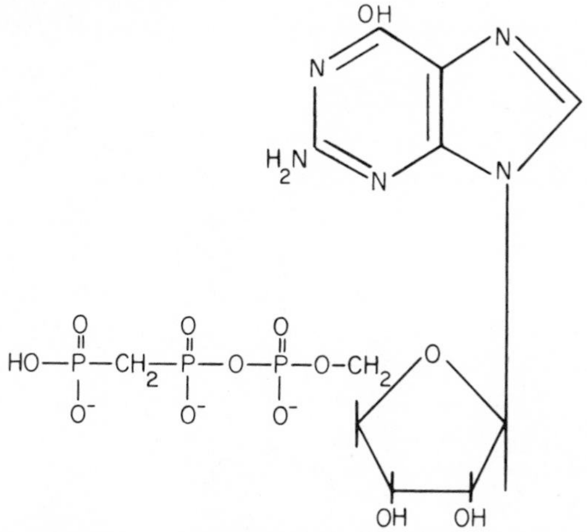

Figure 2.14: The GTP analogue 5′-guanylylmethylene disphosphomate (GMP-PCP)

The amount of inorganic phosphate liberated when GTP is cleaved closely parallels the amount of C^{14}-phe-tRNA bound to ribosomes. Quantitative experiments with poly-U directed poly-phenylalanine synthesis have shown that aminoacyl-tRNA and GTP are present in equimolar amounts in the complex with S3 (Tu) and that the GTP molecule cleaved by the factor is that present in the complex. One GTP molecule is therefore cleaved for every aminoacyl-tRNA molecule bound to the ribosome.

When the transfer complex of aminoacyl-tRNA–Tu–GTP is allowed to react with ribosomes, a C^{14} label in the amino acid and an H^3 label in GTP are retained on nitrocellulose filters, but a γP^{32} label in GTP is lost. As figure 2.15 shows, this implies that the GTP has been hydrolysed to release inorganic phosphate. The Tu–GDP complex which is produced by the hydrolysis is also retained on the filter, however, so these experiments do not show whether it is released from the ribosome or remains attached to it after aminoacyl-tRNA has been inserted at the A site. Collecting ribosomes by centrifugation after they have reacted with the complex of aminoacyl-tRNA–Tu–GTP shows that the hydrolysis of GTP is indeed accompanied by release from the ribosome of a complex of Tu–GDP; most of an H^3 label in GTP is found complexed with

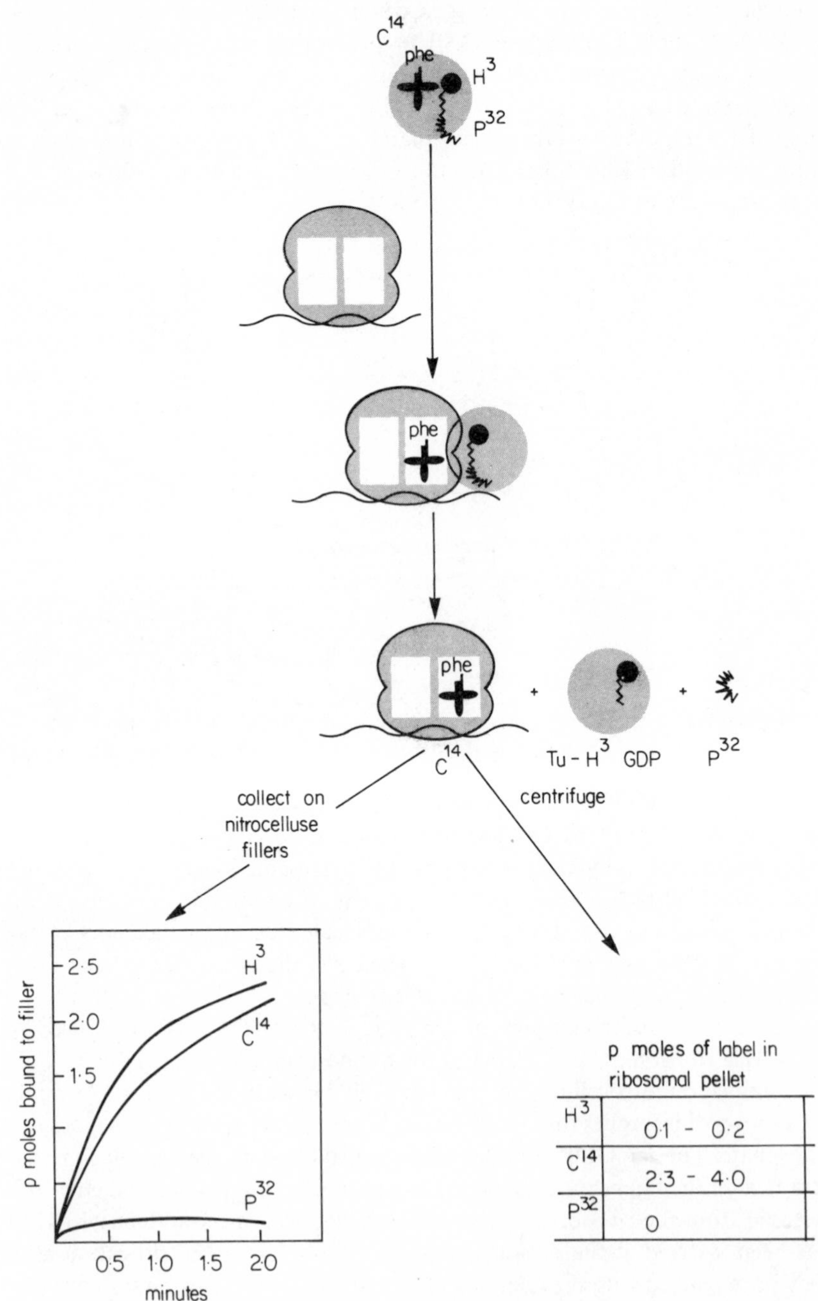
C14
phe
H3
P32
phe
phe
+
+
C14
Tu - H3 GDP
P32
collect on
nitrocelluse
fillers
centrifuge
p moles bound to filler
2·5
2·0
1·5
H3
C14
P32
0·5
1·0
1·5
2·0
minutes
p moles of label in
ribosomal pellet
H3
0·1 - 0·2
C14
2·3 - 4·0
P32
0

Tu in the supernatant after centrifugation, but none of a P^{32} label in γ phosphate is associated with the protein (Shorey et al., 1969; Skoultchi et al., 1969). The release of P^{32} from the ribosomes collected on nitrocellulose filters is therefore often used as a test for the binding reaction. The complex of Tu–GDP released from the ribosome after cleavage of the high energy bond then provides the substrate which Ts uses to regenerate Tu–GTP for binding further aminoacyl-tRNAs to ribosomes.

An intermediate complex of aminoacyl-tRNA–Tu–GTP bound to ribosomes cannot be identified when GTP is used to promote binding, because hydrolysis takes place rapidly and is accompanied by release of Tu–GDP and phosphate. But after binding N-acetyl-phe-tRNA—which mimics peptidyl-tRNA because of its blocked amino group—to ribosomes directed by poly-U, Skoultchi et al. (1969) found that when GMP-PCP is used to promote binding of a second species, phe-tRNA which enters the A site, S3 (Tu) is present on ribosomes isolated by gel filtration; the factor is released in the form of Tu–GDP when GDP is added, presumably because the proper nucleotide can displace its analogue. Figure 2.16 shows the cycle of the T elongation factors in binding aminoacyl-tRNA to ribosomes.

Peptide Bond Synthesis

The synthesis of peptide bonds is catalysed by an enzyme—peptidyl transferase—which is part of the structure of the 50S subunit of the ribosome. Traut and Monro (1964) found that the reaction of polypeptidyl-tRNA with puromycin can take place on the 50S subunit alone, with no requirement for the 30S subunit, mRNA, GTP or supernatant factors. (Polypeptidyl-tRNA appears to enter the P site of the ribosome directly in vitro; aminoacyl-tRNA cannot be used for this reaction because it would be located in the A site.)

In the presence of ethanol or methanol a more extensive reaction can take place on 50S subunits alone. Monro et al. (1969) showed that the formylmethionyl-tRNA initiator species—which enters the P site because it has a blocked amino group—can form a peptide bond with another aminoacyl-tRNA. The peptidyl-tRNA product is active as a donor for further reaction, so that in the presence of a mixture of charged tRNA molecules, recycling takes place to form a variety of di-, tri-, and tetrapeptidyl-tRNA. The amino

---

Figure 2.15: Binding of aminoacyl-tRNA–Tu–GTP to the ribosome. Nitrocellulose filters retain both aminoacyl-tRNA—ribosome and Tu-GDP released upon binding; the filter therefore retains C^{14} label of phe-tRNA and H^3 label of Tu–GDP. Hydrolysis of GTP to GDP upon binding is revealed by loss of γP^{32} label. Centrifugation separates the ribosomes (as a pellet) from the Tu–GDP complex; they therefore retain only the C^{14} label, showing that Tu–GDP must be released from ribosome when aminoacyl-tRNA binds. Data of Brot, Redfield and Weissbach (1970)

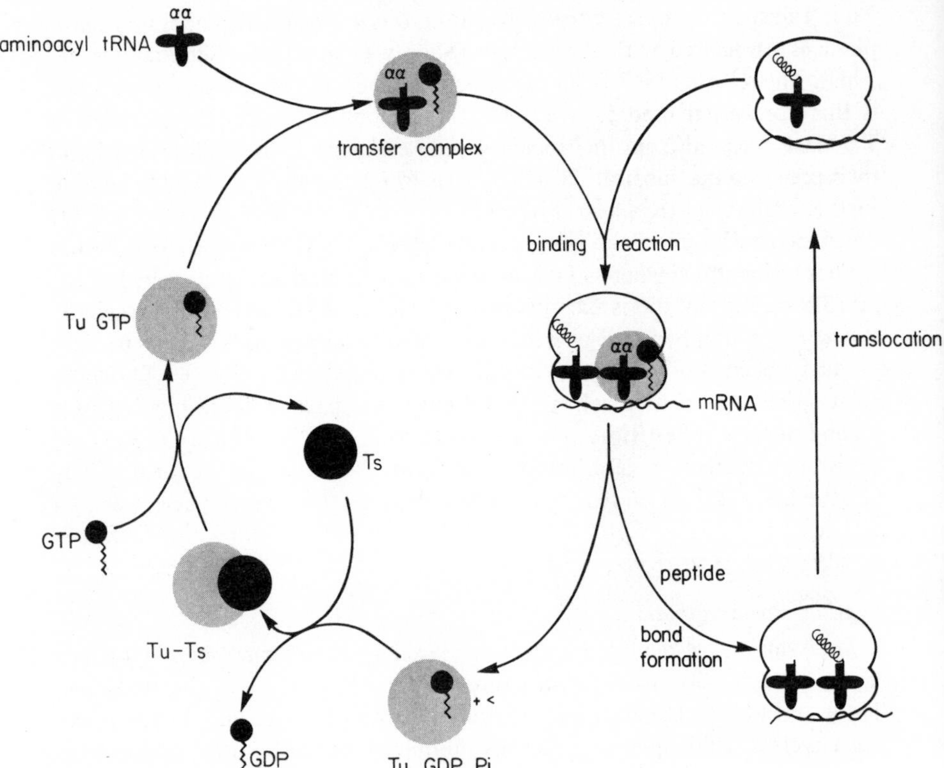

Figure 2.16: the cycle of factor dependent binding of aminoacyl-tRNA to ribosomes

acid sequence synthesised seems to be random and is probably governed by the diffusion of aminoacyl-tRNA molecules to the large subunits and not by direction of a template. Mg^{2+} and K^{+} are needed for the reaction and inhibitors of protein synthesis exert their usual effect; this suggests that peptide bonds are formed in the usual way and supports the use of the alcohol reaction as a model system.

An extension of this reaction is to use fragments of aminoacyl-tRNA as substrates. Variously sized fragments of fmet-$tRNA_f$ can react with puromycin on 50S subunits in the presence of methanol and fmet-ACC and fmet-ACCAAC are each about half as active as the intact molecule. Leu-ACC is much less effective as a donor from the P site than acetyl-leu-ACC, which suggests that a blocked amino group is required. At the entry site, so small a fragment as CA-gly can replace puromycin. Ribosome cores from which all the acidic proteins and some of the basic proteins have been removed by centrifugation in CsCl are also active in the fragment reaction (although they are inactive in protein synthesis). Further loss of proteins inactivates the peptidyl transferase activity and at the same time the ability of the subunits to bind chloramphenicol or lincomycin (antibiotics which inhibit peptide bond synthesis) is lost.

This suggests that there is a localised site within the 50S subunit of the ribosome where the terminal ends of peptidyl-tRNA and aminoacyl-tRNA are recognised and peptide bond formation is promoted. A similar reaction has been found by Vazquez et al. (1969) to take place with the 60S subunits of 80S ribosomes of yeast or human tonsils, which implies that eucaryotic ribosomes also have a catalytic centre on the large ribosome subunit which can be activated by alcohol. Usually, of course, 50S subunits are inactive in the absence of the other components of the synthetic apparatus; indeed, this must be a necessary control in the cell since any independent and undirected catalytic activity would presumably result in random polymerization of free aminoacyl-tRNA molecules.

Movement of the Ribosome along the Message

Role of the G factor Translocase

A ribosome must move three nucleotides further along its messenger after every cycle of peptide bond formation when the peptide attached to the tRNA in the P site has been transferred by peptidyl transferase to the aminoacyl-tRNA in the A site. A ribosome which is ready for translocation therefore carries discharged tRNA in its P site and peptidyl-tRNA in its A site, as shown in figure 2.9. Translocation expels the discharged tRNA from the P site and replaces it with the new peptidyl-tRNA which was in the A site; this frees the A site for entry of the next aminoacyl-tRNA. These concerted reactions are all mediated by the action of the G factor—sometimes termed the translocase protein—and depend upon hydrolysis of a GTP molecule.

The processes involved in ribosome movement have been studied by comparing the conditions needed to form peptide bonds when translocation is necessary with those which will suffice when one aminoacyl-tRNA is introduced directly into the P site so that the need for translocation is obviated. The more stringent conditions of the former situation can only be caused by the need for translocation. Using a poly-U template, Pestka (1968, 1969) compared the conditions required to form diphenylalanine and triphenylalanine. The dipeptide can be formed without addition of either G factor or GTP, whereas both are essential for addition of the third residue. This suggests that one phe-tRNA mistakenly enters the P site directly and a second then enters the A site; so a dipeptide can be formed without translocation. But the phe-phe-tRNA must then be translocated from the A site to the P site before phe_3-tRNA can be produced.

In similar experiments using N-acetyl-phe-tRNA as a peptidyl-tRNA analogue in the P site, Haenni and Lucas-Lenard (1968) showed that G factor is not needed to add one phenylalanine residue to start the chain, but is needed for further extension. And by using a natural phage RNA messenger, Erbe, Nau and Leder (1969) found that the first peptide bond can be formed between the initiator aminoacyl-tRNA—which enters the P site directly—and the next aminoacyl-tRNA without mediation of G factor; but translocation of the product from A site to P site must precede any further peptide bond synthesis.

An antibody to purified factor G protein was prepared by Leder, Skogerson and Roufa (1969) and the anti-translocase specifically inhibits the action of factor G in protein synthesis. The antibody has no effect upon formation of the first peptide bond between an aminoacyl-tRNA species in the P site and one in the A site, but inhibits any further peptide bond synthesis as judged by the availability of peptides to puromycin. The inhibitory effect of the anti-translocase can be overcome by adding increasing amounts of translocase enzyme. E.coli mutants in factor G have been isolated and Felicetti et al. (1969) reported that ribosomes from a strain which is temperature sensitive for protein synthesis can form the first bond in protein synthesis, but cannot form another unless G factor from wild type cells is added. Translocation therefore seems to be the only stage of protein synthesis in which G factor is implicated.

The translation of defined messengers in vitro shows directly that the translocation step catalysed by the G factor moves the ribosomes three nucleotides. The synthetic messenger $AUGGU_{30}$—which directs synthesis of one polypeptide starting from the AUG codon—is useful in experiments of this kind because its poly-U component is amenable to recovery and sizing. Figure 2.17 shows the experiments performed by Thach and Thach (1971). The first two aminoacyl-tRNAs can be bound to the ribosome to form the complex which undertakes peptide bond formation; translocation can take place only after G factor and GTP are added. After treating these pre- and post-translocation complexes with pancreatic ribonuclease, the part of the mRNA covered by the ribosome is protected from degradation and can be recovered.

When the protected segment of RNA is treated with T1 ribonuclease, the AUGG sequence is cleaved to leave a poly-U sequence. The length of the poly-U can be determined by chromatography; as the figure shows, it is three residues longer after translocation has taken place. The length of the messenger protected from the P site to its entry into the ribosome is thus some 16 nucleotides. In the complex containing the first two aminoacyl-tRNAs—but which lacks G factor and GTP and is therefore unable to undertake translocation—the protected mRNA is therefore $AUGGU_{11-12}$. When G factor and GTP are added, translocation moves the ribosome one codon further so that the protected length is $AUGU_{14-15}$. Similar experiments have been performed by Gupta et al. (1971a) with the RNA of phage f2 as template; the addition of one amino acid to the protein synthesised in vitro is coordinated with the

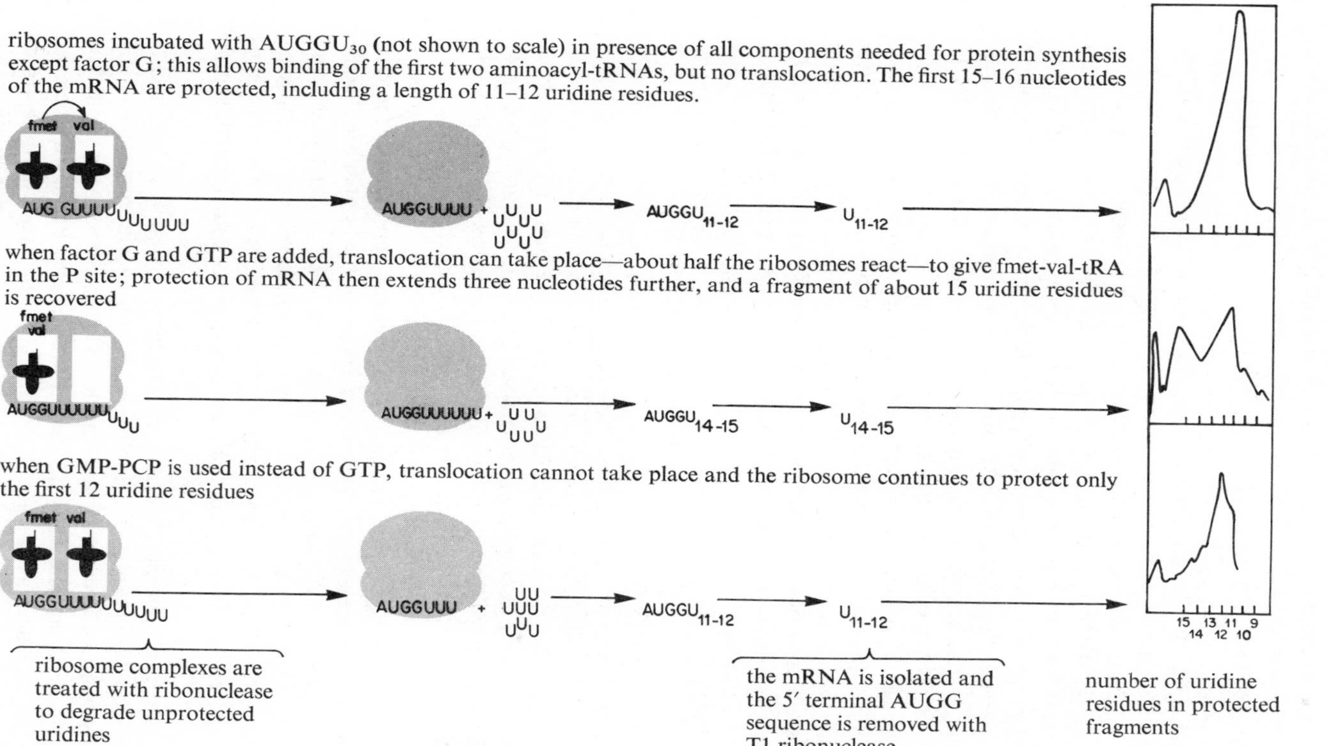

Figure 2.17: Translocation requires factor G and moves the ribosome three nucleotides. The ribosome protects a length of about 15 residues from the P site to the entry point of mRNA. Upper: pretranslocation complexes protect U_{11-12}. Centre: post-translocation complexes protect U_{14-15}. Lower: complexes unable to undertake translocation because GMP-PCP has replaced GTP protect the U_{11-12} length. The ribosome therefore moves three nucleotides for every translocation event and requires factor G and GTP for this action. Data of Thach and Thach (1971)

addition of three nucleotides to the length of the mRNA protected by the ribosome.

Removal of free tRNA from the P site of the ribosome after peptide bond formation appears to be tightly coupled to the translocation step. Lucas-Lenard and Haenni (1969) found that when ribosomes carrying acetyl–C^{14}-phe–H^3-tRNA in the P site are incubated with unlabelled phe-tRNA in the presence of factor G and GTP, reaction takes place to yield ribosomes carrying both H^3-tRNA and acetyl-C^{14}-phe–phe–tRNA. However, the dipeptidyl species cannot react with puromycin; this suggests that although peptide bond formation has occured through transfer of acetyl-C^{14}-phe to the second (unlabelled) phe-tRNA, the acetyl-phe-phe-rRNA product remains in the A site. But when these ribosomes are incubated with factor G and GTP, H^3-tRNA is released stoichiometrically with the amount of C^{14}-acetyl-phe-phe-tRNA which becomes responsive to puromycin. Departure of discharged tRNA from the P site is therefore part of the same translocation reaction which moves the new peptidyl-tRNA from A site to P site. The action of factor G alone may not be sufficient to release discharged tRNA from the P site however, for there are some indications that another protein factor may be needed to help expel tRNA (Ishitsuka and Kaji, 1970; Rudland and Klemperer, 1971).

The GTP Hydrolysis Site of the Ribosome

Factor G is not a part of the intrinsic structure of the ribosome particle but—like the Tu factor—follows a cycle in which it associates with the ribosome to exercise its catalytic function and then dissociates. The G factor has a weight of about 72,000 daltons and comprises more than 2% of the soluble protein of bacteria in exponential growth—a considerable proportion. Its synthesis in E.coli is linked to that of at least some ribosomal proteins (see page 137) so that there is probably one copy of factor G for every ribosome; protein synthesis in cell free systems directed by poly-U depends upon the amount of translocase added, up to a saturation level corresponding to about one molelcule of factor per ribosome. Although the presence of GTP is needed to bind factor G to ribosomes, its hydrolysis is not required since it can be replaced by its analogue GMP-PCP. Translocation cannot take place after the substitution, however, which suggests that the cleavage is needed for ribosome movement.

The EF2 factor of mammalian cells seems to be very similar to factor G in its function (see Moldave et al., 1969); it too has a ribosome-dependent GTPase activity and is needed for the translocation of ribosomes as tested by the puromycin reaction (Hardesty, Culp and McKeehan, 1969). Fusidic acid, which inhibits the action of the bacterial G factor, also inhibits translocation of mammalian ribosomes, although less effectively; this suggests that the complex between EF2 and the 60S subunit may be similar to that between factor G and the 50S subunit, although the association of factor and subunit may be less stable in the mammalian cell. Gill and Dinius (1973) have used the reaction of EF2 with diphtheria toxin—the toxin inactivates the factor by catalysing the attachment to it of an ADP-ribosyl group—to show that the cellular content varies; but there is in general about 1·2 molecules of EF2 per ribosome.

The steroid antibiotic fusidic acid inhibits protein synthesis by acting on the G factor, for cells which are resistant to the drug contain an altered G protein. When fusidic acid is added to in vitro systems for protein synthesis which contain an excess of GTP, it prevents cleavage of the nucleotide and translocation of the ribosome. But when the amount of GTP used to bind G factor to the ribosome is limiting, so that only one translocation event occurs on each ribosome, fusidic acid does not affect the rate or extent of release of γP^{32} from the GTP. The stoichiometry of reaction confirms that fusidic acid allows a translocase molecule to catalyse one round of hydrolysis of GTP on a ribosome, but prevents any subsequent action. Bodley, Zeive and Lin (1970), and Brot, Spears and Weissbach (1971) therefore suggested that fusidic acid stabilises the complex between the ribosome, G factor and GDP which results from translocation, thus preventing dissociation of the G factor and GDP after translocation. Celma, Vazquez and Modolell (1972) showed that the peptidyl-tRNA of fusidic acid treated ribosomes reacts with puromycin; it must therefore be located in the P site.

That fusidic acid "jams" the ribosome in its post-translocation state is demonstrated by the results shown in figure 2.18. When ribosomes which have catalysed peptide bond synthesis and are therefore ready for translocation are incubated with G factor and GTP, cleavage of the GTP and movement of the ribosome take place too rapidly for any intermediate complex to be detected. Neither an H^3 label in the GDP moiety of the GTP nor a γP^{32} label remains associated with the ribosomes. But when fusidic acid is present only the γP^{32} of GTP is released; G factor and GDP are retained by the ribosome. The same results are obtained when 50S subunits are used in place of 70S ribosomes, which suggests that the site of action of the G factor is located on the 50S subunit alone; the activity of this site is influenced by the interaction with the 30S subunit, however, for its GTPase activity is greater when 30S subunits are associated with the 50S subunits (see page 157). McKeehan (1972) has found that mammalian ribosomes display a similar response.

Ribosomes treated with the peptide antibiotic siomycin cannot bind the G factor protein. Modolell et al. (1971) found that siomycin treated ribosomes cannot bind aminoacyl-tRNA to the A site and are unable to participate in the hydrolysis of GTP which is sponsored by factor Tu. This suggests that there may be one site on the 50S subunit, common to both the Tu and G factors, where GTP is hydrolysed whether at the binding of aminoacyl-tRNA or to promote translocation. The related antibiotic thiostrepton has a similar effect

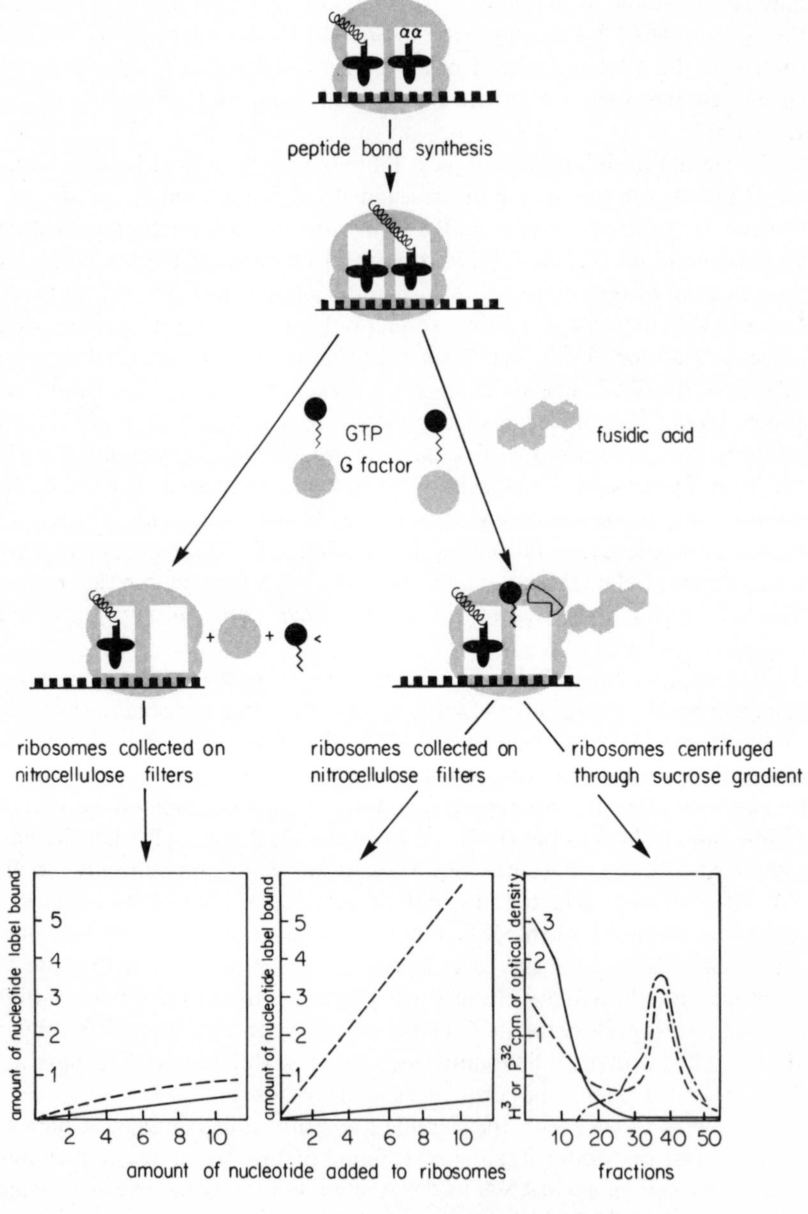
αα
peptide bond synthesis
GTP
G factor
fusidic acid
ribosomes collected on nitrocellulose filters
ribosomes collected on nitrocellulose filters
ribosomes centrifuged through sucrose gradient
amount of nucleotide label bound
amount of nucleotide label bound
H3 or P32 cpm or optical density
amount of nucleotide added to ribosomes
fractions
Optical density = ribosomes
H3 in GDP
P32 from γ-GTP

and Weissbach et al. (1972a) found that similar concentrations of thiostrepton are required to inhibit the hydrolysis of GTP in response to either Tu–aminoacyl-tRNA–GTP binding or G factor translocation.

This discovery suggests that the factors may not themselves catalyse GTP hydrolysis but may act rather to trigger some ribosomal mechanism for cleaving GTP. When 50S subunits are treated with EtOH in the presence of NH_4Cl, a small number of proteins is extracted, including protein L7 (see page 156). As figure 2.19 shows, ribosomes treated in this way are inactive in both Tu and G promoted hydrolysis of GTP; they can neither bind aminoacyl-tRNA to the A site nor undertake translocation. When the L7 protein is added to the EtOH-treated 50S subunits, both activities are restored. This shows that this protein is part of the GTPase hydrolysis site of the 50S subunit and supports the idea that one site hydrolyses GTP in response to the action of either Tu or G factor.

The relationship between the action of the two factors extends beyond their use of a common GTP hydrolysis site, for the ribosome cannot bind both Tu factor and G factor at the same time; their actions are mutually exclusive. When GTP is used to promote binding of factor G to ribosomes, translocation takes place immediately, with the release of GDP and factor G. But when the binding of factor G to the ribosome is stabilised either by the addition of fusidic acid or by the substitution of GMP-PCP for GTP, the intermediate complex between the factor and the ribosome is stable and can be isolated. The ability of aminoacyl-tRNA to bind to these ribosomes can then be measured.

Two criteria have been used to follow the binding of aminoacyl-tRNA to fusidic acid "jammed" ribosomes; after incubation with aminoacyl-tRNA–Tu–GTP, the ribosomes can be examined for their retention of labelled amino acid or for the release of γP^{32} from the GTP of the transfer complex. Both assays give the same result (Cabrer, Vazquez and Modolell, 1972; Miller,

---

Figure 2.18: Fusidic acid stabilises the ribosome–G factor–GDP complex which results from one round of translocation. The stabilised ribosomes cannot undertake any further cycles of translocation. Left: when ribosomes ready for translocation—they carry discharged tRNA in the P site and peptidyl-tRNA in the A site—are incubated with G factor and GTP which has an H^3 label and also a P^{32} phosphate, translocation takes place to release the factor and GDP and Pi. No radioactive label is retained when the ribosomes are collected on nitrocellulose filters. Right: when fusidic acid is also present, one round of translocation takes place to produce ribosomes which are "jammed" with G factor and GDP. P^{32} is released, but the ribosomes retain the H^3 label which is present in the GDP moiety, as shown when they are collected either on nitrocellulose filters or when they are analysed by centrifugation through a sucrose gradient. Data of Bodley et al. (1970) and Bodley, Zeive and Lin (1970)

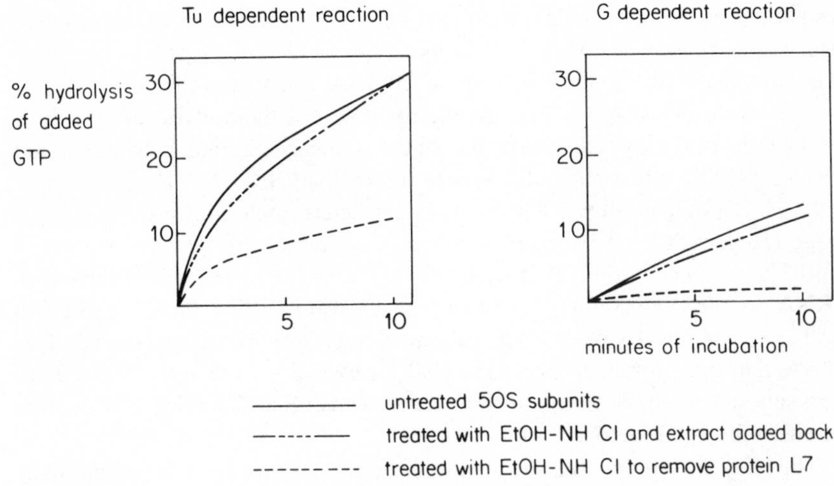

Figure 2.19: Treatment of ribosomes with EtOH-NH_4Cl prevents hydrolysis of GTP in either the Tu dependent aminoacyl-tRNA binding reaction or the G factor catalysed translocation. Ability to perform both reactions is restored when the extracted protein is added back to the 50S subunits. This suggests that ribosomes have only one active centre for GTP hydrolysis. Data of Weissbach et al. (1972a)

1972; Richman and Bodley, 1972; Richter, 1972). As figure 2.20 shows, ribosomes which have undergone normal translocation with G factor and GTP can bind aminoacyl-tRNA and catalyse the hydrolysis of the GTP of the transfer complex. But ribosomes which have been unable to release their G factor from the 50S subunit are inhibited both in binding aminoacyl-tRNA to the A site and in releasing γP^{32} from the transfer complex to about the same extent. Baliga, Schechtman and Munro (1973) showed that a similar inhibition is displayed by mammalian ribosomes which retain EF 2 factor; this interaction may therefore be common to both bacterial and eucaryotic protein synthesis.

A similar inability to bind is found when aminoacyl-tRNA is allowed to interact spontaneously—instead of in the form of a transfer complex with Tu—which implies that entry to the A site as such is inhibited. That the state of the ribosome controls the availability of the A site to aminoacyl-tRNA is suggested also by the observation of Springer and Grunberg-Manago (1972) that under stringent conditions phe-tRNA can bind to ribosomes under direction from poly-U only when the P site is already occupied. Both the states of the GTP hydrolysis site and the P site may therefore influence the entry of aminoacyl-tRNA at the A site.

The binding factor Tu and the translocation factor G thus seem to make use of the same site on the 50S subunit to hydrolyse GTP; G factor must therefore dissociate from the ribosome before Tu can bind. If this one site is alternatively occupied by factors Tu and G, the corollory is that factor G should be unable

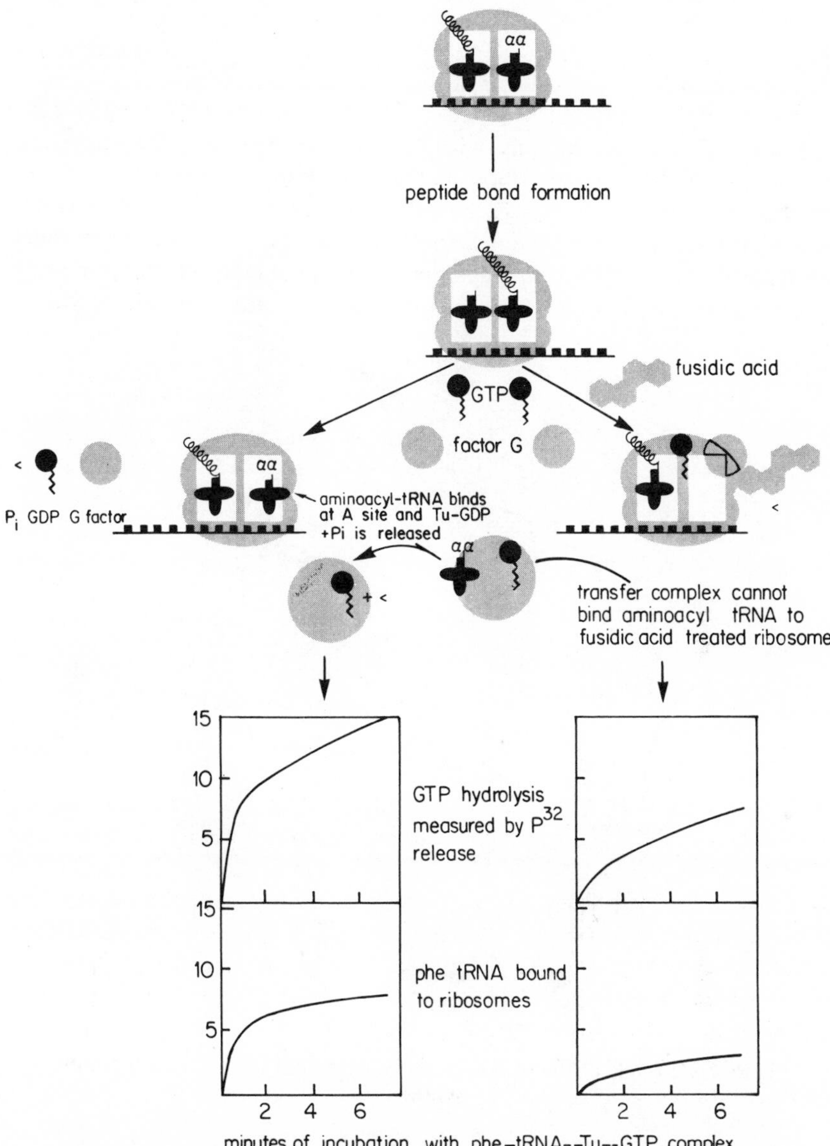

Figure 2.20: Aminoacyl-tRNA cannot bind to the ribosome A site when G factor is present on the 50S subunit. Left: ribosomes which have undergone normal translocation can bind aminoacyl-tRNA from the transfer complex as measured by hydrolysis of GTP to release phosphate (upper curve) or by attachment of radioactive label in amino acid to the ribosomes (lower curve). Right: ribosomes which have been treated with fusidic acid cannot accept aminoacyl-tRNA–Tu–GTP complex because G factor is still bound. Inhibition of aminoacyl-tRNA entry is 60–70% as judged by either hydrolysis of GTP in the transfer complex (upper curve) or attachment of labelled amino acid (lower curve). This implies that G factor and Tu factor must make use of the same site on the ribosome. Data of Cabrer, Vazquez and Modolell (1972)

to bind when Tu is present. Richter (1972) demonstrated that there is some inhibition of G factor binding when the aminoacyl-tRNA–Tu–nucleotide complex is stabilised on the ribosome by substituting GMP-PCP for GTP; but this inhibition is much less than that displayed when bound G factor prevents the access of Tu. Modolell and Vazquez (1973) found that ribosomes which have only phe-tRNA bound in the A site cannot bind G factor; although these ribosomes do not represent one of the usual stages of protein synthesis,

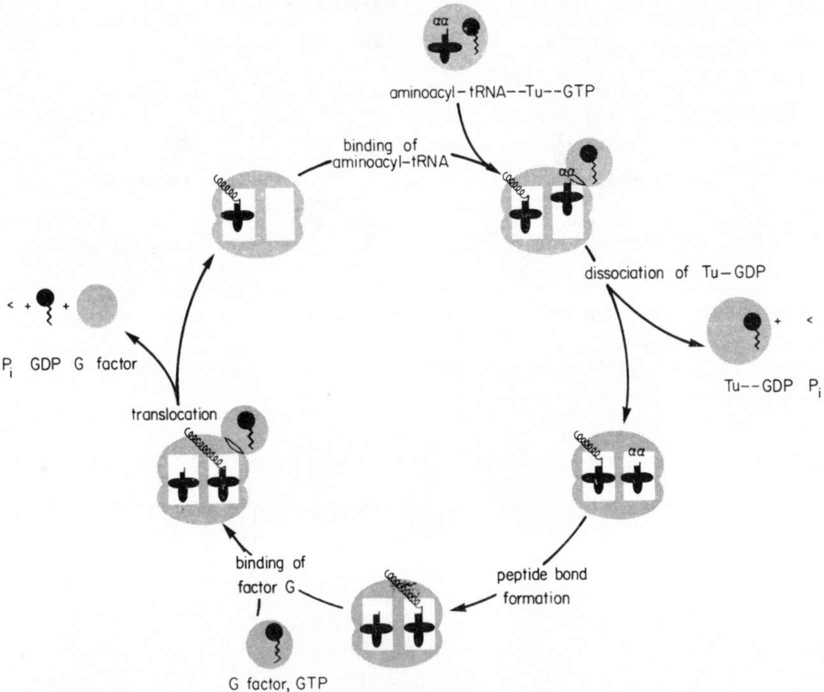

Figure 2.21: Cycle of the ribosome and supernatant factors during protein synthesis

this experiment is consistent with the idea that the ribosome can bind G factor only in the appropriate state.

Figure 2.21 shows the cycle of events through which the association of factors with the ribosome passes and we may imagine that the mutual exclusion of the binding and translocation factors, together with the influence which occupancy of the P site exerts on the A site, ensures that the sequence of events in protein synthesis always proceeds smoothly in the proper order. Factor Tu binds aminoacyl-tRNA only to the ribosomes which possess peptidyl-tRNA in the P site and in so doing hydrolyses GTP; this releases Tu–GDP and frees the GTPase site for G factor and GTP binding. The hydrolysis of

GTP at translocation then releases factor G and GDP, freeing the site for entry of the next aminoacyl-tRNA.

An amino acid therefore passes through several stages in protein synthesis. First of all it is placed on its transfer RNA. Then the aminoacyl-tRNA becomes bound to its codon in the A site of the ribosome, after which the polypeptide chain synthesised so far is transferred to this aminoacyl-tRNA. Translocation moves the new peptidyl-tRNA into the P site. Repetition of this cycle is succeeded by another translocation, which expels the tRNA from the P site, after which it may become charged with another molecule of its amino acid and used again.

The ribosome must possess at least the four active centres shown in figure 2.22. The P site appears to be located largely on the 30S subunit (see next chapter) and the A site mostly on the 50S subunit. This suggests that the model shown in figure 2.10 may indeed account for translocation if we stress the roles of the 30p and 50a subsites and minimise the roles of the 30a and 50p subsites. The peptidyl transferase site is located entirely on the 50S subunit; although we do not know precisely how it relates to the P and A sites, it is reasonable to suppose that it encompasses the 3′ termini of both bound tRNAs. The GTP

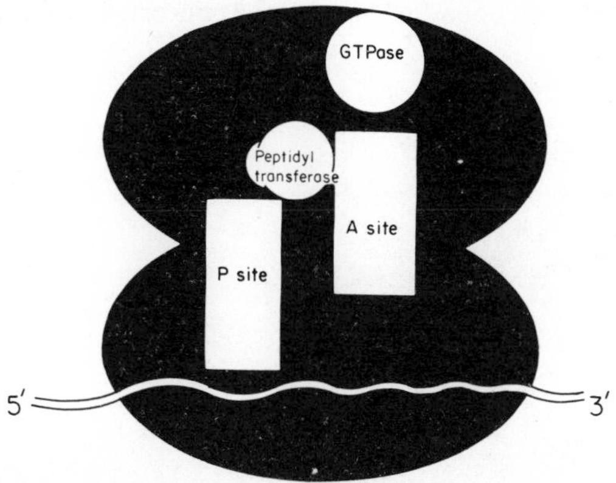

Figure 2.22: The ribosome has at least four active sites. The P site, where peptidyl-tRNA donates the polypeptide chain to the next aminoacyl-tRNA, is located largely within the 30S subunit. The A site, where the Tu transfer complex places incoming aminacyl-tRNA, is located largely within the 50S subunit. About 30 nucleotides of mRNA are protected by the ribosome against degradation; the P site is therefore some five codons distant from the 5′ exit point of the messenger and some five codons distant from the 3′ entry point. The peptidyl transferase centre, which catalyses peptide bond formation, is located entirely on the 50S subunit, presumably in the vicinity of the 3′ terminal ends of the tRNAs bound in the P and A site. The GTPase site, at which both factors Tu and G appear to act, is also located on the 50S subunit, although it is more active when the 30S subunit is associated with the 50S

hydrolysis site is located on the 50S subunit and it is at this centre that both Tu and G factors act to hydrolyse (or trigger the hydrolysis of) GTP in binding and in translocation. The independence of the peptidyl transferase and GTP hydrolysis sites is confirmed by the discovery that they comprise different proteins of the 50S subunit (see page 156).

We do not know how the energy derived from either of the two cleavages of GTP is used in protein synthesis. The energy required for peptide bond formation itself may be provided by cleavage of the energy rich ester link between the peptidyl carboxyl group; Lipmann (1969) has suggested that the energy derived from GTP hydrolysis is probably therefore applied elsewhere. The existence of only one site at which GTP is hydrolysed implies that both the high energy GTP bonds which are cleaved may be utilised in the same way. It is possible that the GTP cleaved when aminoacyl-tRNA is bound is used to provide the energy for some conformational change in ribosome structure. The second GTP cleavage, at least, is presumably used to drive the translocation; and Nishizuka and Lipmann (1966) suggested that there may be an analogy between the ribosome linked GTPase activities and the ATPase linked contraction of muscle. A pulsating ribosomal contraction might occur after peptide bond synthesis to move the ribosome along the messenger and change the status of both bound tRNAs in one concerted movement.

CHAPTER 3

Punctuation in the Message

Signals for Initiation

Incorporation of Formyl-Methionine into Proteins

The phase in which a gene is read is determined by an initiation codon located at its 5′ end recognised by a species of methionyl-tRNA which specifically initiates protein synthesis. The initiator tRNA enters the part of the P site of the ribosome located on a 30S subunit attached to an initiation site on messenger RNA; three initiation factors are also required for the reaction. After the first codon has been recognised by binding of initiator tRNA to the complex, a 50S subunit associates with the 30S subunit to generate a 70S ribosome particle; GTP is hydrolysed at this stage. The 70S ribosome then elongates the polypeptide chain by the processes described in the last chapter and the initiation factors play no further role in protein synthesis.

The idea that some special mechanism might initiate polypeptide chains was first raised by the observation of Waller (1963) that the amino acid composition of E.coli proteins at their N-termini is not random but seems to be restricted to a small number of species, largely:

met	45%
ala	30%
ser	15%

The discovery of N-formyl-methionyl-tRNA (fmet-tRNA) in extracts of E.coli suggested that this species might be involved in initiation; it seemed to be a good candidate for this role because of its blocked amino group, which would prevent it from taking part in chain elongation but would allow it to be the first amino acid of a protein.

The use of RNA phages as messengers to direct protein synthesis in vitro showed that formyl-methionine can indeed be found as the first amino acid in a protein. Adams and Capecchi (1966) and Webster, Engelhardt and Zinder (1966) demonstrated that labelled formyl groups from fmet-tRNA are incorporated into the proteins coded by phage R17 and f2 RNA. The amino acid adjacent to the formyl-methionine initiating the coat protein is alanine, which is the N-terminal group found in phage particles.

When proteins are synthesised in the bacterial cell, extra residues incorporated at the N-terminus must be removed to reveal the proper terminal amino acid. When the usual amino acid in the first position is methionine, only the formyl moiety need be cleaved; when some other amino acid—usually alanine or serine—is to provide the terminus, the methionine must also be removed. By using the peptides fmet-ala and fmet-ala-ser as model systems to test for deformylating activity, Adams (1968) was able to characterize an enzyme in E.coli extracts which removes the formyl group rapidly but does not attack any other bond. Although the system is specific in that it does not attack formyl-methionine alone—there must be a second amino acid attached—the nature of the amino acid adjacent to the methionine does not affect the rate of hydrolysis. Assay for formate shows that the formyl group is released in free form and is not transferred to an acceptor molecule. The deformylase activity is labile under all the experimental conditions which have been used, which probably explains why the formyl group remains attached to proteins synthesized in vitro but not in vivo.

Working with extracts of B.subtilis, Takeda and Webster (1968) detected an aminopeptidase activity which is inactive with the $-NH_2$ terminal analogue formyl-methionyl-puromycin (fmet-puro) but which hydrolyses the met-puro product after deformylase has removed the formyl group. When an $-NH_2$ terminal fragment of coat protein synthesized in vitro by phage f2 RNA was tested with these enzymes, the aminopeptidase by itself proved to be inactive; but on joint incubation with both enzymes methionine can be freed. Some sort of signal must presumably be contained in the amino acid sequence of each protein to tell the aminopeptidase when a protein should lose its N-terminal methionine and when it should retain it.

Formyl-methionyl-tRNA is synthesized by formylation of the met-tRNA produced by the met-tRNA synthetase (rather than by reaction of formyl-methionine with the tRNA). The formylation reaction is catalysed by the enzyme formyl-THF-met-tRNA transformylase and Marcker et al. (1966) and Dickermann et al. (1966) reported the reaction to be:

$$\text{met-tRNA} \xrightarrow{\text{10-formyl-tetrahydrofolate}} \text{N-formyl-met-tRNA}$$

The enzyme is specific in its selection of met-tRNA as substrate and Dickermann and Smith (1971) found that it does not bind non acylated tRNA.

Recognition of AUG by fmet-tRNA$_f$

The methionine accepting activity of E.coli transfer RNA can be fractionated into the two peaks shown in figure 3.1, one of which can be formylated (met-tRNA$_f$) and the other of which cannot (met-tRNA$_m$). When Clark and Marcker (1966) measured the binding of each charged met-tRNA to ribosomes under the direction of various trinucleotides, they found that the methionine codon AUG stimulates binding of both met-tRNA$_f$ and met-tRNA$_m$. The initiator tRNA also responds to the related codon GUG, which usually codes

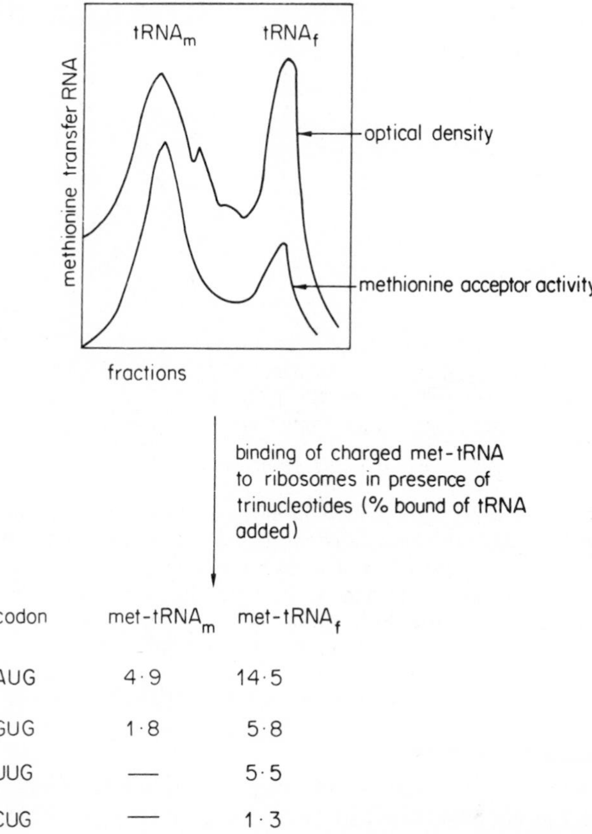

Figure 3.1: Separation of methioine transfer RNAs of E.coli into two species. The first peak from countercurrent chromatography cannot be formylated and responds effectively only to the codon AUG. The second peak, tRNA$_f$, can be formylated and responds in the ribosome binding assay to AUG, CUG and UUG (although it does not respond to UUG used as the first codon of messengers). Data of Ghosh, Soll and Khorana (1967)

for valine; met-tRNA$_m$ responds only to its normal codon, AUG. (Although it was at first thought that UUG is also recognised by met-tRNA$_f$, this seems to occur only in the Nirenberg ribosome binding assay and not when UUG is the initial codon in a message.)

Both of the two methionine transfer species respond to the synthetic polymer poly-A,U,G as shown in figure 3.2. By analysing the incorporation of sulfur from S^{35}-met-tRNA into polypeptide products, Marcker, Clark and Anderson (1966) were able to show that at least 70% of the methionine of met-tRNA$_f$ is incorporated into the N-terminal position whereas, by contrast, incorporation from the non-formylatable species is directed almost exclusively into the

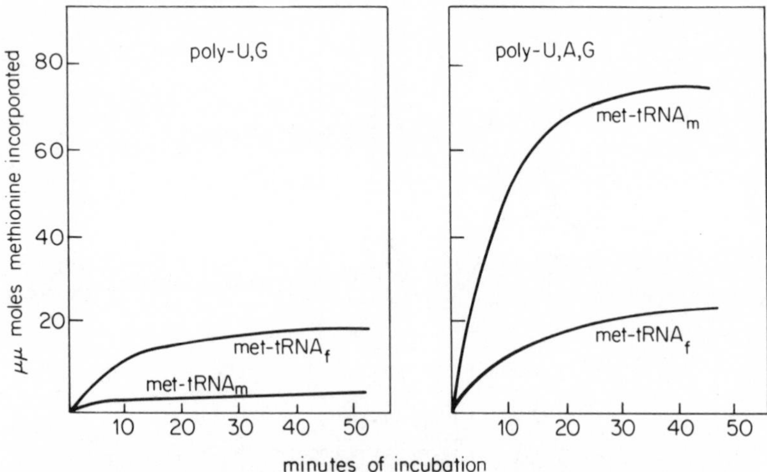

Figure 3.2: Incorporation of methionine into polypeptide from met-tRNA$_f$ (in the charged but not formylated form) and met-tRNA$_m$. Only met-tRNA$_f$ can respond to poly-U,G, which does not contain the AUG codon for methionine but does contain the GUG codon (specifying valine) which can also act as an initiator. Both met-tRNA$_f$ and met-tRNA$_m$ can respond to poly-A,U,G; the met-tRNA$_m$ places methionine internally—and can therefore respond in greater amount—and the met-tRNA$_f$ can donate its methionine only to the NH_2 termini of chains. Data of Clark and Marcker (1966)

internal position. The responses of the two methionine tRNAs to poly-UG show that the codon GUG can be recognised by fmet-tRNA$_f$ when it is located at the 5′ end of a messenger, but is recognised only by val-tRNA in internal positions.

The efficiency of amino acid incorporation in polypeptide synthesis directed by synthetic messengers was at first thought to imply that there might be no need for a special initiation system. But these extracts were generally incubated at high magnesium concentrations (generally above 20 mM) which seem to stabilise artificially the complex between the components of the translation apparatus, abolishing the need for proper initiation. Ghosh, Soll and Khorana (1967) found that at 4 mM Mg^{2+}, poly-UG can be translated efficiently; but poly-UC, poly-AG, poly-AC are all inadequate as templates. But at 11 mM Mg^{2+}, all four templates are translated efficiently. According to Leder and Nau (1967), the optimum concentration for recognition of AUG by fmet-tRNA$_f$ is about 10 mM Mg^{2+}; when the other initiation components—that is protein factors and GTP—are omitted, the magnesium concentration must be raised to about 15 mM Mg^{2+} for efficient binding.

Messengers which lack an initiation codon can therefore be translated only at high magnesium levels when initiation is not necessary. Messengers which commence with an initiation codon are translated most efficiently at low concentrations of magnesium ions, when all the components needed for fmet-tRNA$_f$ to respond to AUG are present. The presence of an initiation codon in a message phases its translation of triplets into amino acids. Ghosh, Soll and Khorana (1967) found that when fmet-tRNA$_f$ is used to initiate synthesis with poly-UG, the message:

U G U G U G U G U G U G U G U G

is read as:

fmet – cys – val – cys – val –

since reading starts in phase with GUG. Sundarajan and Thach (1966) found that if AUG is present at the beginning of a long polynucleotide it suppresses reading of codons which partially overlap it and promotes reading of the subsequent 3′ codons; Thach et al. (1966) found that at low magnesium concentrations the AUG codon exerts a phasing function even when not located at the 5′ terminus itself, but as the ionic level is increased the phase of reading of the message becomes increasingly random.

Interpretation of the code for the triplets AUG and GUG therefore depends upon their position in the message. When at the 5′ end of a gene, both these codons are recognised by the initiator fmet-tRNA$_f$; but within the gene they are recognised by their normal aminoacyl-tRNAs, met-tRNA$_m$ and val-tRNA. Formyl-methionine is probably used to initiate protein synthesis in all bacteria and, apart from E.coli, has been studied largely in B.subtilis (see Migita and Doi, 1970). It is also employed in the organelles of eucaryotic cells, demonstrated by Galper and Darnell (1971) for Hela cell mitochondria and by Marcus et al. (1970) for wheat germ chloroplasts. In general, these organelles synthesise proteins in a manner more akin to that utilised by bacteria than that of the cytoplasm in which they are located (reviewed by Kroon et al., 1972). This further emphasizes the plausibility of the idea that they may have been derived by evolution from bacteria.

Protein synthesis in eucaryotic cells is initiated in a manner very similar to that of bacteria, but with the difference that the met-tRNA initiator is not formylated. Two methionine accepting tRNAs were first separated from mammalian tissues—including mouse ascites tumour cells, mouse liver and yeast—by Smith and Marcker (1970), who found that one of the two species can be formylated in vitro by the transformylase of E.coli. Similar tRNAs have since been found in many other eucaryotic cells, including those of rabbit, rat and wheat. Extract systems of eucaryotic cells share with bacterial systems for protein synthesis in vitro the property that they rely upon met-tRNA$_f$ for translation at low concentrations of magnesium ions (see Smith, 1973).

The coding response of the two eucaryotic met-tRNAs is precisely the same as that of the corresponding E.coli species. Brown and Smith (1970) showed that the molecule which can be formylated—termed met-tRNA$_f$ in analogy with E.coli—directs methionine into terminal positions when poly-AUG or poly-UG are used as messengers; the met-tRNA$_m$ molecule donates methionine only into internal positions in proteins. The met-tRNA$_f$ initiator responds equally well to AUG and GUG as the initiation codons. Under direction from endogenous messengers or added EMC-RNA (encephalomyocarditis virus) only the met-tRNA$_m$ molecule appears to place methionine in proteins. This implies that if methionine is placed at the N-termini of proteins by met-tRNA$_f$ molecules, its removal from the nascent polypeptide chain must be very rapid.

There is no evidence of formylation of eucaryotic met-tRNA$_f$ in vivo in any cell type, which suggests that it functions as an initiator without needing the blocked amino group found in E.coli. Indeed, by using adenovirus RNA extracted from human KB tumour cells, Caffier et al. (1971) showed that yeast met-tRNA$_f$ donates methionine to the N-termini of the proteins synthesized in vitro; but when the initiator is formylated it is inactive. Sufficient similarities with the met-tRNA$_f$ of bacteria have presumably been retained in the evolution of eucaryotes so that the formylating enzyme of E.coli can recognise eucaryotic met-tRNA$_f$ but not met-tRNA$_m$; the features which distinguish the two methionine tRNAs in the cell may therefore be of the same kind in all organisms, although bacteria may in addition rely to some extent on formylation (reviewed by Rudland and Clark, 1972).

That methionine is the eucaryotic initiator has been confirmed by examining proteins very soon after their synthesis has begun. Wigle and Dixon (1970) found that methionine is present at the N-terminus of protamine proteins synthesised in tests cells of the rainbow trout when the polypeptide chain is short, but is removed to reveal the natural N-terminal proline by the time the chain reaches 31 or 32 residues. By examining the nascent peptides found on reticulocyte ribosomes soon after the initiation of haemoglobin synthesis, Wilson and Dintzis (1970) and Jackson and Hunter (1970) found that methionine is located at the N-terminus. Figure 3.3 shows that methionine is present in short chains of α-globin and β-globin. But when longer chains are examined, the usual N-terminal amino acid, valine, is found; the N-terminal methionine which initiates the chain appears to be removed when the nascent polypeptides reach 15–20 amino acid residues in length.

The discovery that E.coli cells contain only met-tRNA$_f$ molecules with an anticodon sequence which should respond only to AUG implies that there must be an unusual degeneracy in its recognition of either A or G as the 5′ nucleotide of the triplet codon; this ability appears to be typical of both bacterial and eucaryotic met-tRNA$_f$ molecules. The lack of species specificity implies that this effect is not due simply to an artefact of the in vitro systems. Since the initiator tRNA enters the P site of the ribosome, in contrast to the other aminoacyl-tRNAs which enter the A site (see later), the 5′ degeneracy of the codon might be a feature of triplet recognition in this site. However, there

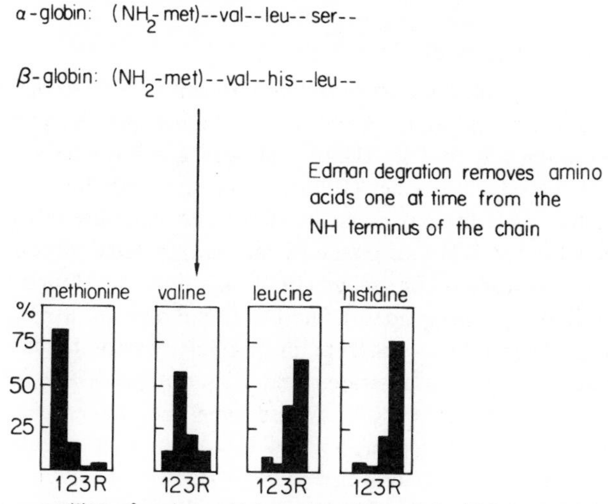

Figure 3.3: Presence of methionine on newly initiated chains of globin. The peptides consist of a mixture of the termini of α-chains (usual sequence NH_2-val-leu-ser) and β-chains (usual sequence NH_2-val-his-leu). When short peptides are analysed by Edman degradation, methionine is found in the first position of the chain (the treatment is only about 75% effective), valine is found primarily in the second position, leucine in the third and also remaining positions; histidine does not occur in the first three positions. This suggests that the normal sequences of the protein chains are preceded by a methionine residue. Data of Jackson and Hunter (1970)

is no evidence to show that GUG is used as an initiation codon within the cell; those bacterial messengers which have been sequenced always appear to utilise AUG and in yeast it seems that GUG cannot function as an initiator. It is therefore possible that although recognition of GUG is a feature of in vitro systems, only AUG is used in vivo.

The only codon used for initiation in yeast seems to be AUG. By isolating yeast variants which completely lack cytochrome c due to mutation at the N-terminal end of the gene, Stewart et al. (1971) characterized revertants able to synthesise the protein. Reversion takes place either by restoration of the original sequence or by the creation of a new AUG codon either at a site soon after the original initiation codon or at a site just before it. The protein made in the double mutant of the last class is longer than usual and includes translation of the mutated (original) initiation site. Its N-terminal amino acid sequence thus reveals the nature of the mutation which prevented initiation. All of the mutants in the initiation codon proved by this analysis to be related to AUG by a single base change; AUG must therefore be the initiation signal. Mutation to give GUG at this site results in failure to synthesise the protein; in yeast, GUG must therefore be unacceptable as an initiation codon.

Initiation Sequences of RNA Phages

One way for a ribosome to undertake its initial attachment to a messenger would be to bind to the free 5′ end and move along to the first AUG codon to form an initiation complex with fmet-$tRNA_f$. However, ribosomes can recognise initiation sites directly. Bretscher (1968c) tested their capability by using a circular messenger; since there is no such mRNA, he made use of the observation that single stranded DNA of phage fd can be translated directly in vitro when neomycin B is added. (Bretscher (1969) has since developed a system in which the antibiotic is not needed.) The DNA can be translated in its circular form; breaking the circles to generate linear ends does not increase the frequency of initiation. This demonstrates that it is not necessary for ribosomes to be "threaded on" the free end of a messenger; they can enter its interior directly. (The specific sequence of $\phi \times 174$ DNA, another circular phage, bound by ribosomes has been determined by Robertson et al., 1973).

This implies that there must be some way for ribosomes to directly distinguish initiation codons from AUG sequences which occur internally in messengers. Some bacterial messengers are translated so rapidly after their synthesis has begun that only one possible initiation site near the free end of the messenger may be accessible in the cell. But this is not necessarily typical of all bacterial messengers and different initiation sites can certainly be distinguished when RNA phages are translated in infected E.coli cells. In eucaryotic cells, the mRNA must be transported from nucleus to cytoplasm for translation, so that the entire messenger must be available and ribosomes must select only the correct initiation sites.

One plausible model is to suppose that some sequence of nucleotides longer than the AUG triplet itself must promote ribosome binding. Ribosome binding sites can be sequenced by binding mRNA to the ribosome, degrading all the unprotected messenger sequences with ribonuclease, and recovering the length of messenger which is protected from enzyme attack by the ribosome. Such experiments are possible, however, only when large amounts of purified messenger are available; the only ribosome binding sites which have been sequenced at present are those of the RNA phages.

As indicated by their name, RNA phages have RNA as genetic material instead of the more usual DNA. The f2 group of phages, whose host is E.coli, consists of the closely related phages f2, R17 and MS2. Their genomes are about 3300 nucleotides long and carry the information for only three genes; Jeppesen et al. (1970) showed that the order of genes is:

5′—maturation protein gene—coat protein gene—RNA replicase gene—3′

Phage, Qβ, serologically unrelated to the f2 group, has a similar organization. Both classes of phage start with a lengthy sequence of at least 50 nucleotides which precedes the first gene and is not translated (Goodman et al., 1970a;

Cory et al., 1972); all the three genes therefore commence with initiation sites located within the genome. Translation of RNA phages has been reviewed by Kozak and Nathans (1972).

The three genes of these phages are translated independently so that ribosomes can be bound to the initiation site of each gene. Ribosomes can discriminate between the three initiation sites, for E.coli ribosomes usually bind only to the maturation protein and coat protein genes of the f2 phages in vitro. This is consistent with the observation of Lodish (1968) that maturation protein and coat protein are synthesised immediately upon infection of E.coli; initiation at the RNA replicase gene takes place in vivo only after translation has proceeded at least part of the way into the coat protein gene. That the initiation site of the RNA replicase gene can be recognised only after some unfolding of the phage structure—Jou et al. (1972) have shown that the RNA phage contains many hairpin loops which are maintained by complementary base pairing—is suggested by the observation of Lodish and Robertson (1969) that reducing the extent of secondary structure by incubation at increased temperature or with formaldehyde allows the RNA replicase gene to be translated without prior translation of the coat protein gene. Progress of the ribosome through the coat protein gene therefore presumably activates the RNA replicase gene by changing the secondary structure of the phage. (These interactions control the initiation of translation; later during infection the coat protein product may bind to the phage RNA at a site between the coat protein and RNA replicase genes and thereby switches off synthesis of replicase—see Bernadi and Spahr, 1972).

Under appropriate conditions, all three of the initiation regions of R17 RNA can be protected from degradation by binding to ribosomes. Steitz (1969) was therefore able to sequence the three initiation sites. The protected RNA fragment extends for about 30 nucleotides and includes about 15 nucleotides prior to the initiation codon—these are not translated—and the first five codons of the gene. The initiation sequences can be identified by equating the sequence of amino acids which these codons represent with the N-terminal sequence of the proteins coded by the phage. Each of the genes of the phage commences with AUG; but there are few other similarities in the sequences of the three protected regions. It is therefore impossible to define any individual sequence which constitutes the recognition site for the ribosome. Nor are there any similar features of secondary structure, since the initiation sequence of the coat protein gene can be written as a hydrogen bonded hairpin loop, whereas no such structure is possible for the initiation sites of the maturation protein and RNA replicase genes.

The three initiation sites of Qβ RNA have also been sequenced; figure 3.4 shows that they too share only the possession of an AUG codon for initiation and possess no extensive homologies with either each other or the recognition sites of R17 RNA. The three Qβ sites can be distinguished by E.coli ribosomes,

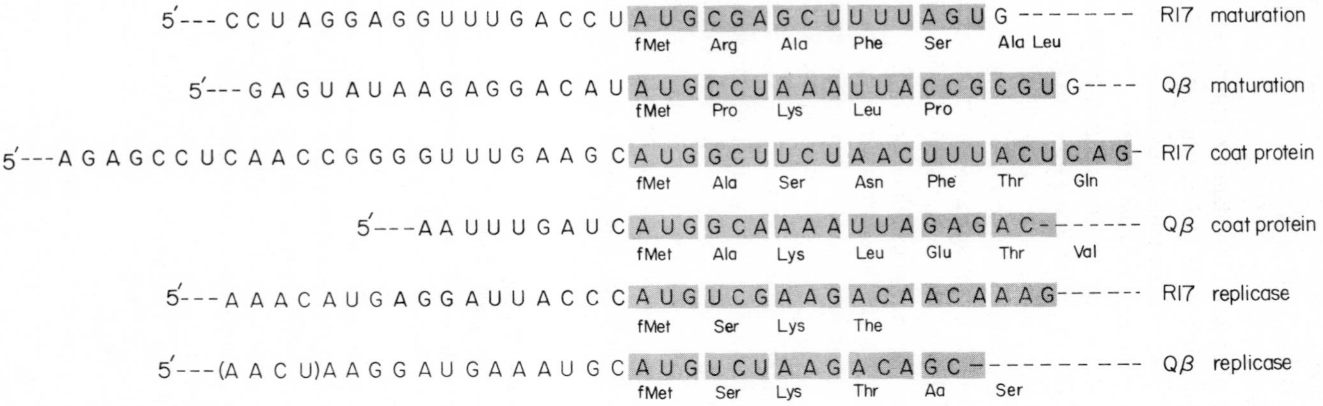

```
5′---CCUAGGAGGUUUGACCU AUG CGA GCU UUU AGU G-------   R17 maturation
                       fMet Arg Ala Phe Ser Ala Leu
5′---GAGUAUAAGAGGACAU AUG CCU AAA UUA CCG CGU G----   Qβ maturation
                      fMet Pro Lys Leu Pro
5′---AGAGCCUCAACCGGGGUUUGAAGC AUG GCU UCU AAC UUU ACU CAG-   R17 coat protein
                              fMet Ala Ser Asn Phe Thr Gln
5′---AAUUUGAUC AUG GCA AAA UUA GAG AC------   Qβ coat protein
               fMet Ala Lys Leu Glu Thr Val
5′---AAACAUGAGGAUUACCC AUG UCG AAG ACA ACA AAG------   R17 replicase
                       fMet Ser Lys The
5′---(AACU)AAGGAUGAAAUGC AUG UCU AAG ACA GC-----------   Qβ replicase
                         fMet Ser Lys Thr Aa Ser
```

Figure 3.4: Initiation sites of phages R17 and Qβ protected by ribosome binding. There appear to be no similarities of sequence or secondary structure. Triplets corresponding to the initial amino acid sequence of each protein are shaded. Data of Gupta et al. (1970). Hindley and Staples (1969), Staples and Hindley (1971), Staples et al. (1971) and Steitz (1969, 1972)

which bind only to the coat protein gene of intact Qβ RNA; Hindley and Staples (1969) sequenced this binding site. Binding to the maturation protein and RNA replicase genes becomes possible when the structure of the phage is reduced or in fragments. Staples and Hindley (1971) and Steitz (1972) sequenced the site bound by a fragment of Qβ RNA containing the RNA replicase gene; Staples et al. (1971) determined the sequence protected in a fragment containing the 5′ end of the phage.

Heterologous ribosomes distinguish between the initiation sites of the phage RNA in a manner different from that shown by the usual host E.coli. Lodish (1970) noted that ribosomes of B. stearothermophilus bind only to the maturation protein gene of f2 RNA. This initiation sequence is therefore recognised by the Bacillus ribosomes as well as by those of E.coli. But Steitz (1973) found that with Qβ RNA, ribosomes of B. stearothermophilus protect a sequence of RNA which corresponds to none of the initiation sites. Since this sequence is specifically bound, it must be related to the initiation sites which are used within the Bacillus system. Because this protected sequence lacks an AUG codon, the initiation site recognised by the ribosome must clearly be longer than the AUG codon itself—although not necessarily as long as the protected sequence—and does not stringently depend upon an AUG triplet.

Recognition of the R17 maturation protein gene site by the Bacillus ribosomes is an intrinsic property, for Lodish (1970) and Steitz (1973) found that in mixed systems in which some components are drawn from E.coli and some from B. stearothermophilus, it is the source of the ribosomes alone which determines the initiation sites protected by the ribosome. Although the use of AUG as initiation codon appears to be universal, the context within which it is recognised by the ribosome may be different in each cell type. In the homologous E.coli system, however, recognition may be influenced by the initiation factors which may control whether a ribosome binds to the maturation protein or coat protein gene (see later). Several sequences may be recognised by ribosomes in E.coli, but we do not know what features of the sequence are pertinent and how different initiation sequences might be distinguished. The idea that some common secondary structure might delineate initiation sequences is clearly excluded by the variety of the sequences which are protected by the ribosome.

Formation of the Initiation Complex

Role of Ribosome Sites in Initiation

The AUG codon must be recognised by fmet-tRNA$_f$ when it initiates a gene and by met-tRNA$_m$ in internal locations. Discrimination between its two roles falls into two parts. First, ribosomes must initially bind to messengers only at AUG codons which function as initiators; second, fmet-tRNA$_f$ must respond to a ribosome engaged in initiation whereas met-tRNA$_m$ must respond to ribosomes engaged in the extension of polypeptide chains. Initiation is mediated by the 30S subunit of the ribosome, which therefore has the responsibility of distinguishing between initiator and internal AUG codons. Selection of the initiation sites presumably depends upon the nucleotide sequence surrounding an AUG codon and this is in part mediated by the subunit itself and in part controlled by the initiation factors. Only fmet-tRNA$_f$ is recognised at initiation because 30S subunits accept fmet-tRNA$_f$ from the initiation factors before they associate with 50S subunits. Once a 70S ribosome has been formed for chain elongation, only met-tRNA$_m$ can be accepted from the elongation factors.

Initiation of protein synthesis therefore depends upon 30S subunits alone, whereas elongation demands the participation of the complete 70S ribosome. Ghosh and Khorana (1967) found that fmet-tRNA$_f$ can bind to poly-UG or poly-AUG templates in the presence of 30S subunits; but this initiation complex must be joined by a 50S subunit before other aminoacyl-tRNAs can bind (see figure 3.6). Free 30S subunits bind fmet-tRNA$_f$ to phage f2 RNA messenger and Nomura, Lowry, and Guthrie (1967) observed that association with 50S subunits inhibits this binding—presumably by removing free 30S subunits—but stimulates the incorporation of other aminoacyl-tRNAs. This inhibition can be overcome by the addition of initiation factors (see later); these assist the disssociation of ribosomes into subunits and such an activity can also account for the apparent ability of 70S ribosome particles to bind to messengers—binding takes place by a dissociation and subsequent reassociation of the ribosome subunits.

A critical test of subunit initiation has been reported by Guthrie and Nomura (1968). Fmet-tRNA$_f$ was allowed to bind to 70S ribosomes which were prepared from cells grown on medium containing heavy isotopes; the reaction mixture contained initiation factors and also a large excess of light 50S subunits derived from cells grown on normal medium. If initiation demands 70S ribosomes, the initiator tRNA should bind directly to the heavy 70S particles. But if the initiation complex is formed only with the 30S subunit, the fmet-tRNA$_f$ can bind only to the product of dissociation of the 70S ribosome. To yield a 70S ribosome for subsequent protein synthesis, the heavy 30S subunit in the initiation complex must reassociate with a 50S subunit; since there is an excess of light 50S subunits, the initiator tRNA must eventually become attached to ribosomes of hybrid density which have a heavy 30S subunit and a light 50S subunit.

A possible flaw in this procedure is that the latter result could also be achieved if a rapid and spontaneous rearrangement to equilibrium were to take place between the heavy ribosomes and the added 50S subunits before formation of the initiation complex—and exchange of subunits has since been shown to be promoted by centrifugation (see page 107). To exclude such an artefact, the binding of normal aminoacyl-tRNA was followed by the same

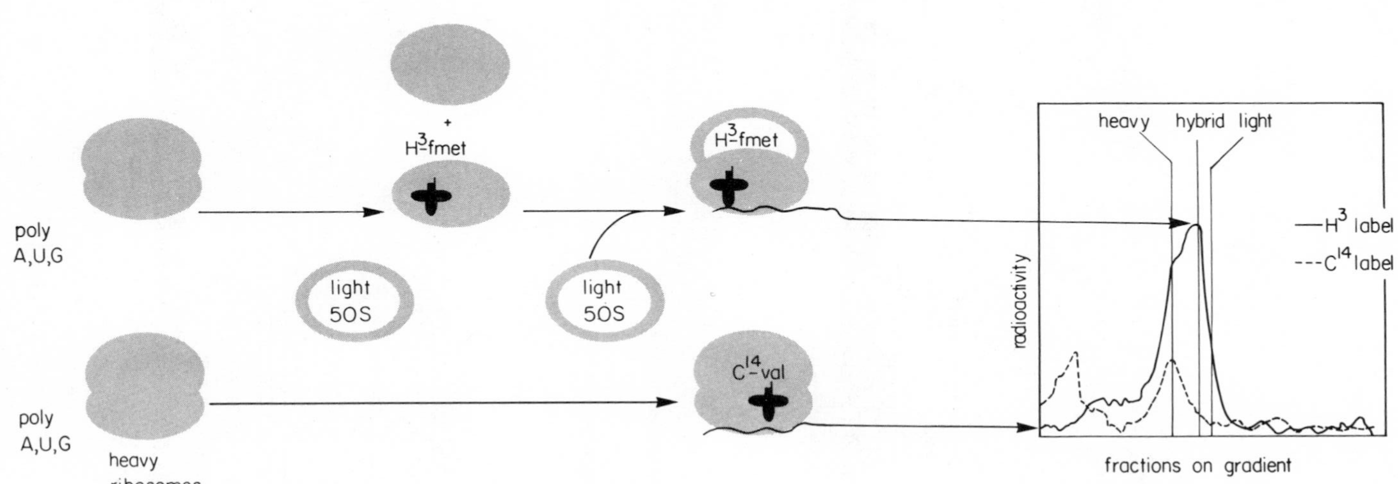

Figure 3.5: Initiation depends on 30S subunits. Aminoacyl-tRNA is bound to "heavy" ribosomes derived from bacteria grown on N^{14} and D_2O; 50S subunits from bacteria grown on "light"—normal—medium are present. Poly-A,U,G is used to direct binding. H^3-fmet-tRNA$_f$ can only bind to 30S subunits, so the heavy ribosome must dissociate before binding the initiator; the light 50S subunits, which are present in excess, may associate with the 30S initiation complex to generate a 70S ribosome of hybrid density. But C^{14}-val tRNA can bind directly to 70S ribosomes; no dissociation need take place so the presence of 50S subunits is irrelevant. The H^3 label is found in a peak of hybrid ribosomes and the C^{14} label remains in the peak of heavy ribosomes. Data of Guthrie and Nomura (1968)

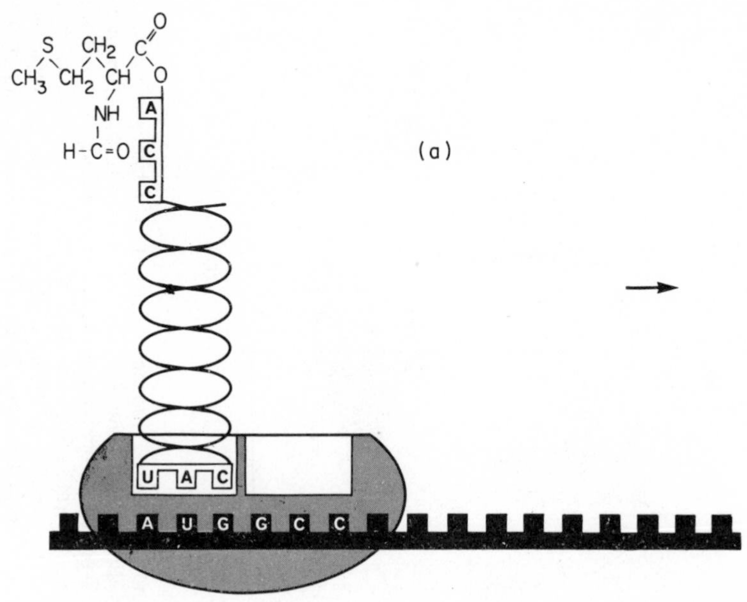
S
CH2
C
O
CH3
CH2
CH
O
NH
H-C=O
A
C
C
U
A
C
A
U
G
G
C
C
(a)

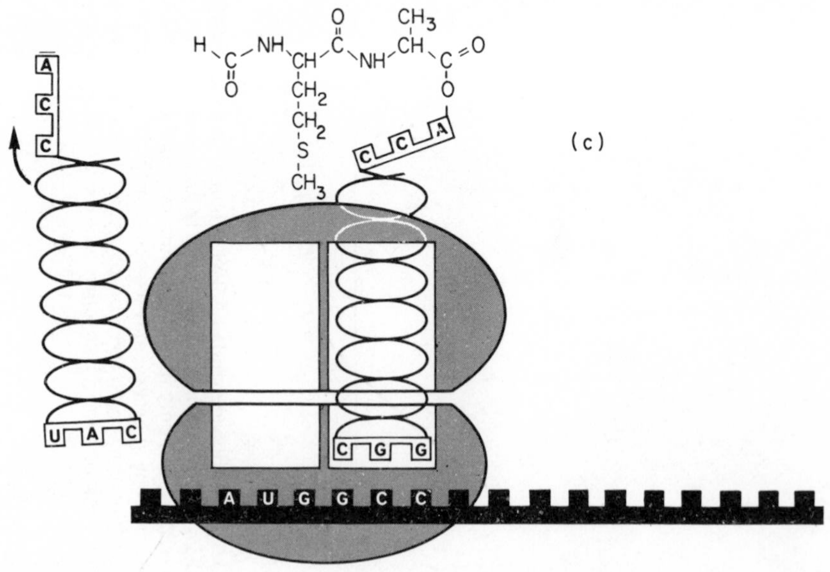
H
C
NH
CH
O
C
NH
CH3
CH
C
=O
O
O
CH2
CH2
S
CH3
A
C
C
C
C
A
U
A
C
C
G
G
A
U
G
G
C
C
(c)

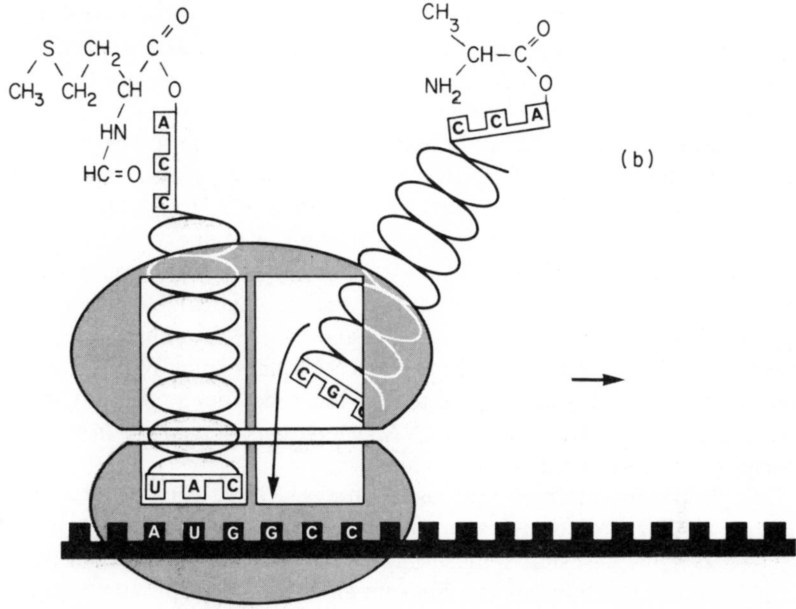

Figure 3.6: Initiation of protein synthesis. (a) An initiation complex is formed between mRNA and a 30S subunit so that the AUG initiation codon is located in the 30p site; fmet-tRNA$_f$ is bound. (b) a 50S subunit attaches and the second aminoacyl-tRNA binds to the A site. (c) After peptide bond formation the free tRNA is released and protein synthesis continues by chain elongation

procedure; this should become attached directly to heavy 70S ribosomes. Figure 3.5 shows that whilst the aminoacyl-tRNA does indeed associate with the heavy 70S particles, initiator fmet-tRNA$_f$ becomes attached to hybrid ribosomes. Initiation probably involves the small ribosome subunit in all organisms; and 40S subunits of eucaryotic cells—including yeast, reticulocytes (Vesco and Colombo, 1970) and rat liver (Leader and Wool, 1972)—can all form complexes of the appropriate nature.

Two major models were proposed to account for the behaviour of the ribosome in initiation when it was thought that complete 70S particles were utilised; and both have since been modified in view of the discovery of the role of 30S subunits in initiation. The single entry model proposed by Hershey and Thach (1967) and Ohta and Thach (1968) envisaged the A site as the only point of entry for all aminoacyl-tRNA species, including the initiator; after an initial binding assisted by initiation factors and GTP, the cleavage of high energy phosphate bonds would translocate the initiator to the P site, leaving the A site available for the next aminoacyl-tRNA.

The double entry model ascribes different roles to the two tRNA binding sites on the ribosome. Peptidyl-tRNA must be bound at the P site before aminoacyl-tRNA can enter the A site. Bretscher (1966) and Bretscher and Marcker (1966) proposed that the initiator fmet-$tRNA_f$ possesses the ability to enter the P site directly; as figure 3.6 shows, the penultimate aminoacyl-rRNA then enters the A site and the first peptide bond is synthesised. There is no ambiguity since only fmet-$tRNA_f$ can respond to codons in the P site of 30S subunits and only met-$tRNA_m$ can respond to AUG codons in the A site of 70S ribosomes. Experiments with puromycin support this idea, since ribosome bound fmet-$tRNA_f$ is sensitive and reacts to give fmet-puro whereas met-$tRNA_m$ is comparatively insensitive under identical conditions. This suggests the fmet-$tRNA_f$ enters the P site and met-$tRNA_m$ enters the A site. That the blocked amino group of fmet-$tRNA_f$ plays some role in ensuring direct entry to the P site is suggested by the behaviour of peptidyl-tRNA analogues such as N-acetyl-phe-tRNA, which can mimic the initiator by entering the P site instead of the A site.

The hybrid ribosome model discussed on page 53 and shown in figure 2.10 accounts for initiation on the supposition that the initiator binds to the 30p site. And as figure 2.22 shows, most of the P site seems in fact to be located on the 30S subunit. Addition of the 50S subunit to the initiation complex could then form a hybrid structure such as the 30p:50a species of part (b) of figure 2.10. The hydrolysis of GTP then converts structure (b) into structure (d), that is converts the fmet-$tRNA_f$ into the reactive state shown by the puromycin reaction. Much data can be interpreted in terms of either the single entry or the double entry models, which are difficult to distinguish clearly by experiments such as those with puromycin; but the double entry model in general provides a simpler explanation and is directly supported by more recent experiments which show there is no translocation at initiation.

The behaviour of formyl-methionine when it joins the initiation complex suggests that the initiator enters the P site directly. Using the $AUGGU_{30}$ messenger described in the last chapter (page 68), Thach and Thach (1971) bound fmet-$tRNA_f$ to the ribosome–mRNA complex and then analysed the length of protected U_n fragment to see whether a translocation is necessary to render the formyl-methionine reactive to puromycin. When GTP is used to bind the initiator tRNA to the ribosome, most (about 80%) of the formyl-methionine is released by puromycin; none reacts when GMP-PCP is used instead. But in each case, the same length of poly-U is protected; this is about 13 nucleotides, which indicates that the AUG codon is in the P site (see figure 2.17).

Although the hydrolysis of GTP is therefore needed to activate the initiator, this reaction does not seem to involve a translocation: we may therefore conclude that fmet-$tRNA_f$ enters the P site directly. Similar results are obtained when fmet-$tRNA_f$ is bound to 30S subunits alone, which supports the idea that the major portion of the ribosome P site may be located on the 30S subunit.

An analogous model may apply to eucaryotic ribosomes, for Levin, Kyner and Acs (1972; 1973) showed that revirus mRNA forms an initiation complex with met-tRNA$_f$ and ribosomal subunits from mouse fibroblasts in which the methionine can be released from the P site by addition of puromycin.

Binding of fmet-tRNA$_f$ to the Initiation Complex

Appropriate recognition of AUG codons requires that only fmet-tRNA$_f$ binds to the 30p site of the initiation complex and only met-tRNA$_m$ enters the A site of 70S ribosomes. Control of the binding of the two met-tRNA species is mediated by the protein factors needed for initiation and elongation. The met-tRNA$_m$ molecule appears to be able to donate methionine only into internal positions in all in vitro systems; Rudland et al. (1969) reported that normal aminoacyl-tRNA molecules—including met-tRNA$_m$—do not bind to the initiation factors, which specifically promote the binding of fmet-tRNA$_f$ alone to the 30S subunit–mRNA initiation complex. Early work on the specificity of initiation has been reviewed by Rudland and Clark (1972).

Although it was at first thought that an analogous mechanism might prevent use of the initiator species within a protein, this situation is now less clear. Ono (1968) observed that fmet-tRNA$_f$ of E.coli is the only aminoacyl-tRNA unable to form a stable complex with Tu-GTP and suggested that it might in this way be prevented from binding to 70S ribosomes. Kerwar, Spears and Weissbach (1970) reported that mammalian factor EF 1 recognises only met-tRNA$_m$ and not met-tRNA$_f$. However, recent experiments have shown that, at least in vitro, met-tRNA$_f$ can donate its amino acid to internal positions of the chain. Drews, Grasmuk and Weil (1972) found that with an E.coli system fmet-tRNA$_f$ is inactive internally but that met-tRNA$_f$ can donate methionine within polypeptides at about 10% of the efficiency of met-tRNA$_m$. This suggests that formylation of the bacterial initiator may be important in reducing recognition by elongation factors.

The met-tRNA$_f$ initiator of mammalian systems can also donate methionine in response to internal AUG codons; Gupta et al. (1971b) and Ghosh and Ghosh (1972) have shown that the internal use of met-tRNA$_f$ relative to met-tRNA$_m$ is greater with E.coli (although formylation rectifies the situation), intermediate with mammalian and plant systems, and very low with yeast which seems to have a particularly efficient discrimination system. Isolation of transfer complexes (under gentle conditions in which they are stable) shows that recognition by the transfer factor of met-tRNA$_f$ relative to met-tRNA$_m$ displays the same ratio as the overall utilizations of the two met-tRNA species within the protein. This suggests that recognition by the transfer factors is the critical step in selecting aminoacyl-tRNAs for entry to the ribosome.

The internal use of met-tRNA$_f$ in vitro depends upon the sequence of the tRNA moiety. Yeast met-tRNA$_f$ retains its characteristic inefficiency in internal use even in heterologous systems, which implies that the differences between the eucaryotic systems may reside in the structures of their initiator tRNAs rather than in the abilities of their elongation factors to discriminate against the initiator. This is consistent with the common evolution of initiator tRNAs implied by their recognition by E.coli transformylase.

Formylation of bacterial met-tRNA$_f$ effectively prevents its use internally and yeast met-tRNA$_f$ does not appear to respond to 70S ribosomes. We do not know whether conditions within mammalian and plant cells are in some way different from those of the in vitro systems, so that their met-tRNA$_f$ species fail to act internally; or whether results in vitro reflect a use of met-tRNA$_f$ in vivo in response to internal AUG codons.

The protein factors required for initiation are distinct from those concerned with chain elongation and comprise three fractions derived by washing 30S subunits with ammonium chloride; they are found only upon the free subunits and not on 70S ribosomes. These factors are concerned only with initiation, for they are essential for translation in vitro of natural messengers (which always require initiation); but although they stimulate protein synthesis by synthetic messengers under conditions demanding initiation, they are dispensable under conditions—such as high levels of magnesium ions—which obviate the need for initiation (Stanley et al., 1966; Salas et al., 1967).

Formation of the initiation complex between a 30S subunit, messenger RNA and initiator fmet-tRNA$_f$ requires the presence of GTP and all three initiation factors (reviewed by Revel, 1972). Figure 3.7 shows that when ribosome subunits are incubated with H^3 MS2 RNA and C^{14} fmet-tRNA$_f$ in the absence of initiation factors, the components remain separate. But when initiation factors are added, the initiation complex is formed and associates with 50S subunits so that both radioactive labels sediment with the 70S ribosomes.

Initiation takes place through the three stages shown in figure 3.8. Factor f3 has two roles; it both assists the dissociation of 70S ribosomes which generates the subunits demanded for initiation and then stimulates the binding of 30S subunits to the appropriate recognition sites on the messenger. Binding of fmet-tRNA$_f$ to the 30S subunit–messenger RNA complex is then undertaken by factor f2. Factor f1 seems both to assist f3 in binding 30S subunits to mRNA and also to stimulate the f2 promoted binding of fmet-tRNA$_f$ to the bound 30S subunits. Its role in these processes is not clear but is probably due to one fundamental interaction with the 30S subunit which is at present undefined. When the initiation complex is complete, 50S subunits attach to it and initiation factors are released from the 70S ribosome which then elongates the protein chain.

The factors f1, f2 and f3 are all derived from uninfected cells of E.coli; some early work also made use of an alternative system in which natural mRNA is produced from phage T4 by infection of host cells and initiation factors needed for translation of the T4 message are isolated from a crude

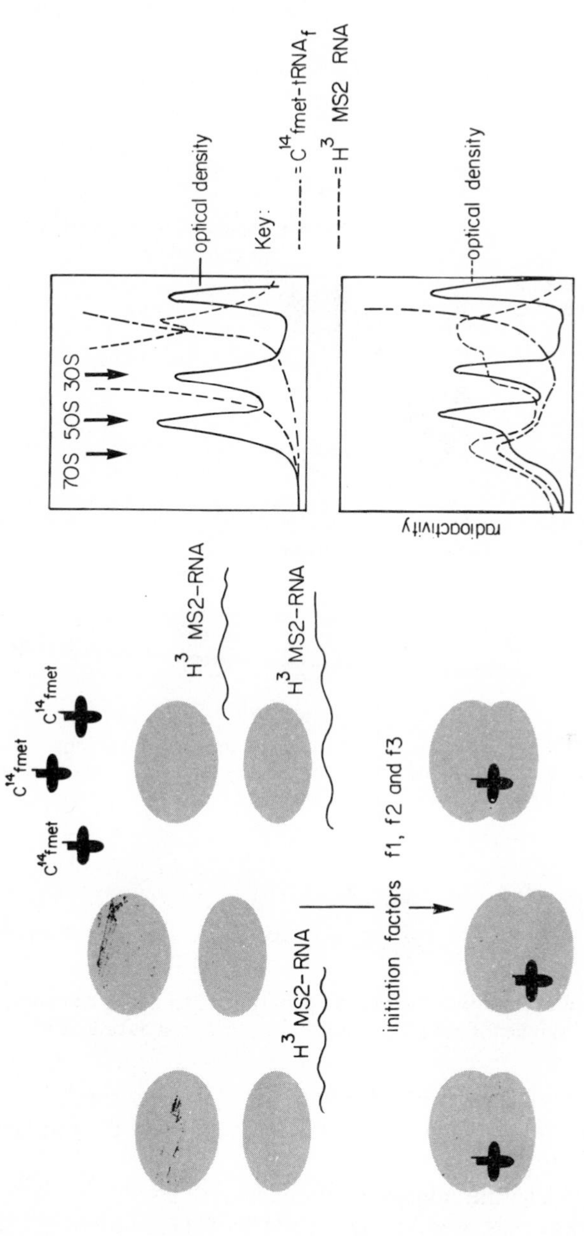

Figure 3.7: Initiation factors are required to bind fmet-tRNA$_f$ and mRNA to ribosomes. Upper: a mixture of 30S and 50S subunits, C^{14}-fmet-tRNA$_f$ and H^3-MS2-RNA does not bind to the ribosomes; the tRNA sediments at the top of a gradient and the MS2 RNA sediments at about 30S. Lower: when initiation factors are added, both radioactive labels sediment at the position characteristic of 70S ribosomes. Data of Sabol et al. (1970)

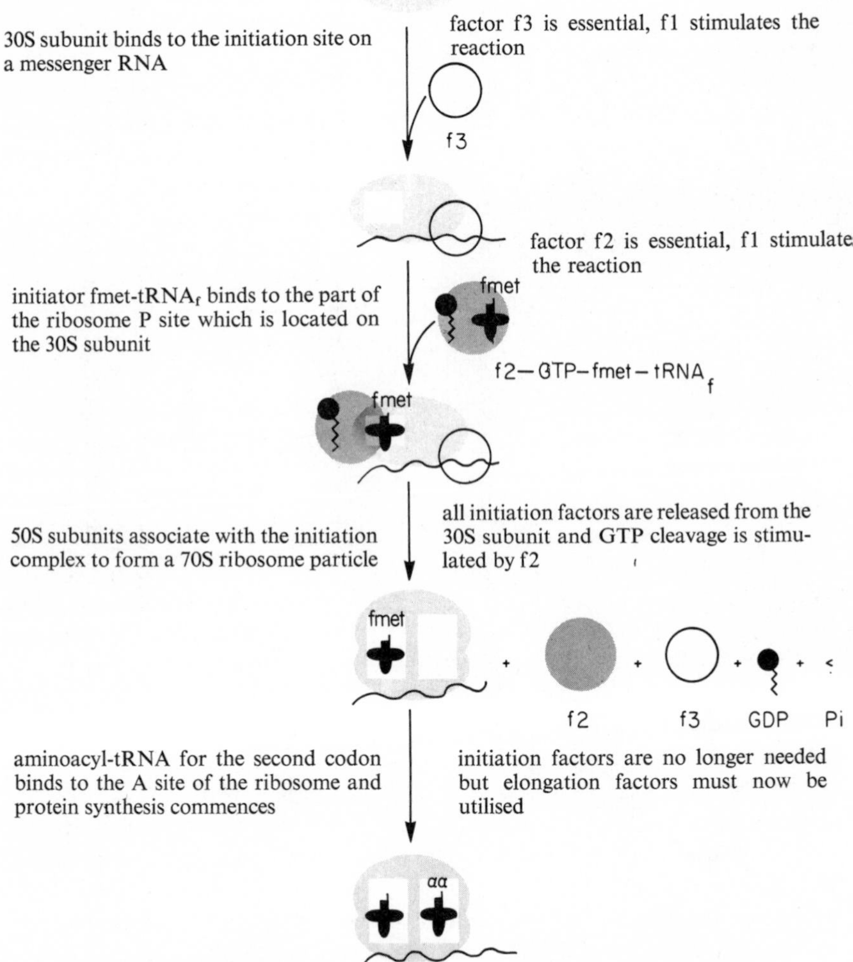

Figure 3.8: Role of initiation factors in formation of an active ribosome–messenger complex. Factor f1 is omitted since its role is not clear

fraction of proteins associated with the ribosomes. Factor A isolated from the phage infected system corresponds to factor f1, factor C to f2, and factor B to f3 (Lelong et al., 1970).

The f2 protein, which is the largest initiation factor of E.coli with a weight of about 70,000 daltons, has a ribosome-dependent GTPase activity and the first step in binding fmet-tRNA$_f$ appears to be the formation of a complex between f2 and GTP—although this complex has not itself been isolated. The activity of f2 is inhibited by reagents which interact with free sulfhydryl groups, but the addition of GTP or 30S subunits confers protection agains the inactivation (Mazumder, Chae and Ochoa, 1969). This suggests that the factor has reactive –SH group(s) which become masked upon formation of a complex with GTP or when it binds to 30S subunits. The idea that its first action is to complex with GTP is supported also by the finding that pre-incubation of f2 with GTP before assay for factor activity results in a faster initial binding of fmet-tRNA$_f$ to ribosomes.

Binding of fmet-tRNA$_f$ to the mRNA–30S subunit complex takes place in two stages, analogous to the interaction of aminoacyl-tRNA with 70S ribosomes. At 0°C, f2 promotes considerable binding of fmet-tRNA$_f$ to 30S subunits; Chae, Mazumder and Ochoa (1969b) suggested that f2 and GTP may interact with fmet-tRNA$_f$ to form a transfer complex which then places the transfer species in the 30p site. This complex is not very stable; Lockwood, Chakraborty and Maitra (1971) found that it dissociates when attempts are made to isolate it on sephadex columns. Groner and Revel (1973) showed that f2 binds fmet-tRNA$_f$ by isolating a complex through centrifugation in glycerol gradients, but under less gentle conditions it is difficult to demonstrate complex formation directly. When f2 is used to bind fmet-tRNA$_f$ to ribosomes at 25°C, its activity is very much reduced; this accords with the idea that the transfer complex is not stable. However, Chae, Mazumder and Ochoa (1969b) found that the addition of f1 increases binding very considerably. This suggests that the transfer complex may in some way be stabilised in the presence of f1.

The role of f1 is not clear, but Sabol et al. (1970) suggested that because it also appears to assist f3 to bind 30S subunits to mRNA its effect on fmet-tRNA$_f$ binding may be a secondary consequence of its direct action. That f1—the smallest initiation factor with a weight of only 9000 daltons—plays a secondary role in formation of the initiation complex is supported by experiments in which Hershey, Dewey and Thach (1969) showed that it binds to 30S subunits only as an integral part of the initiation complex. When f1 is prepared from bacteria grown in a medium containing radioactive isotopes, the radioactive label associates with the ribosomes only when the reaction mixture contains all the components necessary for initiation. That the role of f1 is confined to initiation is shown by its release from the initiation complex when 50S subunits are added to form the 70S ribosome. The function of f1 may therefore be to stabilise the initiation complex as a whole rather than to participate directly in the reactions catalysed by f2 and f3.

The role of GTP in the initiation complex has only recently been resolved. In early experiments the need for GTP seemed to depend upon the conditions of incubation, in particular upon the relative abundances of the different initiation factors, and some results suggested that GTP must be cleaved before fmet-tRNA$_f$ can bind to the 30S subunit. However, more recent work has suggested an alternative that the function of GTP in initiation may parallel its role in binding aminoacyl-tRNAs for elongation; its presence is needed to bind fmet-$tRNA_f$ to 30S subunits but the high energy bond is not cleaved until after association with the 50S subunit. This concept is supported by the observation that maximum expression of the GTPase activity of f2 requires the presence of both 30S and 50S subunits.

One model—for which there is at present no experimental evidence—is to suppose that GTP is hydrolysed at initiation by the same site which is active during elongation; this would imply that the cleavage event must take place on the 50S subunit of a 70S ribosome. Although speculative, this model offers the attraction that it suggests a role for the hydrolysis: to ensure that the ribosome is in the correct conformation to accept aminoacyl-tRNA for the second codon. And in the same way that GTP hydrolysis is linked to the release of transfer and translocation factors during elongation, so might we expect release of the initiation factors to be coupled to GTP release.

In quantitative studies of the role of f2 and GTP, Dubnoff, Lockwood and Maitra (1972) found that f2 is needed in stoichiometric amounts to bind fmet-$tRNA_f$ to a 30S subunit—poly-U,G complex; but binding of the H^3 labelled initiator is much increased when 50S subunits are also present, when f2 appears to act catalytically. This suggests that f2 must remain bound to 30S subunits after placing fmet-$tRNA_f$ in the 30p site, but can be released to act again on other 30S subunits after a 50S subunit has joined the initiation complex. When labelled GTP is used in the reaction mixture, the increase in binding of fmet-$tRNA_f$ to the 30S subunit which f2 promotes is stoichiometric with the binding of GTP. This suggests that one molecule of GTP is implicated in the binding of each molecule of initiator tRNA.

The GTP which is bound in the 30S complex is unstable, for no labelled GTP remains associated with 30S subunits when the initiation complex is isolated on sucrose gradients. About half of the label remains with the complex when it is isolated by gel filtration on sephadex columns. The GTP which is lost, however, is not hydrolysed, for it can be recovered intact. The GTP remaining in the complex is loosely bound for it is displaced by addition of free unlabelled GTP. But, when 50S subunits are added to the 30S initiation complex, any GTP which remains is immediately hydrolysed, as shown by the separation of an H^3 label in the GTP and a γ-P^{32} label. The complex containing GTP and the deficient complex which has lost its GTP during isolation behave in the same manner; both retain a H^3 label in fmet-$tRNA_f$ and generate 70S ribosomes containing the label when 50S subunits are added. All the bound fmet-$tRNA_f$ becomes reactive to puromycin as soon as 50S subunits associate with the complex; this suggests that GTP is not needed once the complex has been formed, although it is hydrolysed if present.

There have been differing reports of the effect of GMP-PCP on formation of the initiation complex, but Dubnoff et al. found that its influence seems to depend on the amount of f2 present. When both 30S and 50S subunits are present and low catalytic amounts of f2 are used, GMP-PCP cannot replace GTP and its substitution does not allow binding of fmet-$tRNA_f$. But at higher levels of f2 when the factor acts stoichiometrically, GMP-PCP can substitute for GTP with about 50% efficiency. Although fmet-$tRNA_f$ then sediments at the 70S position in association with ribosomes, the formyl-methionine moiety is not available to puromycin. When the GMP-PCP is removed by isolating the complex under conditions which cause loss of GTP, the fmet-$tRNA_f$ again becomes reactive to puromycin.

Retention of the nucleotide by the initiation complex may therefore prevent fmet-$tRNA_f$ from reacting with puromycin; GTP hydrolysis may be needed to release factor f2 from the ribosome. When release cannot take place because the GMP-PCP cannot be hydrolysed, the ribosome bound fmet-$tRNA_f$ is unreactive and the f2 protein cannot promote binding of initiator to any further complexes. This reduces the overall binding to a very small level when only catalytic amounts of f2 are present and ensures that the fmet-$tRNA_f$ which is bound cannot interact with puromycin. When larger amounts of f2 are present, of course, each factor can bind fmet-$tRNA_f$ to the initiation complex, although the formyl-methionine is unreactive because of the failure of the complex to release its GMP–PCP and f2.

This model has been supported by experiments in which the location of f2 has been examined after initiation. Lockwood, Sarkar and Maitra (1972) found that when GTP is used to promote binding of fmet-$tRNA_f$ to 30S subunits, the f2 factor is released from the 70S ribosomes as soon as 50S subunits are added. But when GMP-PCP is substituted, the f2 factor remains attached to the 70S ribosomes. This supports the idea that release of initiation factors is associated with a change in the conformation of the ribosome and that both events rely upon the cleavage of GTP. By following the release of labelled f2 from ribosomes after initiation, Fakunding and Hershey (1973) have suggested that the role of GTP hydrolysis is to increase its rate of removal from the 70S particle.

Ribosome Binding to the Messenger

Initiation factor f3 is needed to bind 30S subunits to messenger RNA; the effect of f3, which has a molecular weight of 21,200 daltons and one essential –SH group, is enhanced when f1 is present. One role of this factor is therefore to help 30S subunits to recognise initiation sites and to prevent them from binding securely to other sequences containing AUG codons. Dubnoff, Lockwood and Maitra (1972) have suggested that f3 does not prevent the formation as such of incorrect complexes, but destabilizes the binding of 30S subunits at false locations so that they leave the messenger and are free to seek another binding site.

The first observation which implied that there may be more than one class of initiation site and that translation of messengers may be specifically controlled was that of Hsu and Weiss (1969), who noted that the selectivity of E.coli ribosomes changes after infection with phage T4. Ribosomes of cells infected with phage T4 are much less efficient in translating RNA phage MS2 and bacterial messengers than the ribosomes of uninfected cells; although ribosomes of both uninfected and infected cells can translate the messengers coded by phage T4 DNA. Infection with T4 therefore reduces the ability of the ribosomes to translate messengers other than those coded by the phage itself.

Ribosomes of cells infected with T4 bind less phage R17 RNA than ribosomes of normal, uninfected cells; and the proportions of the different sites recognised within the phage RNA are also changed. Steitz, Dube and Rudland (1970) assayed ribosome binding by characterizing the sequences of RNA which are protected when ribosomes bind to initiation sites. They found that ribosomes from uninfected cells attach to the initiation regions of all three genes; but ribosomes from T4 infected cells discriminate against the coat protein and RNA replicase sites and continue to bind appreciably only to the initiation site of the maturation protein gene.

The selection of initiation sites is controlled by factor f3. Klem, Hsu and Weiss (1970) and Dube and Rudland (1970) identified the components responsible by washing initiation factors from the ribosomes with salt and then testing the characteristics in translation of washed ribosomes directed by the initiation factors of uninfected or infected cells. The source of the washed ribosomes does not control translation, whose specificity depends entirely upon whether the cells from which the initiation factors are derived have been infected. Lee-Huang and Ochoa (1971) showed that factor f3 is responsible; figure 3.9 shows that f3 of uninfected cells directs the translation of phage MS2 and T4 mRNA with equal efficiency but f3 extracted from infected cells shows a ten fold preference for the messengers coded by T4 phage.

The f3 fraction extracted from uninfected bacteria can be resolved into two active fractions by chromatography on phosphocellulose columns; Lee-Huang and Ochoa found that one fraction preferentially stimulates the translation of T4 messengers and the other promotes translation of the RNA phages and of E.coli messengers. Yoshida and Rudland (1972) confirmed that it is the f3 factor which discriminates between the different sites of RNA phages. Using a system in which unfractionated initiation factors promote translation of the maturation protein and coat protein genes of R17 RNA, they found that one class of f3 promotes translation of the coat protein gene whereas the other class directs attachment to the maturation protein gene.

This suggests that E.coli cells contain at least two molecular species of f3. One, f3α, recognises messengers of the bacterial cell and is active with RNA phages, especially with the coat protein gene; it also directs translation of the messengers produced early in infection by phage T4. The second class, f3β, responds efficiently to the messengers synthesised by phage T4 later in infec-

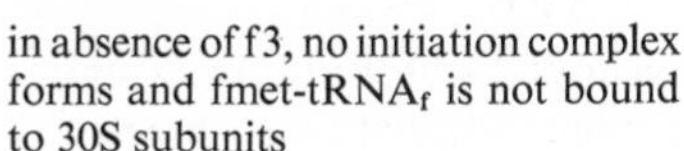

in absence of f3, no initiation complex forms and fmet-tRNA$_f$ is not bound to 30S subunits

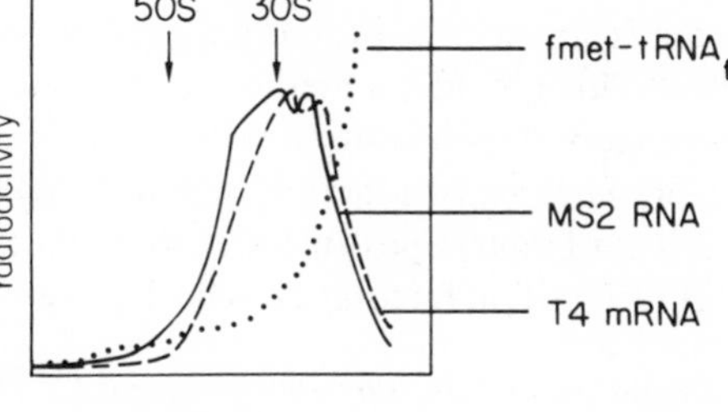

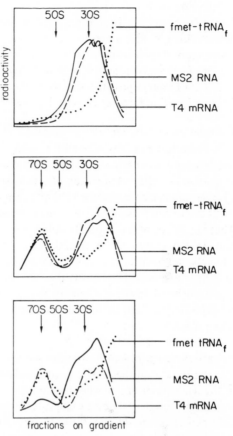

in the presence of f3 from E. coli, an initiation complex is formed, 50S subunits associate with it, and fmet-tRNA$_f$ is bound to the resultant 70S ribosomes. Either T4 mRNA or MS2 RNA is effective in forming the complex

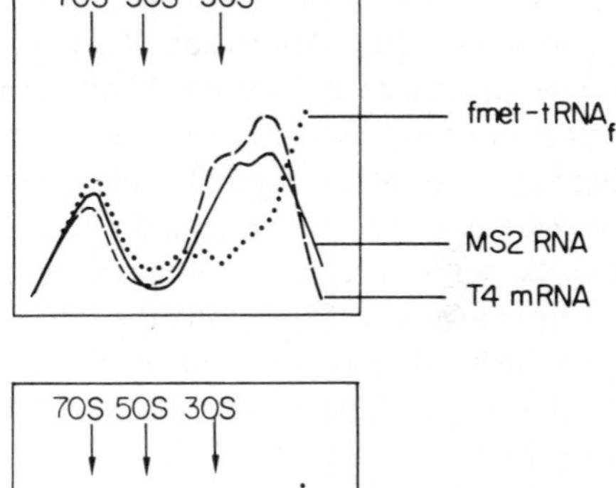

in the presence of f3 from cells of E. coli infected with phage T4, mRNA of the phage forms an initiation complex and binds fmet-tRNA$_f$ to the ribosome, but MS2 RNA is inactive

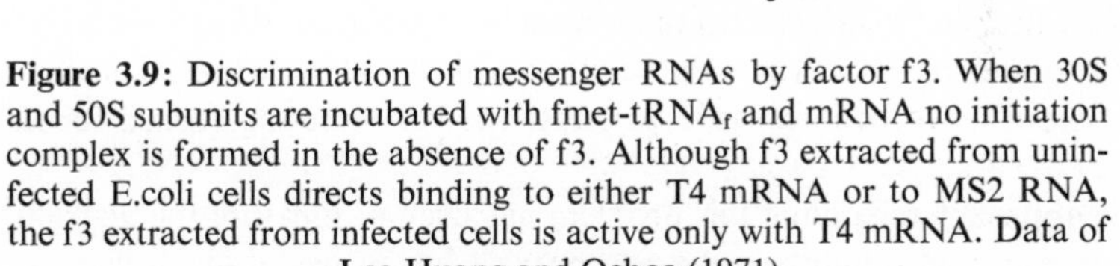

Figure 3.9: Discrimination of messenger RNAs by factor f3. When 30S and 50S subunits are incubated with fmet-tRNA$_f$ and mRNA no initiation complex is formed in the absence of f3. Although f3 extracted from uninfected E.coli cells directs binding to either T4 mRNA or to MS2 RNA, the f3 extracted from infected cells is active only with T4 mRNA. Data of Lee-Huang and Ochoa (1971)

tion; it is less active with RNA phage templates and directs binding only to the maturation protein gene. During the later stages of infection by T4, only f3β is able to initiate synthesis of proteins. Recognition of messengers does not depend solely upon f3, however, for salt washes of E.coli ribosomes contain "interference factors" which inhibit the action of f3. Lee-Huang and Ochoa (1972) fractionated these factors into two classes; the iα factor inhibits the translation promoted by f3α and the iβ factor inhibits the activity of f3β. Interactions between the initiation and interference factors must presumably control the specificity of translation.

Infection of E.coli with phage T4 generates a situation in which the f3α factor is inactive in promoting protein synthesis, whereas the f3β factor is active. This provides a method for the phage to direct protein synthesis to the messengers for which it codes later in infection. We do not know what role the presence of two factors in uninfected bacteria plays in protein synthesis; and it is of course possible that there may be a more sophisticated hierarchy of f3 recognition factors yet to be discovered.

Distribution of Ribosomes in Bacterial Cells

If initiation requires 30S subunits alone, a ribosome which undertakes synthesis of more than one polypeptide chain must dissociate into its subunits at some point after termination of one chain but before initiation of the next. This demands that ribosome subunits cannot remain permanently associated in vivo, but must exchange their partners at each round of protein synthesis. That this happens in E.coli cells and in yeast was shown by Kaempfer, Meselson and Raskas (1968) and Kaempfer and Meselson (1969) in experiments in which bacteria with heavy isotopes were transferred into a medium containing only light isotopes. The density distribution of ribosomes shows a progressive replacement of heavy ribosomes by the hybrid species 50S[L]:30S[H] and 50S[H]:30S[L]. This indicates that dissociation and reassociation must have taken place.

The precise role of the 70S ribosome in cellular metabolism has been controversial; the problem is whether ribosomes dissociate into subunits when they terminate synthesis of a polypeptide chain and can then only associate at the next initiation, or whether ribosomes leave messengers in the 70S form and only subsequently dissociate into subunits. The first model predicts that bacterial cells should contain only free 30S and 50S subunits, with all 70S particles attached to mRNA. In the second case, however, the cell must contain free 70S ribosomes as well as subunits; this implies that some special mechanism dissociates the free ribosomes to generate the 30S subunits needed for initiation.

One approach to solving this problem has been to measure the distribution of ribosomes and subunits in the cell, but this has proved difficult because the results obtained depend heavily on the experimental conditions used to isolate the ribosomes. An approach which has proved more fruitful has been to show that a dissociation factor, the f3 initiation protein, is responsible for dissociating free 70S ribosomes into subunits, presumably at some time after they have been released as such from messengers.

Using a method to prepare polysomes and ribosomes rapidly from E.coli so that there should be no time for changes in the distribution to take place after isolation, Mangiarotti and Schlessinger (1966, 1967) observed that virtually all the free ribosomes are in the form of individual subunits, with less than 2% in the form of free 70S ribosomes. They suggested, therefore, that

70S ribosomes exist only to undertake translation. Direct attempts to measure distributions are not reliable, however, because different preparative conditions yield different profiles. Phillips and Franklin (1969) found that when Na^+ ions are used in the lysing medium, extracts contain only polysomes and 30S and 50S subunits; when K^+ or NH_4^+ is used instead, there is a prominent peak of 70S ribosomes.

It was not clear whether the Na^+ condition represents a false dissociation of 70S ribosomes or the K^+/NH_4^+ a spurious association of subunits, especially as distributions obtained under one condition seemed to be stable when dialysed into the other. Beller and Davis (1971) showed, however, that free ribosomes are dissociated by Na^+ ions in certain conditions, depending particularly on the concentration of magnesium ions; at 5 mM Mg^{2+} free ribosomes are dissociated into subunits, but sodium ions have no effect at 15 mM Mg^{2+}. Figure 3.10 shows that at 10 mM Mg^{2+} free ribosomes are dissociated in Na^+ gradients but ribosomes engaged in protein synthesis on messengers are more stable and do not dissociate; 70S particles obtained by degradation of polysomes (always a possible source of contamination of the 70S peak in extraction procedures) are also stable. This implies that bacterial cells contain free 70S ribosomes which are less stable than those engaged in protein synthesis and are therefore dissociated in vitro by Na^+ ions; however, it is difficult to estimate what proportion they comprise of the total distribution.

Run-off experiments, when ribosomes are allowed to complete synthesis of their current proteins but are not allowed to initiate any new chains, should show directly what happens at termination. If dissociation is automatic, subunits should accumulate; alternatively, ribosomes might be released to swell the peak of 70S ribosomes. Kohler, Ron and Davis (1968) found that depriving cells of amino acids or treating them with puromycin, which causes the release of ribosomes from messengers, yields only 70S particles; the level of subunits remains essentially unaltered. Polysomal and run-off ribosomes behave differently when the magnesium ion concentration is lowered, free ribosomes dissociating at 1 mM Mg^{2+}, but polysomal being more stable and dissociating only at 0·5 mM Mg^{2+}. There is usually a difference also between "native" subunits [those obtained from cells as such] and "derived" subunits [those obtained by dialysis of 70S particles from polysomes]; Schlessinger, Mangiarotti and Apirion (1967) found that derived subunits readily reassociate in 10 mM Mg^{2+} but native subunits cannot do so. The 70S ribosomes found in cells behave in a manner similar to the run-off ribosomes; this suggests that cells contain free 70S ribosomes which are generated when protein synthesis is terminated.

Experiments in vitro, however, have suggested a different conclusion. Using run-off conditions, Kaempfer (1970) incubated together in the ratio $1:10^3$ a mixture of isotopically heavy and light ribosomes extracts; heavy and light ribosomes should run off their messengers together. If ribosomes do not

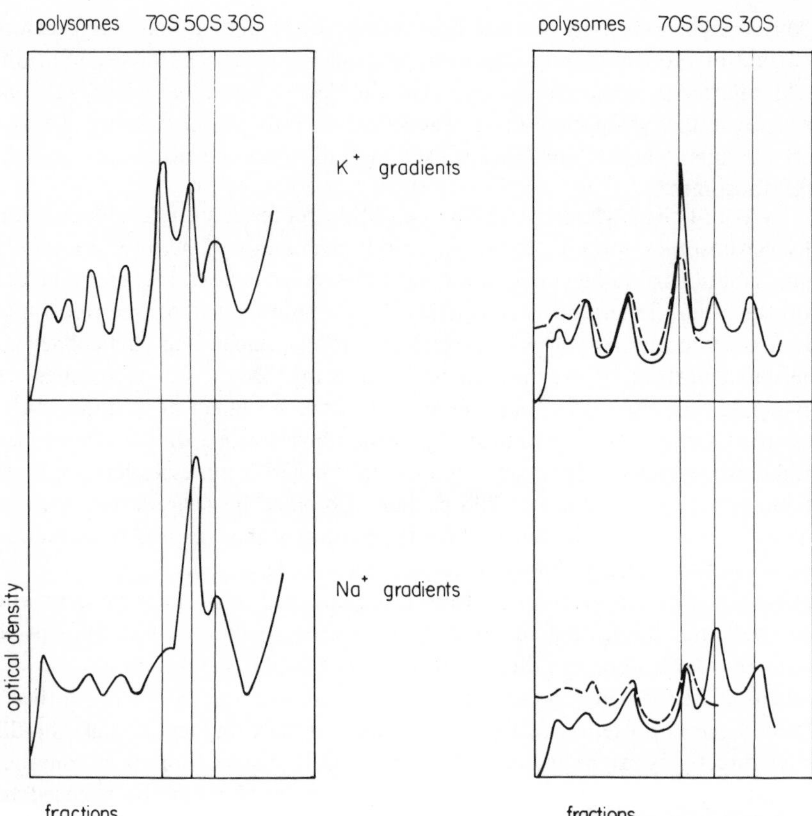

Figure 3.10: Sodium ions dissociate free ribosomes into subunits. The upper two preparations show sedimentation gradients in K^+; the lower two gradients show the ribosome fractions in Na^+ ions. Comparison of the two gradients on the left shows that the Na^+ ions dissociate free ribosomes into subunits (in the presence of 10 mM Mg^{2+}—at lower concentrations of Mg^{2+} sodium ions are even more effective and at higher Mg^{2+} levels they are less effective). The two gradients on the right were prepared from cells labelled with C^{14} amino acids; the label identifies ribosomes engaged in protein synthesis. The 70S peak of the K^+ gradient includes both free and polysomal ribosomes; the 70S peak of the Na^+ gradient retains the C^{14} label and therefore identifies polysomal ribosomes which are stable even in Na^+ ions, although the free ribosomes are lost from this peak and appear as subunits. (—) = absorbance, (---) = radioactivity. Data of Beller and Davis (1971)

dissociate at termination, the heavy 70S particle should remain in this form. If dissociation takes place, however, all the 70S ribosomes should be converted into hybrid form; because of the great excess of light subunits which must be generated when the light ribosomes run-off their polysomes, each heavy subunit should find a light subunit to reassociate with. The conversion of heavy to hybrid ribosomes takes place, at incubation temperatures of either 0°C or 37°C, showing that subunits are exchanged.

An apparent exchange of subunits, however, can be generated by an artefact which makes density centrifugation of ribosomes unreliable. Infante and Baierlein (1971) found by centrifuging ribosomes of sea urchin eggs at high speeds that the distribution depends not only on ionic conditions but also upon the gravitational field and the ribosome concentration. Hydrostatic pressure seems to shift an equilibrium between ribosomes and subunits towards dissociation. The result of this shift is that the apparent sedimentation velocity of the ribosomes is changed; such changes have sometimes been interpreted as evidence that there are changes in the conformation of ribosomes at particular stages of protein synthesis, but these effects are instead an artefact of centrifugation.

The effect of hydrostatic pressure on the heavy ribosomes is to mimic the formation of hybrid ribosomes. Fixation with glutaraldehyde before the ribosomes are centrifuged prevents dissociation in the gradient; for Subramanian and Davis (1971, 1973) showed that when heavy and light ribosomes are allowed to run off polysomes at 0°C and are fixed with glutaraldehyde before centrifugation, only two separate peaks of heavy and light ribosomes are found. The absence of hybrid ribosomes implies that their presence in experiments in which the fixation step is omitted represents an artefact induced by hydrostatic pressure.

But when ribosomes are allowed to run off at 37°C, a peak of hybrid ribosomes is found, even when the fixation step is included. However, this peak can be produced just as readily whether the two sets of ribosomes are allowed to run off together or are run off in separate incubations and are then mixed ten minutes later. This means that although an exchange of subunits takes place, it occurs well after chain termination, presumably because incubation conditions promote a ready dissociation and reassociation of free ribosomes independent of protein synthesis. The temperature dependence of this effect suggests that it may be catalysed by the dissociation factor, whose action is reversible.

In vitro experiments are therefore unable to reveal the state in which ribosomes are released from messengers under physiological conditions. At 0°C, ribosomes are released as 70S particles; and the presence of hybrid ribosomes results only from an artefact of centrifugation. But low temperature conditions may not be typical of those prevailing in the cell. At 37°C, subunits are exchanged to generate hybrid ribosomes in a process which does not depend upon protein synthesis; this therefore provides no information about the state in which they are released from the messenger.

A similar rapid equilibrium between 80S mammalian ribosomes released upon termination of protein synthesis and subunits present in the incubation has been demonstrated by Falvey and Staehelin (1970). The 80S subunits reformed by this exchange continue to exchange subunits but at a much slower rate. This favours a model in which ribosomes are released from the messenger in the form of subunits which immediately associate to reform 80S ribosomes, the state favoured by the equilibrium conditions. But it is clear from distribution profiles of bacterial extracts that 70S ribosomes exist as such in the cell when not engaged in protein synthesis, whether released in this state from messengers or generated by the rapid association of released subunits.

The Ribosome Subunit Cycle

The existence of free 70S ribosomes implies that some event must take place to cause their dissociation before the next initiation can take place. Native subunits obtained from cells are some five to nine times more active with f2 RNA as messenger than are derived subunits obtained from polysomes; as there is no difference between the two types of subunit when poly-U is used as messenger, Eisenstadt and Brawerman (1967) suggested that the difference may be concerned with initiation. They were able to find an initiation factor fraction in the supernatant of a high speed centrifugation of E.coli cell extracts which can abolish the difference between native and derived subunits. They proposed that this factor might be released from the 30S subunit when the ribosome commences translation; and since it is not present in soluble extracts, it must presumably be transferred immediately to a new 30S particle about to undertake initiation.

A more detailed model for the role of such a factor in ribosomal dissociation has been proposed by Kohler, Ron and Davis (1968) and is illustrated in figure 3.11. They suggested that ribosomes dissociate into subunits only when a dissociation factor complexes stoichiometrically with the 30S subunit. When the 30S subunit carrying the factor forms an initiation complex, the factor is released, enabling a 50S subunit to join the complex. The activity of the dissociation factor therefore regulates the cellular balance between free 70S ribosomes and subunits.

A factor which associates only with 30S subunits and is released at initiation is, by definition, an initiation factor. By testing a crude preparation of initiation factors for ability to dissociate 70S ribosomes in vitro, Subramanian, Davis and Beller (1969) identified a dissociation factor with the predicted ability to dissociate ribosomes stoichiometrically. Subramanian and Davis (1970) found that the activity does not reside in a new protein but in the f3 factor previously identified by its role in binding 30S subunits to the messenger. The dissociation activity shown in figure 3.12 has no effect on polysomes and seems to act on 70S ribosomes only after their release from messengers at termination. The activity is not species specific since extracts from Pseudomonas aeruginosa or B. megaterium are just as active as those of E.coli.

If f3 cycles between a subunit bound and a free form at each initiation, all

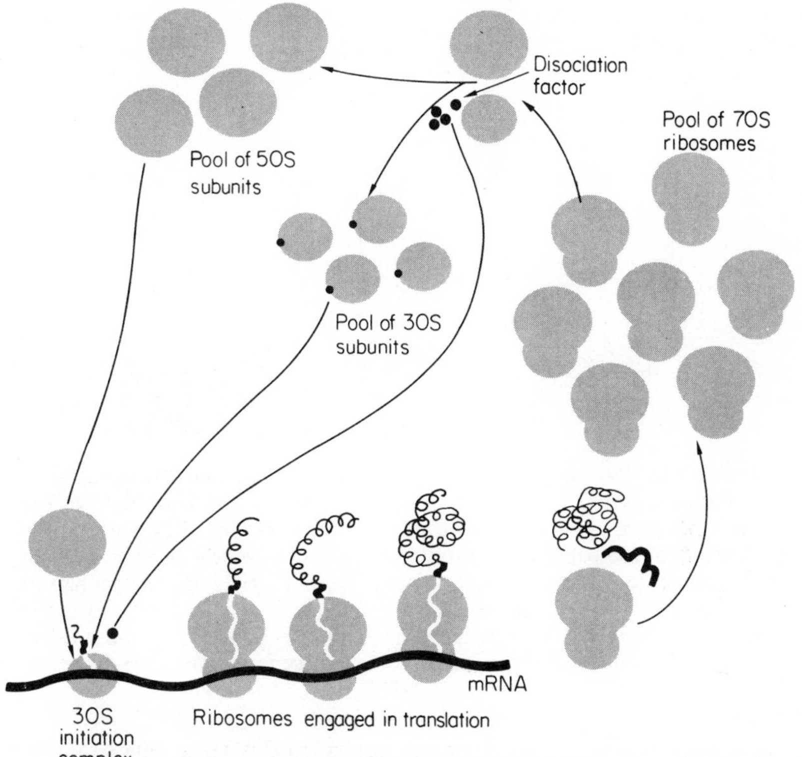

Figure 3.11: Ribosome subunit cycle. When translation is terminated, the 70S ribosome, tRNA and a completed polypeptide chain separate. The ribosome joins a pool of 70S monomers. Ribosomes are withdrawn from the pool when they complex with a dissociation factor to give free 30S and 50S subunits. When a 30S subunit attaches to a messenger to form an initiation complex it releases its dissociation factor so that it can associate with a 50S subunit

70S ribosomes should lack the factor, whether they are engaged in protein synthesis or are free and awaiting dissociation. By using S^{35} labelled f3, Sabol and Ochoa (1971) confirmed that the only particles in the cell which possess f3 are the free 30S subunits. Each 30S subunit appears to have one binding site for f3. When f3 is added to 70S ribosomes, they dissociate into subunits; figure 3.12 shows that the extent of dissociation is proportional to the amount of f3 added.

When a 70S ribosome is generated by association of 30S and 50S subunits (whether by formation of an initiation complex or by increasing the Mg^{2+} concentration), the labelled f3 is released from the 30S subunit. The irrelevance of the reason for association of the subunits suggests that the release of 30S bound f3 is a consequence of subunit association as such, perhaps

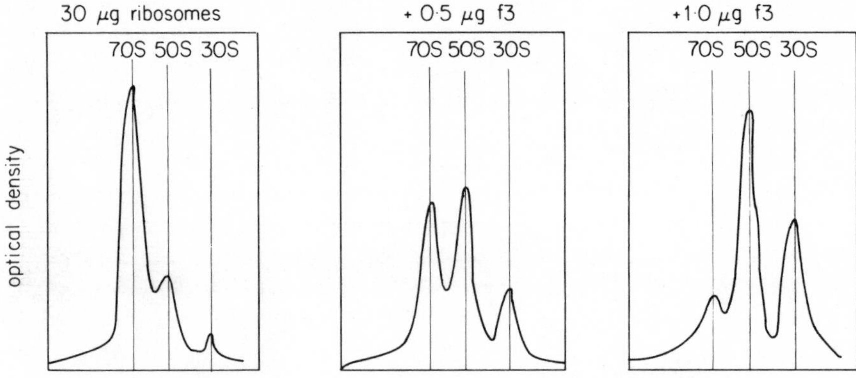

Figure 3.12: Dissociation of 70S ribosomes into 50S and 30S subunits by initiation factor f3. The profile on the left shows the sedimentation pattern on sucrose gradients of a preparation of ribosomes, most of which are in the form of 70S particles. Addition of 0·5 μg of f3 (centre) dissociates about half of the ribosomes; and addition of 1·0 μg of f3 (right) dissociates almost all the ribosomes. These results were obtained at 2 mM Mg^{2+}; f3 is less effective when the concentration of magnesium is raised. Data of Subramanian and Davis (1970); similar results have been obtained by Sabol et al. (1970)

because some conformational change results. In this sense, 60S subunits and factor f3 are antagonists for binding to 30S subunits. The cycle of the dissociation factor has been reviewed by Davis (1971).

Both the forms of f3 identified in E.coli act as dissociation factors, apparently with equal efficiency. When bound to a 30S subunit, an f3 molecule therefore has two effects; it maintains the subunit in a state in which it cannot associate with a 50S subunit and it restricts the binding sites on messenger RNA which can be recognised. After the 30S subunit has complexed with messenger RNA, the f3 protein is released; this allows a 70S ribosome to be generated for translation of the message. The dual functions of f3 are therefore readily reconciled. One model is to suppose that both result from one molecular interaction with the 30S subunit, which both directs it to complex with an appropriate initiation site and at the same time prevents attachment of the 50S subunit; however, this is rendered less likely by the demonstration of Lee-Huang and Ochoa (1972) that the two effects are independent, at least insofar as the interference factors can inhibit messenger recognition without influencing dissociation activity.

An alternative explanation for the action of f3 in dissociating 70S ribosomes has been proposed by Kaempfer (1971, 1972), who suggested that the action of f3 is to prevent the association of 30S and 50S subunits, rather than to promote the dissociation of 70S particles. As with so many aspects of the ribosome cycle, the effect of f3 depends heavily upon the concentration of magnesium ions; it is most effective at 1·5 mM Mg^{2+}, progressively less active at 5 mM and 10 mM and there is no effect at all at 20 mM. Kaempfer argued that the higher concentrations are likely to be the more typical of intracellular conditions, in which case the activity of f3 in dissociating ribosomes in vitro might be caused when it complexes with 30S subunits freed by spontaneous dissociation; by preventing these subunits from reassociating with 50S subunits, an equilibirum between subunits and ribosomes might be shifted in favour of dissociation. But as Davis (1971) has pointed out, whether f3 acts directly to dissociate 70S particles or achieves the same end by removing 30S particles from participation in a reversible equilibrium, its function remains to control the supply of free subunits for initiation.

Termination of the Polypeptide Chain

Factors for Release of Polypeptide from tRNA

During protein synthesis, polypeptide chains are attached to tRNA molecules by the ester bond between the most recently added amino acid and the terminal adenosine of its tRNA. When the last amino acid has been added to the chain, this link must be broken in order to release the completed protein and allow re-use of the tRNA. The spontaneous rate of hydrolysis of polypeptidyl-tRNA is rather low under physiological conditions and protein synthesis in vitro on synthetic templates usually yields only polypeptidyl-tRNA species. Polypeptide is freed from tRNA only when the messengers contain a high proportion of U and A bases; using defined synthetic messengers, Last et al. (1967) were able to show that the UAA codon causes proper release. This implies that chain termination does not take place automatically at the ends of messengers but needs an appropriate signal, that is to say, one of the nonsense codons identified by genetic means as signals for termination (see page 28).

The presence of a sequence which cannot be translated is not sufficient, for when Bretscher (1968b) used f2 RNA in an in vitro system lacking the aminoacyl-tRNA corresponding to its seventh codon, hexapeptidyl-tRNA was formed, not free hexapeptide. When the seventh codon is mutated to an amber sequence, however, hexapeptide is released from tRNA. This implies that there must be some active termination process which separates polypeptide from tRNA in response to nonsense codons.

An early idea that a special tRNA carrying no amino acid might recognise nonsense codons was excluded by experiments using defined systems which contained only known aminoacyl-tRNAs. Rather does chain termination, like initiation and elongation, depend on extra-ribosomal protein factors. An assay system developed by Scolnick et al. (1968) to test for termination activity in cell extracts involves the sequential use of the AUG initiator codon and one of the three terminator triplets, UAA, UAG or UGA. First, f-H^3-met-$tRNA_f$ is bound to 30S subunits or ribosomes in the presence of AUG. Then the nonsense trinucleotide is added together with the cell extract; and the release of radioactivity from the ribosome is measured (reviewed by Beaudet and Caskey, 1972).

Two proteins, which can be separated by chromatography on columns of DEAE-sephadex, have been identified by this assay; both are active as monomers and one is about 44,000 daltons and other some 47,000 daltons. Both catalyse the same release reaction. But R1 is active under direction from either UAA or UAG; and R2 responds to UAA or UGA. About 60% of the UAA-dependent release activity resides in R1. By purifying the factors, Klein and Capecchi (1971) calculated that an E.coli cell contains 500 molecules of R1 and 700 molecules of R2 under conditions in which there are about 30,000 ribosomes in the cell.

The R1 and R2 proteins recognise the sequences of their appropriate nonsense codons directly, for Scolnick and Caskey (1969) found that the factors cause retention of labelled oligonucleotides by ribosomes on nitrocellulose filters; R1 directs binding of UA[H^3]G and UA[H^3]AA whereas R2 promotes retention of only the latter. Although UGA was not tested directly—because of the lack of a tritiated form—it can compete for R2 but not R1. The factor R1 has equal affinity for UAA and UAG; R2 has slight preference for UGA compared with UAA. The R1 directed binding of UAG is inhibited by R2 and UGA, which suggests that there is only one ribosome site at which either R1 or R2 may act. Both ribosome subunits are required for the termination reaction.

Nonsense codons are usually read within the context of the message and an assay for termination using natural mRNA was developed by Capecchi and Klein (1969), who made use of a mutant of R17 RNA which has an amber triplet as its seventh codon in place of the usual CAG coding for glutamine. By omitting the thr-tRNA corresponding to the sixth codon, they were able to isolate a complex in vitro in which the ribosome has translated the first five codons to synthesise the pentapeptide fmet-ala-ser-asn-phe. When thr-tRNA is added, the sixth amino acid is added and the complex can terminate the chain when factor R1 is added. The substrate polypeptide for release must be in the P site of the ribosome, for replacing GTP with GMP-PCP when the thr-tRNA is added blocks the release reaction. Using a similar system, Beaudet and Caskey (1970) showed that when suppressor tRNA which recognises UAG is added, it competes with the R factor for the codon. This suggests that the release factors must enter the ribosome at the A site usually used by aminoacyl-tRNA.

These experiments imply that the triplet sequences of the nonsense codons are sufficient for termination; this is confirmed by experiments to follow the release of coat protein after synthesis in vitro directed by the RNA of phages f2 and R17. The coat protein gene of each phage terminates with two different nonsense codons in immediate succession: UAAUGA. Beaudet and Caskey showed that either R1 or R2 is equally effective in stimulating release of completed coat protein from tRNA (using the assay that coat protein linked to tRNA is retained on DEAE-cellulose whereas free protein is eluted). Both factors fail to influence the rate or extent of protein synthesis but stimulate release from tRNA some three or four fold. (The release which takes place in the absence of added factors is presumably catalysed by endogenous factors.) The effectiveness of either factor implies that the ochre codon located first is adequate for release; the second codon is not needed.

By using antisera to the release factors, Capecchi and Klein (1970) showed that R factor mediated release is the only form of chain termination. Anti-R1 inhibits the release of hexapeptide from the amber mutant in the seventh codon of the coat protein gene of R17 RNA; anti-R2 is ineffective. When either antiserum preparation is added to an in vitro system directed by normal R17 RNA, coat protein is released from the ribosomes as usual to associate with R17 RNA and sediments in the gradient at about 30S. When both antisera preparations are added, however, completed coat protein molecules cannot be released from the ribosomes. Adding an excess of either R1 or R2 to the inhibited preparation allows the coat protein to be released. This supports the idea that the coat protein gene is effectively terminated by the ochre codon, which can use either R1 or R2 to effect chain termination.

R factor proteins have been found in mammalian cells by using the formylmethionine release reaction (reviewed by Beaudet and Caskey, 1972). Although fmet-tRNA$_f$ is not formylated in these cells, their ribosomes have a general "stickiness" for fmet-tRNA$_f$ (whether obtained by formylation of endogenous met-tRNA$_f$ or from E.coli cells) which enables the initiator species to bind even in the absence of any template. Its release can then be measured when protein factors and oligonucleotides are added; the reaction probably works with fmet-tRNA$_f$ and not met-tRNA$_f$ because the formylated species enters the P site whilst the normal aminoacyl-tRNA enters the A site (as judged by reactivity to puromycin).

Using this assay, Beaudet and Caskey (1971) measured the activity of R factor preparations derived from rabbit reticulocytes, guinea pig liver, chinese hamster liver and E.coli; the mammalian factor does not seem to fractionate into more than one species and reticulocyte R factor seems to be a protein of molecular weight about 255,000. Factors from all eucaryotic species show much the same reactivity; UAAA and UGAA are most active, with UAGA showing about 40–60% of this activity. (The reaction does not work with trinucleotides, but with the E.coli system the specificity of recognition is not influenced by adding an extra A to the test oligonucleotide.)

Release Reaction on the Ribosome

In E.coli, a heat labile protein termed the S factor seems to stimulate the release reaction by helping the R factors to recognise their appropriate codons. Its effect depends on the presence of GTP, however, for Goldstein and Caskey (1970) showed that S stimulates formation of the intermediate codon-R factor-ribosome complex in the absence of GTP; but in the presence of GTP or GDP S instead assists dissociation of the complex. Without knowing the precise physiological conditions under which release takes place, it is difficult to decide the role of the S factor, but one possible scheme is to suppose that first of all S stimulates the binding of R factor at the A site of the ribosome; when the ester bond between polypeptide and tRNA is hydrolysed, GTP joins with the complex and S assists its dissociation.

In the presence of ethanol, both R1 and R2 can catalyse the release of formyl-methionine from ribosomes bound to AUG; by analogy with the effect of ethanol on peptide bond formation, it seems likely that the same peptidyl transferase enzyme may catalyse both elongation and termination. Tompkins, Scolnick and Caskey (1970) brought further support to this idea when they found that those antibiotics which inhibit binding of aminoacyl-tRNA to ribosomes also inhibit the recognition part of release, but antibiotics which inhibit peptidyl transferase inhibit the release reaction itself. One possibility is that peptidyl transferase might facilitate a nucleophilic attack on the peptidyl-tRNA ester bond, perhaps with a nucleophilic group on factor R directly participating. An alternative is that R might indirectly encourage peptidyl transferase to cleave the ester bond.

Utilization of Nonsense Codons

Nonsense codons located at the ends of genes have been identified only in the RNA phages. Nichols (1970) found that the sequence at the end of the coat protein gene of R17 RNA consists of two nonsense codons, UAA and UGA, in immediate succession. Studies in vitro suggest that the first nonsense triplet is probably adequate for termination, so that the second is superfluous.

That the maturation protein gene may end in a UAG codon was suggested by the observation of Remaut and Fiers (1972) that when MS2 RNA infects E.coli containing an amber suppressor tRNA—which inserts an amino acid in response to UAG—a protein slightly larger than the maturation protein is synthesized in its place. This suggests that UAG is the termination codon of this gene and that reading past this codon allows translation to continue until a different nonsense codon is met in phase, causing synthesis of a protein longer than usual. This conclusion has been confirmed by the nucleotide sequence determined by Contreras et al. (1973).

Observations of readthrough suggest also that the coat protein gene of phage Qβ may end in UGA; Moore et al. (1971) and Weiner and Weber (1971) found that termination of this protein is leaky in E.coli cells so that a longer protein than usual is produced with appreciable frequency. The characteristics of this readthrough are consistent with those displayed by UGA nonsense codons.

The properties of the RNA phages therefore confirm the conclusion suggested by the isolation of three classes of nonsense mutation in bacterial genes: any one of the codons UAA, UAG, UGA may be used at the end of a gene to cause termination of the synthesis of its protein.

The characteristic patterns of suppression of the various nonsense codons suggest that the ochre UAA codon is largely used in the E.coli cell itself. Whilst amber and UGA suppressor strains of E.coli can be isolated which work with efficiencies of up to 50–60%, even the most efficient ochre suppressors are much weaker, at about 5% only. If a nonsense codon is extensively used in bacteria for chain termination, its suppression would affect not only nonsense mutants, but also the normal chain terminator signals. Extra sequences would thus be added to many proteins. This means that strong suppression of these codons would probably be lethal to the organism, so that only weak suppressors could be tolerated. The strong suppression of UAG and UGA codons suggests, therefore, that they are not usually used as chain terminators. That ochre is used in vivo is supported by the finding of Person and Osborn (1968) that, when an E.coli strain possessing an amber suppressor suffers its conversion to an ochre, the bacterial cells grow more slowly and can support growth of phage T4 only at reduced levels.

This raises the question of why the cell recognises UAG and UGA as terminators, even though it does not appear likely usually to use them as such; it should be more advantageous for these codons to represent amino acids (see page 34). Another query, indeed, is why the termination sequences should be triplets at all since they are recognised by proteins and not by tRNAs. It is interesting that the pattern of recognition shown by R1 is shown by ochre suppressor tRNAs, which must also recognise amber codons because the third base of the codon is read ambiguously (see page 214); there is, however, no known counterpart to the R2 recognition of UAA and UGA amongst transfer species.

Triplets might be used for termination because of the geometry of the ribosome in translation or, more likely perhaps, because of some past function ascribed to these codons during evolution of the code. Suppression of nonsense mutants located at different sites even within the same cistron can vary greatly in efficiency, which led Salser, Fluck and Epstein (1969) to suggest that reading of nonsense codons may depend on the surrounding nucleotide sequence. But since suppression involves competition between the release mechanism and the suppressor tRNA, it is difficult to say whether release recognition or the reaction of suppressor tRNA is influenced. One possible explanation, however, is that the nonsense triplets may direct termination more efficiently in some contexts than in others. In this case, the most effective termination sequences might be any length greater than three nucleotides. Against this idea is the efficiency of the R factors in vitro with triplets and phage mutants.

CHAPTER 4

The Ribosome

Components of the Ribosome Particle

Folded Structure of the Nucleoprotein Chain

The ribosome is a complex cell organelle which carries out the many catalytic activities required to synthesize proteins. We do not know the precise arrangement of the components of each subunit in the sense of defining the position of each protein and the rRNA chains in three dimensions. Nor can we yet resolve the important problems of how messenger RNA passes through the ribosome and of the nature of the interactions which maintain the association of the two subunits.

But a structural view of the ribosome can be derived from studies of the order in which proteins are added to rRNA during assembly of the particle; and in general, the order in which proteins are removed from the completed ribosome by increasing concentrations of salt is the reverse of the order of assembly. This suggests that a small number of proteins binds directly to rRNA and forms the centre of the particle as further proteins are added to the exterior. A picture of the ribosome in functional terms is suggested by experiments in which ribosome particles lacking one of the ribosomal proteins are tested for ability to synthesize proteins. Specific defects can in this way be correlated with the absence of individual proteins. These experiments show that the ribosome has a structure in which groups of proteins interact to create several independent active centres within it.

Models for the general structure of the ribosome have been proposed from experiments which assume that all ribosome particles conform to one nucleoprotein structure. The compact structure of the ribosome is revealed by experiments in which some of cations necessary for its maintenance are removed. Gavrilova, Ivanov and Spirin (1966) found that washing E.coli ribosomes in ammonium chloride dissociates them into subunits which have lost much of their endogenous content of magnesium ions. When the ionic strength of the NH_4Cl is then reduced, the subunits sediment more slowly, although their rRNA remains intact and no proteins are lost. Subunits therefore appear to consist of a single nucleoprotein strand which is folded in a compact manner in part dependent on the presence of magnesium ions. In a study of the ionic requirements of the ribosome, Weiss, Kimes and Morris

(1973) found that 80% of the sites occupied by cations can be filled by any one of several divalent species, but the remaining 20% demand magnesium (the physiological ion) or manganese.

Biophysical studies also show that the components of the ribosome are tightly packed into each subunit. Wittman and Stoffler (1972) reported that the 30S subunit fits the rather flat structure of an oblate spheroid of weight $0{\cdot}55 \times 10^6$ daltons and dimensions $56 \times 224 \times 224$Å. The 50S subunit weighs $1{\cdot}55 \times 10^6$ daltons and has a more symmetrical structure, more nearly spherical in shape. Electron microscopy of negatively stained rat liver ribosomes gives a variety of apparent structures which Nonomura, Blobel and Sabatini (1971)

Figure 4.1: Model for the bacterial ribosome of Nonomura, Blobel and Sabatini (1971)

reconciled to give the model shown in figure 4.1; this visualises the ribosome more as an armchair than a sphere.

Estimates of the extent of secondary structure in ribosomal RNA have varied. Furano et al. (1966) found that more than 90% of the phosphates of rRNA can bind to acridine orange; Cotter and Gratzer (1971) found that when tritiated ribosomes are allowed to exchange their H^3 for unlabelled hydrogen, most of the label is lost from rRNA, although the proteins retain it. Also consistent with the idea that rRNA is accessible to external reagents is the observation of Pinder and Gratzer (1972) that the same oligonucleotides are released upon degradation with pancreatic ribonuclease of either ribosomes or ribosome particles unfolded in EDTA. According to Fellner et al. (1970), a little more than one third of the 16S rRNA is inaccessible to T1 ribonuclease. These experiments suggest that apart from the sequences of rRNA which are directly bound to ribosomal proteins, the polynucleotide chain may be located near to the surface of the ribosome and is in general accessible.

When basic proteins bind to duplex regions of nucleic acid they raise its melting temperature appreciably. But Cotter, McPhie and Gratzer (1967) found that free ribosomal RNA and rRNA within the ribosome have very similar melting curves; the absence of stabilization of the duplex structure suggests that proteins may be bound to the single stranded rather than duplex regions of rRNA. The sequencing studies of Fellner et al. (1972) show that at least some regions of 16S rRNA can form hairpin-like duplex segments when the polynucleotide chain folds back on itself. Melting curves suggest that a little more than half of the rRNA is implicated in duplex formation; but definition of the extent of duplex formation between hairpins and that between different parts of the molecule must await its complete sequence. The sequences of those regions protected from ribonuclease by individual ribosomal proteins may reveal what features of the RNA molecules are recognised directly by ribosomal proteins and which sequences play some other role in ribosome function.

Major RNAs of Bacterial Ribosomes

The 30S subunit of bacterial ribosomes contains an RNA which sediments at 16S; this corresponds to a chain length of about 1500 nucleotides. The 23S rRNA found in 50S subunits is about twice this length. A large number of proteins is found in each subunit, the total in a bacterial ribosome being about fifty-five. In eucaryotic cells, the small 40S subunit contains an RNA which sediments at about 18S and the principal RNA of the large 60S subunit sediments at about 28S. The total number of proteins in both subunits may be as great as eighty. Large subunits of both bacterial and eucaryotic ribosomes also contain one molecule of a small RNA which sediments at about 5S and is some 120 nucleotides long. (The ribosomes of mitochondria and chloroplasts appear to be different both from those of the surrounding cytoplasm and from those of bacteria.)

The principal RNAs of the subunits of bacterial or eucaryotic ribosomes behave as single polynucleotide chains. But chemical estimates of the number of 3′ hydroxyl termini made by Midgley and McIlreavy (1967a, 1967b) suggested that although the 16S rRNA of the small bacterial subunit comprises a single polynucleotide chain, the 23S rRNA may consist of two chains, each about the same size; these might appear to comprise only one larger chain either because their structures sediment more rapidly than would be predicted from their molecular weights alone, or because they are linked in some way other than by the usual covalent bond formation. Although the discrepancy between the number of end groups and the apparent length of 23S rRNA has not been formally resolved, it is commonly assumed that 23S rRNA in fact constitutes only one polynucleotide chain.

An unusual feature of ribosomal RNA, both in bacteria and in eucaryotes, is its high content of modified bases, especially methyl groups. These moieties are added to the ribosomal RNA after its transcription and are part of the process of its maturation. When Brown and Attardi (1965) followed the incorporation of radioactively labelled methyl groups into Hela cell tRNA and rRNA, they found that only 20–24% of the methyl groups in ribosomal RNA are attached to the bases themselves; most are present on the ribose moiety of the nucleotide, probably esterifed as 2′-0-methylribose.

Several cistrons code for ribosomal RNA in the genomes of bacterial cells and there may be several hundred copies of the genes for each rRNA in the cells of higher organisms. Are all the molecules of ribosomal RNA identical or do the genes which code for them vary in sequence? When Fellner and Sanger (1968) and Fellner (1969) studied the nucleotide sequences around the methylated sites in 23S and 16S rRNAs of E.coli, they found that methylation is specific and occurs at only a few sites. The 16S rRNA contains about 14 methyl groups and the 23S chain possesses about 19 groups. Unique fragments containing the methyl groups can be derived from each ribosomal RNA by treatment with T1 ribonuclease, which suggests that both the molecules are homogeneous, at least in their sites of methylation.

The sequence studies of Fellner et al. (1972) showed that there is some variation in the 16S rRNA molecule at individual nucleotides, presumably due to point mutations in some of the genes which code for rRNA. However, the extent of heterogeneity is confined to a very small number of sites. Each major methylated sequence of 23S rRNA appears to be present twice, which suggests that this molecule may consist of two segments which if not identical display appreciable homology. The 23S rRNA may therefore comprise two identical or closely related polynucleotide chains or a single chain with an extensive duplicated sequence; in either situation a gene duplication must have taken place at some point in its evolution.

There seem in general to be similarities between the ribosomal RNAs of species of organisms in spite of differences between their genomes. According to Miura (1962) and Midgley (1962), the nucleotide composition of ribosomal RNAs remain similar in a variety of bacteria whose genomes range in GC/AT ratios from 1·75 to 0·6. Sequences of the genome which code for rRNA must therefore have atypical base compositions relative to the remaining DNA and can be isolated by their different buoyant densities. Cross hybridization experiments between rRNA molecules performed by Moore and McCarthy (1967) suggested that related organisms tend to have rather similar sequences in their ribosomal RNAs. The retention of common characteristics in rRNA in spite of gross differences in the genome presumably reflects some common evolution of the protein synthetic apparatus; some functional capacities of rRNAs have been conserved during evolution of the bacteria, for hybrid ribosome subunits can be reconstituted from the rRNA of one species and the proteins of another (see page 152).

Sequence and Synthesis of 5S Ribosomal RNA

We do not know what function the small 5S rRNA plays in the ribosome. This rRNA was first identified by Rosset and Monier (1963); and Comb and Sarkar (1967) later showed that it is located in the 50S subunit. The presence of 5S rRNA is common to both bacterial and eucaryotic ribosomes and in bacteria is inserted at a late stage of ribosome assembly (see page 129). Attachment is permanent, for Kaempfer and Meselson (1968) found that once 5S rRNA has been inserted in the 50S subunit it cannot exchange with free 5S rRNA or with other ribosomes. The 5S rRNA molecule is not released by some treatments which dissociate many proteins from the ribosome, but is released by unfolding the ribosome with EDTA (see Hosokawa, 1970). This suggests that it may be located within the structure of the 50S subunit.

That 5S rRNA is bound to particular ribosome protein(s) is suggested by the observation of Blobel (1971) that EDTA releases a complex of 5S rRNA with one protein from the ribosomes of rat liver or rabbit reticulocytes. Gray et al. (1972) found that 5S rRNA can form a complex with 23S rRNA in the presence of two or three specific proteins of the 50S subunit of E.coli. The presence of 5S rRNA is essential for the activity of 50S subunits but we do not know whether its role in the subunit is structural or connected with some particular reaction of protein synthesis.

Some 5S rRNAs have been sequenced, but the primary structures do not suggest any obvious conformation for the molecule. E.coli 5S rRNA has no minor bases and Brownlee, Sanger and Barrell (1967, 1968) found that both the strains of E.coli which they used contain two species of 5S rRNA, one of which is the same in each strain and the second of which is related to it by a single base substitution. There must therefore be more than one gene responsible for its synthesis; since there are two types of 5S rRNA and only one 5S rRNA molecule on each ribosome, this implies the presence of two types of ribosome in the cell. But this is unlikely to have any biological significance. Two base sequences—one of eight and one of ten residues—are repeated in the molecule; and when the structure is written so that these are aligned there is enough homology to suggest that the 5S rRNA molecule may have arisen by gene duplication.

The 5S rRNA from human tumour KB cell ribosomes has been sequenced by Forget and Weissman (1969) and, in spite of its similar size, has little homology with the bacterial 5S rRNA. The 5S rRNA isolated by Williamson and Brownlee (1969) from each of two mouse cell lines in culture shows the same T1 and pancreatic ribonuclease digest fingerprints as does that from KB cells; this suggests that its sequence is probably the same in both man and mouse, although the presence of small differences cannot be excluded. Averner and Pace (1972) found the same oligonucleotides in a digest of 5S rRNA of the marsupial rat kangaroo.

The sequence of 5S rRNA may therefore be the same in all mammals. A very large number of genes code for 5S rRNA in the eucaryotic genome (see chapter five of volume 2), so the identification of only one sequence in each species implies that all the 5S genes must be identical or at least so closely related that their oligonucleotide digest patterns remain the same. There must therefore be considerable evolutionary restraint on variation in its sequence both between and within mammalian species. This implies that its function may be the same in all mammalian ribosomes and that most or all of the sequence of the molecule is critical.

But before drawing firm conclusions about the sequence of eucaryotic 5S rRNA, it is important to examine the molecules present in more than one tissue. The apparent conservation of sequence in mammals contrasts with the situation reported by Ford and Southern (1973) in Xenopus. Only one sequence of 5S rRNA appears to be present in kidney cells. But ovary cells possess 5S rRNA which differs principally in seven base substitutions from that of the kidney cells; in addition to this difference, ovary 5S rRNA is heterogeneous and there are some four variations of the ovary sequence which differ in base substitutions. The genes representing 5S rRNA in the X. laevis genome are therefore not identical; and different sets may be expressed in different tissues.

Various secondary structures have been proposed for the 5S rRNA of E.coli since the determination of its sequence. The relative availabilities of adenine residues for oxidation led Cramer and Erdman (1968) to suggest that base pairing is extensive. From the oligonucleotides released upon digestion with ribonuclease, Jordan (1971) suggested that there are 35–45 base pairs; another conformation has been suggested by Mirzabekov and Griffin (1972) from studies of susceptibility to T1 ribonuclease. By testing various tri- and tetranucleotides for ability to bind to 5S rRNA, Lewis and Doty (1970) identified four major single stranded regions of the molecule. But although it seems likely that the 5S molecule has a specific secondary structure (or structures) maintained by complementary base pairing, no model which has been proposed fits all the data; and at present we do not have a satisfactory model for the structure and function of 5S rRNA.

Synthesis and Assembly of Ribosome Subunits

Maturation of Eucaryotic Ribosomal RNAs

The synthesis and maturation of ribomal RNA itself has been most fully characterized in eucaryotic cells, in which a large number of identical genes for rRNA appear to be organised into a cluster at a chromosome site associated with the nucleolus. The activity of the nucleolus is devoted to transcribing these genes into precursor RNAs, modifying the sequences which are to be retained and cleaving excess regions of the precursor, and forming a ribonucleoprotein particle. Much of the maturation process is peculiar to eucaryotic cells and is discussed in full in chapter five of volume 2. However, there are similarities between the bacterial and eucaryotic systems in that the genes for both types of rRNA map close together in bacteria also and may be transcribed into a joint precursor molecule.

The genes which code for both ribosomal RNAs map close together in both

Xenopus laevis and Drosophila melanogaster, in which deletion mutants result in a failure to synthesize both major rRNAs. Wallace and Birnsteil (1966) showed that DNA from homozygous *anucleolate* mutants of Xenopus hybridizes very poorly with rRNA; Ritossa and Spiegelman (1965) showed that the amount of DNA in Drosophila which hybridizes with both rRNAs is proportional to the number of copies of the nucleolar organiser.

Because ribosomal RNAs tend to have a greater content of G and C than the average of the genome, the sequences of DNA which code for rRNA can be isolated by their unusually high buoyant density. By fragmenting preparations of ribosomal DNA to different sizes, Brown and Weber (1968) and Birnsteil et al. (1968) found that any stretch of DNA longer than a 28S rRNA gene can hybridize with both 28S and 18S rRNA. Sequences of DNA corresponding to only one type of rRNA can be produced only by fragmenting the DNA to a size smaller than that of the genes themselves. This suggests that the genes for 28S and 18S rRNA may alternate in their cluster on the genome. Because the buoyant density of the isolated rDNA is greater than would be predicted from the G + C content of rRNA, each pair of 28S–18S genes may be separated from the next by a non-ribosomal region of even greater G + C content.

The precursor from which eucaryotic RNAs are derived is very much larger than the rRNA molecules themselves, although its exact length shows a correlation with the evolution of species. It is at its largest in human cells, in which it sediments at about 45S and has a length of some 14,000 nucleotides. This is about twice the size of the two mature molecules together, for 28S rRNA is about 5000 nucleotides long and 18S rRNA is about 2000 nucleotides long. The excess sequences of the precursor molecule are cleaved in several distinct stages and are then presumably degraded. This feature appears to be characteristic of eucaryotic transcription rather than of ribosomal RNA synthesis; synthesis of large precursor molecules, much of which is degraded so that only part of the total sequences is transported to the cytoplasm, appears also to be used for synthesis of messenger RNA, although a different mechanism is used to select the sequences to be conserved.

By following the fate of a C^{14}-methionine pulse label, Greenberg and Penman (1966) found that Hela cell 45S RNA is methylated; this methylation accounts for all the modified groups in mature ribosomal RNA with the exception of a secondary methylation which takes place later to produce dimethyl-adenine in 18S rRNA. At ten minutes after incorporation of the labelled methyl groups, only 45S RNA is labelled. Its specific activity then declines and the label appears in an RNA sedimenting at 32S; the label then leaves 32S RNA and enters mature 28S rRNA. The 18S rRNA appears to be released for transport to the cytoplasm at an earlier stage than 28S rRNA, for a radioactive label enters cytoplasmic 18S rRNA far more rapidly than the 28S rRNA.

By hybridization experiments, Jeanteur and Attardi (1969) showed that

45S RNA contains the sequences of both 28S and 18S rRNA, the remainder of the molecule being non-ribosomal. The 32S precursor contains the sequence of only 28S rRNA and non-ribosomal regions. By examining the oligonucleotide patterns derived by enzymic cleavage of these molecules, Choi and Busch (1970) have shown that the same 5′ sequence is common to the 45S, 32S and 28S rRNA molecules. This means that the part of the 45S and 32S precursors which ultimately becomes the 28S rRNA must be located at the 5′ end of the molecule.

Although the 45S and 32S precursors are the most prominent when sedimentation on gradients is used to separate the RNA molecules, Weinberg and Penman (1970) have shown by a gel electrophoretic analysis that there are other intermediate species. The first step in maturation is cleavage to produce a 41S RNA; this must involve loss of a small sequence from the 3′ end of the molecule. The 41S RNA is then split into a 32S precursor to the large ribosomal-RNA and a 20S precursor to the small ribosomal RNA. The 32S and 20S precursors are then cleaved into the 28S and 18S rRNAs, which are transported to the cytoplasm.

The 41S RNA has the same number of methyl groups as 45S RNA, which suggests that the cleavage involves loss of a non methylated sequence. The 32S and 20S RNAs together contain this total of methyl groups; and all the methyl groups in each precursor are preserved when it is cleaved to mature rRNA. This suggests that the methylation sites dictate the course of maturation of RNA. Those sequences which have been modified are preserved and ultimately give rise to the mature rRNAs, whilst those sequences which are not methylated are first cleaved from the precursor and later degraded. This implies that the enzymes which cleave the precursors depend upon the methylation sites to distinguish the sequences to be conserved from those which are degraded.

In support of this model, Maden, Salim and Summers (1972) have shown that the fragments produced by T1 ribonuclease digestion of methyl-labelled 45S RNA are identical to the sum of the fragments of 28S and 18S rRNAs. This shows that the 45S precursor contains the sequences of both mature rRNAs and indicates that none of the non-ribosomal regions is methylated. The 41S RNA has the same methyl-labelled oligonucleotide pattern as 45S RNA; 32S RNA has the same pattern as 28S rRNA, and 20S RNA has the same pattern as 18S rRNA. The presence of the additional non-methylated sequences in the precursors can be revealed by labelling the bases themselves; Jeanteur, Amaldi and Attardi (1968) and Birnboim and Coakley (1971) found many nucleotide sequences in 45S RNA in addition to those characteristic of 28S and 18S rRNAs.

This maturation process does not take place on free RNA but in the form of a ribonucleoprotein precursor particle. A label for rRNA first enters nucleolar particles which sediment at about 80S and later enters particles which sediment at about 55S. The 80S particle contains 45S RNA and some of the proteins characteristic of the 60S and 40S ribosomal subunits, although Shepherd and Maden (1972) found that the proteins of the larger subunit appear to be present in greater amount. The 55S precursor contains 32S RNA and some of the proteins characteristic of the 60S subunit, to which it seems to be a direct precursor. Modification and cleavage of the rRNA precursors must therefore take place when they are associated with proteins.

Organization of Bacterial Genes for Ribosomal RNA

The bacterial genome contains several copies of each of the genes which codes for ribosomal RNA; all the copies are identical or very closely related. The proportion of the DNA which is saturated by hybridization with ribosomal RNA corresponds to about six genes for each of the 16S and 23S rRNA molecules in E.coli and about ten in B.subtilis. Both the 16S and 23S rRNA genes appear to be closely linked; and in some bacteria, at least, the 5S rRNA genes may be located in the same region. There has been some controversy about whether all the 16S and 23S rRNA genes of E.coli are clustered at one locus on the chromosome or whether there may be two sets of genes. Matsubaru, Takata and Osawa (1972) reported that there is only one cluster at 77 minutes on the E.coli map.

The arrangement of the two types of gene for rRNA seems to be the alternating array also found in the chromosomes of eucaryotic cells. The genes for all species of rRNA map in the starting region of the B.subtilis chromosome and Colli, Smith and Oishi (1971) have been able to isolate hybrids between B.subtilis DNA and rRNA. Single stranded DNA binds mercuric ions more effectively than DNA in the duplex or DNA-RNA hybrid forms; centrifugation through Cs_2SO_4–$HgCl_2$ can therefore be used to separate DNA hybridized to either 16S or 23S rRNA from the remaining single strands.

By using alkaline hydrolysis to degrade the RNA moiety of the hybrid molecules recovered from the gradient, those DNA sequences which were hybridized with 16S rRNA can be tested for their ability to hybridize with H^3-23S rRNA and P^{32}-5S rRNA; those DNA molecules which were isolated by their ability to bind 23S rRNA can be tested for hybridization to H^3-16S rRNA and P^{32}-5S rRNA. When the DNA fragments used for hybridization are of the order of size of $2–4 \times 10^6$ daltons, those fragment which had first bound 16S rRNA always bind both 23S rRNA and 5S rRNA also; those fragments which had first bound 23S rRNA can also bind both 16S rRNA and 5S rRNA. But when the single stranded DNA is first degraded to a molecular weight of about 1×10^6 daltons, only half of the sequences which first bind 16S rRNA can subsequently bind 23S rRNA and many of the molecules of DNA which had first bound 23S rRNA can no longer bind 16S rRNA. This suggests that the 16S and 23S rRNA genes are located in pairs on the genome; the ability of DNA to bind 5S rRNA seems to depend on its ability to respond to 23S rRNA, which indicates that these two genes are adjacent. An order of 16S-23S-5S rRNA genes would be consistent with these results.

After extracting DNA from E.coli cells in stationary phase—which should contain only one copy of the genome in each cell—Spadari and Ritossa (1970) found that the amount of 16S and 23S rRNA which can hybridize corresponds to between 5·7 and 7·4 genes; this suggests that there are about six or seven genes for each rRNA on the genome. By hybridizing fragments of DNA of different lengths with rRNA, they showed that the extent of clustering is very high, with all the rRNA genes probably located at one site.

The order of genes for 16S and 23S rRNA within each gene pair can be detected by following the synthesis of molecules of ribosomal RNA after transcription is blocked by addition of an inhibitor. Bleyman et al. (1969) used actinomycin, which probably inserts a random block to transcription. This means that the sensitivity of a unit of transcription to the antibiotic should depend on its length and genes which are far from the point where RNA polymerase starts transcription should have a greater risk that a block will prevent RNA synthesis from proceeding through them. In B.subtilis, the synthesis of 23S rRNA is almost twice as sensitive to actinomycin as that of 16S rRNA; since this is the same ratio as the lengths of the genes, synthesis of these two molecules appears to be independent. The synthesis of 5S rRNA is very sensitive to actinomycin, which supports the idea that it is located at the end of a large precursor molecule; one model is to suppose that the 5S rRNA is cleaved from the 23S precursor.

Rifampicin provides a better tool for transcriptional mapping, for this antibiotic specifically inhibits the initiation of transcription without influencing the completion of RNA chains already under synthesis. This means that longer units of transcription can function for greater lengths of time after the addition of rifampicin to bacterial cells, for they possess a greater number of RNA polymerase enzymes already engaged in RNA synthesis. On average, if there are n polymerases transcribing 16S rRNA, there should be $2n$ polymerases transcribing the 23S rRNA gene, so that twice as much 23S rRNA should be produced compared with 16S rRNA after the addition of rifampicin. But if the two genes are transcribed together as one unit with, say, the gene for 16S rRNA preceding that for 23S rRNA, then 23S rRNA will be synthesised by all the polymerase molecules located on the 16S rRNA gene as well as those located on the 23S rRNA gene itself. A similar prediction can be made for the situation in which the 23S rRNA gene precedes the 16S rRNA.

By adding P^{32} together with rifampicin and measuring its incorporation into the 16S and 23S rRNA molecules made in the presence of the antibiotic, Doolittle and Pace (1971) were able to show that the most likely arrangement of genes is one in which 16S rRNA immediately precedes the gene for 23S rRNA, with each molecule of RNA polymerase transcribing both genes. An analysis of the incorporation of H^3-uridine into 5S RNA in similar experiments suggests that the order of the transcription units may be: 16S—23S—5S rRNA.

Synthesis of Ribosomal RNA Precursors

Identifying the various stages through which ribosomes are assembled in the cells of E.coli has proved more difficult than in eucaryotic cells, for the precursor RNAs are only slightly larger than the mature species; no giant precursor has been identified. The precursor RNAs are found in particles of discrete sizes which are smaller than mature subunits, but these proved difficult to identify at first because they are present in such small amounts in normal cells. Mutants which accumulate the precursors are now available, however, although maturation is not always completely blocked. Inhibitors of protein synthesis have been used to block the maturation of precursor particles into mature subunits, but artefacts have resulted during the preparation of the particles on some occasions at least; the concentrations of inhibitor used appear to be critical.

A cell in exponential growth synthesises some 20,000 ribosomes every generation and, according to Mangiarotti et al. (1968), the synthesis of a ribosomal RNA takes some two minutes. This implies that about thirty copies of each rRNA are synthesised each minute from each rRNA gene. After inhibiting RNA synthesis with actinomycin, Adesnik and Levinthal (1969) used an autoradiographic analysis of C^{14} labelled RNA on acrylamide gels to reveal a precursor-product relationship for both the major ribosomal RNAs. The precursor to 16S rRNA appears to sediment at about 17S; the 17S band disappears and is replaced by 16S rRNA within four to ten minutes after the addition of actinomycin. When actinomycin is added to cells, only about twenty per cent of the label in 23S rRNA is found in the mature form, but within a few minutes a larger precursor molecule disappears from cells and is replaced by 23S ribosomal RNA.

The precursor RNAs sediment more rapidly than mature rRNAs and have lower electrophoretic mobilities on gels; this suggests that they represent longer polynucleotide chains. Osawa (1968) has shown that the precursor RNAs are also distinguished from the mature species by differences in the extent of base modification, for the 16S precursor has only 10–20% of its methylated groups and the 23S precursor is about 60% methylated. The content of pseudouridine is also lower in the nascent species.

Four groups of researchers have isolated the "p16" RNA which is the precursor to 16S rRNA and have compared the ribonuclease fingerprints of the two molecules (Brownlee and Cartwright, 1971; Hayes et al., 1971; Lowry and Dahlberg, 1971; Sogin et al., 1971). The p16 RNA appears to be some ten per cent larger than the mature 16S rRNA and contains all the oligonucleotides found in the mature molecule as well as some additional fragments. The number of extra residues in the precursor has been variously estimated as about 40, at least 55 and at least 110. One possible explanation is that several precursors may be generated by sequential cleavages of the immediate product of transcription. There are extra sequences at both the 5′ and 3′ ends of all the precursors studied, so that at least two reactions must be involved in maturation.

The tandem arrangement of genes for 16S and 23S rRNA suggests that they may be transcribed as one unit but cleaved to yield the two precursors before transcription is complete. This would account for the kinetics of their synthesis and also unify the processes of synthesis of rRNA in bacteria and eucaryotes. Such a model explains why no precursor molecule containing both sequences has been identified.

Kinetic experiments suggest that RNA chains begin to associate with ribosomal proteins during their transcription, so that 16S and 23S precursors appear as ribonucleoprotein particles as soon as they are released from DNA. RNA molecules identical to the precursors are accumulated when bacteria are treated with chloramphenicol or starved for some essential amino acid; this suggests that maturation depends upon the synthesis of proteins (see Dalgarno and Gros, 1968a, b). Chang and Irr (1973) reported that the precursors which accumulate when protein synthesis is halted by deprivation of an amino acid turnover rapidly, with a half life of less than 30 minutes.

The precursors accumulated in the cell can be converted to mature ribosomal RNAs when the essential amino acid is restored. These results are consistent with a model in which maturation of ribosomal RNA precursors depends upon synthesis of some of the ribosomal proteins; the absence of these proteins both halts maturation and leaves the precursors vulnerable to degradation by nucleases in the cell. Perhaps the ribosomal proteins which bind to precursors immediately upon their synthesis serve the function of restricting nuclease attack to those regions which must be removed to yield the mature model.

Isolation of Precursor Particles

Precursor particles constitute only a small part of the ribosome population of E.coli and are difficult to identify in growing cells. Blocking ribosome maturation with drugs, isolating the particles accumulated by mutants in ribosome assembly, and identifying the particles found in cells subjected to special growth conditions have been used to follow ribosome synthesis. Assembly of the 50S subunit passes through two discrete stages; it is not clear whether there are two or only one precursors to the 30S subunit.

Addition of chloramphenicol to E.coli cells prevents the synthesis of ribosomes and Kurland, Nomura and Watson (1962) showed that particles sedimenting at 18S and 25S can then be extracted from the cell (reviewed by Vazquez, 1966). One of the problems in such experiments, however, is highlighted by the observation of Yoshida and Osawa (1968) that these particles are artefects produced by the association of rRNA with basic proteins during extraction.

The concentration of chloramphenicol is critical, for Osawa et al. (1969)

found that genuine precursor particles may be isolated when ribosome synthesis is blocked with lower concentrations of the drug. These LCM (low chloramphenicol) particles sediment at 20–22S and 40–43S; the smaller particles contain a 16S precursor rRNA and the larger particles possess a precursor 23S rRNA and also a molecule of 5S rRNA. When the drug is removed, the RNA of the LCM-particles is converted to its mature form and some of the ribosome proteins which are missing from the precursor particles are added.

The same particles can also be found in untreated E.coli cells, which supports the idea that they are genuine precursors to the ribosome subunits. It is difficult to prepare ribosome precursor particles directly by following the fate of a radioactive pulse label in rRNA because of contamination with mRNA. Osawa (1968) overcame this problem by using "shift-up" conditions in which the level of mRNA synthesis is reduced to a low rate relative to that of rRNA. This approach reveals 22S and 26S particles which appear to be precursors to the 30S subunit, and particles sedimenting at 30–32S and 40–43S which are precursors to the 50S subunit. These particles are indistinguishable from the LCM-particles which are produced by treatment with small amounts of chloramphenicol.

Two other approaches have also revealed these precursor particles in treated cells. By using a "fragile" mutant of E.coli in which mRNA can be removed in polysomes after a gentle lysis of the cells, Mangiarotti et al. (1968) found particles containing 16S rRNA which sediment at 26S and particles sedimenting at 32S and 43S which contain 23S rRNA. Forget and Varicchio (1970) found that an unusually high proportion of 30S and 50S subunits are present when E.coli cells are grown at their maximum rate; there is also a small peak of 43S precursor particles. Pulse labelling experiments showed that the rRNA enters these particles from a 26S precursor particle and leaves them for 50S subunits.

Unlike mature subunits, the precursor particles are highly sensitive to degradation by ribonuclease, which suggests that their RNA is probably not fully covered by protein. Matsuura et al. (1970) found that the 30S particles obtained either from untreated cells or after addition of low concentrations of chloramphenicol appear mainly as elongated molecules when observed under the electron microscope. Although these particles are a little more spherical and less filamentous than 23S rRNA itself, their overall structure is similar, which suggests that the association of the first group of proteins with 23S rRNA does not alter its conformation appreciably. The 40S particles are compact spheres which are uniform in size and shape; by this stage, then, the 23S rRNA is folded into the structure which is found in the mature 50S subunit.

The existence of precursor particles of discrete size suggests that the ribosomal proteins are added to the precursor RNAs in groups, starting during their transcription, and that the modification and cleavage of the RNAs must take place in these structures. The precise sequence of events in maturation of the different precursor particles is not known. But both methylation and cleavage are inhibited by the addition of chloramphenicol to cause accumulation of precursor particles. These events therefore presumably take place at about the same time as the addition of further sets of proteins.

Both the precursor particles of the small subunit contain 16S ribosomal RNA in its immature longer form, although we do not know whether there are size reductions in the precursor RNAs present in the intermediate particles compared with the first precursors. The 23S rRNA found in both the 30S and 40S particles is only about sixty per cent methylated, so the remaining modifications to its bases must be made during the final maturation stage. We do not know how the lengths of the 23S rRNA present in the different precursor particles are related and whether they suffer cleavage(s) analogous to the 16S.

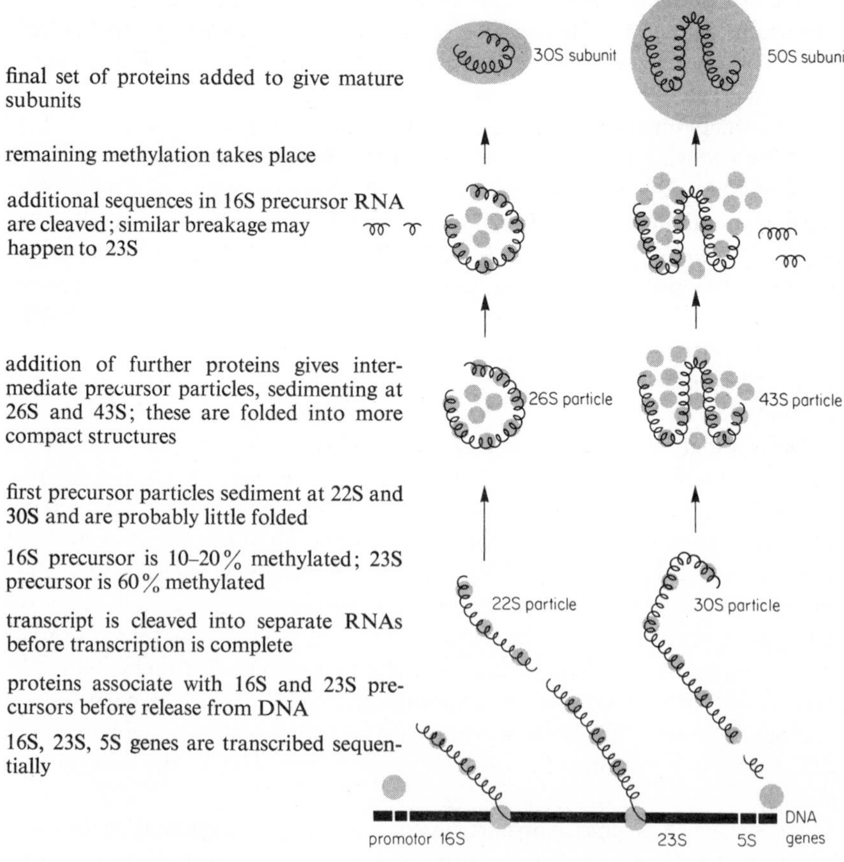
30S subunit
50S subunit
final set of proteins added to give mature subunits
remaining methylation takes place
additional sequences in 16S precursor RNA are cleaved; similar breakage may happen to 23S
addition of further proteins gives intermediate precursor particles, sedimenting at 26S and 43S; these are folded into more compact structures
26S particle
43S particle
first precursor particles sediment at 22S and 30S and are probably little folded
16S precursor is 10–20% methylated; 23S precursor is 60% methylated
22S particle
30S particle
transcript is cleaved into separate RNAs before transcription is complete
proteins associate with 16S and 23S precursors before release from DNA
16S, 23S, 5S genes are transcribed sequentially
promotor 16S
23S
5S
DNA genes

Figure 4.2: Model for maturation of ribosomal RNAs in E.coli

This suggests the scheme for the synthesis of the two subunits shown in figure 4.2. Depending on their conditions of isolation, the 40–43S particles may have some or all of their ultimate content of 5S rRNA, but its association with these precursors is not so stable as in mature subunits, for when present it is easily lost during preparation. This implies that it may be inserted into the ribosomes at the stage of this precursor particle. Feunteun, Jordan and Monier (1972) found that 5S rRNA is inserted in precursor form containing additional nucleotides and matures as part of the particle.

Mutants in Ribosome Assembly

The reconstitution of ribosomes in vitro (see later in this chapter) takes place by a reaction which is strikingly dependent on temperature; Guthrie, Nashimoto and Nomura (1969a, 1969b) therefore argued that mutations which affect the assembly process might be detected as cold temperature-sensitive mutants—assembly might be blocked at low temperature but able to take place at high temperature. Using this rationale, they found the three types of *sad* (subunit assembly defective) mutants shown in figure 4.3.

All these mutants synthesise ribosomes normally at 42°C but are mutant at 20°C. Two classes have a decrease in the amount of 50S subunits when they are grown at 20°C; one, typified by *sad*-68, has an increased peak which sediments at 43S and the other, such as *sad*-19, has an increased peak sedimenting at 32S. When mutant cells are shifted from low to high temperature, both the 43S and 32S peaks disappear and the 50S subunit peak reappears. The ribosomes isolated from these strains are no more cold sensitive than those of the parent strains, so the mutation must represent a defect in assembly and not in function. The abnormality of the third class of mutant, such as *sad*-38, is not restricted to the large subunit, but such cells appear to be defective in the production of both 50S and 30S subunits and to accumulate precursors which sediment at 32S and 21S respectively.

Mutants resistant to various antibiotics which act on the ribosome have been tested to see whether they may have cold sensitive assembly defects. Nashimoto and Nomura (1970) showed that some of the bacterial strains which have reverted from dependence on streptomycin to independence and

---

Figure 4.3: Ribosome precursor particles accumulated at low temperature by three sad mutants. Each graph presents the gradient profile extracted from a mutant with the normal profile of its parent strain shown for comparison. Because different radioactive labels are used, no qualitative conclusions can be drawn. But at 20°C each mutant accumulates precursor(s) instead of subunits. The precursor which sediments at 30–32S occupies the same position on the gradient as mature 30S subunit, but the two types of particle can be distinguished by analysis of their ribosomal RNAs—the precursors have 23S rRNA whereas the mature subunits contain 16S rRNA. Data of Guthrie, Nashimoto and Nomura (1969b)

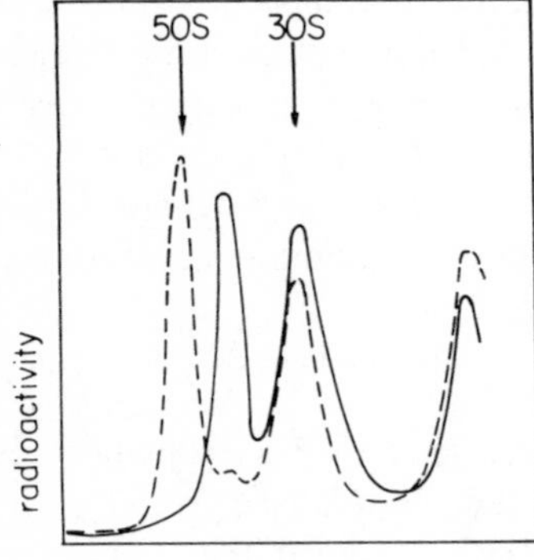

sad-68: fails to produce 50S subunits and instead accumulates 43S precursor particles which contain 23S rRNA. Synthesis of 30S subunits is normal.

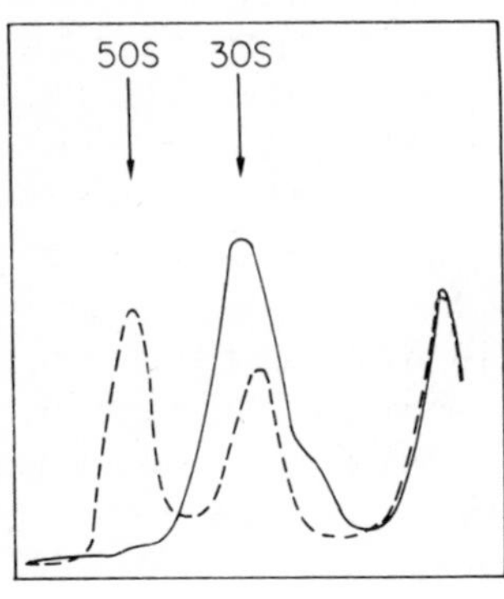

sad-19: fails to produce 50S subunits and instead accumulates 32S precursor particles which contain 23S rRNA. The 32S precursor peak overlaps the peak of normal 30S subunits, but the two particles may be distinguished by their different rRNAs. There is also a small shoulder of 21S precursors.

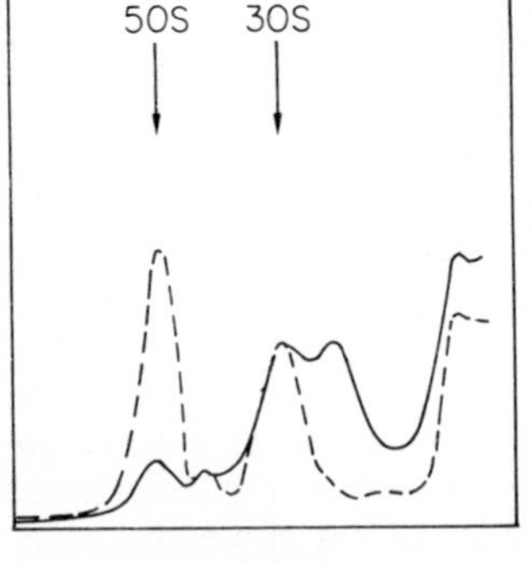

sad-38: fails to synthesize both 50S and 30S subunits; instead accumulates 30S precursors which contain 23S rRNA and 21S precursors which contain 17S rRNA

fractions on gradient

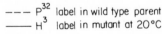

--- P^{32} label in wild type parent
—— H^{3} label in mutant at 20°C

some mutants resistant to spectinomycin show this behaviour (both streptomycin and spectinomycin act on proteins of the 30S subunit, which may be mutated to alter the response of the bacterium to the antibiotic). Some of these mutants show a reduction in synthesis of both 50S and 30S subunit and accumulate all of their precursors. Figure 4.4 shows the presence of 43S, 32S, 26S, 21S particles in *spc*-m24 cells. Others have few 50S subunits and no 30S subunits and accumulate 32S and 21S, and possibly also 26S precursor particles; a third class has reduced synthesis of 50S and probably also 30S subunits and accumulates both the 43S and 26S precursors.

This implies that the assembly of the 50S subunit may depend on the ability of the proteins of the 30S subunit to associate properly with 16S rRNA to form the 30S particle. A general observation which supports this idea is that all *sad* mutants which abolish 30S assembly also inhibit 50S assembly, although

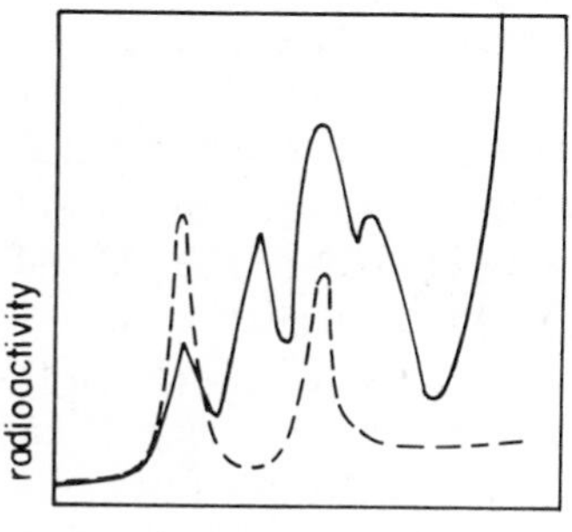

spc-m24: has reduced synthesis of both 50S and 30S subunits; accumulates 43S and 32S precursors to 50S subunits (which contain 23S rRNA) and 26S and 21S precursors to 30S subunits (which contain 17S rRNA). The 32S and 26S precursors form broad peak which overlaps the usual 30S position.

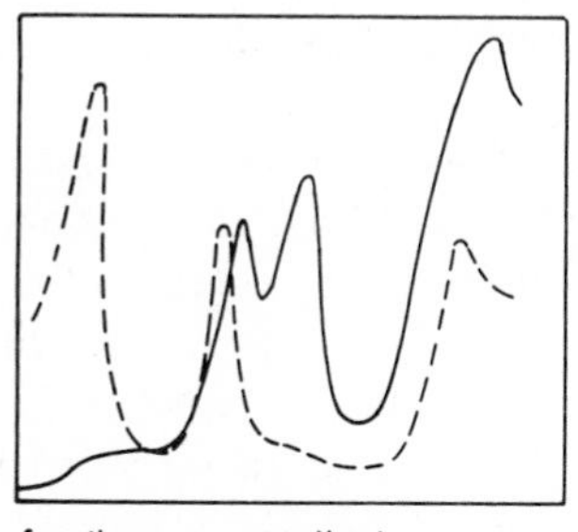

spc–49: fails to synthesise both 50S and 30S subunits and instead accumulates 28S precursors containing 23S rRNA (equivalent to the usual 30–32S particles) and 21S precursors containing 17S rRNA.

fractions on gradient

--- C^{14} reference of 50S and 30S subunits

—— H^3 label of mutant cells at 20°C

Figure 4.4: Ribosome precursor particles accumulated at low temperature by two mutants in the spectinomycin-resistant class. Data of Nashimoto and Nomura (1970)

mutants which interfere with assembly of the 50S subunits do not prevent the production of 30S subunits.

One spectinomycin resistant mutant has been studied in some detail by Nashimoto et al. (1971); this mutant, *spc*-49 which is shown in figure 4.4, has no 50S subunits or 30S subunits when grown at low temperature and accumulates a 30S precursor containing 23S rRNA and a 20S precursor containing 17S rRNA. If the precursors are labelled with H^3 uridine at 20°, and the temperature is then raised to 42°C, the precursor particles disappear and the label enters the mature subunits. Because similar or identical particles are accumulated by the spectinomycin-resistant mutants and independently isolated *sad* mutants which do not map in the spectinomycin gene, it seems likely that mutation in any one of several of the proteins of the 30S subunit can cause the same defect in assembly. This idea is supported by the location of at least one of the *sad* mutations in the same chromosome region as the spectinomycin gene. But it is not easy to see why mutation in the S5 protein, which confers resistance to spectinomycin and is not involved in the early stages of assembly, should cause the *sad* phenotype.

The same particles therefore accumulate in *sad* mutant cells, can be detected in normal cells, and are found in bacteria treated with low concentrations of chloramphenicol. The 43S and 32S particles contain 23S rRNA; the 26S and 21S particles contain 17S rRNA. One reason why the sedimentation characteristics of the particles vary with the method of analysis and, indeed, with particular mutations may be that slightly different complements of proteins are associated with the rRNA in some instances. Nierhaus, Bordasch and Homan (1973) have noted that in most experiments either a 21S or a 26S particle is found as precursor to the 30S subunit; it is rare to find both together. They therefore suggested that there may be only one precursor particle to the 30S subunit, which sediments between 20S and 26S depending upon the experimental conditions.

But all these results are consistent with the concept that ribosome maturation takes place through discrete steps in which several proteins are added at once to the particle. Figure 4.2 shows a pathway based upon the existence of two precursors for each subunit. In the first step, a small number of proteins associates with 23S rRNA to give a 32S particle and with 16S rRNA to give a 21S particle; the rRNA molecules are precursors which are only partly methylated and may be longer than the mature species. In the second step more proteins join the particles, converting the 32S to a 43S particle and increasing the size of the 21S to 26S. Finally, any further steps of maturation of rRNA take place—including insertion of precursor 5S rRNA in the 43S particle—and the last set of proteins is added to yield mature subunits. It is possible, however, that there may be only one precursor to the 30S subunit; and the proteins which may be implicated at each stage of assembly are discussed below (see page 143).

Topology of Ribosome Subunits

Proteins of E.coli Ribosomes

Bacterial ribosomes can be dissociated into their components either partially or completely in vitro; and in appropriate conditions their proteins and RNAs can reassociate to form functional particles. The assembly reaction in vitro appears to be very similar to the processes by which ribosomes are put together in the bacterial cell. *Subribosomal* particles can be formed by using reconstitution mixtures which lack one of the protein components of the ribosome; the inability of these particles to perform some of the reactions of protein synthesis helps to reveal the functions of the missing protein in the ribosome. And the use of heterologous reaction mixtures—in which the various components of the ribosome may be derived from different species of bacteria—shows the extent to which the properties of the different components have been conserved in the evolution of bacteria.

The protein component of ribosomes consists of a large number of different proteins, some of which are involved in the assembly of the ribosome and in maintaining its structure and others of which have specific catalytic roles. In addition to these species, there are also the various protein factors involved in protein synthesis which associate with the ribosome at the appropriate stage of the synthetic cycle. The proteins of E.coli ribosomes have been separated and characterized in some detail since Waller (1964) first isolated the proteins of 70S particles and identified twenty four bands by starch gel electrophoresis. Amino acid analyses, tryptic peptide mapping and responses to antibodies have led Traut et al. (1967, 1970), Fogel and Sypherd (1968), Craven et al. (1969) and Pearson, Delius and Traut (1972) and others also to conclude that the different proteins separated on gels represent distinct molecular species. Completely different sets of proteins are found in each subunit.

Many chromatographic methods have now been used to separate ribosomal proteins and as table 4.1 shows, twenty one different proteins are found in the 30S subunit of E.coli. The two dimensional gel electrophoresis system of Kaltschmidt and Wittman (1970) has been accepted as a standard method for their definition and I shall use the numbering system based on this technique which was agreed by Wittman et al. (1971); its equivalence with the other systems of nomenclature which it has superseded is given in the table. The 50S subunit contains thirty four proteins when analysed on these gels. Methods for isolating and characterizing ribosomal proteins have been reviewed by Wittman and Stoffler (1972).

By collecting nineteen strains of E.coli, Osawa, Takata and Dekio (1970) showed that the same ribosomal proteins are usually present. Strain C has the same 50S subunit proteins as strains B and K but differs in one protein of the 30S subunit; strains B and K also differ by one protein of the 30S subunit. Almost all the E.coli strains tested fall into one of these three categories, although two strains of type R appear to have several differences in their ribosomal proteins. In contrast to the similarities retained between the

Table 4.1: proteins of the 30S subunit of the E.coli ribosome

Berlin	Madison	Uppsala	Geneva	molecular weight	genetic locus	phenotype of mutations
S1	P1	1	13	65,000	*rpxA*	—
S2	P2	4a	11	27,000	*rpxB*	—
S3	P3	9 + 5	10b	28,000	*rpxC*	—
S4	P4a	10	9	26,700	*rpxD* = *ram*	suppresses dependence on streptomycin
S5	P4	3	8a	18,500	*rpxE* = *spc*	resistance to spectinomycin
S6	P3 + P3c	2	10a	17,000	*rpxF*	—
S7	P5	8	7	{26,000* / 23,500}	*rpxG*	K-character
S8	P4b	2a	8b	15,500	*rpxH*	—
S9	P8	12	5	14,500	*rpxI*	—
S10	P6	4	6	18,000	*rpxJ*	—
S11	P7	11	4c	18,300	*rpxK*	—
S12	P10	15	—	16,000	*rpxL* = *str*	resistance to, on dependence on streptomycin
S13	P10a	15b	—	14,000	*rpxM*	—
S14	P11	12b	—	14,000	*rpxN*	—
S15	P10b	14	4b	13,000	*rpxO*	—
S16	{P9	6	4a	13,000	*rpxP*	—
S17	P9}	7	3a	15,000	*rpxQ*	—
S18	P12	12a	2b	10,500	*rpxR*	—
S19	P13	13	2a	14,000	*rpxS*	—
S20	P14	16	1	13,000	*rpxT*	—
S21	P15	15a	0	13,500	*rpxU*	—

The Berlin nomenclature, based upon the two dimensional gel electrophoresis system of Kaltschmidt and Wittman (1971) has been agreed upon—see Wittman et al. (1971)—as a standard code to replace the Madison code of Nomura et al., the Uppsala code of Kurland et al. and the Geneva code of Traut et al.
Molecular weights were determined by Garrett et al. (1972); other estimates have been made by Craven et al. (1969) and by Traut et al. (1970).
The *rpx* nomenclature is based upon the Berlin separation, but some of these loci have previously received other names. Protein S7 of K-strains differs in electrophoretic mobility from that of other strains (Leboy et al., 1964; Birge et al., 1969); and the asterisk indicates the weight of the protein of E.coli K12. Mutations in rpxL (str) were identified by Ozaki et al., 1969; Luzzato et al., 1968a); mutation in rpxE (spc) by Flaks et al., 1969; Funatsu et al., 1972a, c; and mutations in rpxD (ram) by Rosset and Gorini, 1969; Dekio and Takata, 1969; Deusser et al., 1970.

ribosomal RNAs of different bacterial species, however, the E.coli ribosome proteins show little similarity with those of other bacteria. The ribosomal proteins of eucaryotic cells have not been characterized in such detail but also contain a large number of different proteins, about eighty in mammalian cells.

Organization of Genes for Ribosomal Proteins

Mutations have been identified in a small number of the ribosomal proteins and the loci which code for these species have been mapped. By electrophoresing ribosomal proteins from various strains of E.coli on acrylamide gels, Leboy, Cox and Flaks (1964) found that a single protein from the 30S subunit of strain K12 has a lower electrophoretic mobility than its counterpart in other strains. Table 4.1 shows that the K-protein has an increased molecular weight and Birge et al. (1969) have shown that it has three less arginines and one less lysine than the B-protein, which may account for its altered electrophoretic mobility. The locus governing the synthesis of this protein has been mapped by genetic conjugation and transduction with phage P1 and is located close to another site, the streptomycin gene, which also elaborates a 30S protein.

Most of the ribosomal protein genes which have been mapped have been identified by the resistance which mutation in them may confer to the action of antibiotics which inhibit ribosome activity. Mutation in a single gentic locus on the E.coli chromosome confers resistance to streptomycin, which inhibits protein synthesis by interacting with the 30S subunit. The implication is that this locus specifies some protein of the 30S subunit which is necessary for the action of streptomycin and by examining ribosomal proteins of the mutated resistant strains, Ozaki et al. (1969) have shown that the S12 protein is changed by the mutation. This locus, as the first of its class to be discovered, defines the "streptomycin region" of the E.coli chromosome in which several other similar mutations map.

Resistance to spectinomycin maps at a closely linked site (Flaks et al., 1969) and results from amino acid substitutions in the sequence of the S5 protein (Funatsu et al., 1972a, c). A ribosomal ambiguity mutation, *ram*, which affects the accuracy of ribosome function in translation, has been found by Rosset and Gorini (1969) to map close to spectinomycin-resistance; and Zimmermann et al. (1971) have shown that *ram* mutants contain an altered S4 protein. This clustering of mutant activities suggests that this region of the genome may contain the genes for many of the proteins of the 30S subunit.

That genes for many ribosomal proteins may be clustered in the streptomycin region is suggested also by the results of intergeneric crosses, when E.coli genes are introduced into another species of bacteria many of whose ribosome proteins are electrophoretically distinguishable from those of E.coli. By following the synthesis in the recipient cells of ribosome proteins which can be identified as those specified by the E.coli DNA, Sypherd et al. (1969), Dekio,

Takata and Osawa (1970) and Takata (1972) have shown that the streptomycin region includes the genes for at least nine of the 30S proteins and six of the 50S subunit proteins.

Although it is therefore likely that many of the genes which code for ribosome proteins are located in the streptomycin region, one gene at least is located in a different part of the chromosome. Bollen et al. (1973) and Kahan et al. (1973) isolated cells with a mutation in one amino acid of the 30S protein S18 and showed that it maps between 76 and 88 minutes; the streptomycin region is located at 64 minutes on the map. We do not know whether any other ribosome proteins map in this new location.

It is tempting to speculate that the many genes coding ribosome proteins which are located in the streptomycin region may be organised in common units of transcription and translation. This idea is borne out by the experiments of Nomura and Engbaek (1972). One test for coordinate organisation of genes is to show that nonsense mutants in one gene interfere with the expression of other genes (see Chapter 9). No deletion or nonsense mutations have been isolated in the genes coding for ribosomal proteins, however, presumably because such mutations would be lethal. But Nomura and Engbaek have taken advantage of the effect of phage μ on genes of E.coli in order to isolate mutants of this general nature.

Infecting E.coli cells with phage μ causes a high frequency of mutation amongst the surviving lysogenised cells. The reason is that the phage genome is inserted at random sites on the E.coli chromosome; these may be within genes, in which case the cell fails to produce an active gene product. If the insertion takes place in one of the early genes of an operon—that is a gene which is transcribed and translated early in sequence—the expression of the subsequent genes may be reduced or prevented altogether even though they themselves carry no mutations (see page 390). To test for such effects, it is necessary to use a diploid cell—for inactivation of the only ribosomal protein genes in a cell should be fatal—and the strain used had the composition:

chromosome	ery^s spc^s str^s fus^s
episome	ery^r spc^r str^r fus^r

in which the dominant genes which confer sensitivity to erythromycin (acts on the 50S subunit), spectinomycin and streptomycin (act on proteins of the 30S subunit) and fusidic acid (acts on the G translocase supernatant factor) are located on the chromosome of the cell and their resistant alleles are located on an episome.

When these diploids are treated with phage μ, lysogenic survivors which are resistant to the antibiotics can be isolated by plating the cells on appropriate medium; such cells must have inactive sensitive alleles, so that their resistant counterparts can be expressed. All the cells which were isolated for their resistance to erythromycin are also resistant to the other three antibiotics; cells isolated for resistance to spectinomycin may fall into this class, or may be resistant to spectinomycin, streptomycin and fusidic acid. Cells isolated for resistance to streptomycin may fall into either of these classes, or may be resistant to streptomycin and fusidic acid. Cells isolated for resistance to fusidic acid may fall into any of these three classes, or may be resistant to fusidic acid alone.

These results mean that when phage μ inactivates the erythromycin-sensitive gene by insertion in it, the spectinomycin, streptomycin and fusidic acid sensitive genes are also inactivated. When mutants in any of the latter three genes are isolated some of them, of course, may be of this type. When spectinomycin-resistant cells are isolated, for example, some are of this class and resistant to all four antibiotics; others have an insertion in the spectinomycin gene itself, and as a result are resistant also to streptomycin and fusidic acid. A similar situation prevails for the remaining two genes.

The four genes must therefore be located in the sequence: erythromycin—spectinomycin—streptomycin—fusidic acid and must be transcribed in that order into one messenger RNA which is then sequentially translated into proteins. We do not know how the synthesis of this messenger is controlled, but the result of this coordinate clustering is that all four proteins must be synthesised together. Since one of these proteins is part of the 50S subunit, two are part of the 30S subunit, and one is a supernatant factor, this means that the different parts of the ribosome and its associated proteins are probably all under the same control. It seems likely that other ribosomal proteins are also part of this unit of expression; and these results suggest the speculation that a small number of such units may be responsible for the coordinate synthesis of ribosomal proteins. The synthesis of 30S and 50S subunits may therefore be intimately linked in protein as well as rRNA synthesis so that the cell produces one large subunit for every small subunit.

Dissociation and Assembly of 30S Subunits In Vitro

The work which led to our ability to dissociate the ribosome into groups of its components was the finding reported by Meselson et al. (1964) that density gradient centrifugation of 30S (or 50S) subunits in CsCl yields 23S (or 42S) particulate *cores* when a discrete number of proteins, the split proteins, are lost. More detailed studies have since shown that when ribosomes are exposed to high concentrations of CsCl or LiCl, they suffer a series of successive disruptions, each inolving the loss of a group of specific proteins (Itoh, Otaka and Osawa, 1968). This discrete loss of proteins suggests that dissociation does not represent a gradual liberation of various protein components, but supports the picture revealed by the isolation of precursor particles which indicates that the ribosome has groups of cooperatively organised proteins;

disruption of each ordered group results in the more or less cooperative loss of all its proteins together.

The stepwise dissociation of subunits can be reversed in the absence of CsCl, provided that Mg^{2+} ions are present. Ribosomes reconstituted in this way possess appreciable biological activity. Staehelin and Meselson (1966a) showed that the subunits reconstituted by the addition of split proteins to cores are active in cell free systems for protein synthesis; and Nomura and Traub (1968) demonstrated that each reconstituted particle consists of one equivalent of each of the cores and split proteins. The reconstitution reaction is specific for the proteins which have been lost; the addition of 50S split proteins cannot restore activity to 30S cores and vice-versa. The proteins split from the 50S subunit by CsCl centrifugation have been termed the SP50 fraction and the corresponding species lost from the 30S subunit are known as the SP30 fraction.

The 23S cores of the 30S subunit can be further dissociated into their component core proteins, the CP30 fraction, and the 16S rRNA. One prodecure which can be used to isolate ribosomal proteins is extraction in 4M urea and 2M LiCl; ribosomal RNA can be obtained by a phenol extraction. Traub and Nomura (1968b) found that they could reconstitute active subunits by mixing 16S rRNA with the CP30 proteins at 37°C and subsequently adding the SP30 fraction in the cold. High ionic strength is needed for the reconstitution reaction, which suggests that the specific interaction between rRNA and proteins does not involve ionic bonds but requires non-ionic weak bonds such as H-bonds or hydrophobic interactions. Magnesium ions are essential and one possible conjecture about their role is that they help to maintain the proper conformation of rRNA. The reconstituted particles possess all the proteins of the 30S subunit and are active when tested in vitro for their ability to synthesise proteins under direction from poly-U or phage f2 RNA. The ability of the mature ribosomal components to undergo this autonomous reassociation to yield functional subunits suggests that the information required for assembly of the ribosome is contained in the structures of its components.

The kinetics of the reconstitution process have been followed by incubating the reconstitution mixtures at various temperatures, recovering the particles formed, and assaying their activity in an in vitro system at 37°C. Traub and Nomura (1969) and Nomura et al. (1969) found that few active particles result from the incubations at 10°C and 20°C; centrifugation of these reaction mixtures yields inactive species termed RI (reconstitution intermediate) particles. These lack several of the 30S subunit proteins, the S proteins, which can be recovered from the supernatant. The proteins present in the RI particles are similar to those of the 23S core particles and the S proteins resemble those of the SP30 fraction, although there are some minor differences. The S fraction comprises some seven to ten proteins, and although these cannot bind to free 16S rRNA, when they are heated together with RI particles at 40°C for 20 minutes, almost fully active 30S subunits are produced. This suggests that the reconstitution process takes place through discrete steps, the addition of S proteins occurring later in the sequence.

After heating the RI particles alone and cooling to 0°C, the addition of S proteins at the low temperature yields active 30S subunits. Although the overall assembly of the 30S subunit involves the association of some twenty proteins with rRNA, the reaction appears to be uni-molecular; this suggests that the rate limiting reaction may be a structural rearrangement of the RI particle and it is this step which is permitted to take place by the increase of temperature from 0°C to 40°C. The reconstitution process can thus be represented as:

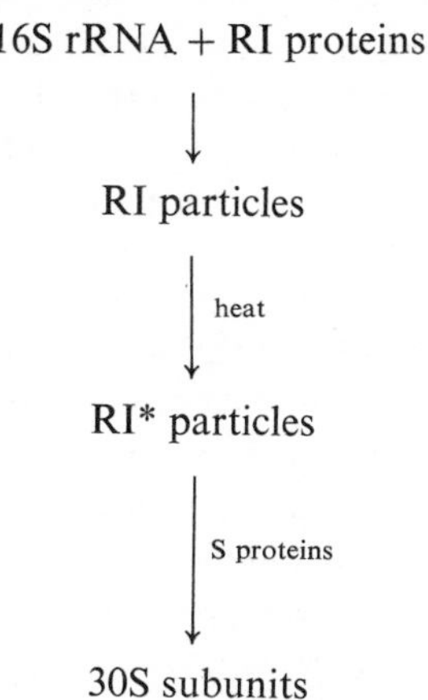

where the RI* particles are the active intermediates whose formation is the rate limiting step.

Dialysis of 30S subunits in the cold for some hours against a solution containing mercaptoethanol and the chelating agent EDTA removes their magnesium ions. The particles which are produced by this treatment sediment at 13S; although they are intact and contain all their components, they have been unfolded. When these unfolded particles are subjected to the same procedure used to reconstitute 30S subunits from 16S rRNA and proteins, they are converted into active 30S subunits. But when they are incubated at low temperature, this conversion is prevented and instead yields particles—termed RI(u) particles—lacking some of the 30S proteins and only weakly active in protein synthesis. The missing protein components can be recovered from the supernatant, and heating these S(u) proteins together with the RI(u) particles gives active 30S subunits by a reaction similar to that of RI particles and S proteins. In fact, the RI and RI(u) particles are functionally identical and the S and S(u) fractions can substitute for each other in the reconstitution process. This suggests that reconstitution involves a step in which a structural conversion is effected by a folding of the particle structure.

It seems likely that the stepwise reconstitution of the 30S subunit which can be achieved in vitro mimics the process of assembly in vivo. This idea is supported by the close relationship between RI particles and 21S precursors (see below). But an important difference between the two processes is that assembly in vivo takes place on precursor RNA, which is both longer and also less modified than the mature rRNA which has been used in vitro. The success of reconstitution experiments in vitro makes it difficult to see what function the precursor RNA serves; it seems unlikely that it contains sequences which must be recognised by ribosomal proteins. Although the maturation of the rRNA takes place as the precursor particles associate with further ribosomal proteins, we do not know whether the maturation of particles depends upon that of the rRNA or whether the two processes, although simultaneous, are independent. One possible speculation is that the precursor rRNA can inhibit the maturation process unless its extra sequences are removed and proper modifications made; this might provide some control over the assembly of ribosome subunits.

Reconstitution of 50S Subunits In Vitro

The reconstitution of 50S subunits in vitro has proved more difficult than that of the smaller subunits. One possible reason is that a higher temperature of incubation is needed but that this would inactivate the the subunits made in vitro. Nomura and Erdmann (1970) and Fahnestock, Erdmann and Nomura (1973) therefore reconstituted 50S particles from Bacillus stearothermophilus, which lives at higher temperatures than E.coli and whose protein synthetic apparatus, including the ribosomes, is stable in these conditions. They found that although this reaction is successful, it is less efficient than the reconstitution of the 30S subunits of E.coli. The small 5S rRNA must be present in the reconstitution mixture, for Erdmann et al. (1971) found that its omission severely restricts all the catalytic activities of the reconstituted subunits. Particles reconstituted without 5S rRNA appear also to be more readily unfolded when their magnesium ions are removed, which suggests that 5S rRNA may play some role in maintaining the compact structure of the large subunit.

By using a gentle procedure to isolate the proteins of E.coli 50S subunits, in which the particles are unfolded by dialysis against 1M NH_4Cl and treated with ribonuclease II to degrade their RNA, Maruta et al. (1971) were able to reconstitute the subunits in vitro. This may open the way to an analysis of 50S subunit assembly. However, the behaviour of the *sad* mutants which inhibit both 50S and 30S subunit maturation in vivo implies that proper assembly of the 50S subunit may depend upon simultaneous assembly of the 30S subunit; in this case reconstitution of the 50S subunit alone may not be typical of its assembly within the cell. But one possible mechanism, which is supported by an analysis of the interaction of ribosomal proteins with rRNA (see below) is that addition of a 30S subunit protein fraction may overcome the problems of 50S assembly in vitro.

Addition of Proteins to Precursor Particles

Reconstitution of ribosome subunits can also be achieved in mixtures in which one of the protein components is omitted. The properties of the resulting particles depend upon which protein is absent; the 30S proteins can be classified in this way into three categories: essential for assembly, partially necessary and dispensable. Nomura et al. (1969) found that when any of the six essential proteins, S4, S7, S8, S9, S16, S17, are omitted, the particles which are formed sediment between 20S and 25S; these particles are defective in assembly because they lack some of the other proteins. The essential proteins must therefore be added to the maturing particle at an early stage of its synthesis to enable other proteins to bind subsequently. Omission of the second group of proteins allows particles which sediment at 27S–29S to form, but these particles lack or have reduced amounts of some of the proteins of 30S subunits, although the loss is not so severe as when essential proteins are omitted. The third group of proteins appears to be dispensable for the formation of subunits, for only the protein omitted is likely to be missing from the reconstituted particle; and particles which sediment as usual at 30S are reconstituted, although they are not functionally active.

Although it was thought at first that all ribosomal proteins are present in all subunits, it now seems likely that there may be some heterogeneity, at least amongst the 30S subunits. Because the 30S subunit appears to contain some 260,000 daltons of protein, but the total weights of the individual proteins amount to some 410,000 daltons, individual subunits may perhaps differ in their contents of some proteins. Estimates of the amount of each protein found in isolated subunits suggest that there may be two classes of 30S protein; there is one copy of each of some proteins, the *molar* proteins, for each 16S rRNA molecule; but others, the *fractional* proteins, are present in smaller amounts. According to Voynow and Kurland (1971), twelve of the 30S subunit proteins are probably present in ratios of one copy per 16S rRNA molecule; as table 4.2 reveals, there is a striking correlation in that all those proteins which are essential for assembly fall into this category.

This suggests that each ribosome contains one copy of each of the proteins essential for assembly and of some other proteins also, but only some, and not all, of the fractional proteins. We do not know whether all ribosomes have the same activity in protein synthesis or whether their individual capacities may depend on which of the fractional proteins they contain. Kurland et al. (1969) have found that adding fractional proteins to isolated 30S subunits allows an exchange of free proteins with the proteins of the subunit and increases their activity in protein synthesis, but it is not clear whether this reaction has any biological significance or is an artefact of work in vitro.

The assembly of ribosomes in vitro can be followed in more detail by adding proteins in a defined sequence. Mizushima and Nomura (1970) found that seven proteins can bind to 16S rRNA itself when sedimented through sucrose with the nucleic acid (see below). Although some of these proteins bind directly to 16S rRNA, others bind more efficiently to the complex which is produced by the association of the first few proteins with the ribosomal RNA. Proteins must therefore be added to the assembling 30S subunit in the appropriate order. When S4, S8 and S20 are bound to 16S rRNA, proteins S7, S13, S16 and S17 can associate quantitatively with the resulting complex. But the binding of S16 and S17, for example, depends upon the prior presence in the complex of either S4 or S20, although not of S8.

Table 4.2: role of proteins in assembly of the 30S subunit of E.coli ribosomes

protein	assembly	21S (*sad*)	21S (*spc*)	RI particle	CsCl	LiCl	proportion	protein
S4	essential	+	+	+	—	—	molar	S4
S8	essential	+	+	+	—	—	molar	S8
S20	dispensable	+	+	+	—	split	fractional	S20
S7	essential	+	+	+	—	—	molar	S7
S15	dispensable	(+)	−	+	—	—	not known	S15
S16	essential	+	+	+	—	—	molar	S16
S17	essential	+	−	−	—	—	molar	S17
S13	dispensable	+	+	+	—	split	fractional?	S13
S9	essential	+	(+)	−	split	split	molar	S9
S19	partial	+	+	−	—	—	fractional	S19
S6	dispensable	+	+	+	—	—	molar?	S6
S5	partial	−	−	−	split	—	molar?	S5
S14	partial	−	−	−	split	split	fractional	S14
S18	dispensable	−	(+)	+	—	split	molar?	S18
S11	partial	−	−	+	—	split	fractional	S11
S10	partial	−	−	−	split	split	molar?	S10
S3	partial	−	−	−	split	split	molar?	S3
S21	dispensable	−	−	−	—	split	fractional	S21
S12*	dispensable	−	−	−	—	split	molar?	S12
S2*	dispensable	−	−	−	split	split	fractional?	S2
S1*	dispensable	−	−	−	split	split	fractional?	S1

Plus indicates that a protein is present; (+) that it is present in reduced amount; minus indicates absence. A query indicates that the proportion is not certain. The positions in the assembly map of the three proteins bearing an asterisk are not known.

The vertical order of proteins follows the assembly map of the 30S subunit shown in figure 4.5. Protein compositions were determined by: 21S (sad) particles—Nashimoto and Nomura (1970); S21 (spc) particles—Nashimoto et al. (1971); RI particles—Nomura et al. (1969); CsCl split proteins—Nomura et al. (1969) and Chang and Flaks (1970); LiCl split proteins—Homann and Nierhaus (1971); proportions of proteins—Voynow and Kurland (1971).

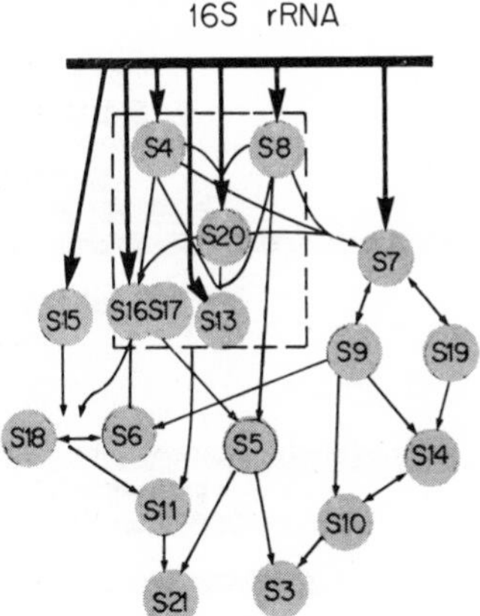

Figure 4.5: Assembly map of the 30S subunit of E.coli ribosomes. Data of Mizushima and Nomura (1970) and Nashimoto et al. (1971)

By following the effect which the omission of one protein has on the ability of others to associate with the particle, Mizushima and Nomura (1970) and Nashimoto et al. (1971) constructed the assembly map shown in figure 4.5, which reflects the order in which proteins bind to the assembling subunit and the interactions which are necessary for them to do so. The proteins which are essential for assembly if particles sedimenting at 30S are to be formed are all located in the early part of the assembly map. But although in general good, the correlation between essential proteins and order of assembly is not absolute, for S13 and S20 both appear to bind at an early stage but are classified as dispensable. In the case of S13, this may happen because the binding of further proteins does not depend upon its presence.

Analysing the proteins included in the 21S precursor particles found in a *sad* mutant and in the spectinomycin-resistant mutant which is defective in subunit assembly shows that there is a good correlation between the position of a protein in the assembly map and its presence or absence in the precursor. Figure 4.6 shows the analysis by two separation methods of the 21S particles accumulated by the *spc-49* mutant. The proteins included in the RI particles are very similar, although not identical, to the proteins of the 21S precursors. In table 4.2 the proteins of the 30S subunit are ordered (approximately) according to their position in the assembly map and it is evident that the proteins which are added earliest during assembly in vitro are in general the same as those present in precursor particles accumulated in mutant cells. Nierhaus, Bordasch and Homann (1973) found that essentially the same proteins are present in a 21S peak isolated as a shoulder of the 30S fraction of E.coli cells.

The similarity between RI particles, the 21S precursors of mutant cells and the 21S peak of normal cells and the evident relationship of their protein

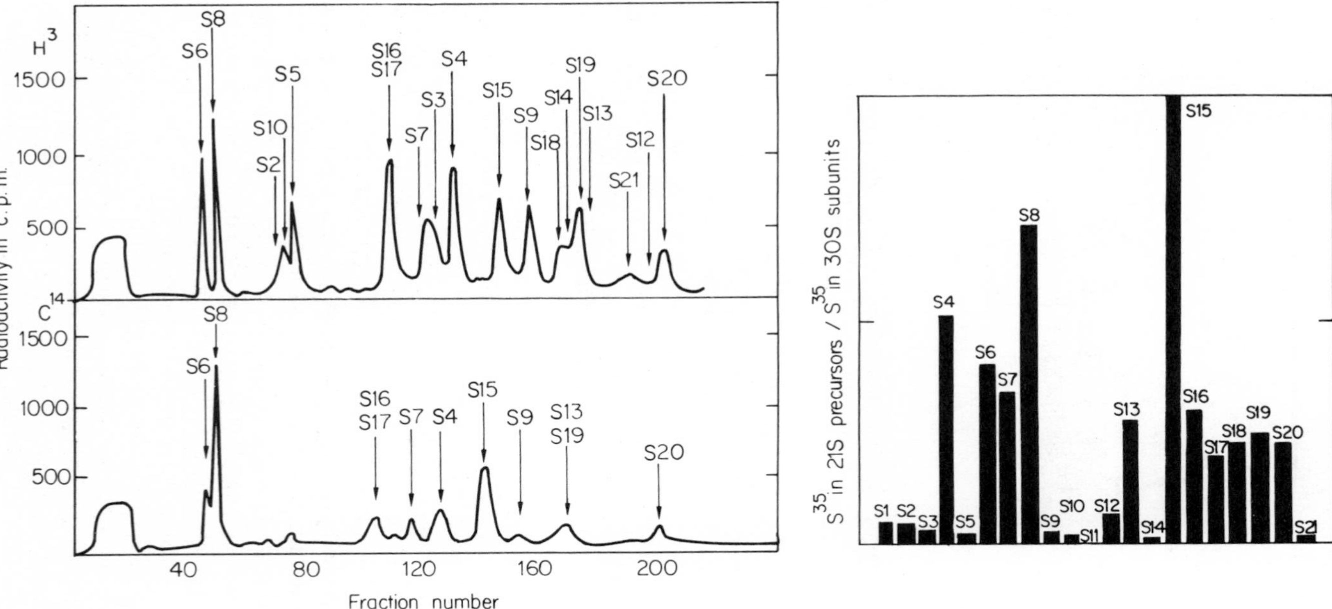

Figure 4.6: Proteins of the 21S precursor particles accumulated by the spectinomycin-resistant assembly defective mutant. Proteins of 30S subunits (upper) and 21S precursors (lower) are compared at the left by chromatography on carboxy-methyl cellulose. The results of an analysis with the two dimensional gel system are shown on the right; protein contents in the precursor are expressed relative to the extent of S15, the protein present in largest amount. Data of Nashimoto and Nomura (1970) and Nashimoto et al. (1971)

contents to the assembly map point to the conclusion that assembly both in vitro and in vivo involves a first step in which a small number of proteins—S4, S8, S20, S7, S15 (?), S16, S17 (?), S13 and perhaps S9, S19 and S6—associate with 16S rRNA to form a precursor particle. It is possible that this may be the only precursor, to which the remaining 30S proteins are then added; or two steps may remain in maturation, forming first a 26S and then a 30S particle.

Organization of Proteins in the 30S Subunit

The order of assembly and the components found in precursors may indicate in an approximate way the topology of the mature subunit. Those proteins which are added first during assembly tend to be those located in the interior of the ribosome and the proteins added later appear to be present at the surface of the ribosome. This idea is supported by the identification of the proteins which are split from 30S subunits by caesium or lithium ions. Table 4.2 shows that both sets of split proteins are similar, although caesium ions are less effective, probably because of their larger ionic radius. The proteins which are split tend to be those which are absent from the precursor particles and are located late in the assembly map. The precursor particles may therefore contain a nucleus of proteins which form part of the interior of the mature subunit when later proteins are added to form the exterior.

Crude maps of the 30S subunit can on this basis be drawn from the relationships of its proteins and as figure 4.7 shows the results are similar according to criteria of either structure or assembly. Proteins have been characterized by their resistance to extraction with salt: the clear outer circle of the salt structure map represents proteins which are removed by CsCl; the shaded intermediate circle shows the additional proteins removed by molar LiCl; and the dark inner nucleus comprises the proteins resistant to both treatments. A very similar arrangement is found in the precursor particle map, in which the inner core consists of proteins found both in the 21S precursors accumulated by the *sad* and spectinomycin-resistant mutants and in the RI particles. The intermediate circle contains the proteins found in some but not all of these particles; and the outer circle marks the proteins which are never found in any of these particles.

Another way to study the topology of the ribosome is to treat 30S subunits with reagents which modify or degrade proteins. Those proteins which are essential for assembly and are found in the precursors tend to be inaccessible to reagents such as iodoacetate or methoxy-nitropone. As table 4.3 shows, the proteins which Chang and Flaks (1971) and Craven and Gupta (1970) found are protected against degradation by trypsin include those present in the precursors. There is an approximately inverse relationship between the sequence of addition of proteins during ribosome assembly and their relative susceptibilities to trypsin or chemical reagents, although there are some exceptions which show that time of addition is not rigorously related to position (reviewed by Wittman and Stoffler, 1972).

The picture of the ribosome suggested by any of these criteria is therefore similar, although the positions of a few proteins may vary. There is a set of proteins, comprising S4, S6, S7, S8, S13, S15, S16, S17 and S20 which provides the nucleus in each case; some of these proteins are bound directly to rRNA and they are in general located within the structure of the subunit. There is a set of proteins including S1, S2, S3, S10 and perhaps S5, S12 and S14 which is added late in assembly and is probably located on the surface of the ribosome from where they are readily removed.

Some proteins show behaviour which is not readily reconciled with too simple a picture of the ribosome; for example, S9 is readily split by CsCl but is essential for assembly although it is added at an intermediate stage and is not included in all precursor particles. Of course, some of the proteins bound to

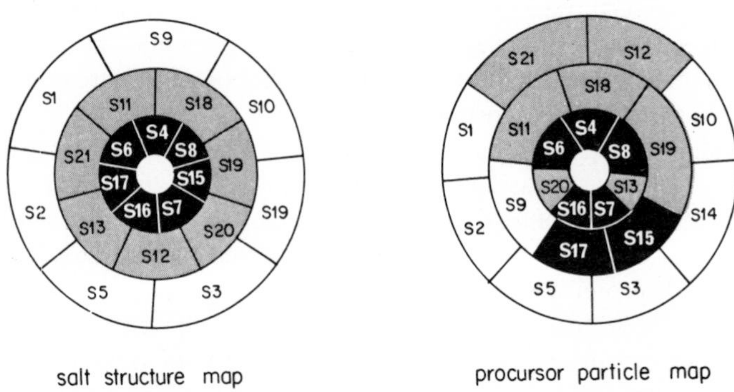

Figure 4.7: Representation of the structure of the 30S subunit of E.coli ribosomes. The salt structure map shows those proteins which are not dissociated by either CsCl or M LiCl in the nucleus (black), those which are dissociated by LiCl form the centre circle (shaded), and those which are dissociated the most readily, by CsCl, are placed in the outer circle (white). The precursor particle map retains the same colour scheme; but the nucleus consists of proteins found in both 21S and RI precursors, the intermediate circle of those proteins present in some but not all of the particles, and the outer circle identifies those proteins never present in any of them. The correspondence between the two maps is appreciable and suggests that the availability of proteins to splitting by salt and their order of assembly reflect an organization of the ribosome in which there is a core of proteins added early in assembly and inaccessible in mature subunits, surrounded by proteins added later which are more accessible. Based on data of Table 4.2

Table 4.3: position of proteins in the 30S subunit of E.coli ribosomes

protein	binding to 16S rRNA	protection against enzyme attack	
S4	+	+	+
S8	+	+	+
S20	+	+	+
S7	+	−	+
S15	+	−	+
S16	(+)	+	+
S17	(+)	+	+
S13	−	+	−
S9	−	−	+
S14	−	−	+
S18	−	+	+
S12	−	+	+

The first column shows the ability of proteins to bind to 16S rRNA, measured by Mizushima and Nomura (1970), Schaup et al. (1970, 1971), Stoffler et al. (1971a, 1972), Garrett et al. (1972). S16 and S17 have only been found to bind weakly in occasional experiments. The last two columns show the measurements of Chang and Flaks (1971) and Craven and Gupta (1970) on the protection of ribosome proteins when 30S subunits are degraded with proteolytic enzymes. There is therefore a reasonable correlation between the ability of a protein to bind to DNA and its protection against degradation.

rRNA may be near the surface of the ribosome but stable to salt or other reagents because of their intimate association with nucleic acid. Proteins such as S15 and S13 which are not needed to help other proteins bind may perhaps fall into this category. Whilst the general nature of the ribosome seems to be an interior which is assembled first and an exterior which is added later, then, this correlation may not necessarily hold for any individual protein.

An important caution in drawing maps of this nature is that the particles which are found as precursors in vivo and in vitro may form a heterogeneous population. Only some 20–40% of the RI particles are the real intermediates of ribosome assembly, as judged by their conversion to 30S subunits; and the 21S particles found in vivo may also vary in their protein content. If individual mature 30S subunits contain different sets of proteins, then the proteins shown in the outer circles of the ribosome maps may on some occasions be absent altogether, as indeed may some of the proteins in the intermediate groups; but it seems probable that the central nucleus, which remains so similar in all these cases, is always present.

Another criterion for estimating the relationships between ribosomal proteins in the mature subunits can be derived from degrading the subunits into smaller particles and seeing which proteins remain together. By briefly exposing 30S subunits to ribonuclease in an insoluble form, Schendl, Maeba and Craven (1972) generated three particles, the largest of which sediments at 22S and contains 15–16 proteins, another which sediments at 15S and contains 12–13 proteins, and the smallest of which sediments at 7S and contains 6–8 proteins. Morgan and Brimacombe (1972) fragmented the 30S subunit by exposure to ribonuclease in the presence of 2M urea. In both sets of fragments the proteins clustered together on the assembly map of figure 4.5 tend to be associated in the fragments, although the relationship does not apply to every protein. This suggests that the order of assembly deduced from the dependence of one protein upon the presence of another is related to the physical interactions in the particle of groups of proteins.

The assembly map does not indicate all the interactions which take place, however, for proteins which are not related on the assembly map can be cross-linked by treating 30S subunits with bifunctional bis-imido esters. The cross-linked proteins can be identified by their slow movement on acrylamide gels, after which the cross links can be destroyed by hydrolysis and the proteins identified by electrophoresis. Bickle, Hershey and Traut (1972) and Lutter et al. (1972) found that dimers may be generated from any two of S6, S7 and S9—which are located in sequence on the assembly map. But S5 and S9 appear also to be related, as do S11–S19, S18–S21, S5–S8 and these pairs are not grouped together on the assembly map. One model to reconcile these results is to suppose that the initial set of proteins bind to different regions of 16S rRNA and then associate with additional proteins along the length of the molecule. But new relationships in addition to these may be generated when the molecule folds to bring into apposition proteins initially concerned with different regions of the rRNA.

Association of Proteins with 16S rRNA

Only those proteins which are located in the early part of the assembly map can bind directly to 16S rRNA. The data of table 4.3 include the results obtained by several research groups; although the amounts bound of each protein vary somewhat, this probably reflects the techniques used for measurement—such as sedimentation through sucrose gradients, electrophoretic separation and immunological assay—rather than biological reality. Proteins S4, S8 and S20 bind to 16S rRNA the most strongly, although S7 and S15 bind in the same total amounts.

One copy of each protein appears able to bind to the RNA, so that proteins which bind are indicated by a plus in the table. Some experiments have suggested also that S13, S16 and S17 can bind to the RNA; these are probably able to do so in the presence of other proteins (see above). Competition experiments between the different proteins show that they all bind to different sites on the rRNA, for the binding of one protein does not inhibit the interaction with rRNA of the others. Similar results are obtained when the 16S rRNA is derived from B.stearothermophilus instead of E.coli, although S20 and S8 then bind less strongly; this suggests that the binding sites in rRNA have been conserved during the evolution of bacteria.

The sequences of 16S rRNA which interact with proteins in the assembly reaction can be identified in either of two ways. After binding protein to rRNA, the complex may be degraded with ribonuclease and the regions which are protected from attack may be recovered and identified; or the rRNA may be fragmented first by the introduction of breaks and different fragments tested for their abilities to bind proteins of the 30S subunit.

Using protein S4, Schaup and Kurland (1972) found that the maximum amount of rRNA is protected by the quantity of protein which saturates the 16S molecule in the assembly reaction; this indicates that the protected region comprises the specific binding site. The protected RNA sequences correspond to about one quarter of the length of the molecule and after isolation will again react with S4 protein in the same way as with 16S rRNA. Protein S4 is not large enough to cover a continuous sequence of rRNA of this length, which implies that it must react with non-contiguous regions of the molecule. This contention is supported by gel electrophoresis of the protected RNA, which behaves as though consisting of at least five fragments of some 60–120 bases each rather than of one continuous length of RNA.

Electron microscopy suggests that S4 binds close to the 5′ end of the polynucleotide chain; Nanninga et al. (1972) showed that complexes of S4 with 16S rRNA show a large aggregate with a tail corresponding to little more than one half of the protein chain. The aggregate appears to contain one end of the chain and several other regions drawn into a complex with the protein.

The 16S rRNA can be split into two fragments. Zimmermann et al. (1972) reported that the larger is a component sedimenting at 12S which contains 900 residues of the 5′ part of the molecule, starting at a point within 50 bases of the 5′ terminus. The smaller fragment sediments at 8S and its 5′ terminus is adjecent to the 3′ terminus of the 12S fragment; this sequence extends to within 50 bases of the 3′ terminus of 16S rRNA. Proteins S4, S8, S15 and S20 can interact with the 12S fragment; S7 binds to the 8S fragment. By degrading the 12S RNA into smaller fragments, the proteins can be ordered by their binding sites along the rRNA. The sites to which they bind are clustered at the 5′ end of the molecule and appear to take up the sequence:

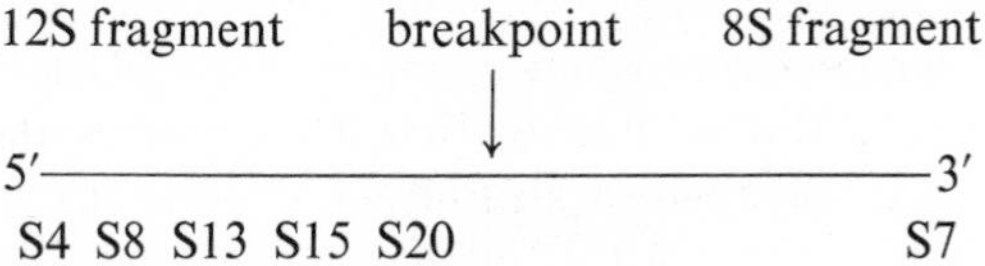

Although S13 does not bind to isolated 12S RNA, it seems to protect from ribonuclease attack a sequence located in this region. Schulte and Garrett (1972) found that the binding of S8 is strongly temperature dependent and suggested that some of the activation energy required for assembly may be needed for this binding reaction.

The clustering of binding sites towards the 5′ end of the molecule suggests that proteins may associate with the rRNA as soon as it is transcribed so that there is unlikely to be any free precursor rRNA in the cell. The assembly of all 30S subunits probably starts with the binding to rRNA of one molecule each of the proteins S4, S8, S15, S20, S7 and perhaps S13, S16 and S17.

That some of the sequences of 16S rRNA have been conserved in evolution is suggested by the success of reconstitution experiments using heterologous components. Nomura, Traub and Bechman (1968) found that the 16S rRNA molecules derived from bacteria which have very different G-C contents in their genomes can nevertheless interact with E.coli proteins to reconstitute functional 30S subunits. The reverse combination of components may also be successful. Hybrid subunits have been constructed from the rRNA and proteins of E.coli, Azobacter vinelandii and various species of Bacillus, although the activity of the hybrids is less than that of subunits reconstituted from homologous components and reverse combinations do not always show the same activity. This implies that the binding sites in the rRNA which interact with proteins have been conserved; and we might expect also to find structural similarities in the proteins of different bacteria which bind to heterologous 16S rRNAs.

As might be expected from the characterization of individual binding sites for each of the 30S proteins, reconstitution experiments reveal a high specificity for the nucleic acid component of the ribosome; 30S particles reconstituted with 16S rRNA treated with heat or alkali are inactive. The 16S rRNA is also highly susceptible to loss of activity by chemical modification; the induction of as few as 6–8 alterations by nitrous acid treatment destroys the activity of reconstituted particles. This may be due to modification in the non-binding regions of the RNA since the modified molecule still seems able to bind its usual proteins.

Apart from its structural function in providing a backbone to which the ribosomal proteins are attached, the role of ribosomal RNA is not at all clear. That its sequence is important for the functions of the ribosome is suggested both by the isolation of mutants which influence its modification and by the effect of mutagens. The antibiotic kasugamycin interacts with the 30S subunit to inhibit initiation. By reconstituting ribosomes in vitro from components derived from kasugamycin resistant or sensitive strains, Helser, Davies and Dahlberg (1971, 1972) showed that the 16S rRNA controls the response to the drug. Analysis of C^{14}-methyl labelled rRNA showed that the change from kasugamycin-sensitivity to resistance depends upon the failure of the cell to methylate two adjacent adenine residues near the 3′ end of the molecule. Sensitive cells possess a methylase activity which is absent from resistant cells and which can catalyse the methylation in vitro. Methylation at this site must therefore in some way control the response of the ribosome to kasugamycin and be important for its function. Lai et al. (1973) showed that the response of staphylococcus aureus to lincomycin, and spiramycin is similarly controlled by a methylation event in 23S rRNA.

The only specific function in which rRNA has been implicated is binding of tRNA. Noller and Chaires (1972) found that kethoxal—which interacts with guanine residues—inactivates 30S subunits by preventing tRNA binding although leaving intact the ability to bind poly-U. Some 10 kethoxal residues are bound to each 30S subunit; but if tRNA is first bound to the ribosomes, only 3–4 ketothoxal residues can be bound and the subunits are not inactivated. This suggests that 6–7 specific guanine residues are needed for the binding of tRNA and are protected by it.

Organisation of the 50S Subunit

Although we know much less about the 50S subunit, its structure also appears to have the ordered features characteristic of the 30S subunit. Table 4.4 shows that its proteins can be organised into groups according to the ease with which they are split from the subunit by CsCl or LiCl. There is a correlation with the presence of proteins in the 43S precursor, for all the proteins most resistant to splitting are in the precursor and very few of the proteins which are easily removed seem to be present.

Eight of the proteins can bind to 23S rRNA and immunological assay of the isolated complexes suggests that they all bind in 1:1 ratios. Four of these proteins are in the group most resistant to extraction by salt and may therefore be located in the interior of the ribosome. None of the 50S proteins binds to 16S rRNA, but one of the 30S proteins, S11, binds specifically to 23S rRNA. This makes this protein an obvious candidate for the role of linking together the assembly of the two subunits and, of course, it is tempting to speculate that it may play some role in the association of the two subunits to form a 70S ribosome.

Functional Sites in Ribosome Subunits

Subribosomal Particles of the 50S Subunit

Although subribosomal particles which sediment normally can be reconstituted in the absence of any one of the proteins which are dispensable for assembly, these particles lack activity when tested for their ability to synthesise proteins. With the 30S subunit, for example, apart from S15 whose function is unknown and S6 which appears to be involved at the stage of initiation only, all the dispensable proteins must be added to the reconstitution mixture if active subunits are to be produced. Of course, the tests for protein synthetic activity with reconstituted particles concern the whole population, so that

Table 4.4: proteins of the 50S subunit of E.coli ribosomes

							proteins remaining after treatment with			
	protein	weight	DNA binding	precursor 32S	43S	CsCl	0·6M LiCl	1·0M LiCl	2·0M LiCl	4·0M LiCl
group I	L17	16,000	+	+	+	+	+	+	+	−
	L19	17,000	+	−	+	+	+	+	+	−
	L23	13,000	+	−	(+)	+	+	+	+	−
	L24	17,000	+	+	+	+	+	+	+	−
	L29	12,000	−	+	+	+	+	+	+	−
	L32	10,500	−	−	−	+	+	+	+	−
group II	L6	20,000	−	+	(+)	+	+	+	−	−
	L9	17,000	−	+	+	+	+	+	−	−
	L18	16,000	−	+	+	+	+	+	−	−
	L25	12,000	−	+	+	(+)	+	+	−	−
	L30	10,000	−	+	+	+	+	+	−	−
group III	L2	29,000	+	−	−	+	+	−	−	−
	L6	22,000	+	−	−	(+)	−	−	−	−
	L11	19,000	−	−	+	+	+	−	−	−
	L15	17,000	−	−	+	+	+	−	−	−
	L27	12,000	−	+	(+)	+	(+)	−	−	−
group IV	L1	23,000	−	+	+	+	+	−	−	−
	L7	15,000	−	−	+	+	−	−	−	−
	L8	18,000	−	+	(+)	+	(+)	−	−	−
	L10	20,000	−	−	(+)	(+)	+	−	−	−
	L25	12,000	−	+	+	(+)	+	+	−	−
	L28	15,000	−	−	−	(+)	−	−	−	−
	L31	10,000	−	−	−	(+)	−	−	−	−
group V	L14	18,000	−	−	+	−	(+)	−	−	−
	L16	20,000	+	−	−	−	−	−	−	−
	L20	17,000	+	+	+	−	−	−	−	−
	L26	12,000	?	?	?	?	?	?	?	?
	L33	10,500	−	−	+	−	−	−	−	−
	L34	9,600	?	?	?	?	?	?	?	?

Plus, +, indicates presence, (+) indicates presence in reduced amounts, and minus, −, indicates absence. Query, ?, shows that the protein was not tested.

Proteins are organised into groups according to their susceptibility to splitting from the subunit from salts. Group V proteins are split by CsCl whereas group I proteins can be split only by 4M LiCl. These groups correlate with the presence of proteins in precursor particles; the group I and II proteins are present in both precursors, group III and IV proteins tend to be added in the 43S precursor. Four of the proteins which bind DNA are in the tightly associated group I. Molecular weights and binding to DNA were determined by Stoffler et al. (1972); proteins present in precursor particles were assayed by Homann and Nierhaus (1971) and Nierhaus et al. (1973).

individual particles might be reconstituted with different proteins and functional capacities; but the need for all proteins to be present in the reconstitution mixture must cast some doubt on the idea that active ribosomes may be heterogeneous in protein content. The functional activities of individual proteins may be deduced by testing the ability in protein synthesis of subribosomal particles which lack them; these particles may be defective in specific reactions of the protein synthetic cycle.

Such experiments can involve only those proteins which are not needed for assembly, for the absence of assembly-essential proteins completely inactivates the reconstituted subparticles. Traub and Nomura (1968a, 1968b) obtained split proteins from both subunits by centrifugation through CsCl density gradients and separated each of the SP50 and SP30 fractions on DEAE-cellulose into an acidic fraction (SP50A or SP30A) and a basic fraction (SP50B and SP30B). SP50A and SP50B each comprise about four proteins, probably drawn from groups IV and V of table 4.4. Subribosomal particles can be reconstituted from the core particles by adding the various split protein fractions both separately and together. By adding the particles to normal subunits of the other size they can be tested for their ability to synthesise poly-phenylalanine under direction from poly-U and for their ability to bind tRNA. Reconstituted 30S particles may also be tested for their ability to bind the poly-U messenger.

The results obtained with the reconstituted 50S particles are shown in table 4.5. Transfer RNA can normally undergo a reversible binding to the 50S subunit in the absence of mRNA—this probably takes place at the P site. SP50B is clearly required for this activity, although since it is not necessary for peptide bond formation, full activity in binding tRNA cannot be a prerequisite for activity of the peptidyl transferase centre. SP50A is not concerned with binding tRNA, but appears to be essential for peptide bond synthesis. This confirms the picture derived from studies of protein synthesis that the 50S subunit may consist of several independent active sites (see page 77).

Table 4:5: activity of partially and fully reconstituted 50S subunits of E.coli ribosomes according to Traub and Nomura (1968a, 1968b)

particle	poly-U system activity	tRNA binding
40S core	almost completely inactive	almost no activity
40S core + SP50A	half the activity of fully reconstituted particles (bottom line)	almost no activity
40S core + SP50B	inactive	half the activity of fully reconstituted particles (bottom line)
40S core + SP50A + SP50B	half activity of native 50S subunits	30% activity of native 50S subunits

The idea that groups of proteins are organised together in the structure of the subunit to make up these active centres is supported by studies of the synthetic capacities of cores derived from 50S subunits by Staehelin, Maglott and Munro (1969). When 50S subunits are centrifuged through CsCl in the presence of 40–50 mM Mg^{2+}, they progressively lose two proteins to give α-cores, another five proteins to yield β-cores, and then some five further to yield γ-cores. We may expect the first two sets of proteins to correspond approximately to group V of table 4.4 and the last set to group IV. The cores should be similar, although not necessarily identical, to those obtained by Traub and Nomura.

The α and β cores retain peptidyl transferase activity as judged by their ability to form N-acetyl-leucyl-puromycin in the puromycin reaction, but the active centre must be removed when the γ cores are formed, for these are quite inactive. The split protein fractions themselves have no peptidyl transferase activity, although this ability is restored to the cores by their addition. Although the β cores are active in peptide bond synthesis, they cannot undertake protein synthesis; this suggests that the group of proteins which constitutes the active site for peptide bond formation may be comparatively independent of the other active sites of the ribosome.

This conclusion is supported by the observation of Bodley and Lin (1972) that CsCl core particles lack the ability to bind G factor and H^3-GTP in the presence of fusidic acid; addition of either the acidic or basic split proteins does not restore much activity, but both together do so. Reconstitution of G factor binding activity is rapid when all the split proteins are restored to the cores, whereas the restoration of peptidyl transferase activity takes place much more slowly. This, together with the different ionic requirements of the G factor binding reaction and peptidyl transferase activity, suggests that two different sites undertake these reaction.

Two proteins in particular seem to be implicated in the GTP hydrolysis site used by both Tu factor and G factor (see page 73). Brot et al. (1971) found that when ribosomes are treated with ethanol and ammonium chloride, 4–5 proteins are released and the particles remaining have a reduced ability to support G factor dependent hydrolysis of GTP and to bind factor G and GTP to the ribosome in the presence of fusidic acid. The activity of the deficient particles is restored by addition of one of the proteins of the extract. Thiostrepton, which inhibits the hydrolysis of GTP, can bind to the deficient particles; this suggests that at least one of the proteins remaining on the particle after EtOH–NH_4Cl treatment is also needed to hydrolyse GTP.

The active protein fraction split by EtOH–NH_4Cl comprises a mixture of L7 and L12, two different forms of one polypeptide chain; L7 is formed by acetylation of L12. Weissbach et al. (1972a) observed that loss of this protein prevents GTP hydrolysis in response to either the binding of Tu–aminoacyl–tRNA–GTP or the binding of G factor and GTP in the manner shown in figure 2.19. That protein L7/L12 is an essential part of the GTP hydrolysis site of the ribosome is suggested also by the observation of Highland et al. (1973) that the only antibodies against ribosomal proteins which prevent formation of the complex with G factor and GDP in the presence of fusidic acid are those responding to L7/L12. Two other 50S proteins, L6 and L10, have been implicated by Schreier et al. (1973) in the GTPase action of the translation factor G. Although the GTPase site cannot yet be fully defined, it therefore comprises only a small number of 50S subunit proteins, presumably located in one restriction region of the particle.

Maximum GTPase activity is displayed only in the presence also of 30S subunits (see page 71). By testing the ability of 30S subunits lacking certain proteins to stimulate the GTPase activity of 50S subunits, Marsh and Parmeggiani (1973) demonstrated that two of the 30S proteins, S5 and S9, are responsible for the stimulation. These two proteins can be cross-linked by bis-imido esters (see page 150), which supports the concept that they are located close together in a region of the 30S subunit which interacts directly with the 50S subunit; they may therefore form part of the interface between the subunits.

Subribosomal Particles of the 30S Subunit

Reconstituted 30S particles may be almost as active as native subunits. As table 4.6 shows, Traub and Nomura found that the SP30B proteins are essential for all the catalytic activities of the subunit, for particles reconstituted without this fraction are inactive. The SP30A proteins do not appear to play any indispensable role, although in the presence of the SP30B proteins they stimulate tRNA binding and poly-U directed incorporation activity. Each of the SP30 fractions has been separated into its individual protein components. SP30B comprises five proteins, S3, S5, S9, S10 and S14; the SP30A includes only two proteins, S1 and S2.

By fractionating the SP30B complement into individual proteins on phosphocellulose columns, Traub, Soll and Nomura (1968) were able to test each

Table 4.6: activities of partially and fully reconstituted 30S subunits of E.coli ribosomes according to Traub and Nomura (1968a, 1968b)

particle	tRNA binding/poly-U activity	poly-U binding
23S core	inactive	inactive
23S core + SP30A	inactive	only weakly active
23S core + SP30B	half of activity of fully reconstituted particles (lowest line)	same as fully reconstituted particles (lowest line)
23S core + SP30A + SP30A	up to 85% of activity of native 30S subunits	highly active

of the five types of reconstructed subparticle lacking only one of these proteins. These subparticles have the deficiencies shown in table 4.7. It is clear that all active ribosomes must possess S10 and S14, so there should be no heterogeneity in the ribosome population with respect to these proteins (although S10 may be molar, however, S14 appears to be fractional in assays of content). The core proteins of the 23S particles contain some components which seem to have the same mobility as S3 and S5 when run on acrylamide gels, so it is possible that the residual activities in the particles lacking these proteins are caused by the presence of small amounts which have not been removed by the CsCl treatment. Protein S9 appears to be implicated in initiation of protein synthesis.

The subparticles which lack the SP50A proteins, S1 and S2, appear to suffer little incapacity in protein synthesis, for they can both initiate and elongate polypeptide chains. But subunits engaged in protein synthesis differ in their content of S1 from free subunits. After incubating 30S subunits with excess poly-U, Van Duin and Kurland (1970) separated the bound subunits—which sediment at 40S—from the subunits which remain free and sediment at 30S. The free subunits contain only half the amount of S1 found in the bound subunits and also have a reduced amount of S21.

When extra S1 is added to the incubation mixtures, the extent of complex formation between 30S subunits and poly-U is increased. Reconstituted subparticles which lack S1 can bind only 10–25% of the amount of poly-U bound by 30S subunits which have previously been incubated with S1; the ability of the particles to bind phe-tRNA depends upon their ability to complex with poly-U. S1 is the most acidic protein of the 30S subunit and its amount per subunit varies more widely than other proteins. It is the only protein which

Table 4.7: activities of subparticles of E.coli 30S subunits reconstituted without one of the basic split proteins

protein missing	activity of reconstituted subparticles
S3	amino acid incorporation reduced to 10–40%
S5	partial activity remains for both tRNA binding and amino acid incorporation
S9	full activity in the poly-U system, but activity reduced to 24–44% in f2 RNA-dependent binding of fmet-$tRNA_f$; this fraction may be required only for initiation and not for elongation
S10 or S14	almost completely inactive in both specific tRNA binding and amino acid incorporation, although normal capacity remains to bind poly-U messenger

These activities are compared to those of particles reconstituted by adding all five proteins, which are comparable in synthetic ability to the particles reconstituted with the unfractionated split proteins or native 30S subunits. Data of Traub, Soll and Nomura (1968).

dissociates spontaneously from the 30S subunit in buffers or high ionic strength. One speculation about its role is to suppose that it stabilises the association of subunits with mRNA after the initiation factors have acted and may later be displaced from the ribosome at chain termination, so that it is not found on free subunits.

Native 30S subunits can also be distinguished from those extracted from polysomes by their reduction in content of S21. Van Duin et al. (1972) found that adding S21 to native subunits inhibits their activity in binding fmet-tRNA$_f$ for initiation. But this effect is only found when 50S subunits are also present in the incubation mixture. This implies that S21 may be implicated in maintaining the association between the subunits of active ribosomes engaged in protein synthesis.

When 30S proteins are added to 30S subunits, some stimulation of their activities is usually found. Randall-Hazelbauer and Kurland (1972) have found that addition of S2, S3 and S14—all three proteins must be present for maximum effect—stimulates the factor dependent binding of tRNA to the 30S subunit. This suggests that these proteins may comprise part of the A site of the ribosome where aminoacyl-tRNAs bind; and this agrees with the role suggested for S3 and S14 by Nomura's reconstitution experiments. Addition of these proteins also stimulates fmet-tRNA$_f$ binding, which suggests that there may be some connection between the ribosome sites used to bind initiator and normal aminoacyl-tRNAs. The effect of the three proteins concerned with aminoacyl-tRNA binding is independent of that of S1 and mRNA binding, which again reinforces the conclusion that each subunit consists of a series of active centres at which the different catalytic activities of the ribosome take place.

Influence of Streptomycin on the Ribosome

The antibiotic streptomycin inhibits protein synthesis and also increases the level of misreading of messengers at translation; although both effects result from its interaction with one of the 30S proteins, the inhibition is not caused by the misreading effect but has a different basis. The addition of streptomycin to systems for protein synthesis in vitro induces distinct patterns of misreading of the messenger. Pestka, Marshall and Nirenberg (1965) and Kaji and Kaji (1965) first observed that streptomycin promotes misreading of poly-U as isoleucine by stimulating the binding of ile-tRNA to the ribosome–messenger complex. Davies, Jones and Khorana (1965) found that streptomycin causes only certain mistakes to be made, with isoleucine and serine responding to poly-U and serine, histidine and threonine responding to poly-C in addition to the usual proline.

The triplets for the misincorporated amino acids are related to the messenger codons by a single base change, which implies that only one base of the codon is misread at a time. These results suggest that streptomycin encourages the misreading of U as C or A and of C as U or A in (at least) the 5′ and internal positions of the code; it is difficult to distinguish misreading in the 3′ position from the effects of wobble. Experiments with other polymers have confirmed the general rule that only the pyrimidines C and U are misread, usually as each other but occasionally as A. The misreading effect is probably not this simple, however, for by using poly-UC, Davies (1966) found that its range of misreading is much more restricted than that of either poly-U or poly-C; this suggests that the misreading of a particular base may be influenced by its neighbours.

Attempts have been made to induce misreading in mammalian systems with factors including streptomycin, ethanol and aliphatic polyamines such as spermine. Although these all enhance ambiguity considerably in bacterial systems, Weinstein et al. (1966) and Friedman et al. (1968) found that they have little or no effect upon systems containing mammalian ribosomes. The difference in the response of bacterial and eucaryotic ribosomes to these agents implies that the role which the ribosome plays in controlling the accuracy of translation may take different forms in these different species.

Although it was thought at first that misreading might provide the basic mechanism for the antibiotic action of streptomycin—by flooding the bacterial cell with non-functional proteins affecting all its activities—Davies (1966) found that streptomycin can inhibit protein synthesis in vitro in the absence of misreading, for example, when poly-AG is used as template. At bacteriacidal concentrations, furthermore, the antibiotic inhibits protein synthesis very rapidly, even before there is time for substantial misreading to occur.

That streptomycin inhibits initiation of protein synthesis is suggested by the observation that its effect is greater in vitro with natural messengers than with those, such as poly-U, which lack proper initiation signals. And the increase in magnesium concentration which abolishes the need for proper initiation also overcomes the inhibition. Streptomycin does not prevent the formation of initiation complexes, but causes release of fmet-$tRNA_f$ from the complex after the 50S subunit has joined it. Release seems to take place from the P site, for Modolell and Davies (1970) and Lelong et al. (1971) noted that the reaction demands the presence of GTP to allow the fmet-$tRNA_f$ to enter a puromycin reactive state.

The distribution of ribosomes in the cell changes when streptomycin is added to inhibit protein synthesis and Luzzato, Apirion and Schlessinger (1968b, 1969) found that the number of polysomes declines rapidly so that most ribosomes are found in the form of monomers. It was at first thought that these monomers represent blocked initiation complexes between the ribosome and mRNA. But Wallace and Davis (1973) suggested an alternative explanation; they found that ribosomes blocked in initiation by streptomycin gradually escape so that they can continue protein synthesis. But upon termination the ribosome remains in the form of a 70S particle with a reduced ability to dissociate into subunits. Wallace, Tai and Davis (1973) showed that streptomycin treated ribosomes have increased resistance to the dissociation promoted by factor f3. Streptomycin therefore acts at two stages; first it inhibits initiation and then when ribosomes escape from this block they exist in a more stable state of association in which it is more difficult to generate the subunits needed for future initiations.

The site of action of streptomycin lies with the S12 protein of the 30S subunit. Davies (1964) and Cox, White and Flaks (1964) first pinned down the action of the antibiotic by forming 70S ribosomes from 30S and 50S subunits of E.coli derived from strains of E.coli either resistant or sensitive to streptomycin; only ribosomes possessing a 30S subunit derived from the sensitive strain are inhibited by the drug. By dissociating 30S subunits into their components, Traub and Nomura (1968c) showed that in reconstituted subunits one of the core proteins determines the response to streptomycin and Ozaki, Mizushima and Nomura (1969) identified this as S12. In a similar manner, Bollen et al. (1969) showed that one of the split proteins of the 30S subunit, later identified as S5, confers resistance to spectinomycin.

Reconstituted subunits in which the S12 protein is derived from bacteria resistant to streptomycin show a ten fold decrease in ambiguity compared to those in which the protein is taken from a sensitive strain. This implies that the S12 protein may control the degree of accuracy with which messengers are read and that mutations which confer resistance to streptomycin reduce the degree of ambiguity and therefore compensate for the increase caused by the drug.

Reconstituted particles which lack S12 appear normal by their sedimentation rates, but are completely resistant to the misreading usually induced by streptomycin. Ribosomes from bacterial strains resistant to streptomycin are usually sensitive to other antibiotics which also cause ambiguity, such as neomycin and kanomycin; but S12 deficient particles are equally resistant to the effects of all these antibiotics. This suggests that the extent of misreading is an intrinsic property of the ribosome and is mediated by S12.

But at least one other protein, S11, also has a role in determining the accuracy of ribosomal reading, for Nomura et al. (1969) found that particles lacking S11 show twice the level of misreading found in subunits which contain it. S12 and S11 therefore seem to act in opposite ways. S12 increases ambiguity, so that its omission prevents the ribosome from responding to streptomycin; and S11 decreases misreading, as revealed by the increase found when it is omitted during reconstitution. None of the other proteins which is dispensable in assembly influences the fidelity of the ribosome when omitted. But the genetic *ram* mutation (see next section) in the gene which codes for the core protein S4 also enhances ambiguity; no mutations in S11 have been found yet.

The interaction of streptomycin with S12 is responsible for its inhibitory effect as well as for its misreading. Using poly-U as template, Ozaki et al. (1969)

have shown that S12-deficient particles possess only half of the activity of fully reconstituted particles, but under direction from f2 RNA their relative activity is decreased even further to 16–20% of the control. The cause of this reduction can be pinned down more precisely, since whereas the binding of normal aminoacyl-tRNA under direction of f2 RNA is reduced to about 80% of its usual activity, the binding of fmet-$tRNA_f$ takes place at only about 25% of control.

Addition of S12 restores full activity. This suggests that particles lacking S12 are deficient in some reaction concerned with the establishment of an initiation complex, which is confirmed by the inability of S12-deficient particles to bind fmet-$tRNA_f$ under direction of AUG triplet. These experiments do not show whether fmet-$tRNA_f$ is bound to and then subsequently released from the initiation complex or whether it fails to bind in the first place; but the S12 protein must play a role in (at least) two of the functions of the ribosome: initiation and translational fidelity. The S12 protein found in strains of bacteria which are sensitive to streptomycin interacts with the drug to inhibit initiation and enhance ambiguity.

Ribosomal Mistranslation

The ability of streptomycin to induce misreading explains its action as a suppressor of mutants in E.coli. After treating bacterial cells with mutagens to create strains which lack the ability to make some of the metabolic intermediates which they usually produce, Gorini and Kataja (1964, 1965) found that addition of streptomycin to the growth medium can be substituted in some one to five per cent of the mutants for the growth factor itself. This means that streptomycin restores the ability of the mutant cells to make the enzyme which synthesises the growth factor.

Such cells depend upon streptomycin for growth in minimal medium, but do not require the drug when they are given the growth factor whose synthesis has been inhibited by the mutation. This phenotype has therefore been described as *conditional streptomycin dependence;* the presence of streptomycin must suppress the mutation and thus overcome the need for growth factor. This mechanism appears to involve misreading of the mutated codon, so that either the original amino acid which was specified or an acceptable substitute for it is placed in the protein only when streptomycin is present; this restores the ability of the cells to make the metabolic intermediate by suppressing the mutation and they can therefore grow.

When cells of E.coli which are sensitive to—that is are killed by—streptomycin are exposed to the antibiotic, they can survive either by becoming resistant to it or by becoming dependent upon it. Resistant cells have ribosomes which no longer respond to the misreading or inhibition of initiation which the drug produces. Most, although not all, of the dependent colonies can be satisfied by addition of either streptomycin or paromycin or, indeed, by ethanol. All these reagents induce misreading in vitro. Such cells require a drug such as streptomycin even when grown on complete medium; its presence is not needed to suppress a mutation, as in conditional dependence, but rather to distort the ribosomes so as to ensure a proper reading of all genes. Without streptomycin, dependent cells cannot translate messengers properly.

Mutations to give the dependent phenotype map at the same locus, *rpxL*, characterized by the wild type *str-s* (sensitive) and the mutant *str-r* (resistant) alleles which dictate whether cells are killed by or survive treatment with streptomycin. Luzzato, Schlessinger and Apirion (1969) have shown that, in fact, one allele, *str-r*, confers either resistance to streptomycin or dependence upon it; when genes of *str-r* E.coli K12 are transduced into a *str-s* strain of E.coli C, both streptomycin resistant and dependent transductants can be recovered. Momose and Gorini (1971) have since mapped a variety of dependent and resistant mutations; they are all sited in the same gene.

The alternative expressions of this gene are determined by the genetic background of the cell in which it is located, for mutations at two other loci determine whether its phenotypic expression is displayed as str-r or str-d. These two loci are closely linked to the *str* (*rpxL*) gene itself and both specify proteins of the 30S subunit. These proteins must interact with the S12 protein coded by *str* (*rpxL*) to influence the extent of misreading induced by streptomycin. In other words, cells which survive exposure to streptomycin by becoming dependent upon it differ from those which become resistant in the structure of one (or more) of the other ribosomal proteins such that their ribosomes require streptomycin if they are to translate mRNA into proteins with the correct amino acid sequence.

The allele at the streptomycin locus, together with these two other genes, can therefore control the accuracy of the translation process. Streptomycin causes the ribosome to misread certain codons and this action may suppress mutations; this explains why conditionally dependent cells require streptomycin. But ribosomes from resistant cells are not distorted into misreading by streptomycin, because the resistant mutations in the *str* (*rpxL*) locus have imposed a *restriction* upon the efficiency of suppression; this restriction applies whether the suppression is mediated by streptomycin or by other means.

Streptomycin itself, therefore, exerts a phenotypic suppression by causing the ribosome to misread codons; restrictive mutations in the streptomycin allele counter this effect and make reading more rigorous. The extent of the restriction varies with the individual mutation and this may play a role in deciding the response of the ribosome to streptomycin. Indeed, it was at first thought that slightly restrictive alleles might confer resistance to streptomycin, whereas alleles causing a greater degree of restriction would require the addition of streptomycin to compensate for this effect; but we now know that one allele may confer either resistance or dependence according to the behaviour of the other ribosomal proteins. In dependent cells, however, the overall structure of the ribosome is such that its restrictive effect will prevent correct translation unless streptomycin is present to counter its influence.

Mutation of the Accuracy of the Ribosome

Alterations in ribosome structure can result not only from the addition of external reagents, such as streptomycin, but also from mutation in the genes, including the streptomycin locus (*rpxL*), which code for ribosomal proteins. If the structure of the ribosome can control the degree of ambiguity with which the code is read, it should be possible to isolate mutants which permit greater ambiguity. We may expect that ambiguous mutants will usually be lethal to the cell, for they must mistranslate many codons; and so Rosset and Gorini (1969)

Table 4.8: extent of misreading in vitro of different E.coli strains

strain	per cent incorporation in response to poly-U of phenylalanine	isoleucine	serine
wild type	90·9	7·3	1·8
*strA*40 mutant	97·4	1·6	1·0
ram-1 mutant	83·0	12·8	4·2
*strA*40, *ram*-1	92·8	4·6	2·6

The *strA* mutation is restrictive and decreases the extent of misincorporation, whereas the *ram*-1 mutant increases it. These two effects counteract each other in the double mutant. Other *str* and *ram* mutants show the same type of effect, but may have a greater or lesser extent of influence. Data of Rosset and Gorini (1969).

have used the phenomenon of restriction to assist in their isolation. Since restriction by streptomycin alleles influences translation in a general manner, they argued that restriction and ambiguity might counteract each other, so that double mutants should show the wild phenotype. Their protocol was therefore to isolate mutants of a restrictive strain which are less restrictive than their parent because their increased ambiguity allows suppression events to take place again.

By following this procedure, they identified a mutant in another gene, the ribosomal ambiguity locus, *ram* (*rpxD*), which neutralizes the effect of restrictive mutations in the *str* (*rpxL*) locus. The *ram* mutation alone helps to suppress nonsense codons, that is it increases misreading. As table 4.8 shows, ribosomes extracted from wild type cells misread poly-U to an extent which is no doubt somewhat greater than that prevailing in the cell; nonetheless, ribosomes from cells bearing a restrictive allele in the *str* (*rpxL*)locus shows a much smaller degree of misreading and ribosomes from strains with the *ram* mutation show an increased degree of misreading. The double mutant shows a degree of misreading similar to the wild type. In vivo, the *ram* mutant can produce a suppressed phenotype similar to that which results from the addition of streptomycin and the effects of the antibiotic and the mutation appear to be roughly additive.

By comparing the proteins of the ribosomes of E.coli strains which are isogenic apart from a *ram* mutation in the *rpxD* gene, Zimmermann, Garvin and Gorini (1971) found that protein S4 is the only one altered. This protein binds to 16S rRNA and is essential for the proper assembly of the ribosome; we do not know how it influences the reading of messengers. Strains of E.coli which are dependent on streptomycin can revert to independence by mutation in streptomycin (*rpxL*) gene itself or, more usually, by mutation at another locus. One of the most common of the reverting loci is the *ram* (*rpxD*) gene for most of the revertants isolated by Deusser et al. (1970) fall into this category. Bjare and Gorini (1971) have demonstrated that the introduction of *ram* mutations into str-d strains releases dependence so that the bacteria can grow in the absence of the drug. Genetic mutations in the *ram* (*rpxD*) gene to change the structure of protein S4 can therefore substitute for the phenotypic effect which is produced by addition of streptomycin; both streptomycin itself and *ram* mutations increase ambiguity and therefore counter restriction.

Such revertants appear to have many changes in the S4 protein. Donner and Kurland (1972) have shown that one such revertant has a large number of amino acid replacements, involving a gain of positive charges, so that the altered protein has a lower affinity for its binding site on 16S rRNA. Funatsu et al. (1972b) have shown that some of the S4 revertants have proteins of increased molecular weight with several alterations located at their –COOH termini; Grosjean et al. (1972) showed that all of these mutant proteins have a lower affinity for 16S rRNA. But mutation in the spectinomycin protein, S5, (coded by *rpxE*) may also enable cells to revert from streptomycin dependence to independence, for Stoffler et al. (1971b) have isolated such a mutant. These genetic experiments identify proteins S12, S4 and S5 as important in setting the accuracy of translation and in vitro reconstitution experiments (previous section) suggest that protein S11 should be included in this category.

What is the mode of action of restrictive alleles at the streptomycin locus? They restrict suppression whether it results from the addition of streptomycin or from mutations in the translation apparatus. Strigini and Gorini (1970) have followed the effect of different alleles on the suppression of nonsense codons which results from the presence in the bacterial cell of tRNA species which can respond to amber (UAG) or to both amber and ochre (UAA) codons. Examining the suppression of amber nonsense codons in the *argF* gene which codes for ornithine transcarbamylase or in the *z* gene of the lactose operon coding β-galactosidase shows that the relative activities of four different mutants are the same no matter which locus is involved, although there are differences between the loci. Table 4.9 shows that the same relative efficiencies in restriction are found no matter whether the codon is read normally, falsely read by an aminoacyl-tRNA, or recognised by a mutated tRNA. This implies that different mutations in the S12 protein impose characteristic degrees of restriction in making translation more rigorous.

Table 4.9: per cent of activity of β-galactosidase synthesised by cells with an amber mutation in the lactose *z* gene

streptomycin allele	su^- cells	su^- cells + streptomycin	su^+ cells
$strA^+$	0·12	0·25	24·6
strA-60	0·12	0·17	22·2
strA-40	0·03	0·15	13·2
strA-2	0·005	0·12	4·1
strA-1	0·003	0·004	3·8

The mutation in the su^- cells (left) may be phenotypically suppressed by addition of streptomycin (centre) or genetically suppressed by introducing a suppressor tRNA (right). Different alleles at the streptomycin locus are ordered according to their degree of restriction; the order is the same no matter whether the parent cell lacks an added suppressor, or is phenotypically or genetically suppressed. This shows that the degree of restriction is a characteristic of the allele and is independent of translational efficiency. Data of Strigini and Gorini (1970).

One complication in using assays of nonsense suppression is that the suppressing tRNA must compete with the protein factors which mediate chain termination. By looking at missense suppression, in which a tRNA has been mutated so that its anticodon recognises the codon for some other amino acid than that with which it is charged—and this recognition may restore insertion of the correct amino acid when a codon has been mutated—Biswas and Gorini (1972) have shown that various streptomycin alleles have the same relative abilities to restrict the action of missense suppressors as they do nonsense suppressors. The genetic effect of the *ram* mutation and the phenotypic effect of streptomycin are the same; they increase the frequency of suppression by either nonsense or missense suppressors, that is they counteract the action of restrictive *str* alleles.

Restriction and ambiguity do not, therefore, depend upon the particular codons being suppressed; they are general effects. Suppression may, in principle, take place in either of two ways. Mispairing between codon and anticodon may enable a tRNA which normally responds to one codon to respond to some other, closely related, codon; this may cause either nonsense or missense suppression. This is the sort of misreading which has been tested in cell free systems, where streptomycin or *ram* cause increases in the amount of isoleucine or serine which respond to poly-U.

But the effect of the restrictive mutations in *str* (*rpxL*) and *ram* (*rpxD*) are much greater in situations involving the second form of suppression, when a mutated tRNA has an anticodon which correctly recognises a nonsense or some other codon; in this case, suppression results because the mutant tRNA, although properly responding to the codon with which its anticodon pairs, carries the amino acid specified by another. Wild type strains of E.coli, with normal species of tRNA, can only use the first form of suppression which demands mispairing in codon-anticodon suppression. But if the *str* and *ram* alleles can control the extent of mispairing to correct a mutation in one gene, they must also control the general level of misreading; any ambiguity which causes suppression at one site should also cause mispairing at other sites which have not been mutated. This should be strongly deleterious to the cell. Wild type cells, however, are relatively insensitive to mutations at the *str* and *ram* loci; the maximum influence of these genes seems to be to cause a 20% decrease in growth rate. This implies that the S12 and S4 proteins act to decrease the specificity of base pairing between codon and anticodon to only a small extent; if they were more effective, they should have a much greater influence on growth rate.

When mutant tRNA missense suppressors are present in the cell, however, mutations at the *str* and *ram* loci may have a drastic effect. A *ram* allele may reduce the growth rate by 4-11 times, presumably by causing missense reading at sites which have not been mutated and so resulting in the synthesis of many useless proteins. The introduction of a restrictive *str* allele may then overcome this effect and restore growth roughly to normal. The effect of *ram* mutations on growth depends on the number of missense suppressors which the cell carries; this supports the idea that the effect of ambiguous *ram* alleles is to aid these suppressors in their (correct) recognition of codons and the effect of restrictive *str* alleles is to inhibit such recognition. The more species of suppressor tRNA are present, the more often an unmutated codon is read by one so that the wrong amino acid is inserted in proteins.

The *str* and *ram* loci therefore seem to exert their principal effect on the action of mutant tRNAs; their influence on codon recognition by normal tRNAs is much less significant in the cell, although such an effect can indeed be displayed in vitro. This idea is supported by experiments which show that the alleles present at these loci have little effect on the suppression of missense mutants in the tryptophan synthetase A protein of E.coli cells containing only normal tRNAs.

What is the molecular action of the S12 and S4 proteins? These results suggest that they do not act to control the accuracy of base pairing, for their effects are greatest when codon-anticodon recognition is correct but the tRNA is a mutant inserting the wrong amino acid. It seems improbable that the ribosome can recognise mutant tRNAs as such, for they may differ from wild type in only a single base of their anticodon. This suggests that the *str* and *ram* loci may control, in a rather general manner, the entry of tRNAs to the A site of the ribosome during chain elongation; changing the characteristics of this interaction will be of much greater importance when mutant tRNAs are present, although it may also have a smaller effect on the extent of mispairing in codon-anticodon recognition.

CHAPTER 5

Transfer RNA

Structure of tRNA

Sequence Determination of RNA

Primary sequences have now been established for a large number of transfer RNA molecules and for some species of 5S ribosomal RNA. The analytical procedure required is analogous to that involved in determining the amino acid sequence of proteins. But since there are only four bases in RNA, compared with twenty amino acids in protein, the diversity of sequences is lower in fragments of nucleic acid. In transfer and ribosomal RNA molecules, however, the presence of modified bases to some extent relieves these limits on diversity. Sequencing has been restricted to small RNAs, for larger molecules have too many repetitions of sequence to yield unique fragments for analysis. But in spite of these difficulties, sequencing techniques have been extended to 16S rRNA of E.coli, about 1600 nucleotides long, and to the small phage RNAs, which are some 3300 nucleotides long. This is probably close to the limits of resolution of the technique (reviewed by Zachau, 1972).

Digesting an RNA molecule with T1 ribonuclease, which attacks internucleotide linkages selectively on the 3′ side of guanosine, usually generates large oligonucleotide fragments. The susceptibility of sites in RNA may depend upon its secondary structure, so that the extent of reaction with the enzyme may be controlled by the conditions of incubation. A different specificity is shown by pancreatic ribonuclease, which attacks RNA on the 3′ side of pyrimidine residues. By using these and other enzymes to cleave the RNA molecule at different sites, it is possible to generate fragments which have regions of partial overlap and which can thus be used to order the sequence of the molecule. (Sequencing single strands of DNA has proved much more difficult because of the lack of enzymes which break chains of deoxynucleotides at specific sites; but see Ziff, Sedat and Galibert, 1973).

The nucleotide content of small fragments of RNA was originally determined by a method developed by Holley et al. (1965) utilising separation on columns of DEAE-sephadex. This technique has since been superseded by electrophoresis in two dimensions, the first on cellulose acetate and the second on

DEAE paper, developed by Sanger et al. (1965) and Sanger and Brownlee (1967). Improvements of the system by Brownlee et al. (1968) led to its ability to separate oligonucleotides of up to twenty five residues in length.

Molecules of RNA may therefore be digested with T1 ribonuclease to yield large fragments, which after their isolation are in turn digested with pancreatic ribonuclease into small fragments whose content is determined. Partial digestion with snake venom phosphodiesterase, which degrades polynucleotides stepwise from the 3′ terminus, can be used to establish most of the sequence, but does not reveal the order of the two or three nucleotides at the 5′ terminus. Micrococcal nuclease may be used to release a dinucleotide from the 5′ terminus.

Transfer RNA is the smallest cellular RNA, sedimenting at about 4S in both bacteria and eucaryotic cells. The molecules at present sequenced vary in length from seventy-five to eighty-five nucleotides. The first tRNA to be sequenced was the alanine accepting species of yeast studied by Holley et al. (1965); transfer molecules for virtually all amino acids in a large number of species have since been determined and many of these sequences are shown in the figures of this chapter.

Clover Leaf Model for Secondary Structure

Although each tRNA has a distinct sequence and there are few bases in identical positions in all tRNA molecules, their overall secondary structures all fall into the same pattern. One feature of all the primary sequences is that no tRNA shows long complementary stretches which could allow base pairing —the greatest extent of duplex formation is usually seven base pairs. To achieve even this length, it is often necessary to include G–U base pairing as well as the more common G–C and A–U pairs. Secondary structures can be written for tRNA sequences by taking into account possible duplex formation between different parts of the molecule. Figure 5.1 shows the sequence of alanine tRNA written as a hairpin or a double hairpin; figure 5.2 shows the structures of several tRNAs, including $tRNA^{ala}$ of yeast, written as clover leafs. The double hairpin and clover leaf structures appear more stable because they rely upon longer segments of duplex RNA.

These models can be distinguished by the different predictions which they make for the susceptibility of the bases at different positions to chemical or enzymic attack. Most ribonucleases attack bonds in single stranded regions preferentially and the points at which the chain should be split are different in the three models. And once bonds have been split, the different models would retain their structures to different extents; the hairpins would tend to separate into fragments rather easily, whereas its hydrogen bonded structure should make the clover leaf more resistant. Another approach is to study the availability of the different parts of the molecule to chemical mutagens; bases in the

Figure 5.1: hairpin and double hairpin models for alanine tRNA of yeast. Data of Holley et al. (1965)

loops of the clover leaf should be more sensitive than those in the central regions of the molecule. According to these criteria, the clover leaf model provides the best fit for the structures of all tRNAs which have been sequenced.

The clover leaf structure shown in figure 5.3 visualises the tRNA molecule in the form of three arms and an amino acid accepting stem. Each arm consists of a loop of unpaired bases projecting beyond a hydrogen bonded duplex stem. The longest hydrogen bonded sequence is located in the amino acid acceptor stem and consists of seven base pairs; this pairing starts with the 5′

end of the molecule and at four bases from the 3′ end, so that the sequence ACCA-OH protrudes as a single strand.

Immediately adjacent to the amino acid acceptor stem—drawn on its right—is the GTΨCG, or pentanucleotide arm, so named because this sequence is found in the same position in virtually all tRNAs. There are a few rare exceptions in which the final G is replaced by an A. The sequence TΨCG marks one end of the looped region, which always consists of seven bases; the 5′ G is the last paired base of the stem of the arm, which consists of five base pairs. Initiator tRNA$^{met}_{f}$ molecules which lack the pentanucleotide sequence have recently been discovered in eucaryotic cytoplasm (see later).

The anticodon arm containing the triplet sequence complementary to the

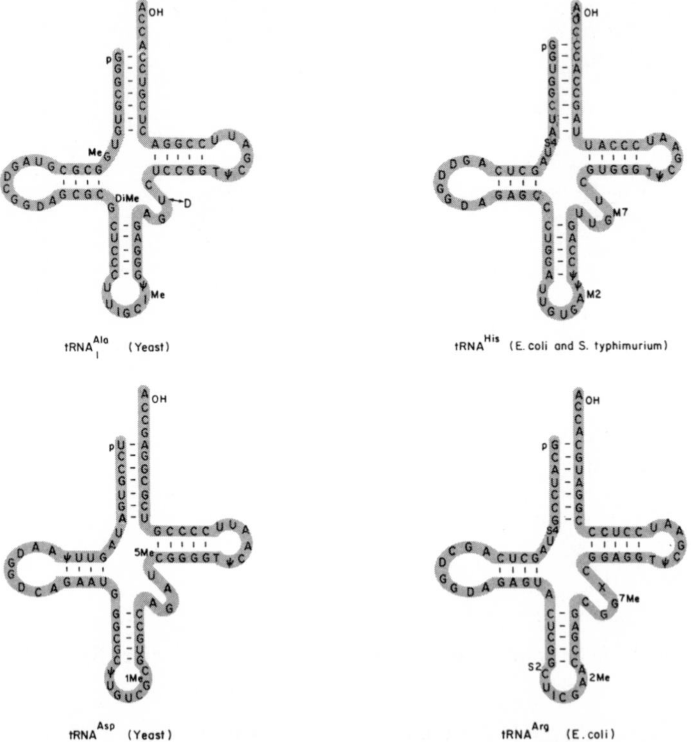

Figure 5.2: clover leaf models for tRNAs. Regions differing in related pairs of tRNA are indicated by boxes. Data of: yeast alanine tRNA$_1$—Holley et al. (1965); histidine tRNA of E.coli and S. typhimurium—Singer and Smith (1972); yeast aspartic acid tRNA—Gangloff et al. (1971); E.coli arginine tRNA—Murao et al. (1972); yeast serine tRNA—Zachau et al. (1966); rat liver serine tRNA—Ginsberg, Rogg and Staehelin (1971); E.coli leucine tRNAs—Blank and Soll (1971b); E.coli glutamine tRNAs—Folk and Yaniv (1972)

Figure 5.2: continued

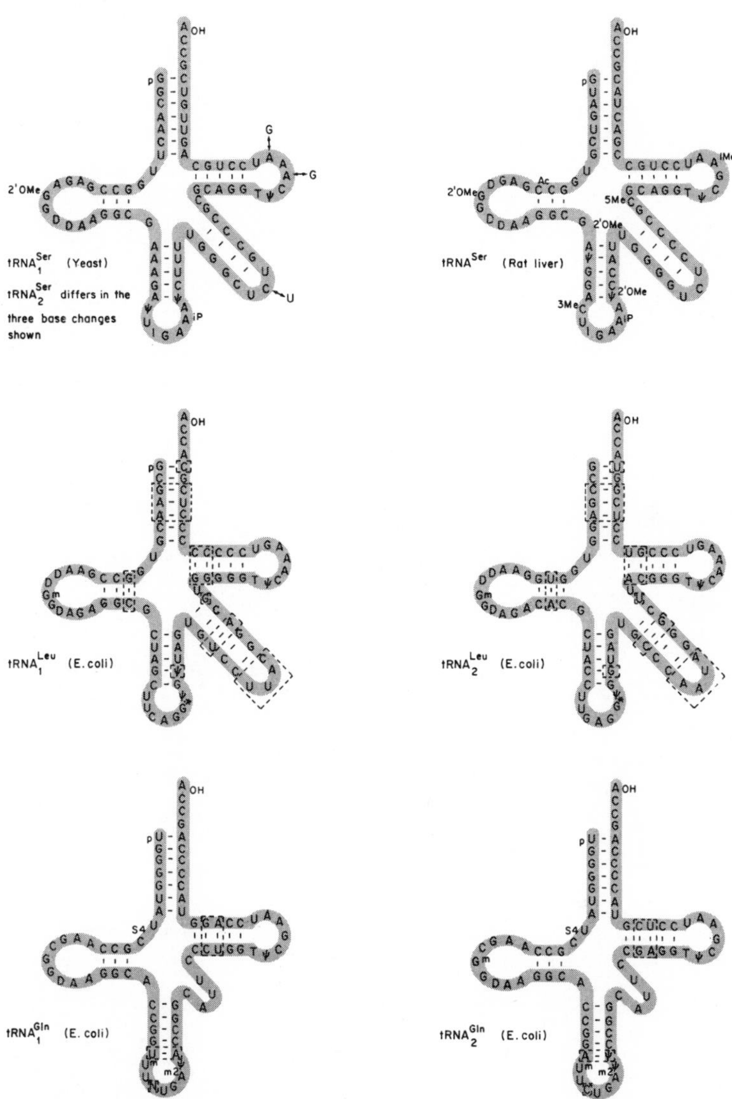

codon(s) for the amino acid carried by the tRNA is located at the far end of the clover leaf from the acceptor stem. The anticodon stem consists of five base pairs, with a G–C pair placed in the penultimate position in either orientation. The loop always consists of seven bases, with the anticodon sequence located as the central three at the furthest projection of this structure. There is always a U residue on the 5′ side of the anticodon and a modified purine, often A, on the 3′ side.

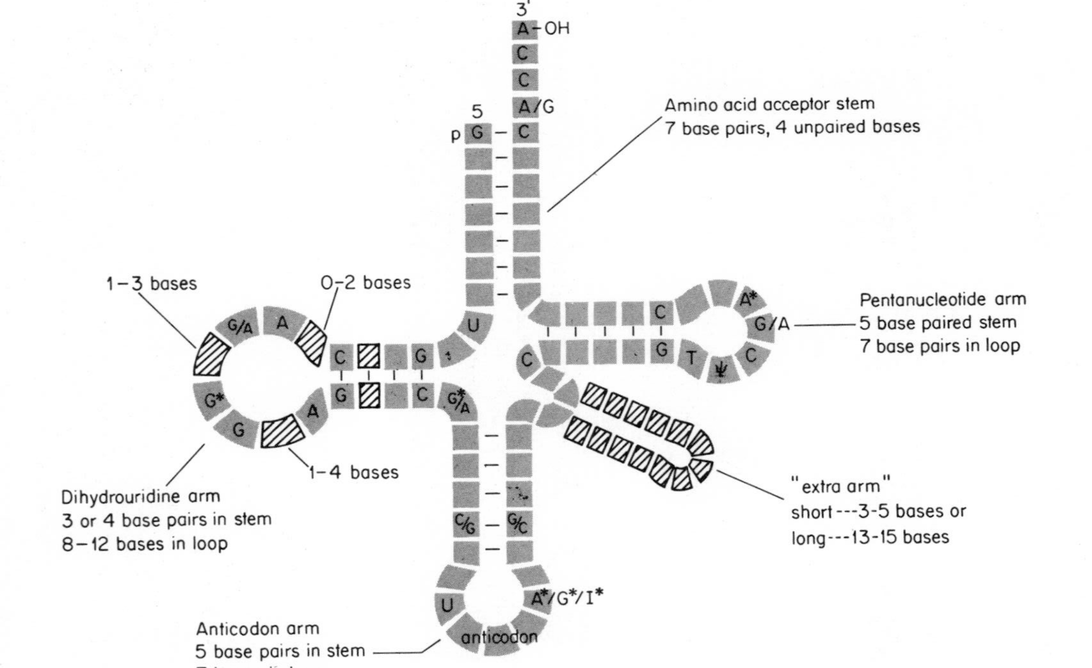

Figure 5.3: generalised clover leaf structure for tRNA. The features shown are found in all tRNAs sequenced (see figures of this chapter). Asterisks indicate bases which may be modified. Exceptions in sequence are: Amino acid acceptor stem: fourth base from end is U in $tRNA^{gly}$ of E.coli. This base is omitted in $tRNA^{his}$ of Salmonella and E.coli. The first base pair is U-A in yeast $tRNA^{asp}$ and E.coli $tRNA^{gln}$; it is occupied by C and A in $tRNA^{met}$ of E.coli. One of the stem positions in $E.coli^{leu}$ is occupied by an A-C mismatch; there are eight base pairs in the stem of $tRNA^{his}$. Pentanucleotide arm: $tRNA^{met}$ of E.coli has central base pair replaced by non pairing A-C match. Extra arm: first base after pentanucleotide arm is U in $tRNA^{leu}$ of E.coli. Anticodon arm: none. Dihydrouridine arm: only the U base instead of two bases separates this arm from the amino acid acceptor stem in $tRNA^{tyr}$ of E.coli which also has U-A pair as first in dihydro uridine stem. Orientation of last base pair is reversed in $tRNA^{leu}$ of E.coli. Only $tRNA^{tyr}$ of E.coli has no dihydrouridine residues in the loop

Located between the anticodon arm and the pentanucleotide arm is the so-called "extra-arm". This varies greatly in length. In some tRNAs, it forms only a slight projection of three, four or five bases; in others it constitutes a complete arm, with thirteen, fourteen or fifteen bases and three, four or five base pairs. Although the first base after the pentanucleotide arm is usually a C residue, there appear to be no other sequence restrictions in this region.

Most tRNAs contain some dihydrouridine residues (D) in the region between sixteen and twenty one nucleotides from the 5′ end of the molecule; this identifies the dihydrouridine arm drawn on the left of the molecule. There are two bases between the amino acid acceptor stem and the dihydrouridine arm; the first is a U, often thiolated to give 4S-U. This arm is the most variable in size and in sequence. The stem may contain either three or four base pairs, the first and last being G–C pairs in the orientations shown. The size of the loop varies from eight to twelve residues and usually contains an AG sequence at its 5′ side and a GG sequence on the 3′ side. Although the loop usually contains D residues and may have as many as five, there may be none at all and their position does not seem to be fixed in any way.

Transfer RNA contains a large number of unusual bases which are not usually found in other RNA chains; amongst the more common of these are dihydrouridine (UH_2, sometimes written as D), pseudo-uridine (ψ), 4-thiouridine, methyl or dimethyl guanine and methyl adenine (see Zachau, 1972), Figure 5.4 reveals that only a limited number of sites in the molecule appear to suffer modification. Dihydrouridine is usually found only in the left arm, and thiolated uridine between this arm and the amino acid acceptor stem. Pseudouridine commonly occurs in the pentanucleotide arm and the anticodon arm also. As the figure shows, the modified bases tend to be located in the loops of arms rather than in the stems, although they may occur also in the last base pair of the stem. We do not know yet how many enzymes are involved in making these modifications or what criteria they use for selecting the sites at which they act; modification takes place on the precursor molecule (see below); this may have a similar secondary structure to tRNA itself so that the loops are more available than the stems of arms. Apart from the few exceptions discussed later in this chapter, we do not know what purpose these modifications serve.

This description of the clover leaf and its sequence restrictions is based upon the structures determined for tRNAs derived from various bacteria, mostly E.coli, and from yeast and mammalian cells. There are no particular features common to the tRNAs of any one species, with the exception that there do appear to be restrictions on the types of modified base which may be introduced into tRNA in any particular cell. It is possible that particular patterns of sequence restriction will emerge when more sequences are known, so that the tRNAs fall into a small number of categories with characteristic homologies

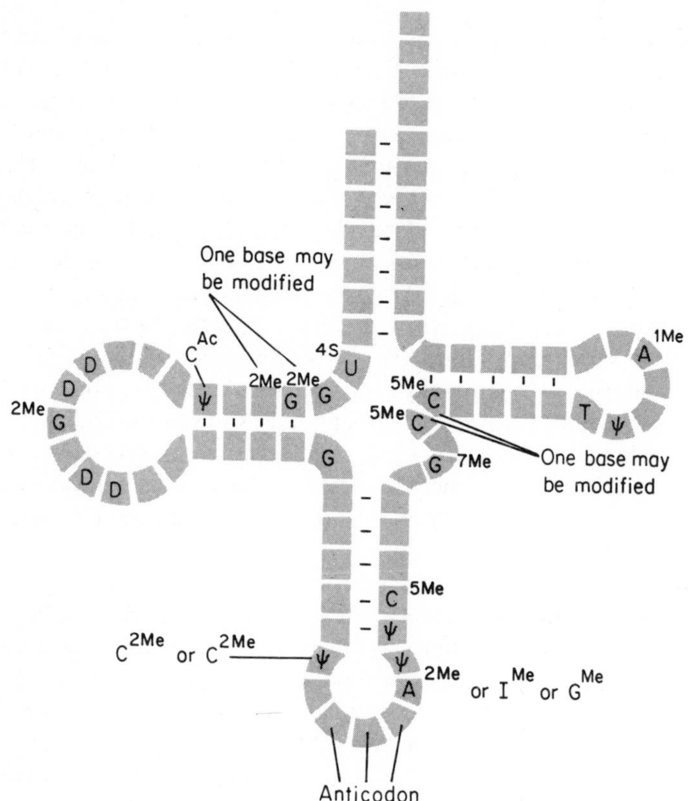

Figure 5.4: sites of modfication in tRNA. All tRNAs have modifications at some of these sites but not at other positions. Based on data of sequences shown in figures of this chapter

between the members of each subset. On the other hand, it is possible that those restrictions in sequence which we have so far found may be reduced in number when more tRNA structures are determined. Not all of the sequences shown in figure 5.3 are common to all tRNAs; the exceptions are mentioned in the legend to the figure and it is too early to say whether they will remain exceptions to a general rule or whether others of each class will be found. The most important point to emphasize, however, is that, by and large, all tRNAs of all organisms fall into a single pattern of secondary structure, although each may contain some particular deviation from the generalised clover leaf of figure 5.3.

Synthesis and Maturation of the Precursor

Genes which code for transfer RNA in bacteria are not clustered in one region of the chromosome but appear to be generally dispersed. The immediate transcription product which gives rise to tRNA is somewhat larger than the mature molecule. By pulse labelling Hela cells for less than thirty minutes, Mowshowitz (1970) found that most of the low molecular weight RNA migrates in gel electrophoresis at a position corresponding to a sedimentation rate of between 4S and 5S; this fraction can be converted to tRNA by incubation with a crude cytoplasmic extract in vitro.

Precursor molecules in E. coli have been detected by Altman and Smith (1971) by pulse labelling cells infected with a ϕ80 phage carrying the gene for a mutant tyrosine tRNA which supresses amber nonsense codons (see later). Upon digestion with ribonuclease, the precursor generates the usual fragments which are found in tyrosine tRNA, although it lacks the oligonucleotides which correspond to the two ends of the mature tRNA molecule. The precursor contains several additional fragments, however, amongst which can be found the sequences of the ends of the tRNA itself. Because ribonuclease appears to attack both precursor and tRNA in the same positions, it seems likely that the tRNA moiety of the precursor molecule has a secondary clover leaf structure which is very similar or identical to its final form. The precursor which has been isolated has three additional nucleotides at the 3′ end of tRNA, although it is of course possible that the immediate product of transcription may have more. The extra 42 residues at its 5′ end presumably represent the first part of the molecule to be transcribed for they commence with the pppG residue characteristic of initiation of transcription.

The structure of the tRNA part of the precursor seems to be important for its proper maturation. Cell extracts contain an endonucleolytic activity which may cleave precursor tRNA to release the 5′ sequence of the precursor. But mutants which insert an adenine at position 2 of the tRNA instead of guanine, or in position 81 in place of cytosine, have low levels of tRNA and do not accumulate precursor. These mutants disrupt base pairs in the amino acid acceptor stem (see figure 5.12) and the A2 mutation can be reverted by a mutation to give uridine at position 80 which restores complementary hydrogen bonding; this suggests that the duplex structure of the amino acid acceptor stem may be necessary if the endonuclease is to recognise the proper cleavage point. Indeed, Abelson et al (1970) noted that most mutants of the su_3^+ tyrosine tRNA are produced in reduced amounts. It seems likely that any mutation which changes the conformation of tRNA may inhibit its synthesis, by reducing its rate of maturation—presumably the modifying enzyme(s) fail to recognise the altered conformation—or because the new structure is unstable and is degraded in the cell.

The sequence of the precursor part of the molecule is also important, for the mutant P, which converts a cytosine to a uridine residue four bases before the beginning of the tRNA, greatly increases the amount of tRNA in the cell. The base located at this position must therefore have some influence on maturation. The two hydrogen bonded structures shown in figure 5.5 can be written for the

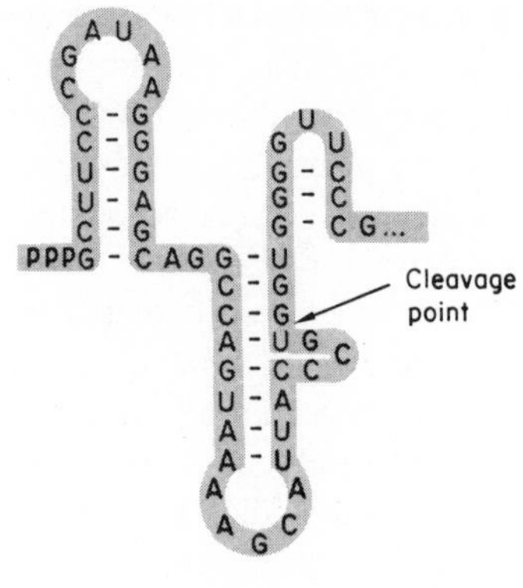

Precursor I

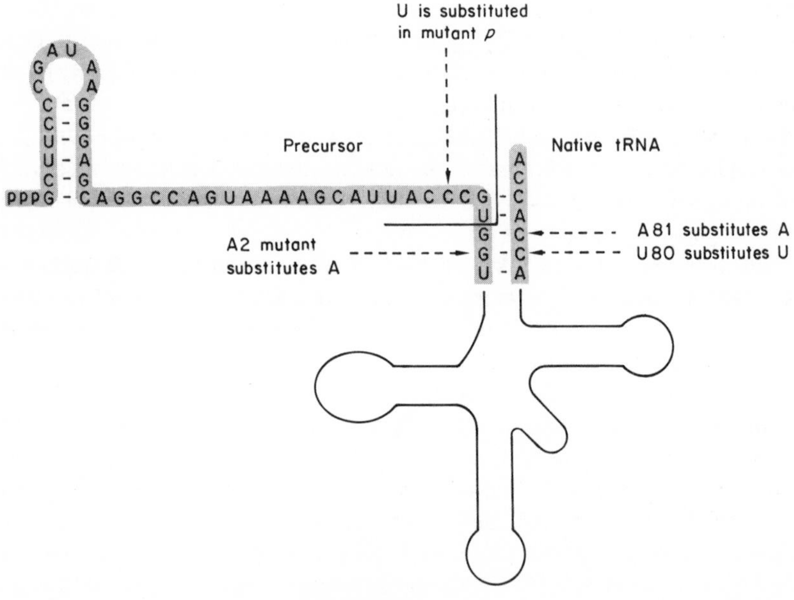

Precursor II

Figure 5.5: possible structures for precursors to tyrosine tRNA of E.coli. Data of Altman and Smith (1971)

precursor; the first may form as transcription proceeds and later be replaced as the sequences of the tRNA moiety are synthesised to provide alternate duplex formations. Transition between the structures might be important in controlling the accessibility of the molecule for cleavage by endonuclease and might be inhibited by the failure to form of certain hydrogen bonds in the tRNA section of the precursor molecule.

The mature form of tyrosine tRNA contains seven modified bases (see figure 5.12), almost all of which are found as the corresponding non-modified bases in the precursor. The differences between the precursor and tRNA are:

position	precursor	tRNA
8 and 9	not known	4-thiouridine
17	guanine	2′-0-methyl-guanine
38	adenine	2-thiomethyl-6-isopentenyl-adenine
40	uridine	pseudouridine
64	10% pseudouridine	
	90% uridine	pseudouridine
63	uridine	thymine riboside

Because cleavage of the precursor takes place before base modification, it seems likely that the modified bases are not needed for maturation; they must therefore be implicated in the functions of the mature molecule.

The unusual bases of tRNA are produced by enzymic modification of the molecule after its transcription. One problem in studying this process is that it is not usually possible to isolate the unmodified precursors to use as substrates. Two types of system have therefore been used. Bacteria may be starved for methyl groups to produce methyl-deficient tRNAs which can be utilised as substrates in vitro (see Baguley and Staehelin, 1969). An alternative is to use heterologous systems and to modify the tRNA of one species with enzymes of another (see Baguley et al., 1970); this has the disadvantage that the final product tRNA contains modified residues which are not present in its native cell. The modifying enzymes in general appear to act on tRNA with a high degree of discrimination, although we do not know what structural features of each tRNA are recognised by the enzymes; nor do we know how many enzymes there are, although at present it seems likely that there may be a large number each of which has a highly specific action.

Alternate Tertiary Conformations of tRNA

The clover leaf model represents the folding upon itself of the linear sequence of tRNA to form base paired regions. But a tRNA molecule is not confined to two dimensions. Completely helical nucleic acids such as DNA exist in a statistical distribution of all possible conformations; they have no tertiary structure unless one is imposed by their interactions with other macromolecules, such as protein. The clover leaf structure of tRNA, however, has many unpaired bases which might interact with other parts of the molecule to organise the clover leaf into a specific three dimensional conformation. Such a structure would accord with Crick's concept (1966) that transfer RNA represents nature's attempt to make a nucleic acid fulfill the type of function more usually exerted by a protein; that is, to activate a specific amino acid through its appropriate synthetase and to recognise only the correct codon(s) on mRNA through mediation of the ribosome. Crystals of tRNA have been formed as a prerequisite for X-ray crystallography (see Kim and Rich, 1969); although no general structure has yet been derived from such studies, the ability of different tRNAs to crystallise together implies that they may have common features of tertiary structure (reviewed by Cramer and Gauss, 1972).

Molecules of transfer RNA appear to be able to exist in more than one conformation and magnesium ions play an important role in transitions between these different structures. Fresco et al. (1966) found that tRNA is denatured by increasing the temperature but that the characteristics of thermal denaturation depend upon the salt form of the tRNA. Changes in structure may be followed by hyperchromicity—which detects the separation of base paired regions, that is disruption of secondary structure—and by changes in viscosity and sedimentation, which depend upon the overall structure of the molcule and may therefore reveal tertiary changes.

With the sodium salt of tRNA a non-cooperative process—one in which different physical properties are affected at different stages—occurs in two steps. The first step involves a slight change in shape with only a small loss of secondary structure; this can be interpreted as the loss of some tertiary structure. The second step involves loss of secondary structure. But the magnesium salt of tRNA—which may be the physiological form—does not show a separate loss of tertiary and secondary structures; tRNA bound to Mg^{2+} is stable to higher temperatures, when a cooperative change occurs in all its properties. This suggests that magnesium ions may stabilise the tertiary structure of tRNA so that it finally collapses at the same time when the secondary structure is lost. The tertiary structure of tRNA revealed by these experiments appears to be biologically significant, for the ability of valine tRNA to be charged with its amino acid is lost at an early stage of thermal denaturation whilst the secondary structure is still intact.

Denaturation of tRNA may be reversible so that active molecules can be reconstituted from the disrupted structure. The denatured form of the molecule is stable and requires a high energy of activation for conversion to its native state. This energy is much too great to be accounted for merely by the imposition of a correct tertiary structure upon the secondary structure; rather must it demand disruption of an alternative tertiary structure before the native conformation can be reformed. By comparing the conformations of different states of $tRNA_3^{leu}$ from yeast, Adams, Lindahl and Fresco (1967) suggested that although the secondary structure remains essentially constant, the denatured form of tRNA has a less compact tertiary structure. If formation of the native tertiary structure depends upon pairing between the unpaired bases of the loops, then in unusual conditions "wrong" associations might lead to the formation of alternative tertiary structures, which must be disrupted before the native structure can be reformed.

Although all transfer RNAs may be able to exist in alternate conformations, the formation and loss of tertiary structure may follow a different series of events in each tRNA. Cole and Crothers (1972) and Yang and Crothers (1972) found that $tRNA^{tyr}$ species (I and II) and $tRNA_f^{met}$ of E.coli display an early melting transition in which tertiary structure alone is lost. But each molecule has a characteristic pattern of renaturation from its partially denatured form which depends upon the "wrong" bonds which may form and have to be disrupted during acquisition of the tertiary conformation.

Magnesium ions probably play a role in maintaining the structures of at least some tRNAs. Phenylalanine tRNAs of several organisms contain an unusual base Y, a modified adenine, adjacent to the 3′ side of the anticodon (see figure 5.10). This base fluoresces naturally and Beardsley, Tao and Cantor (1970) noted that its intensity is enhanced when magnesium ions are added to the tRNA. The Y base itself does not interact directly with magnesium ions, for in isolation its fluorescence is independent of their concentration. Circular dichroism spectra of the tRNA show that a small change in the conformation of tRNA accompanies the increase in fluorescence when magnesium ions are added. In the presence of magnesium ions the Y base thus seems be to protected from solvent by the surrounding tRNA structure; removal of the cations changes the structure so that the base suffers an increase in exposure to the solvent which quenches its fluorescence. The structure of the anticodon loop must therefore depend upon magnesium ions.

The influence of magnesium ions upon structure appears to vary for each tRNA molecule. Reeves, Cantor and Chambers (1970), for example, found that magnesium ions cause the two alanine tRNA species of yeast to fall into compact ordered structures by increasing the extent of base pairing. On the other hand, Yarus and Rashbaum (1972) suggested that isoleucine tRNA of E.coli does not rely at all upon magnesium ions to maintain its structure. Levy and Biltonen (1972) have presented a model for the effect of magnesium ions on the thermal unfolding of yeast $tRNA^{phe}$. Whatever the precise nature of tRNA tertiary structure, it seems likely that magnesium ions are implicated in its maintenance in many molecules.

One idea which has commonly been proposed is that tRNA may change its conformation when charged with an amino acid; such a change might help the elongation factors to distinguish aminoacyl-tRNA molecules from uncharged tRNA. Different techniques—including optical rotatory dispersion, circular dichroism, sedimentation or chromatographic elution—have been used to detect these changes in structure, which always appear to be subtle (see Cramer et al., 1968; Stern, Zutra and Littauer, 1969; Watanabe and Imahori, 1971). It is clear, however, that there can be no gross changes of structure, for Hangii and Zachau (1971) found that subjecting tRNA to very gentle degradation with a variety of exonucleases and endonucleases of differing specificities yields the same results whether tRNA is charged or free. It is difficult to attribute biological significance to those changes which have been detected, for there is no pattern of changes consistent for all tRNAs which might distinguish charged and uncharged molecules.

Models for Tertiary Folding of the Clover Leaf

Although the properties of unfractionated mixtures of tRNA suggest that all transfer molecules may have some features of secondary structure in common, the behaviour of individual molecules usually proves to be distinctive. There may therefore be differences between the conformations of each tRNA and—as more clover leaf structures and their inter-arm reactions are defined—it may ultimately prove impossible to formulate one structure to satisfy them all. The X-ray scattering data obtained by Lake and Beeman (1968) argue strongly against a general open arrangement of the clover leaf but are consistent with a compact structure in which the arms are folded on each other.

Many different classes of tertiary conformation can be constructed by folding the arms of the clover leaf; some of the models which have been built are illustrated in the schematic representation of figure 5.6. The structures of these and other models have been discussed in detail by Cramer and Gauss (1972). The general principles upon which tertiary structures are based are similar to those used to derive the clover leaf secondary structure: stability is achieved by allowing unpaired bases to hydrogen bond and also by stacking the bases of one arm upon those of another. All the models have some base stacking and (d) and (e) rely upon extra base pairs. All have the helical axes of their ordered regions nearly parallel. A general feature of all structures apart from (b) is therefore that they possess a long axis and small cross-section; this is consistent with the physical properties of tRNA molecules in general. These models usually have the amino acid acceptor stem at one end of the molecule and the anticodon arm at the other end, which means that recognition of amino acids and codons is well separated in space.

Each model makes certain predictions about which bases should be protected by the tertiary structure from chemical modification; the susceptibility of particular regions to chemical reagents can therefore help to distinguish between the models. Nucleotide sequences which are not base paired can be assayed directly by their ability to interact with short complementary oligonucleotides in a method developed by Uhlenbeck et al (1970); in general these results are consistent with the structure implied by responses to chemical agents (Uhlenbeck, 1972). This technique has proved particularly useful for defining the structure of the anticodon arm (see below). Restrictions on the bases available for chemical reaction or base pairing support the clover leaf model and suggest that further bases are removed from reaction by interactions additional to those forming the secondary structure.

Carboiimides react with bases U, Ʉ, G and I; by treating yeast alanine tRNA with this reagent, Brostoff and Ingram (1968) found that the anticodon loop is particularly well exposed; the extra loop found in some tRNAs is also reactive. That the structure of each tRNA may be different is implied by the observation of Chang (1973) that although the sequences of the anticodon loops of $tRNA_f^{met}$ and a tyrosine suppressor of E.coli are very similar, residues in $tRNA_f^{met}$ are less accessible to carboiimides and to methoxyamine (which reacts with cytosines).

That there must be other base pairs as well as those of the clover leaf itself is suggested by the failure of certain bases—which are not paired in the clover leaf itself—to interact with reagents which attack unpaired bases. Monoperthalic acid, for example, oxidises unpaired adenine residues in a reaction which can be followed by the change it causes in the ultraviolet spectrum. Using this technique, Cramer et al (1968) found that an average of only four adenines in each tRNA of a bulk preparation remains unpaired; in yeast phenylalanine tRNA they are located at positions 35, 36, 38 and 76 (see figure 5.10). All the models of figure 5.6 except (d) can account for this result.

That the arms of the clover leaf interact to form the overall tertiary structure is shown by experiments in which modification of a residue in one part of the molecule influences the structure of another region. Cameron and Uhlenbeck (1973) demonstrated that removal of the Y base from yeast phenylalanine tRNA, for example, changes the ability of the loop of the dihydrouridine arm to bind to a complementary oligonucleotide; the change in conformation of the molecule induced by this excision is therefore not confined to the anticodon loop in which it is located.

Methoxyamine reacts with unpaired cytosine residues and Cashmore (1971) compared the reactivities of mutant suppressor tyrosine tRNAs of E.coli (see figure 5.12). Although the residue at position 57 is usually inactive, it becomes reactive in a mutant in which the guanine at position 15 is mutated to an adenine. The cytosines located at positions 20, 48 and 54 also become more reactive in this mutant. Cashmore suggested that the G15 residue of the dihydrouridine arm base pairs with the C57 residue adjacent to the pentanucleotide arm; when this base pair cannot form, the structure of the tRNA is changed so that the cytosines at positions 20, 48 and 54 are also less protected in the structure. In some tRNAs, the common G15 residue is replaced by an A; all these molecules have a U instead of C at the position adjacent to the 5′ stem of the pentanucleotide arm. Base pairing between these two sites may therefore be important in maintaining the structure of the molecule. Model (e) allows for this interaction.

Direct evidence that two bases separated in the clover leaf are located close to each other in the tertiary structure has been provided by irradiation of tRNAs which contain 4-thiouridine. This base absorbs light at longer wavelengths than any other and because a sulfur atom is present in the pyrimidine ring can be modified by irradiation to produce a characteristic change in its spectrum. The base is always located adjacent to the amino acid acceptor stem

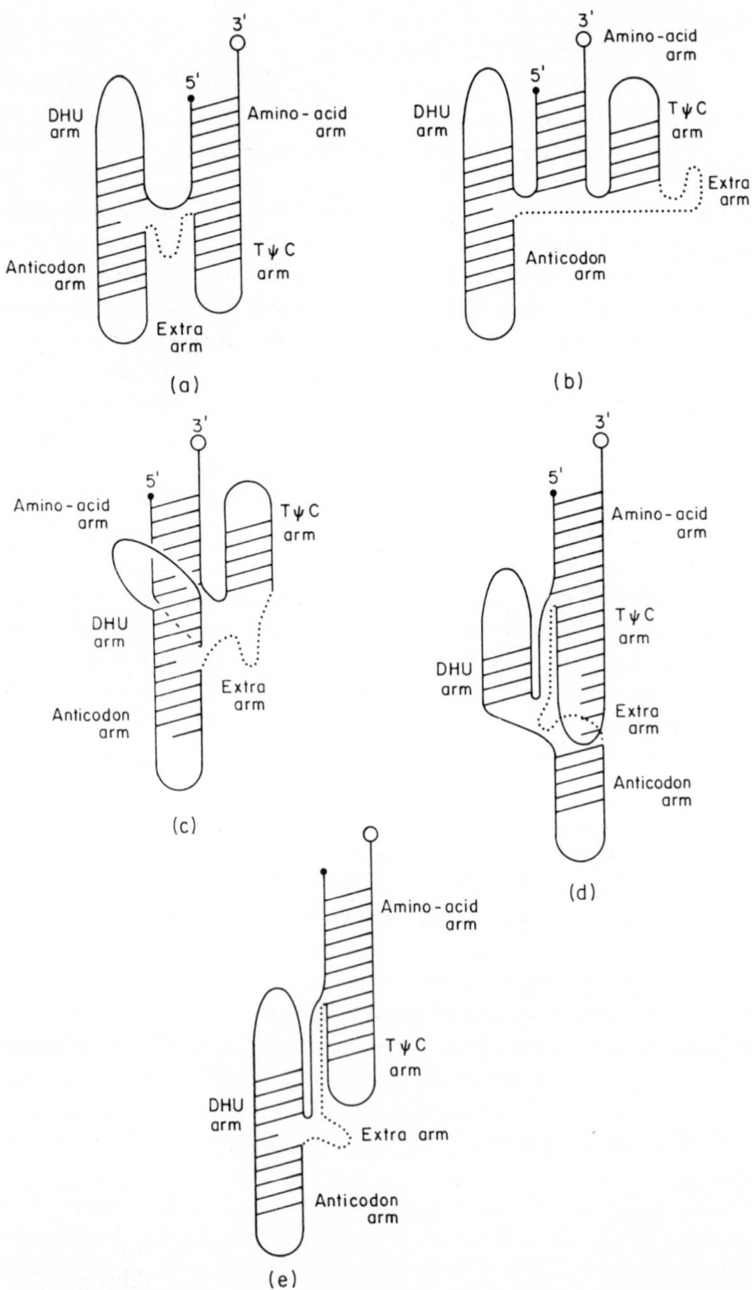

Figure 5.6: models for the tertiary structure of tRNA proposed by: (a) Lake and Beeman (1968); (b) Cramer et al. (1968); (c) Ninio et al. (1969); (d) Fuller and Hodgson (1967); (e) Levitt (1969)

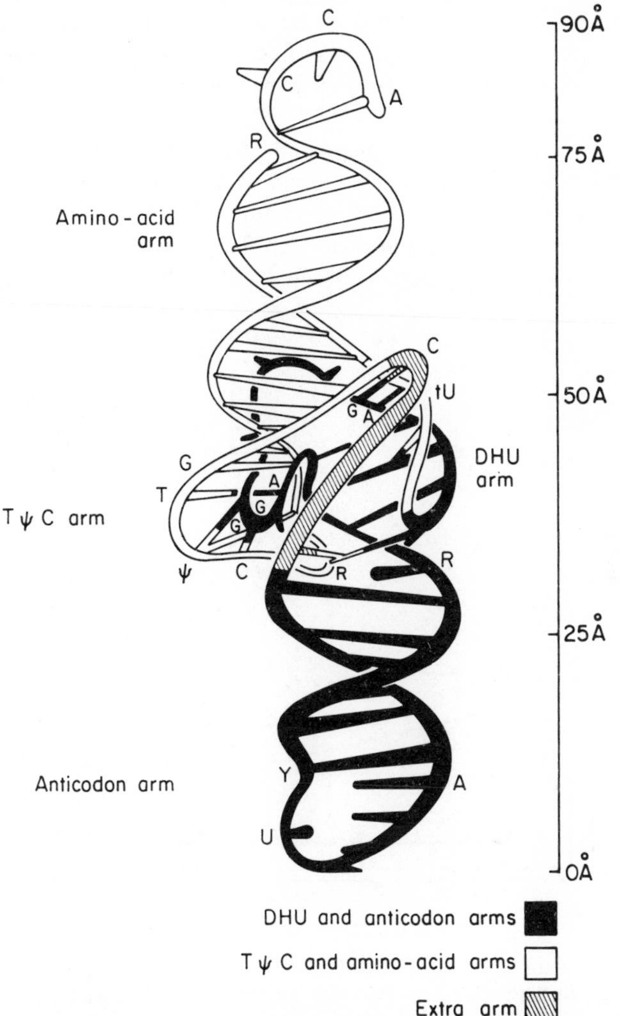

Figure 5.7: tertiary structure of $tRNA_f^{met}$ of E.coli according to model of Levitt (1969) also shown as part (e) of figure 5.5

(see figure 5.4). When Favre, Michelson and Yaniv (1971) degraded valine tRNA of E.coli—one of the tRNAs which contains this base—with ribonuclease after its degradation, they found that most of the thiouridine can be recovered in a fragment which takes the structure:

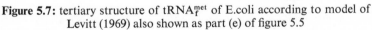

$$\begin{array}{ccccc} 6 & 7 & 8 & 9 & 10 \\ \text{A}- & \text{U}- & \text{U}_s- & \text{A}- & \text{G} \\ & & | & & \\ \text{C}- & \text{U}- & \text{C}- & \text{A}- & \text{G} \\ 11 & 12 & 13 & 14 & 15 \end{array}$$

The numbers indicate the positions of the bases in the tRNA: U_s represents 4-thiouridine (see figure 5.17). The cross-link suggests that positions 8 and 13 are close to each other in the tertiary structure although separate in the clover leaf.

The photoproduct can be obtained only from the "native" form of the tRNA; partial destruction of the secondary and/or tertiary structure by temperature or pH treatment therefore changes the relative positions of the two interacting bases. But irradiated tRNA differs very little from native tRNA in its physical properties; indeed, Yaniv et al. (1971) showed that the irradiated

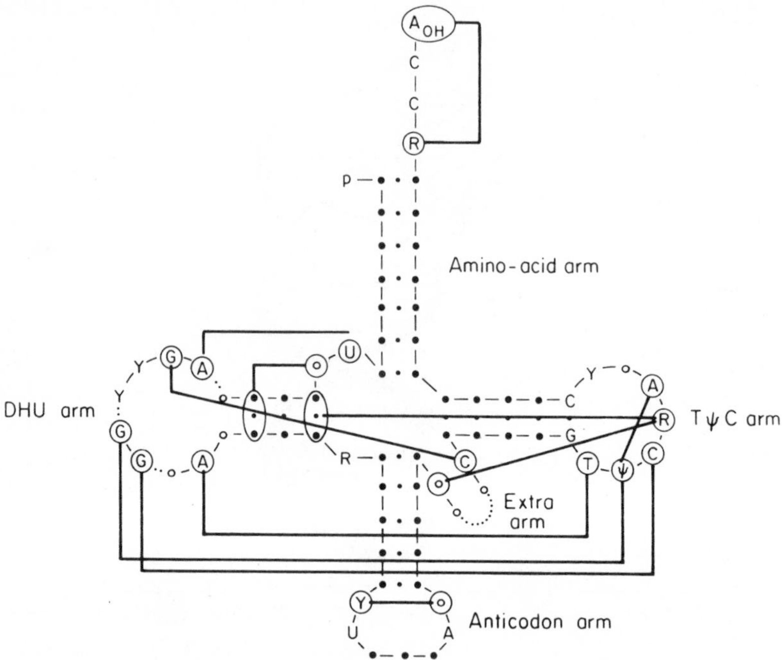

Figure 5.8: folding of arms of tRNA to give the tertiary structure proposed by Levitt (1969) and shown also in figure 5.6 part (e) and figure 5.7

tRNA can donate valine for protein synthesis in vitro, although less efficiently than the native molecule. Several other tRNAs which contain 4-thiouridine also give rise to the photodimer, so the juxtaposition of bases 8 and 13 may be a common feature in tRNA conformations. This interaction is accounted for by models (c) and (e).

Observations of a variety of transfer molecules therefore show that bases in different parts of the clover leaf may be located close together in space because of the folding of the arms. The interactions detected in each individual tRNA are usually possible with some, but not all, of the other tRNAs. Model (e) proposed by Levitt (1969) can account for some tRNAs in the form shown in figure 5.7, which is a model of the initiator methionine $tRNA_f$ of E.coli.

Because this model draws upon the different structural features found in several tRNAs, only slight modifications to it are needed to conform with the structures of all known transfer molecules. Figure 5.8 shows the interactions postulated to determine the folding of the arms. Two regular duplex regions are achieved by stacking the amino acid stem on the pentanucleotide arm and by stacking the dihydrouridine arm on the anticodon. The molecule should be very stable, because all but eight pyrimidines are stacked in this model. The centre of the molecule is formed by interacting nucleotides in the pentanucleotide, dihydrouridine and extra arms.

It would however be premature to conclude that all transfer RNAs take this form; other interactions between bases may yet be discovered and it is possible that tRNA conformations may differ significantly between each molecule. If the structural features revealed with individual tRNAs prove not to apply to all other tRNA molecules, then it may become necessary to build models for each tRNA, or perhaps class of tRNAs. But the Levitt model is consistent with what we know at present about the tertiary structure of tRNAs.

Functions of the Transfer Molecule

Modification of Function In Vitro

All transfer RNAs except the initiator are bound by the elongation factor Tu and enter the A site of the ribosome; they should therefore presumably display common features of structure. Each of the aminoacyl-tRNA synthetase enzymes charges only the transfer molecules corresponding to one amino acid; these tRNAs should possess common features which distinguish them from all other tRNA molecules. Intuitively, of course, it seems probable that recurrence of certain bases at the same positions of all or sets of related tRNAs must be important in their structure or function to have been conserved in evolution; such bases are therefore obvious candidates for the role of providing common sites of recognition.

But although many attempts have been made to equate particular sequences of the clover leaf with specific functions, recognition seems to be the prerogative of tertiary conformation rather than base sequence as such. Experiments in which certain base positions are chemically or physically modified to show whether they are implicated in particular reactions have produced different results with various tRNA species; there is no consistent pattern to suggest that any arm of the clover leaf has responsibility for corresponding reactions in all tRNAs, or that common bases play the same role in all transfer molecules. We are therefore at present unable to identify general characteristics of tRNAs which are recognised by synthetases, factor Tu or the ribosome.

Experiments which rely upon chemical modification of tRNA have in fact led to many different—and sometimes inconsistent—suggestions about the roles of the different regions of the molecule. One general difficulty with experiments which attempt to modify specific regions of the molecule is that the effect of the modification may not be restricted to the bases which react; local modifications may alter the conformation of other regions and it may be the conformational changes and not the sequence modification itself which influences the response of the tRNA. Another difficulty lies in obtaining reaction conditions which allow modification of only a single site. Of course, these objections do not apply to experiments which show that some base is not involved in some function.

That different features of the transfer molecule are recognised in its various reactions is shown by modification events which change only one response. The amino acid acceptor terminus of phenylalanine tRNA of yeast, for example, appears to be implicated in binding to Tu but not in charging by synthetase. Factor Tu binds only charged tRNAs and the terminal ribose ring must be intact; Chen and Ofengand (1970) found that treating $tRNA^{phe}$ with periodate and borohydride—which breaks the 2′—3′ carbon–carbon bond—permits the tRNA to be charged by its synthetase but prevents its interaction with the Tu–GTP complex.

One prominent sequence common to all tRNAs is the terminus of the amino acid acceptor stem. The use of fragments representing only a part of the tRNA molecule suggests that this sequence may be recognised by the ribosome. Pestka et al. (1970) found that sequences such as —CACCA-phe bind directly to the ribosome in a reaction which is inhibited by the same antibiotics which prevent aminoacyl-tRNA from binding. But this region must comprise only part of the elements recognised, for Pestka (1970) showed that progressive removal of the —CCA terminus of tRNA by phosphodiesterase reduces, but does not eliminate, its ability to compete with aminoacyl-tRNA for ribosome binding sites. The requirements for binding directly to the ribosome therefore appear less stringent than those of the binding reaction mediated by Tu, which demands intact tRNA.

A second feature found in almost all tRNAs is the common five base order, GTΨCG, of the pentanucleotide arm. Direct experiments have failed to implicate this arm in any common reaction of all tRNAs. But Simsek et al. (1973) and Petrissant (1973) showed that the initiator $tRNA_f^{met}$ molecules of wheat germ, sheep mammary gland and rabbit liver lack GTΨC and may instead contain GAUC, the sequence identified in the yeast initiator by Simsek and RajBhandary (1972). The absence of the usual pentanucleotide sequence cannot therefore be due simply to lack of modification. It is tempting to speculate that this difference may help to distinguish the eucaryotic initiators from all other cellular tRNAs. (But both $tRNA_m^{met}$ and $tRNA_f^{met}$ of E.coli possess the common sequence; figure 5.9 shows that they differ appreciably in sequence but the initiator has no particular features to distinguish it from other tRNAs; and the initiators of yeast and E.coli are not well related.)

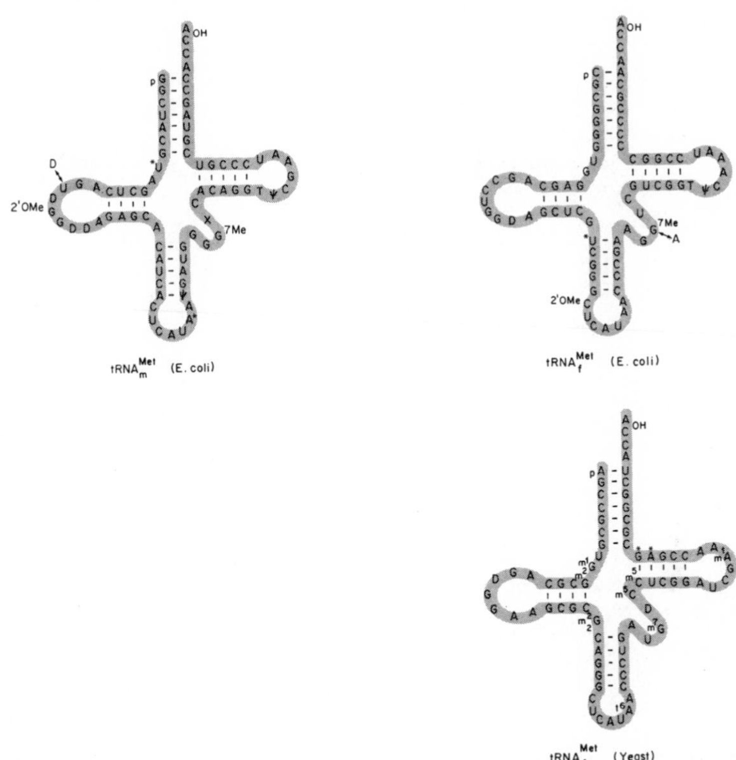

Figure 5.9: clover leaf structures for methionine tRNAs. The two species of E.coli show little similarity. The initiator of yeast is not closely related to that of E.coli and lacks the common pentanucleotide sequence. Data of Cory and Marcker (1970) and Simsek and RajBhandary (1972)

Phenylalanine tRNA of yeast contains the unusual fluorescent base Y on the 3′ side of the anticodon; according to Fairfield and Barnett (1971), this base is a marker for "eucaryotic" $tRNA^{phe}$, for it is found in the cytoplasmic species of Euglena and Neurospora but not in their chloroplasts or mitochondria. Different $tRNA^{phe}$ species display different fluorescent spectra, so the form of the base may vary; Thiebe et al. (1971) have characterized the Y base of yeast $tRNA^{phe}$ and Blobstein et al (1973) showed that the Y base of bovine phenylalanine tRNA differs in the presence of a hydroperoxide group on a side chain. Thiebe and Zachau (1968a) found that mild treatment of $tRNA^{phe}$ with acid excises the base without breaking the phosphate-sugar backbone; the polynucleotide chain therefore retains its integrity. Treated $tRNA^{phe}_{HCl}$ does not bind to poly-U in the ribosome binding test. Its inactivity probably results from a change in the conformation of the anticodon loop caused by the base excision.

But in spite of its deficiency, the $tRNA^{phe}_{HCl}$ can be charged with its amino acid

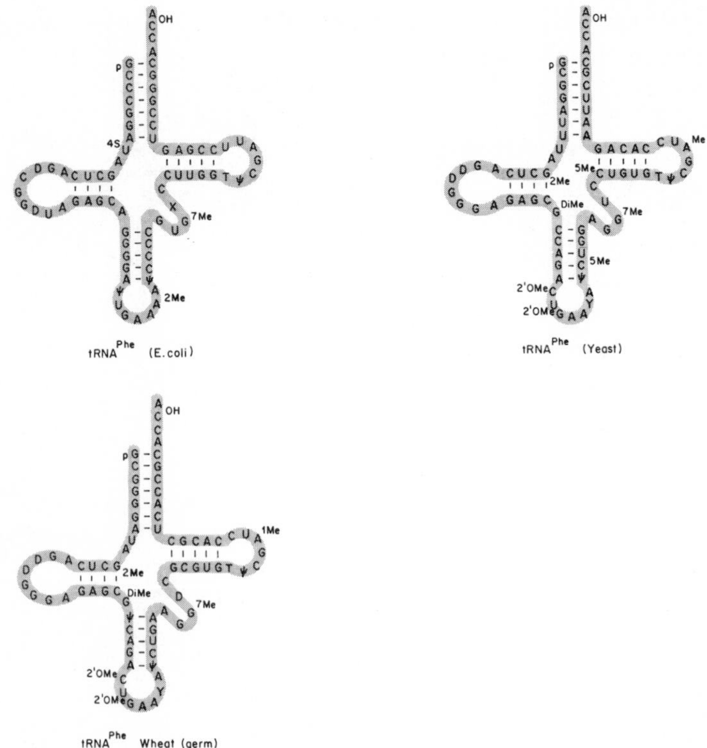

Figure 5.10: clover leaf structures for phenylalanine tRNAs. The three species do not show extensive homologies. Data of RajBhandary et al. (1967) and Dudock et al. (1971)

just as efficiently as the native phenylalanine tRNA. This suggests that Y itself and the anticodon loop in general are not essential parts of tRNA for recognition by synthetase. Thiebe and Zachau (1968b, 1969) subsequently used the synthetase of E.coli to charge the $tRNA^{phe}$ of yeast; although native tRNA can accept phenylalanine, $tRNA^{phe}_{HCl}$ does not react. The requirements for heterologous charging therefore appear to differ in that they rely upon the anticodon region.

The phenylalanine tRNAs of yeast and wheat germ—shown in figure 5.10—differ in 16 of their 76 nucleotides but both can be charged by the yeast enzyme. After excision of the Y base, each molecule can be split at this site by chemical means. Thiebe and Zachau (1969) found that both species of phenylalanine tRNA continue to accept their amino acid after this cleavage. The "half-molecules" produced by the chain scission can be separated on columns of DEAE-Sephadex; neither half alone can accept phenylalanine. But both the homologous and heterologous combinations of half molecules can accept an appreciable proportion of their usual load of phenylalanine. Because of the differences in base sequence between the molecules, the heterologous combinations must have disruptions of the usual structure, and recognition must presumably rely on localised structural features which are retained.

This approach has been taken a stage further by Thiebe et al. (1972), who made partial enzyme digests of yeast phenylalanine tRNA under various conditions and then recombined the fragments remaining to reconstitute tRNA molecules lacking a specific part of their sequence. These combinations can be charged by the homologous synthetase of yeast. The extent of charging is always lower with fragment combinations than with intact phenylalanine tRNA, probably because the mixture contains a variety of active and inactive reconstituted molecules which differ in their tertiary structures. The reconstituted molecules retain some activity when short sequences of the anticodon or dihydrouridine loops are missing which implies that specific bases in these loops are not used for recognition; but removal of more extensive lengths of the molecule greatly reduces acceptor activity, again probably due to the destruction of features of tertiary conformation rather than to the loss of particular bases.

Mutants of su_3 tyrosine tRNA

Point mutations in the sequence of tRNA offer another opportunity to analyse the influence on function of the base at some particular site. Point mutations have the advantage that only one base is altered; and just as a mutant may be selected by some criterion, so may its reversion by further mutation be followed. Of course, depending on selection techniques restricts the range of mutations to those which influence the transfer molecule sufficiently to generate cells of mutant phenotype. The simplest changes to select are mutations in the anticodons of those tRNAs which carry amino acids whose codons are related to nonsense triplets by only one base change; mutant tRNAs with altered coding properties can be detected by their ability to suppress nonsense mutations (see page 211). Other mutations can then in turn be selected by their influence on the ability of the tRNA molecule to suppress nonsense triplets.

The su_3^+ tyrosine tRNA of E.coli is a mutant which responds to the amber codon UAG whereas the wild type su_3^- tRNA responds to the UAC and UAU codons which represent tyrosine. Figure 5.11 shows the results of selecting su_3^+ cells for loss of their ability to suppress nonsense codons. Reversions which prevent su_3^+ from responding to amber codons may, of course, take place in the anticodon so that the tRNA recognises either its original codons or the codon(s) for some other amino acid. But mutations can also occur in other parts of the molecule; any mutation which generates a defective tRNA—whatever the cause of the defect—abolishes its ability to suppress nonsense mutants by inserting tyrosine in response to UAG.

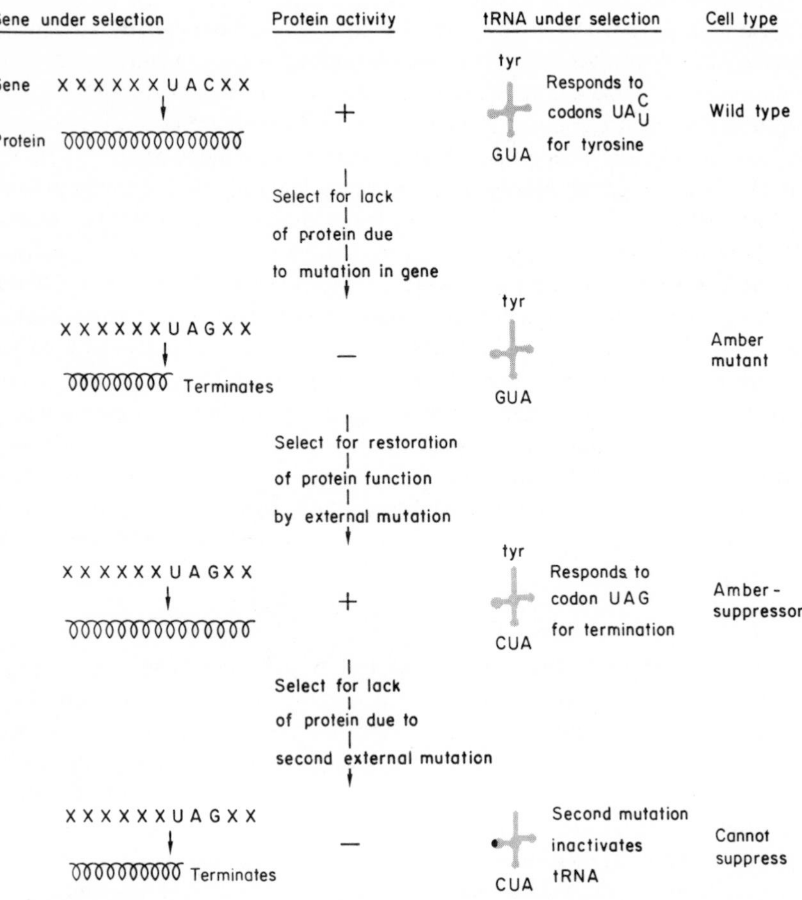

Figure 5.11: protocol for isolating nonsense suppressor tRNAs and revertants which have lost the suppressor activity. Only some of the mutants isolated at each stage fall into the desired class; others—including direct revertants of the original mutation—are omitted for clarity

Several mutations in which an A replaces a G at some point in the molecule have just this effect; these tRNAs carry the anticodon sequence which recognises UAG but have the su^- characteristic because they are deficient in some stage of protein synthesis. Two of the mutants isolated by Abelson et al. (1970) are described as A15 and A31: A15 has an A instead of the G which is usually found in position 15 of the dihydrouridine loop; A31 has an A in place of the G usually found at position 31 of the stem of the anticodon arm (see figure 5.12). Because in appropriate conditions the mutant molecules can bind efficiently to ribosomes in the presence of poly-A,U,G, codon recognition itself must be unimpaired by the second site of mutation.

Although A31 binds poorly to ribosomes in the absence of the T transfer factors, it can bind efficiently when either factor T is added or the magnesium concentration is lowered. This suggests that the mutation changes the conformation of the anticodon loop, probably by disrupting the base pair G31-C41, so that the molecule fits less well into the ribosome A site; but transfer factor or lower magnesium ion concentrations allow the anticodon loop to regain its proper conformation. An interaction between the anticodon arm of tRNA and the transfer factor may therefore be important in placing the tRNA molecule on the ribosome.

But although both mutants can recognise the UAG codon, they are only half as active as the parental su_3^+ tRNA when tested for their ability to insert tyrosine in response to an amber mutation in the coat protein gene of phage f2 RNA. The A31 mutant displays a ten fold rise in its Km for synthetase—so it must have a much reduced affinity for the charging enzyme. Either the anticodon stem itself must interact with the enzyme, therefore, or its duplex structure must be required to maintain the conformation of some other part of the molecule involved in this recognition. Anderson and Smith (1972) showed that conversion of the C at position 41 to yield U restores suppressor activity to the A31 mutant; the A31.U41 double mutant has an A-U base pair in place of the G–C pair of su_3^+ tRNA. The presence of a base pair at this site is therefore essential, but its nature does not matter. Conversion of C to U at positions 16 and 45 can also restore activity of the A31 mutant to suppress amber codons; which emphasizes that bases located at several different positions may influence whatever features of conformation are recognised in tRNA.

Another factor which may contribute to the failure of cells to suppress nonsense codons when su_3^+ tRNA is mutated to A15 or A31 is extrinsic to their role in protein synthesis; both the mutant tRNAs are synthesised in greatly reduced amounts by the cell. In fact, this failure of synthesis may be the critical feature which establishes the level at which mutants and their revertants can suppress amber codons; presumably, failure to synthesize the tRNA results from conformational changes altering recognition by maturation enzymes. Some mutants may therefore change the production instead of or as well as reducing the activity of tRNA in protein synthesis.

By selecting mutants which can suppress amber codons at 32°C but not at 42°C, Smith et al. (1970) isolated temperature sensitive variants of su_3^+ tRNA. The A2 and A81 mutants both have single base substitutions which disrupt pairs in the duplex region of the amino acid acceptor stem. Although they can insert tyrosine in response to UAG codons at low temperature, they fail to do so at high temperature. The A2 mutant has been reverted to the su_3^+ form which can insert tyrosine at UAG codons at both temperatures by further mutation in which a U replaces the C normally present at position 80. The A2.U80 double mutant—which has an A–U base pair in the second position of the duplex stem instead of the usual G–C pair—appears to behave as su_3^+. This implies that the base sequence itself does not matter in the duplex region of the amino acid acceptor stem; it is the ability to form hydrogen bonded base pairs and thus to generate a duplex structure which is important. That base pairs as such, of either type, are critical in maintaining the structure and function of tRNA is suggested also by the reversion of another temperature sensitive mutant, A25; the double mutant A25.U11, achieved when U replaces C at position 11, has the su_3^+ activity. Once again, the reverting mutation restores base pairing by generating an A–U pair in place of a G–C pair.

Interaction of tRNA with Synthetase

Transfer RNA molecules display two classes of multiplicity. First, the existence of several codons which specify one amino acid may mean that tRNA molecules with different anticodons must be charged with the same amino acid. And second, even where one tRNA might in theory suffice, a cell may contain different tRNA molecules which have the same anticodon but differ elsewhere in the molecule. The multiplicity of tRNA molecules (see figure 5.15) prompts the question: is there a similar repetition of synthetases, with one enzyme for each tRNA, or can one synthetase charge more than one of the transfer molecules representing its amino acid? If multiple charging is possible, does it include redundant tRNAs only with the same anticodon or can tRNAs with different anticodons be charged by the same synthetase?

No definite answer can yet be given, but it seems likely that bacterial cells at least—and probably rat liver also—contain only one synthetase for each amino acid; this one enzyme must charge all the tRNAs which respond to that amino acid. Reports of multiple synthetases for one amino acid have in general proved to be the result of confusion with the subunit structure of the enzyme. In eucaryotic cells, of course, the cell organelles—mitochondria and chloroplasts—may contain a different set of enzymes from those of the cytoplasm. Several cases are known of enzymes which charge all the tRNAs responding to one amino acid; there do not at present appear to be any examples of redundant synthetases.

Cells of E.coli, for example, contain no less than five leucine tRNAs which differ in coding specificity and structure, but Blank and Soll (1971a) found that all are charged by the same leucyl-tRNA synthetase. Five serine tRNAs are also all charged by one synthetase (Roy and Soll, 1970). The affinities of the different tRNAs which are charged by one enzyme may vary; for example, two classes of valine tRNA in E.coli differ in twenty two bases but both are charged by the same enzyme. However, Helene, Brun and Yaniv (1971) found that one species, which is found in the cell in much smaller quantities than the other, has a much higher association constant for the enzyme; this presumably ensures that it will become charged and is not excluded by competition with the more frequent species.

The ability of one synthetase enzyme to charge tRNA molecules which have different anticodons (and therefore recognise the different codons for an amino acid) excludes the possibility that the anticodon sequence is in general used to identify tRNAs by the enzyme. Further evidence against this idea is provided by the unimpaired charging of the nonsense suppressor tRNAs which bear only a single base change in the anticodon; although they respond to nonsense codons instead of their usual codons, they continue to be charged by the same synthetase. But in particular instances, mutation in the anticodon sequence may influence the affinity for synthetase, for three missense mutant glycine tRNAs—which recognise codons for amino acids other than their usual ones—are charged by the gly-tRNA synthetase less efficiently than their wild type counterparts. This emphasizes the difficulty of deducing general rules which may dictate how transfer molecules are recognised by the various proteins with which they interact.

When one synthetase charges several different tRNAs which respond to the same amino acid, we may expect to find similarities between the molecules which account for their common recognition. Two problems prevent the deduction from their sequences of these common features. It is necessary to sequence all the tRNAs recognised by the enzyme and also to view them in the perspective of all those sequences which are not recognised; in effect this demands sequencing the entire cellular complement of tRNAs, a difficult task because some of them may be present in rather small amounts. Even the clover leaf structures may not reveal the relevant common features, however, which may depend upon similarities of tertiary confirmation produced by different combinations of base sequence.

That the discrimination of tRNAs by synthetases must be set against the cellular complement of transfer molecules is shown by experiments in which tRNAs of one species are charged by the synthetase of another. Recognition of the appropriate tRNAs is often retained across species barriers. For example, Anderson (1969) fractionated the alanine accepting activity of human spleen tRNA into six peaks, that of rabbit liver into four and that of E.coli into two; the alanyl-tRNA synthetase derived from any of these organisms can charge any of these fractions. Common features of tRNA—synthetase recognition must therefore have been conserved in evolution.

But the specificity of recognition is altered in heterologous charging reactions. For example, Dudock et al. (1971) and Roe and Dudock (1972) reported that yeast phenylalanyl-tRNA synthetase can charge phenylalanine tRNA species of yeast, wheat and E.coli; and can also recognise $tRNA_1^{val}$ and $tRNA_1^{ala}$ of E.coli. This implies that features of recognition which in yeast are confined to phenylalanine tRNA may be found in other tRNAs in other species of cell. Although the criteria for recognition exercised by the synthetase for each amino acid may therefore have been conserved to some extent in evolution, they suffer the imposition within the cell of the demand that all classes of recognition are mutually exclusive; the evolution of any tRNA sequence within a species may therefore be restricted by the sequences of the other tRNAs.

By determining the sequences of the tRNAs recognised by one synthetase—whether they derive from homologous or heterologous cells—it might be possible to find similarities of sequence responsible for common recognition. The sequences of the tRNAs recognised by yeast phenylalanine tRNA have in common only the base paired sequence

```
C—U—C—G
|  |  |  |
G—A—G—C
```

in the stem of the dihydrouridine arm and the presence of A as the fourth base from the 3′ end (see figures 5.10 and 5.17). The other similarities between these molecules are those found in all tRNAs. One interpretation of these results is that the dihydrouridine stem is recognised by the synthetase, a contrast with the more common postulate that sequences of loops are likely to be recognised. An alternative is to suppose that there are resemblences in their tertiary structures which are not revealed by the clover leaf sequence; this contention is supported also by the observation that although all these tRNAs are charged by one enzyme, they react at different rates.

The difficulty of making deductions from the primary sequence is emphasized by the remarkable degree of similarity displayed between the E.coli tRNA molecules valine$_{2B}$ and glycine$_3$ which, as figure 5.17 shows, differ only in two positions of the dihydrouridine loop and in six base pairs in the stems of arms. This similarity is in fact much greater than is often found between different tRNAs coding for the *same* amino acid. It implies that the criteria used by the val-tRNA and gly-tRNA synthetases of E.coli to distinguish these two tRNA molecules must be very precise. In contrast to situations in which the sequences of loops appear more decisive in tRNA function, this distinction must involve the sequences of base paired regions. It would be interesting to know the frequencies of errors in recognition of the codons to which these tRNAs respond.

That fine distinctions are drawn between tRNA structures within the cell is emphasized by the isolation of point mutations of tRNA which change the amino acid with which it is charged. Some amber nonsense mutants cannot be suppressed by the su_3^+ tRNA because insertion of tyrosine at the mutant position produces an inactive protein. However, these nonsense mutants can be suppressed by the insertion of another amino acid, for example glutamine in cells possessing the su_2^+ tRNA gene. By selecting su_3^+ cells for their ability to suppress such a nonsense mutant, Smith and Celis (1973) isolated mutants in the tRNA which, although they continue to recognise UAG, insert glutamine

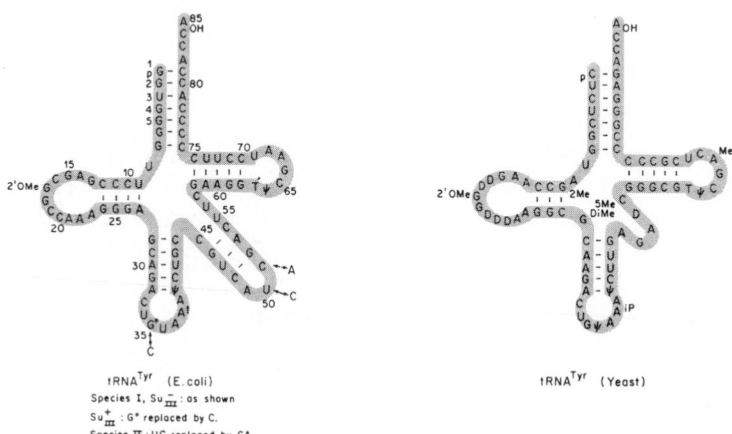

Figure 5.12: clover leaf structures for tyrosine tRNAs. The bases of the molecule are numbered from the 5′-P end to the 3′-OH end. There is little homology between the molecules. Data of Abelson et al. (1970) and Madison et al. (1966)

instead of the usual tyrosine. All four of the mutations isolated are located in the amino acid acceptor stem of the clover leaf.

Determination of the sequence of one of these mutants shows that it suffers only one change in its nucleotide sequence; the G present at position 1 of su_3^+ tRNA (see figure 5.12) is replaced by an A. Isolation of the protein produced when cells carrying this tRNA are infected with an amber mutant of phage T4 confirmed that glutamine is inserted in response to UAG.

The su_3^+ A1 mutant tRNA can be charged with glutamine in vitro by the glutaminyl-tRNA synthetase of E.coli; and its charging with tyrosine by tyr-tRNA synthetase is greatly reduced. The two glutamine tRNAs which are usually charged by this synthetase differ from each other by the seven nucleotides indicated in figure 5.2; and neither has any marked homology with the su_3^+ A1 mutant sequence. Forty-five of the seventy-five bases of the glutamine tRNA are the same as those in the corresponding positions in tyrosine tRNA; but many of these homologies are those common also to other tRNAs and there is no extensive sequence which might serve as a recognition site for the enzyme. And it is not clear why the A1 mutation in su_3^+ tRNA should confer recognition by gln-tRNA synthetase, for the base in position 1 of glutamine tRNA is a U residue.

In spite of the common function of all aminoacyl-tRNA synthetases—to charge tRNA only with its appropriate amino acid—there seems to have been little conservation in evolution even of their overall structures. A common class of activating enzymes consists of a single polypeptide chain of about 100,000 daltons. However, not all synthetases fall into this category since some comprise identical subunits and others consists of non-identical subunits. There is at present insufficient evidence to decide to what extent the structure of the synthetase for a given amino acid has been conserved in evolution.

The Synthetase Charging Reaction

Each aminoacyl-tRNA synthetase can bind three substrates: its specific amino acid, ATP and the transfer RNA which it charges. The behaviour of those enzymes which comprise a single polypeptide chain suggests that there is one binding site for each substrate. Although it has a subunit structure, phenylalanyl-tRNA synthetase of E.coli also appears to follow this rule (Kosakowski and Bock, 1971). Such an organization is the basis of most models for synthetase action. The charging reaction has been reviewed by Chapeville and Rouget (1972).

Models for the action of the synthetase enzymes have been based on the observation that cooperative effects exist between the different sites. Yarus and Berg (1969) found that when isoleucine binds to its synthetase it promotes an increase in both the rates of dissociation and association of its tRNA; since both are raised about six fold, the equilibrium constant remains unaltered. This suggests that the availability of the site to tRNA—both for entry and exit—is increased by the binding of amino acid. Because the maximum velocity of the charging reaction itself is greater than the maximum rate at which the charged ile-tRNA can leave its site on the enzyme, the release step may be rate limiting. Measurements obtained by Yaniv and Gros (1969) on the affinity of association between valine tRNA and its synthetase suggest that dissociation of the enzyme—aminoacyl-tRNA complex is likely to be the rate limiting step in forming other charged tRNAs also.

Although their data formally require only two conformational states for the synthetase—one with slow entry and exit and one with fast—Yarus and Berg suggested the model of figure 5.13 in which the enzyme passes through four states in its catalytic cycle. Each moiety which binds to the enzyme causes a conformational change. The binding of isoleucine—reaction 1—"opens" the tRNA binding site, which is then filled by tRNA—reaction 3. Binding of the tRNA in turn affects the amino acid binding site so that aminoacyl-tRNA can be formed. When the amino acid is transferred onto the tRNA and leaves its site—reaction 5—the binding of another isoleucine and AMP is necessary to reconvert the tRNA site into its "open" form—reaction 6. A second isoleucyl-adenylate complex is formed before the first ile-tRNA is released, assisting the rate limiting step—reaction 8—in which the ile-tRNA is lost. The cycle is then repeated. Yaniv and Gros (1969) found that the valine enzyme has two distinct and relatively independent sites, one which is responsible for the specific recognition of the acceptor tRNA and one which activates the amino acid and transfers it to the tRNA. Since the Yarus and Berg model does not require the ATP binding site to change its conformation, it also contains two sites which are actively involved in the charging reaction, so that it is consistent with these data on the activation of valine.

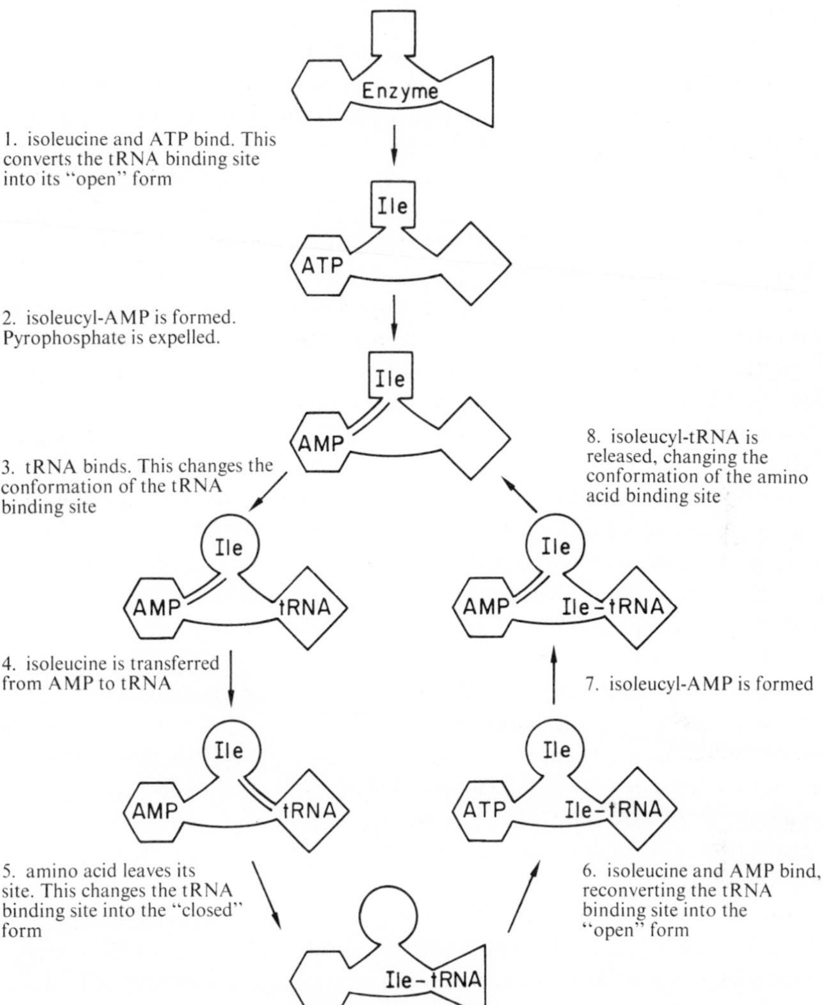

Enzyme
1. isoleucine and ATP bind. This converts the tRNA binding site into its "open" form
Ile
ATP
2. isoleucyl-AMP is formed. Pyrophosphate is expelled.
Ile
AMP
3. tRNA binds. This changes the conformation of the tRNA binding site
8. isoleucyl-tRNA is released, changing the conformation of the amino acid binding site
Ile
AMP
tRNA
Ile
AMP
Ile-tRNA
4. isoleucine is transferred from AMP to tRNA
7. isoleucyl-AMP is formed
Ile
AMP
tRNA
Ile
ATP
Ile-tRNA
5. amino acid leaves its site. This changes the tRNA binding site into the "closed" form
6. isoleucine and AMP bind, reconverting the tRNA binding site into the "open" form
Ile-tRNA

Figure 5.13: catalytic cycle of isoleucyl-tRNA synthetase. This model suggests that the enzyme can exist in any of four conformations. When isoleucine is bound, the tRNA binding site is in "open" state, permitting rapid entry and exit. Because the model permits a second ile-AMP complex to be formed before the first ile-tRNA has been released, the amino acid binding site is occupied at the time of ile-tRNA release, thus increasing its rate of dissociation which is probably the rate limiting step in production of aminoacyl-tRNA. Data of Yarus and Berg (1969)

Although the behaviour of the ile-tRNA synthetase of E.coli reflects a general property of many synthetases—reactions at one binding site may influence the other—the form which this effect takes depends upon the individual synthetase molecule. Rouget and Chapeville (1971a) found that a specific order must be followed when substrates bind to the leu-tRNA synthetase of E.coli; binding of ATP is the first step of the reaction and leucine binds only after the enzyme-ATP complex has been formed. When the synthetase is incubated with C^{14}-leucine and H^3-ATP, both radioactive labels are recovered with the enzyme after passage through columns of sephadex; when ATP is omitted, leucine cannot bind to the enzyme.

When leu-tRNA is incubated with the synthetase enzyme, it binds to it and can exchange with free $tRNA^{leu}$. The rate of this exchange is independent of the concentration of $tRNA^{leu}$, which suggests that the rate limiting step is dissociation of the complex of leu-tRNA with synthetase. In this respect, the leucine activating enzyme shows the same behaviour as the ile-tRNA synthetase. A five fold increase in the rate of exchange is achieved, however, when leu-tRNA is bound to the complex of enzyme with ATP and leucine. Because ATP is needed for the binding of leucine, both ATP and leucine must be present to increase the association/dissociation rate of the leu-tRNA—synthetase complex (whereas the amino acid alone is sufficient with ile-tRNA synthetase). In spite of these differences of detail, however, the general mechanism of reaction may be similar.

That the different activities of the enzyme are separate, as shown in figure 5.13, is supported by the effect on leu-tRNA synthetase of reagents which interact with free –SH groups. Rouget and Chapeville (1971b) found that the activity of the enzyme is influenced by amounts of N-ethyl-maleimide or p-chloro-mercuribenzoate which inactivate only one sulfhydryl group. Inactivating this group does not change the affinity of the enzyme for its substrates, but means that the presence of ATP and leucine no longer enhances the exchange reaction with $tRNA^{leu}$, that is, does not promote dissociation of charged leu-tRNA. This suggests that one of the –SH groups of the enzyme is essential for interaction between the different sites and is needed to transfer leucine to tRNA; but it is involved only weakly, if at all, in the other catalytic activities of the protein.

Most of the synthetases of E.coli, at least, require the activity of some –SH group(s) in their function. Ile-tRNA synthetase loses its activity when stored in the presence of reducing agents and Iacarino and Berg (1969) found that the uptake of a single mole of N-ethyl-maleimide inhibits the reaction of the synthetase with ATP and isoleucine and its subsequent binding of $tRNA^{ile}$; but the reducing agent does not change the intrinsic ability of the enzyme—that in the absence of stimulation by amino acid—to bind ile-tRNA. In this situation, then, it seems likely that the –SH group is involved in catalysing the synthesis and/or use of the synthetase–adenylate–amino acid complex.

The role of –SH groups in the actions of the leucine and isoleucine charging enzymes therefore appears to be similar but not identical.

Interaction between the tRNA and the amino acid binding sites appears to be a two way reaction—as shown in figures 5.13—and the influence of the tRNA site on amino acid and/or ATP binding is revealed by its influence on the exchange between bound ATP and free pyrophosphate. Charlier (1972) found that tRNA binding stimulates the ATP/PP_i exchange reaction of the ile-tRNA synthetase of Bacillus stearothermophilus, which is the most common effect found. However, the influence of tRNA may be the reverse, for Buonocare and Schlessinger (1972) found an inhibition of this exchange when $tRNA^{tyr}$ binds to the tyrosine activating enzyme of E.coli. In this instance, at least, therefore, the model developed for isoleucine and leucine charging cannot apply.

An alternative model for synthetase action has been proposed by Loftfield (1972), who suggested that instead of the sequential series of actions usually postulated, a "concerted" reaction may take place; according to this model, all three substrates bind independently and only then does a single reaction take place to form aminoacyl-tRNA. (This model implies that the ATP/PP_i exchange reaction stimulated by amino acids is a separate property of the enzyme, not representing its charging of tRNA.) The arginyl-tRNA synthetases of E.coli and B. stearothermophilus may follow a reaction of this nature. Papas and Peterkofsky (1972) found that the E.coli enzyme follows a random sequential mechanism in which its substrates bind in random order, after which reaction is initiated; Parfait and Grosjean (1972) found that a particular order of binding must take place with the Bacillus enzyme, but that a concerted reaction is initiated only after all three substrates are bound.

The interactions of synthetase with amino acid and with tRNA occur with different specificities, for Bergmann et al. (1961) showed that isoleucyl-tRNA synthetase of E.coli can convert either isoleucine or the related amino acid valine to the aminoacyl-adenylate; but transfers only the correct amino acid to isoleucine tRNA. Loftfield (1963) calculated that valine is incorrectly recognised and converted to valyl-AMP with a frequency of about 1/50 but that the error rate for charging of amino acid is less than 1/5000. Loftfield and Vanderjagt (1972) demonstrated that during synthesis of globin protein in vitro, the substitution of valine for isoleucine is about 3/1000. Since these amino acids are coded by closely related codons, the error rate of other substitutions may be even lower.

By comparing the properties of the two aminoacyl-adenylate complexes, Baldwin and Berg (1966) demonstrated that the ile-tRNA synthetase has a mechanism for preventing transfer of the wrong amino acid to tRNA. When the aminoacyl-adenylate complexes formed with the enzyme are isolated by gel filtration and incubated with $tRNA^{ile}$, only the ile-AMP–$enzyme^{ile}$ complex transfers its amino acid to $tRNA^{ile}$; the val-AMP–$enzyme^{ile}$ complex instead breaks down to release valine.

Breakdown of the wrong complex is caused specifically by $tRNA^{ile}$, for no other tRNA can be substituted. Digestion of the isoleucine tRNA with ribonucleases abolishes its ability to breakdown the complex; and there is a good correlation between the ability of the partially degraded transfer molecule to accept its amino acid from a properly charged synthetase and to breakdown an improperly formed enzyme complex. That a special mechanism—rather than some general conformational change caused in the enzyme by tRNA binding—is responsible for causing breakdown is suggested by the failure of several chemically modified $tRNA^{ile}$ species which can bind to the enzyme to cause breakdown. A requirement for a free 3′ hydroxyl group on the $tRNA^{ile}$ suggests that the enzyme may start the procedure for transferring valine to the $tRNA^{ile}$ but somewhere during or after its formation the complex val-$tRNA^{ile}$ is destroyed.

A further safeguard against mistakes in protein synthesis allows incorrectly charged tRNAs to be hydrolysed. Yarus (1972) prepared ile-$tRNA^{phe}$ by using the ile-tRNA synthetase of E.coli to charge the $tRNA^{phe}$ under abnormal conditions which stimulate false recognition. The ile-$tRNA^{phe}$ may be hydrolysed by the phe-tRNA synthetase of E.coli; and when ATP and phenylalanine are added to the incubation, the isoleucine is replaced by phenylalanine. The maximum rate of hydrolysis of the mischarged tRNA is close to the maximum rate of charging; both reactions may rely upon the same sites of the enzyme.

Accuracy of Translation

Codon-Anticodon Recognition

One prominent feature of the genetic code is that whilst a change in the first or second base of a triplet usually produces a codon which specifies a different amino acid, a change in the third base very often gives another triplet representing the same amino acid. The third base is therefore not so rigidly defined as the first two and the patterns of third base degeneracy are:

in 8 groups,	any of U, C, A or G codes the same amino acid, that is,	32 codons
1	any of U, C, A	3
7	either U, C	14
6	either A, G (including the UAA/UAG terminators)	12
2	only G	3
1	only A (this is the UGA terminator)	1

This raises the question: does one transfer molecule respond to all the codons related by third base degeneracy for some particular amino acid, or does each triplet interact with a different tRNA? If only one transfer molecule is required for the related triplets, it follows that the base pairing between the third base of the codon and its complementary base on the anticodon cannot always follow the normal pattern of A = U, G = C.

The pattern of third base degeneracy shows that it is not possible for codon meanings to rely upon a unique third base of C or of A; any codon ending in C represents the same amino acid if U is substituted and, similarly, any amino acid codon ending in A is not affected by the substitution of a G. (The sole exception to this rule, UGA, is not recognised by tRNA.) Crick (1966) proposed that there may be a certain amount of "wobble" in the third base position so that the anticodon on any one tRNA can respond to more than one codon. The observed degeneracy patterns can be accounted for if a certain amount of unusual pairing takes place between the third base of the codon and the first base of the anticodon such that:

U in the anticodon recognises	either A or G in the codon
C	only G
A	only U
G	either C or U
I (inosine)	all of U, C, A

Although the pairing of C and A in the first position of the anticodon is restricted to recognition of their usual partners, U and G can recognise each other as well as forming their usual base pairs. This unusual recognition demands the slight distortion of bond angles shown in figure 5.14; this is presumably promoted by the topology of the codon-anticodon interaction in the ribosome. Inosine, one of the unusual bases found in tRNA, has the

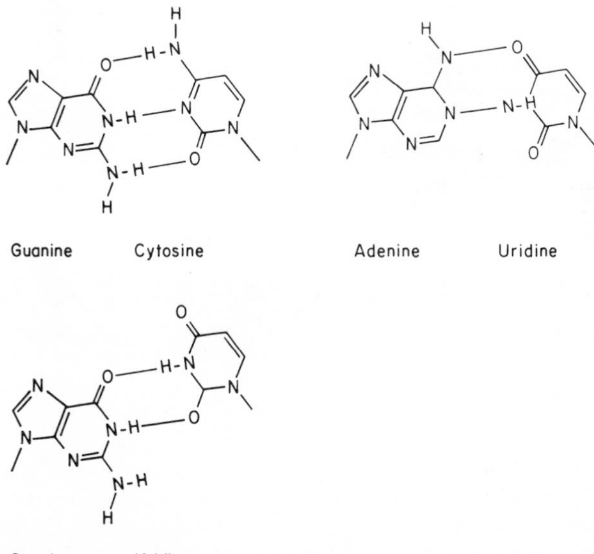

Figure 5.14: base pairing between tRNA and mRNA. The usual rules governing base pairing allow only G–C and A–U pairs (upper). With a slight distortion of bond angles, G–U pairs (lower) can be formed; this is permitted only at the third base of the codon. Model of Crick (1966)

base hypoxanthine which pairs equally well with all the usual bases except guanine.

The isolation of individual species of transfer molecule—whose response to triplets in the ribosome binding assay or to codons in synthetic messengers has been measured—has confirmed this scheme. Soll, Cherayil and Bock (1967), for example, found that when tRNA species exhibit multiple codon recogni-

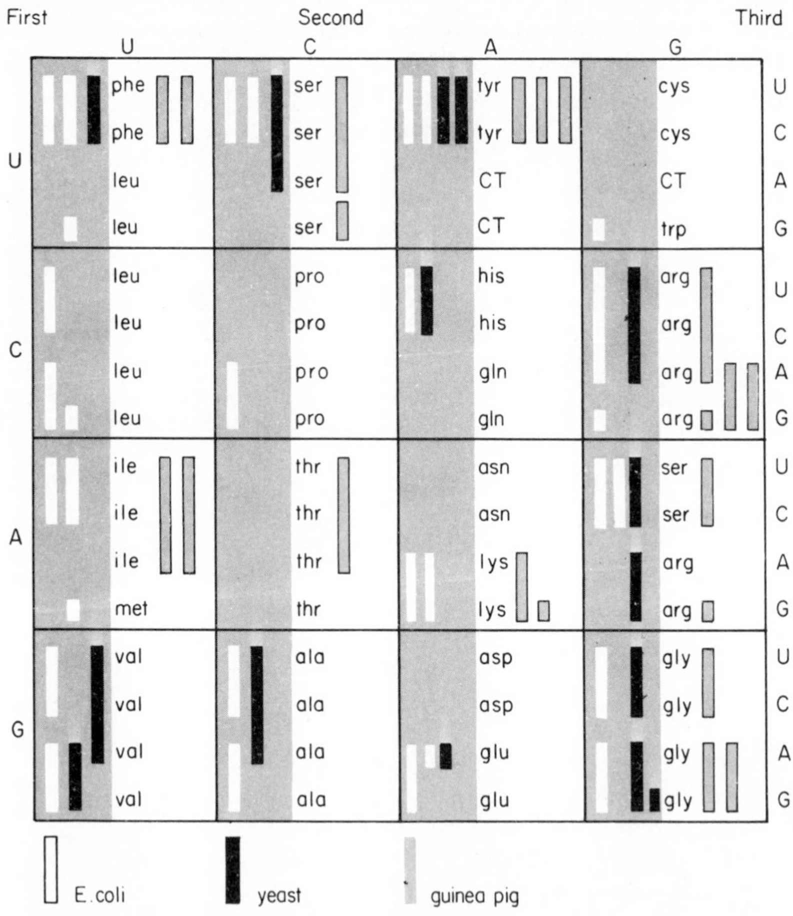

Figure 5.15: patterns of tRNA codon recognition in E.coli, yeast and guinea pig (based on properties of tRNAs in vitro). No particular mode of synonym recognition predominates in any cell. Where more than one tRNA responds to the same codons, additional vertical lines are drawn. Data of Caskey et al (1968), Ishikura and Nishimura (1968), Kruppa and Zachau (1972), Mirzabekov et al. (1968), Nishimura and Weinstein (1969), Ohashi et al. (1970), Roy and Soll (1970), Soll et al. (1967), Soll and RajBhandary (1967), Staehelin et al. (1968) Takeishi et al (1972), Yoshida et al. (1970)

tion, the synonym codons—codons representing the same amino acid—to which one tRNA responds are related according to the rules established by the wobble hypothesis. In the several organisms whose tRNAs have been isolated, as figure 5.15 shows, recognition of three codons by one tRNA always involves U, C and A as the terminal bases, recognition of two codons occurs when either both pyrimidines (U and C) or both purines (A and G) occupy the terminal positions, and when only one codon is recognised its terminal base is always guanine. The only predicted type of tRNA which has not been observed is one which responds only to a codon terminating in U; this failure may be due to a ready deamination of adenosine in the anticodon during preparation, or may reflect the absence of such tRNAs from the cell, perhaps because there are no codons which end in U where C cannot be substituted—the cell may find it more economical to produce a tRNA with G in the anticodon which can respond to both U and C. The validity of wobble pairing is directly confirmed by the recognition properties of those tRNAs whose anticodon sequences are known; table 5.1 shows that the presence of U, C, G or I in the anticodon confers recognition of the synonym codons predicted by wobble.

An apparent exception to the wobble hypothesis is provided by $tRNA^{glu}_2$ of E.coli and $tRNA^{glu}_3$ of yeast, both of which respond only to GAA and not to GAG. But Ohashi et al. (1970) and Yoshida et al. (1970) found that the first

Table 5.1: codons recognised in vitro by tRNAs with known anticodon sequences

organism	tRNA	anticodon	codons	wobble pattern
yeast	alanine	IGC	U GCC A	I recognises U, C, A
yeast rat liver	serine	IGA	U UCC A	
yeast	phenylalanine	Me–GAA	UU^{U}_{C}	
E.coli	phenylalanine	GAA	UU^{U}_{C}	
E.coli S. typhimurium	histidine	$\overset{*}{G}$UG	CA^{U}_{C}	G recognises U and C
E.coli	$leucine_1$	GAG	CU^{U}_{C}	
E.coli	$glycine_3$	GCC	GG^{C}_{U}	
E.coli	$valine_2$	GAC	GUC	

Table 5.1:—*continued*

organism	tRNA	anticodon	codons	wobble pattern
E.coli	tyrosine	$\overset{*}{G}$UA	UAU > UAC	modified G recognises U more readily than C in codons which have A as second base
E.coli	histidine	$\overset{*}{G}$UG	CAU > CAC	
E.coli	aspartic acid$_1$	$\overset{*}{G}$UC	GAU > GAC	
E.coli	asparagine	$\overset{*}{G}$UU	AAU > AAC	
E.coli	methionine$_f$	CAU	AUG(GUG)	C recognises only G
E.coli	methionine$_m$	CAU	AUG	
E.coli	tyrosine su_3^+	CUA	UAG	
E.coli	glutamine$_2$	CUG	CAG	
E.coli	tyrosine su_{oc}	UUA	UA^A_G	U recognises A or G
E.coli	glycine$_{ins}$	U′CC′	GG^A_G	
E.coli	valine$_1$	VAC	GU^A_G > GUU	V = uridine-5-oxyacetic acid, recognises A and G more efficiently than U
E.coli	serine$_1$	VGA	UC^A_G > UCU	
E.coli	glutamic acid$_2$	NUC	GAA	N is a modified 2-thiouridine which recognises A alone
yeast	glutamic acid$_3$	NUC	GAA	
E.coli	glutamine$_2$	NUG	CAA	

All the predicted patterns of wobble recognition between the usual four bases, U,C,A,G have been confirmed, except the recognition of U alone by A. Exceptions to these patterns are caused by modification of bases in the anticodon which changes the coding response. For references see figures of clover leaf structures in this chapter and text

base of the anticodon—which responds to the third, wobble base of the codon —is a derivative of 2-thiouridine, probably 5-methyl-amino-methyl-2-thiouridine in E.coli and 5-acetyl-methyl-2-thiouridine in yeast. The sulfur atom which replaces oxygen at position 2 of the uridine ring should be unable to hydrogen bond with a G residue, although, as figure 5.16 shows, a thiouridine-adenine pair remains possible. In this instance, then, recognition of a third A alone is possible. A similar situation may exist in rat liver and rabbit reticulocytes, where one tRNA for lysine responds to only AAG and the other recognises only AAA (Rudloff and Hilse, 1971; Liu and Ortwerth, 1972).

Another unusual form of recognition is found in tRNAs which recognise as the third base of the codon A or G—as predicted for U in the anticodon—but

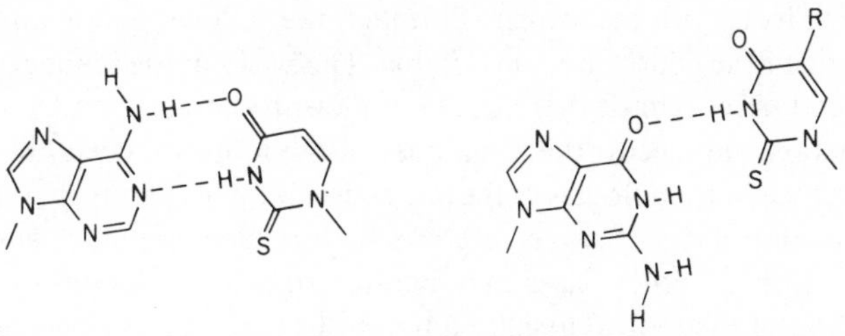

Adenine 2-Thio-uridine Guanine 2-Thio-uridine

Figure 5.16: unusual hydrogen binding properties of 2-thiouridine which recognises adenine but can no longer undertake wobble pairing with guanine. Data of Yoshida et al. (1970)

respond as well, although much less efficiently, to a third base of U. Both $tRNA_1^{val}$ and $tRNA_1^{ser}$ of E.coli contain an unusual residue at the first position of the anticodon, which Ishikura et al. (1971) and Kimura et al. (1971) found to be uridine-5-oxy-acetic acid. This modification of uridine appears to allow it to recognise U in the codon at some 20% of the level of its recognition of A and G; this conclusion is based on vitro assays and we do not know what significance it may have for translation in the bacterial cell. A residue of G or dimethyl-G in the anticodon usually recognises U or C in the codon, but several tRNAs of E.coli contain a modified guanosine called G* or Q, whose structure is not known, which Harada and Nishimura (1972) found to have more affinity for U in binding assays than for C; all these tRNAs recognise A in the second position of the codon and once again we do not know whether the differential binding observed in vitro is important in vivo.

Neighbouring bases appear to influence codon–anticodon recognition, for base pairing between the two triplet sequences seems to depend on provision of the proper environment, which may implicate bases in the anticodon loop of tRNA other than the anticodon sequence itself. Uhlenbeck, Baller and Doty (1970) made use of the reaction of tRNA with small oligonucleotides to detect unpaired regions in the molecule, for sequences as short as three or four bases can interact with available complementary regions. Association is measured by equilibrium dialysis, in which a concentrated solution of tRNA is loaded on one side of a dialysis membrane and radioactively labelled nucleotides on the other side; trinucleotides reach equilibrium in 24 hours and tetranucleotides in forty eight hours. This method can detect unpaired sequences of three or more nucleotides, but cannot detect any sequence which, although unpaired, is held in a rigid conformation incompatible with duplex formation.

The triplet AUG binds to the anticodon of $tRNA_f^{met}$, but its binding is greatly increased when it is tested in the form of the tetranucleotide AUGA. This implies that the U residue on the 5′ side of the anticodon must be able to pair with the A, for no other fourth base will stimulate the binding of AUG. The second base on the 5′ side of the anticodon also seems to be able to undergo some base pairing with messenger; although the following base and the bases on the 3′ side of the codon are unavailable. The bases of the sequence CUCAU in the anticodon loop must therefore be sufficiently single stranded to bind the complementary sequences; the remainder of the loop is unavailable.

The existence of homologies in the anticodon loop—apart from the sequence of the anticodon itself—between tRNAs with related responses suggests that the structure of the other bases may be important in maintaining the proper conformation of the loop. The base adjacent to the 5′ side of the codon is a U residue and the base adjacent to the 3′ side is usually an A or modified A (sometimes an I or G). The A residue always seems to be modified in tRNAs which must recognise codons beginning with a U or A; the modification may perhaps be important in ensuring that the anticodon takes the proper conformation for pairing with the codon.

That wobble pairing depends upon the structure of the anticodon loop is shown by the properties of the $tRNA^{phe}_{HCl}$ which has lost the Y base—a modified A—adjacent to the anticodon. Ghosh and Ghosh (1970, 1972) found that excision of Y prevents phe-tRNA from responding to UUU, although it continues to respond to UUC, under the usual conditions of assay. The response to UUU can be restored by increasing the magnesium concentration, a condition which is known to influence the structure of the molecule. That the excision changes the conformation of the anticodon so that it loses its ability to recognise U as well as C was confirmed by the observation of Pongs and Reinwald (1973) that excision changes the ability of the tRNA to bind oligonucleotides complementary to the anticodon loop.

Although the meaning of codons is the same in all organisms, they make different uses of the available patterns of codon-anticodon recognition. A third base degeneracy of U,C,A,G, for example, may be provided in one species by tRNAs which respond to U,C and A,G and in another by tRNAs which recognise U,C,A and only G. Figure 5.15 shows that valine recognition in E.coli and yeast is accommodated by just these alternatives. As the figure shows, none of E.coli, yeast and guinea pig tRNA complements shows any restriction to particular classes of recognition; all possible modes of recognition of synonym codons may be utilised in any one cell type and there need be no correlation with the recognition patterns of other species. Caskey, Beaudet and Nirenberg (1968) for example, compared the responses in the ribosome binding assay of E.coli and guinea pig aminoacyl-tRNA preparations corresponding to twelve codon sets representing six amino acids; only about half of the synonym recognitions are the same in the two species. Marshall and Nirenberg (1969) found that thirty seven codons in Xenopus laevis represent the same twelve amino acids as in E.coli, but there are many differences in how synonym codons are recognised.

Wobble pairing would enable considerably fewer than sixty-one transfer molecules to respond to all codons, but all organisms probably contain more tRNA species than required simply to recognise all codons. The multiplicity of tRNA molecules representing a single amino acid arises from two causes. As well as the *degenerate* tRNAs which recognise the different synonym codons for any one amino acid, there are *redundant* tRNAs which respond to the same anticodon(s) but are structurally different in other parts of the molecule. According to Muench and Safille (1968) at least fifty-six tRNAs can be separated from E.coli extracts by partition gradient chromatography.

The number of tRNA molecules representing each amino acid appears to have no correlation with its number of codons and varies from one to five; it is notable that most amino acids have more than one tRNA molecule—this may confer an important advantage in evolution in helping the organism to withstand deleterious mutations changing the coding response of a tRNA which might otherwise cause the substitution of one amino acid by another at all its codons (but see discussion below of suppressor tRNAs).

Redundant tRNAs may be closely related. Identical redundant tRNAs may be synthesized when a cell possesses two genes coding for tRNA with the same sequence; closely related tRNAs may be synthesized by two genes which have evolved by duplication but have since suffered different mutations. The E.coli genome, for example, possesses two identical genes which represent one of the two tyrosine tRNA species and a third, almost identical gene codes for a second tyrosine tRNA (see figure 5.12; and also below). As figure 5.17 shows, valine tRNAs 2a and 2b of E.coli differ only in three base pairs in the amino acid and pentanucleotide stems.

Degenerate tRNAs may be less similar in sequence to each other, but may retain sufficient homology to indicate their evolution from a common ancestor. Figure 5.17 shows that the resemblance between the two $tRNA_2^{val}$ molecules and $tRNA_1^{val}$ of E.coli is more limited. Figure 5.2 shows that there is a greater degree of resemblance between two leucine tRNAs; and two degenerate tRNAs for glutamine are as closely related as many redundant tRNAs. Since all the tRNAs for one amino acid are recognised by a single synthetase, they must have common structural features in spite of any differences of sequence (see page 194).

Insufficient tRNAs have at present been sequenced to reveal the relationships between tRNAs coding for the same amino acid in different organisms; but there is at present no evidence to suggest any conservation of sequence. Although figure 5.2 shows some resemblance between the serine tRNAs of yeast and rat liver, figures 5.10, 5.12 and 5.17 show few homologies amongst the phenylalanine, tyrosine, valine or glycine tRNAs which have been sequenced in more than one species. Specific recognition by heterologous enzymes implies some conservation of structure between, for example, the phenylalanine tRNAs, but these need not be apparent from the sequence of the clover leafs (see page 195).

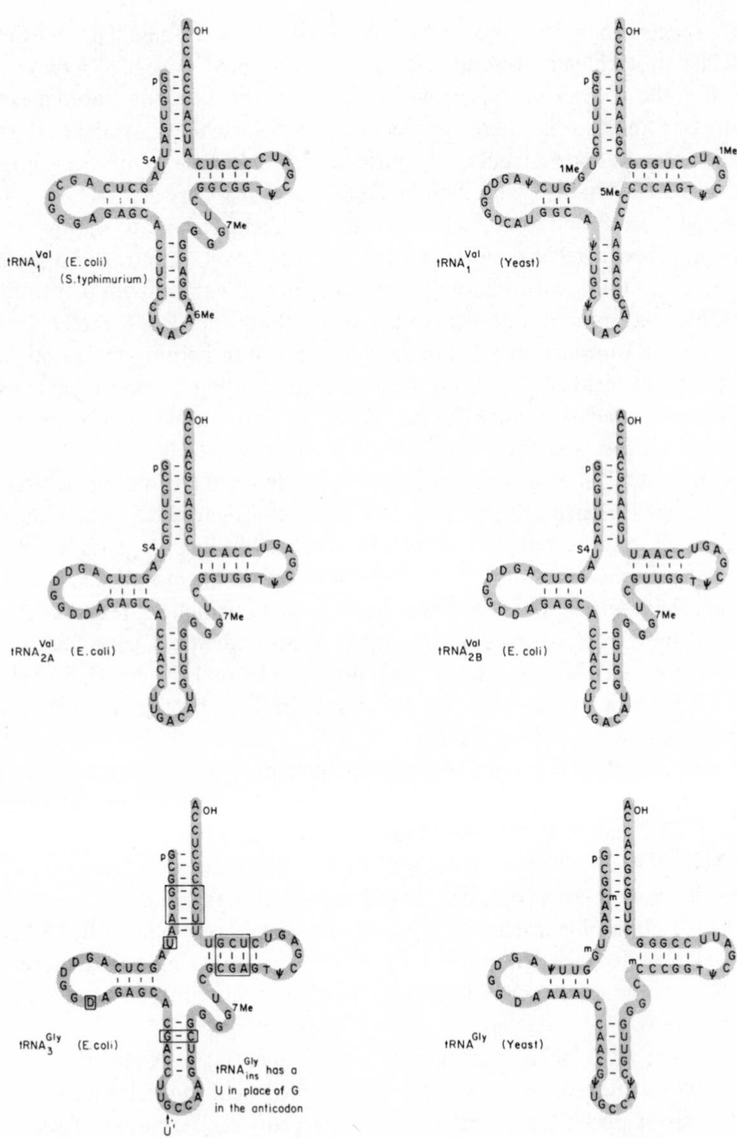

Figure 5.17: clover leaf structures of valine and glycine tRNAs. The tRNA$_2^{val}$ species of E.coli are closely related to each other and poorly related to tRNA$_1^{val}$ which recognises different codons. None are related to tRNA^{val} of yeast. The tRNA$_3^{gly}$ of E.coli differs from the tRNA$_{2b}^{val}$ only in the positions enclosed in the boxes; it is not well related to tRNA^{gly} of yeast. Data of Kimura et al. (1971), Squires and Carbon (1971), Yaniv and Barell (1971) and Yoshida (1973)

Suppression of Amber and Ochre Mutations

Mutation in the anticodon of a transfer molecule can change its coding properties so that, although it continues to be charged with the same amino acid, it recognises a new codon(s) instead of the original codon(s) to which it responded. When the codon to which the mutated tRNA responds is a nonsense triplet, the insertion of an amino acid means that protein synthesis is not terminated but can continue. If a nonsense codon has arisen by mutation of a codon specifying an amino acid, recognition of the codon—even by some amino acid other than that originally responding in the wild type cell—may suppress the mutant phenotype by allowing completion of the protein. As figure 5.11 shows, tRNAs which respond to nonsense codons can therefore be isolated by selecting cells with a nonsense mutation in some protein for ability to synthesize the protein again. This is nonsense suppression.

When a codon has been mutated to missense, so that a different amino acid from usual is incorporated into protein, mutation in the anticodon of a tRNA may allow the original amino acid, or some acceptable substitute, to be placed in the protein in response to the mutant codon. This is missense suppression.

Another class of suppression—discovered more recently—applies to frameshift mutations where the reading frame can be restored to normal by a mutant tRNA. Nonsense, missense and frameshift suppressor tRNAs are all isolated by application of the same principle; a cell mutated in some protein is selected for revertants, some of which are located at the original site but others of which identify tRNAs mutated in such a way as to restore at the mutant codon the original (or an acceptable substitute) reading of the genetic code.

A transfer RNA molecule may suffer two classes of mutation which change its coding response; either the codons to which the tRNA responds may change or the amino acid with which it is charged may be altered. All the mutant tRNAs isolated as nonsense suppressors arise from mutations which change their nucleotide sequence so that they respond to a nonsense codon(s) instead of to their usual codons; but in spite of this change the transfer molecule continues to be charged with its usual amino acid by synthetase. Missense suppressor tRNAs also can result from such a mutation, in this case the new codon comprising that representing an amino acid rather than termination.

Missense suppression may also result from mutations which change the amino acid recognised by a tRNA without changing the codon(s) to which it responds. This demands that the mutant tRNA fails to be recognised by its proper synthetase and is instead charged by the synthetase for some other amino acid. Such mutations can generate only missense suppressors and cannot account for the suppression of nonsense codons since there are no tRNAs in the wild type cell which correspond to nonsense triplets. (Of course, cells which contain a nonsense suppressor may suffer a second mutation which changes the amino acid recognised by the tRNA; the double mutant tRNA is then a nonsense suppressor which has both altered codon and amino acid recognitions; see page 197).

That nonsense codons might be suppressed by the production of transfer species which recognise them as amino acid code words was first suggested by Benzer and Champe (1962) and Garen and Siddiqui (1962)—see page 28. Many such suppressors, both of nonsense codons and of the missense type, have since been isolated in E.coli; evidence that suppression is mediated by transfer RNAs has so far been obtained only in bacteria (reviewed by Garen, 1968).

The first three suppressors identified in E.coli insert amino acids in response to amber (UAG) mutations. Weigert and Garen (1965a, 1965b) and Weigert, Lanka and Garen (1965) showed that there are marked differences in the alkaline phosphatase molecules synthesised when these different suppressors act on nonsense mutations in the protein; this suggests that each suppressor inserts a different amino acid at the position of the nonsense mutant. As table 5.2 shows, *su*1, *su*2 and *su*3 insert serine, glutamine and tyrosine respectively. Stretton and Brenner (1965) demonstrated that they insert the same amino acids at nonsense sites in mutants of phage T4.

Each suppressor allows propagation of chain synthesis of any particular mutation to a characteristic extent; as the table shows, the absolute efficiencies of suppression depend on the mutant site, but the efficiencies of the suppressors relative to each other are usually similar, *su*3 being slightly more efficient than *su*1 and *su*2 being much less efficient (see Garen, 1968). These variations may in part reflect the acceptibility of the amino acid which is inserted into protein—assays often involve measurement of protein activity rather than synthesis as such—but may also depend upon an influence of neighbouring bases on the reading of nonsense triplets.

Each suppressor locus can exist in an active form (su^+) and as an inactive allele (su^-). The su^- allele is found in wild type cells and codes for the tRNA which recognises the proper codon(s) for the amino acid which it carries. The su^+ allele directs synthesis of the mutant suppressor which recognises a nonsense codon. Capecchi and Gussin (1965) showed that the component active in suppressing amber (UAG) mutants in the alkaline phsophatase gene of $su1^+$ E.coli cells is a serine accepting tRNA not found in extracts of the $su1^-$ parent strain. When a mutant of R17 phage RNA which has a nonsense triplet in its coat protein gene was used to direct protein synthesis in vitro, an extract derived from the $su1^-$ parent strain could not make active protein, but an extract from the $su1^+$ suppressor strain could do so by inserting serine at the mutant site. Soll (1968) showed that $su1^+$ but not $su1^-$ cells contain a tRNA which inserts serine in response to UAG.

Cells of $su6^+$ bacteria insert leucine in place of amber codons and Gopinathan and Garen (1970) demonstrated that the leucine accepting tRNA of $su6^-$

Table 5.2: coding responses of nonsense suppressors of E.coli and their derivation

	suppressor tRNA		wild type parent tRNA			suppression efficiency			
gene	codons recognised	anticodon	anticodon	codons recognised	amino acid	a	b	c	d
su-1	UAG (amber)	$\underline{CUA}$	$\underline{CGA}$	$\underline{UCG}$	serine	63	28	41	36
su-2	UAG	$\underline{CUA}$	$\underline{CUG}$	$\underline{CAG}$	glutamine	30	14	24	5
su-3	UAG	$\underline{\underline{CUA}}$	$\underline{\underline{GUA}}$	UA^{C}_{U}	tyrosine	51	55	40	42
su-6	UAG	$\underline{CUA^*}$	$\underline{CAA^*}$	UUG	leucine				
su-7	UAG	CUA	CUG	CAG	glutamine		77 (recessive lethal)		
su-B	UAG	CUA	CUU	AAG	lysine	8			
su-F	UA^{A}_{G} A (ochre) G (amber)	UUA	CUA	(*su*-2) UAG	glutamine			2	3
su-G	UA^{A}_{G}	UUA	CUA	(*su*-3) UAG	tyrosine			4	19
su-oc	UA^{A}_{G}	$\underline{\underline{UUA}}$	$\underline{\underline{CUA}}$	(*su*-3)UAG	tyrosine				
su-4	UA^{A}_{G}	UUA	GUA	UA^{U}_{C}	tyrosine		16 12		
su-5	UA^{A}_{G}	UUA	UUU	AA^{A}_{G}	lysine		5 6		
su-8	UA^{A}_{G}				—		4		
su-9	UG^{A}_{G} A (nonsense) G (tryptophan)	UCA	CCA	UGG	tryptophan				
su-UAG	UG^{A}_{G}	$\underline{\underline{CCA}}$	$\underline{\underline{CCA}}$	UGG	tryptophan				

Codon and anticodon sequences have been deduced from the amino acids inserted in response to nonsense codons in vivo, taking into account the predictions of wobble recognition. Those sequences underlined once have been confirmed by in vitro assays of tRNA responses and those sequences underlined twice have been determined directly. Suppression efficiencies have been measured in (a) T4 head protein, (b) alkaline phosphatase, (c) β-galactosidase, (d) ornithine transcarbamylase. The ochre suppressors *su*-F, *su*-G, *su*-oc are all derived by single step mutation of amber suppressors; *su*-G and *su*-oc may be the same. Data of Altman, Brenner and Smith (1971), Chan and Garen (1970), Garen (1968), Goodman et al. (1968), Gopinathan and Garen (1970), Hayashi and Soll (1971), Hirsh and Gold (1971), Kaplan (1971), Soll and Berg (1969a, 1969b), Stretton and Brenner (1965), Strigini and Gorini (1970).

E.coli which responds to UUG can be separated into two fractions on columns of benzoylated DEAE-cellulose. In *su*6$^+$ cells, only one fraction responds to UUG; the other instead responds to UAG. Hayashi and Soll (1971) separated the suppressor tRNA species into two fractions, one of which is probably an unmodified precursor of the other. The suppressor species has the same sequence as the leucine tRNA-4 except for a single base change in its anticodon, which probably represents the replacement of CAA by CUA.

A bacterial strain which suppresses amber mutations therefore possesses a tRNA which recognises UAG as a signal for an amino acid; this species is absent in the parent strain which does not suppress nonsense mutations. The idea that suppressor tRNAs always result from a change in the coding properties of a tRNA in the parent strain is supported by the relationships of the amino acids which can be inserted by different suppressors in response to nonsense codons; all the amino acids noted in table 5.2 have codons (amongst their degeneracies) which are related to the UAG codon by a single base change. The activities of suppressor tRNAs of E.coli and S.typhimurium have been reviewed by Whitfield (1972).

Ochre suppressors have been identified by the same criteria used to isolate amber suppressors (see Brenner et al., 1965); because they are usually much less efficient than amber suppressors, it is more difficult to determine the amino acid which is inserted in response to the nonsense triplet. As table 5.2 shows, although several ochre suppressors have now been isolated, they are much less well characterized than the amber suppressors.

Ochre suppressors in E.coli always respond to both ochre (UAA) and amber (UAG) codons, usually with the same efficiency. The *su*2 and *su*3 amber suppressors have been converted to ochre suppressors by mutations in a single base (Ohlsson et al., 1968; Strigini and Gorini, 1970); if amber suppressor tRNAs possess a mutated anticodon, CUA, which recognises UAG, the conversion to ochre suppression can be accounted for by a further change to generate the related anticodon UUA. Altman et al. (1972) confirmed by sequence analysis that the *su*3 amber suppressor suffers this change in its anticodon upon mutation to become an ochre suppressor. According to the wobble hypothesis, the UUA anticodon must recognise both the codons UAA and UAG; the inability of tRNAs to recognise only codons ending in A—unless unusual bases are present in the anticodon—explains why ochre suppressors also recognise amber codons.

Suppressors which insert tyrosine in response to ochre (UAA) although not to amber (UAG) codons have been isolated in yeast by Gilmore, Stewart and Sherman (1971); and Sherman et al. (1973) isolated suppressors which insert tyrosine in response to amber codons. Although there is no evidence on the molecular nature of this suppression, it is reasonable to speculate that these suppressor strains possess changed tRNAs; however, because the ochre suppressors do not follow the predictions of the wobble hypothesis, suppression of nonsense codons in yeast is not identical to nonsense suppression in E.coli.

Suppressor Mutations in the Anticodon

Suppressor tRNAs with changed codon recognitions might result from either of two classes of mutation. The *su* gene might code for a modifying enzyme—for example a methylase—so that the mutant enzyme activity fails to make some modification essential for proper codon recognition. Such mutants should be recessive. (It is of course possible that mutation in an activating enzyme might alter rather than abolish its activity, but this is less likely to occur.) Alternatively, the *su* locus may comprise a structural gene for the tRNA, in which case the mutation usually directly changes one of the bases in the anticodon. These mutants should be dominant. The existence of allelic amber and ochre suppressors which may be converted by single base changes suggests that these *su* loci at least represent structural genes.

It is difficult to isolate suppressor tRNAs from cells in amounts sufficient for sequencing because they are usually derived from minor tRNA species which are only synthesised to a small extent. The *su*3 suppressor locus, however, is located close to the attachment site of phage ϕ80; Smith et al. (1966) therefore used lysogenization of prophage ϕ80 followed by induction to obtain transducing phages; these ϕ80d*su*3 phages are defective (indicated by the "d") because they contain certain bacterial genes—including the *su*3 locus—in place of phage functions which are needed to complete the reproductive cycle of the phage.

Two genes which code for the *su*3$^-$ species of tRNA are located close together —probably adjacent—on the bacterial chromosome; both these genes are gained by the transducing phage. Bacteria with the *su*3$^+$ characteristic are mutated in one of these genes but retain a wild type *su*3$^-$ copy of the other; the structure of the phages prepared from *su*3$^+$ bacteria is therefore ϕ80d*su*3$^+$-*su*3$^-$, although—when bacteria were thought to contain only one locus coding for *su*3 tRNA—it was originally designated as ϕ80d*su*3$^+$.

When the ϕ80d*su*3$^+$*su*3$^-$ phage infects bacteria, large amounts of suppressor su3$^+$ tRNA and wild type su3$^-$ tRNA are synthesised from the two bacterial genes which it carries. When cells are infected with the phage, lysis is delayed because it is defective in late functions but the reproductive cycle can be even further prolonged by the addition of chloramphenicol to block protein synthesis. Transfer RNA can then be extracted from the infected cells with phenol. Comparison of the accepting levels of tRNA preparations for tyrosine and other amino acids in cells infected with ϕ80d*su*3$^+$*su*3$^-$ shows that the relative proportion of tyrosine tRNA is greatly increased by infection whereas the accepting capacity for other amino acids remains unaltered. In cells infected with ϕ80d*su*3$^+$*su*3$^-$, the tyrosine accepting tRNA responds both to its usual codons UAU and UAC and also to the amber triplet UAG. When similar experiments are performed with the phage ϕ80d*su*3$^-$*su*3$^-$ prepared from *su*3$^-$ bacteria, the same increase in tyrosine accepting activity is found, but the extra tRNA molecules continue to respond only to the normal codons UAU and UAC.

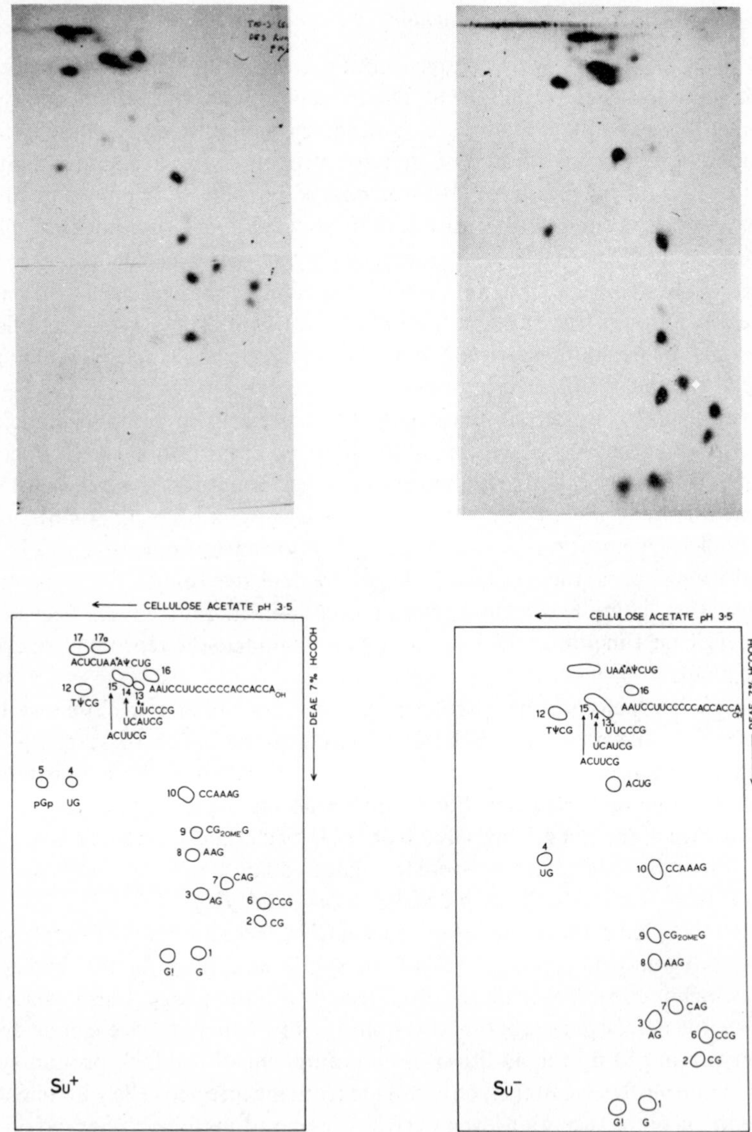

Figure 5.18: comparison of su$_3^+$ (left) and su$_3^-$ (right) tyrosine tRNAs of E.coli. Two dimensional electrophoresis separates the digest products of T1 ribonuclease; results are shown above and a drawing of the spots below. The molecules differ only in the oligonucleotide containing the anticodon sequence. The su$^+$ molecule has two spots 17 and 17a (representing different extents of modification of a base in the fragment) of sequence: ACUCUAA*AψCUG (anticodon underlined). The su$_3^-$ molecule has two spots instead, cleaved at the G of the anticodon: ACUG UAA*AψCUG. Data of Goodman et al. (1968)

The nucleotide sequences of the *su*3$^+$ and *su*3$^-$ tRNAs have been determined by using the large quantities of tRNA which can be extracted from cells infected with the ϕ80 transducing phages and Goodman et al. (1968, 1970b) found that the two species differ in only one fragment in their ribonuclease fingerprints, shown in figure 5.18. The wild type tRNA coded by *su*3$^-$ gene has the anticodon sequence GUA shown in figure 5.12, which recognises the codons UAU and UAC representing tyrosine; the suppressor tRNA coded by *su*3$^+$ has the mutant anticodon sequence CUA which recognises only the amber UAG triplet. These experiments provided the first proof of the molecular basis of suppression by mutant tRNAs and strongly suggest—although they do not formally prove—that the *su*3 locus is the structural gene for tyrosine tRNA. This conclusion has been confirmed by hybridization experiments which show that the transducing ϕ80 phages which carry *su*3 genes can anneal with tyrosine tRNA, whereas the DNA of phage ϕ80 itself cannot do so.

When a tRNA is mutated so that it recognises an amber or ochre nonsense triplet, it loses its ability to recognise its former codons. This implies that tRNAs which can be mutated to suppressors must be dispensable; for unless the cell retains the ability to recognise the codons to which the mutated tRNAs formerly responded, the mutation should be lethal by virtue of preventing synthesis of any proteins which rely upon the translation of these codon(s). In other words, tRNAs which are essential to the cell, because they provide the only capacity to translate some codons, cannot be mutated.

Species of tRNA for only three—serine, glutamine and tyrosine—of the seven amino acids related to the ochre and amber triplets by one base change were at first identified as suppressors; leucine and lysine have since also been identified, although they comprise less efficient suppressor tRNAs, but glutamic acid and tryptophan (which is related only to the amber codon) have not been found as suppressors. This implies that there may be only one transfer RNA each for glutamic acid and tryptophan (see also below) whereas there must be multiple tRNAs which respond to the codons for the suppressor amino acids, only one of which is mutated in each case.

The su_3^+ mutation which yields amber suppressor tRNA does not deprive the cell of the ability to recognise the codons for tyrosine, for it identifies only one of three genes which code for tyrosine tRNA. That the genome of E.coli must contain at least two genes for tyrosine tRNA is implied by the presence of two species of tRNA which differ slightly in nucleotide sequence (see figure 5.12). These two molecules are synthesized in the proportions 60:40 and it is the minor species which is changed by the su_3 mutation. But only a part of the minor tRNA in su_3^+ cells has the suppressor characteristic; most of the mole-ecules remain responsive to the usual tyrosine codons. This suggests that (at least) two identical genes must code for the minor tRNA, only one of which is mutated in suppressor strains.

The minor tRNA (species I) appears to be the product of two genes, both of which lie close to the phage $\phi 80$ attachment site and can be placed on the phage; these genes are probably adjacent. Russell et al. (1971) confirmed this model by testing the predictions which it makes for the genomes which should be produced by unequal recombination between the two adjacent genes. The three species of tyrosine tRNA found in su_3^+ cells are synthesized in the proportions 60:25:15, the last representing the suppressor mutant. (Of course, the major tRNA species II respresenting 60% of the tyrosine accepting capacity of the cell may be coded by more than one gene.)

When a cell is made diploid for a tRNA gene, the objection to its mutation must be overcome, for even if one copy is mutated to yield a suppressor, the other continues to specify wild type tRNA which can recognise the codons for its amino acid. Such mutations should be recessive-lethals, for if the wild type gene is lost the cell loses its capacity to respond to some codon(s) and must die. Using this technique, Soll and Berg (1969a, 1969b) found two amber suppressors, *su*7 of high efficiency which inserts glutamine, and *su*8 which is of much lower efficiency. Miller and Roth (1971) have found recessive-lethal UAG and UGA suppressors in Salmonella typhimurium by this means. The isolation of a glutamine recessive-lethal suppressor is curious, for we already know that glutamine tRNAs are redundant because of the isolation of the *su*2 suppressor. The recessive-lethal glutamine tRNA may perhaps have some unusual importance for the cell; or it may provide most of the capacity to recognise CAG whereas the *su*2 suppressor tRNA may be only a minor species whose mutation has less effect.

Only the tyrosine suppressor tRNA is related to the amber and ochre triplets by a change in the third base position. The other suppressors are related by changes in the first base (glutamine and lysine) or by second base changes (leucine and serine); suppressor tRNAs derived by single base mutations from these tRNAs must therefore retain their usual pattern of third base-recognition. All these four amino acids are specified by codons which may have as their third base either A or G. According to the wobble hypothesis, a U in the anticodon could recognise either base; but in this situation, the suppressor would be compelled to recognise both ochre (UAA) and amber (UAG) codons. A suppressor which recognises only amber codons must have C as the first base of its anticodon so that only G is recognised in the third position of the codon.

The existence of the serine, glutamine, leucine and lysine amber suppressors therefore implies that these amino acids must be represented by at least two tRNA species; one which responds only to the codon ending in G (and can be mutated to give an amber suppressor) and one which responds to both A and G (because without special modification it is impossible to respond to A alone; only glutamine is represented by such a molecule). Since tRNAs responding to the third base G alone have not been isolated from wild type E.coli, they may comprise minor species whose existence is revealed only by the nonsense suppressor mutation.

Suppression of UGA Codons

Two suppressors have been found which act upon UGA triplets. Because only tRNAs especially modified can recognise a third base A alone, they must respond also to other codons. According to the wobble hypothesis, the anticodon UCA should recognise the tryptophan codon UGG as well as the UGA nonsense triplet; and the anticodon ICA must recognise the cysteine codons UGU and UGC as well as UGA. This means that it may be possible to mutate only the tRNAs coding for tryptophan and cysteine to suppress UGA codons; for mutation of any other tRNA to create a UGA suppressor would mean that either the tryptophan or the cysteine codons would become ambiguous—although they would be read with their proper meaning by their normal tRNAs, they must also be recognised (falsely) by the suppressor tRNA. Such mutations would probably be lethal. Mutation of the cysteine or tryptophan tRNAs, however, would allow suppression of UGA; although the tRNA should continue to recognise its former codons, unlike the amber and ochre suppressors which lose this ability.

Both the UGA suppressors, *su*9 and *su*-UGA, insert tryptophan in response to the nonsense codon. Both also respond to the normal tryptophan codon, UGG. The failure to find an amber UAG suppressor which inserts tryptophan—which could be achieved by single step mutation of the tryptophan anticodon CCA to the suppressor CUA—implies that this tRNA may be the only transfer molecule in E.coli which can respond to the tryptophan codon UGG. Its mutation to recognise both UGG and UGA, however, does not seem to impair the ability of the cell to insert tryptophan into proteins.

We might expect that the anticodon CCA of the tryptophan tRNA should be mutated to UCA in the suppressor; this confers the ability to recognise UGA as well as UGG. This may be the cause of the *su*9 suppressor mutation found by Chan and Garen (1970). But by sequencing fragments of the *su*-UGA suppressor tRNA, Hirsh (1971) found that its mutation is not located in the anticodon of the molecule but lies at position 24 in the stem of the dihydrouridine loop; the su^- tRNA has the structure G24 whereas the su^+ molecule has instead A24. The anticodon sequence remains CCA.

This implies that the UGA triplet is read by a CCA anticodon, which must demand pairing between A and C in the third position. This pairing need not be very efficient, however, to explain the suppressor activity of the molecule and it may be much less active in recognising UGA triplets than its normal UGG triplets. That the suppressor tRNA can indeed recognise UGA triplets has been confirmed by Hirsh and Gold (1971), who characterized its suppressor activity in a cell free protein synthetic system. The base change from G24 to A24 changes the structure of the tryptophan tRNA molecules by replacing a G–U base pair in the dihydrouridine stem with an A–U pair; this increases the stability of the helix and must in turn influence the conformation of the anticodon loop.

Strains of E.coli which have UGA mutations are usually leaky; they can suppress this nonsense mutation, to some extent at least, without a special suppressor. Model, Webster and Zinder (1969), for example, found that whereas a UAG mutation in the RNA replicase gene of phage f2 completely prevents the synthesis of replicase in vitro, a UGA mutation in this gene allows 20% of wild type synthesis using extracts from an su^- strain. This suggests that the su^- tryptophan tRNA may respond to UGA codons, although weakly, as well as to its usual UGG codon; the mutation to create the *su*-UGA suppressor may then enhance this ability rather than create it anew. Such suppression of UGA may be common, for Roth (1970) noted that UGA mutants are found only rarely in the histidine operon of S.typhimurium, perhaps because they are often, if not always, read as sense. Ferretti (1971) found that several strains of Salmonella contain weak UGA suppressors.

Recognition of Natural Chain Termination Signals by Mutant tRNAs

When a mutant tRNA gains the ability to recognise nonsense triplets it may act as a suppressor of nonsense mutations; but it must also recognise other nonsense codons where they are used as signals for the termination of protein synthesis. Nonsense suppressor tRNAs may therefore cause new damage to the cell by preventing natural termination signals from being read as such. The same argument applies to missense suppressors; although a tRNA which recognises the "wrong" codon may suppress a mutation at one site, this pattern of recognition must create misreading at other sites where the codon which it recognises is intended to specify its normal amino acid.

Suppressors which have a widespread effect of this nature would be lethal to the cell by introducing so many errors in translation that many protein functions would be inhibited. This explains why strong missense suppressors cannot be isolated. Strong suppressors for both amber (UAG) and UGA nonsense triplets can be isolated; this may mean that the cell does not usually use these codons to terminate protein synthesis. As table 5.2 reveals, suppressors for ochre (UAA) mutations are much weaker and no strong ochre suppressors have been isolated. This suggests that the cell may rely upon ochre (UAA) signals to terminate protein synthesis.

Although this explanation for the difference between the ochre suppressors and the amber and UGA suppressors is attractive, it cannot completely account for their relationship. Ohlsson, Strigini and Beckwith (1968) have shown that allelic amber and ochre suppressors—which differ in only one base in the anticodon of tRNA—have very different efficiencies even when they act upon the same mutation; in such instances, the difference in efficiencies of suppression can be due only to a lower ability of the ochre suppressor to recognise its codon. For example, Strigini and Gorini (1970) found that *su*F suppresses an amber nonsense mutant in the lactose operon at 2·5% efficiency; *su*2 acts on the same mutant with 24·2% efficiency. Similarly, *su*G and *su*3 have suppression efficiencies at this site of 4·1% and 40·0%.

The absolute efficiencies vary, for they are different when measured in the argF locus, but the ochre suppressors remain less efficient than the amber suppressors, even when the same amber mutant is involved. That the ochre and amber suppressors are allelic eliminates the possibility that there may be different amounts of the two types of suppressor tRNA; and that they suppress the same mutation excludes any effects of neighbouring base sequences or the action of the termination factors.

It is, of course, possible that the cell does not rely upon nonsense codons alone; the protein termination factors might read nonsense triplets with greater efficiency within the context of some longer signal, so that the cell can discriminate between nonsense mutations and genuine terminators (see page 114). Such mechanisms would weaken the argument that amber (UAG) and UGA codons cannot be used as termination signals because they are suppressed strongly when found as mutations. However, whatever the absolute extent of use of the different signals, the difference between ochre suppressors and amber and UGA suppressors implies that the cell is likely to place much greater reliance, if not sole dependence, on the UAA sequence.

Missense Suppression by Mutant tRNAs

Missense suppression is more difficult to detect than nonsense suppression and only three missense suppressors have been studied in detail in E.coli. All involve mutations in glycine tRNAs; glycine is represented by four codons of the form GGX—the last base is irrelevant—and there are three species of glycine tRNA. The glycine missense suppressors have been isolated by their ability to suppress a mutation in the tryptophan synthetase A protein which replaces the glycine present at position 210 with arginine (coded by AG_{G}^{A}); the mutation changes a GGA triplet into AGA and the subsequent amino acid substitution inactivates the enzyme.

Missense suppressors restore activity to the enzyme by inserting glycine in response to the mutant codon. Hill, Squires and Carbon (1970) found that these suppressors map at one of two loci, *gly*T and *gly*U. Suppressor cells of either class contain tRNA molecules which can respond to the mutant AGA codon by inserting glycine instead of arginine. One assay for the suppressor tRNAs in vitro is to test their response to the alternating polymer poly-AG, which usually codes for the sequence arg–glu–arg–glu–. Suppressor cells contain tRNA which can insert glycine into this polypeptide chain in response to the triplet AGA.

Fractionation on benzoylated DEAE-cellulose resolves glycine tRNA into three species, with the coding responses shown in table 5.3. *Gly*U su^+ cells lack glycine $tRNA_1$ and contain instead a suppressor tRNA; the synthesis of suppressor depends directly upon the number of su^+ *gly*U genes in the cell, which suggests that *gly*U is the structural gene for glycine $tRNA_1$. Su^+ mutants of *gly*T lack glycine $tRNA_2$ and instead possess a suppressor tRNA. The amount of this suppressor depends on the number of su^+ *gly*T alleles in the cell, so that *gly*T appears to be the structural gene for glycine $tRNA_2$.

Normal cells possess one copy of the *gly*U and one copy of the *gly*T gene. The $tRNA_1$ coded by *gly*U is dispensable, for the ability to recognise the glycine GGG codons is also provided by the glycine $tRNA_2$ molecule. Cells in which *gly*U has been mutated from the wild type su^- to the suppressor su^+ therefore retain the ability to insert glycine into proteins, although they will, of course,

Table 5.3: codons recognised by the three glycine tRNAs of E.coli and their suppressor mutants

wild type alleles				mutant suppressor alleles		
codons recognised	anticodon	tRNA	locus	tRNA	anticodon	codons recognised
GGG	CCC	gly_1,su^-	*gly*U	gly_1,su^+	UCU	AG^A_G (arg)
GG^A_G	UCC	gly_2,su^-	*gly*T	gly_2,su^+	UCU	AG^A_G (arg)
GG^U_C	GCC	gly_3,su^-	*gly*V	gly_{ins}	UCC	GG^A_G (gly)
GG^U_C	GCC	gly_3,su^-	unknown	none		

The gly_3,su^- and gly_{ins} tRNA molecules have been sequenced; the other anticodon sequences are based on the codons recognised by the tRNAs. Data of Carbon, Squires and Hill (1970) and Squires and Carbon (1971).

on occasion place glycine where AGA or AGG codons demand the insertion of arginine. It is not possible to recognise a third base A alone—relying upon the usual four bases, to which anticodon sequences are probably restricted in mutant tRNAs—so that the suppressor must recognise both the arginine codons; this demands the anticodon UCU in the suppressor tRNA. However, as table 5.3 shows, this would demand more than one base change in the anticodon sequence of the tRNA. *Su*$^+$ mutants at the *gly*U locus are, indeed, rare as this double transition suggests and can be isolated readily only after irradiation with ultraviolet light; their nature is not clear.

The su^+ mutation of the *gly*T locus to change the coding properties of glycine $tRNA_2$ can be accounted for by a single change in the anticodon of the tRNA, from UCC to UCU. Mutagenesis with hydroxylamine, which causes transitions in DNA from G–C base pairs to A–U base pairs, increases the frequency of occurrence of suppressors at the *gly*T locus by more than one hundred fold compared with the spontaneous rate. This supports the idea that this mutation is due the substitution of one base in the tRNA.

Suppressor mutations in glycine $tRNA_2$ abolish the ability of the cell to respond to the GGA codon, although glycine $tRNA_1$ allows recognition of

GGG as glycine. Carbon, Squires and Hill (1970) found that su^+ mutants in *gly*T retain only 8% residual activity in recognising GGA; the origin of this recognition is unknown. Loss of the ability to recognise GGA confers a general disadvantage on the cell which is relieved when a normal, su^- *gly*T allele is introduced. The *gly*T suppressor is therefore closely analogous to the recessive—lethal nonsense suppressors, except that it is recessive-deleterious rather than lethal. Its failure to be lethal is presumably due to the residual GGA recognition.

The ability of the cell to recognise GGA may be restored by a second mutation, at a locus *gly*V, which converts about one third of the glycine $tRNA_3$ molecules into a form, $tRNA^{gly}_{ins}$, which responds to the codons GGA and GGG—although with only about a fifth of the efficiency of the glycine $tRNA_2$, in vitro at least. (The sequences of the $glycine_3$ and $glycine_{ins}$ tRNAs are shown in figure 5.17). At least two genes must therefore code for glycine $tRNA_3$, only one of which (*gly*V) is mutated to restore recognition of GGA; Squires and Carbon (1971) showed that this mutation converts the GCC anti-codon of $tRNA^{gly}_3$ into the UCC sequence expected of wild type glycine $tRNA_2$. Because most of the glycine $tRNA_3$ molecules remain unaltered, the cell retains its normal capacity to translate the GGU and GGC codons. When cells are made diploid for the region in which *gly*V maps, their synthesis of $tRNA^{gly}_3$ increases some three fold; this suggests that this locus is the structural gene for some of the glycine $tRNA_3$ molecules.

The two missense tRNAs which recognise arginine codons in place of glycine codons do not in themselves appear to be deleterious to the cell, perhaps because their efficiencies may be low and they must compete with the arginine tRNA itself to place glycine in codons demanding arginine. The loss of ability of the cell to respond to one of its glycine codons, however, confers the severe pleiotropic inhibition of cell functions expected of recessive-lethal (or recessive-deleterious) mutations. The glycine $tRNA_{ins}$ species can be considered in the same class as the other two missense suppressors, for it too recognises a different codon from wild type; because this codon also specifies glycine, however, no misreading occurs. Nonetheless, GGA and GGG codons in su^+ *gly*T, *gly*V_{ins} cells must be read by an aminoacyl-tRNA species different from that utilised in wild type cells.

Suppression of Frameshift Mutations

Frameshift mutations are created in DNA by certain mutagens which cause the insertion or deletion of an extra base pair in DNA; this changes the reading frame of the message so that triplets are translated in the wrong sets beyond the mutation (see page 19). When Riddle and Roth (1970) examined frameshift mutants of the histidine operon of Salmonella typhimurium, they were able to isolate external suppressor mutations which appear to restore the correct reading frame; these suppressors do not act upon nonsense codons nor do they appear to be of the type which substitute one amino acid for another. Most of the mutants which are suppressed are in the single insertion class.

The suppressors and mutants fall into two groups. Each suppressor acts on some or all of the mutants in one group, but does not act at all on any of the mutants in the other group. The efficiencies of suppression vary from 1% to 15%; Riddle and Roth (1972a) have characterized these mutations into the six classes shown in table 5.4. *SufA*, *sufB* and *sufC* comprise one group; *sufD*, *sufE* and *sufF* constitute the other. The *sufD* mutants include three phenotypic classes and the *sufE* class includes two phenotypic categories; these may be different alleles at the one locus or due to mutations in closely linked genes.

Table 5.4: frameshift suppressors of Salmonella typhimurium

locus	% suppression enzyme synthesis	efficiency polarity	genetic type	tRNA altered
sufA	2	5	dominant	proline 1
sufB	1	5	dominant	proline 2
sufC	2	3	—	—
sufD	11	15	dominant	glycine 1
sufE	2	1	dominant	—
sufF	3	10	recessive	glycine 2

Suppression efficiencies have been measured either by the extent permitted of enzyme synthesis or by the influence of the suppressor mutation on the synthesis of subsequent proteins of the histidine operon (polarity measure). The dominant loci appear to the structural genes for tRNAs; the recessive locus may specify a modifying enzyme Changes in the properties of tRNAs have been detected by their elution from benzoylated DEAE-cellulose columns. Data of Riddle and Roth (1970, 1972a, 1972b).

Frameshift mutants seem to be particularly prone to occur in sequences of repeated bases (which may perhaps slightly distort the structure of the double helix). The suppressors might therefore lead to the production of tRNAs which correct the reading phase. Two of the frameshift mutants suppressed by *sufA*, *sufB* and *sufC* insert an extra C residue into a sequence of repeated cytosines. *SufB* suppressors insert a proline residue (CCC codes for proline) at the point where the reading frame becomes normal. By using columns of benzoylated DEAE-cellulose to separate the proline tRNAs of Salmonella, Riddle and Roth (1972b) have identified three fractions. Proline $tRNA_1$ is missing from *sufA* strains, which contain instead a new peak of elution. *SufB* mutations appear to influence a minor proline tRNA, $tRNA_2^{pro}$.

The *sufD* mutation maps at a locus corresponding to the *glyU* site of E.coli. *SufD* mutations change the elution position of glycine $tRNA_1$ of Salmonella; the loci *glyU* in E.coli and *sufD* in Salmonella therefore seem to direct the synthesis of homologous tRNAs. The *sufF* mutation affects synthesis of some of the glycine $tRNA_2$ molecules—the peak of glycine $tRNA_2$ is decreased in size and a new peak appears instead. Mutations *sufA*, *sufB*, *sufD* and *sufE* are all dominant, which suggests that they are the structural genes for proline and glycine tRNA molecules. *SufF* is recessive and this, together with its partial rather than complete influence on glycine $tRNA_2$, suggests that it codes for a modifying enzyme which influences the coding response of the tRNA.

How can a transfer RNA alter the reading frame of the message? Because one of the codons for proline is CCC and one of the codons for glycine is GGG, a transfer RNA which recognises a sequence of four bases (CCCC or GGGG) instead of the usual triplet could correct the reading frame in a sequence of repeated bases. This model has been supported by the isolation by Riddle and Carbon (1973) of the glycine tRNA coded by *sufD*; the tRNA possesses an additional cytosine residue in its anticodon loop, suggesting that a new anticodon sequence of CCCC may be able to respond to the quadruplet GGGG. This tRNA can therefore suppress frameshift mutations created by the insertion of an additional G into a sequence of repeated guanines coding for glycine. A prediction of this model is that two further classes of frameshift suppressors should exist, one concerning phenylalanine (codon UUU) and one concerning lysine tRNA (codon AAA).

Control of Transcription

CHAPTER 6

Transcription of Phage Genomes

Synthesis of RNA by RNA Polymerase

Transcription of Duplex DNA Templates

Transcription of DNA into RNA represents the first stage of expression of those genes which code for protein and the reaction responsible for production of the RNA molecules needed for protein synthesis. The enzyme RNA polymerase is implicated in all three stages of transcription: initiation, elongation and termination. One of its components, the aggregate of four polypeptide chains comprising the core enzyme, is sufficient for elongation; its other component, the single polypeptide chain of the sigma factor, is implicated only in initiation and is not associated with the core enzyme during elongation. Termination of the chain synthesized by the core enzyme demands the mediation of a protein, the rho factor, which is implicated only at this stage.

Both synthesis of proteins and replication of DNA demand the participation of many proteins; the ribosome represents the central component of the apparatus of protein synthesis and a membrane bound complex of many enzymes may define an apparatus necessary for replication. Transcription of RNA for many years appeared to be the responsibility of RNA polymerase alone; but the more recent discoveries of factors necessary for transcription suggest that a more complex apparatus may be necessary for initiation and for termination, although not for elongation.

Initiation of transcription demands the selection by RNA polymerase of specific binding sites on DNA. Identification of these sites is accomplished by the sigma factor, which prevents the enzyme from binding firmly to other regions of DNA. The ability of RNA polymerase to initiate transcription at certain loci of the bacterial chromosome, however, may be controlled by extraneous factors, whose interaction with the enzyme and DNA is not yet defined (see page 304). During phage infection, changes may be introduced in the structure of RNA polymerase or further factors may associate with it to alter its transcriptional specificity to that demanded by the phage DNA.

Termination of RNA synthesis does not appear to be an intrinsic property of the enzyme in vitro at low ionic strength but depends upon at least one extraneous protein, the rho factor, which causes transcription to cease at specific sites on DNA. However, a second class of site may be recognised by the enzyme alone in conditions of high ionic strength. Factor mediated termination may be used to control the selection of genes for transcription, since an interaction preventing the termination allows the polymerase to read the additional sequences of its template beyond its former site of termination. Anti-terminator proteins, which may inactivate the termination factor or prevent RNA polymerase from responding to it, are synthesized during phage infection and may represent additions to the transcription apparatus.

The interaction with DNA of core polymerase during chain elongation is not well defined but probably represents a fundamental synthetic activity of the cell which remains invariable irrespective of the sequences of DNA used as template. Site specific changes in the transcription apparatus may be confined to the stages of initiation and termination, so that the same mechanisms are responsible for the elongation on bacterial DNA of messenger RNA (which is subsequently translated into protein) and transfer and ribosomal RNAs. All RNA chains grow by the sequential addition of nucleoside triphosphate precursors in a reaction in which pyrophosphate is lost.

Each RNA transcript is an exact copy of one of the strands of its duplex DNA template. It is possible to construct models such as those of Stent (1958) and Zubay (1962) which allow RNA chains to grow in the wide groove of the double helix of DNA, the addition of bases being directed by base pairing between the added ribonucleotides and the base pairs of DNA. The most likely scheme for transcription, however, is to suppose that the two strands of the DNA duplex unwind so that the RNA chain can be formed by complementary base pairing with one of them. One unresolved problem of this model is the topology of DNA unwinding (this is also a problem in DNA replication). Although not yet directly proven by experiment, the assumption that template DNA unwinds for transcription is consistent with many data concerning transcription; and most models for the action of RNA polymerase are predicated upon this basis (see for example Krakow and Von der Helm, 1970).

Complexes comprising the DNA template, RNA polymerase enzyme and nascent RNA have been isolated from bacterial cells; and it is clear that the transcription reaction does not proceed through formation of a long hybrid RNA-DNA such as results from hybridization (Chamberlin and Berg, 1962; Bremer and Konrad, 1964). Rather does it seem that transcription involves a *local* unwinding of the DNA duplex just at the head of the RNA under synthesis. The simplest explanation for the behaviour of DNA—RNA polymerase—nascent RNA complexes is to suppose that this region moves along DNA as synthesis proceeds.

Complexes obtained in different preparations have different physicochemical properties, but part of the nascent RNA is usually resistant to degradation with ribonuclease. (The extent of protection of DNA from attack by

DNAase seems to depend much more heavily on the preparation.) When Hayashi and Hayashi (1968) examined the RNAase resistant part of the RNA bound in a complex with ϕX174 duplex DNA, they found that the size of the RNA—measured by its release with formamide after different periods of incubation—increases as transcription proceeds, but that the extent of the RNAase resistant region remains constant.

Chasing a radioactive label through RNA shows that the resistant region becomes sensitive to RNAase attack as transcription proceeds; this indicates that it is located at the growing point of the RNA chain (and not at the tail where transcription was initiated). A resistant region of about fifty nucleotides therefore seems to move along the DNA template as transcription proceeds. Using a complex obtained from E.coli, Tongur et al. (1968) found that the resistant RNA becomes sensitive when the DNA is degraded with DNAase. When the protein present is degraded by pronase, about half of the RNA becomes sensitive. This leaves a picture of RNA at the growing point protected in part by its association with DNA—most probably because it is hydrogen bonded to one strand of the duplex—and also, to some extent, by the surrounding protein.

The ability to unwind DNA for transcription must be intrinsic to RNA polymerase, for transcription can take place in purified in vitro systems without any special unwinding components. One general picture to view transcription is illustrated in figure 6.1. This attributes to RNA polymerase the functions both of unwinding the DNA duplex for transcription and of reforming it afterward, for RNA-DNA duplexes are stable and would presumably remain in this form unless their milieu dictates otherwise. Indeed, one of the difficulties in defining the mechanism of transcription is that the local environment of DNA engaged in RNA synthesis may depend heavily upon its interaction with polymerase.

That DNA engaged in RNA synthesis may exist in an altered state is suggested by electron microscopic visualization of T7 DNA under transcription. Phage T7 is a linear duplex of DNA of some 26×10^6 daltons, about 12–13 μ long; RNA polymerase of E.coli initiates transcription close to its left end and continues for a distance which depends upon the provision of termination factors (see Millette et al., 1970; Summers and Siegel, 1970; Dunn and Studier, 1973). Because this template is a reasonable size and its products have been characterized, it is especially suitable for *heteroduplex mapping*.

Heteroduplex mapping allows double stranded regions of DNA or DNA-RNA hybrids to be distinguished from those which are single stranded; as its name implies, the technique is usually used to characterize preparations of denatured DNA which have been renatured with either another preparation of DNA or one of RNA. Davis, Simon and Davidson (1971) have reviewed the two variations of the technique, which depend upon the medium in which the preparation is spread for electron microscopy. A protein, usually cytochrome

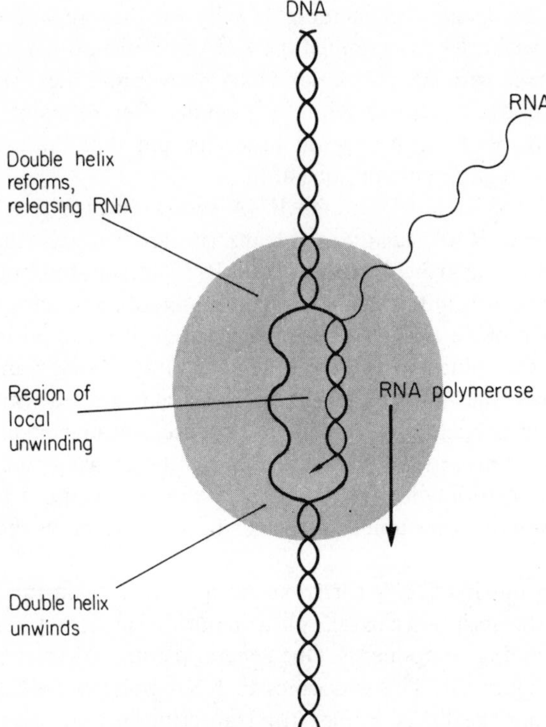

Figure 6.1: model for transcription

c, is included in the medium so that the structure which is visualized in the electron microscope comprises a column of basic protein collapsed around nucleic acid.

After aqueous spreading, duplex DNA or DNA–RNA appears as an elongated thread; single strands of nucleic acid, however, are condensed into bushes because of random base interactions. This distinguishes clearly between duplex and single strands and allows the position of the single strands to be established relative to the duplex, whose length can be accurately measured.

When the preparation is instead mounted in the presence of an appropriate concentration of formamide and salt, the duplex regions remain stable but the random interactions which generate the bushes of single stranded nucleic acid are prevented. Single stranded sequences therefore also appear as elongated threads, although thinner and more twisted than the duplex sequences. Both single stranded and duplex contour lengths can then be measured.

When two closely related but not entirely homologous DNA molecules are denatured and renatured together, the heteroduplex products of cross-renaturation are double stranded in the complementary regions but fail to anneal in regions of non-homology. (The homoduplex products of self-

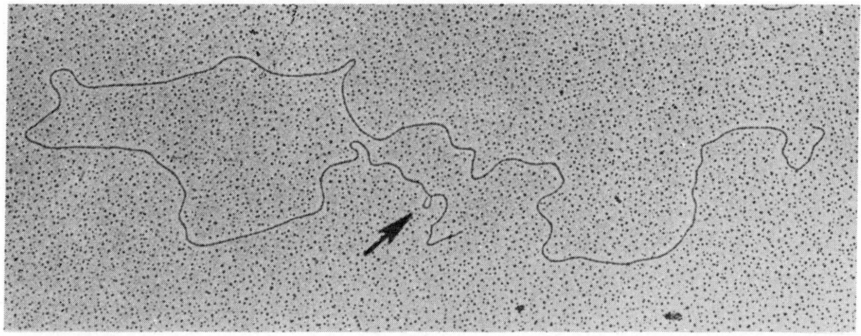

Figure 6.2: heteroduplex formed between complementary strands of T7 and T3 DNA and spread in formamide. The arrow identifies a region in one phage which has been deleted in the other, so that it generates a single stranded loop. Data of Davis and Hyman (1971)

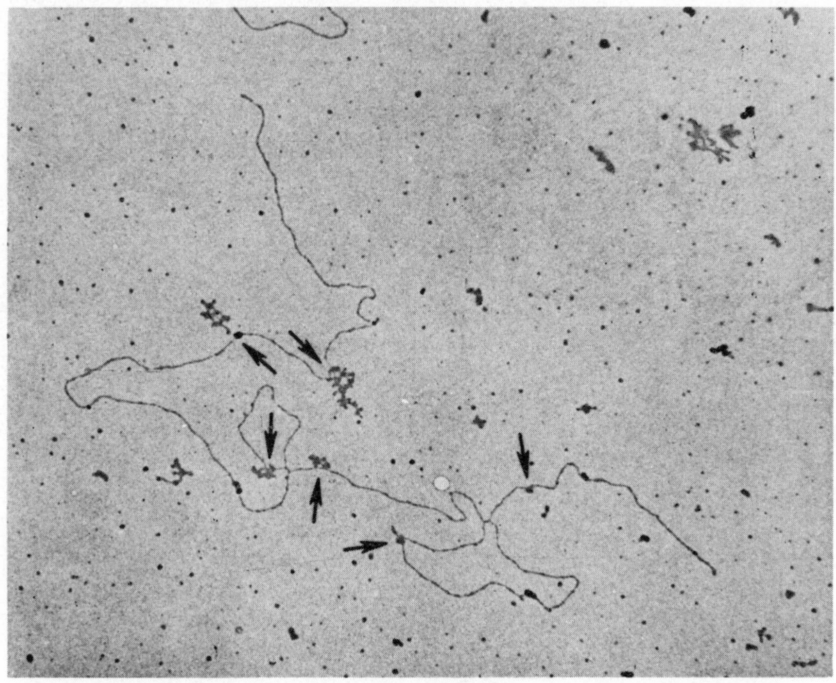

Figure 6.3: aqueous spreading of T7 DNA engaged in transcription. The bushes marked by arrows indicate aggregates of growing RNA chains, which increase in size proceeding from one end of T7 DNA. Data of Davis and Hyman (1970)

renaturation should comprise perfect duplex molecules.) Figure 6.2 shows formamide spreading of the heteroduplex formed between one strand of phage T7 DNA and the complementary strand of phage T3 DNA. Davis and Hyman (1971) found that the two phages differ in a deletion near to one end which is revealed by the single stranded loop marked by the arrow; only this short sequence in one strand cannot find a complement with which to anneal in the other strand.

Native T7 DNA engaged in transcription can be mounted for electron microscopy in the same way. Using the aqueous spreading technique, Davis and Hyman (1970) showed that when only a very short time is allowed for transcription, the condensed bushes of RNA are located close to one end of the template. This maps the sites of initiation. Figure 6.3 shows T7 DNA which has been allowed to continue transcription for 30 minutes; although RNA polymerases are not visible, the position of each enzyme is marked by the condensed bush of its product. The bushes grow progressively larger from one end to the other of the template, consistent with the increase in the length of RNA as the polymerase continues.

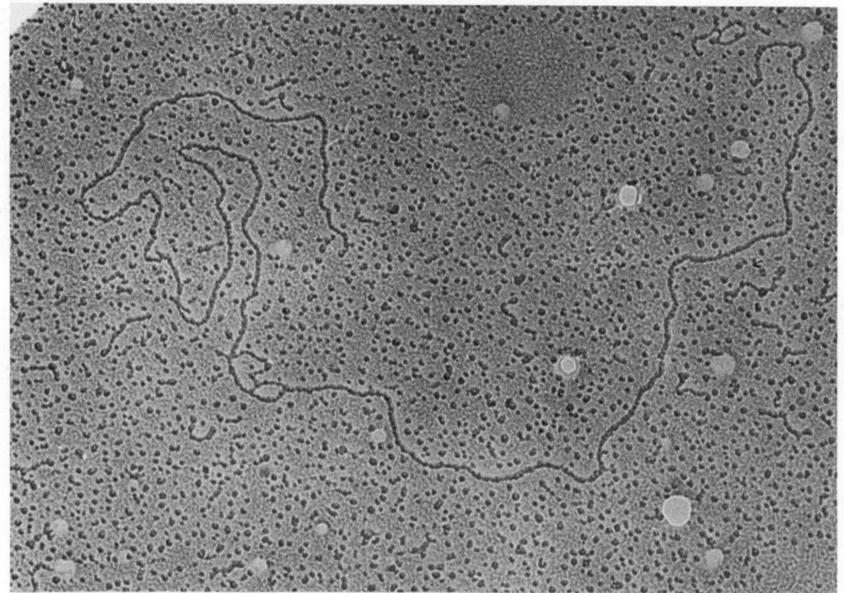

Figure 6.4: transcription bubble formed in visualization of T7 DNA engaged in RNA transcription. After 2.5 minutes of RNA synthesis, the template was spread in formamide. The arrow indicates a transcription bubble with a short attached RNA chain (much of the RNA chain is probably lost by breakage during preparation). Magnification × 50,000. Photograph kindly provided by Dr. M. D. Bick

In a variation of the formamide spreading technique, Wolfson, Dressler and Magazin (1972) found that it is possible to construct a denaturation map of T7 DNA. When exposed to increasing concentrations of formamide, certain restricted regions of the molecule progressively become denatured to single strands. These sequences may identify regions rich in A–T base pairs.

By applying the formamide spreading technique to T7 DNA engaged in transcription, Bick, Lee and Thomas (1972) observed that sequences close to the polymerase are denatured very readily. Concentrations of formamide well below those needed for denaturation mapping cause the strands of DNA to separate to form *transcription bubbles.* Growing RNA chains extend from the duplex template as single strands. Figure 6.4 shows a transcription bubble accompanied by a growing RNA chain.

Although RNA polymerase cannot be located within the bubble, the most likely model to explain its production is to suppose that DNA is locally unwound in a short region within or close to the polymerase, probably located at the head of the bubble. The bubble may be generated by a transient interaction between the growing RNA and the DNA immediately behind the polymerase. Of course, the presence of these structures in preparations spread in formamide does not imply that they exist in solution or in the cell. However, the effect of the RNA chain upon the structure of DNA in its immediate environment is consistent with the idea that the local milieu of DNA within and close to the polymerase may be different from that prevailing elsewhere, in part because of the protein itself and in part because of the presence of the polynucleotide chain of RNA.

Separation of RNA Polymerase into Core Enzyme and Sigma Factor

A single RNA polymerase enzyme species undertakes synthesis of all RNA in bacteria. Eucaryotic cells contain at least two distinct enzymes, one located in the nucleoplasm and the other in the nucleolus; and in addition to these and any other nuclear RNA polymerases also possess a mitochondrial RNA polymerase (see chapter six of volume two). All RNA polymerase activities catalyse the synthesis of RNA directed by a DNA template—RNA replicase enzymes may synthesize RNA under direction from an RNA phage template—drawing upon nucleoside triphosphates as precursors and cleaving pyrophosphate with each addition to the growing chain.

Bacterial RNA polymerase has been purified and its catalytic activities analysed in detail. The ionic strength of the incubation medium is a critical parameter of enzyme activity, for E.coli RNA polymerase exists as a dimer, or greater aggregate, at low ionic strength but dissociates into monomers when the ionic concentration is raised (for review see Richardson, 1969). After an initial period of confusion, it is now clear that the monomer is the active form of the enzyme, which Travers and Burgess (1969) found to comprise four types of polypeptide chain. The total weight of this form of the enzyme, which sediments at 13–14S, is 495,000 daltons organized in the form:

subunit	number in enzyme	molecular weight
α	2	40,000
β	1	155,000
β'	1	165,000
σ	1	95,000

In medium of low ionic strength, each enzyme molecule can initiate synthesis of only one chain of RNA; thus measurement of the number of RNA chains initiated by a known amount of enzyme gives the molecular weight of the active enzyme molecule involved in initiation. This proves to be 480,000—in good enough agreement with the total of the $\alpha_2\beta\beta'\sigma$ molecule.

The growth of an RNA chain proceeds from 5′ phosphate to 3′ hydroxyl terminus and Maitra et al. (1967) were able to show that the first nucleoside triphosphate incorporated retains all its phosphate groups (α, β and γ) during the subsequent chain elongation. By contrast, only the α phosphate is retained by nucleotides incorporated internally. The incorporation of a γ-P^{32} radioactive label into DNA can therefore be used to follow the chain initiation reaction, whilst C^{14} or an α-P^{32} label follows the overall synthesis of RNA. RNA chains are preferentially initiated with purine triphosphates (ATP and GTP) when RNA polymerase from E.coli is used in vitro and the general characteristic of the reaction seems to be the same when the enzyme from Azotobacter vinelandii is employed instead.

RNA polymerase of E.coli has been separated into two components by Burgess et al (1969) who found that chromatography on phosphocellulose removes the *sigma factor* (the σ polypeptide) from the *complete enzyme* ($\alpha_2\beta\beta'\sigma$) to give a *core enzyme* ($\alpha_2\beta\beta'$). With a heterologous template, such as calf thymus DNA, the core enzyme is as active as the complete enzyme. But when phage T4 DNA is provided as template, the sigma factor greatly enhances RNA synthesis, although it has no catalytic activity itself. This suggests that the core enzyme possesses the catalytic activity of synthesising phosphodiester bonds and that sigma plays some other role specific for the usual template.

The functions of the polypeptides of the core enzyme are not well defined. Mutants in the β subunit have been identified by selecting strains of E.coli resistant to the antibiotics rifampicin (which inhibits initiation) and streptolydigin (which inhibits elongation); both mutations reside in the same gene (see Iwakura et al., 1973). This implicates the β subunit in both the stages of initiation and elongation. Zillig et al. (1970) found that the β' subunit alone can bind heparin; because heparin inhibits transcription by competing with RNA polymerase for DNA, this suggests that β' may be involved in binding the enzyme to its template. The genes coding for the α, β' and σ subunits have not been identified.

Release of Sigma Factor by Core Enzyme at Initiation

Sigma plays no role in chain elongation for its action is confined to promoting initiation of new chains. (In a formal sense, it plays the same role for RNA polymerase core enzyme that initiation factors play for ribosomes.) That sigma functions in initiation was suggested by experiments in which Travers and Burgess (1969) followed the effect of the factor by measuring the incorporation of a γ-P^{32} label into RNA. Its overall stimulation of RNA synthesis is paralleled by an increase in initiation.

Two incubation protocols show that each molecule of sigma factor can be re-utilised several times by different molecules of core enzyme. When the complete enzyme is added to a reaction mixture at low ionic strength, a complex with T4 DNA is formed in which only one RNA chain is started for each complete enzyme molecule added. But when a large excess of core enzyme is added, there is a renewed burst of initiation. This means that the sigma factor originally present must have been re-used by the added core enzyme molecules many times to enable them to initiate new chains.

The second protocol uses the antibiotic rifampicin, which inhibits RNA synthesis by interfering with the ability of the core enyzme to initiate new chains, but does not inhibit the extension of chains already under synthesis. Rifampicin inhibits bacterial RNA polymerase in vitro or in vivo and does not inhibit any of the enzymes found in the cells of higher organisms. Rifampicin-resistant mutants of E.coli have an altered RNA polymerase and Di Mauro et al. (1969) found by using C^{14} derivatives of the antibiotic that it binds to normal RNA polymerase, but cannot do so when the enzyme is prepared from a resistant bacterial strain. Rifampicin acts on the β-polypeptide of the core enzyme, which has an altered electrophoretic mobility in strains of E.coli resistant to the antibiotic.

Using conditions of low ionic strength, the addition of rifampicin inhibits the burst of initiation which takes place when an excess of core enzyme is added to the complex of complete enzyme with T4 DNA. But when core enzyme used for the second round of initiation is prepared from a strain of E.coli resistant to rifampicin instead of from the usual sensitive cells, the burst of renewed initiation becomes possible again. This implies that the sigma present in the original complex between complete enzyme and T4 DNA can be used again with the second set of molecules of core enzyme.

Sigma factor might act in either of two ways to assist several molecules of core enzyme to initiate RNA synthesis. One possibility is that the sigma polypeptide remains attached to DNA and assists further core enzyme molecules to bind and initiate new RNA chains there. This model demands that the sigma polypeptide itself recognises DNA. An alternative is that the sigma

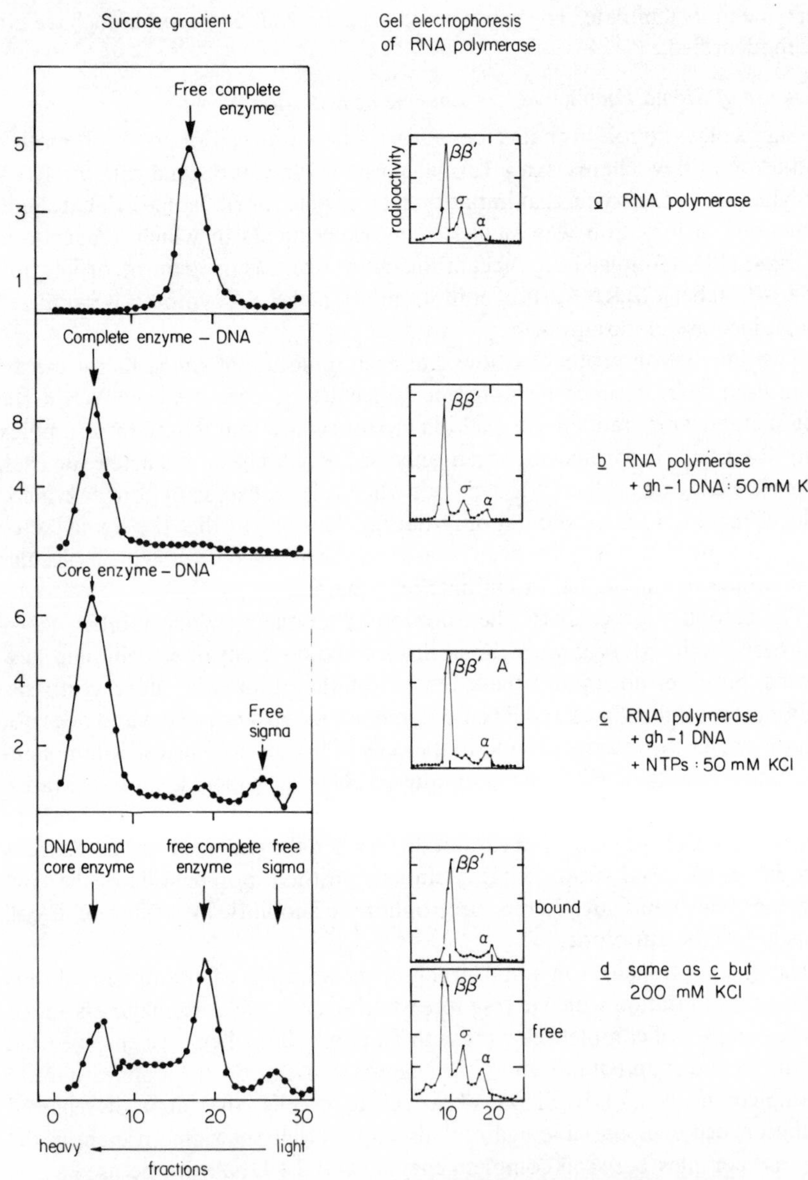

Figure 6.5: sigma is released from RNA polymerase when transcription is initiated on phage gh-1 DNA. Left: sucrose gradients of RNA polymerase of P. putida. Right: analysis on acrylamide gels of the enzyme recovered from the main peak of the sucrose gradient. The gel system resolves the enzyme into three peaks, one containing both β and β', one containing α, one corresponding to σ. (a) free RNA polymerase contains all its polypeptide subunits, (b) polymerase bound to DNA in 50 mM KCl retains all subunits, (c) addition of nucleoside triphosphates forms an initiation

factor may be released after initiation so that it is free to combine with another core enzyme molecule to yield a complete enzyme able to undertake initiation elsewhere. In this case, sigma might promote initiation either by binding to DNA directly or by an indirect action in which it enables the core enzyme to do so.

Whether sigma is released after initiation can be determined by making an initiation complex between complete enzyme and phage ϕ80 DNA and then adding an excess of T4 DNA together with extra core enzyme. If sigma is released at initiation it should combine with a new core enzyme and initiate synthesis on the T4 DNA which is in excess. But if sigma remains bound to the ϕ80 DNA, the new molecules of core enzyme should attach to the ϕ80 DNA through the sigma polypeptide so that only ϕ80 specific DNA is synthesised. Travers and Burgess (1969) found that some 83% of the RNA is complementary to the T4 DNA. Since specific RNA synthesis on phage T4 DNA demands the presence of sigma factor in these conditions, this implies that sigma must have been released from the ϕ80 template to which it was first bound and then re-used by further molecules of the core enzyme to transcribe T4 DNA.

This experiment shows that sigma factor is released from the core enzyme bound to ϕ80 DNA when an excess of free core enzyme is added, but does not reveal the stage of reaction at which the sigma polypeptide is released. Indeed, in such experiments it is possible that its release is caused by the excess of added core enzyme—the free core enzyme might compete for the sigma factor and so cause its release through an exchange reaction.

The constitution of RNA polymerase is similar in many bacteria. In Azotobacter vinelandii the sigma factor appears to have about the same affinity for core polymerase as in E.coli. In Pseudomonas putida, however, core enzyme and sigma factor are more tightly associated for they are not separated by elution on phosphocellulose. Gerard, Johnson and Boezi (1972) have used the interaction of the enzyme of P. putida with the DNA of its phage gh-1 to show that sigma is released from the complete polymerase—DNA complex only after transcription has been initiated.

When complete enzyme labelled with S^{35} is allowed to bind to gh-1 DNA, it sediments more rapidly through a sucrose gradient as shown by parts *a* and *b* of figure 6.5. The bound polymerase retains its sigma factor, for analysis on SDS acrylamide gels identifies all the polypeptides, $\alpha,\beta,\beta',\sigma$. Part *c* of the figure,

---

complex and sigma is released from enzyme to form peak at fractions 24–28 on gradient. Gel electrophoresis shows that the polymerase has lost about 60% of its sigma subunits. (d) some polymerase enzymes form initiation complex in 200 mM KCl and start transcription; those which are inactive remain free in gradient. Enzyme active on DNA completely lacks sigma; enzyme which remains free on gradient retains it. Data of Gerard, Johnson and Boezi (1972)

however, shows that about 60% of the sigma factor of the complete enzyme is released to form a separate band when nucleoside triphosphates are added to start transcription.

The most probable explanation for the fractional release of sigma is that only 60% of the RNA polymerase molecules commence transcription; the others may fail to transcribe RNA, although remaining bound to the DNA template. Complexes of RNA polymerase and DNA engaged in transcription can be distinguished from those which are inactive by incubation in a suitable

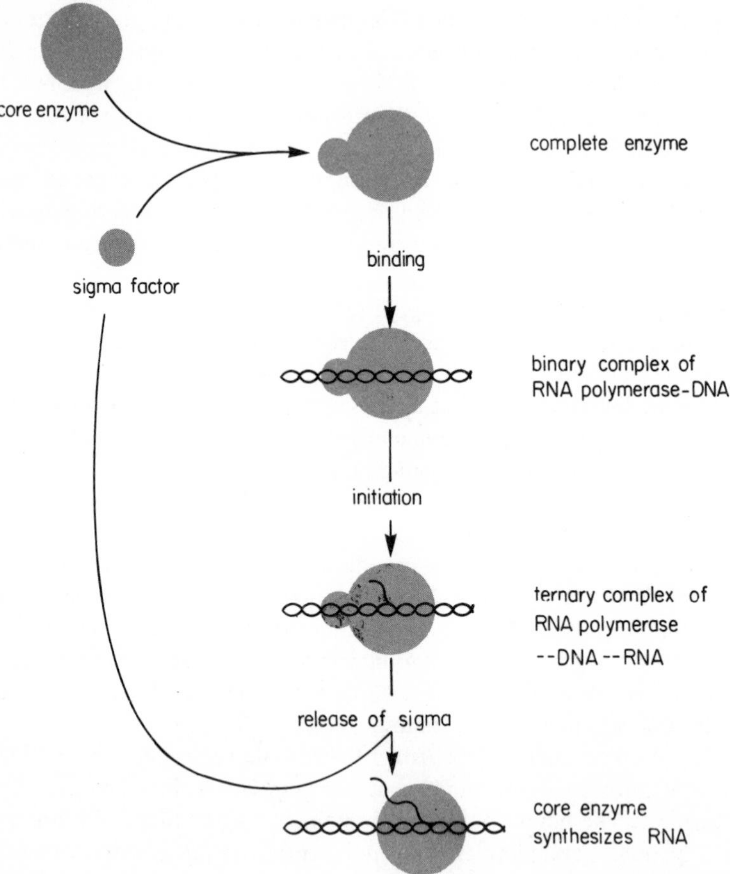

Figure 6.6: sigma factor cycle. A free sigma factor associates with core enzyme to yield a complete enzyme which binds to DNA to form the RNA polymerase–DNA binary complex. Nucleoside triphosphates are needed for initiation, forming a ternary complex of RNA polymerase–DNA–RNA. Sigma factor is released after initiation and the core enzyme continues to synthesize RNA. Because sigma factor has a high affinity for core enzyme, it probably remains free for only a short time before associating with another molecule of core enzyme

concentration of KCl. The affinity of RNA polymerase for DNA is reduced as the ionic strength of KCl is increased; at the 50 mM concentration used in part *c* of the figure, RNA polymerase bound to DNA—whether engaged in transcription or not—forms a stable complex. But if the concentration is raised to 200 mM, RNA polymerase can bind to DNA only when undertaking transcription. Part *d* of the figure shows that in these circumstances all polymerase molecules bound to DNA lack sigma whereas the inactive enzyme now free in solution retains it.

Sigma factor therefore remains part of the polymerase upon binding to DNA but is released whenever the enzyme engages in transcription. This suggests the cycle for the sigma factor shown in figure 6.6. Release of sigma from core enzyme must take place at an early stage of transcription and Krakow and Von der Helm (1970) suggested that in A. vinelandii it may occur when the nascent polynucleotide chain reaches a certain length, of the order of 10 bases. The sigma factor cycle is probably essentially similar in all bacteria, although it is of course possible that the precise step at which it is released may vary. One caution implicit in this formulation of the cycle, however, is that the release and re-use of sigma has been demonstrated only in vitro although it is presumably the same in vivo.

Binding Reaction and Initiation

Restriction of Strand Transcription by Sigma

Two interactions must take place to start RNA synthesis: polymerase must first bind to DNA and then initiate transcription. One of the most important questions about these interactions is whether they take place at the same point in the genome or whether two overlapping or separate sites are utilised. We may define the sequences of DNA at which the two reactions occur as the *binding site* and *initiation site*. Binding sites comprise specific sequences of base pairs which are recognised by the polymerase and provide the only loci at which enzyme and DNA can associate to form a stable complex.

Our original model for initiation was to suppose that RNA polymerase also commences transcription at this site. More recently, however, transcription of DNA phage templates has suggested that a second sequence, the initiation site, may be used by the enzyme to direct the start of RNA synthesis. We do not know whether initiation sites exist as independent entities in bacterial DNA or whether initiation takes place at the binding site. *Site* is used in this book to describe a length of DNA whose base pair sequence is specifically recognised by proteins, contrasted with *genes* which code for proteins or RNA and must therefore be transcribed to fulfill their function.

Sites utilised by RNA polymerase may be defined by either genetic or biochemical studies. Genetic definition of mutants which fail to transcribe certain bacterial or phage genes has identified *promotor sites*; even a single base mutation in a promotor sequence may prevent or greatly reduce transcription of the genes adjacent to it. A promotor must include a binding site; one model for these mutations is therefore to suppose that they may reduce its affinity for polymerase and so prevent transcription. However, since any mutation which prevents transcription of adjacent genes is characterized as a promotor, such mutants might also be located in initiation sites or indeed in any other contiguous control element (see pages 307 and 350). To define the molecular nature of promotor mutations therefore requires their biochemical characterization.

Binding of RNA polymerase to DNA in vitro shows that each template possesses a restricted number of binding sites. The role of sigma factor is to ensure that RNA polymerase recognises only these sites on DNA and thus transcribes only the appropriate strand of DNA; the binding site may therefore be described as a sequence of DNA which is stably bound by complete enzyme but not by core enzyme (but see below). The initiation site can in principle be located by the sequence of DNA to which the first few bases of the RNA chain correspond.

Since it has not yet proved possible to isolate the sequences of DNA bound by polymerase and/or corresponding to the start of the RNA chain, we lack experimental evidence on whether binding site and initiation site coincide or are separate. Nor has it been possible to characterize the binding of RNA polymerase to templates in which promotor mutants have been mapped. Although promotor mutations are often equated with binding sites, this concept has therefore yet to be confirmed by experiment (but see page 356).

That sigma does not itself recognise the nucleotide sequence of the binding site but enables core polymerase to do so is suggested by observations that core enzyme alone can bind to DNA but that its association at appropriate binding sites is enhanced by sigma whilst binding to other regions is depressed. Comparison of the extent of binding to DNA of RNA polymerase in the absence and presence of sigma suggests that complete enzyme can bind to only a restricted number of sites (presumably including those utilised in vivo) whereas core enzyme binds at random. Indeed, in appropriate conditions, core enzyme has a general affinity for DNA and its binding is limited only by the space available on the template.

Hybridization experiments with T4 DNA have been used by Bautz et al. (1969) to show that the RNA molecules synthesized in vitro when sigma is present correspond to the sequences which are usually synthesized during the early part of infection in vivo (see later). When sigma is omitted, there is no such restriction and RNA synthesis seems to be initiated at random sites along the T4 molecule. When phage DNA is incubated with complete RNA polymerase enzyme for several minutes before nucleotides and rifampicin are added, some of the enzyme molecules become resistant to rifampicin. The number of resistant complexes formed with each DNA molecule is quite small.

Bautz and Bautz (1970) found that these resistant complexes can only be formed when sigma is present—they must be a consequence of proper initiation.

If each molecule of RNA polymerase initiates one chain of RNA under these conditions—which is probably approximately, although not precisely true—measuring the incorporation of γ-P^{32} into RNA should provide an estimate of the number of initiation sites. ATP is more commonly used in initiation than GTP, and the sum of the two types of initiation event depends on the template used:

T7 DNA	1 ATP starter	1 GTP starter =	2 initiation sites
lambda DNA	2	1	3
T4	22	3	25
T5	32	3	35

Although it is not possible to say whether these initiation events take place at the same sites which are used in vivo, the total number of initiation sites is in each case about the number of promotors which we expect to be present. This supports the idea that RNA polymerase can form two types of complex with DNA. Core enzyme can bind at random and any ensuing initiation events do not take place at proper initiation sites. Complete enzyme can bind efficiently and tightly to a small number of sites, which may correspond to those used in vivo.

The double stranded replicative form of phage fd is a useful template for following transcriptional specificity because only one strand is transcribed in vivo. The DNA of mature phage particles consists of a single strand which is converted into a duplex by synthesis of its complement shortly after infection. The RNA extracted from infected cells cannot hybridize with the mature single strand but hybridizes with the strands produced by denaturing the replicative duplex. This implies that only the complementary strand synthesized after infection is used as a template for RNA synthesis during phage development.

Core enzyme alone can transcribe duplex fd DNA, but Sugiura, Okamoto and Takanami (1970) found that sigma greatly increases the extent of RNA synthesis. However, the RNA sequences synthesized by core enzyme and complete enzyme are very different. RNA transcribed by core enzyme can hybridize with either denatured duplex DNA or with the single strands from mature phages; RNA transcribed by complete enzyme finds complements only with the denatured DNA of the duplex. This indicates that core enzyme transcribes both strands of the duplex fd DNA template whereas complete enzyme transcribes only the strand expressed in vivo.

When complete enzyme is used to transcribe fd duplex DNA in vitro, the RNA product comprises largely one class of molecule, which sediments on a sucrose density gradient at about 26S. Core enzyme yields only RNA molecules which are much lighter and distributed throughout the gradient. Addition of sigma restores the original sedimentation profile. In the absence of sigma the core enzyme is therefore free to initiate RNA synthesis at random sites on either strand of the DNA template; the RNA products are dispersed in size distribution and hybridize with both DNA strands. Sigma stimulates initiation from starting points on one strand and depresses transcription from other starting points. The 26S RNA is about the size expected of a transcript of the complete length of fd DNA.

The triphosphates used for initiation also suggest that core enzyme starts randomly whereas complete enzyme uses specific initiation sites. Only 45% of the RNA molecules synthesized by core enzyme start with either ATP or GTP whereas most of the starting sequences transcribed by complete enzyme do so. Takanami, Okamoto and Sugiura (1970) showed that the heterogeneous starting sequences produced by core enzyme are replaced by only three starters—pppApUpG, pppGpUpA and pppGpUpU—when sigma is added. Sigma therefore restricts initiation to a small number of sites; we do not know how the three starting sequences are related to each other or to transcription in vivo.

Binding of RNA Polymerase to DNA

The stability of complex formation between RNA polymerase and DNA depends on several parameters, including temperature and ionic strength. Although both core enzyme and complete enzyme bind to DNA, under physiological conditions binding in the presence of sigma factor is more secure. An assay for measuring the binding of enzyme to DNA has been developed by Hinckle and Chamberlin (1970, 1972a). RNA polymerase is mixed with H^3-labelled DNA and filtered through a nitrocellulose membrane; RNA polymerase and any DNA bound to it are retained by the membrane, but any free DNA is eluted. The sigma polypeptide is not needed to form this complex, although DNA is retained with a two fold greater efficiency when it is present.

The reason for this increased efficiency is that the complex formed by core enzyme is unstable and dissociates rapidly, releasing DNA. The complex formed by complete enzyme is more stable and is retained longer by the filter. The stability of these complexes can be measured by binding polymerase to H^3-labelled DNA and then incubating the complex with an excess of unlabelled DNA. When a polymerase molecule dissociates from the H^3-DNA to which it is first bound, it is free to bind to another template; it usually finds an unlabelled DNA since this is now present in excess. The H^3-DNA which has lost its polymerase is then no longer retained by the filter. Incubating a mixture under these conditions and measuring the decline with time of the H^3 label which is retained when samples are filtered therefore estimates the stability of the complex. Figure 6.7 shows that the complex formed between complete enzyme and T7 DNA is stable at 37°C, whereas the complex formed with core enzyme is unstable and dissociates rapidly. Similar results have been obtained by Dausse et al (1972a, b).

The effect of temperature shows that the complexes formed with T7 DNA by complete enzyme and core enzyme are different in nature. The complex formed by complete enzyme has a half life of at least 50 hours at 37°C, of 15 hours at 25°C and only 30 minutes at 15°C. There is thus a drastic increase in the stability of the complex as the temperature is raised. The complex formed by core enzyme behaves in precisely the reverse manner. The half life of the complex at 37°C is 20 minutes; but it is increased to 60 minutes when the temperature is reduced to 15°C. This complex therefore decreases in stability as the temperature is raised.

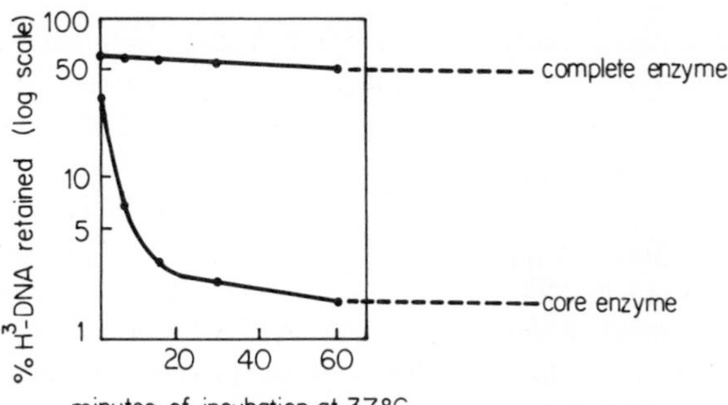

Figure 6.7: dissociation of RNA polymerase from H^3 labelled T7 DNA. About 60% of the DNA binds to complete enzyme and there is little dissociation during 60 minutes incubation. Only 40% of the DNA binds to core enzyme and very little is retained 60 minutes later. Data of Hinckle and Chamberlain (1972a)

This suggests that the complex formed by core enzyme involves a simple binding of polymerase to DNA, and dissociates more readily with increasing temperature. The presence of sigma factor, however, allows a very stable complex to be formed at higher temperatures, so that the dependence of complex formation on temperature is reversed. The temperature dependent effect must involve some change in state of the complex. The most likely change is a melting of the double helix to provide a localised region of separated strands which can move along the template during transcription. When the temperature is sufficient to allow this melting, the binding of polymerase to DNA becomes more secure. Since sigma does not itself bind to DNA, it presumably enables the core enzyme to undertake the strand separation reaction.

Another reaction which supports the idea that sigma is implicated in melting DNA is the development of complexes of RNA polymerase and DNA resistant to heparin. Zillig et al. (1970) found that complete enzyme can form a complex with T4 or T3 DNA which is resistant to inactivation by heparin; core enzyme cannot do so. Formation of the resistant complex depends upon temperature. When sigma is added to the complex of core enzyme and DNA at temperatures above 20°C, a heparin-resistant complex is formed; but at temperatures below 12°C sigma has no effect and the complex remains sensitive.

Dependence of formation of a heparin-resistant initiation complex on temperature follows a transition curve very like that of the melting of DNA, although the transition temperature depends on the enzyme and not merely on DNA. With RNA polymerase of E.coli it is 17·5°C; but with the enzyme of B. stearothermophilus it is much higher at 30–35°C. If the role of sigma is to assist the core enzyme to melt DNA to start transcription, then, the unwinding reaction cannot depend on DNA alone, but must rely upon the milieu provided by the enzyme.

When single strand breaks are made in DNA, the binding activity of core enzyme is stimulated but that of complete enzyme is inhibited. But Hinckle, Ring and Chamberlin (1972) found that, on average, every single strand break introduced into T7 DNA promotes the formation of a stable complex with 1–2 molecules of complete enyzme. However, most of these bound enzyme molecules are inactive and cannot initiate RNA synthesis. This suggests that single strand breaks act as sites of binding activity for core polymerase, perhaps because they provide sites at which the DNA strands are readily separated. Because core enzyme is less able (or unable) to undertake strand separation, it is comparatively inactive on templates of perfect duplex DNA; the introduction of single strand breaks substitutes for the action of sigma in melting DNA (although, of course, the breaks do not occur at physiological binding sites). When sigma is present, however, polymerase is prevented from initiating transcription at the breaks, although it continues to bind. This shows that melting the two strands of DNA, although necessary, is not sufficient to allow complete enzyme to start transcription.

Core polymerase binds with a very similar affinity to all regions of DNA. At saturation, binding seems limited only by space available on the DNA; there are some 1300 of these "loose binding" sites on T7 DNA. Hinckle and Chamberlin (1972a) have shown that all of the core polymerase molecules bound to T7 DNA dissociate in the same way, so that all loose binding sites are indistinguishable to the core. In this experiment, core polymerase is bound to unlabelled T7 DNA and the mixture is then diluted with excess H^3-labelled T7 DNA. The amount of label retained by nitrocellulose filters with increasing time of incubation shows how many of the core polymerase molecules have dissociated from their original template. The binding of core enzyme to H^3-T7 DNA alone—that is without a previous incubation with unlabelled DNA—acts as a control.

Figure 6.8 shows that none of the core polymerase molecules originally bound to the unlabelled DNA is retained by this template, for within 60 minutes the retention of H^3-DNA is the same as the control. And all of the core polymerase molecules are bound to their original template with a half life in the range 10–20 minutes. This indicates that core polymerase—when there are on average 27 molecules of enzyme for each T7 DNA—binds equally poorly to all loose binding sites.

Different results are obtained with complete enzyme. As the figure shows, complete enzyme shows a biphasic dissociation curve. When 13 molecules of polymerase are present for each unlabelled T7 DNA molecule, about one third (five) of the bound molecules rapidly leave this template and associate with one of the H^3-labelled T7 DNAs present in excess. This reaction takes

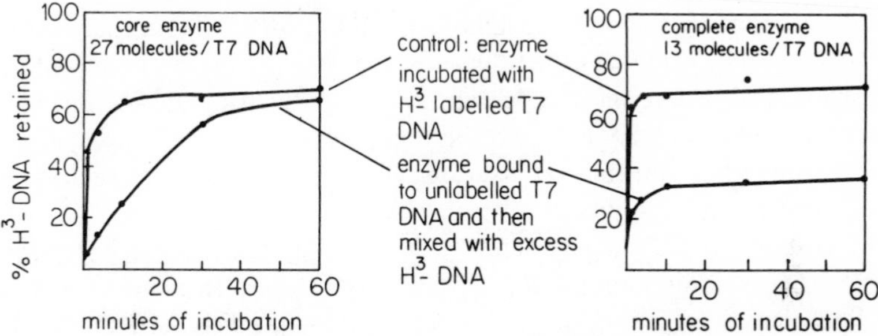

Figure 6.8: binding sites on T7 DNA for RNA polymerase. In control experiments, both core enzyme and complete enzyme bind rapidly to H^3-labelled T7 DNA so that about 70% is retained on nitrocellulose filters. Left: when a complex of core enzyme bound to unlabelled DNA is mixed with excess H^3-labelled DNA, the core molecules which dissociate from their first template associate with labelled DNA. The half life of dissociation is 10–20 minutes; by 60 minutes retention of label equals that of the control so there can be no tight binding sites for core enzyme. Right: with complete enzyme about five of the thirteen molecules bound to unlabelled DNA dissociate rapidly and bind to the labelled DNA. The remaining eight enzyme molecules remain bound to unlabelled DNA so that the plateau of H^3-DNA binding remains at this low level. Data of Hinckle and Chamberlain (1972a)

place as rapidly as the association of free RNA polymerase with DNA. This indicates that these loosely bound complete enzyme molecules dissociate from their template very quickly, more rapidly than do core enzyme molecules bound to loose binding sites.

The remaining eight complete enzyme molecules bound to the unlabelled DNA do not dissociate from it during the experiment, since there is no further increase in the retention of H^3-labelled DNA. Similar experiments with different concentrations of RNA polymerase for each T7 DNA molecule confirm that there are about eight "tight binding" sites on T7 DNA at which complete enzyme can form a stable complex. Once these sites are occupied, excess molecules of complete enzyme must bind to loose binding sites, and they do so less securely than do molecules of core enzyme.

If core enzyme is bound to H^3-labelled T7 DNA, it dissociates steadily from it. But if sigma factor is added, the rate of dissociation is increased. This indicates that the affinity of core enzyme for loose binding sites is reduced when sigma factor associates with it. Complete enzyme therefore has a decreased affinity for loose binding sites as well as the ability to recognise tight binding sites. The association constants for the different sites are:

complete enzyme recognises 8 tight binding sites with		K = 10^{12}-10^{14} moles^{-1}
complete enzyme	1300 loose	10^8-10^9
core enzyme	1300 loose	2×10^{11}

When core enzyme is bound to DNA, the addition of sigma therefore has two effects. First, it reduces the affinity of the core for loose binding sites, so that the complete enzyme is released more rapidly from DNA. Second, it confers an ability to recognise tight binding sites. The overall effect of sigma factor is thus to ensure that polymerase ceases to bind DNA at random and instead attaches only to proper promotor sites.

The stability of RNA-polymerase—DNA complexes is much greater in solutions of low ionic strength than at higher ionic levels. In 50 mM NaCl—the standard incubation condition used in the experiments discussed above—little dissociation of the T7 DNA—complete enzyme complex occurs. But at 100 mM NaCl the half life of the complex is reduced to 2 hours and at 150 mM NaCl to about 10 minutes. The dissociation data do not produce first order linear plots, which suggests that RNA polymerase may bind with different affinities to the various tight binding sites present on T7 DNA.

Binding of RNA polymerase to DNA takes place very rapidly. Hinckle and Chamberlin (1972b) found that the half life of tight complex formation between complete enzyme and DNA is about 20 seconds. Binding of core enzyme to form a loose complex takes place more rapidly; the reaction is too rapid to allow its half life of formation to be measured, but is 80% complete within 5 seconds. Loose complex formation must therefore take place at least ten times more rapidly than tight complex formation.

These interactions suggest that RNA polymerase may find its tight binding sites on DNA by the process of trial and error illustrated in figure 6.9. Loose complexes are formed very rapidly by collision of the polymerase with random segments of DNA; but in the presence of sigma their dissociation is very rapid. This allows the enzyme to undertake many associations and dissociations within a short time; this process ends when a tight binding site is encountered. It is the speed of the loose complex formation and release which ensures that the enzyme does not spend a long period of time engaged in attempts to find its proper binding site.

Core enzyme must presumably have a high affinity for DNA in order to exercise its catalytic functions. But if complete enzyme were to have a high affinity for all DNA sequences, the associations and dissociations needed to find a tight binding site could take many hours. By reducing the stability of the

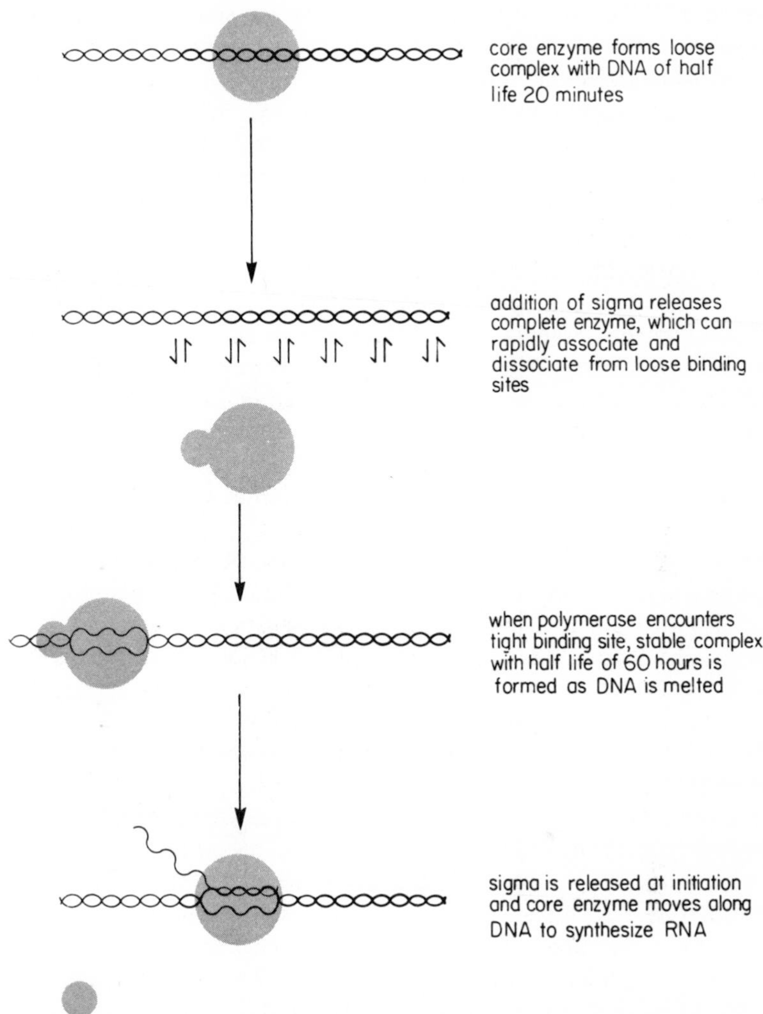

core enzyme forms loose complex with DNA of half life 20 minutes
addition of sigma releases complete enzyme, which can rapidly associate and dissociate from loose binding sites
when polymerase encounters tight binding site, stable complex with half life of 60 hours is formed as DNA is melted
sigma is released at initiation and core enzyme moves along DNA to synthesize RNA

Figure 6.9: model for interaction of RNA polymerase with DNA

loose complexes, sigma ensures that dissociation as well as association is rapid. When the complete enzyme encounters a tight binding site, a stable complex is formed, transcription is initiated and sigma is released; this leaves the core enzyme, with its general affinity for DNA, able to continue transcription.

Initiation of RNA Synthesis

The effect of rifampicin on the interaction of RNA polymerase with T7 DNA suggests that more molecules of enzyme form tight binding complexes than initiate transcription. Hinckle, Mangel and Chamberlin (1972) found that rifampicin has no effect on the binding of complete enzyme to DNA, nor does it change the rate of dissociation of the enzyme. Since polymerase becomes completely resistant to inactivation once transcription has started, rifampicin must block one of the later stages in initiation of RNA synthesis. Because loose complexes formed by core enzyme are slow to initiate RNA synthesis, they are more sensitive to inactivation than tight complexes with complete enzyme—there is time for rifampicin to inactivate the core enzyme before it initiates transcription. Rifampicin can therefore distinguish between loose and tight enzyme binding complexes.

When nucleoside triphosphates are added to a complex of complete enzyme with T7 DNA, transcription is initiated with GTP and ATP molecules. Initiation increases rapidly until a plateau is reached after 1–3 minutes and then increases slowly. When rifampicin is added together with the nucleoside triphosphates, the second, slow phase of initiation does not take place. This suggests that it corresponds to reinitiation by molecules of RNA polymerase which have completed synthesis of the chains they originally initiated. Rifampicin therefore allows only a single round of initiation by those polymerase molecules previously bound at tight binding sites.

Using this incubation system, Chamberlin and Ring (1972) found that in the absence of rifampicin one mole of complete enzyme initiates 0·24 moles of GTP starter and 0·42 moles of ATP starter chains; these yields per enzyme remain constant over a range of enzyme : DNA ratios of 1 to 20. This indicates that at least 13 RNA chains ($0{\cdot}66 \times 20$) can be initiated at the eight tight binding sites of T7 DNA during the two minute incubation period.

When rifampicin is added to the incubation mixture together with the nucleoside triphosphates, its effect depends upon the ratio of RNA polymerase molecules to DNA genomes. When there are less than eight molecules of complete enzyme for each T7 DNA, rifampicin has little effect, even at high concentration. But at higher enzyme:DNA ratios, the additional polymerases are progressively inhibited as the concentration of the drug is increased. The resistant level of initiation achieved in the presence of eight or more enzyme molecules per genome corresponds to about 1 GTP starter and 2 ATP starter chains per genome; initiation events in excess of these rely upon polymerase molecules which first bind at a loose site, only later moving to a tight binding site, so that they are inhibited by rifampicin.

Because binding to tight sites is irreversible, initiation must take place without release from DNA; the initiation site must be the same as the binding site or polymerase must be able to move between them without dissociating from DNA. But although there are eight tight binding sites, there appear to be only three initiation sites. One possible explanation is to suppose that more than one polymerase binds to each tight binding location, so that the eight polymerases are bound at only three loci on T7 DNA. The binding of all these molecules is not equal, for only three can initiate transcription in the presence of rifampicin.

Another important discrepancy concerns the use of these sites in vivo. Immediately after infection of E.coli by T7 DNA, the host RNA polymerase transcribes the first 20% or so of the genome from a single site which appears to be located at the left end of the molecule. Only the *r* strand of the DNA is transcribed. Of the RNA made in vitro, most of the ATP chains appear to represent transcription of the *r* strand, although the nature of the GTP starter is not clear and may comprise more than one molecule. Considering only the ATP starters, then, two chains are initiated simultaneously in vitro whereas there seems to be only one such event in vivo.

One possible model is to suppose that the promotor is a complex structure, consisting of more than one tight binding site so that several polymerase molecules may bind, and including more than one site at which initiation may take place. Because promotor mutants of T7 have not been isolated, however, we cannot define the site(s) used in vivo; an alternative approach would be to compare the terminal sequences of the RNAs synthesized in vitro and in vivo. Until transcription in vivo is better characterized, it remains possible that the apparently complex interactions observed in vitro do not represent the situation prevailing within an infected cell.

Calculation of how often any particular nucleotide sequence would occur by chance in E.coli DNA suggests that the minimum length needed to establish a recognition sequence which is not fortuitously repeated elsewhere is 12 base pairs (see page 289). The polymerase enzyme molecule probably covers about 30 base pairs when bound to DNA. Little is known about how specific sequences of nucleotide base pairs in DNA might act as recognition sites for protein, although models have been built to show that active groups in the wide groove of DNA might distinguish the bases and allow a specific interaction with protein (and see page 495).

One possible way to explain the affinity of binding sites on DNA for proteins is to suppose that their difference from the rest of the genome is rather gross and may distort the structure of the duplex. Regions of DNA rich in A–T base pairs might be used as recognition sites, for X-ray scattering experiments have suggested that stretches of A-T base pairs may have a structure distinct from that of the normal duplex (Bram, 1971). Such regions would also be less stable in duplex form that G–C pairs and would therefore be easier to unwind. Another idea is that there might be a cluster of pyrimidines alone on one strand, matched by a corresponding cluster of purines on the other strand. Unique sequences of up to 20 successive pyrimidine residues have been found by Ling (1972) in the DNA of phages fd, f1 and ϕX174, supporting the earlier ideas of Szybalski et al. (1966) who detected clusters of deoxycytosine in a variety of organisms by their reaction in complexing rapidly with poly-G.

Signals for Termination of Transcription

Termination Sites Recognised by Rho Factor

Synthesis of RNA on most DNA templates in vitro proceeds for a distance depending on the incubation conditions but does not usually terminate at unique sites; instead, random termination produces a very dispersed distribution of RNA molecules of various sizes. Termination at specific sites can take place in vitro under two types of condition. At high salt concentration, termination seems to occur at specific sites on some templates; in low salt concentration, a protein factor, *rho*, can promote termination at what appears to be a different set of sites. Almost all work on termination has made use of templates of phage DNA and there are few data to say how termination takes place on bacterial DNA.

Rho was first prepared by Roberts (1969, 1970) as a protein factor from E. coli which enables RNA polymerase to complete transcription at certain sites on phage lambda DNA. The factor was initially identified by its ability to depress to a plateau value the incorporation of precursors into RNA. Only a limited part of in vitro transcription of lambda DNA is depressed and it is the propagation of RNA chains which is affected; rho has no effect on initiation (as measured by incorporation itno RNA of a γ-P^{32} label in GTP).

The RNA transcribed from λ DNA remains associated with its template,

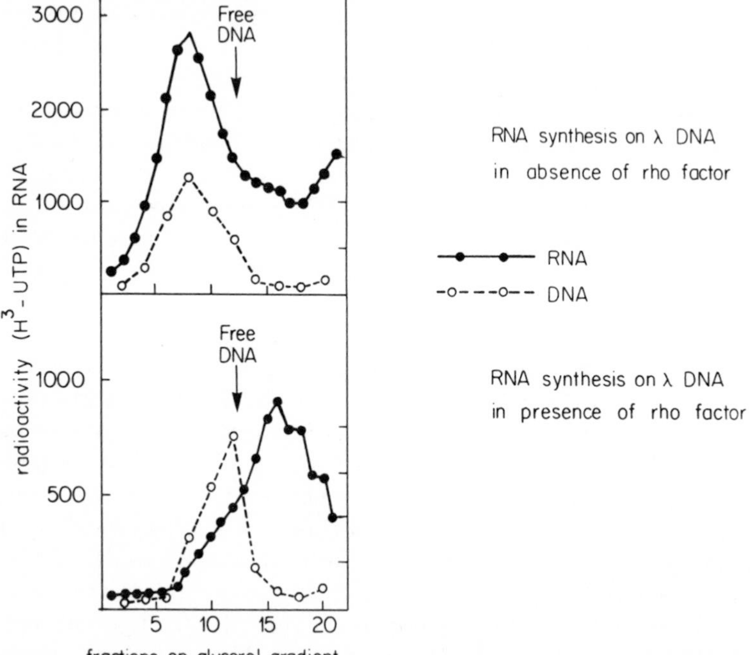

Figure 6.10: release of newly synthesized RNA from λ DNA by rho factor. Upper: when RNA polymerase alone is used, newly synthesized RNA sediments in a complex with λ DNA (amount of DNA is in arbitrary units, not to same scale as RNA). Lower: addition of rho factor to incubation mixture reduces synthesis of RNA (to 1000 c.p.m. instead of 3000 c.p.m.) and transcripts are released from DNA to sediment at lighter position. Data of Roberts (1969)

unless rho factor is present to cause its release as shown in figure 6.10. In the absence of rho, RNA and DNA sediment together on a glycerol gradient in a heavy complex at more than 50S, but after rho has been added, the RNA sediments at a much lighter value and the DNA sediments at the usual position of 32S characteristic of lambda. The addition of rho causes a large decrease in the size of the RNA extracted from the incubation mixture after transcription. Rho must exert its effect when the RNA is transcribed and not at a later stage because, when assayed in the usual reaction conditions, it has no effect on the RNA made in its absence.

If rho factor terminates RNA synthesis at specific sites on the template, the RNA molecules produced should be of discrete sizes. In accordance with this idea, when rho is present during transcription in vitro, the RNA product comprises the two classes of molecules shown in figure 6.11, one sedimenting

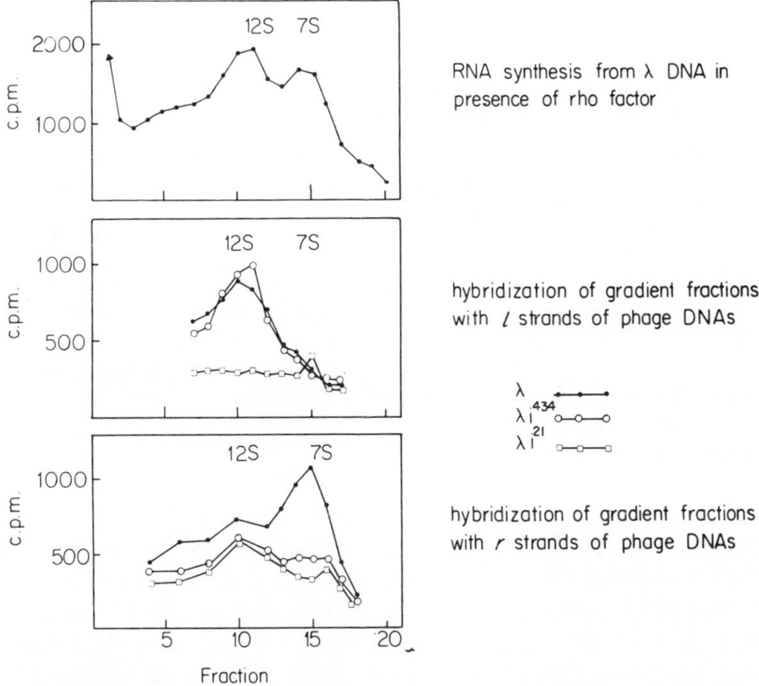

Figure 6.11: characterization of RNA synthesized by phage λ in presence of rho factor. Upper: the two principal species of RNA transcript sediment at 12S and 7S. Centre: 12S RNA hybridizes with *l* strands of λ and λi^{434} but not with λi^{21}; it must therefore correspond to the N region of λ (see figures 6.12 and 6.13). The 7S RNA does not hybridize with any *l* strands. Lower: 7S RNA hybridizes only with the *r* strand of λ; it must therefore correspond to the tof region (see figures 6.12 and 6.13). The 12S RNA does not correspond to any *r* strand regions. Data of Roberts (1969)

at 12S and the other at 7S. In the absence of rho, there is no trace of either of these peaks and little material in this light part of the gradient. What part of the template these RNAs represent can be determined by hybridization experiments with the DNA of λ and two related phages, shown in figure 6.12, which possesses the immunity region of another phage in place of the immunity region of lambda.

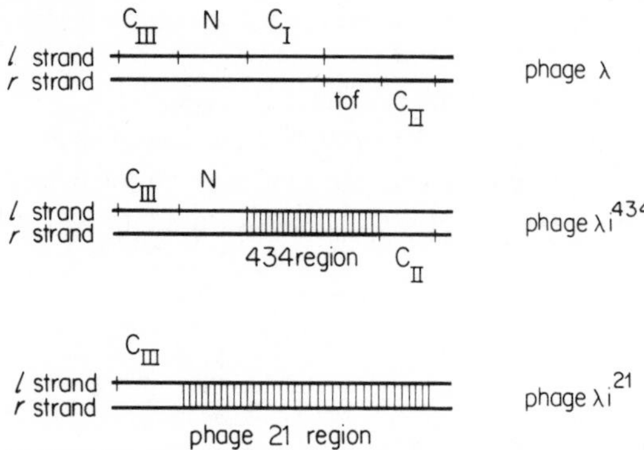

Figure 6.12: immunity region of phage lambda. In phage λi^{434} (derived from a cross between phages λ and 434), genes cI and tof are replaced by DNA of phage 434. In phage λi^{21} a larger region of λ DNA is replaced by 21 DNA, including genes N, cI, tof and cII. When hybridized with RNA transcribed from lambda, the DNAs of λi^{434} and λi^{21} behave as though the replaced parts of the immunity region have been deleted. Any RNAs hybridizing with λ but not λi^{434} must correspond to cI or tof; any RNAs hybridizing with λ but not λi^{21} must correspond to the N-cI-tof-cII region. The two strands of each phage can be separated and assayed for hybridization with RNA; since N and cI lie on the *l* strand and tof and cII on the *r* strand, these phage DNAs allow each gene to be specifically assayed

Because there is no homology between the different immunity regions, the effect of the substitution in λi^{434} is to delete the region of genes *cI* and *tof* of λ for hybridization purposes; in phage λi^{21}, gene N of λ is also deleted. Figure 6.11 shows that the 12S RNA is homologous to the *l* strands of both λ and λi^{434}, but not λi^{21}, which implies that it is synthesised to the left from a region in the vicinity of the N gene. The 7S RNA is homologous to the *r* strand of λ DNA, but not that of λi^{434} or λi^{21}, which implies that it must be synthesised to the right from an area entirely within the immunity region.

Transcription of lambda DNA is initiated in vivo at two promotors, one on each strand of DNA as shown in figure 6.13. Mutants of lambda mapping at these sites abolish or greatly reduce transcription in vivo to the right or to the left from the immunity region. In vitro, λ*sex* mutant DNA synthesises 7S RNA but not the 12S RNA so that the *sex* mutation identifies the left strand promotor p_l. Mutants mapping at the promotor in the *x* region considerably reduce 7S RNA synthesis in vitro without inhibiting synthesis of 12S RNA; *x* mutants thus identify the right strand promotor, p_r.

This suggests that the 7S and 12S RNA peaks represent RNA molecules transcribed from the promotors which are used in vivo. It is difficult to compare the 12S and 7S RNA species directly with those made in vivo, however, because one of the functions of the *N* gene protein (which is probably coded by 12S RNA) is to allow transcription to proceed further than is allowed in vitro; the mechanisms by which this may take place and the way in which rho may

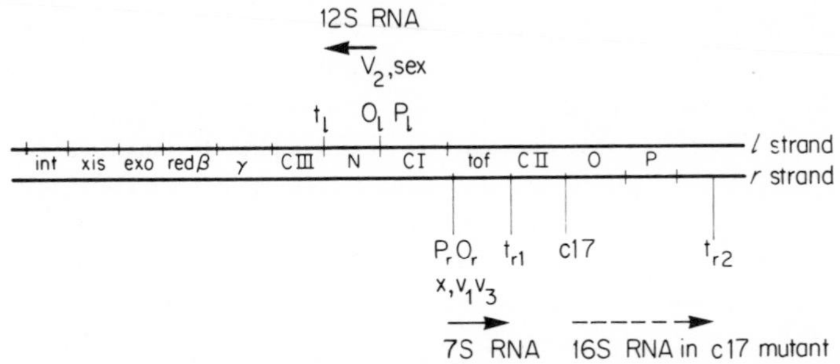

Figure 6.13: transcription of phage lambda. Left strand transcription commences at the control site o_1p_1 identified by mutations v2 in the o_1 operator and sex in the p_1 promotor. When rho is present transcription terminates at t_1 to give 12S RNA. Right strand transcription starts at the control site o_rp_r identified by mutations v1, v3 in the operator o_r and of the type x in the promotor p_r. When rho is present transcription terminates at t_{r1} to give 7S RNA. The c17 mutation creates a new promotor for initiation of transcription on the r strand; in the presence of rho the polymerase continues to the termination site t_{r2} to give an RNA product sedimenting at 16S

be used during development of lambda are discussed below (page 269). However, these results demonstrate that rho terminates transcription only at certain sites of the genome, presumably by itself recognising or causing core polymerase to recognise particular nucleotide sequences.

Termination in Different Ionic Environments

One of the problems in working with transcription systems in vitro is that they depend very heavily upon the ionic milieu; rho factor does not act on phage lambda templates, for example, when the KCl concentration is raised significantly above 0·1 M. Indeed, before the existence of various factors to initiate and terminate chains was realised, the effect of ionic strength led to much confusion, for the RNA polymerase enzyme appeared to show different types of activity in different ionic environments (see Richardson, 1969, 1970). Under conditions of low ionic strength, the initiation of RNA chains ceases soon after addition of the enzyme, which starts only one RNA chain and remains attached to its template in an enzyme—nascent RNA—DNA complex. When the salt concentration is greater, reducing the affinity of the polymerase for DNA, RNA synthesis continues for longer and the polymerase and nascent RNA chains are released from the template.

But there are two types of termination reaction, depending on the ionic conditions. In medium of low ionic strength, rho acts to terminate transcription at specific sites. But when the ionic strength is increased, rho is less effective or even completely inactive, depending on the template; and termination can then take place at other sites, also apparently specific although less well identified, in a reaction which does not need rho. Transcription in vitro of the duplex DNA of phage fd has distinguished more clearly between the two types of termination reactions. By using nucleotide precursors labelled with γ-P^{32}, Takanami, Okamoto and Sugiura (1970, 1971) have identified the RNAs transcribed from specific starting points. Only ATP and GTP are used in genuine initiation events. (When RNA chains have been initiated at random starting points, it is difficult to detect discrete species even if termination is specific, and it would not be easy to know what significance to ascribe to the termination even if it could be detected.)

The addition of rho causes the same sort of decrease to a plateau value of RNA synthesis from fd duplex DNA that is found with phage lambda, the final amount of RNA synthesis depending upon the concentration of KCl. Rho is more effective at levels below 0·1 M KCl, when the plateau synthesis is reduced to 35% of its previous value. But with this template, rho also functions in 0·2 M KCl, although the plateau level of RNA synthesis is 65% of that in the absence of rho. The cause of the reduction is that rho decreases the size of the RNA released from the template and a similar effect is found also with phage ϕ80 DNA as template.

The RNA initiated with ATP grows to a molecule of size 26S in the absence of rho factor at either 0·05 M or 0·2 M KCl. This size of RNA is about that expected from transcription of the complete length of fd DNA, which suggests that transcription may be able to proceed once around the template, but is then terminated at some site in an action which does not need rho. The GTP starters have two types of initiation sequence and comprise three molecular sizes, 17S, 13S, 10S in the absence of rho. It is reasonable to suppose that these molecules represent initiation at two sites closer to the "end" of the template; they must be terminated at the same site as the ATP starters.

Both ATP and GTP starters are reduced in size when rho is present. Figure 6.14 shows that in 0·05 M KCl the presence of rho reduces the size of the ATP starter to 10S; in 0·2 M KCl rho confines the ATP starter to 17S. The GTP starters are also reduced in size by the action of rho; in 0·05 M KCl they sediment at 10S and in 0·2 M KCl fall into two classes, sedimenting at 10S and 13S.

Completion of the ATP starter takes about 5 minutes in vitro. If rho is added at various times after transcription has started, the polymerase is allowed to move a certain distance along the template before any rho-dependent termination can take place. To ensure a synchronous start to the movement of polymerase, initiation can be allowed to take place in the presence of low amounts of nucleoside triphosphates; when the amount of the precursors is increased, elongation starts from the pre-initiated complex.

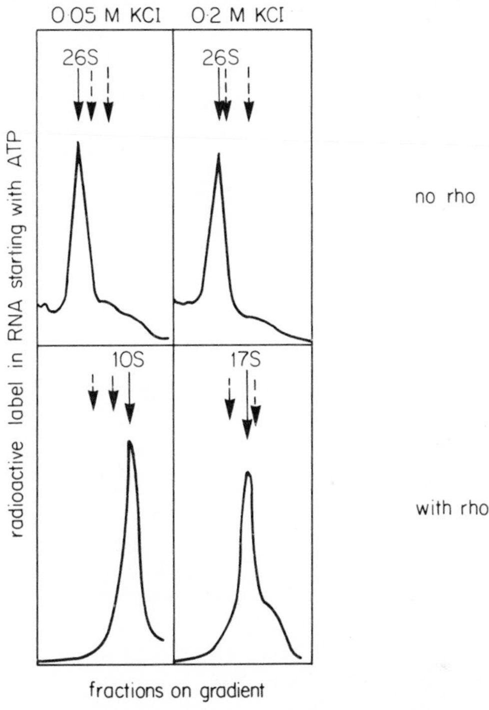

Figure 6.14: transcription in vitro of duplex fd DNA. Upper: in the absence of rho a 26S RNA is synthesized in either 0.05 M KCl (left) or 0.2 M KCl (right). Lower: when rho is added the size of the RNA product is reduced, to 10S in 0.05 M KCl (left) and to 17S in 0.2 M KCl (right). Data of Takanami, Okamoto and Sugiura (1970)

The size of the ATP starter chain depends upon the time at which rho is added. Figure 6.15 shows that 10S RNA is produced when rho is present from the beginning of transcription. But when rho is added after transcription has proceeded for 1 minute, a peak is also found at 13S. This implies that some polymerase molecules have progressed past the point at which rho can act to terminate the chain at the size of 10S and they are therefore able to continue to the next site at which rho acts, which generates a chain of 13S. Similarly, if rho is added only after two minutes of transcription, 17S RNA is produced by RNA polymerase molecules which must have passed the 13S termination site and can continue until they encounter the next site at which rho acts. These

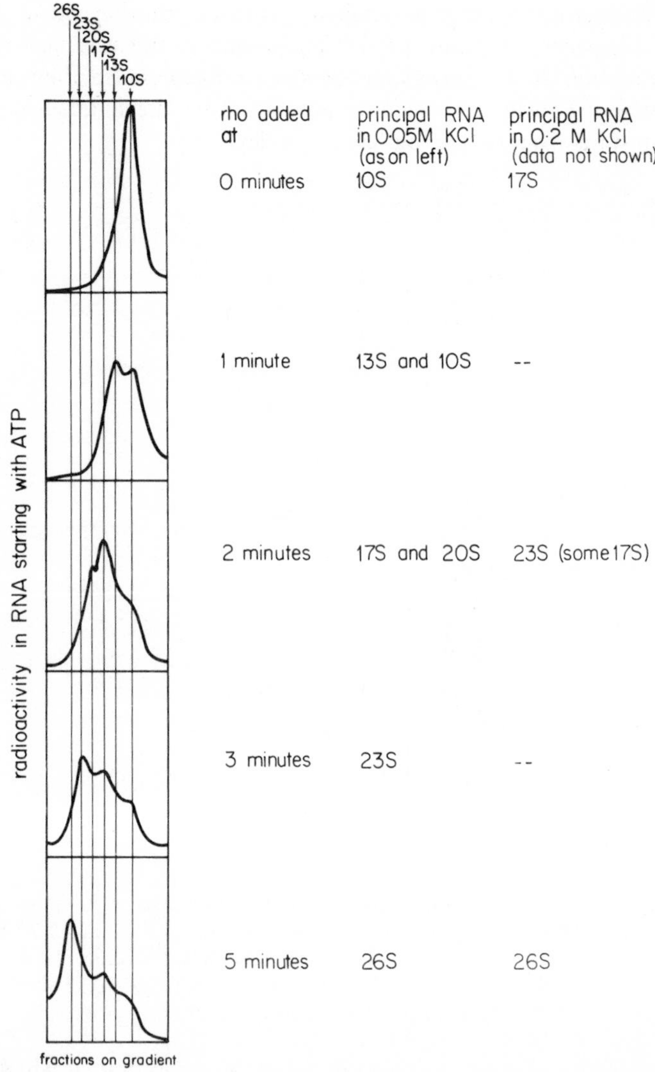

Figure 6.15: sites on fd DNA which respond to rho factor in vitro. Gradients on the left show the RNA species synthesized in 0.05 M KCl; the table on the right identifies the major RNA products in both 0.05M KCl and 0.2 M KCl. When rho is added after the start of transcription, some enzyme molecules have time to progress beyond termination sites to produce longer RNAs. In 0.05 M KCl, RNA molecules of 10S, 13S, 17S, 20S, 23S and 26S are produced depending upon how far the polymerase proceeds before rho is added to terminate transcription at the next available site. In 0.2M KCl, the sites which produce 10S, 13S and 20S RNA are not recognised by rho, so that only 17S, 23S or 26S RNAs are produced. Termination to yield 26S RNA is independent of rho and can take place in its absence. Data of Takanami, Okamoto and Sugiura (1971)

results are obtained at a KCl concentration of 0·05 M, in which a total of 6 discrete classes of RNA can be found, sedimenting at 10S, 13S, 17S, 20S, 23S and 26S. The action of rho can be confined to any one of the first five sites by selecting the time at which it is added; termination at the last site, to give 26S RNA, can of course take place in the absence of rho.

The length of the chain depends upon the KCl concentration as well as upon the time at which rho is added. In 0·2 M KCl, there are fewer discrete classes of RNA. A peak at 17S is generated when rho is present from the beginning of transcription; some chains are allowed to grow to 23S when rho is not added until 2 minutes after transcription has started; and the usual 26S RNA is given when rho is added at 5 minutes.

There therefore seem to be some five sites on the DNA where rho can terminate transcription, with an efficiency determined by the ionic milieu. In low salt, rho works at all the sites. If it is present before RNA polymerase reaches the first, all the chains are released at the 10S size and RNA polymerase continues no further along the template. If rho is added after polymerase has passed this point, the factor acts at the next termination site to give 13S RNA. Addition of rho at successively later times produces 17S, 20S, 23S RNA species. But when the salt concentration is raised, rho is active only at the 17S and 23S sites and fails to act at the sites generating 10S, 13S and 20S RNA. And, of course, in the absence of rho the polymerase can transcribe past all these termination sites until it reaches the final region where 26S RNA is terminated irrespective of the presence of rho.

Two classes of model may explain the effect of KCl on the activity of rho. There may be different types of rho present in the protein preparation and these may possess different sensitivities to salt. Or there may be one type of rho, but more than one class of base sequence which it recognises, some of these sites having an affinity for rho even in high salt when others are inhibited. Termination in the cell presumably depends on the ionic milieu. We do not know whether this is such that all five sites function in vivo but three fail in vitro when the KCl is too high, or whether only two usually function in vivo and recognition of the three additional sites is an artefact induced by low salt concentration.

Rho is usually assayed in vitro in low salt, for with other phage templates its sensitivity to KCl may be greater and it may not work at all with 0·2M KCl, which would argue that all the sites which it recognises are genuine termination signals. But, of course, these are two extreme theories and different sites may function with different efficiencies in vivo as well as in vitro. There must be differences between the DNA base pair sequence of the rho sites and the final recognition site, however, since termination to yield 26S RNA can take place in purified systems in which no factors are added to assist RNA polymerase. This suggests that there may be two classes of termination site, one responding to rho and one which does not need the factor; these classes may of course be further subdivided.

Interaction of Rho Factor with RNA Polymerase

With most templates of phage DNA, the concentration of KCl appears to have a greater effect than with fd DNA. With these templates, rho appears to be inactive in 0·2 M KCl and active only at lower concentrations. But termination events can take place in the high salt concentration which cannot take place in the condition of low salt under which rho must be assayed. Transcription of phage T4 DNA in the absence of rho is greatly stimulated when the KCl concentration is increased to 0·2 M; and Maitra et al. (1970) and Richardson (1970) showed that the increase in RNA synthesis is almost entirely due to reinitiation events taking place after termination. Experiments in which transcription is allowed to take place on one template after which a different phage DNA is added to compete for the enzyme polymerase reveal that the enzyme is released at termination before it reinitiates at another site. In the absence of rho factor, then, a high concentration of KCl is needed for termination and the termination reaction releases RNA polymerase from its template.

A more specific examination of the RNA species synthesised is possible with smaller DNA templates and Millette et al. (1970) and Maitra et al. (1970) have shown that with T7 DNA as template, KCl promoted transcription again appears to generate an RNA species corresponding to transcription of the complete length of the template. This RNA sediments at about 28–30S and its termination seems to be specific for 76% of the RNA molecules terminate in uridine; and essentially all of these molecules have the same terminal oligonucleotide fraction.

The basis for the action of rho factor is probably an interaction with RNA polymerase rather than with DNA itself. Goldberg and Hurwitz (1972) found that although rho factor does bind to DNA, very large amounts of protein are needed for the reaction. Varying the proportion of rho to DNA template does not influence the activity of the factor, which also suggests that it does not recognise DNA termination sites directly. However, the ratio of RNA polymerase to rho is critical; at high levels the termination factor is comparatively ineffective, but its activity increases as the amount of RNA polymerase is reduced.

Rho exerts a half maximal effect at ratios of 9 polymerase molecules for every molecule of rho factor. This indicates that the factor acts catalytically and must presumably be released after one termination event so that it can undertake another with a different polymerase. The action of rho is restricted to E.coli polymerase, for its effect upon the transcription of phage T7 DNA depends on whether the transcribing enzyme is the host cell polymerase or that coded by the phage itself. Rho reduces the size of the RNA product when E.coli polymerase is used at a KCl concentration of 0·01 M, although it is completely inactive at 0·2 M KCl. The specificity of its action is indicated by the termination of almost all RNAs in uridine. The factor has no effect on RNA synthesis conducted by the phage T7 RNA polymerase.

The action of rho does not appear to dissociate RNA polymerase from its template, although it terminates RNA synthesis. When poly-dAT is added to an incubation of RNA polymerase with T7 DNA, the incorporation of GTP into RNA—this is a measure of transcription of T7—is reduced to 10% of its previous value. But when polymerase is first incubated with T7 DNA, either in the presence or absence of rho, the subsequent addition of poly-dAT has little effect on the incorporation of GTP in the low ionic conditions needed for rho activity. This shows that the action of rho does not release RNA polymerase molecules from their template—if it did, many of them would transcribe the poly-dAT. Rho does not irreversibly attach polymerase to the template, however, because an increase in ionic strength after rho has acted allows resumption of RNA synthesis and transcription of any poly-dAT present.

Whether T4 or T7 DNA is used as template, and whatever criteria are used to follow transcription, then, it seems that rho can function only in conditions of low ionic strength in which the termination reaction leaves RNA polymerase attached to its template and unable to initiate synthesis of new chains. This failure to re-initiate explains why rho causes an overall reduction in the extent of RNA synthesis. In the T phage systems, the two types of termination reaction appear to be mutually exclusive in vitro, for rho can only function at low KCl concentrations and termination (and re-initiation) in its absence can only take place at high KCl. This must presumably reflect the conditions of incubation in vitro rather than comprising a genuine dilemma for transcription in vivo. The KCl promoted termination must be analogous to the termination event which generates 26S RNA at the final termination site on fd DNA; the rho-dependent termination seems to be the same as that shown with fd DNA, although it is affected to a greater extent by the salt concentration. Although it is not possible to arrange a hierarchy of sites on T phage DNAs with respect to their response to rho, it is of course possible that different sites may have different characteristics.

How does termination take place in vivo? The results obtained in vitro may mean that there are two classes of termination site. One class is sensitive to rho, perhaps with varying efficiencies; and the other type of termination event does not need rho but is a function only of the interaction of RNA polymerase with DNA. The extent to which these different classes of termination sites are used is difficult to decide, for it is unlikely that the lack of re-initiation when rho is active in vitro reflects its action in vivo. One model to explain the need for two types of site is to suppose that RNA polymerase transcribes rather long segments of DNA, more or less continuously, until it reaches a "KCl" termination site at which it is released from the template. Rho factor might act either to "chop" the RNA after transcription has passed certain sites or to cause termination and immediate re-initiation of transcription there. In either case, however, although RNA polymerase should remain attached to the template after rho has acted, it should be able to continue transcription instead of halting.

It is difficult to decide what role rho plays in terminating phage transcription in vivo because its action may be antagonised during development of a phage (see below). This allows RNA polymerase to read through the rho-dependent termination sites and means that the species of RNA synthesised in vitro and in vivo cannot be compared to see whether they terminate at the same site. Another important problem is the lack of defined templates of bacterial DNA, so that there is little evidence to say whether the two types of termination detected on phage templates in vitro are used in the uninfected bacterial cell (but see page 406). The role of rho in termination of transcription from bacterial DNA is therefore unclear.

Changes in RNA Polymerase During Phage Development

Switches in Early Transcription of Phage T4

After infection of a bacterial cell by phage T4, a temporal sequence of events is set in train so that different phage RNA molecules are synthesised at appropriate times during the development of the phage. The phage messengers can be divided into four different classes by the ability of the RNA sequences found in the cell at any particular time to compete in hybridization for the phage DNA with the sequences found in infected cells at other times.

There are two general classes of phage mRNA, each of which is further divided (see Guha et al., 1971a). The early sequences are those synthesised before replication of the phage DNA, which occurs at about 11 minutes after infection; the (*true*) *late* sequences are synthesised after replication. An exception to this pattern is that some sequences are synthesised in small amounts before replication although their transcription is not fully switched on until after replication; these are known as the *post-replicative early*, or sometimes as the *quasi-late*.

When RNA made during the first five minutes of infection was labelled and hybridized to DNA in the presence of unlabelled competing RNA extracted from cells at different times during development, Salser, Bolle and Epstein (1970) obtained the results shown in figure 6.16. Some sequences appear before 1·25 minutes after infection, others between 1·25 and 2·50 minutes, and a few more between 2·50 and 3·75 minutes, when all the sequences found at five minutes are present. The sequences present before 1·25 minutes comprise the *immediate early* messengers; they can be distinguished from the *delayed early* messengers transcribed between 1·25 and 3·75 minutes by the effect of chloramphenicol. When protein synthesis is inhibited in infected cells, only the immediate early sequences can be synthesized and the delayed early sequences fail to be transcribed.

The immediate early messengers must therefore be transcribed by the host RNA polymerase, but some phage coded protein must be synthesized (by translation of one of the immediate early mRNAs) before the delayed early messengers can be transcribed. Competitive hybridization experiments show that both the immediate and delayed early messengers are present in much smaller amounts in infected cells at 20 minutes than at 5 minutes, so their synthesis must later be turned off. The post-replicative early (quasi-late) sequences, however, increase in amount by this time; and the true late messengers, which do not compete with any of the early sequences, are also switched on by 20 minutes after infection. Each class of RNA synthesis seems to involve initiation events at several promotors at different loci on the T4 genome.

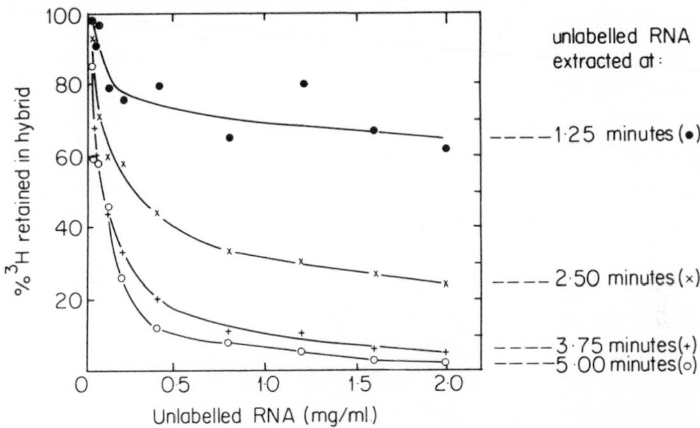

Figure 6.16: characterization of the early messengers of T4 development. RNA extracted from cells labelled for the first 5 minutes of infection is hybridized to phage T4 DNA in the presence of increasing amounts of unlabelled RNA extracted at 1.25, 2.50, 3.75 or 5.0 minutes after infection; the plateau level of label remaining hybridized shows the difference between the unlabelled and labelled preparations. The RNA synthesized by 1.25 minutes (the immediate early messengers) competes with 30% of the labelled RNA; RNA synthesized by 2.50 minutes competes with 70%; and RNA synthesized by 3.75 or 5.0 minutes competes with virtually 100%. The immediate early sequences therefore provide 30% of the total early sequences; and the delayed early sequences are synthesized between 1.25 minutes and 3.75 minutes. Data of Salser, Bolle and Epstein (1970)

Thesc temporal changes in the pattern of RNA synthesis may be achieved by the introduction of sequential changes in the specificity of RNA polymerase. The host sigma factor can direct core enzyme to recognise only immediate early genes of the phage and the rho factor terminates transcription at the ends of these genes. Two models have been proposed for the transition from transcription of immediate early to delayed early genes. One model is to suppose that a new sigma factor is coded by one of the phage immediate early genes and that this protein replaces the host sigma so that the core enzyme recognises delayed early genes. An alternative model is that host polymerase is unaltered and continues to recognise immediate promotors, but that its rho-mediated termination of transcription at the ends of the immediate early genes is prevented by an anti-termination factor coded by one of the phage immediate early genes; this allows the polymerase to continue transcription past its usual termination site and thus to transcribe the delayed early genes.

Extraction of T4 infected cells early in infection led to the isolation of a protein factor which appeared to direct core polymerase to transcribe delayed early as well as immediate genes. Travers (1970b) suggested that this protein might represent a T-early sigma factor which directs core polymerase to recognise the promotors of both the immediate and delayed early classes of gene, perhaps with different affinities. However, more recent research has supported the alternative model that the initiation specificity of RNA polymerase remains unaltered during early infection but that its termination specificity is changed.

The proportion of delayed early sequences present in a population of RNA molecules can be estimated by comparing its levels of competition in hybridization to DNA with the RNA molecules extracted from cells before 1·25 minutes and at 5 minutes after infection; the difference must be due to the presence of delayed early sequences. When E.coli polymerase is allowed to transcribe T4 DNA in vitro for only a short period of time, only immediate early RNA molecules are produced. But Milanesi et al. (1970) found that when incubation is allowed to proceed for longer, delayed early messengers are also synthesised. The addition of rho factor, however, restricts termination to the immediate early genes only.

A model which accounts for this finding is to suppose that the segments of immediate early genes are separated from succeeding delayed early regions only by rho-dependent termination sites. When polymerase is given sufficient time in vitro, in the absence of rho, transcription can continue past the end of the immediate early genes and into the delayed early genes. When rho is added, transcription must cease at the end of the immediate early genes. Readthrough, and not termination and reinitiation, must be involved in vitro, for the delayed early messengers continue to be made even when rifampicin is added after transcription of the immediate early genes has started.

The effect of rho on transcription in vitro is therefore very similar to the effect of inhibiting protein synthesis in vivo. The relative time of production of delayed early messengers in infected cells is similar to that in vitro and could be accounted for by a similar readthrough. This suggests the model shown in figure 6.17 in which the protein(s) coded by the phage which is needed to transcribe delayed early genes might function by antagonising termination by rho.

One implication of this model is that there must be at least two types of termination site on T4 DNA. One, at the ends of immediate early genes, is recognised by rho and inactivated by some anti-rho factor. The other, located at the ends of delayed early genes, must function in the presence of anti-rho, since otherwise transcription of these genes would not be terminated. These sites might therefore either respond to rho and anti-rho differently from immediate early terminators, or may perhaps not rely on rho for termination.

The idea that delayed early genes are transcribed by readthrough from immediate early genes has an important implication for the character of the delayed early RNA molecules. According to this model, initiation of delayed early messengers takes place at the promotors of the immediate early genes; so delayed early mRNA sequences should be carried at the 3′ ends of messengers which have immediate early sequences at their 5′ ends. But if new initiation events are demanded for delayed early transcription, the two types of RNA messenger should be independent, since each will commence at its own promotor. Most attempts to distinguish the two types of situation have made use of this prediction.

Delayed early sequences should be in larger RNA molecules if they are transcribed as the result of readthrough, although it is difficult to draw firm conclusions from data on the size of RNA because transcription produces a spectrum of molecules. However, Brody et al. (1970) have suggested that most delayed early mRNAs are too large to be the result of initiation at special promotors, but are more likely to carry immediate early sequences as well. And Witmer (1971a, b) found that in the presence of rho, the RNAs synthesised from T4 DNA by E. coli RNA polymerase compete to about 40% in hybridization with the longer RNAs made in the absence of rho. The large molecules contain all the sequences found in the small molecules. This suggests that rho prevents polymerase from reading through immediate early into delayed early genes.

Following the expression of individual T4 genes allows immediate early and delayed early messengers to be characterized. Schmidt et al. (1970) identified the messengers of the *rIIA* and *rIIB* genes transcribed in vivo by hybridization with phage DNAs carrying specific deletions in this region. These two genes are contiguous delayed early loci; *rIIA* starts to be transcribed about 1·25 minutes after infection and although it would take about 1 minute for RNA polymerase to reach the end of the gene, *rIIB* commences transcription at about 1·8 minutes. By testing to see whether *rIIA* and *rIIB* messengers might be part of one molecule, however, Schmidt et al. found some messengers carrying both sequences. This suggests that some of the transcription of *rIIB* results from readthrough past the end of *rIIA* but that some also results from new initiation events.

By hybridizing the RNA transcribed in vitro with denatured T4 DNA and degrading unhybridized regions of DNA with a nuclease, Jayamaran (1972) isolated the DNA corresponding to transcripts; the genes included can be characterized by using the DNA in a phage transformation assay. This technique defines three classes of early gene. The immediate early genes *42* and *30* are transcribed in vitro by RNA polymerase irrespective of the inclusion of rho factor. The delayed early genes *43*, *e*, *rIIA* and *rIIB* can be transcribed only if rho is omitted; this suggests they are usually read by a readthrough mechanism which antagonises rho. The delayed early genes 41 and 40 are not read at all in this system; their transcription may require modification of initiation specificity of the host polymerase (although a negative result of this nature may, of course, be due to some other defect peculiar to the in vitro system).

Both readthrough and new initiation may therefore be implicated in activation of the delayed early genes and some genes, such as *rIIB*, may utilise both mechanisms. To determine whether this apparent diversity prevails in vivo, however, requires the development of systems for the transcription of delayed early genes which depend upon purified T-early sigma or anti-rho proteins; and the genes which code for these proteins must be identified in the phage by mutations which influence their activity.

Changes in Template and RNA Polymerase During Late T4 Development

Transcription of true late messengers demands new initiations whose occurrence depends upon two classes of event. First, certain phage functions, in particular genes *55* and *33* must be expressed, for mutants at these loci are defective in late development. An obvious role for such late functions would be to code for proteins modifying the transcriptional specificity of RNA polymerase. One model supposes that a T-late sigma must be produced to direct the host core enzyme to recognise the promotors for late genes. The second event necessary for late development is a change in the state of the DNA template; only DNA in a certain condition appears to be able to bind the late polymerase.

By following the infection of E.coli with a T4 phage bearing a temperature sensitive mutant in DNA polymerase, Riva, Cascino and Geiduschek (1970a) have been able to show that transcription of true late genes can take place only at the low permissive temperature; raising the temperature to stop replication also prevents transcription. The quasi-late genes are excluded from this conclusion, because the assay for late mRNA involves measuring the difference in competition of an RNA preparation with 5 minute RNA (containing early and quasi-late sequences) with 20 minute RNA (containing also the true late).

This result can be explained if the parental phage DNA is incompetent as a template for late transcription until it is replicated to convert it to a competent form. Continual replication is needed to generate competent templates, which suggests that the replicated DNA is soon removed from the pool of competent templates, presumably when it matures ready for its incorporation into new phage particles. When Riva, Cascino and Geiduschek (1970b) repeated their experiments with phages bearing other mutations, however, they found that they could uncouple the relationship between replication and transcription. In phages bearing both an inactive ligase and exonuclease (this latter mutation is necessary because otherwise the exonuclease degrades the DNA of a ligase deficient phage), late transcription can take place in the absence of replication.

The RNA polymerase enzyme of late infected cells therefore appears to recognise only templates bearing single strand breaks. During the course of infection, these breaks are continually introduced during replication and sealed by ligase; in the mutant lacking ligase, breaks cannot be repaired and so DNA not suffering replication can be used as template. This concept is supported by experiments in which Bruner and Cape (1970) introduced a second phage template (bearing specific gene markers) into cells at a late stage of infection with T4. Although the appropriate RNA polymerase activity must be present in the cells, the late genes of the unreplicated second DNA cannot be transcribed. Figure 6.17 illustrates a model in which additional phage coded proteins change the behaviour of host RNA polymerase so that it recognises single strand nicks.

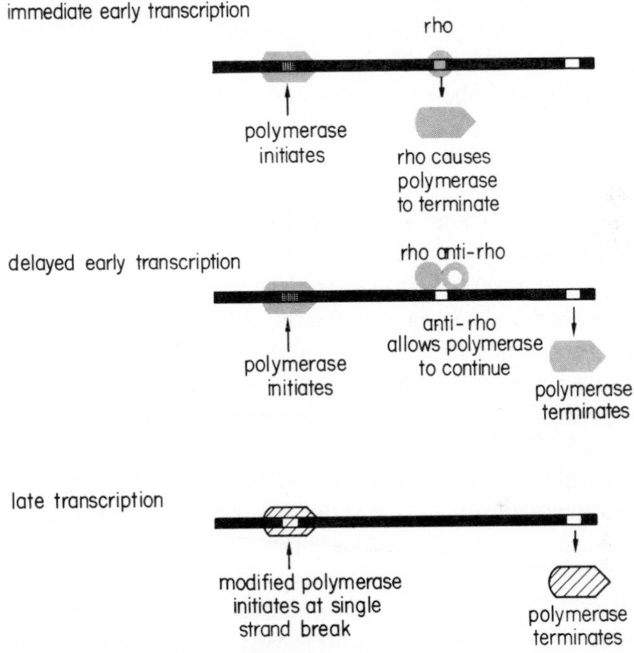

Figure 6.17: model for the development of phage T4. Upper: immediate early genes are transcribed by the RNA polymerase of the host cell. Transcription is terminated at specific sites in response to the rho factor. Centre: one of the proteins coded by the immediate early genes is an anti-rho factor which prevents this termination; this allows RNA polymerase to read through from the immediate early genes into the adjacent delayed early genes. Termination must take place at sites different in nature from those used by the immediate early genes. Although the anti-rho factor is depicted as interacting with rho and DNA, this model is speculative and the molecular action of rho and anti-rho has not been defined. Lower: transcription of late gene depends on the production of single strand breaks in DNA by replication and on the modification and association of host polymerase with new proteins, coded by early genes, which may act like the sigma factor in specifying the promotors at which polymerase can bind

Two types of change may take place in the host polymerase enzyme to redirect its transcriptional specificity: the subunits of the core polymerase are modified during phage development and we may speculate that sigma factor also is either modified or replaced. About two minutes after infection, the α subunit gains a P^{32} label which Goff and Weber (1970) identified as an AMP moiety bound to peptide through a 5′ phosphodiester link. Polymerase isolated at 18 minutes after infection contains a β' subunit of greater electrophoretic mobility than that of uninfected cells and Travers (1970a) observed that this change takes place between 10 and 15 minutes after infection; we do not know whether the change is a replacement or modification.

Changes in the subunits of the core are probably not themselves concerned with regulation but may be necessary to allow it to associate with new regulator molecules. For example, the affinity of the core for host sigma factor might be reduced so that it could be replaced by a phage coded sigma factor specific for late genes. Another reason for these changes may be that the core must be modified so that it can initiate transcription at single strand breaks instead of on intact duplex DNA.

These changes in the core enzyme and the need for a template with single strand breaks make it difficult to interpret experiments using an apparent T-late sigma activity in vitro. Travers (1969) reported that a protein is synthesized between 5 and 15 minutes after infection which changes the specificity of transcription by normal host enzyme with T4 DNA in vitro; this protein appeared to promote synthesis of both quasi-late and true-late messengers from unmodified T4 DNA. Because the template and core protein in these experiments were in the condition characteristic of early infection, we cannot conclude that this protein represents a T-late sigma; it is necessary to utilise late core and template and also to show that the protein is coded by one of the T4 genes whose expression is necessary for late development.

That changes in the activity of the host polymerase are controlled by the phage is suggested by the identification of additional components in the enzyme during the late stages of infection. Stevens (1972) observed that one of these polypeptides, of 22,000 daltons, is coded by gene *55* of the phage since it is absent in cells infected with mutants in this locus. Another polypeptide, of 12,000 daltons, is coded by gene *33*. Two other additional components have not been identified. We do not know whether these proteins comprise a T-late sigma activity or act in some other way to change the properties of the host polymerase so that it transcribes only replicating DNA.

Transcription of Delayed Early Genes of Phage λ

The development of phage lambda is similar to that of phage T4 in that it may be divided into stages of transcription. It is less complicated, however, in that the genes to be transcribed are arranged in compact blocks and there is no formal connection between replication and transcription later in the cycle.

The lytic development of lambda may be prompted in either of two ways. Phage DNA may infect a cell and enter the lytic cycle directly, in the same way as other bacteriophages. Alternatively, the λ DNA may be integrated into the bacterial chromosome, where it may exist in a stable *prophage* state (see page 345), until it is later induced into excision from the bacterial chromosome and enters the lytic cycle.

When protein synthesis is blocked in cells suffering induction of phage lambda, only two genes are expressed, *N* and *tof*. These are analogous to the immediate early genes of phage T4. In this situation, transcription starts at the two promotors from which the 12S and 7S RNAs are made in vitro in the presence of rho and does not proceed beyond the end of the *N* gene to the left and the *tof* gene to the right (see figure 6.13). The product of gene *N* is required for synthesis of the proteins coded by the genes on the left which participate in integration, excision and recombination and for expression of the *O* and *P* genes on the right which are needed for replication of λ DNA (reviewed by Echols, 1971, and Szybalski et al., 1970). When an active *N* gene protein is not produced, either because translation is inhibited or because the gene is mutated, transcription ceases at the end of the *N* and *tof* genes, shown by experiments in which Kumar, Calef and Szybalski (1970) hybridized the RNA of infected cells with λ DNA bearing certain deletions. Some transcription takes place past these points, presumably by readthrough at the termination sites, but extensive transcription of the delayed early genes can take place only when *N* protein is present.

Two types of molecular interaction might provide the positive control mediated by the *N* gene. The *N* protein might direct RNA polymerase to initiate transcription at new promotors, perhaps by acting as a new sigma factor, or it might prevent the rho-dependent termination discovered by Roberts (1969) at the ends of the *N* and *tof* genes. The action of the *N* protein is specific for λ DNA since only the λ *N* gene (and not 434 *N*) supports transcription beyond λ termination sites. Isolation of the *N* protein may allow its action to be determined in vitro; because it appears to be extremely unstable, any cessation of *N* gene transcription should rapidly halt expression of the delayed early genes (see page 366).

That *N* protein acts upon transcription by RNA polymerase is suggested by the observation of Georgopoulos (1971) that *groN* mutants of E.coli, upon which lambda fails to grow because *N* protein is unable to function, have an altered RNA polymerase. That *N* protein functions as an anti-terminator is suggested by the observation of Luzzati (1970) that it is active only under conditions in which RNA polymerase can initiate transcription at the immediate early promotors; and that RNA polymerase is enabled by *N* action to continue transcription is suggested also by the hybridization experiments of Portier et al. (1970) which showed that the leftward delayed early messengers contain immediate early sequences.

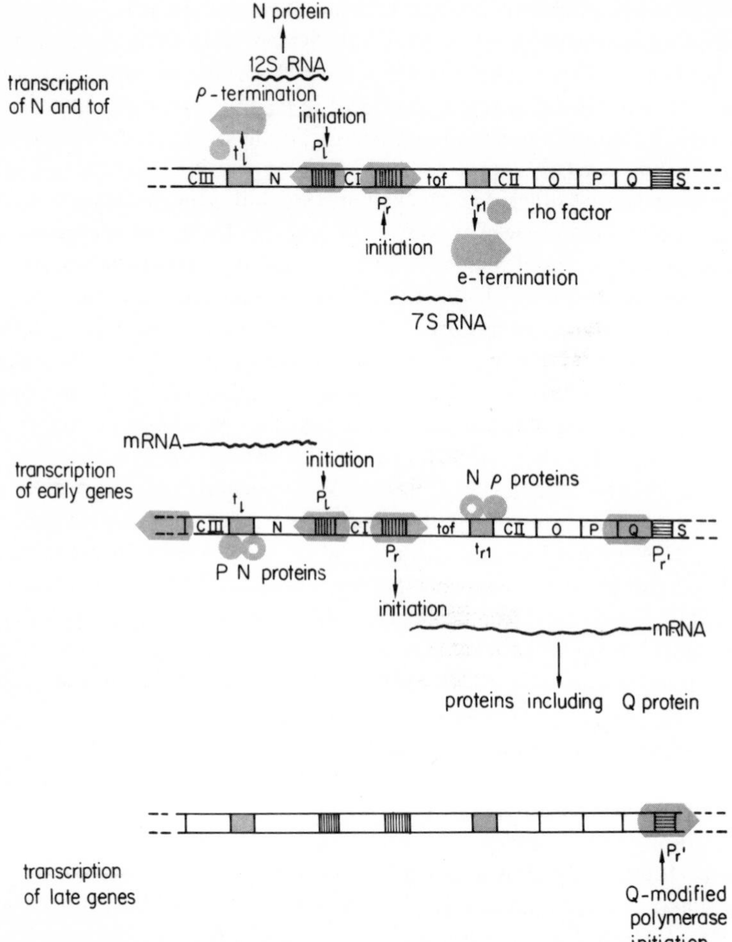

Figure 6.18: model for development of phage lambda. Upper: Only genes N and tof are transcribed by the host RNA polymerase, giving 12S RNA coding for N protein and 7S RNA coding for tof protein. Termination depends upon rho factor. Centre: When the N protein is synthesised, it acts as an anti-rho factor and allows transcription to continue past the rho-responsive termination sites t_1 and t_{r1}. The interaction between polymerase, rho and anti-rho has not been defined. The sites at which delayed early transcription is terminated are not known, but the recombination genes to the left and cII, O, P, Q to the right are all transcribed. Lower: When the Q protein is synthesised, it associates with RNA polymerase and directs it to transcribe the late genes, starting at the $p_{r'}$ promotor. The phage is in a circular form so that transcription continues past all the late functions and may progress into the recobmination functions. Its site of termination is not known.

The action of *N* is not sufficient to allow expression of the late genes, which are silent unless the phage bears an active *Q* gene. The late genes of λ defective in gene *Q* can be *trans*-induced, however, by the presence of a second infecting phage which bears an active *Q* gene. By studying the ability of this second active *Q* gene to switch on the late genes of a Q^- phage also carrying one of a series of deletions, Herskowitz and Signer (1970) mapped the site of action of the *Q* gene product at a locus between genes *Q* and *S* on λ DNA. This one promotor serves to initiate transcription from the *r* strand of all the late genes, for in the circular molecules of the phage formed from the linear chains released by excision, the two blocks of the late genes fuse into one. When this late promotor site, p_r' of figure 6.18, is deleted, there is some inefficient transcription of the late genes which shows the same control features characteristic of the delayed early genes; this suggests that it can take place by an inefficient readthrough past the normal termination sites by RNA polymerase enzymes which started at the p_r promotor.

The action of the *Q* gene product therefore seems to be to enable RNA polymerase to commence transcription at a promotor site which cannot be recognised by the host enzyme of its own accord. One way for Q^- mutants to revert is by gaining a secondary mutation which creates a new promotor, enabling the late genes to be transcribed by host enzyme without mediation of the *Q* gene. This supports the idea that phage λ DNA contains promotor sites which have different specificities for RNA polymerase. That *Q* gene acts directly on RNA polymerase is suggested by the finding of Naono and Tokuyama (1970) that RNA polymerase isolated from cells induced for λ during the late period contains two peaks of enzyme activity; one is present in uninduced cells but the second is extra. The extra peak seems to represent some late RNA polymerase activity which has a preference for different regions of λ DNA from the normal enzyme. Efforts to identify the molecular basis for this activity, however, have been hampered because the late-λ enzyme is extremely unstable, especially after it has been dissociated into subunits. It remains likely nonetheless that its function is to behave as a new sigma factor which recognises late promotors, although it is possible that it may instead effect some change in the core enzyme (or act as a new enzyme itself).

CHAPTER 7

The Lactose Operon

Organization of Gene Clusters

Induction and Repression

It was demonstrated at the beginning of the century that certain enzymes of yeast are formed only in the presence of their specific substrates; this effect was subsequently termed enzyme adaptation and is now known as *induction*. Most studies of induction have been carried out in bacterial systems, in which the effect is considerable; E.coli grown in the absence of a β-galactoside contain only about five molecules per cell of the enzyme β-galactosidase, whereas cells grown in the presence of the substrate may contain up to 5000 molecules of enzyme. Enzyme activity can be detected very rapidly (within 2–3 minutes) after addition of substrate; and removal of substrate results in an equally rapid cessation of activity. The rate of appearance of enzyme activity varies according to the particular substrate employed, reaching different saturation values for different substrates.

Induction might be caused either by activation of an inactive precursor form of the enzyme already present in the cell or by the de novo synthesis of new enzyme molecules. Immunological studies have shown that the induced enzyme is antigenically distinct from any protein previously present in the cell, whilst isotopic studies have confirmed that it is not derived from a precursor molecule synthesised before induction. Inhibition of bacterial protein synthesis with antibiotics has shown that new enzyme molecules must be synthesised if induction is to take place; experiments with precursors of RNA, or using direct measurements of messengers, have shown that induction takes place by transcription of a new mRNA which is then translated into protein.

The phenomenon of repression—specific inhibition of enzyme synthesis—was discovered by Monod and Cohen-Bazire (1953) when they found that synthesis of the tryptophan synthetase protein of E. coli is inhibited by tryptophan. Isotope incorporation experiments have shown that the effect involves repression of enzyme synthesis and not merely inhibition of activity. When synthesis of an enzyme is repressed, its mRNA is no longer synthesised and is soon degraded; after this, no further protein molecules are synthesised and those present in the bacterial cell are diluted out with division and eventually decay.

Coordinate Activity of Clustered Genes

Induction or repression is not usually confined to a single enzyme. In bacteria, the genes governing the synthesis of the various enzymes sequentially linked in a metabolic pathway are often closely linked to form a cluster on the genome. In such cases, all the enzymes coded by the cluster of genes are usually induced or repressed together. The first system of this type to be defined was the *lactose* (*lac*) genes of E. coli, used for much of the research which has worked out the method by which induction is controlled in the cell.

Genetic mapping has shown that the order of the three adjacent genes of the lactose system is *z*, *y* and *a*, shown in figure 7.1 (reviewed by Beckwith, 1970). The *z* gene codes for the enzyme β-galactosidase, which catalyses the hydrolysis of β-galactosides, such as that of lactose, to glucose and galactose. β-galactosidase is a tetramer composed of identical subunits which are unusually large with a molecular weight of some 135,000 daltons (reviewed by Zabin and Fowler, 1970). The colinearity of the *z* gene with β-galactosidase shows that it is transcribed and translated from left to right.

The protein specified by the *y* gene is the β-galactoside permease, a membrane

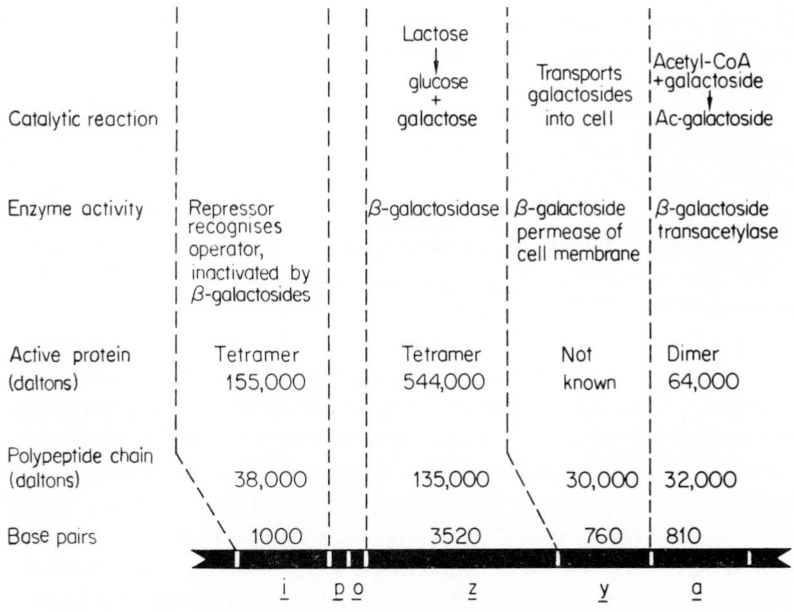

Figure 7.1: the lactose operon. The order of genes and control sites has been determined by genetic mapping; p is the promotor and o is the operator. The length of the genes is estimated from the size of the proteins for which they code and therefore does not include any possible intercistronic regions. The length of the p-o control segment is not known but may be less than 100 base pairs. Data on the products of the operon are taken from studies of purified proteins

bound protein which is part of the transport system responsible for taking up β-galactosides into the bacterial cell. The permease protein itself is poorly characterized, but seems to comprise a fundamental subunit of about 30,000 daltons (reviewed by Kennedy, 1970). Cells with mutations in either the *z* or *y* genes are lac^-, for the z^- cannot metabolise lactose and the y^- cannot take it up from the medium.

The function of the β-galactoside transacetylase specified by the *a* gene is quite mysterious; its catalytic function in vitro is to transfer an acetyl group from acetyl-CoA to galactosides (reviewed by Zabin and Fowler, 1970); but a^- mutants do not seem to suffer any disadvantage from lacking this ability and the transacetylase protein does not, so far as is known, play any role in the metabolism of β-galactosides in E. coli. Most assays for the products of the lactose genes have used β-galactosidase activity, although transacetylase is sometimes assayed as well; the *z* gene is therefore better characterized than the *y* and *a* genes.

Induction of any gene system is extremely specific and can be achieved only by the substrate for its enzymes, or other molecules closely related to it. The small molecules which cause induction are termed *inducers* (or *co-repressors* for systems of repression). Only molecules with an intact unsubstituted galactoside residue can induce the lactose enzymes. But the ability to induce does not depend on the ability to act as substrate, since thiogalactosides such as IPTG (isopropylthiogalactoside), which cannot be metabolised by β-galactosidase, are excellent inducers. Molecules which induce the system but are not metabolised by its enzymes are known as *gratuitous inducers* and are particularly useful because they remain in the cell in their original form, even when the system is active.

That there is no inherent correlation between the molecular structure of the inducer and the structure of the active centre of the enzyme is indicated by the lack of any quantitative correlation between inducing capacity and substrate activity of various galactosides. Thus melibiose, an α-galactoside lacking affinity for β-galactosidase, can act as an inducer. And certain mutants in which the β-galactosidase protein is immunologically normal but lacks its normal catalytic centre can nonetheless induce the enzyme as normal.

There is complete correlation between the induction of each of the three enzymes of the lactose system. Not only are the same compounds active or inactive as inducers of each enzyme, but the relative amounts of the enzymes synthesised in the presence of different inducers or various concentrations of one inducer are constant, even though the absolute amounts may vary considerably. This suggests that the inducer governs their coordinate synthesis by interacting with some common controlling element. That the specificity of induction does not depend on the catalytic activity of the enzymes and that the rates of synthesis of different enzymes are under common control suggests that this controlling element is not elaborated by the genes coding for the enzymes themselves.

Structural and Regulator Genes

These observations led Jacob and Monod (1961) to propose the concept that there are two different classes of gene: structural genes and regulator genes. Whilst *structural genes* are concerned solely with elaborating proteins required by the cell, *regulator genes* are responsible for controlling the synthetic activity of these structural genes (although this control is exercised through synthesis of regulator proteins). Mutants of the controlling system should not behave as alleles of the structural genes and the regulator elements of the lactose system were detected by the discovery of *constitutive* mutants.

Mutations in the structural genes cause loss of activity of their respective enzymes, but mutation in two other regions, termed o and i, can produce bacteria with the abnormal ability to synthesise large amounts of the *lac* enzymes in the absence of an inducer. The o region maps immediately to the left of the structural genes and the i region a little further along the genome. The normal alleles at these loci are termed o^+ and i^+ and the unusually active constitutive mutants o^c and i^-. Most constitutive mutants synthesize more enzyme in the absence of inducer than is produced by wild type bacteria after induction, but the relative proportions of the enzymes made are the same as in normal cells; this suggests that the regulator loci identified by these mutations constitute the control system which responds to inducer in wild type cells.

The chromosomal location of the lactose system is near a site of attachment of the F factor and F′ episomes which carry these bacterial genes can be obtained by sexduction. Bacteria which are normally halpoid can therefore be made diploid for the lactose region by introducing an F′ *lac* episome and this allows the effects of different alleles of the regulator genes to be compared in the same cell. In the reciprocal double heterozygotes

bacterial chromosome—	$i^+\ z^-$	and	$i^-\ z^+$
episome————————	$i^-\ z^+$		$i^+\ z^-$

the normal inducible allele i^+ is dominant to the constitutive i^- and is active in the *trans* position with respect to z.

Any i^+ gene present in a cell therefore controls any z^+ genes in the usual way —judged by their production of active β-galactosidase—even if the structural gene is present on a different DNA from the regulator. The same pattern of control is shown over the other structural genes, y and a. This indicates that the i mutations must belong to an independent cistron which governs the expression of all three structural genes by directing the synthesis of some component which can diffuse through the cytoplasm. The dominance of inducible over constitutive shows that the former is the active state of the gene; this conclusion has been confirmed by the isolation of strains carrying deletions of the i gene, which behave as i^- mutants.

Jacob and Monod suggested that the i^+ gene governs the synthesis of a *repressor* protein. The specificity of action of the *i* gene in affecting only the lactose system and its pleiotropy in controlling the synthesis of all three enzymes implies that the repressor acts on some feature of the system common to all the structural genes. This is the *operator* (*o*) region defined by the class of o^c mutations. In the absence of inducer, the repressor protein combines with the operator and this interaction switches off the complete *zya* segment. Addition of inducer inactivates the repressor so that it can no longer combine with the operator; in the absence of such interaction, the system resumes protein synthesis—it has been induced, as shown in figure 7.2.

In i^- mutants the repressor protein is either inactive or absent so that the operator cannot be switched off and the three structural genes synthesize their

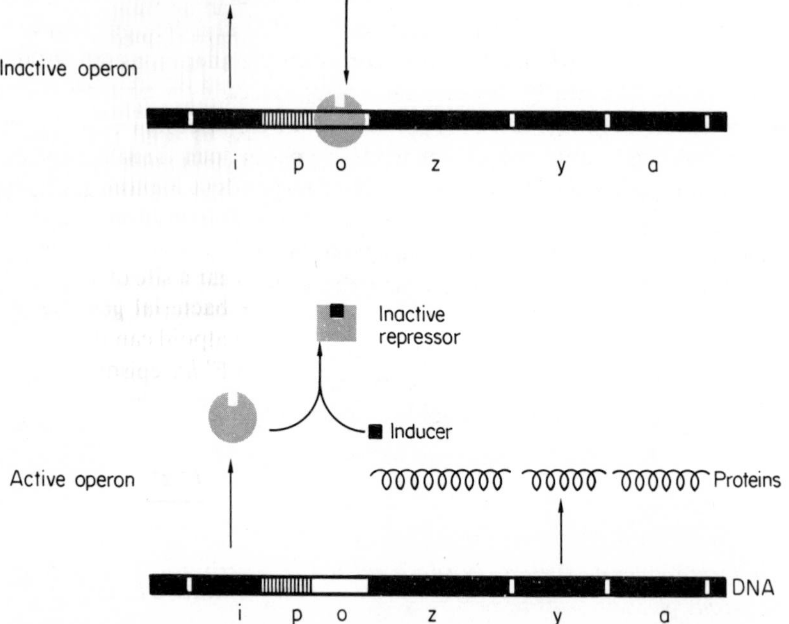

Figure 7.2: regulation of the lactose operon. The regulator gene i synthesises a repressor protein which binds to the operator (upper) and prevents the structural genes from functioning. The active form of the repressor is represented by a circle, with an insertion for the site to which inducer binds. When inducer is added (lower) it binds to the repressor protein converting it to an inactive form, represented by a square. The change in activity probably results from a change in conformation and means that the repressor can no longer bind to the operator. This allows the structural genes to direct synthesis of their protein products. The size of the p and o elements is exaggerated in this and subsequent figures

enzymes continuously. The normal i^+ allele is dominant to i^- because its presence restores normal repressor protein able to switch off the system, but responsive to inactivation by inducer. This relationship is illustrated in figure 7.3.

Other classes of mutant in the *i* regulator gene have also been isolated, some of which specify repressor proteins with unusual properties. Uninducible i^s mutants cannot synthesize the lactose enzymes, even in the presence of an inducer. This defect does not reside in the structural genes, but can be explained

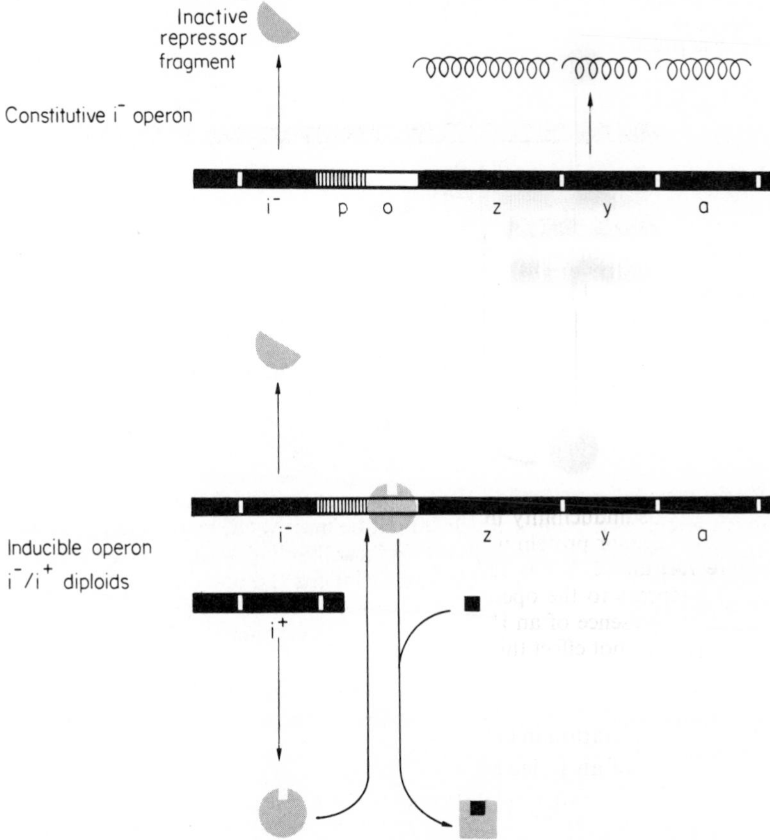

Figure 7.3: constitutive regulator mutants in the lactose operon. Inactive repressor produced by i^- mutants cannot bind to the operator. This means that the operon functions constitutively in making proteins (upper). The same result is achieved if no repressor is produced. Because any i^+ gene in the cell can make active repressor which binds to operator and is released in response to inducer, diploids containing an i^+ gene as well as the i^- mutant (lower) are normal in lactose induction; they are switched off unless inducer is added to inactivate the repressor protein. This means that i^+ is dominant over i^-

if the i^s gene produces an altered repressor, which although still active in turning off the operator is no longer subject to inactivation by inducer. The i^s mutation is dominant in i^s/i^+ diploids, because the altered repressor protein switches off the operator as shown in figure 7.4, no matter whether a normal repressor is present or not (see Willson et al., 1964).

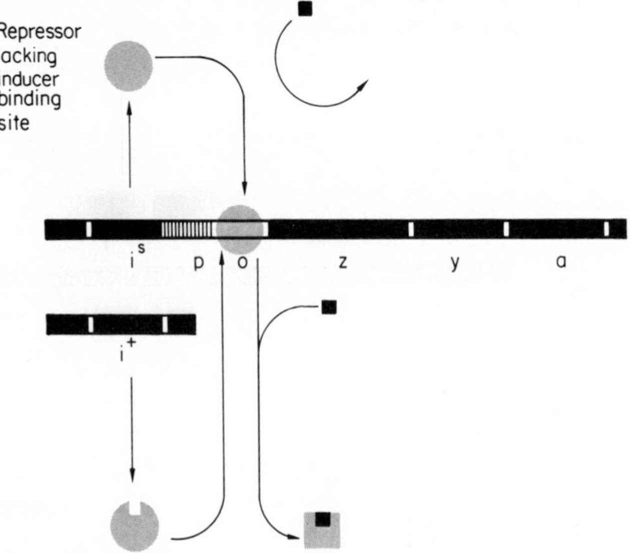

Figure 7.4: uninducibility in the lactose system. The i^s mutant gene produces a repressor protein which cannot be inactivated by the inducer (for example, because it has lost the inducer binding site). This repressor therefore binds to the operator and maintains the operon in an inactive state. The presence of an i^+ gene, coding for a normal repressor protein, cannot effect this situation; so i^s is dominant over i^+

The o^c mutants identifying the operator were originally isolated by selecting for constitutive mutation in cells made diploid for the normal lactose genes by the introduction of an F′ lac episome:

bacterial chromosome—$i^+ \ z^+ \ y^+ \ a^+$

F factor episome ———$i^+ \ z^+ \ y^+ \ a^+$

Because the i^- constitutive mutation is recessive, mutation in both i^+ alleles would be required to produce constitutivity dependent on the i gene; and such double mutation is very rare. Constitutive operator mutants could arise from an impairment of the affinity of the o region for repressor protein, in which case the structural genes could not be switched off. Figure 7.5 shows that such mutants must be dominant since the presence of a second operator sensitive to repressor cannot influence the inability of the mutated operator to respond to control. Another prediction is that if mutation from o^+ to o^c results from the loss of sensitivity of the operator to repressor, the o^c cells must be insensitive to the presence of the altered repressor synthesized by the i^s mutant; and it has been shown that o^c constitutivity holds over i^s uninducibility.

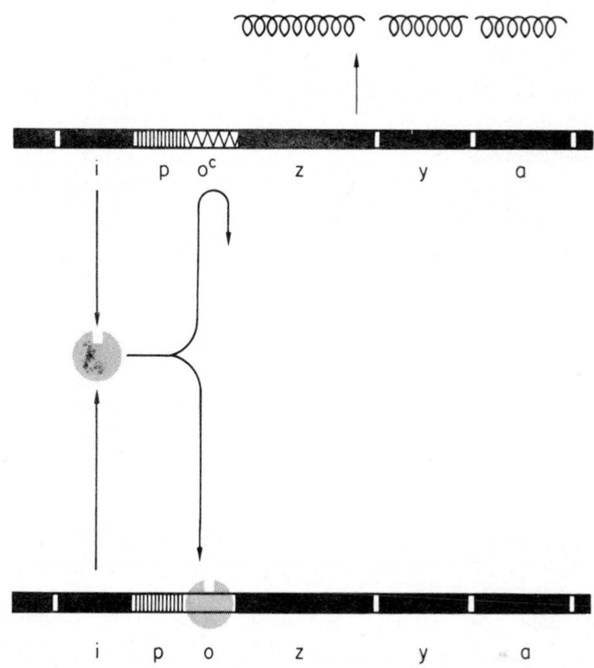

Figure 7.5: cis-dominance of operator constitutivity in the lactose operon. An o^c operator cannot bind repressor protein, so the genes adjacent to it—that is in cis arrangement—function constitutively in the absence of inducer. If a second, normal operon is present in the cell, its operator binds repressor protein normally and responds to inducer. The two operons function independently, for the o^c mutant can only control the genes contiguous with it on DNA. Cells which have an o^c operator linked to a wild type set of structural genes are therefore constitutive for expression of the lactose genes

The o^c mutation is pleiotropic, affecting all three structural genes of the system. It is *cis* dominant but *trans* recessive; lactose genes on the chromosome carrying the o^c mutation are constitutive, but those on a different chromosome remain under the control of their own operator. The operator therefore controls the adjacent genes of its own segment of DNA but has no effect on structural genes of a different DNA. According to the complementation test for the gene (see figure 1.1), the operator locus can be regarded as part of the same cistron as *z* or, indeed, by a similar argument as part of the same cistron as *y* or *a*.

These characteristics demonstrate that the operator does not code for some product which can diffuse through the cytoplasm to govern the activity of the structural genes. It must instead control some integral property of the adjacent *zya* segment by serving as a recognition element; Jacob and Monod proposed that the interaction of the operator sequence with repressor might prevent transcription of the adjacent genes. Whereas both structural and regulator *genes* require transcription and translation to fulfil their functions, regulator *sites* such as the operator function by constituting a nucleotide sequence whose recognition by a protein controls the expression of adjacent genes. All such sites may be defined by their *cis* dominant, *trans* recessive control of structural genes.

The term *operon* describes the system as a whole, including both structural genes and regulator elements. Operons are usually organized in bacteria so that the structural genes which code for a series of metabolically related proteins comprise a cluster on the chromosome. Their expression is under coordinate control and is governed by the interaction of an operator situated at one end of the gene cluster with a protein synthesized by a regulator gene (which need not necessarily map near the cluster of structural genes). Although the operon contains structural genes elaborating independent proteins, it therefore behaves as a single unit in the transfer of genetic information from DNA to protein. Some operons are controlled by a set of interactions taking the same form as those of the lactose operon, but control circuits may also be constructed from other sets of interactions; these are discussed in chapter eight.

Transcription of the Structural Genes

The model proposed by Jacob and Monod allows for the interaction between repressor protein and operator to take place either at the level of transcription of DNA into messenger RNA or (formally analogous) at the translation of the messenger into proteins. Figure 7.6 shows that control of transcription implies that the repressor must interact with the segment of DNA genetically defined as the operator; if control were at translation, this segment would have to be transcribed into mRNA and the repressor would act on a transcript representing all three structural genes.

All the systems of induction and repression which have been studied appear to be controlled at transcription, so that recognition sites always comprise base pair sequences of DNA. Whether translation of messengers is used as an additional control step in some bacterial operons is controversial. Control at the subsequent level of protein activity is well characterized and in many instances the same inducer or co-repressor influences both the production of an enzyme and its activity.

If control of the operon is exercised at transcription, inhibiting or interfering with RNA synthesis should prevent induction of the structural genes; and messengers for these genes should be present only in induced cells. Both these predictions have been confirmed. One early demonstration of transcriptional control was the observation of Nakada and Magasanik (1964) that addition of 5-fluoruracil just before addition of inducer prevents synthesis of active β-galactosidase. Incorporation of 5-fluoruracil into RNA causes miscoding,

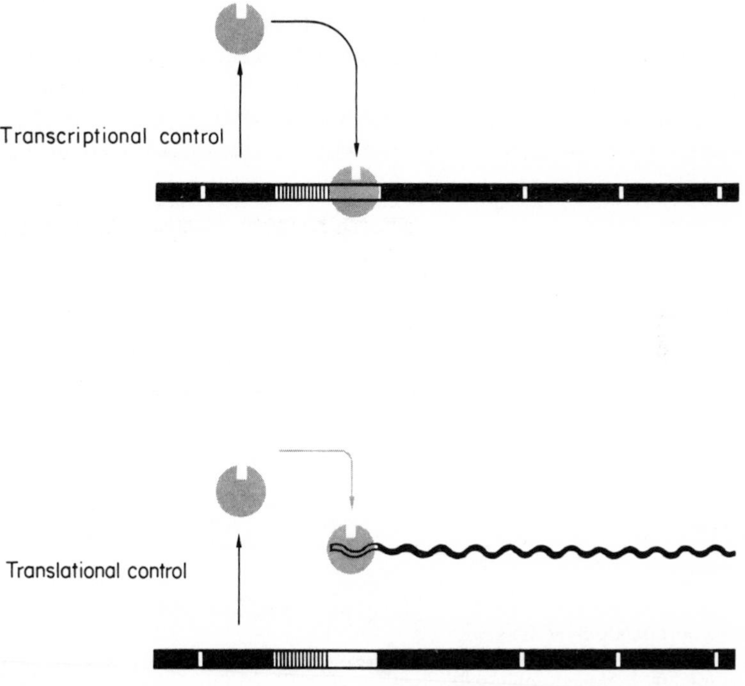

Figure 7.6: level of control of the lactose operon. If repressor protein interacts with DNA (upper) it will prevent transcription of mRNA until released by inducer. If repressor binds to the transcript on mRNA of the operator region (lower), the operon must always make messengers but can only translate them into proteins when the repressor is released by inducer. Measurements of specific RNAs in bacterial cells show that inducible and repressible operons are usually controlled at transcription

so that newly synthesized messengers are translated into altered, enzymically inactive proteins; the β-galactosidase messenger must therefore be transcribed only after induction. And by comparing the RNAs of induced and non-induced cultures, Guttman and Novick (1963) showed directly that the only change in cellular RNA following induction of β-galactosidase is the synthesis of a new RNA species sedimenting at about 30S.

RNA complementary to the lactose operon can be detected by hybridization with the DNA of phages carrying the lactose genes. In some strains of E.coli, the lactose operon has been transduced from its usual position on the genetic map to a locus near the attachment site of phage $\phi 80$ and $\phi 80$d*lac* phages have been obtained from these cells. (The phage is defective, as signified by the "d" in $\phi 80$d, because some of the phage genes have been replaced by the lactose genes.) There is little homology between the DNA of the phage and bacterial DNA apart from the lactose genes which are present (and any other adjacent regions of the bacterial chromosome which may also have become attached to the phage). RNA extracted from bacterial cells which have not been induced therefore anneals only poorly with the phage DNA; this homology probably represents non-lactose regions of the bacterial chromosome which are also carried on the phage. Addition of IPTG to the bacterial culture to induce the lactose enzymes causes a much greater extent of hybridization

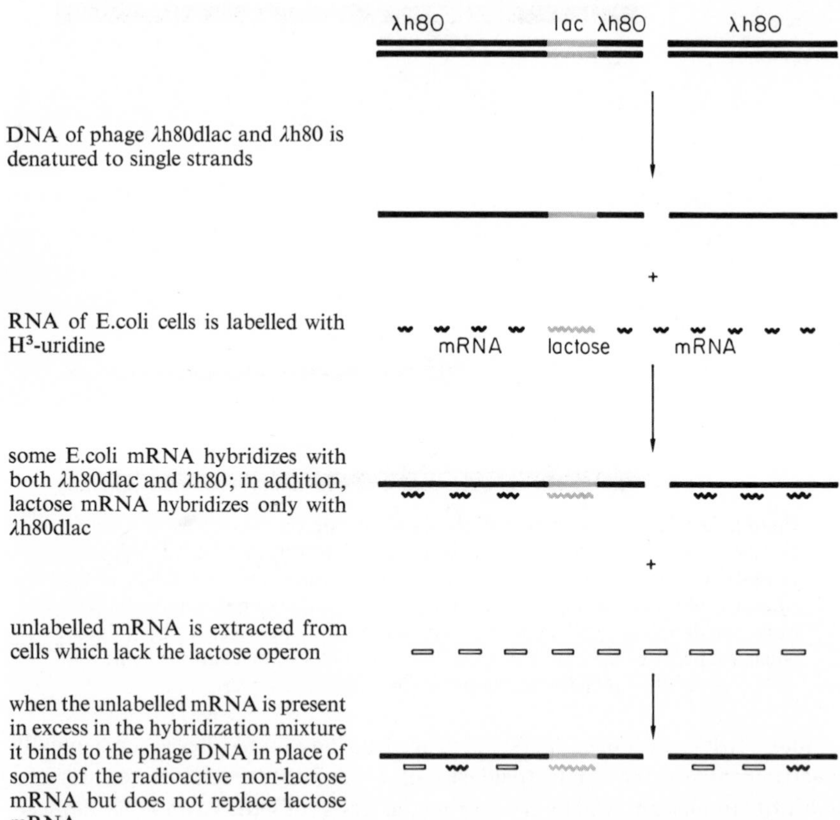

λh80
lac λh80
λh80
DNA of phage λh80dlac and λh80 is denatured to single strands
+
RNA of E.coli cells is labelled with H³-uridine
mRNA
lactose
mRNA
some E.coli mRNA hybridizes with both λh80dlac and λh80; in addition, lactose mRNA hybridizes only with λh80dlac
+
unlabelled mRNA is extracted from cells which lack the lactose operon
when the unlabelled mRNA is present in excess in the hybridization mixture it binds to the phage DNA in place of some of the radioactive non-lactose mRNA but does not replace lactose mRNA

Figure 7.7: hybridization assay for lactose mRNA. The difference in hybridization of RNA with λh80 dlac and the λh80 control gives the amount of lactose mRNA. Background hybridization of labelled non-lactose mRNA is reduced by including excess unlabelled RNA from E.coli cells deleted for the lactose genes

(reviewed by Contesse, Crepin and Gros, 1970). Such experiments show clearly that transcription of the lactose genes precedes the synthesis of β-galactosidase and the other enzymes of the operon.

Another phage which has been used for these experiments is λh80, a derivative from recombinants containing a λ immunity region but a ϕ80 host range and attachment site. Both direct and competition hybridization assays for reaction of lactose RNA with the DNA of λh80d*lac* phages have been developed by Varmus, Perlman and Pastan (1970a, b). The protocol for the direct assay, shown in figure 7.7, is to hybridize RNA extracted from E. coli cells labelled with H^3-uridine with control λh80 DNA (which does not carry the lactose genes) and with the transducing phage λh80d*lac*. Lactose mRNA binds only to λh80d*lac* DNA.

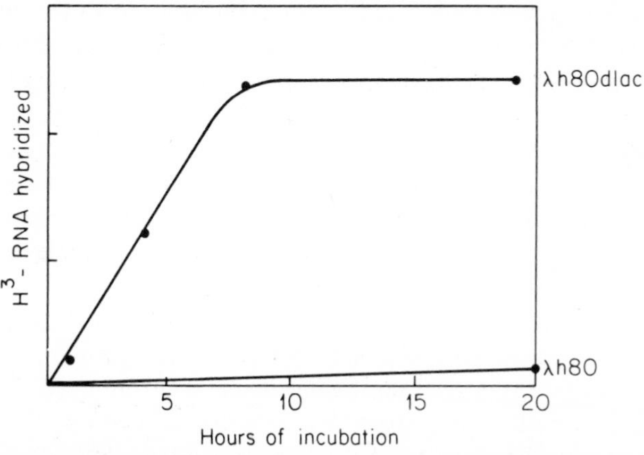

Figure 7.8: measurement of lactose mRNA by hybridization with the DNA of λh80dlac. The low level of hybridization with the λh80 DNA used as control shows that the assay is specific for lactose messengers. Data of Varmus, Perlman and Pastan (1970a)

There is some binding of other RNAs to the phage part of the DNA, but this can be compensated by subtracting the amount of RNA hybridized to the λh80 DNA itself. Interactions between the phage DNA and bacterial RNA which are not due to lactose RNA can be reduced by adding a large excess of unlabelled RNA extracted from a lac deletion strain. The unlabelled RNA competes with all the radioactive RNAs except lactose mRNA, so that the lactose-specific hybridization forms a greater proportion of the total. Figure 7.8 shows that efficient hybridization with the lactose genes on the phage is found only after induction of the lactose operon.

The direct assay measures the rate of lac mRNA synthesis relative to the rate of total RNA synthesis and does not estimate its concentration in the cell. This can be deduced, however, from competition hybridization experiments in which λh80d*lac* DNA is hybridized with enough labelled RNA from induced cultures to saturate the lac binding sites on the DNA molecules. Unlabelled RNA preparations extracted from cells grown under different experimental conditions can then be tested for their ability to compete for the lactose DNA (see figure 7.15). Addition of increasing concentrations of IPTG to cells induces increasing amounts of competing RNA, whereas RNA from uninduced cells competes hardly at all. And because competition experiments compare unlabelled RNA preparations, they do not suffer the disadvantage of falling subject to variations in the rate of H^3-uridine incorporation. These results therefore allow the number of lac mRNA molecules in an induced cell to be calculated; this is about ten on average.

The kinetics of expression of several operons, including lactose, galactose and tryptophan, show that their proteins are synthesized sequentially following induction, starting with that coded by the gene adjacent to the operator (see chapter nine). This suggests that transcription of messenger RNA can start only at this end of the gene cluster, the polymerase proceeding through the subsequent genes in order and thus sequentially generating the template for translation. The immediate product of transcription is therefore polycistronic, comprising an RNA chain representing all the genes of the operon. The existence of a single site adjacent to the operator where transcription can be initiated explains how the interaction of repressor at a single operator site controls all the structural genes (see later).

Experiments to detect transcription of particular regions of the operon have been possible with the tryptophan gene cluster of E.coli, for which a series of ϕ80 phages carrying the structural genes have been prepared from various bacterial strains with well defined deletions in the operon. By testing the hybridization with these DNAs of RNA extracted from cells synthesizing the tryptophan enzymes, Imamoto et al. (1965a, b) demonstrated that the messengers include the sequences of all the structural genes; these RNA sequences are absent from cells with an inactive tryptophan system. When Imamoto (1968a, b) synchronised the start of transcription of the operon, he was able to show that RNA synthesis starts at the end proximal to the promotor and operator regulator sites and proceeds sequentially along the structural genes to the far end (see chapter nine).

These experiments confirm that messenger RNA is sequentially transcribed from several cistrons, but do not say whether it is translated in this form or is first broken down into smaller units of one gene each before protein synthesis. Polarity effects, in which a mutation in one gene influences the translation of genes beyond it in the cluster, strongly suggest that the polycistronic messenger is translated into successive proteins as it is transcribed (see chapter nine). The isolation of polysomes synthesizing specific proteins also suggests that translation takes place on polycistronic mRNAs. Kiho and Rich (1965) found polysomes carrying nascent β-galactosidase and observed that genetic modification in the lactose structural genes is associated with changes in the size of these polysomes; bacterial strains with deletions in either the *z* or *y* genes have smaller polysomes. Bagdasarian et al. (1970) have detected the phosphohistidine phosphatase enzyme of the histidine operon on polysomes too large to code for one protein alone; they are probably polycistronic messengers representing the entire operon. Polycistronic messengers are therefore the units of expression of the operon for both transcription and translation.

The kinetics of induction show that the addition of inducer causes synthesis of protein at the maximum rate within a very short period. This induction lag is the time required to transcribe mRNA, translate it into polypeptide chains, and assemble the polypeptide monomer subunits into active proteins. The equally rapid cessation of synthesis when inducer is removed implies that the messenger RNA cannot be stable; either it must be inherently unstable or it must be destroyed by enzymes very soon after its synthesis. Indeed, the instability of mRNA is an integral part of the control of the operon. Otherwise an operon would remain functional after it has been switched off, because the stable mRNAs would continue to synthesize proteins even though no further RNA were to be made.

The instability of mRNA was inferred from experiments in which Jacob and Monod (1961) introduced a chromosome loaded with P^{32} from z^+ cells into z^- recipients by conjugation; the radioactive decay of the P^{32} inactivates the z^+ gene by mutation. β-galactosidase is made in the progeny cells only so long as the integrity of the z^+ gene is maintained. This result implies that new messengers must be synthesised continually from DNA because those previously made do not survive for long. The decay of mRNA can be measured directly by the decrease in synthesis of β-galactosidase or the reduction of hybridizable mRNA in cells when inducer is removed.

The half life of lactose mRNA is about 2·0–2·5 minutes and the rate of decay is independent of the presence of inducer or other factors such as the rate of protein synthesis or energy production (Leive and Kollin, 1967). This short half life seems to result from degradation of the messenger by nucleases very soon after its transcription has commenced (see page 384) and the rate of decay of lactose mRNA can be altered by mutations at a locus which may code for one of the enzymes involved (Varmus, Perlman and Pastan, 1971b). Thus the intracellular level of lactose messenger—and presumably of other messengers also—depends only on its rate of synthesis and on the activity of the enzymes responsible for its degradation.

Interaction of Repressor and Operator

Promotor for the Structural Genes

In their original model, Jacob and Monod suggested that the operator might possess the two properties of acting as the recognition site for repressor and initiating the transcription (or possibly translation) of RNA. At this time it was difficult to say whether the operator might be a separate element or comprise the starting part of the *z* gene. Fine structure mapping, however, localizes the operator to the left of all known *z* gene mutants (see Reznikoff and Beckwith, 1969); and the β-galactosidase enzyme produced in operator mutant strains is always wild type. All recognition sites concerned with the control of transcription appear to share this characteristic; they lie adjacent to but are not part of the gene sequence coding for protein.

The idea that the operator might mediate both the initiation of transcription and its control was supported by the isolation of mutants in which the ability of the operon to produce its enzymes is completely lost. These are recessive to o^c and o^+ and map very close to the o^c mutants; thus they were at first interpreted as representing some form of inactivation of the operator and were termed o^o. However, Beckwith (1964) showed that "o^o" mutants can be restored to normal activity by external nonsense suppressors; this means that they result from chain terminating nonsense mutations very early in the *z* gene which act by preventing translation rather than transcription. Accurate mapping experiments have since confirmed the conclusion that "o^o" mutants are really a form of extreme polar mutation in the *z* gene (see page 394).

The ability of cells to make lactose enzymes is clearly distinct from that of regulating gene expression, for o^c mutants which cannot respond to repressor protein are nevertheless transcribed into RNA and translated into the proteins of the operon. The existence of a promotor region, distinct from the operator, where RNA polymerase might bind and initiate transcription was first inferred from two lines of evidence. Jacob, Ullman and Monod (1964) found that they could not induce constitutive operator mutants with 2-aminopurine or ultraviolet irradiation and noted also that o^c mutants did not seem to revert to wild type. They suggested that o^c mutations are the result of deletions in the operator region; this idea was lent support by the finding that many o^c mutants also appeared to be i^-, which would suggest that they arose by deletions extending across both the *i* gene and operator site. The ability of these cells to transcribe the operon was thought to mean that the deleted region is not needed for binding RNA polymerase.

A site apparently indispensable for the activity of the operon was isolated by examining revertants to *lac* y^+ activity from diploid i^s uninducible mutants. Because i^s is dominant over i^+ or i^-, these revertants should arise almost entirely from deletion of the operator to yield o^c mutants. But when they were mapped, it appeared that none of these deletions extended over the operator into the *z* gene. Because neither the absence of the operator nor the deletion of part of the *z* gene alone prevents expression of the *y* gene, this was taken to imply that there must be some site between *o* and *z* which is essential for operon function; its deletion would prevent expression of all three structural genes. This could have been the promotor site where RNA polymerase is recognised and commences transcription.

Support for the concept that the promotor comprises a site on the genome separate from the operator was lent by the isolation of strains with the properties which might be expected to result from mutation at this locus. Such mutants are *cis* dominant and *trans* recessive—this means that they influence only the structural genes on their own DNA molecule—and reduce the maximum rate of operon expression without influencing the mechanism of regulation. Certain base transitions lower the activity of the lactose operon some fifteen fold, presumably by reducing the affinity of the promotor for proteins needed for transcription (see Scaife and Beckwith, 1966; Arditti, Scaife and Beckwith, 1968). (Mutations in the promotor of the *i* gene may have an even more striking effect.) By using the deletion strains across *o* and *i*, the structural promotor mutants were mapped at the expected site between *o* and *z*.

The original arguments for the existence of the promotor have proved to be misleading, however (reviewed by Miller, 1970). Most of the putative deletions across *i* and *o* have in fact proved to be point mutations of the i^{-d} type (which synthesises the dominant i^- repressor discussed below); others are double mutants. And although o^c mutation can result from deletions, the operator has now been characterised by the isolation of many o^c mutants which result from point mutation. Deletions extending from the operator into the *z* gene have also now been isolated, so that both the results taken to argue for the idea of a promotor between *o* and *z* have proved invalid. Eron, Beckwith and Jacob (1970) showed that transacetylase can be synthesized in mutants with deletions of *o* and the beginning of *z*; this region is therefore essential neither for transcription nor for translation.

When Ippen et al. (1968) reexamined the mapping data obtained with the apparent *i-o* deletions, they found that by taking the real nature of the mutants into account they could place the mutants in the promotor site between *i* and *o*, instead of between *o* and *z*. Although the original arguments used to support the idea of the promotor do not in fact identify it, therefore, this site exists nevertheless and is located before the operator with regard to the structural genes of the operon. That the promotor site is essential for transcription of the operon has been confirmed by hybridization studies to show that the promotor mutant strains producing little β-galactosidase also produce little mRNA on induction (reviewed by Contesse, Crepin and Gros, 1970). Studies on the transcription of the lactose operon in vitro from phage DNA templates carrying the lactose genes have also shown a correlation between ability to transcribe lactose RNA and the genetic constitution of the promotor (see page 306). The promotor is therefore probably the site at which RNA polymerase binds to DNA.

The order of sites in the operon, promotor–operator–structural genes, suggests the model for control of the lactose system depicted in figure 7.9. By binding to the operator, repressor protein prevents RNA polymerase from entering the structural genes. Addition of inducer releases the repressor and

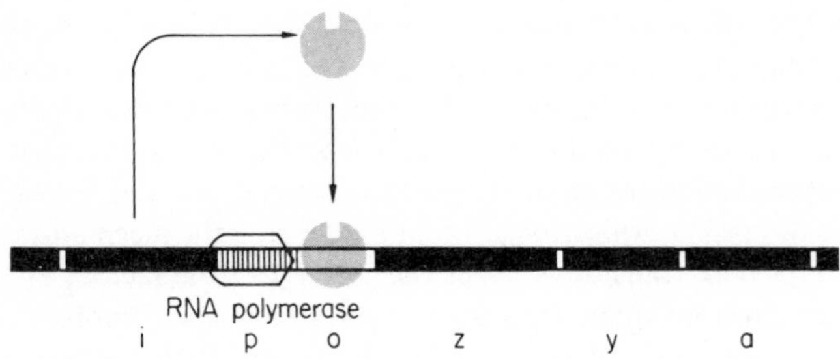

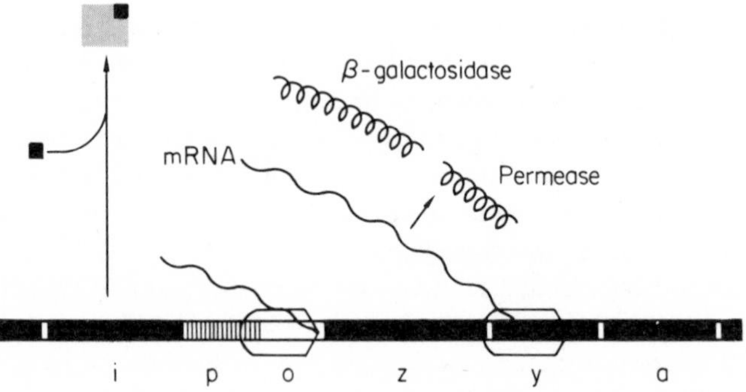

Figure 7.9: model for the action of repressor protein. In repressing conditions (upper) the repressor binds to the operator sequence and prevents progress of the RNA polymerase enzyme into the operon. There is therefore no transcription. When inducer is added (lower), the repressor is converted into an inactive form which no longer binds at the operator. This removes the impediment to the progress of the polymerase and transcription produces a messenger RNA for the complete operon which is translated as such. More than one polymerase molecule may be active on the operon at any given moment

allows the polymerase to continue, thus switching on the operon. Experiments using lactose DNA in vitro support this model (see) later). If transcription commences at the polymerase binding site, the operator as well as the structural genes must be transcribed into RNA and should be represented by the sequence at the starting (5′) end of the messenger.

Sequence of the Operator

It is difficult to estimate the size of the regulator regions of the operon because genetic mapping results are not reliable over small distances. Recombination is subject to high negative interference—the occurrence of one recombination event stimulates others nearby—and to marker specific effects, when the presence of a mutant influences the recombination frequencies. Deletion mapping has proved more successful for ordering mutants, but again does not give accurate estimates of distances. However, it is possible to deduce the *minimum* size of recognition regions.

An important characteristic of the operator should be that the same sequence of base pairs does not occur by chance elsewhere in DNA coding for proteins. The possibility that a given sequence will be present in the genome can be calculated from the base composition of DNA, for although base sequences are not random over short distances, they are effectively random over lengths of ten base pairs or more. For the operator to possess a high probability of comprising a unique sequence, it must be some 12 base pairs long. Sadler and Smith (1971) have pointed out, however, that this demand is not sufficient, for point mutations in the operator can generate sequences which still have an appreciable affinity for the lactose repressor. If the fortuitous occurrence of sequences only one base different from the operator is to be avoided, the minimum number of base pairs in the operator becomes 15; if sequences only two bases different also present a threat, the length of the operator must be greater than 18 base pairs. This minimum sequence need not be continuous, but might be enclosed in a longer sequence containing spacer elements whose sequence is not preserved. This argument applies not only to operator regions, of course, but to any recognition site (such as a promotor) whose role is to be recognised by a protein.

Another way to deduce the minimum size of a regulator element is to try to mutate every base in it and to order the mutants; the number of different sites which can be mutated identifies a minimum number of base pairs for the region. By inducing a large number of o^c mutants with mutagens of different specificities, Sadler and Smith have shown that both AT and GC base pairs are present in the operator. Ordering 26 o^c mutations identified 16 sites of point mutation (the rest were deletions covering more than one site). The operator must therefore consist of at least 16 base pairs; there may, of course, be other base pairs which are not mutated.

The classification of these o^c point mutants suggests that there may be symmetry in the sequence of the operator, for an equivalent set of mutations seems to be found in each of two segments. One possible explanation is that this reflects a symmetry in the way repressor binds to protein; each half of the operator might bind one subunit of the repressor. The repressor could certainly cover a larger region than this, for each subunit could bind to between 10 and 20 base pairs, making up one or two turns of the double helix, depending on the tertiary conformation of the protein. It is possible, of course, that only part of the repressor protein recognises and covers DNA whilst the rest remains accessible to the inducer.

One surprising result which Sadler and Smith found when analysing their o^c mutants was that only one class of mutant could be found at each site of the operator, whereas three classes should be possible (corresponding to all three possible base pair substitutions). This implies that there may be some peculiar restraint on the sequences which can be present at the operator; it is difficult to see why certain changes in the operator sequence should be lethal to the cell, but one possible explanation is that the structure of the operon is important in some way for cellular functions other than expression of the lactose genes. Another pointer that events at operator may not be autonomous is the observation that, although operator and promotor are distinct, there may be some sort of interaction between them, for many of the o^c mutations have some effect on the activity of the promotor.

Isolation of Repressor Protein

The role demanded of the repressor suggests that it should be protein since it can exist in active or inactive forms as the result of its interaction with a small molecule inducer. This is most readily explained as a conformational effect; the protein must possess a binding site for the operator whose activity is controlled by the binding of inducer at a second site. This idea provides a molecular explanation for the behaviour of repressor variants characterised by the properties of the *i* gene mutants. The i^- mutants must have lost their binding site for operator, so that they can no longer switch off the operon; and the i^s mutants must contain repressors which can no longer bind inducer, and so cannot respond to induction. The formal model of the operon, however, remains the same irrespective of the molecular nature of the repressor; for indeed, when the idea of the repressor was first proposed it seemed probable that RNA would be implicated rather than protein, as no way could be seen at the time for protein to interact specifically with DNA base pair sequences.

But the *i* gene shows all the properties of a gene coding for protein. Some i^- constitutive regulator mutants can be suppressed by external nonsense suppressors (Bourgeios, Cohn and Orgel, 1965); this implies that the active gene product must be made by translation of mRNA into protein. Complementation between different *i* mutants can take place in diploid cells; for example, certain i^-/i^+ cells show the uninducible i^s phenotype; this suggests that the repressor protein is polymeric and that the subunits specified by these particular i^- and i^+ alleles can interact to form a multimer which has the properties of the protein usually made by assembly of the i^s subunits. By using intragenic complementation between certain pairs of temperature sensitive mutants, Lieb (1969) has been able to show that the phage λ repressor also is polymeric.

The amount of repressor in wild type cells is very small but mutants have been obtained which make more repressor. The technique used by Gilbert and Muller-Hill (1970) was to revert to wild phenotype a temperature sensitive repressor, i^{tss}, by growing cells on a medium in which constitutive bacteria die. To survive, cells must produce sufficient molecules of active repressor. This might happen in either of two ways. A second mutation in the regulator gene might restore activity to the repressor protein; or cells might gain a mutation to cause the synthesis of more repressor, so that there would be enough active molecules at the high temperature. This second type of mutation generated the double mutant $i^q i^{tss}$.

The i^q mutant has been recovered separately by genetic recombination. Cells bearing i^q make about ten times more repressor than wild type and the mutation maps at the far end of the *i* region; it is probably a mutant in the promotor of the *i* gene. The amount of repressor made in bacterial cells can be increased yet further by placing the genes on a defective phage, which makes many hundreds of copies of DNA carrying the lactose genes when it infects bacteria, but does not lyse the cells. In cells infected with λ dlac carrying an i^q mutation, as much as 0·5% of the cellular protein is lactose repressor.

The repressor has been isolated from such cells by purifying an extract for its ability to bind IPTG. It is a protein tetramer of identical subunits of 38,000 daltons each. IPTG-binding protein can only be isolated from i^+ cells and cannot be found in cells carrying mutant *i* regulators which do not possess a repressor with the ability to bind inducer according to their genetic definition. Temperature sensitive mutants in the *i* gene, however, produce a IPTG binding protein of altered temperature stability. This provides clear evidence that the IPTG binding activity identifies the protein product of the *i* gene. There are only about ten copies of repressor made in a normal cell, although its synthesis is subject to gene dosage and is increased appropriately in diploid and triploid cells. Repressor proteins have also been isolated for other systems; their properties are discussed in the next chapter.

Binding of Repressor to Operator DNA

If the repressor acts on DNA to prevent transcription, it should bind in vitro to DNA containing the lactose operator, but not to DNA lacking this recognition site. Two types of assay have been used to measure the binding of repressor to DNA. Gilbert and Muller-Hill (1967) sedimented labelled repressor protein together with λDNA carrying the lactose genes on a glycerol gradient; figure 7.10 shows that the IPTG binding protein associates with DNA containing the operator region of the lactose operator but does not bind to the DNA when the genome of an o^c mutant is substituted. The addition of the IPTG inducer releases the repressor from DNA. There are four IPTG binding sites on the repressor for each DNA binding site, which accords well with the tetrameric structure.

A more sophisticated technique for measuring the binding of repressor to DNA has been developed by Riggs, Newby and Bourgeois (1970) by using nitrocellulose filters. Free DNA is washed through the filters, but DNA complexed with repressor protein is retained. When increasing amounts of purified repressor are added to a fixed amount of P^{32} DNA, increasing amounts

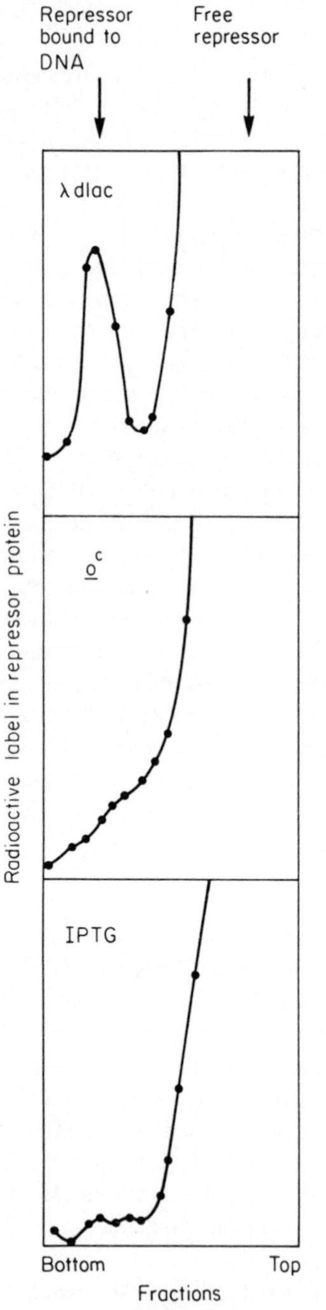

when a mixture of radioactive repressor protein and λ dlac DNA is sedimented through a glycerol gradient, repressor protein binds to the DNA and sediments rapidly with it.

binding is specific for DNA containing a lactose operator sequence of wild type, because when the phage carries a lactose operon derived from a cell with an o^c mutation, little repressor protein binds to DNA. (The extent of binding to DNAs with mutant operators varies with the individual mutation).

the addition of $1{\cdot}2 \times 10^{-4}$ M IPTG releases the repressor from dlac DNA.

Figure 7.10: binding of repressor protein to lactose DNA. Free repressor sediments slowly, at about 7S; the DNA of phage λdlac (and any repressor protein bound to it) sediments more rapidly, in excess of 40S. Data of Gilbert and Muller-Hill (1967)

of DNA are retained on the filter until a plateau is reached when the operator is saturated with repressor. DNA binds to the filter only when it carries the lactose operator; there is little or no binding in operator mutants and IPTG is a very effective inhibitor of operator-repressor binding. Repressor binds efficiently only to duplex DNA; it is one hundred times less effective with denatured DNA. Figure 7.11 shows that the relative efficiencies of various inducers as inhibitors of binding in vitro are the same as their activities in promoting induction in vivo, which suggests that the binding assay and its inhibition by IPTG reflects the manner of control of induction in the bacterial cell (reviewed by Bourgeois, 1971).

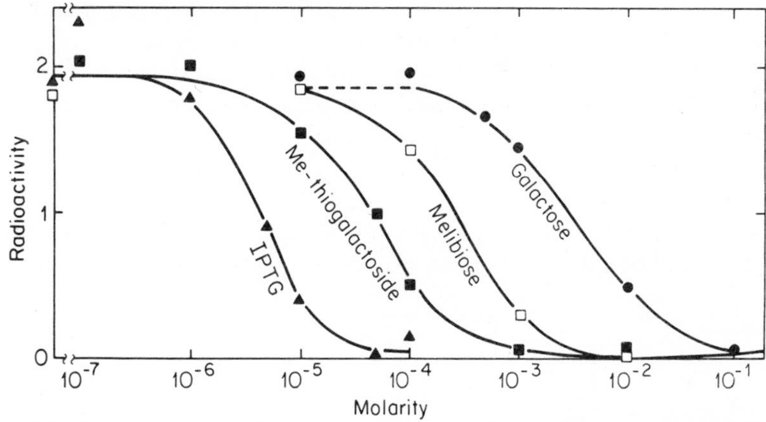

Figure 7.11: effect of inducers on binding of lactose repressor to operator. Labelled DNA containing the operator is incubated with repressor in the presence of an inducer for five minutes and then filtered. Only DNA bound to repressor is retained by a nitrocellulose membrane. All inducers cause increased reduction of repressor–operator binding as their concentration is increased, in the order of effectiveness IPTG, methyl-thiogalactoside, melibiose, galactose. Data of Riggs, Newby and Bourgeois (1970)

Lactose, which is the substrate for β-galactosidase, is not itself the natural inducer of the operon. By growing cells on lactose, Jobe and Bourgeois (1972) were able to isolate the repressor protein bound to its natural inducer; the sugar eluted from the repressor is allolactose. Digestion of lactose by β-galactosidase gives glucose and galactose as the main reaction products. However, the enzyme also transfers the galactose moiety to certain acceptor molecules; the main products of this side reaction are allolactose and galactobiose. (Allolactose is produced by transfer of the galactose to glucose, so that the reaction comprises a molecular rearrangement.) When lactose is digested with β-galactosidase and the products are separated and tested in vitro for their ability to release DNA from its complex with repressor protein, only allolactose is effective. Allolactose is as good as IPTG as an inducer in vitro and is more effective in vivo.

Cells must therefore contain a basal level of β-galactosidase which produces allolactose when cells grow on lactose; this allolactose then induces the operon. Jobe and Bourgeois (1973) observed that lactose itself is an anti-inducer and can block the effect of IPTG on repressor-operator binding. The parameters of the reaction with repressor of lactose and allolactose suggest that within the cell the inducing action of the content of allolactose is sufficient to overcome the anti-induction of lactose and therefore to ensure that growth on lactose induces the operon.

Repressor protein binds very tightly to DNA and the rate of dissociation of the complex can be measured by a challenge technique with a second template. After repressor has been bound to labelled DNA, a large excess of unlabelled DNA is added. The unlabelled DNA can bind repressor only after the protein has been released by dissociation from its pre-existing complex with labelled DNA. Following the decline in the radioactive counts retained on the nitrocellulose filter as the mixture is incubated therefore gives the rate of dissociation of the complex. Dissociation depends on the ionic conditions, but is rather slow. In 0·2M salt the half life of the repressor-operator complex is 5-6 minutes; in 0·05M salt it is 19 minutes.

But the addition of IPTG greatly reduces the stability of the complex and

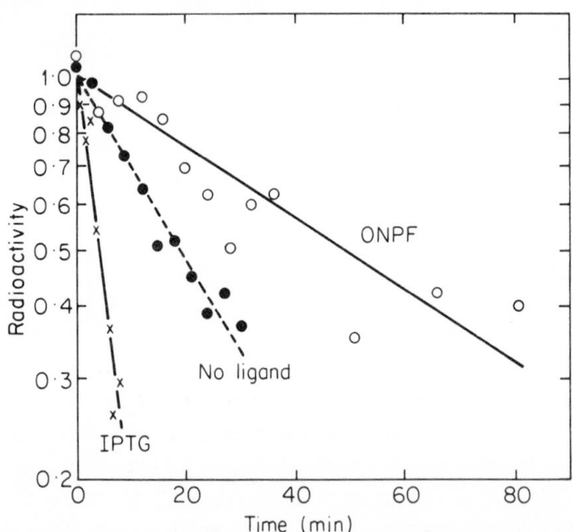

Figure 7.12: release of repressor from lactose operator DNA. DNA containing the lactose operator is mixed with sufficient repressor to reach saturation; after incubation for five minutes in 0.05 M salt, IPTG is added to reach a concentration of 3×10^{-6} M no ligand is added, or ONPF is added to reach 1.8×10^{-3} M. Samples are removed at increasing time periods and filtered through nitrocellulose membranes to assay the retention of labelled DNA. IPTG increases and ONPF decreases the rate of dissociation. Data of Riggs, Newby and Bourgeois (1970)

dissociates repressor from DNA much more readily. Figure 7.12 shows that in 0·05 M salt, a 3×10^{-6} M concentration of IPTG reduces the half life to about 4 minutes. Correspondingly, the anti-inducer ONPF (o-nitrophenyl-fucoside) increases the stability of the complex to a half life of about 50 minutes. These results imply that the action of repressor is not mediated by allostery. Allosteric interactions would demand that the repressor is in *rapid* equilibrium between a free and a DNA bound form. Figure 7.13 shows that according to an allosteric model, inducer would free DNA of repressor by binding to the free form of the protein to upset the equilibrium.

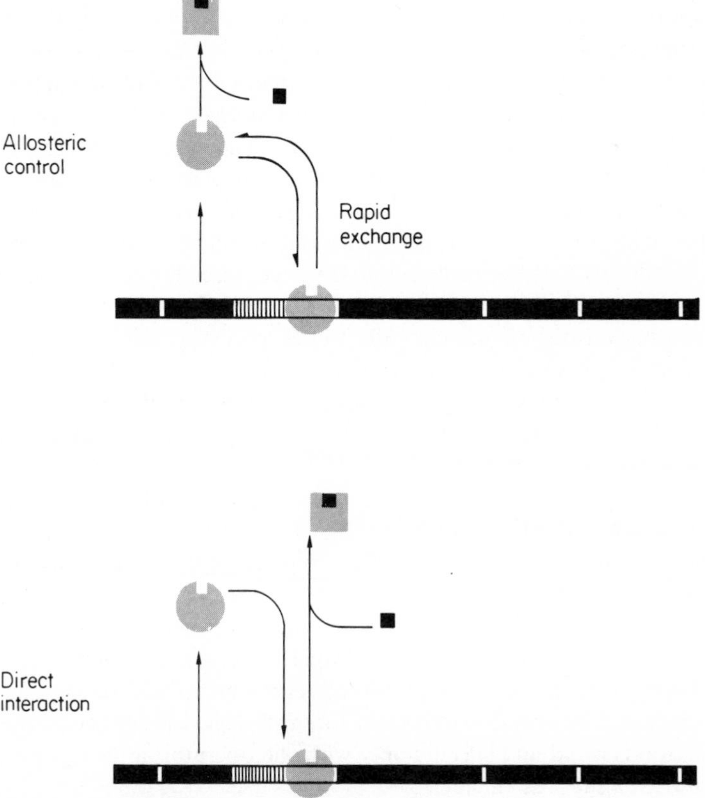

Figure 7.13: models for release of repressor from lactose DNA by inducer. Allosteric control (upper) demands that the repressor protein is in rapid equilibrium with operator DNA; the inducer binds only to free repressor, thus upsetting the equilibrium and ensuring that molecules of repressor do not return to the operator once released from it. Direct interaction (lower) supposes that inducer may bind to repressor protein complexed with the operator, thus reducing its affinity for the DNA and releasing it. This model is supported by the slow exchange of free and bound repressor in vitro and the effect of inducer on repressor–operator dissociation rates

But the dissociation of repressor from the operator is a slow process compared with the speed of induction. Instead of an allosteric interaction, IPTG seems able to react with repressor protein already bound to DNA to change its affinity for the operator; figure 7.13 shows that, in effect, this prises the repressor protein away from the DNA by increasing its rate of dissociation. That this interaction is caused by a conformational change in the protein is supported by the demonstration of Laiken, Gros and Von Hippel (1972) that the fluorescence spectrum of the repressor is changed when IPTG binds to it.

The overall equilibrium constant for binding between repressor and operator shows that complex formation is very stable, for the rate of association with repressor is very fast. Electrostatic interactions are implicated by the effect of increasing salt concentration and an important influence is probably the reaction between negatively charged phosphates on the operator and positively charged groups on the protein. Although the repressor can be characterised as an acidic protein by its isoelectric point, that it has basic groups able to react with DNA is implied by its ability to displace the phosphate groups of phosphocellulose columns. The speed with which the repressor recognises the operator implies that forces other than random diffusion must be involved; one possibility is that the protein initially recognises DNA itself and moves along the genome until it reaches its stable binding site at the operator.

By binding various oligonucleotides to the repressor, Lin and Riggs (1971) found that poly-dAT competes very effectively with the operator DNA. The nature of the substituent at the 5′ position of the pyrimidine ring is important in determining repressor affinity, however, for poly-dAU is much less effective than poly-dAT whereas poly-dA-BRdU is much more effective. This conclusion has been confirmed in vivo by experiments in which Lin and Riggs (1972a) substituted BRdU for thymidine in E.coli cells and prepared P^{32}-labelled λh80d*lac* DNA with about 90% replacement of thymidine. The loss of the P^{32} label bound to a filter is ten times slower than when normal DNA is used. The strong influence of BRdU on repressor binding implies that the operator may have a comparatively simple structure very rich in A-T base pairs. In measurements of the ability of repressor to bind to various DNA species lacking the operator sequence, Lin and Riggs (1972b) confirmed that the reaction is rapid and specific only with the operator, although poly-dAT is recognised almost as efficiently.

The affinity of repressor for DNA can vary even in cells which show normal induction, for Jobe, Riggs and Bourgeios (1972) found that repressors from different E.coli "i^+" strains are not identical in their binding properties. Figure 7.14 shows that in such strains the half lives of the repressor operator complexes (measured in 0·05 M salt) are 20, 45 and 75 minutes. This implies that the operon can be controlled in the same manner even when the precise affinity of repressor for DNA is altered; presumably, these variations in the repressor do not make enough difference to the protein to alter the pattern of

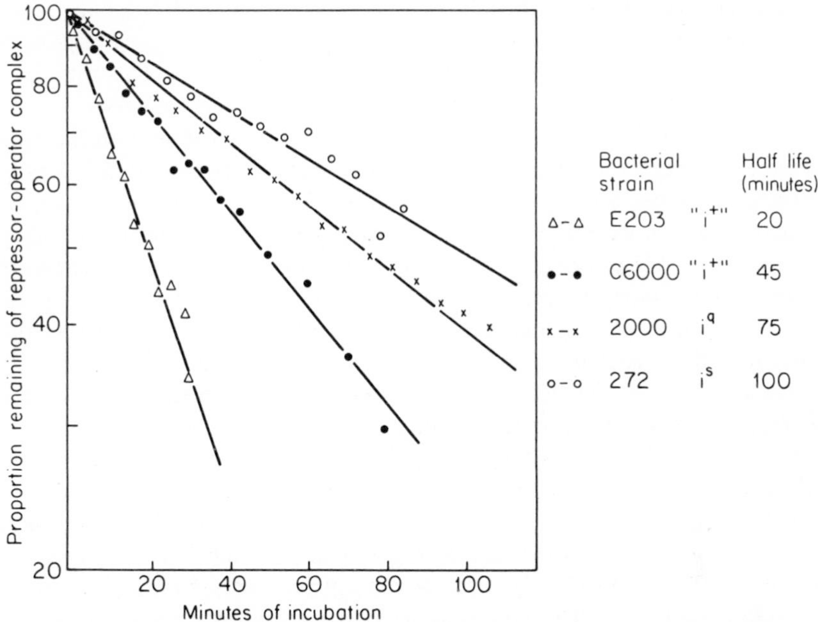

Figure 7.14: dissociation of lactose repressor–operator complex. Repressor isolated from three strains with "wild type" protein (the i^q mutation affects only the promotor and not the structure of the i gene) varies in affinity for the operator (measured in 0.05 M salt). Repressor from an i^s uninducible mutant strain has an increased affinity for operator. Data of Jobe, Riggs and Bourgeois (1972)

control of the operon. The prediction that i^s mutants should not bind inducer has been confirmed in eleven strains of E. coli; eight of these repressor proteins have normal operator binding and three an even tighter binding than usual, which implies that the two sites of the repressor protein can be mutated independently.

Repression of Transcription

How does the binding of repressor protein at the operator inhibit transcription? One possible model is to suppose that repressor and RNA polymerase compete for binding to the same site or to overlapping sites on DNA. This seemed likely when it appeared that the operator was the recognition site of both proteins, but is less plausible now that two distinct sites have been identified (see also page 306). It would be possible, however, for repressor binding at the operator to exert some steric interference with the binding of RNA polymerase to promotor.

In the lactose operon, it seems that the action of repressor may be simply to bind to the operator DNA located between the promotor and the first structural gene. As figure 7.9 shows, this places a block in the way of RNA polymerase molecules binding to the promotor, which cannot pass the repressor and are therefore prevented from transcribing the structural genes. This may be a general mechanism for the action of repressors on bacterial DNA, for the kinetics of repression of the tryptophan operon are consistent with the same scheme (see page 379); when the tryptophan operon is repressed, RNA polymerase molecules already engaged in transcription can continue along the structural genes, except for enzyme molecules at the very beginning of the operon which are prevented from commencing transcription. The repression of phage lambda, however, seems to be mediated by a repressor which may compete with RNA polymerase for binding to DNA (see page 356), so that there may be more than one way for repression to be established.

Transcription of the lactose operon in vitro has shown that repressor and RNA polymerase bind independently to DNA, but that the binding of repressor prevents the RNA polymerase from undertaking transcription. But this effect depends on the template used. With the DNA of phage λh80d*lac*, which carries a wild type lactose operon, Chen et al. (1971) found that a complex can be formed between RNA polymerase and DNA, as measured by the resistance of RNA polymerase to rifampicin. This is presumably achieved by stable binding of RNA polymerase at the promotor. When repressor is added to this complex, it can bind to DNA and prevent transcription of the lactose genes. When the components are added in reverse order, repressor before RNA polymerase, the enzyme can still bind to DNA to form the rifampicin resistant initiation complex, although the presence of repressor inhibits transcription. This implies that the binding of the two proteins is independent, for each can bind to DNA already complexed with the other. Repressor can prevent transcription, therefore, no matter whether it is added before or after RNA polymerase, which supports the model of figure 7.9.

But different results were obtained both by Chen et al. and by Eron and Block (1971) when the lactose genes carried on phage DNA possessed a mutation which greatly increases the transcription of the operon; this has been interpreted as a super-promotor, p^s, which is more active than the usual promotor. In this case, addition of repressor after formation of the initiation complex with RNA polymerase allows only partial inhibition of transcription. The addition of repressor before formation of the complex produces complete repression of RNA synthesis and must partially prevent the binding of RNA polymerase, for subsequent addition of IPTG to release repressor from the complex can only partly reverse the inhibition—this implies that fewer RNA molecules are able to initiate transcription.

There are two plausible explanations for the differences in behaviour between the wild type and the p^s promotors. The p^s mutation may be a deletion between operator and promotor which has the effect both of increasing recognition of promotor by RNA polymerase and of bringing the two sites closer together so that the binding of their prospective proteins overlaps. Another, perhaps more likely, explanation is that the p^s mutation has created a new promotor site altogether, close to the other side of the operator, so that RNA polymerase bound at p^s is no longer subject to proper repression, although its binding is inhibited by the presence of repressor bound to the operator.

Strains of E.coli have been isolated with deletions which remove some or all of the control elements of the lactose system and thus fuse the remaining and intact structural genes of the operon to the tryptophan system adjacent on the chromosome. Reznikoff et al. (1969) found that the activity of the lactose genes is then controlled by the tryptophan system, so that both operons are repressed when tryptophan is added. The expression of the lactose genes shows that transcription can readthrough from one operon into the next, presumably because the usual termination signals have been deleted. Eron et al. (1971) have shown by hybridization assays with ϕ80p*trp* and ϕ80p*lac* phage DNAs that some, at least, of the lactose mRNA is attached to RNA specifying the tryptophan genes.

Readthrough depends on the nature of the deletion involved; when the deletion does not remove the end of the *i* gene, readthrough into lactose does not take place, which implies that there is some barrier to transcription at the junction between *i* and *p*. When *p* and the remaining part of *i* are deleted, however, this barrier is removed and readthrough takes place. But the introduction of an i^+ gene on an F′ lac episome can then prevent transcription of the lactose structural genes. As this transcription must start some distance away from the lactose operator, this observation provides a strong argument that the repressor protein binds to the operator to provide a physical block to the progress of polymerase.

Synthesis and Structure of the Repressor

How is the synthesis of repressor protein itself regulated? The *i* gene seems to be transcribed from a promotor at the end farther from the structural genes of the operon at a level which is a function only of the affinity between the *i* gene promotor and the RNA polymerase enzyme. The initiation point for *i* gene transcription has been identified in two ways. The mutants i^q and i^{sq} which make much more repressor than wild type cells map at the left end of *i* and produce repressor protein which seems to be normal; they are almost certainly mutations in the promotor for the *i* gene (reviewed by Miller, 1970).

The *i* gene is transcribed from the same strand of DNA as the structural genes of the operon; this second line of evidence also implies that its promotor must be at the left end. Kumar and Szybalski (1969) have hybridized RNA from i^q mutants with the strand of phage DNA carrying the lactose genes to which the lactose mRNA usually hybridizes. When this RNA fraction is then eluted from the DNA, it can be hybridized anew with the denatured DNA of an E.coli strain which has the *i* gene but lacks the rest of the lactose operon. That the RNA which hybridizes is the *i* gene messenger can be confirmed by elution and assay in another hybridization, with DNA of bacterial strains which contain or lack the *i* gene. The identification of this RNA as *i* gene messenger shows that the coding sequence of the lactose operon lies on one strand for both regulator and structural genes. Also in support of this contention is the effect of the presence of the end of the *i* gene in the deletion strains fusing lactose to tryptophan genes; the signal for terminating transcription appears to lie at the right end of the gene.

That the amount of repressor protein found in wild type cells depends on the number of *i* genes present is consistent with constitutive transcription of its promotor. The presence of about 10 copies of active repressor in wild type cells implies that there must be 40 translations in toto of *i* gene mRNA; but we do not know how many copies of the *i* messenger are present in the cell and how often each is translated.

That the monomeric subunits of the repressor associate at random in the cell to form the active tetramer is revealed by the complementation which takes place in diploid cells to generate phenotypes which would not be predicted from the presence of two independent *i* genes. Diploid i^-/i^+ mutants may show the uninducible i^s phenotype; and some i^+/i^s diploids may suffer mixing of subunits to generate repressors which are partially inducible, alleviating the normal dominance of the i^s gene. Certain i^- mutants are dominant; these i^{-d} alleles specify a repressor which is wild type apart from its inability to bind to the operator. As Gilbert and Muller-Hill (1970) pointed out, the dominance of the i^{-d} mutants implies that the i^{-d} gene codes for a "bad" subunit which combines with the "good" subunits produced by the i^+ gene and distorts them so that none of the subunits of the mixed tetramer can bind to the operator; this reverses the usual dominance relationship (see also Weber et al., 1972). The activity of the repressor is therefore a function of its proper tertiary conformation.

Some of the i^{-d} mutations are mis-sense, resulting in substitution of one amino acid for another; since they all map in the early part of the *i* gene, this suggests that the NH_2-terminal region of the repressor protein may be responsible for binding to the operator. Some i^{-d} mutants are ambers located at the beginning of the gene which can restart protein synthesis spontaneously later in the gene. Their translation should produce two polypeptides. One should comprise an NH_2-terminal fragment, ending at the site of nonsense mutation. The other should be a protein which contains the C-terminal part of the repressor, from the site of restart to the end. Platt et al. (1972) characterized some of the early amber mutations and observed that because they are located in the very beginning of the gene, the protein active in negative complementation is probably the C-terminal restart fragment and not the NH_2-amber fragment.

To obtain more material of one such mutant, i^{100}, a double mutant with i^q

was made and the operon was then placed on a defective λ h80 phage. Extracts from cells infected with this phage contain material which binds IPTG and cross reacts with anti-repressor antibody. The affinity of this protein for IPTG is the same as that of wild type repressor; since the extracts are ten times less active in binding IPTG than the extracts of wild type cells, the mutant protein must be produced in ten fold smaller amounts.

The i^{100} repressor protein has a subunit with a molecular weight of 34,000 daltons. Its N-terminal amino acid sequence corresponds to a start at the methionine residue in position 43 of the wild type repressor. This implies that the absence of the first 42 N-terminal amino acids does not change the ability of the repressor protein to bind the IPTG inducer. The defective protein subunit can also aggregate to form a tetramer; its activity in negative complementation in diploids implies that it can also interact with monomers of the wild type repressor to form mixed tetramers. The inability of the mutant repressor to repress the operon supports the concept that the binding site for the operator may be located in the N-terminal region of the protein which has been lost. A model for the action of repressor based upon its N-terminal amino acid sequence has been proposed by Adler et al. (1972).

One intriguing mutant which has been isolated by Myers and Sadler (1971) has an inverted control of the lactose operon; it shows constitutive synthesis of the lactose enzymes in the absence of inducers, but synthesis is repressed in their presence. This repressible constitutive mutant maps in the *i* gene (i^{rc}) and has the usual coordinate effect on the expression of all lactose genes; its maximum level of synthesis is that of fully induced i^+ cells, which suggests that only the repressor protein is altered.

Two classes of model may be postulated to account for this unusual behaviour. The binding properties of the repressor might be altered so that the interaction of protein with inducer promotes instead of inhibiting its binding to the operator. Or the repressor protein may have normal activity, but there may be some defect in its synthesis which is remedied only when inducer is added. In this case, repressor should be absent from uninduced cells, leading to constitutivity, but made in the presence of IPTG, which might allow some partial repression.

The kinetics of gene expression can distinguish these two situations. It is difficult to measure the introduction of repression when IPTG is added to i^{rc} cells, but it seems to take at least half a doubling time of the cells to reduce the expression of the lactose genes. When IPTG is washed out of the cells which are partially repressed, the repression takes 2–3 doublings of the cells to relax. The control of induction in i^+ cells, of course, is very much more rapid.

The dependence of establishing or relieving repression on the growth rate of the i^{rc} cells suggests that repressor synthesis in the mutant may be linked to the cell cycle. IPTG binding activity is present in extracts only when the cells have grown repressed in the presence of the inducer, which suggests that the presence of IPTG is needed if repressor monomers are to be synthesised. The model which Myers and Sadler proposed to account for these results is to suppose that repressor monomers are synthesised in bursts during the cell cycle and that IPTG influences the synthesis of a whole burst in the i^{rc} mutant. The function of the inducer is probably exercised at an early stage in repressor synthesis, possibly its completion and release from the ribosome.

Induction of Transcription at the Promotor

Glucose Repression by Cyclic AMP

The rate of synthesis of β-galactosidase (and also of other inducible enzymes) is reduced when cultures of E.coli are grown in the presence of glucose. When the lactose genes are induced, the rate of β-galactosidase synthesis is greatly reduced in cultures growing on glucose, compared with cells for which some other metabolite has been provided as carbon source. This has been termed *catabolite repression* (or *permanent repression*). The absolute extent of the repression varies with the particular bacterial strain, but the rate of β-galactosidase synthesis may be reduced to about a third of that shown on non glucose medium.

Transient repression is exhibited when glucose is added to induced cultures of E.coli growing on some other carbon source, such as glycerol. In this case, there is a severe repression of β-galactosidase synthesis and no enzyme is produced for roughly the next half generation. After this period, the enzyme is synthesized at the rate characteristic of normal growth on glucose cultures.

Both catabolite and transient repression result from the effect which glucose exerts on lowering the level of 3′–5′ cyclic AMP in the cell. At least one way in which this may happen is for the addition of glucose to promote the excretion of cellular cyclic AMP into the incubation medium. The addition of cyclic AMP to cultures suffering glucose repression overcomes the repression effect and stimulates the production of β-galactosidase. (A third effect of glucose on inducible enzyme synthesis is to reduce the uptake of inducer from the medium and thus its intracellular level; this is not related to the level of cyclic AMP).

The influence of cyclic AMP on the synthesis of β-galactosidase is exerted at transcription by increasing the synthesis of lactose messenger RNA. However, the effect is not mediated through the operator-repressor control system since the addition of cyclic AMP is as effective with mutants in these control loci as it is with wild type strains. Response to glucose repression and cyclic AMP stimulation is altered, however, in promotor mutants. Using a promotor mutant which permits the lactose system to make only about five per cent of its usual level of enzymes when induced, Pastan and Perlman (1968) found that glucose does not cause transient repression in such cells.

Mutants mapping at the promotor vary in their effect on the response of the lactose operon to cyclic AMP; Silverstone et al. (1969, 1970) observed that promotors can be mutated to affect either or both of the activities of setting the level at which the operon may be transcribed and responding to cyclic AMP. Silverstone and Magasanik (1972) found that these promotor mutations influence catabolite repression of transacetylase synthesis as well as that of β-galactosidase and must therefore identify the only site in the operon at which this system acts.

Glucose repression works directly on transcription, for when Varmus, Perlman and Pastan (1970a, 1970b) measured the level of lactose mRNA in induced cells, they found that the amount of RNA hybridizing with λh80dlac

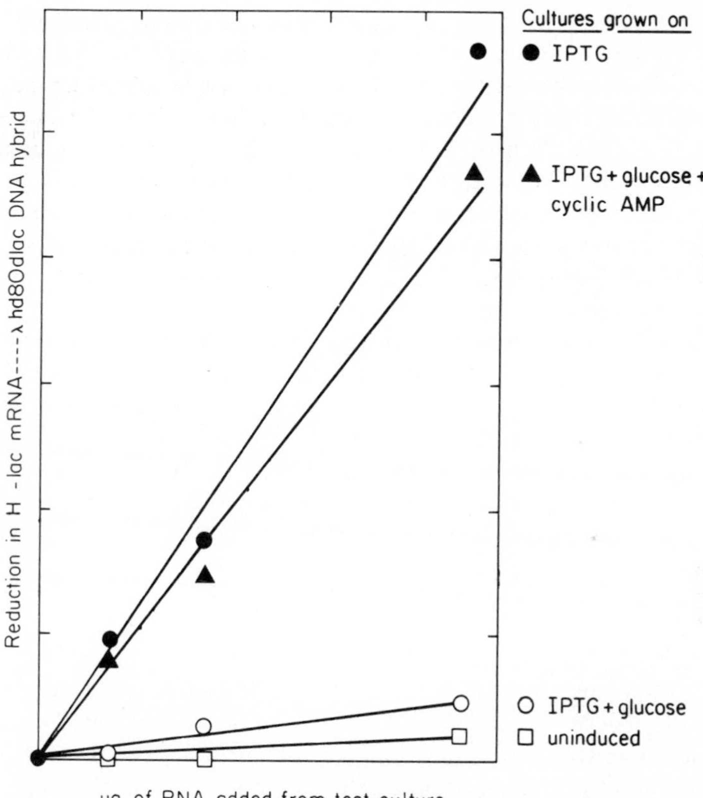

Figure 7.15: detection of lactose mRNA in cultures of E.coli grown under different conditions. H^3-labelled lactose mRNA is annealed to filters containing λh80dlac DNA in the presence of unlabelled RNA extracted from the experimental culture. If lactose mRNA is present in the unlabelled preparation, it competes with the H^3-lactose mRNA and reduces the radioactivity retained on the filter. Addition of IPTG induces lactose messengers. When cells are grown on glucose the level declines almost to that of uninduced cells; but the addition of cyclic AMP restores synthesis of lactose mRNA

DNA depends on whether the cells are grown in the presence or absence of glucose. There are about 10 molecules of lac mRNA in induced cells grown without glucose; but only one molecule per cell when glucose is added. Figure 7.15 shows that the addition of cyclic AMP overcomes the effect of glucose and restores the level of lactose mRNA. Glucose repression must act on the synthesis of mRNA, for the rate of degradation of lactose messengers, measured by the disappearance of a radioactive label when induction is prevented by addition of the anti-inducer ONPF, is the same irrespective of the presence or absence of glucose.

Cyclic AMP is essential if the lactose operon is to be switched on; this conclusion follows from the identification of two classes of mutant which cannot synthesise the enzymes of various inducible operons. One such class of mutant lacks cyclic AMP because the adenyl cyclase enzyme which makes the cyclic nucleotide has been mutated; and such cells have no more β-galactosidase or lactose mRNA when IPTG is added than they do when uninduced. The addition of cyclic AMP restores the normal response to induction. This shows that the cyclic AMP system does not function through a repressor; rather is it a positive control system whose function is an essential as that of RNA polymerase if the lactose operon is to be switched on when inducer removes repressor protein from the operator.

Glucose does not itself repress the expression of the operon, therefore, but acts by reducing the level of cyclic AMP and thus inactivates the system which is needed for proper transcription. The cyclic AMP control system acts not only on the lactose operon, but is needed for expression of many inducible operons; this implies that it may provide a way for the bacterial cell to balance the amount of energy devoted to synthesising certain inducible messengers vis a vis the synthesis of other RNA species.

Cyclic AMP Control Protein

Cyclic AMP seems unlikely to react directly on the DNA of the promotor and the protein which mediates its action has been identified by the existence of the second class of mutants, which has normal levels of cyclic AMP but nevertheless fails to induce many operons. Cells of this *crp*$^-$ mutant genotype contain little lactose mRNA, even when induced with IPTG. Two approaches have been used to identify the protein inactivated in these mutants. Emmer et al. (1970) have compared the ability to bind cyclic AMP of extracts from wild type and mutant cells. Most of the cyclic AMP binding activity of wild type cells is present in the supernatant, can be bound to phosphocellulose columns and elutes from them when the ionic strength is increased. One *crp*$^-$ mutant lacks this protein activity; another possesses a protein which has a much lower affinity for cyclic AMP.

By using a DNA-dependent cell free system for synthesising β-galactosidase, Zubay, Schwartz and Beckwith (1970) have shown that extracts from *crp*$^-$

mutants have only about 5% of the activity of wild type extracts. Cyclic AMP stimulates transcription of lactose mRNA in the presence of the proteins extracted from wild type cells, but cannot do so when the proteins are derived from mutant cells. By using this system, they were able to purify the cyclic AMP binding protein for its activity in stimulating transcription of the lactose operon. As Riggs, Reiness and Zubay (1971) and Anderson et al. (1971) have reported, the CRP factor (cyclic AMP receptor protein—sometimes described as CAP) is a dimer of subunits of 22,300 daltons and has an isoelectric point of just above pH 9.

Transcription of the lactose operon in vitro has confirmed that cyclic AMP and its binding protein are essential for RNA synthesis. Proper transcription of the lactose genes in vitro seems to depend on the DNA to which they are attached, for Eron and Block (1971) found that transcription of ϕ80plac DNA is asymmetric and the synthesis of lactose mRNA is decreased by the addition of rho factor; this suggests that, with this template, the lactose operon is transcribed only when RNA polymerase reads through from the adjacent phage genes at which it initiates transcription. With ϕ80dlac DNA, however, the lactose genes are transcribed only in the presence of CRP factor and cyclic AMP. Transcription with this template seems to be initiated at the proper site for its characteristics are changed when promotor mutants in the lactose operon are substituted for wild type promotor; and it is inhibited by the addition of lactose repressor. (The "p" in λplac or ϕ80p phages indicates that these strains form plaques, that is they are not defective. The advantage of using such phages for hybridization studies is that mature particles are produced from which DNA may readily be isolated.)

A purified system in which all the components necessary for transcription have been identified has been developed by de Crombrugghe et al. (1971a, 1971b). Their assay for transcription of lactose genes is to hybridize the RNA made in vitro from λh80dlac DNA templates with phage λ DNA to remove any phage sequences; lactose sequences can then be identified by hybridization with λplac DNA. There is little homology between the DNAs of phages ϕ80 and λ, so only lactose sequences should hybridize in this second reaction. The precision of the hybridization can be increased by using isolated single strands of the DNA of the phages. These experiments have shown that transcription takes place from only one strand of the template and is inhibited by the addition of the lactose repressor. Using the p^s mutant operon, which synthesises more lactose mRNA in vitro, they found that transcription requires RNA polymerase, including sigma factor, cyclic AMP and the CRP factor. Transcription is inhibited 80% by the addition of lactose repressor and the inhibition is lifted when IPTG is added.

The need for sigma factor shows that the CRP protein is not a sigma-like factor which directs the attention of core polymerase to the lactose promotor. Rather is its binding to DNA required to allow the complete RNA polymerase enzyme to transcribe the lactose genes. The CRP factor binds to DNA as might be expected from its ability to exchange with phosphocellulose columns. Nissley et al (1972) found that when C^{14}-labelled CRP factor is incubated with poly-dAT, λ*pgal* or calf thymus DNA and centrifuged through a sucrose gradient, the protein and the DNA sediment together only when cyclic AMP is added. The binding depends specifically upon cyclic AMP; cyclic GMP, for example, is ineffective. The same characteristics of reaction are found if a nitrocellulose filter is used to trap the complex of CRP factor and DNA.

Since the protein does not bind to RNA polymerase, its action probably implicates direct recognition of a DNA site; however, its lack of specificity in binding to DNA means that we cannot fully define its molecular action. But by testing the ability of RNA polymerase to form a rifampicin-resistant initiation complex with the DNA of the galactose and lactose operons in vitro, Nissley et al. (1971) and de Crombrugghe et al. (1971c) have shown that cyclic AMP and the CRP protein act at the stage when RNA polymerase must bind to DNA. This suggests that the cyclic AMP system and RNA polymerase may interact at adjacent sites on DNA.

All the promotor mutants which have been found in the lactose operon appear to be concerned with the action of the cyclic AMP system. Beckwith, Grodzicker and Arditti (1972) noted that although the L1 and L8 deletions remove the initial part of the promotor, the lactose operon can still function at about 2% of the level of wild type cells. This synthesis depends upon the remaining part of the promotor and is insensitive to catabolite repression. Mutants which eliminate CRP factor activity also show the residual 2% level of expression; this is found in cells which are mutant in both the adenyl cyclase and *crp* genes—so that they cannot be leaky in the cyclic AMP system—whether they have wild type, L1 or L8 deletion promotors. These deletions map

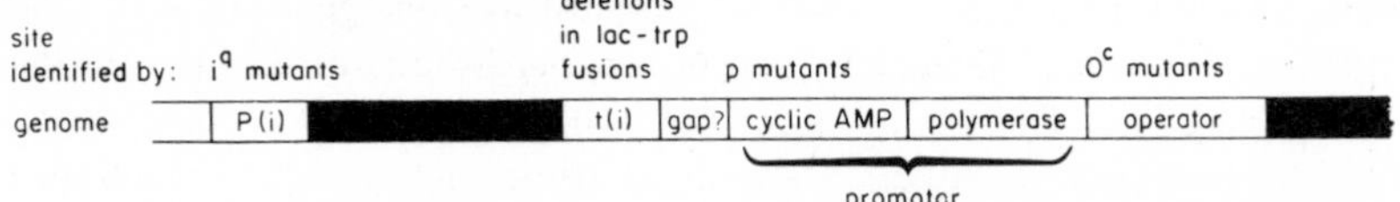

Figure 7.16: control sites in the lactose operon. The i gene has a promotor at the left and a termination site at the right. We do not know whether the control sites for the structural genes lie immediately adjacent to the i gene region or whether there is a gap between them. All "promotor" mutants so far identified appear to fall into the left part of the region and may identify the cyclic AMP protein binding site; the RNA polymerase binding site may be located immediately to the right. Both sites must be occupied before transcription is initiated. We do not know where RNA polymerase initiates transcription. The operator site where repressor protein binds lies to the right of the promotor region but to the left of the structural genes; it is possible that there may be some overlap between the polymerase binding site and the operator

in the left end of the gene; a revertant of L8 which makes 70% of the usual level of β-galactosidase, and is insensitive to cyclic AMP, maps to the right of L8.

The "promotor" region may therefore be divided into the two sites shown in figure 7.16, a cyclic AMP protein recognition site at the left and an RNA polymerase binding site at the right. Both these sites must be occupied by the proteins, presumably so that CRP protein interacts with polymerase, before transcription can be initiated. Since the promotor mutants so far isolated fall into the cyclic AMP region, we have no genetic information on whether the RNA polymerase binding site overlaps with the operator (although biochemical experiments suggest that the binding of polymerase and repressor is independent). Only the isolation of mutants in the polymerase binding site can decide this point. It is important also to realise that although we may expect the RNA polymerase binding site to lie adjacent to the cyclic AMP site, this does not imply that transcription is initiated at the same point. It is possible that after attachment to DNA, the polymerase may move to another site to initiate transcription. But in any case, the promotor is clearly a structure more complex than just a binding site for polymerase.

Isolation of the Lactose Genes

It is the existence of the different types of phages carrying the lactose genes which has made possible the establishment of the in vitro systems for expression of the operon. The only homologies between different transducing phage DNAs may lie in the lactose genes, so that hybridization of the RNA transcribed from one phage with the DNA of another provides a specific assay for lactose mRNA. Phages ϕ80 and λ, for example, have less than 0·5% homology with each other, so that almost the only common sequences between ϕ80d*lac* and λ*lac* are those of the lactose genes. By using DNA-DNA hybridization techniques instead of DNA-RNA annealing, Shapiro et al. (1969) have used these phage strains to isolate the DNA corresponding to much of the lactose operon.

Two particularly useful phages are derivatives of λ and ϕ80 which carry the lactose genes in opposite orientation. These phages were obtained by preparation from two bacterial strains in which the lactose operon has been transposed to near the attachment site of the phages, but is inserted in the opposite direction from usual at the ϕ80 site. Because both λ and ϕ80 integrate into the bacterial chromosome with the same overall polarity, the resultant transducing phages have lactose DNA inserted into their genomes in opposite directions.

After denaturation of the phages, the single strands can be separated by their different densities into the heavy (H) and light (L) strands. Because the lactose operon is present in the ϕ80p*lac*-1 and λp*lac*-5 phages with reversed polarities, the coding strand of the lactose gene is carried on the H strand of the ϕ80p*lac*

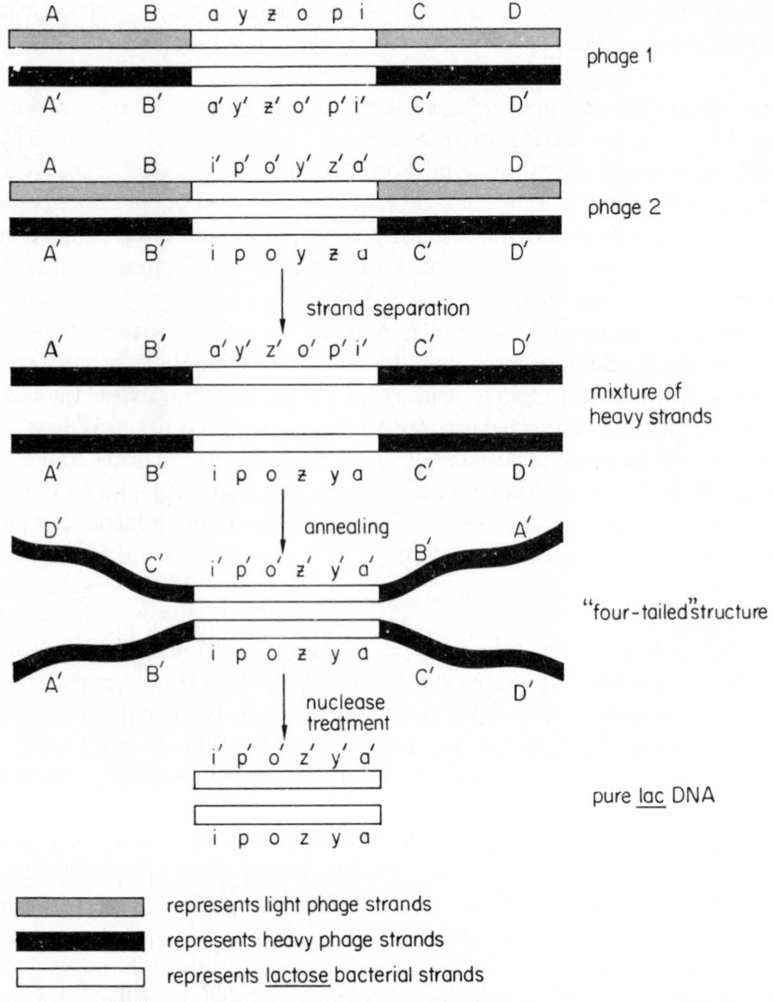

Figure 7.17: isolation of lactose operon DNA. Each transducing phage preparation has the lactose genes inserted in opposite orientation. After denaturation and centrifugation, the two separated H strands are allowed to anneal together. Only the lactose sequences are complementary and can form a duplex; the phage sequences remain single stranded. Treatment with a nuclease specific for single stranded DNA leaves intact only the duplex region of the lactose operon. After Shapiro et al. (1969)

DNA but on the L strand of the λp*lac* DNA. Figure 7.17 shows that if the separated H strands of the two phages are then hybridized, one carries the coding strand of lactose and the other its complement; they should therefore renature to form duplex lactose DNA. However, the phage regions contained in the DNA preparations are not complementary and remain single stranded. This yields the four tailed structure in which the only duplex region should be the lactose genes. When the single stranded regions are removed with a nuclease specific for denatured DNA, only duplex DNA of the lactose operon should remain.

This isolation relies on absolute non complementarity between the H strands of the two phages as otherwise duplex regions would also be produced for phage genes. The small extent of homology between $\phi80$ and λ DNA is reduced even further by using the isolated H strands instead of an H strand and an L strand and there is, in fact, no detectable homology in the phage regions of the DNA. But contamination of the duplex DNA could still arise from the presence in the phages of other bacterial genes able to behave in the same manner as the lactose operon. Indeed, this is a real difficulty, for the tranducing phage preparations almost always contain other bacterial genes as well as those of the lactose system.

To avoid this, and to ensure that the two phage strains have in common only chromosomal material derived from the lactose operon, one of the phages was prepared from strains which contain two deletions, one extending from the middle of the *y* gene to past the end of the operon and the other extending into the operon from the other end and terminating in the *i* gene. This means that any extra bacterial DNA carried by this phage corresponds to the new regions adjacent to the lactose DNA in the deletion strain and must be different from those present in the other transducing phage. The isolated duplex segment does not, therefore, contain the whole lactose operon, but has intact only the region *p-o-z*.

Electron microscopy on the preparation of annealed DNA shows the expected "four-tailed" structures and, after digestion with nuclease, the presence of only pure duplex regions. These are some 1·4–1·5 μ long, which corresponds to about 4250 base pairs; this is in good agremeent with the size of the lactose operon which can be estimated from the size of the proteins coded to be about 4700 base pairs from *i* to *y* (see figure 7.1).

CHAPTER 8

Operon Control Circuits

Bacterial Operons

Positive and Negative Control Systems

Four simple types of control circuit can be designed to show how either positive or negative control may, in principle, be used to achieve either induction or repression of operons in response to the external environment. Systems of induction and repression are defined by their response to the small molecule inducer or co-repressor; just as it may be advantageous for a bacterium to induce a set of enzymes only after addition of the inducer substrate which they metabolize, so also the enzymes which synthesize some metabolic intermediate may be repressed when this co-repressor product of the pathway is added to the cell. Inducible operons, therefore, are active only when the inducer is present; whereas operons which can be repressed fail to show activity in the presence of co-repressor and are active only in its absence.

Regulator mutants in systems of repression can give the phenotypic states of *derepression* and *super-repression*, just as mutants in inducible operons may be constitutive or uninducible. Derepressed operons cannot be switched off and show the same characteristic of continual function as constitutive mutants of inducible operons; super-repressed operons remain repressed even in the absence of co-repressor, showing the same failure to function which is found in the uninducible mutants of induction systems.

Positive and negative control systems may be distinguished by the response of the operon when no regulator protein is present. For genes under positive control, expression is possible only when an active inducer protein is present; but genes under negative control function unless they are switched off by interaction with an active repressor protein. Either type of control could be used to achieve either induction or repression by allowing an appropriate interaction with the inducer or co-repressor molecule to determine the activity of the regulator protein.

If the repressor protein characteristic of negative systems is active unless inactivated by an inducer, the pattern of induction shown by the lactose operon (figures 7.2, 7.3 and 7.4) is the result. If the repressor protein lacks activity until activated by a co-repressor, the repression of figure 8.1 is achieved; this may be the mode of control of the tryptophan operon (see page 379).

But in either induction or repression, the regulator gene synthesizes a repressor protein which must interact with the operator if the operon is to be switched off.

In systems of negative control, deletion of the operator thus causes constitutive (or derepressed) synthesis of the enzymes of the operon, because there

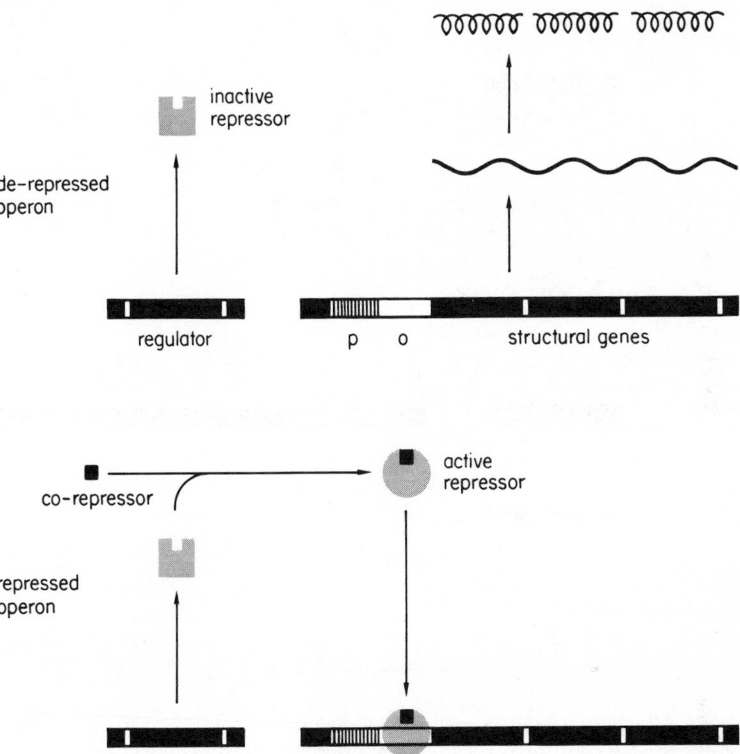

Figure 8.1: negative control of repression. The regulator gene synthesizes a repressor protein which in free form cannot bind to the operator to switch off the operon (upper). Addition of co-repressor converts the repressor protein to an active form which can bind at the operator to inactivate the operon (lower)

is no longer any site at which the repressor protein can act. Operator mutants are always *cis*-dominant, *trans* recessive. An inactive regulator gene (produced, for example, by deletion or nonsense mutation) causes the same continual synthesis of the enzymes of the operon, for there is never any repressor protein available to switch off the structural genes. Figures 7.3 and 8.2 show that constitutive or derepressed regulator gene mutants are recessive, because the introduction of a normal repressor protein restores the response to inducer or co-repressor.

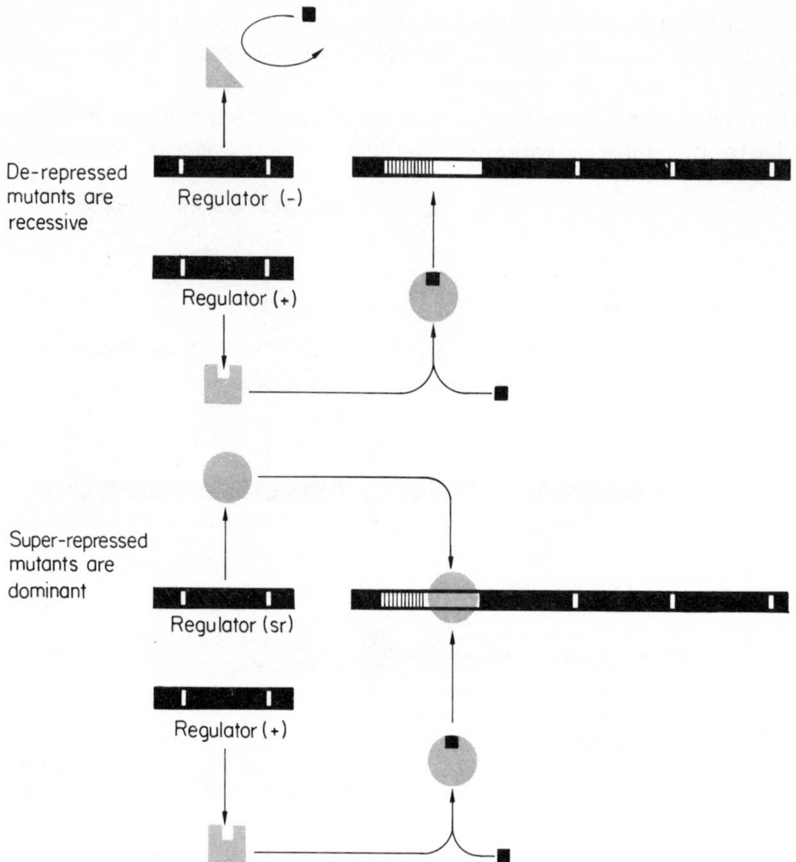

Figure 8.2: mutants in repression systems under negative control. Upper: derepressed (constitutive) mutants synthesize a repressor protein which cannot bind the operator, for example because it has lost its operator binding site or co-repressor binding site. Introduction of a wild type regulator gene to make a diploid synthesizes a repressor protein which can be activated by co-repressor to recognise the operator and therefore restores normal control. Derepression is recessive to wild type. Lower: super-repressed (uninducible) mutants synthesize a repressor which takes up an active conformation without needing to interact with co-repressor. This switches the operon off. Although the introduction of a wild type gene produces a repressor protein amenable to normal control, this cannot prevent binding of the super-repressor to the operator; super-repression is dominant to wild type

Uninducible or super-repressed mutants in the regulator gene are dominant as illustrated in figures 7.4 and 8.2. Uninducible mutants can result when the regulator gene produces a repressor protein which can recognise the operator but can no longer bind the inducer which would inactivate it. Super-repressed mutants can result from mutation to produce a repressor protein which no longer needs to be activated by co-repressor but is itself active. These mutants are dominant because the presence or absence of a normal repressor protein cannot influence the mutant repressor (except by intragenic complementation or gene dosage effects).

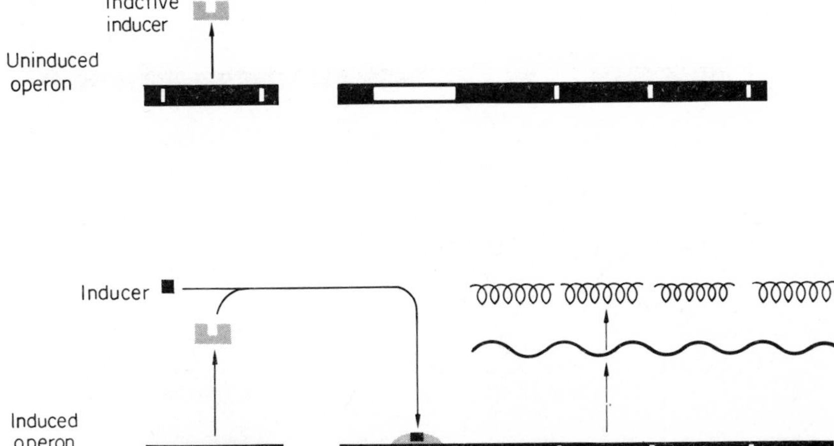

Figure 8.3: positive control of induction. The regulator gene synthesizes an inducer protein which is inactive in free form (upper), so that the operon remains non-functional in the absence of inducer. When inducer is added (lower) the inducer protein is converted to an active form and interacts with a recognition site in a reaction necessary to switch on the operon

These dominance relationships should be reversed in positive control systems in which an active inducer protein must be present to enable the operator to function. Inducer proteins might, in theory at least, be used to achieve induction if the protein were to be activated by a small molecule inducer to allow interaction with the operon as shown in figure 8.3. And repression control would result if the inducer protein were active unless inactivated by a co-repressor, as shown in figure 8.4. The opposite nature of the influence of inducer and repressor proteins means that positive and negative control systems can be distinguished by their different responses to mutations in the regulator gene; table 8.1 summarizes the relationships of such mutants to wild type.

Deletion of the region which is recognised by the active form of an inducer protein must abolish expression of the operon, giving the uninducible phenotype in induction systems and the super-repressed characteristic in repression systems; such mutants should be *cis* dominant, *trans* recessive and should share many of the properties of mutants in the promotor region of negative

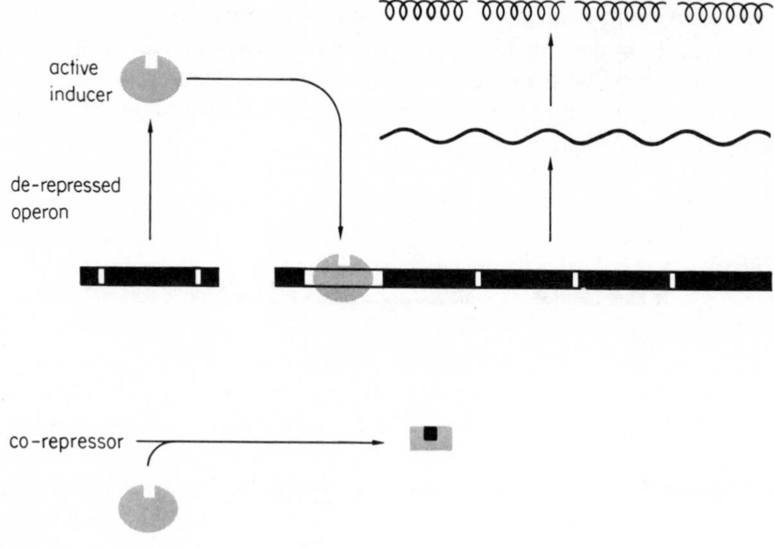

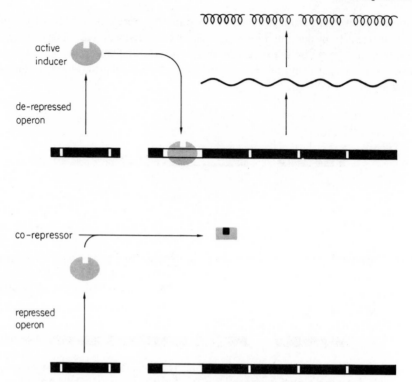

Figure 8.4: positive control of repression. The regulator gene synthesizes an active inducer protein which maintains the operon in a functional state (upper). Addition of co-repressor converts the inducer protein to an inactive form, switching off the operon (lower)

Table 8.1: characteristics of positive and negative control systems

induction		repression
negative: *operon functions unless operator is switched off by repressor protein*		
inducer inactivates repressor protein		co-repressor activates repressor protein
	regulator gene mutants	
constitutive	recessive	derepressed
uninducible	dominant	super-repressed
positive: *operon only functions after inducer protein switches it on*		
inducer activates inducer protein		co-repressor inactivates inducer protein
	regulator gene mutants	
constitutive	dominant	derepressed
uninducible	recessive	super-repressed

systems such as the lactose operon. A defective regulator gene should produce the same phenotypes, for there can be no protein present to switch on the operon; these mutants must be recessive because the introduction of a normal inducer protein would restore the normal response to the inducer or co-repressor molecule.

Not all these simple control circuits are used in bacteria. Most operons, whether they are inducible by their substrate or repressible by their product, are under negative control. The cyclic AMP system, however, which acts on many of the inducible operons, is an example of positive control—the behaviour of the cyclic AMP binding protein is formally analogous to the situation shown in figure 8.3, when the inducer protein (CRP) must be activated by an inducer (cyclic AMP). A difference between this system and the substrate-induction or product-repression control of operons is that cyclic AMP is not a substrate or product of the metabolic pathway catalysed by the enzymes of the operon; and, of course, the level of cyclic AMP influences many operons, not just one. However, operons on which cyclic AMP acts fall under two types of control; they show (usually) negative control by virtue of their response to the inducer or co-repressor molecule related to the metabolic activities of the operon and display positive control in response to cyclic AMP.

An assumption usually made in early theories about the control of induction and repression was that negative and positive control can be treated in formal terms as different aspects of the same fundamental type of control mechanism. This implies that the molecular basis of each type of control must be similar. If negative control is exerted by the binding to DNA of a protein which prevents transcription, the analogous action for positive control must be the binding to DNA of a protein whose presence is necessary for transcription. This is just the role played by the cyclic AMP binding protein, of course, and if it recognises some part of the DNA of the promotor rather than the RNA polymerase enzyme, its action may not be dissimilar to that of the repressor protein (with the reservation that CRP may interact with polymerase as well as DNA, whereas repressor may interact specifically with only DNA).

No other simple positive control system of this nature has been found, although in principle a similar interaction could be used to govern induction or repression (see page 344 for discussion of the advantages of negative systems). The implication of a positive control mechanism would be that the promotor site must play a role corresponding to that of both operator and promotor in a negative control system (and akin to that originally envisaged for the operator of the lactose operon); for the promotor would be a site essential for the transcription of the operon at which the control protein would act. The recognition element of figures 8.3 and 8.4 would in effect comprise more than one region—its DNA sequence would contain the information needed to respond to inducer protein, RNA polymerase, and perhaps the cyclic AMP system also.

Another positive control mechanism would be the use of hierarchies of sigma factors to recognise different classes of promotors. This mode of control is formally analogous to the existence of different classes of f3 initiation factor to direct translation of different populations of messengers (see page 102), although the regulator role of the f3 factors within the E.coli cell itself is not clear. Such a mechanism does not seem to be used to control transcription in bacteria, however, where the sigma polypeptide is simply part of the RNA polymerase complex enzyme which recognises all promotors and initiates transcription. There is no evidence to support the idea that there may be sigma factors specific for particular operons or even particular classes of operon.

But the introduction of new sigma factors may be used by phages to turn on a general class of genes, which presumably share in common the same promotor sequences (which must be distinct from those identifying the other classes of genes). It is likely that such a mechanism is used to express the late genes of phages, although a different positive control system—the synthesis of anti-terminator factor to allow RNA polymerase to readthrough past terminator sites—may be used to express the delayed early genes. (There is no evidence to suggest that anti-termination is used as a control mechanism in uninfected E.coli.) Changes in the specificity of B. subtilis RNA polymerase occur when the cells make a transition from vegetative growth to sporulate; it is possible that this may rely upon alterations in the sigma factor as well as the core enzyme (see Greenleaf et al. 1973; Linn et al. 1973).

The general rule, then, seems to be that negative control systems are used in bacteria to regulate the expression of individual operons, although the systems discussed below show that the interactions constituting the control system comprise several variations on the theme of negative control. Positive control systems seem to be utilised to achieve more wide ranging effects, such as the control by cyclic AMP of many bacterial operons. And the introduction of changes in RNA polymerase appears to be a drastic, perhaps irreversible, step for the bacterial cell (whether infected with phages or not), which is reserved for situations of temporal change in which a large number of genes of one class must be switched off and those of another class turned on.

The Histidine Operon

The cluster of nine structural genes shown in figure 8.5 codes for the enzymes which synthesise histidine from PRPP (phosphoribosyl-pyrophosphate) and ATP in Salmonella typhimurium. The gene cluster comprises a coordinate unit of control for the enzymes are synthesised only in the absence of histidine; when cells are starved for histidine there is an increase of about 25 fold in the level of histidine enzymes. Hybridization experiments between the RNA of derepressed bacteria and the DNA of either normal Salmonella or strains deleted for the histidine region suggest that the increase in histidine mRNA is about 10 fold upon derepression.

The histidine operon is probably under negative control of repression, but exhibits two unusual features; no regulator gene has been identified and the co-repressor which controls the structural genes is not their product, histidine, but his-tRNA. Regulator mutants can be isolated by their resistance to the histidine analogue triazole-alanine (TRA) which acts as a false co-repressor of the operon. When TRA is incorporated into proteins, it damages the cell; selection of mutants resistant to TRA gives two classes, one of which is mutant in the trivial inability to take up the analogue from the medium. The other class comprises bacteria which are derepressed for the histidine enzymes; in spite of the presence of the substitute for histidine, they have a supply of normal histidine which can be used in protein synthesis, so that they resist the damaging effects of incorporating TRA into proteins.

Six classes of regulator mutants, termed *hisO*, *hisR*, *hisS*, *hisT*, *hisU*, *hisW* have been identified in this way by Roth, Anton and Hartman (1966) and Anton (1968); all can be derepressed yet further to give even higher levels of the histidine starvation, which suggests that the mutations represent alteration in, but not destruction of, these genes. The mutants in *hisO* are located at the right end of the operon and Fink and Roth (1968) showed that they have the *cis*-dominant, *trans*-recessive characteristic expected of an operator.

Amber mutants in genes at this end of the operon reduce the expression of the subsequent genes, which suggests that the operon is transcribed into a polycistronic messenger, starting at *hisG*. The order of the genes does not correspond to the order in which the enzymes are used. By measuring the enzyme activities resulting from derepression, Whitfield et al. (1970) have shown that the *hisG* polypeptide is synthesised in amounts about three times greater than the *hisD*, *hisC* and *hisA* products, which seem to be made in about equal amounts. We may speculate that more of the *hisG* product may be needed because the active form of the enzyme is an aggregate of six polypeptide chains, whereas the other enzymes which have been isolated appear to be active as monomers or dimers. Figure 8.5 describes the proteins which have been characterized.

All the regulator genes other than *hisO* are dispersed on the chromosome and do not map near the histidine operon; they appear to be concerned with the synthesis and modification of the tRNA which accepts histidine or with its charging of amino acid. The *hisS* gene codes for histidyl-tRNA synthetase and *hisS* mutants have an altered activating enzyme with a reduced affinity for histidine. Roth and Ames (1966) showed that addition of histidine to these mutants can overcome the effects of mutation; and this led to the suggestion that growth of the mutants and repression of the histidine operon are limited by the supply of charged histidyl-tRNA. There seems to be only one his-tRNA synthetase in S. typhimurium, which de Lorenzo and Ames (1970) found to

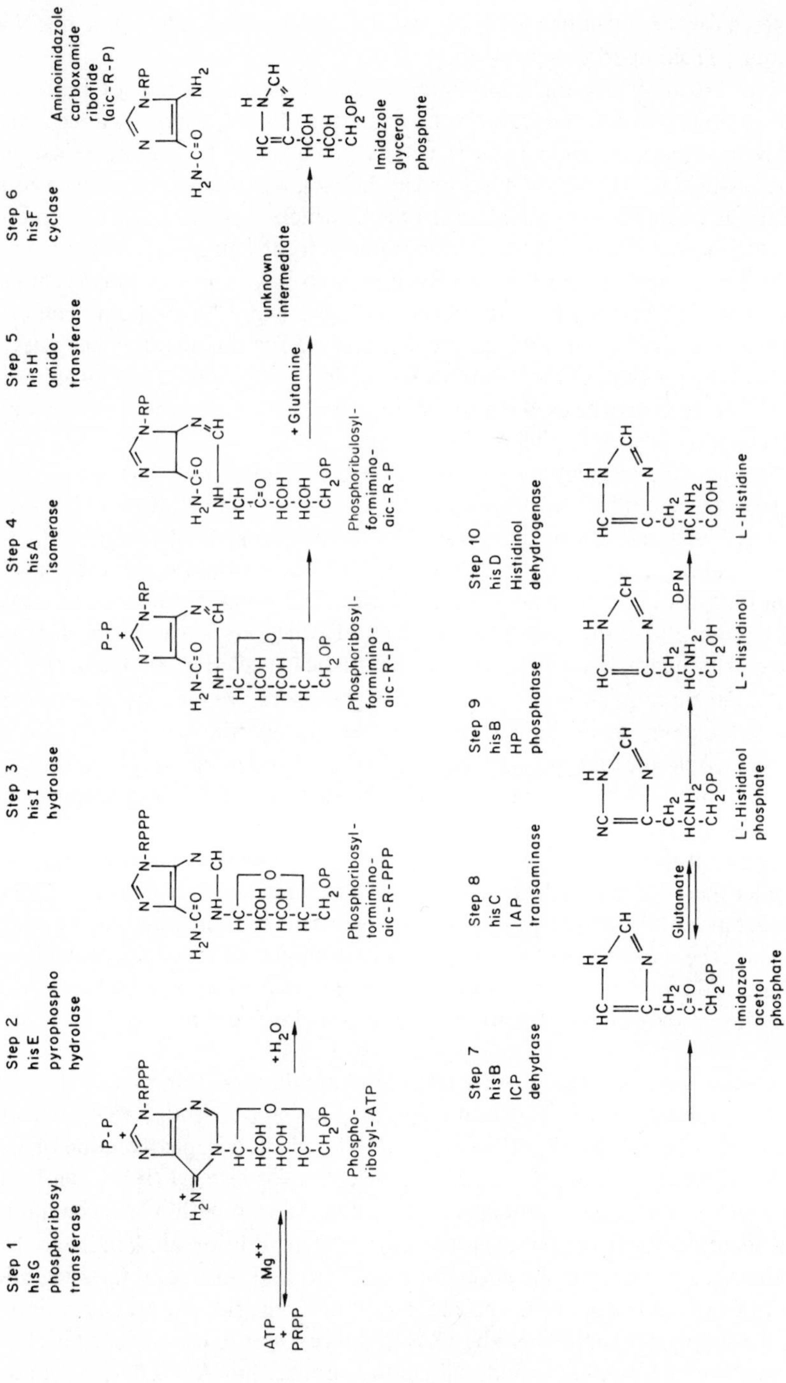

Step 1
hisG
phosphoribosyl
transferase

Step 2
hisE
pyrophospho
hydrolase

Step 3
hisI
hydrolase

Step 4
hisA
isomerase

Step 5
hisH
amido-
transferase

Step 6
hisF
cyclase

ATP + PRPP ⇌ (Mg^{++}) Phospho-ribosyl-ATP → ($+H_2O$) Phosphoribosyl-formimino-aic-R-PPP → Phosphoribosyl-formimino-aic-R-P → Phosphoribulosyl-formimino-aic-R-P → (+ Glutamine) unknown intermediate → Imidazole glycerol phosphate

P-P + Aminoimidazole carboxamide ribotide (aic-R-P)

Step 7
hisB
ICP
dehydrase

Step 8
hisC
IAP
transaminase

Step 9
hisB
HP
phosphatase

Step 10
hisD
Histidinol
dehydrogenase

→ Imidazole acetol phosphate ⇌ (Glutamate) L-Histidinol phosphate → L-Histidinol → (DPN) L-Histidine

step catalysed	2	3	6	4	5	7,9	8	10	1	
enzyme activity	pyrophospho hydrolase	hydrolase	cyclase	isomerase	amido-transferase	dehydrase phosphatase	transaminase	dehydrogenase	phosphoribosyl-transferase	
active form	—	—	—	—	—	—	monomer	dimer	hexamer	
protein monomer	—	—	—	—	—	—	59,000	40,000	35,000	
base pairs							1600	1100	1000	
gene	E	I	F	A	H	B	C	D	G	O

Figure 8.5: the histidine biosynthetic system. Data of Ames and Hartman (1963), Roth et al. (1966)

have an affinity for uncharged $tRNA^{His}$ great enough to imply that much of the tRNA and enzyme may be complexed together.

Mutants in *hisR* have only about half of their usual activity in forming his-tRNA; and this gene is the structural locus coding the tRNA for histidine. Attempts to find more than one type of histidine tRNA in the cell have failed, although it is possible that there may be some minor species as well as the one major species. And, of course, there may be more than one gene coding for the same $tRNA^{His}$, only one copy of which is mutated in the *hisR* mutants. According to Brenner and Ames (1972), the *hisT*, *hisU* and *hisW* genes all seem to be concerned with making modifications to the tRNA after its synthesis. All the mutants which have been tested are recessive to their wild type alleles, which suggests that the mutations prevent proper completion of the $tRNA^{His}$ molecule.

Another finding which has implicated his-tRNA in repression is the finding of Schlessinger and Magasanik (1964) that α-methyl-histidine derepresses the histidine operon; this analogue inhibits the rate of formation of histidyl-tRNA by competing with histidine for its activating enzyme, although the α-methyl-histidine cannot be transferred to the tRNA. Roth et al. (1966) therefore suggested that the histidine operon is under negative control, his-tRNA acting as the co-repressor which activates the repressor protein in the

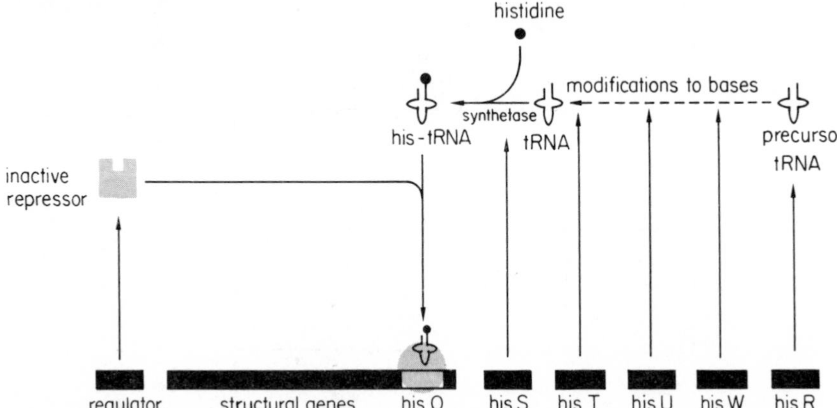

Figure 8.6: model for control of the histidine operon. The structural genes are controlled by the his0 operator at their right end. His-tRNA acts as the co-repressor which activates the repressor protein which binds the operator. Any mutation preventing the synthesis or changing the structure of his-tRNA may therefore reduce the level of co-repressor and thus derepress the operon. Mutations of this nature have been isolated in hisR (the structural gene for $tRNA^{his}$), in hisT, hisU and hisW (which modify bases of the tRNA) and hisS (which codes for the his-tRNA synthetase). Since no regulator gene has been identified, the part of the model showing the synthesis and activation of repressor protein is speculative; it has been suggested that the product of the hisG structural gene may act as the repressor

manner of figure 8.1. The various mutations which appear to have a regulatory effect thus represent changes affecting the ability of the cell to make co-repressor and are not mutations in regulator genes as such. Figure 8.6 illustrates this model.

This idea is supported by the experiments of Lewis and Ames (1972) demonstrating that the extent of derepression of the histidine operon is correlated with the amount of his-tRNA found in Salmonella cells. Wild type cells growing with a doubling time of 47 minutes have 77% of their tRNA^{His} charged with the amino acid: addition of histidine increases the level of charged his-tRNA to 88%. In the absence of added histidine, *hisS* mutants have 8.5% charged his-tRNA and are derepressed relative to the wild type some 18 times. Addition of histidine increases the amount of his-tRNA and the operon becomes less derepressed (more repressed) in parallel.

The extent of derepression in *hisR*, *hisU* and *hisW* mutants shows that the absolute amount of his-tRNA in the cell, not the proportion of tRNA^{His} which is charged, appears to set the level of expression of the operon. In these mutants, the proportion of tRNA^{His} which is charged is normal, but synthesis of the transfer species is reduced so that there are fewer molecules in total in the cell. Figure 8.7 shows that the correlation between cellular content of

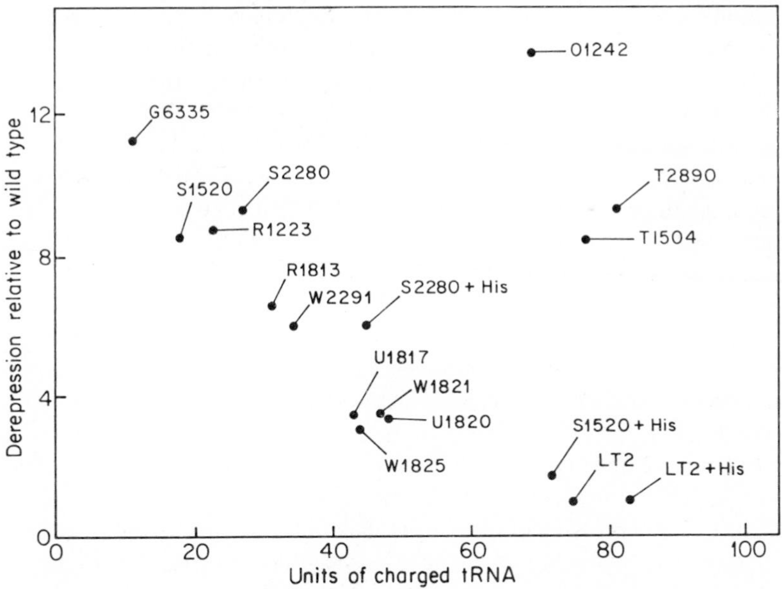

Figure 8.7: correlation of levels of charged his-tRNA and repression of the histidine operon. The extent of derepression relative to wild type (LT2) is directly related to the amount of his-tRNA in each mutant Derepression in his0 and hisT mutants cannot be accounted for by lack of charged his-tRNA, Data of Lewis and Ames (1972)

his-tRNA and derepression (expressed relative to wild type cells) accounts for the derepressed behaviour of mutants in *hisS*, *hisR*, *hisU* and *hisW*. It does not explain the behaviour of *hisO* or *hisT* mutants. Since *hisO* mutants identify the operator, they should not depend upon changes in co-repressor concentration. *HisT* mutants have an altered $tRNA^{His}$ which lacks two pseudouridine modifications in the anticodon loop and we must suppose that although this molecule is active in protein synthesis it is defective as a co-repressor (see also Brenner et al. 1972).

One problem in this model for control of the histidine operon is that amino-acyl-tRNA is somewhat larger than the substrate molecules usually providing inducers or co-repressors. And if his-tRNA acts as co-repressor, this control system cannot show an all or nothing response but must titrate the *amount* of his-tRNA in the cell instead of responding to its absence or presence. (Of course, other operons, such as the lactose genes, respond in this way to increases in concentration of inducer, but in the concentrations in which the cell finds itself this probably causes an on/off switch of gene expression; because his-tRNA is an internal co-repressor which is always needed for protein synthesis, its basal level within the cell must always be significant.) There are about 1500 molecules of $tRNA^{His}$ per cell and the rate of turnover of his-tRNA in protein synthesis is some 5800 molecules per second for cells grown on minimal medium. The level of charged his-tRNA in wild type cells seems to be well above that required to sustain protein synthesis; the excess his-tRNA may act as a buffer so that a decline in its level derepresses the histidine operon before protein synthesis is inhibited by the lack of his-tRNA.

A more serious problem with this model is that no regulator gene has been found which may specify the repressor protein with which his-tRNA must presumably interact. One explanation for this failure may be that the repressor protein also exercises some other function. One possible candidate which has been suggested for this role is the synthetase protein itself, which might act as repressor when bound to his-tRNA. But such a role would impose two quite different functions on this protein—forming his-tRNA for protein synthesis and recognising the *hisO* operator—and also encounters the difficulty that production of synthetase may itself lie under the control of his-tRNA.

The first enzyme of the operon, phosphoribosyl-transferase, coded by *hisG*, has been implicated in repression of the histidine structural genes by the finding of Kovach et al. (1969, 1970) that the state of its feedback sensitive site influences the kinetics of repression of the operon. Histidine inhibits the catalytic activity of this enzyme by binding at a feedback site which is distinct from the active catalytic centre. The analogue thiazole–alanine binds to the enzyme at the feedback site and inhibits its catalytic activity; when the enzyme has been inhibited in this way, the analogue triazole-alanine can no longer repress the operon. Feedback resistant mutants of the *hisG* protein mimic the effect of thiazole-alanine—they are not repressed by addition of TRA.

These results have led to the suggestion that the *hisG* protein may be the repressor of the operon. The "physiological derepression" seen in *hisG* mutants whose catalytic activity is lost may be a consequence of their inability to synthesize histidine, thus depleting the cell of his-tRNA; figure 8.7 shows that the derepression of at least one *hisG* mutant can be explained in this way. But this model does not explain why inhibition of the feedback sensitive site of the enzyme prevents repression, since in these mutants the enzyme has normal catalytic activity; failure in feedback inhibition may therefore exert some other, direct influence upon repression, for example inactivating the ability of the protein to act as a repressor by binding to *hisO*. Vogel et al. (1972) reported that the *hisG* enzyme has a specific affinity for $tRNA^{his}$ and preferentially binds the charged form of the enzyme.

This model implies that the *hisG* enzyme should (in response to his-tRNA) control its own synthesis and should therefore establish a feedback cycle which is auto-regulatory. It may be doubtful, however, whether evolutionary constraints would allow the enzyme to fulfill two such widely divergent functions as acting as phosphoribosyl-transferase and repressor protein. Whatever the protein which acts as repressor, however, it seems probable that it must have more than one function to explain the absence of regulator protein mutants; it is possible that the second function is essential for the cell so that mutation in the protein is lethal. To demonstrate that any protein acts as the histidine repressor, of course, requires its specific binding to *hisO* DNA and elution by his-tRNA.

Positive-Negative Control of the Arabinose Operon

The three enzymes of E.coli which convert L-arabinose to D-xylulose-5-phosphate are regulated as a unit of coordinate control by induction. The *ara*$^-$ mutants which cannot utilise the pentose as a carbon source can be grouped into six genes; genes *araA-D* are closely linked in the sequence *araDABC* shown in figure 8.8, the first three representing the structural genes for the enzyme of the metabolic pathway, epimerase, isomerase and kinase. E.coli has two transport systems which are responsible for the uptake of arabinose. *AraE*, the first characterized genetically, specifies a permease which acts as a system of low affinity for arabinose; *araF* specifies an arabinose binding protein of high affinity which also takes up arabinose (Brown and Hogg, 1972). Both *araE* and *araF* are unlinked to the *araDABC* cluster and both appear to respond to the same control system which acts on the cluster.

The operon appears to be transcribed as a polycistronic messenger from *araB* to *araD*; nonsense mutations in the first gene inhibit synthesis of the subsequent two enzymes. Hybridization experiments with *λdara* show that the messenger RNA for the cluster is found only in induced cells (Wilcox, Singer and Heffernan, 1971); Schleif (1971) has found that strains of bacteria which are constitutive for the arabinose operon have the same level of mRNA as induced

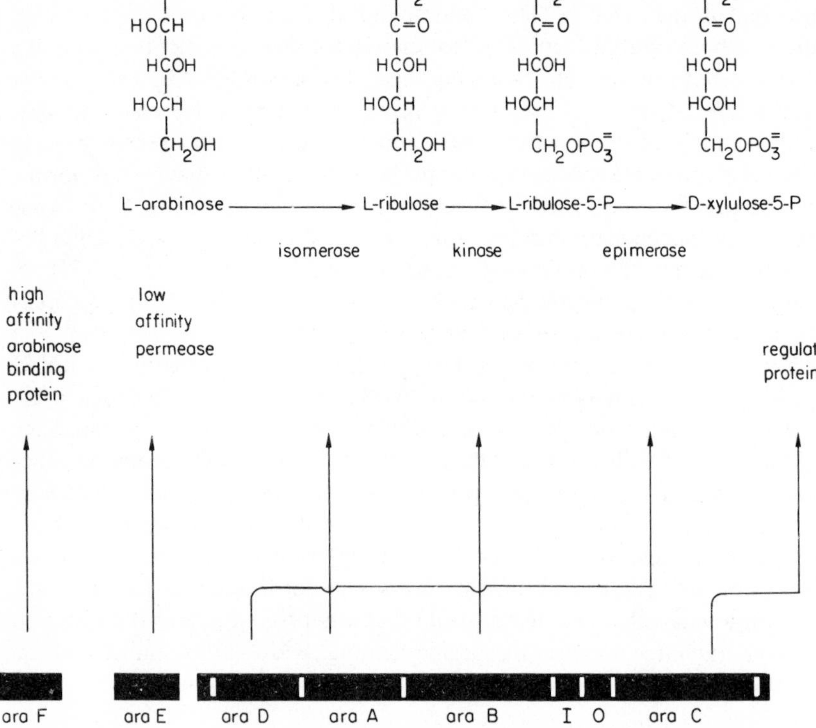

Figure 8.8: the L-arabinose system of E.coli. Data of Englesberg et al. (1965), Schleif (1971), Wilcox et al. (1971)

cells. The efficiency with which the mRNA is translated into protein seems to be similar to that of the lactose operon.

Mutations in *araC* are pleiotropic, affecting all three enzymes and both transport systems; *araC*$^-$ mutants cannot synthesize any of the enzymes whilst *araC*^{c} mutants show constitutive synthesis. Its control of the unlinked genes *araE* and *araF* excludes the possibility that *araC* provides a recognition site and suggests that it may code for a regulator protein. Deletions or nonsense mutants in the *araC* gene give the *araC*$^-$ uninducible phenotype, recessive to *araC*$^+$ or *araC*^{c} alleles in diploids for this region, which is the response demanded of a positive control system. This suggests that the *araC*$^+$ gene might code for an inducer protein which is inactive until converted by arabinose into a form able to switch on the operon; this is the positive induction system of figure 8.3.

But according to this model *araC*^{c} mutants should produce an inducer protein which is active even in the absence of arabinose and should therefore be dominant over *araC*$^+$; whereas the reverse relationship has been found.

Dominance of inducible over constitutive regulator alleles suggests that negative control is exerted through the production of a repressor protein which is inactivated by arabinose; or that subunit interactions between the *araC*^{c} and *araC*$^{+}$ proteins always lead to the loss of the constitutive character. To account for the ability of the *araC*$^{+}$ product to prevent the action of the *araC*^{c} product, Englesberg et al. (1969a, 1969b) proposed that the *araC*$^{+}$ gene synthesizes a repressor protein P1 which switches off the structural genes. Its interaction with arabinose, however, does not merely inactivate P1, but converts it into an alternate form, P2, which acts as an inducer protein whose activity is required to switch on the gene cluster. The interactions of the regulator protein with the unlinked genes *araE* and *araF* have not been defined.

Expression of the *araBAD* segment is therefore determined by the relative concentrations of the P1 repressor and P2 inducer forms of the *araC* regulator protein. Figure 8.9 postulates that in the absence of arabinose the protein is largely or entirely in the form of P1, so that the operon is inactive. Addition of arabinose converts the protein to the P2 state and thus induces the structural genes. One model for these interactions is to suppose that P1 and P2 exist in an equilibrium which is biased towards P1 in uninduced cells, but which is displaced towards P2 in induced cells. This control system therefore has features characteristic of both negative and positive control, for the interaction of repressor protein with inducer does not inactivate the regulator protein and allow gene expression by default; instead the P2 inducer form of the *araC* protein must act to ensure that the structural genes function.

Further dominance relationships have been determined by the isolation of temperature sensitive mutants of the *araC* gene, which fall into the two classes *araC*^{ts} and *araC*^{cts}. The *araC*^{ts} mutants are inducible at 28°C but show the araC^{-} phenotype at 42°C; the *araC*^{cts} class of mutants is partially constitutive at 28°C, showing 10% of the inducible level of the arabinose enzymes, but behaves as araC^{-} at 42°C. According to the negative-positive control model, both the *araC*^{ts} and the *araC*^{cts} proteins are inactive at high temperature, although the *araC*^{ts} protein is normal at low temperature and the *araC*^{cts} protein has an altered equilibrium at 28°C which favours P2 inducer protein more than does the equilibrium prevailing in wild type *araC*$^{+}$ cells.

By constructing diploids, Irr and Englesberg (1971) have shown that *araC*$^{+}$ is dominant to either *araC*^{ts} or *araC*^{cts} at either 28°C or 42°C; under all circumstances the operon is switched off unless arabinose is added to induce it. Introduction of a wild type *araC*$^{+}$ gene restores repressor protein to the cell and thus the ability to respond to arabinose to relieve repression. At low temperature, *araC*^{ts} is dominant to *araC*^{c}, for the P1 repressor protein can control the operon; at high temperature, *araC*^{c} is dominant over the then inactive *araC*^{ts} or *araC*^{cts} alleles.

Both the *araDAB* cluster of structural genes and the regulator gene *araC* appear to read from right to left. By placing the arabinose operon in the

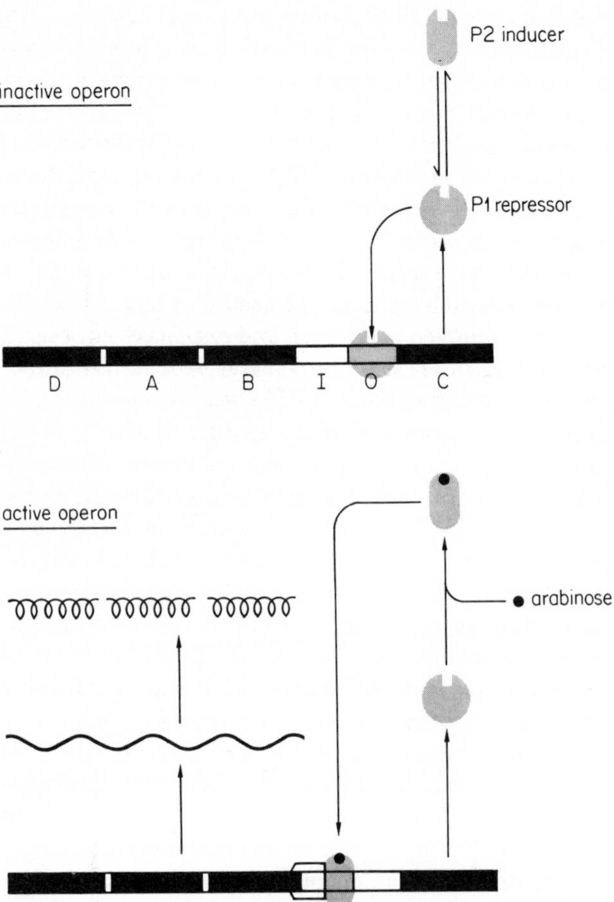

Figure 8.9: repressor-inducer model for control of the arabinose operon. The regulator gene araC codes for a repressor protein, Pl, which binds at the operator (ara0) to switch off the operon. The repressor is in equilibrium with an alternative protein conformation, P2, which acts as an inducer; but in the absence of arabinose P1 is the predominant state so the operon is inactive (upper). Addition of arabinose converts P1 repressor to P2 inducer; the inducer acts at the initiator site araI to switch on the operon. The molecular basis for its interaction is not defined but it is shown here as interacting with RNA polymerase to assist the initiation of transcription. Model of Englesberg, Squires and Meronk (1969)

late region of a hybrid λ-ϕ80 phage, Schleif et al. (1971) showed that both *araDAB* and *araC* can be expressed as the result of transcription by an RNA polymerase starting at a phage promotor; the polymerase reads through the intervening genes into the bacterial operon. Because the direction of transcription of the phage genes is known, that of the arabinose operon can be deduced. A promotor must therefore lie on bacterial DNA to the right of *araC*; and if *araDAB* is under separate control—that is, the structural genes are not read by readthrough from *araC*—a second promotor must lie to the immediate right of *araB*.

Two sites between *araB* and *araC* have been inferred from the properties of deletion strains to comprise the loci where the P1 repressor protein and P2 inducer protein act. Deletions which extend from *araC* into *araB* display the ara$^-$ phenotype, with a pleiotropic loss of expression which is *cis* dominant, *trans* recessive for all three of the *araDAB* structural genes. This locates the control sites for the *araDAB* cluster in the region between *araB* and *araC*.

By comparing the properties of two deletion strains, *Δ719* and *Δ766*, Englesberg et al. (1969a,b) suggested that an operator site, *araO*, to which P1 repressor binds may be located at the right side of this region and that an initiation site, *araI*, at which P2 inducer protein acts may occupy the left side of the intercistronic divide. The deletion *Δ719* covers all known sites in *araC* but does not extend to *araB*; its left end is therefore located in the intercistronic region. Deletion *Δ766* covers most of *araC*, but its left end is located within the gene so that the intercistronic region remains intact.

Both deletions display the uninducible phenotype of *araC*$^-$ cells; but their responses differ when an *araC*$^+$ gene is introduced into the cell in *trans* array to restore the presence of regulator protein. The *araC*$^+$ gene causes a 35 fold increase in the synthesis of isomerase from the *araA* gene in cells with *Δ719*, but causes only a two fold increase in cells with the *Δ766* deletion. The excess induction in the *Δ719* strain has been interpreted to show that this deletion covers *araO*, so that repression of *araDAB* by the P1 molecules in the cell is impossible; but because the deletion does not extend over *araI*, induction can be achieved by any molecules in the form of P2 protein. Because *Δ766* retains both *araO* and *araI*, introduction of a normal *araC*$^+$ gene restores normal control (that is repression in the absence of arabinose).

The site from which transcription is promoted has also been identified by the isolation of *araI*^{c} revertants of *Δ719* able to synthesize isomerase. The *araI*^{c} mutation confers *cis* dominant, *trans* recessive constitutive expression of the *araDAB* gene cluster at about four times its usual base level. Gielow, Largen and Englesberg (1971) found that when an *araC*$^+$ gene is introduced into *araI*^{c} *Δ719* cells, the synthesis of isomerase doubles, consistent with the idea that the *Δ719* deletion allows induction but not repression. But when an *araC*$^+$ gene is introduced into *araI*^{c} *Δ766* cells, the constitutive level of expression is reduced to the usual basal level, consistent with the idea that the *Δ766* deletion leaves intact an *araO* locus for P1 binding.

The nature of the site identified by the *araI*^{c} mutation is not well defined; the mutation maps close to *araB* and may either locate the site at which P2 inducer acts, the site at which polymerase binds, or may perhaps create a new promotor in a site formerly possessing some other function. Eleuterio et al. (1972) isolated point mutations mapping in the left part of the intercistronic region which may identify a site utilised by RNA polymerase, for they produce only low levels of isomerase when induced. Using deletion mapping, Schleif (1972) estimated the length of the intercistronic region as about 500 base pairs, probably somewhat larger than the control region of, for example, the lactose operon; we may speculate that this region contains several control sites, binding P1 repressor, P2 inducer, RNA polymerase, the cyclic AMP receptor protein (to which the gene cluster responds); how the proteins bound at these sites may interact with each other to control transcription requires a more precise definition of the system.

That the regulator protein of the arabinose operon is similar to the repressor proteins of other operons in recognising specific control sites on DNA has been shown by experiments in vitro. Wilcox et al. (1971) prepared a derivative of fucose—a very efficient anti-inducer of the arabinose genes—attached by covalent linkage to a column of sepharose; extracts of *araC*$^+$ but not *araC*$^-$ cells contain a protein which binds to the column. This protein binds specifically to $\phi 80\lambda$d*ara* phage DNA on nitrocellulose filters, although the location of the binding site has not yet been identified.

Using a cell free system with ϕ80d*ara* DNA as template, Zubay, Gielow and Englesberg (1971) were able to transcribe and translate the arabinose genes in vitro to yield ribulokinase, although it is not yet clear whether transcription from this template takes place from the arabinose promotor or by readthrough from a phage promotor. The DNA of λd*ara* can also be expressed to yield ribulokinase; and Greenblatt and Schleif (1971) reported that *araC* protein and arabinose are needed for transcription of this template, although they are not necessary when ϕ80d*ara* DNA is used. Addition of fucose inhibits the production of ribulokinase from the λd*ara* template; but the fucose has no effect when the *araC* protein is extracted from an *araC*^{c} strain instead of from *araC*$^+$ cells. Production of ribulokinase from either template requires mediation of the cyclic AMP system. The molecular action of the *araC* protein has not yet been defined, which is necessary to confirm the positive-negative control model.

Dual Control of the Galactose Operon

The three enzymes coded by the galactose operon metabolize galactose to glucose-1-phosphate by the series of reactions shown in figure 8.10. The three genes *galETK* are contiguous and when galactose is added to E.coli the enzymes for which they code appear in the sequential order epimerase, transferase, kinase. Mutations in any one of the three structural genes create the gal$^-$ phenotype of cells which cannot metabolize galactose; a fourth, unlinked gene, *galU*, which codes for the enzyme UDPG-pyrophosphorylase, functions constitutively, so that mutants at this locus cannot supply UDP-glucose and are therefore also gal$^-$.

The operon is transcribed as a unit from the right end at which an operator has been identified by o^c *cis*-dominant, *trans*-recessive mutations; mutation at an adjacent promotor site reduces the maximum level of expression of the operon. The gene which codes for the repressor protein of the operon, *galR*, is not linked to the structural genes. Adhya and Shapiro (1969) found that *galR*$^-$ mutants—which include deletions in the gene—are constitutive and recessive; Saedler et al. (1968) isolated uninducible *galR*^{s} mutants which are dominant. This suggests that the galactose operon is under a negative control similar to that which applies in the lactose operon.

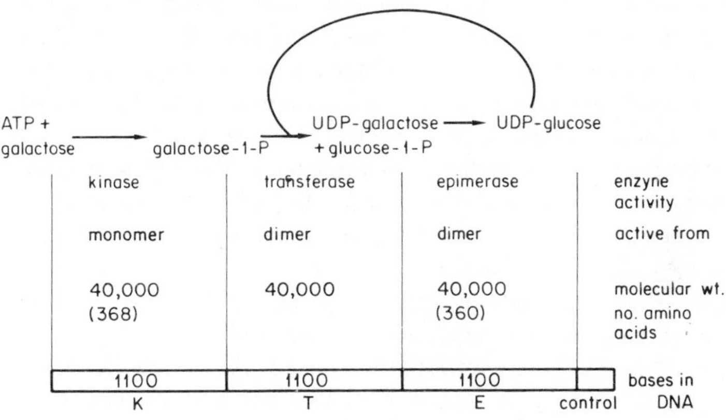

Figure 8.10: the galactose operon of E.coli. Data of Saedler et al. (1968) and Wilson and Hogness (1969)

That a polycistronic messenger is made for all three structural genes is suggested by the influence of nonsense mutations in *galE* upon the expression of *galT* and *galK*; proper translation of the first gene is necessary for synthesis of the proteins of the subsequent genes. And by comparing the activities of the epimerase and kinase enzymes in crude extracts prepared from induced cells with the catalytic activities of the purified enzymes, Wilson and Hogness (1969) were able to show that the same number of molecules of each is synthesized; between 5600 and 5900 polypeptide chains are made from each of the *galE* and *galK* genes on induction. This corresponds to about 43–50 translations of each genome per minute.

The galactose operon is located near the attachment site of phage lambda in E.coli so that λ*gal* transducing phages are available. By using a hybridization assay with λp*gal* DNA, Miller et al. (1971) have shown that the addition to cells of fucose, a gratuitous inducer of the operon, causes the synthesis of galactose mRNA. The operon is subject to glucose repression, for synthesis of its messenger RNA is greatly reduced when cells are grown in glucose; however, the enzymes can be induced again at normal levels when cyclic AMP is added. Using a strain of E.coli deficient in adenyl cyclase, Miller et al. showed that the presence of cyclic AMP is required if the operon is to be fully induced. In the absence of cyclic AMP, fucose cannot induce the operon, but induction is considerable when cyclic AMP is present.

The galactose operon can be transcribed in vitro when it is carried on a lambda phage DNA and the RNA made can then direct the synthesis of at least the galactokinase enzyme. Parks et al. (1971a) and Wetekam, Staack and Ehring (1971) found that synthesis of galactose mRNA in vitro is stimulated by cyclic AMP, although not absolutely dependent upon it; Nissley et al. (1971) showed that cyclic AMP and CRP protein act at the stage of initiation of transcription by RNA polymerase (see page 306). By taking advantage of its affinity for analogues of galactose, Parks et al. (1971b) purified the repressor; the protein binds the DNA of the operon to nitrocellulose filters and the dissociation constant of the complex is the same as that displayed by the components of the lactose operon. Fucose and galactose are equally effective in preventing the repressor from associating with DNA. That this interaction controls transcription in vitro has been confirmed by Nakanishi et al. (1973) who found also that the galactose repressor cannot prevent transcription of DNA bearing of *gal* o^c mutation.

Lactose DNA does not compete with galactose DNA for binding to the galactose repressor so that the *lac* and *gal* operators must comprise unrelated sequences of DNA. This supports the concept that repressor proteins can recognise specific base pair sequences and do not demand only some simple feature of DNA secondary structure for binding. Control of the galactose operon by the galactose inducer is thus very similar to the control of the lactose operon; a repressor protein recognises a specific nucleotide sequence at the operator but is released by the small molecule inducer. And cyclic AMP and CRP protein must be present for RNA polymerase to act at the promotor, the same situation prevailing in the lactose operon.

But in addition to the negative-induction control mediated by galactose and the positive control exerted by cyclic AMP, the galactose operon falls under control of a third system. This appears to relate to the function of the epimerase in the production of capsular polysaccharide. Synthesis of capsular polysaccharide—needed for the cell wall of E.coli—is controlled by the unlinked regulator genes *capR*, *capS* and *capT*. Mutation in any one of these genes causes overproduction of the polysaccharide, which is revealed by a mucoid phenotype. Figure 8.11 shows that at least ten enzymes are implicated in the production from glucose of the sugar precursors utilized for assembly of the capsular polysaccharide; and synthesis of many of these enzymes, whose genes are unlinked, is derepressed in mucoid (cap^-) strains of E.coli.

Differences between the three mucoid strains show that each of the *cap* genes controls a characteristic spectrum of structural genes. The enzyme UDP-galactose epimerase, coded by *galE*, is derepressed 4-6 times in $capR^-$ mutants and 3-4 times in $capT^-$ mutants; since there is only an increase of

50% in the enzyme level in *capS*⁻ mutants the *galE* locus must be controlled only by *capR* and *capT*. The *capR* locus has been identified in some E.coli strains by the *lon*⁻ mutation; *lon* and *capR* are synonyms for the same gene.

All three enzymes of the galactose operon are constitutively expressed in *capR*⁻ cells which suggests that the *capR*/*capT* system may act upon the transcription or translation of the polycistronic messenger. By using hybridization with λp*gal* DNA to assay galactose messengers, Mackie and Wilson (1972a) and Buchanan et al. (1973) confirmed that the synthesis of *gal* mRNA is increased by the *capR*⁻ and *capT*⁻ mutations.

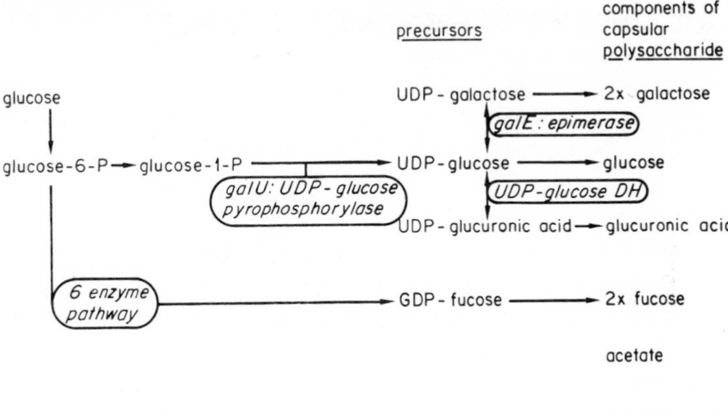

Figure 8.11: synthesis of the capsular polysaccharide of E.coli. Enzymes implicated in the later stages of synthesis and the genes which code for them are in italics. Data of Hua and Markowitz (1972)

The effects of the *capR*/*capT* control system and the normal induction exercised by galactose appear to be independent. Hua and Markowitz (1972) observed that the addition of fucose induces the galactose enzymes to the same relative extent—a 10-25 fold increase—when either wild type or *cap*⁻ mutant cells are grown on glycerol (growth on glycerol avoids glucose repression). Although mutant *galR*ˢ cells are not induced by fucose, the introduction of a *cap*⁻ mutation causes the low (3–6 fold) level of constitutivity characteristic of the particular *cap* mutant. Although constitutive expression is considerable in *galR*⁻ cells, a further two fold increase in enzyme synthesis is caused by *cap*⁻ mutations. The *capR*⁻ and *capT*⁻ loci also cause two fold increases in expression of *gal* o^c mutants. The reason why the effect of the *cap* system is limited to a doubling of expression in *galR*⁻ and o^c cells may be that this level of enzyme synthesis is the maximum possible for the operon. But below this limit, the effects of the galactose induction (allowing for the control of this response by cyclic AMP) and the *cap* induction appear to be additive. When all three controls are turned from off to on, the operon is induced about 1000 fold (the same as the extent of induction of the *lac* genes).

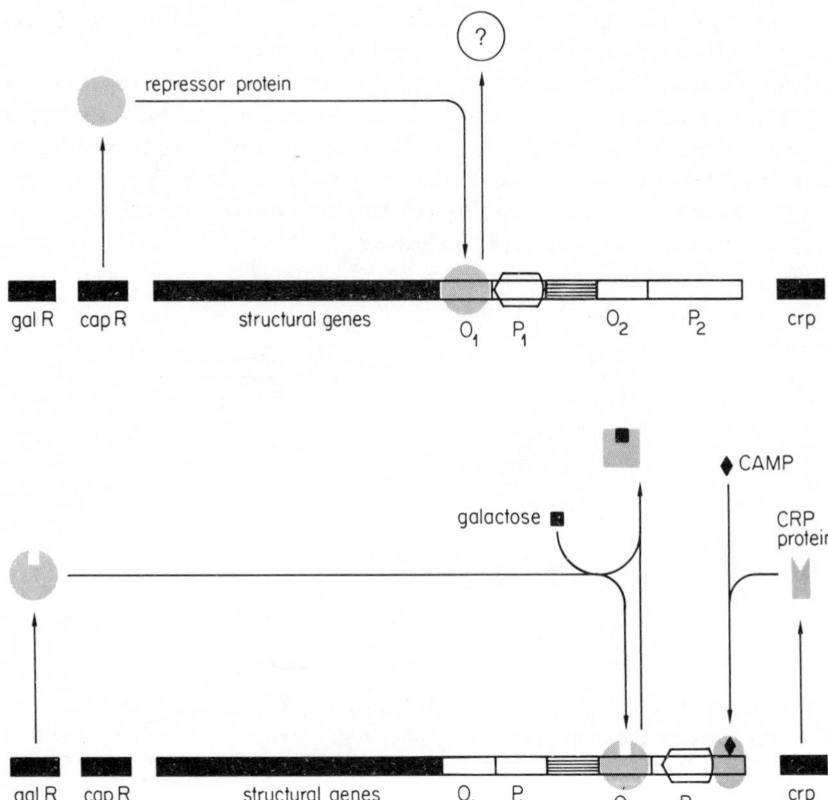

Figure 8.12: control of the galactose operon. Transcription of the structural genes may be initiated at either or both of the two control sites, each of which consists of an operator and a promotor. Upper: at site 1, the capR repressor may bind to the operator to prevent transcription by RNA polymerase binding at the adjacent promotor. We do not know how its release from the operator is achieved. Lower: at site 2, the galR repressor binds to the operator unless inactivated by galactose. RNA polymerase can initiate transcription at the adjacent promotor only with the participation of the cyclic AMP binding protein. The order of the control sites o_1/p_1 and o_2/p_2 shown in the figure is arbitrary. There is genetic evidence only for the existence of the o_2/p_2 type site at the right end of the gene cluster. We do not know how the sites interact, but their action seems to be independent so that binding of repressor at one site does not prevent transcription from the other site; although the molecular nature of this region is not defined the two sites are illustrated as separated by a gap. Model of Hua and Markowitz (1972)

The *cap*⁻ mutations are pleiotropic in their effects, causing the mucoid phenotype, the increased synthesis of the enzymes specified by at least four spatially separated operons; and also generating an increased sensitivity to ultraviolet and ionising radiation (shown by formation of non-septate filaments and subsequent cell death). Hua and Markowitz have suggested that the product of the $capR^+$ gene is a repressor protein which binds to the operators of the several structural genes which it controls. $CapR^-$ mutants must produce only an inactive repressor protein, leading to expression of the operons. Since $capR^-$ $capT^-$ double mutants show the same constitutive expression as $capR^-$, both loci may function in the same pathway; $capT^+$ may therefore usually function to produce some enzyme which catalyses production of a co-repressor needed to activate the *capR* protein.

The independence of the galactose/cyclic AMP and *capR* control systems suggests that the galactose operon may have two sets of control elements. Figure 8.12 illustrates a model in which one operator binds to the repressor protein inactivated by galactose and its adjacent promotor responds to cyclic AMP. A second operator site binds the *capR* repressor and its adjacent promotor does not respond to cyclic AMP. One problem raised by this model is how polymerase might read through from the outside operator/promotor locus when the appropriate repressor is bound at the inner operator. Since the existence of these different control sites cannot be confirmed until mutants have been isolated and mapped in all the control functions which they exercise, the order of the two operator/promotor loci is arbitrary and we can at present only speculate on how their functions may relate to each other. The complex nature of the control of the galactose operon presumably results from the utilization of the sugar for two different purposes, catabolism and cell wall synthesis.

Control of Dispersed Genes in the Arginine System

Eight enzymes are involved in the metabolic conversion of glutamic acid to arginine, but their genes map at the separate loci on the E.coli chromosome shown in figure 8.13. One locus comprises the four genes *argECBH* and each of the remaining genes maps at a distinct site. In spite of their separation, however, the expression of all eight enzymes is subject to a common repression by arginine. Repression of the structural genes is mediated by one regulator gene, *argR* which is unlinked to any of them.

The RNA produced by the *argECBH* gene cluster can be assayed by hybridization with a ϕ80d*arg* phage DNA; and Rogers et al. (1971) and Krzyzek and Rogers (1972) observed that the level of hybridizing mRNA is reduced 7–24 times when arginine is added to E.coli cells. Repression is established within two minutes after arginine has been added; full derepression is achieved within one minute of starvation for arginine. Repression of transcription follows the same kinetics when either arginine or rifampicin is added, which is

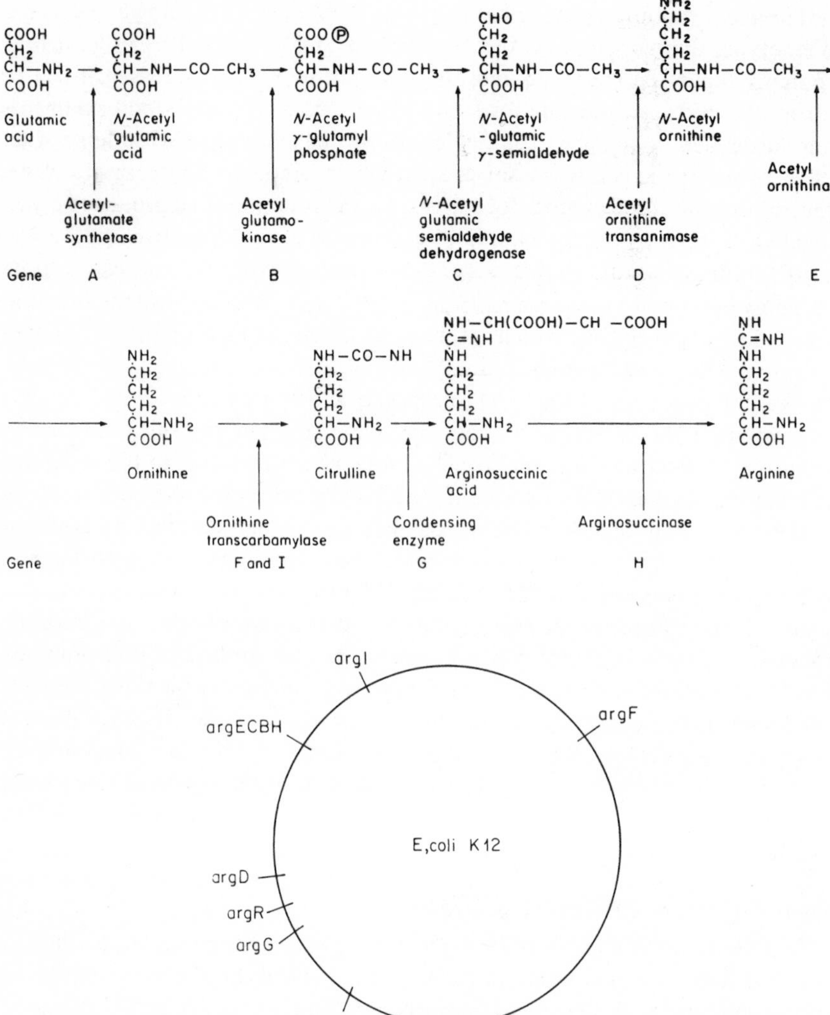

Figure 8.13: the arginine biosynthetic system of E.coli K12. Strains B and W are similar but lack argF, so that OTC enzyme is produced only by argI

consistent with the idea that arginine acts as a co-repressor to prevent new initiations of transcription.

The enzyme ornithine transcarbamylase (OTC) is the simplest to assay because its specific activity varies most widely with changes in the arginine concentration. Strains of E.coli differ in the genes which code for this enzyme. E.coli K12 contains two genes which specify ornithine transcarbamylase, *argF* and *argI*. Legrain et al. (1972) reported that each gene produces a protein chain of 35,000 daltons. The active form of the enzyme is a trimer and these E.coli cells contain all four possible forms, FFF, FFI, IIF, III. This situation must presumably have arisen by a gene duplication. But strains B and W of E.coli have only one gene coding for OTC; Jacoby (1971) showed that this locus maps in the position of *argI*, not *argF* as had previously been thought.

ArgR$^-$ mutants produce about the same level of *argECBH* mRNA—either in the presence or absence of arginine—as bacteria derepressed by starvation for arginine. By studying the repression of OTC in response to arginine in strains diploid for the *argR* locus, Maas et al. (1964) and Mass and Clark (1964) showed that *argR*$^+$ is dominant to *argR*$^-$; amber mutants have the *argR*$^-$ genotype and this suggests a negative control system in which the *argR* gene produces a repressor protein which is activated by arginine (see figure 8.1).

The repressor protein has been isolated by Urm et al. (1972) by its ability to bind to a phage carrying the *argECBH* cluster. There are about 200 molecules of arginine repressor in an E.coli cell, some 20 times more than the amount of lactose or tryptophan repressor; the increase is presumably due to the dispersion of arginine genes, with the consequent need for repressor to bind to more than one operator.

Although most strains of E.coli, including K, C and W, display repression of all eight enzymes when arginine is added to the medium, strain B is an exception. Cells of E.coli B have a slightly repressed level of the arginine enzymes in the absence of arginine, but addition of arginine induces instead of further repressing the structural genes. Jacoby and Gorini (1969) showed that insertion by recombination of the K cell *argR*$^+$ gene into B cells introduces repression in response to arginine; and replacement of the usual allele of K cells by the *argR*^{B} allele introduces induction in place of repression. The mode of regulation of the arginine structural genes thus depends solely on the *argR* allele present.

Regulator mutants in strain B have been isolated by growing bacteria under conditions in which they must acquire the K-type repression control to survive. Using different selective criteria, Jacoby and Gorini (1969) and Kadner and Maas (1971) showed that single step mutations can convert *argR*^{B} control to a response similar, although not identical, to that of the K cells. Intermediate classes of control can also be produced by mutation, including mutants only partially repressed by addition of arginine (whose repressor must therefore have a lower affinity for the operator sequences) and mutants which are partially derepressed irrespective of the addition of arginine (their repressor must have a lower affinity for operator but fail to recognise arginine). The precise affinity of the *argR* repressor for its operator sequences and its response to arginine therefore depend upon the amino acid sequence of the protein in a very precise manner.

Both the *argR*^{B} and *argR*$^+$ alleles are dominant to the *argR*$^-$ derepressed mutants. In $argR^B/argR^+$ diploids, control is dominated by $argR^B$ in the absence of arginine so that the structural genes function at the partly repressed level characteristic of strain B cells; but addition of excess arginine achieves the full repression characteristic of $argR^+$ cells. A model to account for these results has been proposed by Jacoby and Gorini (1969) and Karlstrom and Gorini (1969), who suggested that E.coli K cells possess a negative control system in which the $argR^+$ gene synthesizes a repressor protein activated by arginine. In cells containing the $argR^B$ allele, the repressor protein has a greater affinity for its operators in the absence of arginine than its counterpart in K strains; this means that the repressor protein itself is partially active, so there is some repression in the absence of arginine. The interaction of added arginine with the $argR^B$ repressor protein, however, must reduce the affinity of the regulator protein for its operators; this induces rather than represses the operon.

Control of the dispersed arginine structural genes shows that it is possible to exercise control of the synthesis of the enzymes constituting a metabolic pathway even when the genes which code for them are not gathered in a cluster. Each of the loci representing the arginine structural genes must possess an operator sequence which is recognised by the *argR* regulator protein. The structural genes of the system are therefore under *parallel* control since all respond to the same repressor; but this control is non-coordinate since the levels of expression of the genes at the separate loci may be quite different.

Two types of variation in the control sites may explain how the level of expression of each locus is established. Although the various operators must all respond to the same repressor, each may constitute a different (although related) base pair sequence; each operator may therefore have a characteristic affinity for repressor which sets its binding of protein at a certain level. An alternative model is to suppose that the promotors of the different loci function with varying efficiencies in recognising RNA polymerase and initiating transcription. Of course, these models are not mutually exclusive.

The operator at the *argI* locus has been identified by Jacoby and Gorini (1969) by the isolation of mutations in B cells which cause *cis* dominant synthesis of the OTC enzyme at partially derepressed levels in the absence of arginine. The mutation in $argO_I$ also causes partial derepression in cells possessing the $argR^+$ regulator allele; the level of derepression is reduced by addition of arginine, so that some response to the amino acid remains. Cells possessing the $argO_I$ mutation in combination with $argR^B$ also show this response, the addition of arginine reducing the extent of derepression. The mutation therefore reverses the usual relationship between operator and B repressor protein, so that the protein bound by arginine has a greater affinity for the operator than the free protein; the other arginine genes in the B cell retain their usual response of induction by arginine. The behaviour of the $argO_I$ mutants therefore demonstrates that recognition of operator by repressor protein depends critically upon both their sequences and is consistent with the idea that differences in operator sequences at the separate arginine loci might establish their levels of expression.

The four gene *argECBH* cluster appears to comprise two operons rather than one, with the *argCBH* genes expressed coordinately and the *argE* gene under parallel but non-coordinate control. According to Baumberg, Bacon and Vogel (1965) and Cunin et al (1969), although both *argCBH* and *argE* respond to the same repression control, the enzymes specified by the *argCBH* genes may be repressed by fifty or seventy fold, whereas the *argE* gene can be repressed only twenty fold. The *argCBH* segment seems to be read from *argC* towards *argH*, for nonsense mutants in *argB* and *argC* reduce the expression of *argH*. But amber mutants in *argE* do not influence the expression of the other three genes. This suggests that *argE* and *argCBH* are expressed by transcription from different control points.

Deletions which abolish the boundary region between *argE* and *argC* reduce the expression of *argH* to a low level which is similar to that of *argE*. This observation was originally taken to mean that there is a control site for the *argCBH* genes located to the left of *argC*, with a control site for *argE* situated at the left end of *argE*; deletions of the region between *argE* and *argC* might place *argH* under the control of *argE* by allowing a readthrough which is usually prevented at the boundary.

But the expression of *argH* in the deletion strains cannot be repressed by arginine and appears to take place from an inefficient additional promotor located to the left of the *argH* gene. Elseviers et al. (1972) have shown that a deletion which removes all known mutant sites in *argC* but does not appear to enter *argE* has two effects. It reduces the expression of *argE*; and although *argH* continues to be expressed at the same level in constitutive cells, its activity can be repressed only 12–15 times in the deletion strain. This suggests that *argE* may be controlled by a site in this region, that is at the right end of the gene, so that it is transcribed in the opposite direction from the *argCBH* cluster.

The deletion may remove part of the promotor and operator for *argE* for mutations located between the deletion and *argE* itself can restore activity of the gene. The deletion appears also to remove part of the operator for *argCBH* but leaves its promotor intact. One model to explain these results is shown in figure 8.14; the promotor for *argCBH* must be independent of the other sites, which must overlap at least in part because the deletion affects all their functions. The idea that these sites overlap is supported by the behaviour of a mutant isolated from the deletion strain which is active for *argE*; *argH* and *argE* are both expressed constitutively in these cells. This implies that the mutations generates promotor activity for *argE* but abolishes operator function for both *argE* and *argH*. Another observation which supports this model is the behaviour of operator mutants which Jacoby (1972) isolated by selecting *argR*^{B} diploids for constitutive expression of *argC*. The o^{c} mutation maps

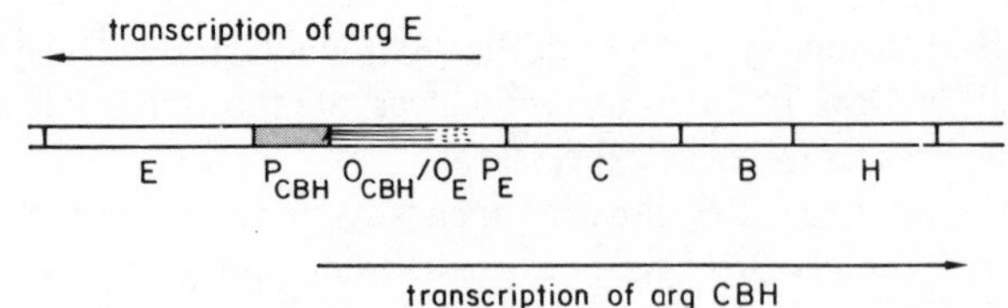

Figure 8.14: the argECBH cluster. Both argE and argECBH are controlled from sites between argE and argC. The promotor for argCBH (p_{CBH}) is independent of the other sites. The operators o_{CBH} and o_E may overlap and the promotor p_E may also overlap with them in part. The order of sites shown is hypothetical

between *argE* and *argC* and increases the expression of argE as well as that of the *argCBH* cluster.

Isolation and fine structure mapping of more regulator mutants will be needed to decide the details of the structure of this control region. However, one problem is that if repression takes place by a mechanism similar to that of the lactose operon, binding of repressor protein to either the *argE* or *argCBH* operator regions should prevent RNA polymerase from progressing from either promotor to the structural genes. Yet *argE* is expressed to a greater extent under repressing conditions than the *argCBH* cluster. And if the promotors are situated on the far side of the control region from their respective structural genes, RNA polymerase must transcribe—or must at least move along—both strands of DNA in the region between them. At present we can only speculate on how the cell might solve these topological problems.

Complex Gene Clusters

That a repressor protein may act at more than one operator is shown by the parallel control of the arginine system and, indeed, as the small cluster of *argECBH* genes reveals, two operator/promotor control elements may be located close to each other—perhaps reflecting the evolution of the system. At least two other systems in E.coli, the histidine utilization (*hut*) and the biotin (*bio*) genes also seem to comprise two operons located at adjacent sites on the chromosomes under parallel but non-coordinate control.

The four enzymes of the histidine utilization pathway are induced by urocanic acid, the product of the first enzyme of the metabolic pathway shown in figure 8.15. After entry into the cell, histidine is converted into urocanate by the small basal amount of histidase present; that the urocanic acid produced by this reaction is the inducer, is suggested by the failure of histidine to induce the *hut* enzymes in strains which lack the histidase enzyme. Bacterial mutants which lack urocanase suffer constitutive synthesis of the other enzymes, for the histidine made endogenously gives rise to urocanic acid, which cannot be metabolised and therefore induces the histidine utilization operon.

The histidase and urocanase genes are under coordinate control as are the two hydrolase enzymes; but although the qualitative effect of induction is the same for all four enzymes, the proportions made are different and each pair of enzymes suffers glucose repression to a different extent. Control mutations in the regulator elements may affect the synthesis of either pair of enzymes without influencing the other. By constructing an F′ factor bearing the *hut* genes, Smith and Magasanik (1971) have tested the dominance relationships of the different regulator mutants.

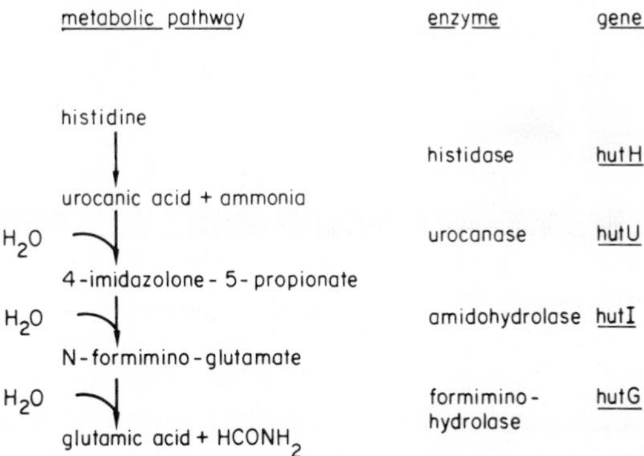

Figure 8.15: the histidine utilization (hut) system of E.coli. Data of Smith and Magasanik (1971)

Mutants with an inactive *hutC* gene exhibit constitutive synthesis of all four enzymes, but $hutC^+$ is dominant to these $hutC^-$ alleles; this suggests that *hutC* specifies a repressor protein which is inactivated by the urocanic acid inducer. Figure 8.16 shows a model for its action. Mutants *hutP* and *hutR* are *cis* dominant, *trans* recessive and seem to define a promotor region which controls the expression of the *hutH* and *hutU* genes. Analogous mutants called *hutQ* identify the operator for these two genes, although they tend to affect promotor functions also; one possible explanation is that there is some overlap between operator and promotor. Mutants in *hutM* identify the promotor locus for the two hydrolase enzymes; an adjacent operator site has not yet been identified by mutation. The action of the *hutC* regulator, therefore, is to control the activity of the genes on either side of it through independent interactions at their separate operator sites.

By using the DNA of a λ*phut* phage, Hagen and Magasanik (1973) confirmed that a protein present only in $hutC^+$ cells binds to the operon and is released by urocanate. The model of figure 8.16 implies that the *hutC* protein repressor regulates its own synthesis; that it is under control of the *hutM* promotor is suggested by the observation that deletion of *hutM* causes failure to synthesize the hydrolase enzymes but results in constitutive synthesis of histidase and urocanase (presumably because the *hutC* repressor is not synthesized). Mutants of the p^s class in *hutM* increase synthesis of the hydrolases

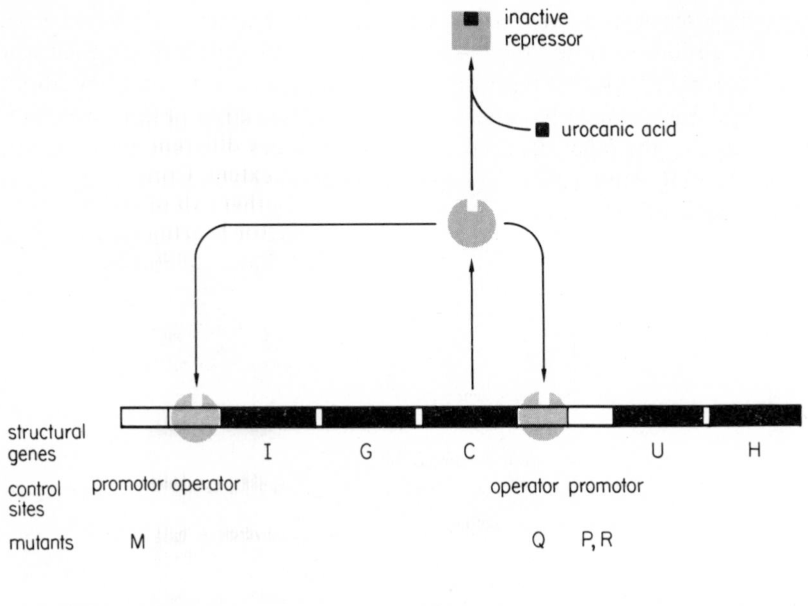

Figure 8.16: model for control of the histidine utilization system. HutC species a repressor protein which acts at two separate sites, one to the left of hutIG and the other to the left of hutUH. The promotor locus for hutIGC is identified by hutM mutants; hutC is therefore repressed by its own product. The operator shown adjacent to hutM has not been identified by mutation. The promotor for hutUH is identified by hutP and hutR mutations; the operator is identified by hutQ. Urocanic acid, the product of action of the hutH enzyme on histidine, acts as inducer which inactivates the hutC repressor protein. Data of Smith and Magasanik (1971) and Hagen and Magasanik (1972)

but depress synthesis of histidase and urocanase; the depression can be relieved by mutation in *hutC*.

The biotin locus of E.coli has a structure analogous to that of the *argECBH* cluster; figure 8.17 shows that most of its genes are transcribed from one strand of DNA, but that one gene is transcribed from the other strand. By hybridizing RNA from E.coli pulse labelled in conditions either of repression or derepression for the biotin genes, Guha et al. (1971) showed that 40% of the biotin mRNA hybridizes with the *l* strand and 60% hybridizes with the *r* strand of λ phage DNAs carrying the biotin locus. By using phages carrying various deletions within the biotin region, they have shown that *bioA* appears to be transcribed from one strand whereas the other genes are transcribed as a cluster in the opposite direction.

Escape synthesis of a bacterial operon can take place by two different mechanisms when cells are infected with a phage carrying the operon. Replication

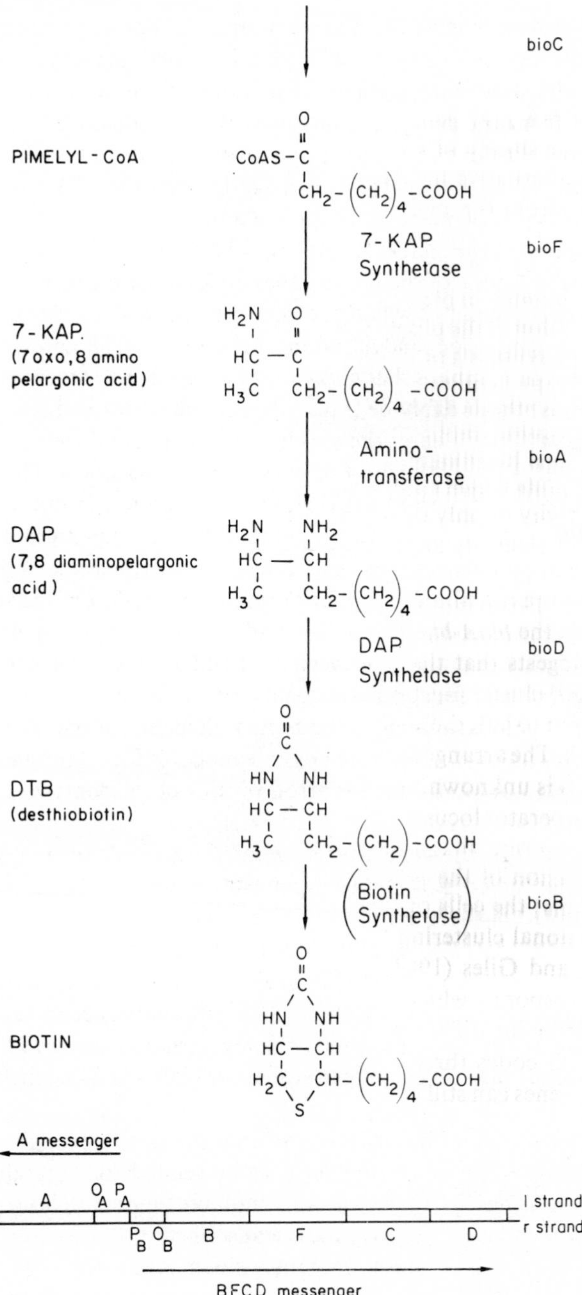

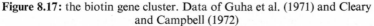

Figure 8.17: the biotin gene cluster. Data of Guha et al. (1971) and Cleary and Campbell (1972)

of the phage DNA may generate a large number of copies of the structural genes so that the supply of repressor protein is inadequate to repress all the operators. An alternative mechanism is for transcription to proceed into the bacterial genes from the phage regions immediately adjacent to them; this transcription does not depend upon the operator/promotor sites of the operon.

The two types of escape synthesis can be distinguished because the first is inhibited by mutants in phage replication and the second by mutations which inhibit transcription of the phage. Krell et al. (1972) made use of this distinction to permit escape synthesis of the biotin genes under these different conditions. *BioA* shows escape synthesis when the phage is allowed to replicate; *bioD* displays escape synthesis dependent upon transcription. The known direction of phage transcription implies that *bioD* must be transcribed from left to right, whereas *bioA* must presumably be transcribed in the opposite direction.

Deletion mutants which enter the *bio* cluster from either its left end or right end abolish activity of only the *bioA* or *bioD* genes respectively. This suggests that the control elements are located within the operon and not at its ends. Cleary, Campbell and Chany (1972) found that no deletion entering from the right end of the operon diminishes its activity, but deletions entering from the left and passing the *bioA-bioB* boundary prevent expression of the remaining genes. This suggests that the promotor, and presumably the operator also, for the *bioBFCD* cluster must lie between *bioA* and *bioB*. If *bioA* is to be transcribed from right to left, the control elements for its activity must be located in the same region. The arrangement of the control elements for the two directions of transcription is unknown, but one possible model is for a repressor protein to bind at one operator locus which controls transcription from two promotors orientated in opposite directions.

The phenomenon of the gene cluster seems, in general, to be confined to bacterial systems; the cells of higher organisms do not appear to possess the extent of functional clustering found in bacteria. For example, according to Ahmed, Case and Giles (1964) and Fink (1966), the genes of the histidine system of Neurospora—which is metabolically similar to that of Salmonella—are scattered over the different chromosomes, although in each case there is one locus which codes three enzymes. However, as the dispersed bacterial systems show, genes can still be under parallel control even when they are sited apart.

The gene clusters identified in certain fungi do not appear to represent operons; rather does the clustering seem to be related to the ability of the proteins of these genes to aggregate into multienzyme complexes. The tricistronic histidine locus of Neurospora is transcribed into a messenger RNA which directs synthesis of three proteins which Ahmed (1968) found to aggregate into a multifunctional complex. A cluster of five structural genes in Neurospora codes for the enzymes of the polyaromatic pathway—which converts hydroquinolic acid to chorismic acid; and Case and Giles (1968,

1971) found that the products of these genes form an aggregate of all five enzymes. No loci corresponding to operator/promotor control elements have been found in these fungal systems and it appears probable that the gene clustering is needed for some function concerned with formation of a multiprotein aggregate, such as the ability of ribosomes to translate a polycistronic messenger so that the enzymes can aggregate as they are synthesised.

Evolution of Control Systems

Induction and repression of bacterial operons is controlled by the inhibition of transcription caused by binding of an active repressor protein to the operator at the starting end of the sequence of structural genes. The ability of the repressor to respond to the presence or absence of a small inducer or co-repressor molecule may perhaps comprise a specialized system which has evolved to meet the particular needs of bacteria; such unicellular organisms are likely to suffer sudden fluctuations in their external environment and may therefore require the ability to rapidly activate or inactivate particular gene systems.

Models for gene control in eucaryotic cells which have a formal similarity to the Jacob-Monod model are discussed in chapter six of volume two; but whatever form the eucaryotic control systems take, there are probably substantial differences between the molecular interactions on which they rely and those of bacteria. Both the production of messenger RNA and the organization of the genome are very different in bacterial and higher eucaryotic cells. Whilst bacteria synthesize short lived messengers so that operons can be switched off as rapidly as on, much of eucaryotic mRNA appears more stable; and its synthesis requires maturation from giant precursor molecules in the nucleus. The lack of functional gene clustering, the much increased quantity and the more complex state of DNA in the nucleoprotein structure of eucaryotic chromatin raise questions about how eucaryotic regulator molecules recognise their sites of action in DNA.

Experiments in selection have shown that when a bacterial strain loses its proper control mechanism, it is at a selective disadvantage. Zamenhof and Eichoran (1967) showed that regulator mutants of the tryptophan system of B. subtilis which cannot be repressed as usual by tryptophan suffer a selective disadvantage, presumably because of their wasteful production of the amino acid. Baich and Johnson (1968) compared the growth of wild type E.coli with a strain lacking the ability to control proline synthesis by end product inhibition. In a mixed culture, there was a steady state increase in the number of wild type cells at the expense of the mutant.

A bacterial strain may gain a selective advantage by acquiring a mechanism to repress the production of an enzyme when it is not required or to induce it only when it is needed. Protein synthesis is energetically expensive—three high energy bonds are used for every amino acid added to the chain—so that it is wasteful for the cell to produce unnecessary proteins. RNA synthesis also consumes energy, so the most economical control system for cellular functions is to halt the transcription of genes whose products are not needed at any particular time. The evolution of negative control systems—whether used for induction or repression—in which the regulator gene synthesises a repressor protein is therefore easy to explain; a cell which acquires a way to switch off functions when they are not needed can spend less energy in order to achieve the same growth and so presumably gains a selective advantage.

But it is more difficult to see how positive control systems, in which the regulator gene synthesizes an inducer protein, might evolve; since in this situation the cell must originally have had the capacity to synthesize the enzyme concerned no matter what the external conditions. This may explain the apparent predominance of negative control systems where individual operons are concerned. And an advantage of negative control may be that it provides a failsafe system in the sense that failure of the function leaves the genes which it controls in operation. This must presumably be less harmful to the cell than the inactivation of a positive control system, which leaves the cell completely unable to utilise the gene functions concerned.

One possible way for positive control to evolve may be through a change in the function of a regulator protein in evolution; systems such as that of the arabinose operon in which arabinose converts the repressor protein into an inducer instead of simply inactivating it might reflect an intermediate stage in the evolution of positive control systems. We can imagine that a repressor protein whose activity resides in competing with RNA polymerase for a DNA site might evolve an interaction with the enzyme which becomes necessary for initiation of transcription.

Development of Phage λ

Repression of Phage DNA Expression

The small bacteriophage lambda comprises a circular duplex DNA of about 31×10^6 daltons—about 46,500 base pairs. As a *temperate* phage, two options are open to lambda on infection of its host bacterium, E.coli. The λ DNA may behave in a *virulent* manner and enter the *lytic cycle*; that is to say, phage genes are expressed and phage DNA replicated to reproduce many more phage particles which are eventually released by lysis of the cell. But instead, lambda may exist in harmony with its host cell in a non-infectious manner known as *lysogeny*. As shown in figure 8.18, lysogenic bacteria carry a phage DNA which exists in a form known as prophage, when λ DNA is integrated as a linear molecule into the bacterial chromosome. The choice of which option is followed depends on the conditions of infection and the genotypes of phage and bacterium. As part of the bacterial chromosome, prophage is inherited as are bacterial genes; but prophages may be *induced*—for example, by ultraviolet irradiation—to enter the lytic cycle, in which case the linear DNA is excised from the bacterial chromosome to form an infectious circle.

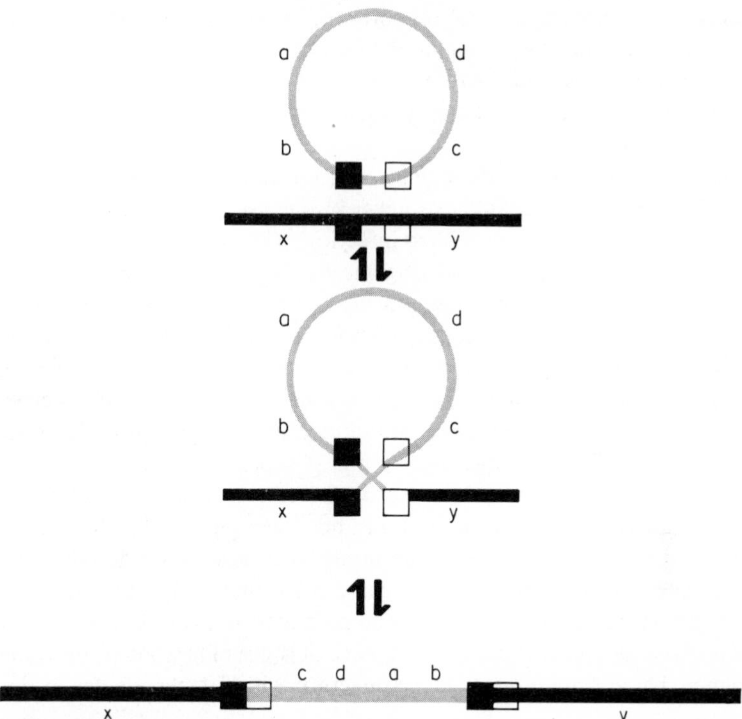

Figure 8.18: the Campbell model for lysogenic integration of bacteriophage into the bacterial chromosome to form prophage. Recombination events take place at the specific regions marked by squares. Lysogeny is reversible; the same sequence of events is followed in reverse during induction of prophage

Lysogenic bacteria are *immune* to infection by further phage particles—called *super-infection*—of the same type as the prophage. Lysogeny and immunity result from the operation of the same control system in which a repressor protein prevents expression of the lambda genes which code the proteins needed by the phage for lytic development (reviewed by Ptashne, 1971). Lambda DNA includes at least 33 genes, 18 concerned with manufacturing the head and tail proteins of the mature phage particle, 9 with recombination, replication and lysis, and 6 which are regulators. Four of the regulators are concerned with establishing or maintaining lysogeny; three are implicated in controlling the proper sequence of lytic development (one regulator is used in both lysogeny and the lytic cycle).

Phage lambda utilises both negative and positive control systems during its infective cycle. One negative control system synthesises a repressor protein which prevents the expression of all λ genes except its own. The genes which are repressed include the *N* gene, which is a positive regulator required to express the delayed early genes and set lytic development in train, and the *tof* regulator, which acts to stop the transcription of repressor and switches off some delayed early genes later in development. By inhibiting these two regulator genes, the λ repressor maintains λ DNA in an inert state; compared with bacterial operons, therefore, the control of lysogeny and lytic development of the phage is very economical.

Mutation in any one of three genes, *cI*, *cII* and *cIII* inhibits the establishment of lysogeny; *cI* mutants can never lysogenise and *cII* and *cIII* mutants manage to do so only rarely. Mixed infection of a host cell with wild type and *c* mutant λ phages yields cells lysogenic for both phages; so the wild type alleles are dominant. After infection of bacteria with pairs of mutants, it is possible to isolate lysogenic cells carrying *cII* or *cIII* mutants alone, but not *cI* mutants; this suggests that *cII* and *cIII* are needed only to help establish lysogeny, but that continual action of *cI* is necessary to maintain it (see below). The *cI* gene codes for a repressor protein which switches off all the genes of lambda, either of a prophage or of a phage super-infecting a lysogenic cell.

That immunity can be maintained by the *cI* gene alone is shown by the properties of hybrids between lambda and other phages which consist of λ DNA apart from a small area, the immunity region, which is derived from the other phage. As shown in figures 6.12 and 8.19 λimm^{434} has the immunity region of phage 434 in place of the *cI* and *tof* genes of λ; phage λimm^{21} has the *N*, *cI*, *tof* and *cII* genes of phage 21 in place of those of lambda. Cells lysogenic for phage 434 are immune to super-infection with further phages of the 434

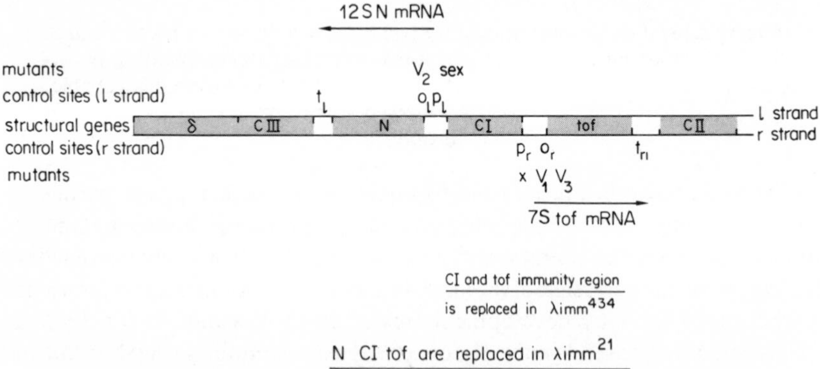

Figure 8.19: control sites for early expression of phage lambda. The *cI* gene is the only locus transcribed in lysogenic cells and codes for a repressor protein which acts at two sites, o_1 (identified by the v2 mutation) and o_r (identified by the v1, v3 mutations). The two promotors corresponding to these operators are p_1 (identified by the sex mutation) and p_r (identified by mutants of the x type). If repressor fails to bind to o_1, transcription of the left strand takes place from p_1 to give 12S mRNA coding for N protein. If repressor does not bind to o_r, transcription of the right strand takes place from p_r to produce 7S mRNA for tof. Sites t_1 and t_{r1} mark the points where these transcripts cease

type, just as lysogenic λ cells are immune to super-infection with λ. But the immunity regions of the two phages are different, for lysogenic λ cells are not immune to super-infection with phage 434; nor are cells lysogenic for phage 434 immune to super-infection with λ. The hybrid λimm^{434} behaves as does phage 434 in conferring immunity, although all its DNA except the immunity region is derived from λ; this implies that both the *cI* gene of phage 434 and the sites at which it acts are located within the immunity region and are sufficient to maintain immunity; other phage functions are not involved.

In cells lysogenic for λ, only the *cI* gene of the phage is transcribed and all its other functions are quiescent. On induction, messenger RNA is made which will hybridize with λ DNA; and experiments using the separated strands of λ DNA have suggested that transcription starts off from two sites, each located in the immunity region, one on each strand. Virulent mutants of lambda, called λ*vir*, cannot lysogenise and are immune to the action of repressor protein for they can super-infect and reproduce in cells which are already lysogenic for λ. The λ*vir* characteristic is conferred by the presence of three mutations in λ DNA. According to Hopkins and Ptashne (1971), the mutation *v2* maps between the *cI* and *N* genes; and the two closely linked mutations *v1* and *v3* map just to the right of *cI*. The position of these mutations is shown in figure 8.19.

The virulent phages behave as operator-constitutive mutants which can no longer recognise the *cI* repressor protein. Mutant *v2* identifies the leftward operator, o_1, which controls the expression of λ to the left of the immunity region; *v1* and *v3* identify the rightward operator, o_r, which controls transcription to the right. This means that transcription to the left cannot be prevented in λ*v2* phages and transcription to the right is not inhibited in λ*v1v3* mutants. This enables the phage to proceed with lytic development even in the presence of repressor protein.

Isolation and Function of Repressor Protein

Some mutants in the *cI* gene respond to external nonsense suppressors, others are temperature sensitive; and chloramphenicol prevents synthesis of repressor. These results suggest that the repressor is a protein. Ptashne (1967a, 1967b) has isolated the λ repressor by using conditions in which the relative rate of synthesis from the *cI* gene should be greatly increased. Certain mutants in the *cI* gene, known as ind^-, cannot be induced by ultraviolet irradiation; other mutants, ind^s, are unusually sensitive to ultraviolet. These phenotypes probably result from an altered ability of repressor protein to interact with some inducer produced by the ultraviolet treatment.

Cells which are lysogenic for phage λ carrying the *cI* ind^- mutation are immune to lytic infection by further phage particles. Ptashne irradiated such cells with ultraviolet light to block bacterial protein synthesis and super-infected the cells with many wild type phage particles. But the ultraviolet irradiation leaves the synthesis of *cI* protein relatively unscathed; and the presence of the *cI* protein prevents the super-infecting phages from synthesising any protein other than the phage repressor itself. The repressor can then be isolated by comparing the H^3/C^{14} ratios in the proteins of cells infected either with λcI^+ or with mutants in the *cI* gene which should make no protein. The repressor protein isolated by this method is missing from all cells infected with *cI* nonsense mutants and is produced in a modified form by temperature sensitive and missense mutations. Mutations which map at the extreme ends of the *cI* gene modify the same polypeptide chain, which suggests that *cI* codes for a single protein. The phage 434 repressor was subsequently isolated by Pirrotta and Ptashne (1969) by a similar experimental protocol.

The polypeptide specified by the *cI* gene of λ consists of a chain of about 27,000 daltons; the phage 434 repressor monomer is slightly smaller at about 26,000 daltons. The repressors bind specifically to their appropriate phage DNAs and Ptashne and Hopkins (1968) showed that the DNA of λ*vir* binds λ

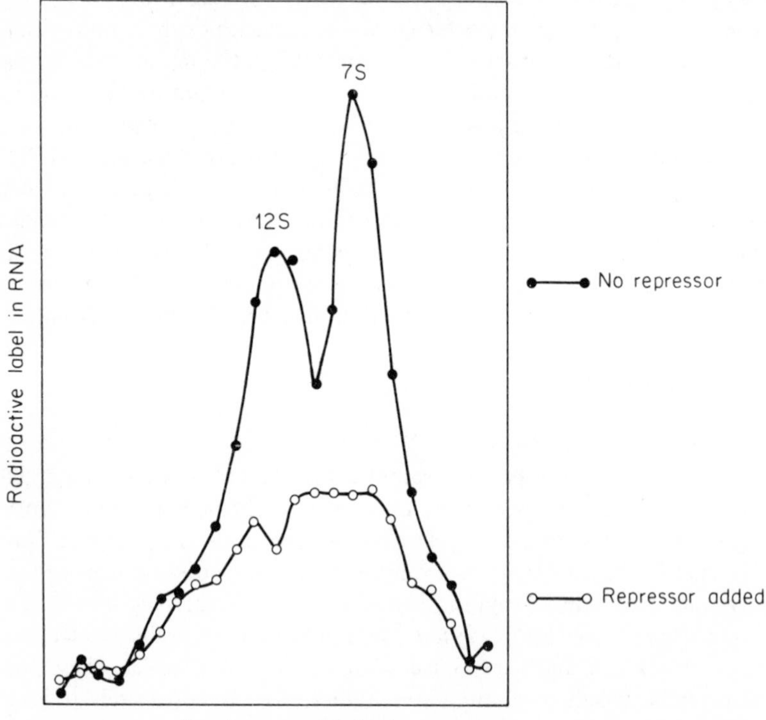

Figure 8.20: repression of transcription of phage λ DNA by purified λ repressor protein in vitro. In the absence of repressor, RNA polymerase synthesises 12S RNA, corresponding to transcription of the left strand of gene N, and 7S RNA corresponding to the right strand of gene tof. Addition of repressor protein inhibits transcription of both RNAs. Data of Steinberg and Ptashne (1971)

repressor much less efficiently than does wild type λ DNA. The mutant DNAs of $\lambda v2$ and $\lambda v3$ each have a lowered affinity for repressor and the $\lambda v1$ $v3$ double mutant responds to repressor only at exceptionally high concentrations of the protein, which supports the idea that repressor protein binds independently at the two sites o_1 and o_r.

By using the membrane filter technique to measure the binding of λ DNA to repressor protein, Pirrotta, Chadwick and Ptashne (1970) were able to show that the active form of the repressor is probably a dimer. The binding of λ repressor to DNA follows a sigmoid curve as the concentration of repressor is increased; and the binding of labelled repressor to DNA is greatly increased

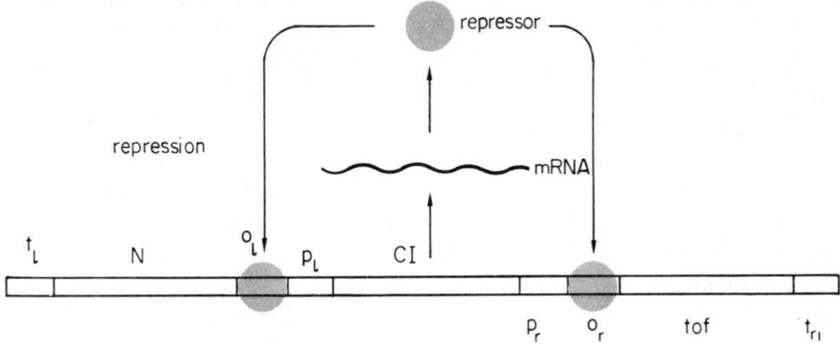

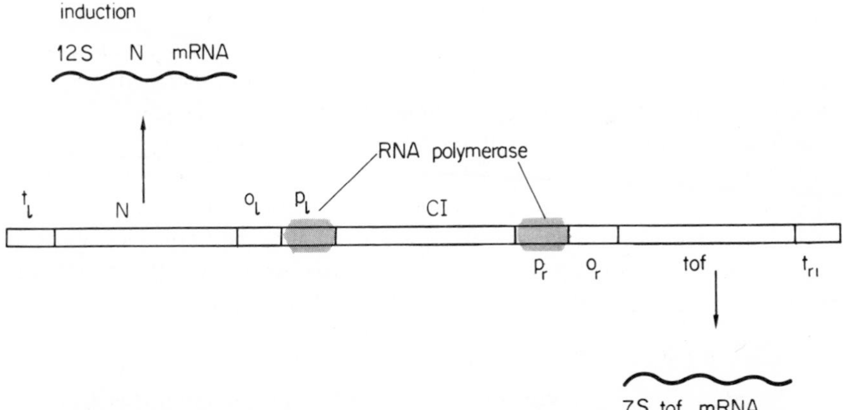

Figure 8.21: control of λ development. The genome is repressed in the prophage state when repressor protein coded by the cI gene binds to the o_1 and o_r operator sites (upper). Transcription of cI itself is not inhibited. If repressor fails to bind, RNA polymerase is allowed to transcribe the N gene into 12S RNA (starting at p_1) and the tof gene into 7S RNA (starting at p_r). Transcription is terminated at the t_1 and t_{r1} sites (lower). The order and interaction of operator and promotor sites is not known but is shown here as p-o-structural gene for convenience

by the addition of unlabelled repressor. This implies that tight binding of repressor to operator requires the interaction of two or more subunits with the DNA. The affinity of the λ repressor for its operators is about the same as that shown between the lactose repressor and its operator; and as with other operator-repressor interactions, the rate of association of repressor and operator is too fast to allow the assumption that the repressor finds its binding sequence on DNA by free diffusion.

When λ DNA is transcribed in vitro in the presence of rho factor, the two species of RNA, sedimenting at 12S and 7S, which are made (see page 253) seem to correspond, as judged by hybridization assays, to the *N* and *tof* genes respectively. The 12S RNA hybridizes to the *l* strand of phage λ or λimm^{434}, but not to that of λimm^{21}; the difference between the first two phages and the last lies in the substitution of the *N* gene of phage 21 for that of λ. This shows that 12S RNA is coded by the N gene of phage λ. Synthesis of the 12S RNA is greatly reduced by a mutation called *sex*, which probably identifies the p_l promotor, adjacent to the o_l operator, where RNA polymerase binds. The 7S RNA hybridizes with the *r* strand of λ DNA but not with that of λimm^{434} DNA, which suggests that it is transcribed rightward from within the immunity region. This means that it must correspond to gene *tof*.

As figure 8.20 shows, Steinberg and Ptashne (1971) found that addition of purified repressor to this in vitro system blocks the synthesis of both 12S and 7S RNAs. The λ repressor is specific, for it has no effect on transcription of the corresponding messengers directed by phage λimm^{434} DNA; the operator sites on which λ repressor acts must therefore be different in the two phages, so that o_l and o_r must be located within the immunity region. Mutant $\lambda v2$ continues to make 12S RNA in the presence of repressor although synthesis of 7S RNA is repressed as usual; $\lambda v1\ v3$ shows the reverse property, with 7S RNA synthesis constitutive and 12S RNA repressed as in the wild type. Repressor therefore appears to act independently at o_l and o_r in vitro as well as in vivo. This suggests the model for control of λ development shown in figure 8.21. Because higher concentrations of repressor seem to be needed to block 7S RNA synthesis than 12S RNA synthesis, it is possible that the o_r operator may have a lower affinity for repressor than does the o_l operator.

Structure of the Operator/Promotor Sites

The interaction of λ repressor with its operators is not defined well enough to say whether the binding of repressor to o_l and o_r prevents movement of RNA polymerase molecules bound at p_l and p_r, or whether the overlap of operator and promotor sites means that repressor and polymerase compete for DNA. That there may be interpenetration of the operator with its associated promotor is suggested both by biochemical studies of protein binding to the o_l/p_l control site and by the genetic map locations of mutants in the o_r/p_r control site. The structure of both the o_l/p_l and o_r/p_r operator/promotor sites may be different from those of bacterial operons, for in both the repressor-binding sequence appears to extend over an appreciable distance, containing the promotor sequence within it.

That there may be competition between repressor and polymerase for λ DNA is suggested by the observation of Steinberg and Ptashne (1971) that repressor is without effect on transcription when RNA polymerase has previously been bound to λ DNA. (In the lactose operon, repressor is effective whether added before or after polymerase; see page 298). It is not possible to test whether RNA polymerase can bind to λ DNA already complexed with repressor because the conditions of binding of RNA polymerase are not stringent enough; in complexes of RNA polymerase and λ DNA, the enzyme seems to bind at many sites other than the two promotors. However, Chadwick et al. (1970) were able to test whether binding of polymerase to λ DNA inhibits binding of repressor by comparing binding to wild type λ DNA with binding to the λ*sex* mutant in p_1.

In the absence of RNA polymerase, λ*sex* and wild type λ bind equal amounts of repressor, as judged by sedimenting the complex made between DNA and radioactive repressor through density gradients. But when RNA polymerase is added before addition of repressor, the λ*sex* DNA binds more repressor than the λ DNA. This suggests that the *sex* mutation lowers the affinity of p_1 for RNA polymerase and that as a result the binding of repressor to o_1 becomes more efficient. In other words, there may be competition between binding of polymerase and repressor such that reduction in polymerase binding by the *sex* mutation allows increased binding of repressor to the adjacent o_1 site.

Mapping experiments have suggested that the o_r operator and the p_r promotor may overlap each other. Ordal (1971) has isolated *super-virulent* lambda mutants (λ*vs*) by their ability to grow on bacteria carrying the plasmid λ*dv*. The DNA comprising λ*dv* constitutes some 15% of the λ genome, including the immunity region, and it can behave as an episome in infected cells, which usually carry about 10 copies of it. Such cells are immune to superinfection although the mechanism of this immunity is not certain; it may depend on other factors besides synthesis of repressor protein. But supervirulent mutants can grow on such cells and the *vs* mutation maps to the right of *v3* but to the left of mutants in *tof*. The λ*vs* phage DNA can transcribe gene *O* even in the presence of repressor, which suggests that *vs* might be a mutation in the o_r operator. The alternative that the *vs* mutation creates a new promotor seems less likely because *vs* lowers the affinity of λ DNA for repressor in vitro. However, mutants of the class *x*, which block transcription in vivo from the *r* strand and seem to locate the p_r promotor, lie between *v3* and *vs*. This suggests that there may be some overlap between the two regulator elements.

By using the DNA of λ*b2* as a template for RNA polymerase in vitro—the *b2* region seems irrelevant for reproduction of λ and λ*b2* has a deletion which removes this region and therefore eliminates its transcription—

Blattner and Dahlberg (1972) found that RNA chains start and grow to about 175 nucleotides in 0.4 minutes at 25°C. Four different RNA species are produced by this reaction. By testing the ability of these molecules to hybridize with the DNA of various deletion strains of λ, each RNA has been located on the λ genome. More than 85% of the RNA originates from the promotors p_1 and p_r; a small amount of RNA synthesis takes place from two other promotors which seem to be located to the right of the immunity region. One minor species is transcribed to the right from a starting point which may be the late promotor, p_r'. The other is transcribed to the left, that is toward the immunity region, from a starting point which seems to be in the region where replication of λ DNA is initiated. What the function of this last mRNA might be is not known. There does not seem to be any transcription from the *cI* gene which codes for the repressor protein.

The 5′ terminal sequence of each of these RNA species was determined by labelling with γP^{32}-ATP or γP^{32}-GTP. The minor RNA which is transcribed to the left from the right arm of the λ chromosome starts with GTP but the other three RNAs all start with ATP. Only the messenger starting from p_r has an initial sequence of AUG which might start a protein chain.

Does RNA synthesis start at the promotor site itself or elswhere on the chromosome? The startpoint, s_1, for the major leftward synthesis which responds to the promotor p_1—this is the 12S N messenger RNA—has been mapped by using deletion strains which end in the region of p_1. As figure 8.22 shows, the two strains of λ*bioN2* and λimm^{434} share a short overlapping region of λ homology located to the left of p_1. The length of the homologous region can be estimated in base pairs by measuring the formation of the heteroduplexes between the phages. In this technique, the two DNAs are annealed and the regions of homology show up as duplexes, whereas regions which are not complementary must remain single stranded. When two DNAs are largely complementary apart from some regions of substitution or deletion, formamide spreading causes the non-homologous regions to protrude as loops from the axis of the duplex (see page 232). Measuring the length of duplex formed between the two derivatives of λ DNA shows that the homology extends for about 215 base pairs (for review see Fiandt et al. 1971; Simon et al. 1971). The p_1 RNA molecule hybridizes just as efficiently with the *l* strands of each of these phages as with λ itself, which shows that its sequence of 175 nucleotides must lie within the region of overlap.

Using the λ*bio30* deletion strain, which has a shorter overlap with λimm^{434} of only about 120 nucleotides, reduces the degree of hybridization of the p_1 promoted RNA to only 60% of its previous value. In this strain, the biotin deletion extends further towards the right and therefore cuts into the 3′ end of the RNA molecule. After the p_1 RNA has been hybridized with the DNA of this deletion, treatment with ribonuclease degrades any unpaired regions of RNA. The hybridized part of the RNA remains resistant to this treatment and can later be eluted from its DNA partner.

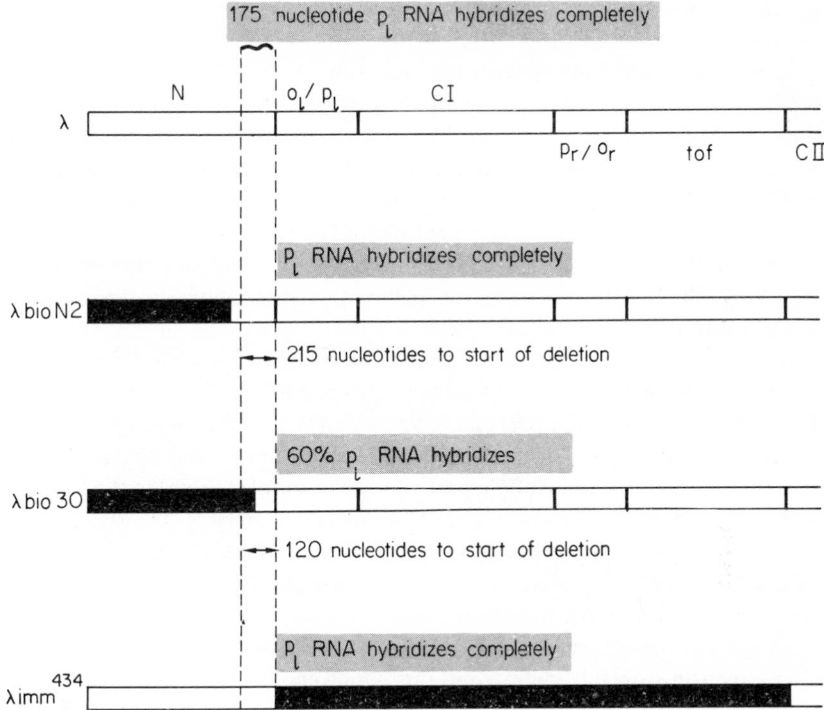

Figure 8.22: mapping the p_1 RNA of phage λ. Black indicates regions which are deleted or replaced in mutant strains. When the p_1 RNA is hybridized with three strains of λ bearing deletions extending into the region from which the mRNA is transcribed, it anneals just as well with λbioN2 and λimm^{434} as with λ itself; since these three phages have in common only a 215 length of nucleotides located to the left of the o_1/p_1 region, the 175 nucleotide long RNA must be transcribed from within this region of overlap. This agrees with the reduction to 60% in its hybridization with λbio30, which has only 120 nucleotides of homology in this region. Data of Blattner and Dahlberg (1972)

Characterizing the fingerprints of this RNA showed that they belong to the 5′ end of the molecule and stretch for about 120 nucleotides to the right of the end of the biotin deletion. As this is about the same size as the overlap of the deletion λ strain with λimm^{434}, the startpoint s_1 must be very close to the region of non-homology of the λimm^{434} DNA. The precision of this technique demands that s_1 must be within 20 nucleotides of the boundary between the λ DNA and the phage 434 DNA in the hybrid λimm^{434}. The p_1 RNA made from λimm^{434} DNA has the same sequence as that made under direction of λ DNA itself; and the 5′ end of the p_1 RNA remains protected from degradation by ribonuclease when it is hybridized with the DNA of λimm^{434}. The startpoint for the RNA must therefore lie to the left of the phage 434 immunity region in λimm^{434}.

But the promotor and operator regions at which RNA polymerase must bind to transcribe p_l RNA can be defined by the *sex* mutation which reduces the extent of transcription to the left both in vitro and in vivo and by the *v2* mutation which has a reduced affinity for the λ repressor. In the in vitro system of Blattner and Dahlberg, the *sex* mutation reduces the synthesis of p_l RNA to 10% of its previous level, although the RNA which is made has the usual sequences characteristic of p_l RNA. This shows that the mutant site itself is not transcribed into RNA and that its effect must be to reduce the frequency of initiation from some adjacent site.

But the *sex* and *v2* mutations cannot recombine with λimm^{434} to give recombinants with the control typical of wild type λ DNA. This indicates that the p_l/o_l region lies within the part of λ DNA which is replaced by phage 434 in the λimm^{434} hybrid; instead of λ control sequences, λimm^{434} contains the different control sites characteristic of phage 434, which cannot respond to the regulator proteins of lambda. The p_l/o_l site must therefore map to the right of the λ-434 boundary, although the s_l initiation site maps to its left. A length of DNA thus separates the site where RNA polymerase must bind to DNA, p_l, from the site where RNA synthesis starts, s_l.

The physical distance between p_l and s_l has been measured by Blattner et al. (1972) by using several deletion mutants which end in this region. The distance from the *sex* mutation identifying p_l to the end of two biotin deletions which end at different points between s_l and p_l was measured by recombination. The distance between the ends of the two biotin deletions can be deduced by measuring the length of heteroduplex formation by electron microscopy and gives a calibration of recombination units in base pairs. The use of heteroduplex mapping and its combination with genetic mapping is illustrated in figure 8.23. These experiments show that the distance from *sex* to the boundary between phage 434 and λ in λimm^{434} is some 195 ± 80 nucleotide base pairs. Since the s_l site must lie within 20 nucleotides to the left of this boundary, the distance between p_l and s_l must be at least 195 nucleotides.

But more recent experiments have produced results which suggest that the RNA polymerase binding site identified by the *sex* mutation is located very near the λ/imm^{434} boundary. The position of *sex* is very close to the site in o_l where repressor protein first binds; and the distance between the polymerase binding site and the start of *N* appears to be rather small. The mismapping of *sex* which suggested a greater separation may have been due to a false definition of one of the deletion strains.

That the o_l operator may itself comprise more than a single binding site is suggested by the lengths of DNA isolated by Maniatis and Ptashne (1973) by their resistance to degradation when bound to repressor. The fragments protected by λ repressor protein are present in λDNA but not in λimm^{434} DNA; and no fragments are recovered if the DNA is denatured before repressor

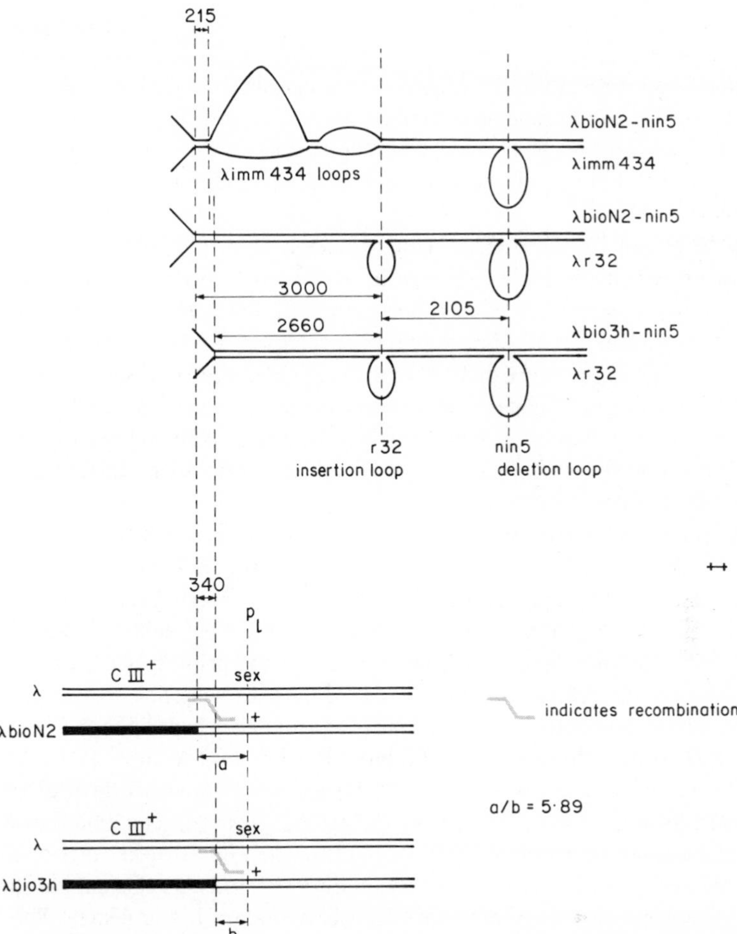

Figure 8.23: mapping of s_1 of λ by combination of heteroduplex technique and genetic recombination. Upper: denatured single strands of different λ deletion strains are hybridized with each other. Homologous regions form a duplex axis, but insertions or deletions cause single stranded loops to protrude from the axis. The first hybridization shows that the overlap between λbioN2 and λimm^{434} is 215 nucleotides; the p_1 RNA starts within 20 nucleotides of the 434/λ boundary marking the right end of the region of homology. The ends of the λbioN2 and λbio3h deletions are shown in the nin5 deletion, which are 2105 bases apart and which can be used as a calibration to adjust for any variations in individual experiments. The distance between the ends of the λbioN2 and λbio3h deletions is 3000–2660 = 340. Lower: recombination between $cIII^+$ sex and bio p_1^+ strains to give $cIII^+$ p_1^+ can only take place in the interval between the end of the deletion and the sex mutation. The ratio of the proportion of wild type recombinants achieved when λbioN2 or λbio3h is used is therefore a/b, which equals 5.89. But the heteroduplex mapping shows that $a - b = 340$. We can therefore deduce that $a = 410$. The end of λbioN2 is 215 nucleotides from s_1, so that the distance from s_1 to p_1 must be 410–215 = 195 (±80) nucleotides. Data of Blattner et al. (1972)

binding. Similar results are obtained with substrates of *λv1v3* (repressor binds only to o_l) and *λv2* (repressor binds only to o_r), although more repressor protein is needed to protect the o_r fragments; the two operators may therefore possess similar but not identical structures in which o_r has a lower affinity for repressor.

The length of the DNA fragment protected from degradation by the repressor protein depends upon the ratio of repressor to DNA. With DNA of *λv1v3*, at low concentrations of repressor (about 0·05 repressor dimers per o_l operator), a sequence of 35 base pairs is protected. As the concentrations of repressor is increased towards 10–30 dimers per o_l operator, discrete fragments of 45, 65, 75, 85 and 100 base pairs become protected. Pyrimidine tract analysis suggests that the sequence of the smallest fragment is included in the larger fragments; and shows also that these o_l fragments differ from the fragments of o_r protected when *λv2* is used as substrate.

The model suggested by Maniatis and Ptashne to account for these results supposes that at each operator a dimer of repressor first binds to a primary sequence some 30 base pairs long; as more repressor protein becomes available, monomers bind one at a time to further, secondary segments of DNA each about 15 nucleotides long which lie adjacent to the primary sequence. The λ repressor may therefore recognise several different nucleotide sequences.

The operator/promotor sites can also be isolated when λ DNA is cleaved with restriction endonucleases. Maniatis, Ptashne and Maurer (*Cold Spring Harbor Symp. Quant. Biol.*, **38**, pp. 857–868, 1973) showed that four of the fragments produced when λ DNA is split by the Hin enzyme can bind repressor. One of these contains the o_l primary binding site and also the start of the *N* gene; another contains the o_r primary binding site and also the start of the *tof* gene. The other two fragments contain the secondary binding sites. The Hin enzyme therefore cleaves both operators between their primary and secondary binding sites; and the primary binding site of each operator must be adjacent to the structural gene which it controls. Examining the repressor-binding abilities of fragments derived from mutant λ strains showed that the *v2* mutation maps in the primary binding site of o_l; and one of the double mutations *vlv3* maps in the primary site of o_r, the other residing in one of the adjacent secondary sites.

Two of the fragments produced by cleaving λ DNA with another enzyme, the Hpa restriction endonuclease, can bind repressor protein; one of 350 base pairs contains o_l and the start of the *N* gene, and another of 2000 base pairs contains o_r and the start of the *tof* gene. The 350 base pair fragment is split into pieces of 150 and 200 base pairs by the Hin enzyme. The 150 base pair fragment corresponds to a length of 120 (± 20) base pairs representing the start of the *N* gene; the remaining 30 base pairs comprise the primary repressor binding site, which must therefore lie within 20 base pairs (the precision of the technique) of the start of the *N* gene.

The 350 base pair Hpa fragment can bind RNA polymerase as well as repressor, but polymerase binding is much reduced when the fragment is derived from λ*sex* mutant DNA. The Hpa 350 fragment derived from λ*sex* DNA is not cleaved by Hin endonuclease; and binding of polymerase to Hpa 350 fragments of wild type λ DNA protects them against Hin attack. The *sex* mutation must therefore lie within the site recognised by the Hin enzyme and this same sequence must be part of the site recognised by RNA polymerase. *Sex* therefore identifies a polymerase binding site located within o_1, very close to the start of *N* where transcription is initiated.

Since the operator and promotor sequences at the o_1/p_1 site clearly interpenetrate, repressor binding and polymerase binding may be competitive. The organization of the o_r/p_r site seems to be very similar, although comprising nucleotide sequences which are not identical with those of the o_1/p_1 site. We do not yet know how these sites function in controlling the repression of λ.

Control of Synthesis of cI Repressor

Expression of the *cI* gene is not constitutive, as is transcription of the *i* regulator gene of the lactose operon (see page 299), but is itself under control of other λ genes. The *cII* and *cIII* genes have been implicated in switching on repressor synthesis after infection of a bacterial cell, although we do not know how their protein products act. In cells which have already established lysogeny, however, the transcription of *cI* seems to depend only on the presence of active repressor itself, for when the repressor protein is inactivated, production of further molecules of repressor may cease. Different control sites on λ DNA seem to be used by the two systems, the first for establishment and the second for maintenance of lysogeny.

A temperature sensitive mutant in the *cI* gene (cI_{857}) produces a repressor protein which is active at 30°C but inactive at 42°C. (The host cell is not lysed at the higher temperature if the phage also carries other mutations which block lytic development.) When cells lysogenic for λcI_{857} are grown at the higher temperature they gradually lose the ability to regain immunity for subsequent growth at 30°C. One way to explain this result is to argue that active repressor is itself needed for expression of the *cI* gene; when the temperature is raised in cI_{857} lysogens, the amount of active repressor decreases and this hampers the restoration of immunity when the temperature is later lowered because by then the cells lack the active repressor protein needed to allow transcription of *cI*.

Measurements of repressor activity in λcI_{857} lysogens show a decline with growth at 42°C. But such experiments do not show whether activity alone is reduced or whether synthesis of further molecules of repressor is inhibited. The amount of repressor protein present in lysogenic cells grown under various conditions has been measured directly through an antigenic assay developed by Reichardt and Kaiser (1971). Although λcI^+ strains have the same amount

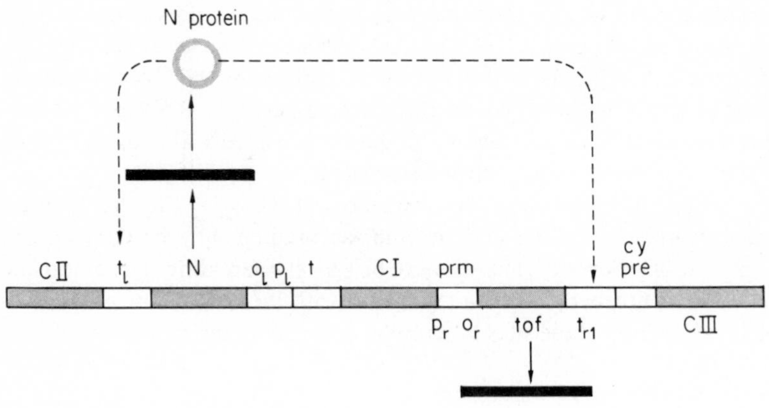

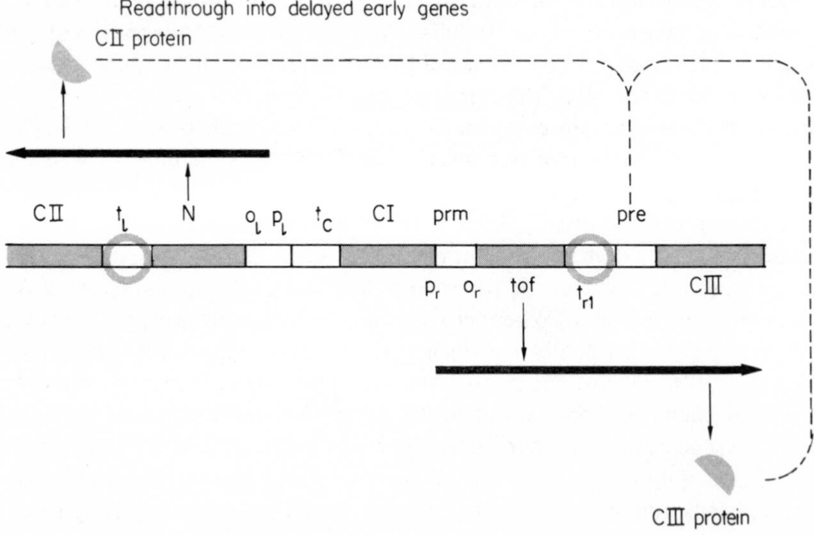

Figure 8.24: establishment of lysogeny by phage λ. Infection of host cell; the host polymerase recognises p_1 and p_r and transcribes the N and tof genes, terminating transcription at t_1 and t_{r1}. Gene cI cannot be transcribed from prm because the cell contains no active repressor. Readthrough into the delayed early genes: the N protein synthesized from the mRNA made upon

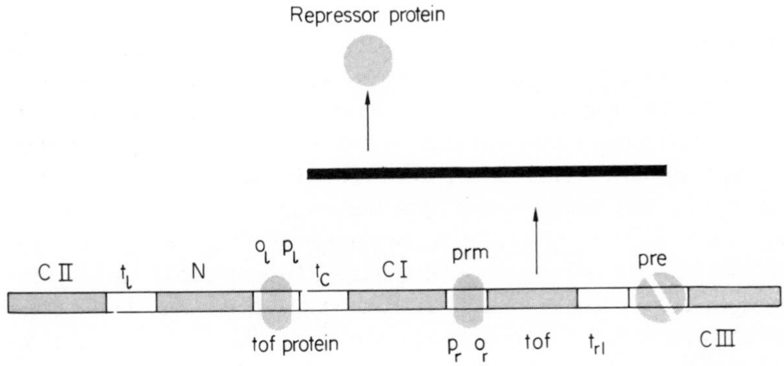

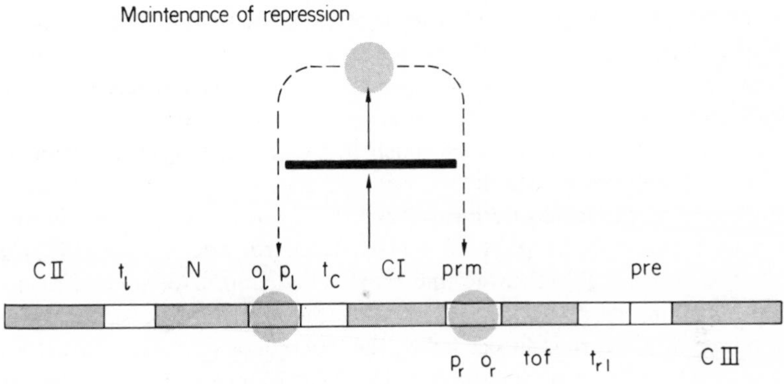

infection allows polymerase to continue transcription past the t_1 and t_{r1} sites into the delayed early genes. These include cII and cIII (needed to establish repression) and other functions such as int, O, P (needed for integration of phage DNA into the host chromosome). Establishment of repression: the cII and cIII proteins act at the pre promotor to encourage transcription into gene cI; polymerase must terminate at a site t_c located before the N gene. Tof protein acts at o_1 (and perhaps also at o_r) to halt transcription of the early genes. Maintenance of repression: repressor protein binds to o_1 and to o_r, preventing transcription of λ genes. Binding to o_r also serves the purpose of promoting transcription of cI from prm. Transcription from pre is switched off as cII and cIII proteins are no longer made

of repressor at either 30°C or 42°C, lysogens of λcI_{857} contain little repressor protein after growth at 42°C. This supports the idea that the absence of active repressor protein leads to a decline in synthesis of further molecules of repressor.

Hybridization experiments with the λ RNA present in lysogenic cells have shown that it is the left strand of the *cI* gene which is transcribed (Spiegelman et al. 1970). The promotor site for transcription of the *cI* gene during maintenance of lysogeny (*prm*) must therefore be located to the right of *cI* gene. This sequence may overlap or be identical with the rightward operator o_r, for Reichardt and Kaiser found that the mutants *v1v3* which identify o_r cannot make repressor according to the antigen binding assay. This suggests that o_r may have a dual function, so that it is used not only to control rightward transcription but also to control expression of *cI*. To maintain lysogeny, it must be necessary for repressor to bind to o_r and this binding must in some way stimulate transcription of the *cI* gene (see figure 8.24).

Although the maintenance of lysogeny appears to require only the presence of repressor protein, establishment of lysogeny when λ DNA infects a host cell requires in addition the functions *cII*, *cIII* and *cY*. Mutants in *cII* and *cIII* show the complementation behaviour to be expected of genes which code for proteins and the cII^- and $cIII^-$ alleles are recessive to the wild type; this suggests that the cII^+ and $cIII^+$ genes specify proteins whose activity is needed to switch on transcription of *cI* in the first place. The activity of these proteins is not needed once lysogeny has been established, for the rare lysogens formed by cII^- and $cIII^-$ phages are stable.

This model is supported by measurements of repressor synthesis upon infection of E.coli cells by phage λ DNA. Reichardt and Kaiser (1971) and Echols and Green (1971) showed that there is little repressor synthesis during the first five minutes, after which there is a rapid rate of synthesis for the next 10 minutes; this in turn is succeeded by the low rate of synthesis characteristic of cells in which lysogeny has been established. Figure 8.25 shows that cII^-, $cIII^-$ or cY^- mutants accumulate greatly reduced amounts of repressor upon infection. That these functions are needed only to establish lysogeny is shown by the synthesis of repressor at the same low rate in cells lysogenic for wild type λ, cII^-, $cIII^-$ or cY^- mutants.

Mutants in *cY* do not show the complementation behaviour typical of genes which code for proteins, but are *cis*-dominant, *trans*-recessive as expected for a control element of DNA which is recognised by regulator proteins. The *cY* site maps in the position shown in figure 8.24, to the right of the *tof* gene. The behaviour of cY^- mutants suggests that this site may identify the promotor where transcription of *cI* is initiated when lysogeny is being established. One model for its use is to suppose that the *cII* and *cIII* proteins form an oligomer which acts at this site to allow RNA polymerase to initiate transcription. This means that the RNA made from this promotor (the *pre* site) should contain a transcript of the *l* strand of *tof* at its start and a transcript of the *l* strand of *cI* at its end. Since *tof* is usually transcribed from the p_r/o_r—that is from the *r* strand of λ—this means that the first part of the messenger would be a transcript of its anti-coding strand and would, presumably, be meaningless and therefore not translated.

The prediction that *tof* anti-coding strand RNA should be transcribed during establishment of lysogeny has been tested by Spiegelman et al. (1972), who have made use of a prophage deletion which retains a sequence of λ DNA extending only over the *cIII—tof* region. This prophage transcribes RNA from the *r* strand of *tof*. When labelled RNA from this source is hybridized with unlabelled RNA from cells newly infected with λ DNA, some of the *r* strand *tof* RNA forms a double stranded RNA molecule—it must have hybridized with a transcript of the *l* strand of *tof*. Using this assay shows that the synthesis of *l* strand *tof* RNA follows the same kinetics as that of repressor protein; it first increases and then declines. We may therefore assume that the *l* strand *tof* RNA is at the starting end of the same RNA molecules which carry transcripts of *cI*. That the *cII* and *cIII* proteins act at *c Y* to stimulate transcription of this RNA is supported by the observation that mutants at these loci produce much decreased amounts of *l* strand *tof*. That transcription of *cI* starts at a site between *tof* and the *cI* gene itself during maintenance of lysogeny is shown by the failure of lysogenic bacteria to synthesise *l* strand *tof* RNA.

When λ DNA first infects the bacterial cell, no repressor is present, of course, so that host RNA polymerase can transcribe genes *N* and *tof* from the promotors p_l and p_r. The enzyme is unable to transcribe *cI* from promotor *prm* because of the lack of repressor and cannot act at *pre* because the *cII* and *cIII* proteins have not yet been made. Once the *N* mRNA has been translated into protein, however, transcription can continue into the *cII* and *cIII* genes. Their protein products can then act at the *pre* promotor identified by the *c Y* mutation. The lag after infection before repressor is synthesised must be the time required for this series of events, shown in figure 8.24, to take place.

Repressor synthesis proceeds at a rapid rate when the *pre* promotor is active. The active repressor then binds at o_l and o_r to prevent transcription of the genes of λ and thus to maintain the phage in an inert state. Its binding at or near to o_r also has the effect of stimulating further transcription of *cI*, but in this instance from the *prm* promotor immediately adjacent to the *cI*, gene itself. Transcription from the *pre* promotor is switched off at this time; because the *pre* promotor is some 7–8 times more efficient than the *prm* promotor, this means that the rate of repressor synthesis drops. The maintenance of lysogeny therefore relies upon a slower rate of repressor synthesis than does its establishment.

Transcription from the *pre* promotor must be switched off when lysogeny has been established. Since the products of the *cII* and *cIII* genes have not been isolated, we do not know how stable these proteins are. If they are unstable however, some inactivation of *pre*-transcription might be caused by the action of repressor in switching off all the lambda genes except its own. But it must take time for active repressor to be made and the inactivation of *pre* probably takes place before this can happen. Repressor synthesis at the high rate characteristic of *pre* promoted transcription appears to depend upon the absence of the *tof* protein; for as figure 8.25 shows, synthesis continues at this high rate in *tof*$^-$ mutants where it is not reduced to the low level typical of lysogeny. On infection with *tof*$^+$ phage, however, repressor synthesis is reduced at about 20 minutes after infection.

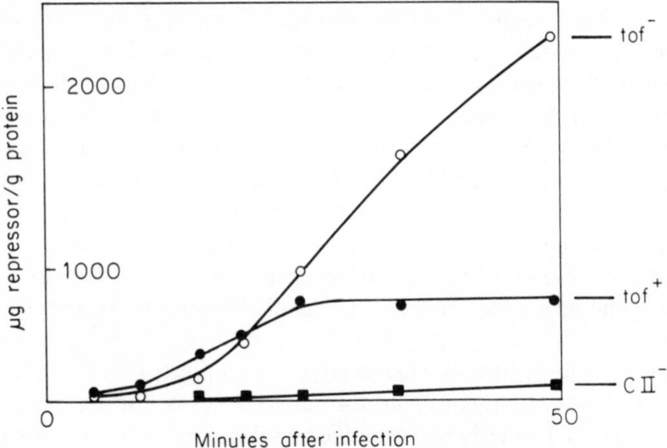

Figure 8.25: synthesis of repressor protein after infection. With both tof$^+$ and tof$^-$ phages repressor synthesis commences after a short lag period at a rapid rate. Synthesis later ceases in tof$^+$ phages, but can continue in tof$^-$ mutants. Mutants in cII synthesize very little repressor protein. Data of Reichardt and Kaiser (1972)

Since *tof* is one of the first proteins made upon infection, its late effect argues that its action is indirect. One site of *tof* action is the o_l operator (see below); another may be the o_r operator although it does not appear to switch off the system for maintaining repression via *prm*. This suggests that *tof* may prevent transcription from *pre* by binding to o_l (and perhaps also o_r) to prevent transcription of *cII* (and perhaps also *cIII*). If these proteins are unstable and then decay, *pre* will become inactive. The maintenance system must be able to function when *pre* is switched off, in spite of the presence of *tof* at *prm*, because by this time sufficient repressor has been synthesised to ensure that *prm* is activated.

The effect of *tof* on repressor synthesis was discovered because although many strains of the temperature sensitive mutant repressor λcI_{857} lose their ability to regain immunity at 30°C when they are grown for extended periods at 42°C, certain mutants at other loci can recover immunity at the lower temperature. These fall into two classes. Mutants of the *x* class in the p_r promotor are able to regain immunity, presumably because some gene under the control of this promotor must usually function to prevent the cells from regaining immunity. The second class of mutants identify this gene, which has been called *cro*. Mutants isolated in other experiments and called *tof* or *fed* identify the same function; I shall use the name *tof* (turn off function) here.

The cI_{857} tof^- double mutants regain immunity when the temperature is lowered, then, whereas cells lysogenic for cI_{857} tof^+ do not. The wild type tof^+ function is dominant over the mutant tof^- which suggests that the product of the tof^+ gene is a protein which prevents the expression of *cI* and therefore synthesis of repressor. In cells in which the temperature sensitive mutant repressor has been inactivated by incubation at 42°C, transcription of *tof* takes place from the o_r/p_r site at which repressor now fails to bind, and the protein which is translated from this messenger prevents the *cI* gene from functioning. When the *tof* protein is absent—as in p_r or tof^- mutants—the *cI* gene can function, so that immunity can be regained when the temperature is lowered to 30°C to activate the repressor protein.

This model implies that immunity is regained in this situation by use of the establishment system. Immunity is lost in the first place because the increase of temperature inactivates the repressor protein and therefore prevents transcription of *cI* from *prm*. That the *tof* function is not implicated in this loss is shown by the fall in repressor synthesis in cI_{857} lysogens whether or not a tof^+ or tof^- allele is present. That the ability to regain immunity when the temperature is lowered depends upon transcription from *pre* is suggested by the measurements of *cI* transcription made on cI_{857} tof^- lysogens by Szybalski et al. (1970). Upon incubation at high temperature, there is a rapid cessation of transcription of *cI* (presumably due to inactivation of transcription from *prm*); but transcription is resumed after some 10–20 minutes and appears to start at a point beyond the right end of the *cI* gene. That the *tof* protein acts as a repressor of the *cI* gene is also shown by the results of Kumar, Calef and Szybalski (1970) and Calef et al. (1971), who have measured the transcription of *cI* from wild type tof^+ and mutant tof^- prophages upon induction. The mutants can transcribe the *cI* gene under conditions in which the wild type cells cannot.

The mutual repression of the *tof* and *cI* genes by the protein product of the other gene ensures the proper temporal functioning of lambda in both lysogeny and lytic development. In lysogeny, the repressor maintains the phage in an inactive state by binding to o_l and o_r to repress transcription of *N* (needed to induce the delayed early messengers) and *tof*. The binding of repressor at o_r maintains the continued transcription of *cI*. When the repressor already existing in the cell is inactivated upon induction, or when phage DNA infects a new bacterial cell, the two immediate products of transcription are N and tof proteins.

The *N* protein acts to promote transcription of the delayed early functions. The *tof* protein acts to turn off transcription of the early genes. During establishment of lysogeny, this turn off allows the switch to be made from *pre* to *prm* promoted transcription of *cI*. If the cell is entering the lytic cycle, the turn off of the early genes means that *cII* and *cIII* cannot function to produce repressor protein and nor can the other early functions continue to be transcribed when they are no longer needed. The control of phage functions is finely balanced, so that in some situations at least the selection of which genes are to be transcribed may depend on quantitative and temporal effects—such as the relative timing and extents of synthesis of *tof* and repressor proteins.

The *tof* gene is transcribed in different directions at different times; upon infection or induction it is transcribed to the right from the p_r promotor but as soon as *cII* and *cIII* proteins have been made it is transcribed to the left from the *pre* promotor. We do not know how the topological problems raised by the transcription of *tof* in both directions are solved by the phage. It is possible that there is in fact no conflict and that DNA may be transcribed simultaneously in opposite directions. Alternatively, the two transcriptional events may be mutually exclusive on a single phage DNA (demonstrations that both directions are transcribed rely upon a population of infected cells); and this unusual situation may even in itself constitute some form of control over λ development.

Sequence of Lytic Development

Measurement of the RNA species made in lysogenic and lytic cells shows that the transcription of lambda is an ordered process which depends upon the sequential actions of several regulator functions. One reason for the apparent complexity of these interactions may be that it is necessary to ensure a precise temporal sequence of events both in lysogeny and in lytic development. By employing a hybridization assay with the DNAs of the three phages, λ, λimm^{434} and λimm^{21}, Kourilsky et al. (1970, 1971) have followed the progress of transcription when prophage is induced into lytic development.

Hybridization with λimm^{21} measures all λ RNA except that corresponding to the *N*, *cI* and *tof* genes; annealing with λimm^{434} DNA measures all lambda RNA except that corresponding to *cI* and *tof*. As figure 8.19 shows, the difference in hybridization between λ and the other two phages therefore measures the transcription of *cI*, which is present only in λ DNA. Transcription of the *N* gene is revealed by the difference in hybridization with the DNAs of λ or λimm^{434} (which contain *N*) and the DNA of λimm^{21} (which lacks *N*). And the extent of hybridization with λimm^{21} measures the transcription of the delayed early regions outside the immunity and N regions.

By extracting the RNA synthesized after heat induction of prophages, Kourilsky et al. found that the RNA made during the first 10 seconds hybridizes only with λ DNA; this shows that it corresponds only to the *cI* gene previously transcribed in lysogenic cells—so this time represents an induction lag period. During 10–25 seconds after induction, the RNA hybridizes with λ and also with λimm^{434} but not with λimm^{21}; this shows that genes *N* on the *l* strand and *tof* on the *r* strand are being transcribed. And from 25–150 seconds, hybridization takes place with all three templates as transcription proceeds into the delayed early genes which all have in common. Using the separated single strands of the phage DNAs shows that the induced transcription starts simultaneously from the promotors on each strand of λ DNA.

The patterns of transcription are different, however, in regulator gene mutants of λ. Comparison of *tof*$^+$ and *tof*$^-$ phages shows that the *tof* product acts to turn off the early genes after they have been expressed. The transcription of *cI* mRNA seems to be reduced at a time before *tof* acts to reduce the levels of the early mRNAs; this implies that the system for maintenance of lysogeny in the prophage is switched off upon induction without assistance from *tof* (presumably because inactivation of repressor prevents transcription from *prm*). Transcription of the *l* strand of λ is usually turned off at about 5 minutes after infection, when the transcription of the late genes is turned on. But in *tof*$^-$ mutants, *l* strand transcription is enhanced at early times and is not turned off at all.

This suggests a sequence of events in which relief of repression—that is inactivation of repressor upon induction—leads to transcription from p_l and p_r. The same situation is found upon infection of a host cell when the lytic cycle is followed. After the *tof* gene product has been transcribed from the messenger which starts at p_r, it acts at o_l to turn off any further transcription of the *N* and subsequent genes. The level of early *r* strand RNA is slightly enhanced in *tof*$^-$ mutants, which suggests that *tof* may also act on the o_r operator. We do not know why *tof*$^-$ phages are deficient in the late transcription of the *r* strand genes. These experiments have also confirmed that transcription of the delayed early genes requires the product of the *N* gene, for mutation in the λ *N* gene allows the RNA which is transcribed to hybridize with λ but not at all well with λimm^{21}; this implies that the *N* gene itself is transcribed but that transcription proceeds no further.

Another way to follow the action of the *tof* protein in turning off the transcription of the early *l* strand genes is to assay the level of exonuclease produced after infection. The exonuclease gene is located shortly after *N* on the *l* strand and synthesis of exonuclease usually takes place early in infection, after which it is turned off. But phages mutated in either the p_r promotor or which are *tof*$^-$ do not turn off synthesis of exonuclease. A similar situation is found with λimm^{434}; just as the λ *tof* product can act only on the λ immunity region, so the λimm^{434} *tof* product can act only on the phage 434 immunity region. This again agrees with the idea that *tof* codes for a repressor protein which acts at a site within the immunity region to turn off transcription of the *l* strand.

By assaying the effect of a series of deletions on the turnoff of exonuclease synthesis, Pero (1971) mapped the site of action of *tof* between *cI* and *N*. This, together with the finding of Sly et al. (1971) that the *v2* mutation which identifies the o_l operator, decreases the effect of *tof* on expression of the *l* strand genes, supports the idea that *tof* protein acts on the p_l/o_l control site. This site is therefore recognised by both repressor protein coded by the *cI* gene and by the *tof* protein; the o_r/p_r site may similarly respond to *tof* as well as to repressor. The action of the two proteins on *prm*, however, must be different, for the repressor encourages transcription to the left whereas *tof* does not. That the *v2* mutation affects both repressor and *tof* protein action suggests that the same operator, or different operators which overlap at the *v2* mutant site, are recognised by the two proteins; recognition of different regulator proteins by one region of DNA may provide a role for the rather long sequences of the two lambda operators.

When lambda is induced, therefore, the repressor ceases to bind to o_l and o_r. As a result, host RNA polymerase can commence transcription from the p_l and p_r promotors (the sequence of events has been reviewed by Szybalski et al., 1970). In *N* gene mutants, this transcription continues only to the ends of the *N* and *tof* genes, where the termination sites t_l and t_{rl} are located. In wild type phages, however, the *N* gene product is translated from the leftward mRNA and acts to allow transcription to continue beyond these points. The action of N protein is specific for the λ phage DNA (λimm^{21} cannot substitute for λ) so its action cannot be to generate a simple inactivation of rho but must be concerned to polymerase to continue beyond the λ termination sites (see page 269).

By transcribing λ DNA in an in vitro system containing rho in which RNA synthesis is initiated at p_l and can therefore continue beyond the t_l site only if N protein activity is present, Greenblatt (1973) and Dottin and Pearson (1973) developed an assay for *N* protein. Synthesis of proteins coded by genes to the left of *N* depends upon the addition of *N* protein activity; extracts containing *N* protein can therefore be purified for the anti-termination activity by following their ability to support transcription and translation of these genes when added to systems lacking *N* protein. Assays of the production of *N* protein during infection confirm that it is synthesized during the very early stages of infective development; but its synthesis is later switched off in *tof*$^+$ but not in *tof*$^-$ phages. Because *N* protein is unstable and decays with a half life of about 2 minutes, the action of *tof* in abolishing transcription of *N* must very rapidly cause cessation of expression of the delayed early genes.

Production of N protein allows transcription to the left to continue into the galactose genes of the host chromosome when the prophage remains in its integrated state. But the phage is usually excised by this time to give a circular form in which transcription continues into the *b2* region. No functions have yet been located in the *b2* region, which takes its name from the *b2* deletion which covers this area; *b2* phages appear to possess all the functions necessary for growth and lysogeny. Transcription to the left presumably terminates somewhere in the *b2* region, but we do not know where.

Transcription to the right passes the t_{r1} termination site in the presence of *N* protein. Analysis of the mutation which creates a new promotor at a site beyond t_{r1} has shown that there is a second site, t_{r2}, at which termination can take place unless N protein acts (see figure 6.13). Transcription then proceeds through the head (*A–F*) and tail (*Z–J*) functions until it enters the *b2* region (in a circular phage) in the opposite direction from leftward transcription. No termination signal for this rightward transcription has yet been detected.

The transcription of the *Q* protein allows the late products to be transcribed from their genes, starting at the p_r' promotor at an efficiency some 10–20 fold greater than that of p_r. The *Q* protein, which may act as a sigma factor to redirect the activity of RNA polymerase, enables the enzyme to start transcription at p_r'. We do not know where the transcription which starts at this point is terminated. By the time these late genes are under transcription, the *tof* product has acted to switch off the leftward early genes and perhaps also the rightward early genes.

The temporal expression of lambda genes in lytic development thus lies under the control of a series of regulator genes whose interactions with each other ensure the proper sequential expression of structural genes. By stopping transcription of the *N* and *tof* products, the repressor coded by *cI* keeps the phage genome in an inert state. When repressor is removed from DNA, the first genes to be transcribed are *N* and *tof*. The *N* gene product permits transcription to proceed beyond the *N* and *tof* regulator genes into the delayed early phage genes. The *tof* gene acts to stop synthesis of the early genes once they are no longer needed. One of the delayed early genes, *Q*, in turn makes a protein product which allows expression of the remaining, late genes.

Control of Stable RNA Synthesis

Synthesis of Ribosomal RNA

Synthesis of the messengers of inducible and repressible operons of E.coli is controlled by the response of their repressor proteins to the milieu in which the bacteria grow. Other messengers seem to be synthesized at constitutive rates which are set by the affinities of their promotors for RNA polymerase; these genes probably code for enzymes which are vital to the cell and must always be made. But the synthesis of stable RNA species does not appear to be merely constitutive; rather it is under a specific control. We know more about the synthesis of ribosomal RNA, whose genes seem to be clustered and for which accurate hybridization assays are available than about the synthesis of transfer RNAs, which are specified by cistrons dispersed on the genome.

By following the synthesis of ribosomal RNA during the cell cycle of E.coli,

Dennis (1971) has shown that there are no discontinuous jumps in the production of rRNA, such as might be expected at the time of replication of genes which are expressed continually at a level depending only on the efficiency of their promotors. The cellular content of ribosomal RNA increases steadily during the cell cycle, which implies that some control mechanism must act to establish its level.

The proportion of the bacterial genome which is devoted to coding for ribosomal and transfer RNAs is very small. Hybridization assays show that only about 0·3% of the denatured single strands of E.coli DNA corresponds to ribosomal RNA and only about 0·06% to transfer RNA (see Yanofsky and Spiegelman, 1962; Birnsteil, Sells and Purdom, 1972). Since only the coding strand of the denatured DNA can hybridize, this means that of the $4{\cdot}6 \times 10^6$ base pairs of the E.coli genome, there are about 6 sets of genes for 16S rRNA (about 1500 nucleotides long) and 23S rRNA (about 3000 nucleotides long); and there must be about 60 genes coding for tRNA precursors of about 120 nucleotides in length.

But in spite of their small representation in the genome, stable RNAs form by far the greater part of the total RNA synthesised in the bacterial cell. Norris and Koch (1972) have found that only 3–4·5% of the RNA of E.coli cells comprises messengers; 70–80% is ribosomal and 15–25% is tRNA, the exact proportions depending on the growth rate. Since messenger RNA is unstable and turns over continually, its rate of synthesis is of course much greater than the cellular content would imply. But synthesis of rRNA may occupy 40% of the total production of RNA of E.coli cells in exponential growth.

Some synthesis of ribosomal RNA takes place in vitro from the DNA templates of most bacteria, but the proportion does not reach the in vivo level under conditions in which new chains of ribosomal RNA must be started in vitro. But using a complex of DNA and protein extracted from Bacillus megaterium—or from E.coli—which contains most (95%) of the RNA polymerase activity of the cell, Pettijohn et al. (1970) found much greater amounts of rRNA synthesis. The complex completes chains already under synthesis; it does not start new ones. The discrepancy between the extent of initiation of ribosomal RNA synthesis in vitro (about 10%) and synthesis in vivo (about 40%) or in vitro when chains are completed (also about 40%) supports the idea that the high level of ribosomal RNA transcription in the cell is due to some mechanism which acts at initiation to direct the attention of RBA polymerase to the ribosomal RNA genes.

One way to account for the discrepancy between gene content and extent of transcription is therefore to suppose that some specific protein factor directs RNA polymerase to transcribe the genes which produce ribosomal RNA. Indeed, Travers, Kamen and Schleif (1970) found that almost no ribosomal RNA appears to result from transcription in vitro of E.coli DNA, but that the amount of synthesis appears to be greatly increased by the addition of a protein factor, the ψr factor, isolated from bacterial cells. The ψr activity appeared to reside in two proteins which are found in E.coli cells but which are also used by phage Qβ RNA to make up the RNA replicase of the phage, together with another host protein and a protein coded by the phage. A preparation of these two proteins—which are readily obtained from the Qβ replicase—stimulated synthesis of ribosomal RNA in vitro. Blumenthal et al. (1972) have since shown that these two proteins are Tu and Ts, the factors which are used in elongation of polypeptide chains, although their function as part of the Qβ replicase is mysterious.

But in contrast with these results, Haseltine (1972) found that in appropriate conditions ribosomal RNA comprises a constant 7–14% of the RNA made in vitro from E.coli DNA. By using a competition hybridization assay between H^3 RNA made in vitro from bacterial DNA and purified P^{32} ribosomal RNA, he demonstrated that the amount of total RNA synthesis in vitro is indeed increased by addition of the two components of Qβ replicase with apparent ψr activity; but this stimulation applies equally to all RNA synthesis and the proportion of ribosomal RNA remains constant.

That incubation conditions may be critical in establishing the level of ribosomal RNA synthesis has been suggested by Travers et al. (1973), who found that the proportion of ribosomal RNA synthesis is greatly increased if the DNA template is first incubated at 37°C and the reaction is performed in a KCl concentration greater than 0.075M. They therefore suggested that the promotors at which rRNA synthesis commences may differ from those representing other genes in requiring a conformational change, perhaps a melting of the double helix, before they can support rRNA synthesis.

A role for the ψr (Tu–Ts) and other factors in assisting RNA polymerase was suggested by Travers and Buckland (1973) on the basis of their observations that crude preparations of RNA polymerase sediment over a wide range on a gradient, with peaks at 16S and 21S, in contrast to the value of 14S displayed by purified enzyme. They found that the 16S polymerase fraction supports synthesis of rRNA in vitro at 10–15% of total RNA synthesis, but the 21S RNA polymerase synthesizes little (less than 2%) ribosomal RNA. They suggested that the 16S fraction represents a complex of RNA polymerase with Tu–Ts; and the 21S fraction represents a complex with other factors. The disappearance of 16S polymerase when cells are grown in conditions in which ribosomal RNA synthesis is greatly reduced is consistent with the concept that the formation of this form of the polymerase is responsible for rRNA synthesis.

Stringent/Relaxed Control of Transcription

Cells of E.coli possess a system, identified by mutations which abolish one of its activities, which adjusts the level of ribosomal RNA synthesis in response to environmental conditions. In wild type E.coli cells, the synthesis of RNA appears to depend upon an adequate supply of amino acids; starvation of a bacterial strain for an amino acid which it cannot make for itself—because it is mutated in the synthetic pathway—not only blocks protein synthesis but also causes a large reduction in RNA synthesis. In such cells, amino acid starvation reduces the rate of RNA synthesis to about 30% of the rate before starvation.

But in certain mutant strains, RNA synthesis continues normally during amino acid starvation. This ability is conferred by mutation at a single locus in which the wild type *stringent* gene rel^+ (previously known as RC^{str}) is replaced by its *relaxed* mutant allele, rel^- (previously known as RC^{rel}). When the effect of amino acid starvation on RNA synthesis was first discovered, it seemed natural to attribute it to some form of feedback from the protein synthetic system to the transcription process, mediated by the product of the *rel* gene, which must be active in stringent rel^+ strains but absent or inactive in relaxed rel^- mutants (reviewed by Edlin and Broda, 1968).

It does not appear to be the lack of amino acids as such which inhibits RNA synthesis. Cells with a temperature sensitive mutation in valyl-tRNA synthetase, for example, can show the stringent response at high temperature. At 30°C, these cells can charge tRNA with valine, but they cannot do so when the temperature is raised to 42°C to inhibit protein synthesis. According to Neidhardt (1966) stringent strains are prevented from synthesising RNA by an increase in temperature, but relaxed strains do not show this response. Translation itself is implicated by other experiments also, for Shih et al. (1966) blocked protein synthesis with trimethoprim, which inhibits the enzyme dihydrofolate reductase and thus prevents synthesis of fmet-$tRNA_f$ (although all other tRNA species remain fully charged). Trimethoprim prevents RNA synthesis in stringent but not in relaxed strains. This suggests that the critical step in establishing the stringent control of RNA synthesis is concerned with some reaction of protein synthesis subsequent to charging of tRNA.

The stringent control system does not affect RNA synthesis alone, for other metabolic activities such as the uptake of nucleotide precursors into the cell are also inhibited during amino acid starvation. Synthesis of all four nucleoside triphosphates is subject to stringent control, although purines respond more effectively than pyrimidines. Gallant and Cashel (1967) and Cashel and Gallant (1968) found that amino acid starvation reduces the rate of conversion of UMP into UTP and CTP by about half in rel^+ cells but has no effect in rel^- cells. The use of actinomycin to inhibit transcription does not prevent the formation of UTP and CTP, so that synthesis of the triphosphates is not inhibited merely as a consequence of inhibition of transcription but is an independent effect.

The idea that inhibition of RNA synthesis might be a secondary effect of some other metabolic control exerted by the stringent system is excluded by the observation that the stringent response does not exercise a coordinate control over all RNA synthesis. Although there is an overall reduction in RNA synthesis when stringent strains are starved for amino acids, some RNA is transcribed even in these conditions. This synthesis was not detected at first, probably because of the restriction which amino acid starvation imposes on the entry of labelled nucleotides to intracellular pools. But an examination of the RNA which *is* synthesised during starvation of stringent bacteria led Sarkar and Moldave (1968) to suggest that it comprises messenger species. According to Lazzarini and Dahlberg (1971), ribosomal and transfer RNA comprise only about 12% of the RNA which is synthesised during amino acid deprivation; the major effect of the starvation is therefore to curtail rRNA and tRNA synthesis but to leave uninhibited the transcription of messengers.

Following the production of specific messengers in both stringent and relaxed strains has confirmed this conclusion. Morris and Kjelgaard (1968) examined the rates of synthesis and breakdown of β-galactosidase mRNA after induction of amino acid starved bacteria for the lactose enzymes; although the net rate of RNA synthesis in the stringent strain is reduced to 10% of that found in relaxed cells, the synthesis of lactose messenger remains unaffected. Other experiments have suggested that the same conclusion applies to the tryptophan operon (Edlin et al., 1968; Lavalle and de Hauwer, 1968). By using a competition hybridization assay for detecting tryptophan messengers, Stubbs and Hall (1968a, 1968b) determined the level of transcription of the tryptophan operon in relation to the level of total RNA. They obtained the same results with either stringent or relaxed strains; the state of the *rel* allele does not influence the level of tryptophan mRNA.

That there is no coordinate control over all RNA synthesis suggests that the stringent response does not reflect any coupling—whether direct or indirect—between translation and transcription. Rather does it imply that the purpose of stringent control may be to adjust the synthesis of ribosomal and transfer RNAs to correspond with the rate at which amino acids enter protein. That is to say, the level at which the protein synthetic apparatus is itself made responds to the availability of amino acids. The observation that starvation for any one of several amino acids results in essentially the same overall reduction in the rate of RNA synthesis suggests that this deprivation activates some general mechanism for restricting the synthesis of ribosomal and transfer RNAs, but does not influence messenger synthesis.

That the effect of the stringent response is exerted directly on the initiation of transcription is suggested by the results of Stamato and Pettijohn (1971), who found that the complex of DNA and protein which usually makes some 40% of ribosomal RNA in vitro has much less polymerase activity when extracted from stringent cells which have been starved. But the state of this complex in relaxed cells has not been determined. That restoration of ribosomal RNA synthesis in stringent cells is blocked by rifampicin also indicates that the stringent system exerts its control at the initiation of transcription.

As well as responding to the halt in protein synthesis caused, for example, by deprivation of amino acids, cells adjust the ratio of their stable: unstable RNA synthesis to their growth rate. By measuring the rates of RNA synthesis, Nierlich (1972a, b) and Bremer et al. (1973) showed that the synthesis of stable RNA molecules increases from 29% at doubling times of 87 minutes to 66% of total RNA synthesis at a generation time of 29 minutes. When cells are "shifted up" from slower growth medium in which 50% of RNA synthesis represents stable RNA to a richer medium, the rate of stable RNA synthesis transiently rises to 75% before establishing a level of 66% of total synthesis.

The level of stable RNA synthesis is therefore set by the growth rate of the cell and may be rapidly adjusted in response to changes in growth rate. Response either to deprivation of amino acids or to changes in growth conditions is usually attributed to control of the level of initiation of ribosomal and transfer RNA synthesis. We should note, however, that an alternative explanation has been proposed by Donini (1972), in which the stringent response is mediated by control of the rate of degradation of ribosomal RNA; this model postulates a constant rate of synthesis but variable rate of turnover.

Accumulation of Guanosine Tetraphosphate in Stringent Cells

Stringent cells which are starved for amino acids contain two unusual nucleotides, which Cashel and Gallant (1969) termed MSI and MSII (MS stands for magic spot). After labelling E.coli cells with P^{32} phosphate, phosphorylated nucleotides can be separated by thin layer chromatography and identified by autoradiography; an extra spot is found when stringent cultures are starved but is absent in relaxed cells. This spot has been resolved into the nucleotides MSI and MSII. The structure of MSI is a guanosine tetraphosphate, ppGpp, with one disphosphate group attached to the 5′ position and the other diphosphate attached to either the 2′ or 3′ position (we do not know which). The structure of MSII has not been investigated in detail, but it may be a guanosine pentaphosphate, with a triphosphate group attached at the 5′ position and a diphosphate attached to either the 2′ or 3′ position.

Although ppGpp is observed in appreciable amounts in stringent cells but is absent from relaxed cells starved for amino acids, the unusual nucleotide can be produced by relaxed cells under other conditions. (The tetraphosphate is more readily identified and so its appearance has been followed in some detail, although MSII is produced as well.) Relaxed cells contain a base level of ppGpp which, although not increased by amino acid starvation, can respond to other environmental changes. During the transition from fast to slow growth, both relaxed and stringent cells restrict their synthesis of ribosomal RNA, although protein synthesis is not inhibited. When the RNA/protein ratio characteristic of the new medium is achieved, ribosomal RNA synthesis resumes at the rate characteristic of slow growth. During the transition, ppGpp accumulates in both stringent and relaxed cells in amounts approaching those found in cells starved for amino acids.

The base levels of ppGpp found in relaxed and stringent cells during balanced growth vary inversely with the growth rate and the RNA content of the cells. Lazzarini, Cashel and Gallant (1971) found that the level of RNA synthesis seems to be closely correlated with the amount of ppGpp in the cell, both during normal growth and in the occurrence of the stringent response to conditions of change. Figure 8.26 shows that this correlation is good at various growth rates, with one exception which is not understood. A similar correlation has been observed by Fiil et al. (1972) by growing the temperature sensitive mutant in valyl-tRNA synthetase at a range of increasing temperatures.

The level of ppGpp (and pppGpp) therefore determines the extent of stable RNA synthesis; the amount of these nucleotides in the cell is increased in response to conditions such as amino acid starvation or shift to slower growth rates and as a result ribosomal RNA synthesis is inhibited. Relaxed mutants have lost the ability to produce ppGpp in response to amino acid starvation, although they can still synthesize the nucleotide in response to other environmental changes; this means that they fail to exhibit the stringent response when starved for amino acids, but continue to do so when transferred from conditions of rapid to slow growth.

The occurrence of ppGpp (and its companion spot MSII) in bacterial cells appears to be the only unusual event to which we can look for an explanation of the effect of environmental changes on stable RNA synthesis. The tetraphosphate may cause at least some, and probably all, of the inhibition of nucleoside triphosphate synthesis which stringent cells show in response to amino acid starvation, for Gallant, Irr and Cashel (1971) found that ppGpp can inhibit the enzymes which synthesize AMP and XMP (the immediate precursor of GMP).

How ppGpp (and pppGpp) may inhibit ribosomal RNA synthesis is not clear. The tetraphosphate inhibited the action of Tu-Ts in promoting rRNA synthesis reported by Travers et al. (1970); but Haseltine (1972) showed that although ppGpp depresses RNA synthesis in vitro, it does so whether or not Tu-Ts is present and has an equal effect on production of all classes of RNA. Travers (1973) reported that the effect of ppGpp on RNA synthesis depends upon the conditions of incubation in vitro. He found that the nucleotide tetraphosphate depresses synthesis of all RNA by purified RNA polymerase; but when a crude polymerase–Tu-Ts complex undertakes transcription, synthesis of ribosomal RNA may be preferentially inhibited in certain conditions.

Because these conditions correspond to those when the ribosomal RNA promotors appear to be in an inactive state (that is below 37°C), it is not clear that an interaction of ppGpp with Tu-Ts bound to polymerase would be suffi-

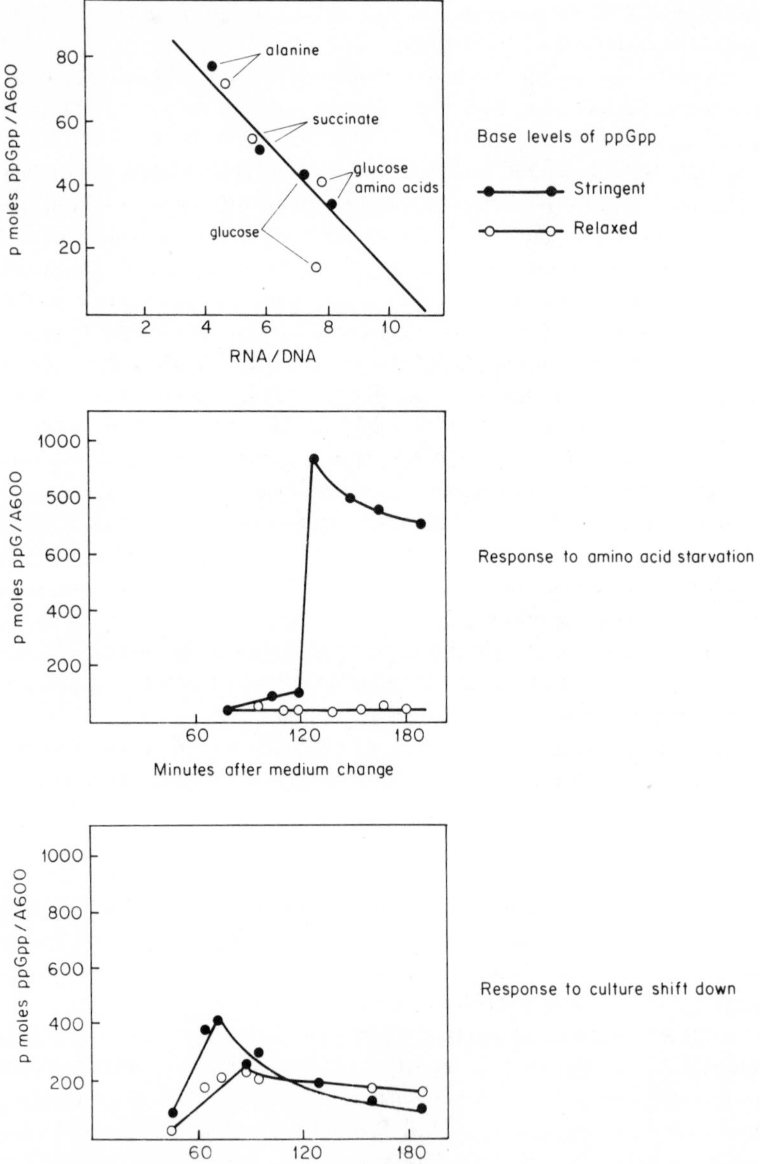

Figure 8.26: level of ppGpp in stringent and relaxed E.coli cells. The upper panel shows that (with one exception) the level of ppGpp correlates with the cellular content of RNA in both stringent and relaxed cells grown on minimal medium plus alanine, succinate, glucose or glucose + amino acids. The centre panel shows that stringent cells accumulate ppGpp upon starvation for amino acids whereas relaxed strains do not. After

cient to control rRNA synthesis in vivo. Although the most plausible model for the action of ppGpp and pppGpp is to suppose that the unusual nucleotides inhibit some protein factor needed by RNA polymerase to transcribe stable RNA genes, we cannot yet conclude that an interaction with Tu-Ts alone is responsible.

What event causes the production of ppGpp? Since the stringent response is exerted whenever protein synthesis is inhibited in rel^+ cells, we may suppose that the production of ppGpp is a signal from the non-utilised synthetic apparatus to prevent the synthesis of further components. The correlation between inhibition of protein synthesis in rel^+ cells and production of ppGpp applies no matter whether protein synthesis is inhibited by amino acid starvation, increase in temperature, or addition of trimethoprim, an inhibitor of the initiation of protein synthesis. In rel^- cells, all these treatments fail to cause accumulation of ppGpp.

The maximum level of ppGpp in rel^+ cells is reached within about 10 minutes of inhibition of protein synthesis. The ppGpp present in the cell turns over rapidly, so that its level responds almost immediately to environmental changes. Lund and Kjelgaard (1972) found that addition of amino acids to a starved culture of rel^+ cells causes the ppGpp which has accumulated to disappear with a half life of about 30 seconds, and a consequent resumption of RNA synthesis. In starved rel^+ cells, about 2×10^6 molecules of ppGpp accumulate per cell, so that the rate of synthesis must be about 50,000 molecules each second in the cell if its half life is only 30 seconds.

When chloramphenicol is added to rel^+ cells which have been starved for amino acids, the ppGpp which has accumulated decays at about this same rate, and the stringent response is abolished. Tetracycline, fusidic acid and puromycin all cause a similar effect. This supports the concept originally proposed by Cashel and Gallant (1969) that ppGpp is produced as the result of some "idling reaction" when protein synthesis is prevented. In order for the stringent reaction to occur, the synthetic apparatus must be functional; and the effect of these antibiotics is to prevent proper ribosomal function. They must therefore abolish the stringent response.

This idea has been supported by a detailed study of the effect of chloramphenicol made by Gallant, Margason and Finch (1972). If chloramphenicol directly inhibits the synthesis of ppGpp by interfering with ribosome function, it should do so regardless of the *rel* allele present. (An alternative idea is that the

---

transfer to a medium containing only very little methionine, the cells use up their supply of the amino acid and then show a sudden accumulation of ppGpp. The lower panel shows that stringent and relaxed cells share the same response to a shift from succinate to glucose medium; when the remaining succinate has been utilised there is a sudden accumulation of ppGpp. Data of Lazzarini, Cashel and Gallant (1971)

effect of chloramphenicol is only indirect, occuring because it inhibits residual protein synthesis in starved *rel*$^+$ cells, thus allowing recharging of tRNA with amino acids; in this case, it should affect only the stringent response of starved *rel*$^+$ cells.) Chloramphenicol causes a rapid disappearance of the basal amount of ppGpp present in unstarved cells of either *rel*$^+$ or *rel*$^-$ genotype. During a transition to conditions of slower growth, the antibiotic greatly reduces the rate of accumulation of ppGpp in either *rel*$^+$ or *rel*$^-$ cells.

The rapid disappearance of ppGpp whenever chloramphenicol is added to cells suggests that the unusual nucleotide usually suffers a rapid rate of turnover and that chloramphenicol reveals this situation by interfering with the production of ppGpp. The kinetics of accumulation of ppGpp when uninhibited cells suffer a step down in growth conditions suggest that the rate of degradation is decreased to cause ppGpp to accumulate. In *rel*$^+$ cells starved for amino acids the rate of synthesis of ppGpp appears to be adjusted. This explains why *rel*$^-$ cells shown normal accumulation of ppGpp in response to step down, but cannot respond to amino acid starvation; their mechanism for degradation of the nucleotide is not affected by the mutation, but their synthetic pathway is inhibited. Whether the cell achieves its final level of ppGpp by adjusting its rate of synthesis or degradation, however, the effect of accumulation is to switch off ribosomal RNA synthesis.

The implication of the ribosome in ppGpp formation has been supported by the discovery that both the magic spot nucleotides can be produced by an in vitro protein synthetic system. Haseltine et al. (1972) found that both nucleotides can be synthesized by a reaction mixture containing GTP or GDP, ATP, salt washed ribosomes, tRNA, messenger RNA and a factor present in the 0·5M NH_4Cl wash of ribosomes of stringent cells. The wash factor taken from relaxed cells is inactive. Since labelled products result when labelled ATP^{32} is used, and MSI is the predominant product if GDP is provided whereas MSII is predominant if GTP is provided, it seems likely that the two unusual nucleotides are synthesised by the reactions:

$$5' \text{ ppG} + 5' \text{ pppA} \rightarrow 5' \text{ ppGpp } (2' \text{ or } 3') + \text{pA}$$
$$5' \text{ pppG} + 5' \text{ pppA} \rightarrow 5' \text{ pppGpp } (2' \text{ or } 3') + \text{pA}$$

The signal which triggers the idling reaction appears to be the presence of an uncharged tRNA in the A site of the ribosome. Haseltine and Block (1973) observed that accumulation of ppGpp and pppGpp in vitro is absolutely dependent upon addition of an uncharged tRNA able to recognise the messenger used as template by the ribosomes. Provision of $tRNA^{phe}$ is effective with templates of poly-U, poly-U,C or poly-U,A to which it can respond, but not with templates such as poly-A, poly-C or poly-A,G which it cannot recognise. When poly-U is used as template, ppGpp and pppGpp accumulate as soon as the supply of charged phe-tRNA has been utilised (see also Pederson, Lund Kjelgaard, 1973).

That the uncharged tRNA must enter the A site in response to the messenger has been shown by using the initiation complex which fmet-$tRNA_f$ forms with phage R17 RNA (see page 94); the initiator tRNA is located in the ribosome P site, leaving the A site available for the tRNA corresponding to the second codon of each gene. Addition of uncharged $tRNA^{ser}$ or $tRNA^{ala}$—which respond to the second codons of coat and replicase genes—causes ppGpp and pppGpp to accumulate. These tRNAs cause this reaction only in response to the specific complex in which their codons are located in the A site.

Synthesis of the magic spot nucleotides does not require the presence of the elongation factors; but when present G factor hydrolyses the pppGpp product to ppGpp. Addition of Ts-Tu and charged tRNA abolishes the reaction. Inhibition of translocation by addition of fusidic acid prevents the idling reaction, probably because the ribosome is jammed in a state in which the A site is not available to transfer RNA. Rabbani and Srinivasan (1973) obtained analogous results by showing that incubation at high temperature of cells with a temperature sensitive mutation in the G factor abolishes the stringent response.

Ribosomes engaged in protein synthesis do not accumulate ppGpp or pppGpp; one model is to suppose that the idling reaction substitutes for translocation in stringent cells, causing synthesis of the unusual nucleotides. According to this model, the rel^+ factor can cause synthesis of ppGpp and pppGpp but the rel^- factor is inactive in mediating this response to amino acid deprivation. The small basal level of ppGpp and pppGpp found in both rel^+ and rel^- cells is presumably produced by the same reaction in both; this implies that it is only the increased production in response to amino acid starvation which is prevented by the rel^- mutation.

That the *rel* protein plays some role in protein synthesis is suggested by the observation of Hall and Gallant (1972) that rel^- cells make inactive β-galactosidase molecules when starved for amino acids. The total amount of synthesis of this protein is reduced by about one third by the rel^- mutation; and the protein molecules which are made have only about one third of the usual activity. These two effects therefore account for the reduction to 10% of the usual activity of β-galactosidase in rel^- cells. These results imply that the rel^- mutation has two effects on the protein which mediates the effect of this system at protein synthesis. The protein is defective in its response to amino acid starvation so that it both fails to produce ppGpp and also causes the synthesis of abnormal proteins. Large amounts of small molecular protein material are found in starved rel^- cells, which suggests that this latter effect may be concerned with premature terminations of protein synthesis. The wild type rel^+ protein must presumably therefore play some role in normal protein synthesis as well as mediating the stringent response.

CHAPTER 9

Expression of the Operon

Kinetics of Synthesis of the Tryptophan Enzymes

Transcription and Translation of mRNA

The tryptophan operon comprises the five structural genes shown in figure 9.1, which are regulated as a single unit of expression by an unlinked regulator locus, *trpR*. An operator locus has been identified at the left end of the gene cluster by its *cis*-dominant derepression of all the enzymes. Transcription commences at this end of the operon from a promotor site adjacent to the operator.

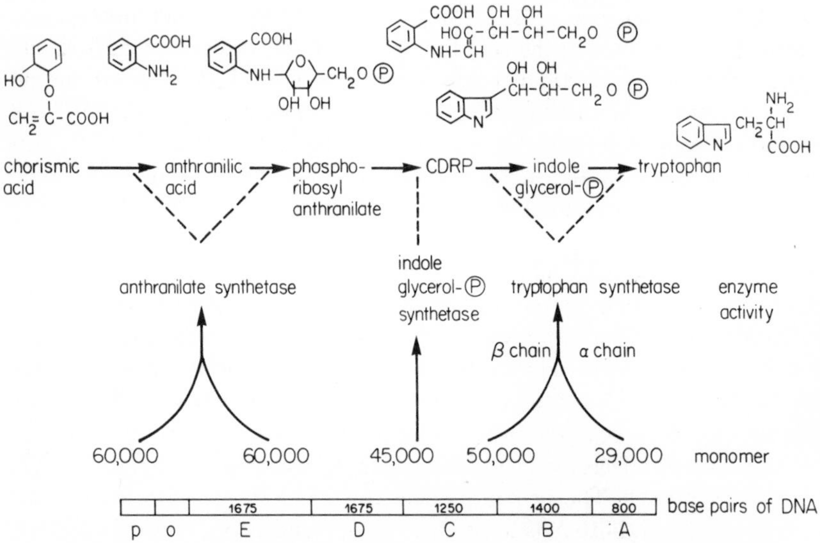

Figure 9.1: the tryptophan operon of E.coli. The order and function of genes is the same in Salmonella, although the nomenclature is different with the genes named trpABEDC from left to right. The position of the operator has been determined by mapping of mutations which show cis-dominance in E.coli; the promotor has been identified by cis-dominant mutations in S. typhimurium. The length of each gene has been calculated from the size of its protein product in E.coli. CDRP is an abbreviation for carboxy-phenylamino-1-deoxyribolose-5′ phosphate. Data of Imamoto and Yanofsky (1967), Callahan et al. (1970) and Yanofsky et al. (1971)

The system is essentially the same in E.coli and in S. typhimurium, the only difference being an unfortunate change in nomenclature. I shall refer here only to the genes as named for E.coli.

Amber nonsense mutants in the regulator gene create the *trpR*⁻ genotype in which cells are derepressed for all the tryptophan enzymes; *trpR*⁺ is dominant over *trpR*⁻, which suggests that the regulator gene codes for a repressor protein. This genetic conclusion has been supported by biochemical studies; Zubay et al. (1972) found that the synthetic activity of a DNA-dependent cell free system is repressed by a protein extracted from *trpR*⁺ but not from *trpR*⁻ cells. Each cell seems to contain about 10 copies of the protein, which is about 58,000 daltons in weight. It is not yet clear whether tryptophan itself acts as the physiological co-repressor; however, the behaviour of mutants in *trpS* (the gene coding for trp-tRNA synthetase) has contradicted earlier suggestions that trp-tRNA might be implicated.

The availability of a series of strains of E.coli bearing deletions which remove different regions of the tryptophan operon has allowed derivatives of phage $\phi 80$ to be prepared—the tryptophan genes map close to the attachment site of this phage—which carry only some of the structural genes of the operon. The $\phi 80$*trp* phages enable transcription of each structural gene to be followed by hybridization. Imamoto et al. (1965a,b) found that the messengers for the operon are formed only under conditions of derepression. Mosteller and Yanofsky (1970) have shown that rifampicin has the same effect on the expression of the structural genes as does tryptophan itself; transcription of the early genes of the operon halts very rapidly, although the later regions continue to be transcribed by those polymerase molecules which have initiated transcription before addition of the antibiotic or amino acid.

That transcription of the operon may start some distance before the beginning of the *trpE* gene is suggested by two sets of experiments. Imamoto (1970) suggested that synthesis of mRNA representing about the first 4% of the tryptophan operon is halted when repression is established, although polymerases which have progressed past this region can continue transcription. This implies that the activation of repressor protein blocks the movement of RNA polymerase past an operator site when the polymerase has initiated transcription at some previous site.

And by hybridizing tryptophan mRNA with the DNA of phages derived from strains with deletions in the early region of the operon, Hiraga and Yanofsky (1972) showed that a sequence of less than 200 nucleotides preceding the beginning of the *trpE* is transcribed. Cohen, Yaniv and Yanofsky (1973) have isolated the early *trp* mRNA by hybridizing RNA made by a strain possessing only the first 15% of the *trpE* gene with the DNA of a phage carrying only *trpE*. The discrete fragments which they separated on acrylamide gels may in part be the products of breakage in vivo of the messenger sequence preceding the *trpE* gene.

That the region between the operator and the start of the *trpE* gene may have some role in regulation is suggested by the observation of Jackson and Yanofsky (1973) that deletions of it may increase the maximum rate of expression of the operon. This effect is exercised at the level of transcription.

Movement of RNA polymerase along the operon can be followed by derepressing the tryptophan genes, allowing RNA to be synthesised in the

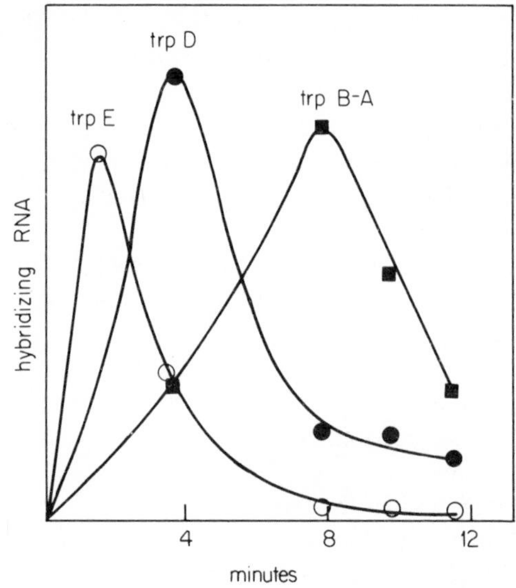

Figure 9.2: production of tryptophan mRNA in E.coli. Cells are derepressed for 1.5 minutes in the presence of H^3-uridine and then repressed by addition of tryptophan. RNA corresponding to different parts of the operon can be assayed by hybridization with DNA of ϕ80 carrying only those regions. Transcription proceeds sequentially long the operon from trpE to trpA. The messengers are degraded very soon after their synthesis as shown by the rapid decline in hybridizable mRNA. Data of Morse, Mosteller and Yanofsky (1969)

presence of H^3-uridine, and then adding tryptophan to repress the operon after only a very short time. Only polymerases which have initiated transcription in the brief period of derepression can proceed along the operon. Figure 9.2 shows the appearance and disappearance of mRNA for different regions of the operon in such an experiment. As polymerases proceed along the operon, mRNA is transcribed first from the *trpE* gene, then from *trpD*, *trpC* (not shown) and finally from *trpB* and *trpA*. Degradation of the mRNA follows very soon upon its synthesis, for *trpE* mRNA starts to suffer breakdown before synthesis of *trpD* is completed; similarly, degradation of *trpD* commences before transcription has reached *trpB-A* at the end of the operon.

When rifampicin is added to cultures which are derepressed because they carry a *trpR*$^-$ allele, new initiation events are blocked although the polymerases presently transcribing the operon can continue to the end. The decline in the rate of incorporation of an H^3-uridine label into mRNA for the *trp* genes in effect measures the progress of the last polymerase along the operon. By using this technique, Rose, Mosteller and Yanofsky (1970) have shown that the rate of transcription remains the same throughout the operon. The rate of transcription seems to depend only on the temperature of the cells, for their growth rate can be varied at each temperature but the transcription time remains set at a characteristic level. Since we know the approximate length of the operon, the rate of transcription can be calculated from the time taken to transcribe all five structural genes. This gives the results:

temperature	transcription time	rate of transcription
25°C	6·5–7·0 minutes	1000 nucleotides/minute
30°C	4·0–4·5	1600
37°C	2·5–3·0	2200–2700

The times required to transcribe the first quarter, first half and whole operon are simply proportional to the distance which RNA polymerase must travel.

The appearance of enzyme activity corresponding to each gene lags slightly behind the time of transcription of the gene, usually by 1-2 minutes. This means that the first gene of the operon has been translated into protein before the last one has been transcribed into RNA; this agrees with the value obtained by Imamato and Ito (1968), who found that under growth conditions in which transcription of the operon takes 8 minutes, the mRNA of *trpE* appears within 2.5 minutes and the enzyme of the *trpE* gene within about 0·5 minutes of completion of its transcription.

The kinetics of expression of the lactose operon also are consistent with a scheme in which translation starts before transcription is completed, for Leive and Kollin (1967) found that 2.5 minutes are required to transcribe the β-galactosidase gene upon induction. If transcription of the remaining genes proceeds at the same pace, another 1.5 minutes should be required to complete transcription of the gene cluster. The time required for translation and activation of the enzymes from the messenger is some 1·5 to 2·0 minutes for each protein. But β-galactosidase is available in the cell before the time which these processes would take if consecutive, so that translation and activation of the first enzyme must occur before the far end of the operon has been translated into RNA. And according to Schwartz, Craig and Kennell (1970) the messengers transcribed from the lactose operon after induction start to be degraded even before their synthesis is complete.

The inference drawn from kinetic studies of individual operons that transcription and translation takes place simultaneously is supported by the electron microscopic observations of Miller, Hamkalo and Thomas (1970). Figure 9.3 shows that when cells of a fragile mutant of E.coli are burst osmotically

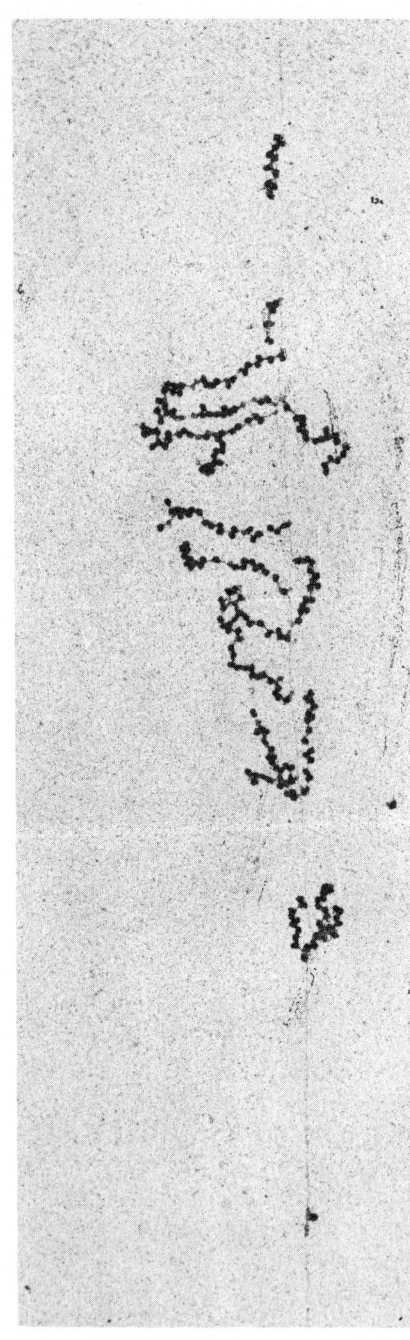

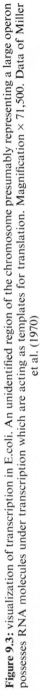

Figure 9.3: visualization of transcription in E.coli. An unidentified region of the chromosome presumably representing a large operon possesses RNA molecules under transcription which are acting as templates for translation. Magnification × 71,500. Data of Miller et al. (1970)

by rapid dilution into water, masses of thin fibres appear with attached strings of granules. Treatment with DNAase destroys the central fibre and treatment with RNAase removes the granular strings from the fibres. The granules are about the size of E.coli ribosomes, so that the granular strings represent polysomes attached to DNA (the central fibre).

Since few free polysomes have been observed in these preparations, translation of messenger RNA as soon as it is transcribed appears to be a general phenomenon. That the polysomes represent messengers still under transcription is suggested by the attachment of the granular strings to the central fibre at an irregularly shaped granule of about 75Å diameter, presumably RNA polymerase. Ribosomes are closely spaced in the polysomes and the ribosomes at the newly synthesized end of the messenger are usually very close to the granule where they join the central fibre. The longest polysomes observed possess about 40 ribosomes. Because the spacing of polysomes along the central fibre is irregular, transcription must be initiated at statistical rather than exact intervals.

The rate of movement of ribosomes along an mRNA can be estimated by taking the difference between the times at which the enzymes of two of the genes first appear; this measures how long it takes for the ribosome to move from the end of the first gene to the end of the second gene. Using this criterion, Morse, Baker and Yanofsky (1968) found that under conditions in which transcription of the tryptophan operon proceeds at about 1000 nucleotides per minute, translation seems to take place at about 1200 nucleotides per minute; the rates of transcription and translation are very similar, if not identical. The time at which the last ribosome passes the end of a gene is revealed by the moment when the increase in activity of the enzyme which it specifies comes to a halt; the rate of ribosome movement estimated in this way is the same as that estimated from the time of first appearance of an enzyme, which shows that the last ribosome to translate the messenger proceeds at the same pace as the first.

By using a pyrimidine requiring mutant strain of E.coli in which all the CTP and UTP is derived from the medium—so that an equilibrium is reached in which the specific activity of the cellular RNA is the same as that of the nucleotide pools—Baker and Yanofsky (1970, 1972) determined the number of molecules of tryptophan messenger present in derepressed cells. About 0·12% of the total RNA of derepressed cells represents tryptophan messengers. After derepression for one minute with the tryptophan analogue 3-indoleacrylic acid, there appear to be about 5.3 equivalents of mRNA for *trpE* per cell. At this time, the average RNA polymerase probably has about 0.6 of a chain of *trpE* mRNA attached to it—it takes about 50 seconds to transcribe *trpE* in these conditions—which suggests that about 9 polymerases are engaged in transcribing the tryptophan genes. Since on average there are probably about 1.8 tryptophan operons present in each cell (through chromosome replication), there seem to be about 5 initiation events every minute for each tryptophan operon.

As the operon can be transcribed in between 2·75 and 3·0 minutes under these conditions, the entire cluster of structural genes might accommodate about 15 polymerase molecules at once. E.coli cells can synthesise about 460 molecules of tryptophan synthetase A protein per cell through one minute of derepression, which corresponds to about 50 ribosomes per messenger. The number of initiations appears to be less in cells which have established steady state derepression; *trpR*$^-$ cells exhibit about 2.6 initiations per operon every minute; about 30 ribosomes seem to translate each messenger to produce some 140 molecules of tryptophan synthetase B each minute. In the steady state, there are some 7–8000 molecules of enzyme per operon.

Using *trpR*$^-$ cells which are continually derepressed, Mosteller, Rose and Yanofsky (1970) found that the rate of synthesis of the enzymes of the tryptophan operon varies in direct proportion to the growth rate, although the rates of transcription and translation remain constant, as does the rate of degradation of messenger RNA. But there can be as much as 8–11 fold more tryptophan mRNA in rapidly growing cells compared with those at slow growth rates. This implies that the initiation of transcription takes place more frequently at the tryptophan promotor when cells grow faster. This might be achieved by some special control mechanism or, perhaps more likely, may simply be the result of changes in the availability of RNA polymerase molecules to undertake transcription; the rate of initiation must presumably depend not only on the affinity between polymerase and on promotor but on the relative number of enzyme molecules available in the cell. Rose and Yanofsky (1972) found that the frequency of initiations at the tryptophan promotor decreases when cells are shifted from poor to rich medium; such a shift probably causes polymerases to bind preferentially to the genes for stable RNA, thereby reducing their availability for messenger synthesis.

Degradation of Messenger RNA

The rapid kinetics of transcription and translation of messenger RNA and its low stability imply that the several events which take place during the lifetime of a messenger may occur simultaneously. A sequence of events must take place in rapid succession in which transcription is initiated at the promotor, a cluster of ribosomes attaches to the free end of the messenger as soon as it becomes available, and a nuclease commences degradation of the messenger shortly after attachment of the last ribosome. As figure 9.4 shows, all these events may take place at the 5′ end of the message before RNA polymerase has transcribed its 3′ end.

In experiments with the tryptophan operon, Morikawa and Imamoto (1969) and Morse et al. (1969) have shown that degradation proceeds along the messenger from the 5′ end to the 3′ end, in effect following the last ribosome along the message. After the tryptophan operon has been derepressed for eight minutes, the first round of transcription has been completed and a steady state of transcription is achieved. Administration of a short pulse label of H^3-uridine during this subsequent period labels all the segments of tryptophan mRNA; and the later addition of unlabelled uridine then rapidly dilutes the radioactivity in the RNA precursor pools so that little further label is transferred to RNA. The rate of degradation of the mRNA can be measured by following the disappearance of the radioactive label from the various parts of the messenger.

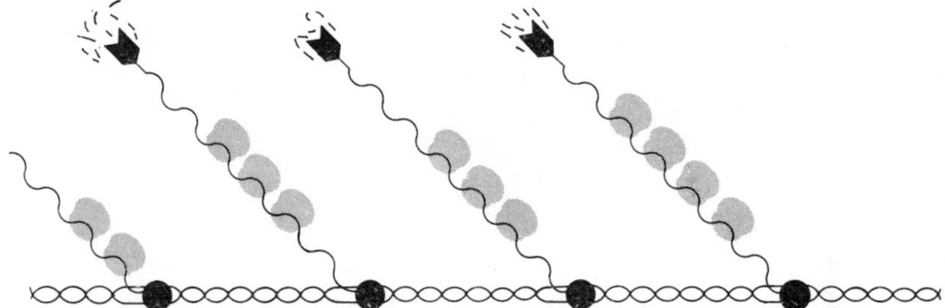

Figure 9.4: the operon shortly after induction. Ribosomes attach to a messenger and commence translation immediately after its synthesis has begun. A nuclease attaches to the messenger after the last ribosome and commences to degrade the RNA sequentially even before transcription is complete. This means that each messenger under transcription bears a cluster of ribosomes located shortly after the polymerase, which in turn is followed by a nuclease. Several polymerases may transcribe the operon at the same time

Another way to follow the synthesis and degradation of successive regions of the tryptophan operon, used by Mosteller, Rose and Yanofsky (1970), is to treat the derepressed cells with rifampicin to block new initiation events; RNA polymerases already engaged in transcription continue to the end of the operon. Hybridization with transducing phages carrying particular parts of the operon shows when these regions cease to be transcribed, for at this point the increase in incorporation of a radioactive label into mRNA stops. The disappearance of these regions can then be measured as they are degraded.

The label in the mRNA corresponding to the *trpE* gene begins to disappear very soon after dilution with unlabelled uridine (about 60 seconds) or after addition of rifampicin (about 70 seconds). The label in the *trpD* region begins to be lost after about 120–130 seconds and that of the *trpC-A* region at about 5 minutes. This means that degradation of the messenger proceeds in the same 5′ to 3′ sequence as its synthesis and translation. But degradation must begin at the 5′ end before the 3′ end has been transcribed; indeed, the rapid degradation of the *trpE* region implies that the enzymes responsible must set to work very soon after the last ribosome attaches to translate the messenger.

Degradation seems to proceed more slowly than transcription or translation however, which excludes the idea that a nuclease might travel along the messenger a constant distance behind the last ribosome. Almost all the radioactive label is lost from *trpE* in 10 minutes, from *trpD* mRNA in 15 minutes, and from *trpC-A* in 25 minutes. This indicates that degradation of individual messengers may start at varying times after their transcription has commenced and takes about 15 minutes to complete—this is about twice the time required for transcription under these conditions.

One implication of these results is that the operator distal regions of the mRNA must remain intact for longer than the regions coded by genes nearer to the operator. Forchhammer, Jackson and Yanofsky (1972) have confirmed this prediction by measuring the half lives with which different parts of tryptophan mRNA decay when repression is established. *TrpE* mRNA has a half life of 58 seconds, *trpD-C* of 75 seconds, *trpB-A* of 95 seconds. When the region between the operator and *trpB-A* is deleted, the *trpB-A* mRNA is degraded at the rate usually characteristic of *trpE*. This means that it is position in the messenger which determines stability, not some feature of the structure of mRNA. These results are consistent with a model in which *trp* mRNA is attacked at its 5′ end at random times after initiation, degradation then proceeding along the messenger more slowly than the ribosomes.

The kinetics of messenger disappearance seem to result from sequential degradation in the direction 5′ to 3′. An obvious mechanism to achieve this result would be for a 5′→3′ exonuclease enzyme to attach to the free end of the messenger and move along behind the ribosomes. Indeed, Kuwano, Schlessinger and Apirion (1970a, b) and Kuwano, Apirion and Schlessinger (1970) detected such a ribonuclease activity, called ribonuclease V, which appeared to be associated with ribosomes and which might therefore travel with the last ribosome of the cluster to degrade the message. If a certain proportion of ribosomes were to carry ribonuclease V, only a certain number of rounds of translation could start on a messenger before the attachment of one of these ribosomes; this would explain why a constant number of ribosomes, on average, translate each messenger and also how degradation follows the last ribosome engaged in translation. But this ribonuclease has proved to be an artefact, for Bothwell and Apirion (1971) and Holmes and Singer (1971) have shown that the apparent RNAase V activity of ribosomes is in fact a contaminant of the enzyme ribonuclease II, which has a 3′ to 5′ exonuclease activity. There is, therefore, no known enzyme with the ability to degrade RNA in the direction 5′ to 3′.

The only known ribonucleases are exonuclease enzymes which degrade RNA from 3′ to 5′ and endonucleases which make internal scissions in the chain. The apparent lack of any ribonuclease able to degrade RNA from 5′ to 3′ suggests that the apparently sequential 5′ to 3′ degradation of messenger RNA might in fact be achieved by a two part degradation reaction. First, an endonuclease might cleave the messenger, presumably initially at a point near the 5′ end and than at a series of sites progressively closer to the 3′ end. After the introduction of these internal breaks, an exonuclease might degrade the fragments released in this way from their 3′ ends towards their 5′ ends. Once ribosomes have ceased to attach to the 5′ end of the mRNA, therefore, this end would become exposed as ribosomes proceed along the message and might become susceptible to attack by the endonuclease. This would explain why the overall degradation of messenger proceeds from the 5′ end to the 3′ end.

But the drawback of this model is that it does not explain why ribosomes should stop translating the message and demands some mechanism to set the number of ribosomes which attach to a molecule of mRNA; the advantage of schemes which invoke an exonuclease to proceed from 5′ to 3′ along mRNA is that such an enzyme might compete with ribosomes for the free end of the messenger so that the frequency of translation could be established by the ratio of enzyme molecules to the number of ribosomes. At present, none of the known ribonuclease enzymes of E.coli have been implicated in the degradation of mRNA, so there is no evidence to say whether degradation is really sequential—involving a 5′ to 3′ exonuclease which has not yet been discovered—or whether the sequential kinetics result from the action of two or more enzymes, some of which may be amongst those already identified (see also below).

Internal Initiation of Transcription and Translation

Despite the coordinate control over expression of their cistrons, transcription and translation of the tryptophan operons of both E.coli and Salmonella can be initiated at internal sites. Bauerle and Margolin (1967) examined enzyme synthesis by the tryptophan operon of Salmonella in a group of deletion mutants which extend into the operon from the operator end and terminate at various points within it. They found that the pleiotropy exerted by deletion of the promotor adjacent to the operator site does not extend to all five cistrons. All strains with a deletion ending in either of the first two genes lack expression of both enzymes, but the remaining three genes can function at a low level.

The ability to express the last three genes is lost whenever the deletion extends beyond the boundary between the second and third cistrons. The last three genes can be expressed at this low level independently whenever they are not expressed together with the first two genes, for although nonsense mutants in the first gene reduce the expression of all the remaining genes, the effect is not coordinate; expression of the second gene is more severely limited than are the last three.

The tryptophan operon of E.coli also shows non-coordinate expression of its last three genes. According to Morse and Yanofsky (1968), the basal level of synthesis of the last three enzymes is about five times that of the *trpE* and *trpD* products, so that the *trpCBA* segment shows a low level synthesis of enzymes, even under conditions of repression, at about 2% of the derepressed rate. After derepression, however, there is coordinate synthesis of all five enzymes, because a polycistronic messenger is synthesised to represent the whole operon and is translated to give (generally) equimolar yields of its protein products.

This suggests that the non-coordinate synthesis is unlikely to result from translation of just the last three genes of a messenger representing the whole operon, but more probably results from the independent transcription of *trpCBA*. This demands the presence of a second promotor, located before the third gene of the operon, where transcription can be initiated; this results in the synthesis at a low level of messenger RNA for the last three genes and so causes repressor-resistant synthesis of their enzymes. Jackson and Yanofsky (1972a) mapped the internal promotor by isolating deletions which lack it; these can be identified by their effect in reducing the basal level of the last three enzymes (although they do not change the derepressed level of expression). This locates the internal promotor within *trpD*; it must therefore comprise some sequence needed for the function of the *trpD* enzyme which, by chance, resembles the sequence of the promotor of the operon. If missense mutations in this region can be found, it may therefore prove possible to deduce the nucleotide sequence of the internal promotor from the wild type and mutant amino acid sequences.

New promotor sites may also be created by mutation, for mutants which have deletions extending from the operator into the first gene do not express the second gene. But revertants able to express this gene can arise spontaneously or may be induced with the mutagens 2-aminopurine or nitroso-guanidine. The mutational events which confer reversion may be as simple as a single base pair transition; Wuesthoff and Bauerle (1970) found that such mutations can occur in either of two particular regions of the first gene, where they function as low level constitutive initiators for the subsequent genes. The most likely explanation for this behaviour is to suppose that these regions of the first gene resemble a promotor sequence and may be mutated to give new sites at which RNA polymerase can initiate transcription.

Similar mutants have been found in the histidine operon, which is usually expressed only as a coordinate unit. Atkins and Loper (1970) tested the activity of the various genes of the operon by seeing whether a Salmonella host bacterium could supply the functions missing from F'*his* episomes derived from strains of E.coli carrying mutations in the histidine genes. The two sets of genes do not recombine, so any inter-generic complementation of functions missing from the operon carried by the episome must be caused by synthesis of the active enzyme from genes on the bacterial host chromosome.

The first three genes (*hisG,D,C*) cannot be expressed in any of the bacteria where the chromosome has a deletion which removes the operator, but revertants able to produce some of the other enzymes can be divided into two classes. One class has all the enzymes of the operon except the first three; the other has only the enzymes coded by the last two genes. This implies that reversion can result from a mutation which creates a new promotor either before the fourth gene or before the penultimate gene of the operon. Although there is no evidence to say whether the mutant initiators act at transcription or translation, it seems likely that they function by creating new promotors which direct the binding of RNA polymerase rather inefficiently when the proper promotor is absent—the new initiators do not seem to function in cells which have the first promotor intact.

Under certain abnormal conditions, both transcription and translation can be initiated at several sites within the tryptophan operon. Imamoto and Ito (1968) and Imamoto (1969) found that treating derepressed bacteria with DNP (dinitrophenol) inhibits transcription; but RNA synthesis regains its normal rate within one minute when the inhibitor is removed. If tryptophan is added to the cells at the same time when the DNP inhibitor is removed, the amino acid prevents the initiation of new rounds of transcription at the promotor adjacent to the operator.

After derepression, the tryptophan mRNA synthesised during the first minute or so usually corresponds to only the *trpE* gene. But the tryptophan mRNA synthesised after DNP treatment may contain molecules which correspond to other regions of the operon. After treatment with DNP for 1·5 minutes, the mRNA synthesised on its removal corresponds only to genes *trpE* and *trpD*; after 4–9 minutes of treatment the messenger made during the first minute of derepression corresponds to the segment *trpEDC* and to part of *trpB*; and after treatment with DNP for 11–29 minutes, the mRNA found in these cells corresponds to the whole operon. That this is due to simultaneous initiations of transcription within the operon is supported by the finding that RNA molecules containing the operator-distal regions are smaller than would be expected if their transcription was initiated at the beginning of the operon.

After treatment with DNP, the proteins coded by the *trpC* and *trpA* genes may appear simultaneously, which indicates that there are (at least) two sites within the operon at which ribosomes can attach to tryptophan messengers lacking the normal 5′ sequence. What seems to happen in the cells treated with DNP is that RNA polymerase molecules may move along the DNA template to which they are bound, but without catalysing the synthesis of RNA. When the inhibitor is removed, the polymerases can resume RNA synthesis and therefore start new chains at the points within the operon to which they have progressed.

These results, together with the existence of active internal initiators (see above) and the successful translation of the remaining genes in strains deleted for the start of an operon (see page 299), indicate that ribosomes can commence translation at sites within messengers lacking the usual operator-proximal regions. This implies that no special sequence of nucleotides at the 5′ end of the message is needed to direct the attachment of ribosomes. Rather must each gene have its own signal for initiating protein synthesis—whether this is solely an AUG codon or, more probably, a more complex sequence—which can be used whenever it is available to ribosomes.

Polarity of Translation

Influence of Nonsense Mutations on Operon Function

Because the messengers coded by operons are polycistronic and are sequentially translated into proteins, events in an early gene of the operon may affect not only that gene but also the subsequent cistrons. Ames and Hartman (1963) found that some half of the point mutations isolated in cistrons of the histidine operon of Salmonella have dual effects; as well as inactivating the enzyme specified by the mutated gene (by premature termination), the *amounts* of all the enzymes coded by genes on the side of the mutated cistron distal from the operator are reduced, as figure 9.5 shows. The amounts made of the enzymes coded by the genes on the operator-proximal side of the mutated gene remain unaffected. (The operator-proximal part of an operon or of a gene within it describes the region closer to the operator/promotor control site; the operator-distal region is that further from the starting sites).

All mutational events which produce these *polar* effects—with the exception of the insertion discussed below (page 406)—do so through the introduction of a nonsense codon; and amber, ochre and UGA triplets are all equally effective (reviewed by Zipser, 1969). Suppressor genes which act on nonsense codons

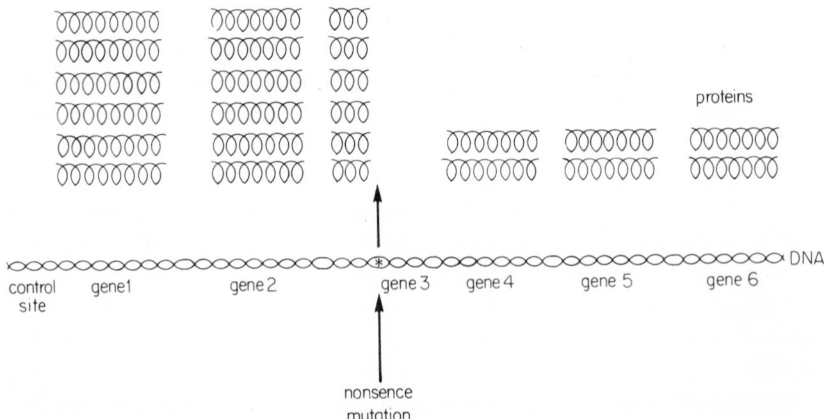

Figure 9.5: polarity in bacterial operons. A nonsense mutation in the third structural gene causes premature termination of synthesis in that gene. However, it also reduces the *numbers* of proteins synthesized by each of the subsequent genes. Each of genes 4–6 suffers the same extent of reduction

partially relieve polarity, and Imamoto, Ito and Yanofsky (1966) found that the extent of relief is about equivalent to the extent of suppression characteristic of the suppressor used. Since the suppression of nonsense codons takes place at translation, this implies that the polar effect results from events in protein synthesis. The degree of polarity is usually expressed as the per cent synthesis of enzymes succeeding the mutated gene compared with the extent of synthesis in the wild type operon; zero synthesis of the subsequent enzymes is equivalent to complete polarity.

By following the effect of nonsense mutations in the z gene upon translation of the lactose operon of E.coli, Newton et al. (1965) were able to show that there is *conservation* of polarity; for nonsense mutants in *z* create the same degree of polarity in the expression of *a* as they do of *y*. The effect of nonsense codons in creating polarity is not limited to the first gene; amber mutants in *y* reduce the amount of transacetylase synthesised by the *a* gene. When the positions of nonsense z^- mutants within the gene were located by deletion mapping, Newton et al. found that the extent of the reduction in synthesis of permease and transacetylase depends on the position of the mutation in the *z* gene; mutants near the operator are completely polar and allow virtually no synthesis of the subsequent enzymes. But mutations at the operator-distal end of the gene display little polarity and allow production of the second two enzymes at levels approaching that of the wild type strains.

One possibility, suggested by Fink and Martin (1967) from results obtained with the histidine operon, is that the first gene of an operon might have a steep gradient of polarity with position of mutation, whilst mutation in subsequent genes might produce a gradient of more shallow slope. This idea has been excluded by Michels and Reznikoff (1971), who have examined the polarity of nonsense mutants in the *z* gene in three strains of E.coli. In normal cells, *z* is the operator proximal gene of the lac operon; in some deletion strains which fuse the tryptophan and lactose operons it remains the operator proximal gene of its unit of expression; but in others it becomes a distal cistron of the fused operon. The same gradient of polarity is found in all three situations—it is a characteristic of the gene itself and not a consequence of its position in the operon. Figure 9.6 shows the gradients of polarity of the *trpC*, *trpD* and *trpE* genes, in which Yanofsky et al. (1971) have measured polarity by the level of synthesis of the *trpA* or *trpB* genes.

The deletion of a large segment of *z* lying between the site of mutation and the end of the gene restores permease activity to hitherto strongly polar mutants in *z*. This suggests that the crtical distance in creating the gradient of polarity is that from the mutation to the next gene, rather than that from the beginning of the gene to the mutant site. This conclusion has been confirmed by Zipser and Newton (1967), who constructed double mutants in *z* which possess a deletion either operator-proximal or operator-distal to a nonsense mutant triplet. Deletions between the mutant site and the beginning of the gene do

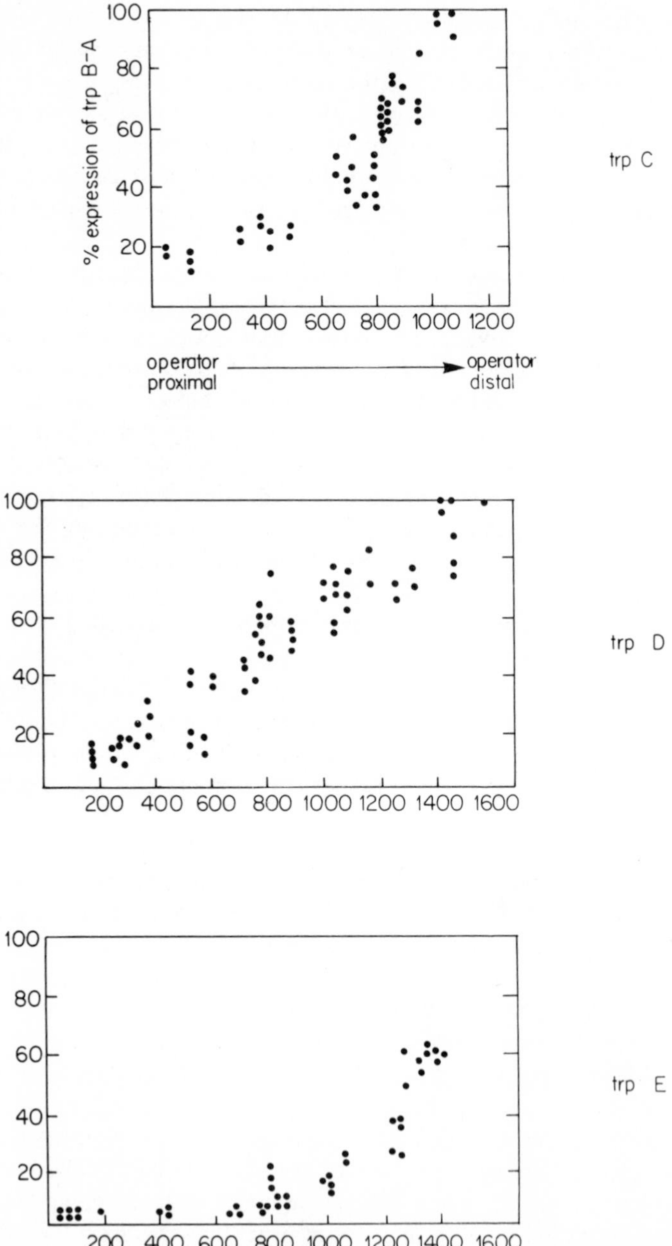

Figure 9.6: gradients of polarity in the tryptophan operon. The expression relative to wild type of the last two genes (measured by their synthesis of the enzyme tryptophan synthetase) is shown as a function of the map position of nonsense mutants in the first three genes. Each gene has a characteristic

not change the degree of polarity, but the removal of a sequence of DNA beyond the site of mutation partially suppresses the polar effect; essentially, such deletions "move" the polar mutant to a new point on the gradient.

Role of Reinitiation Sites in Polarity

The behaviour of double mutants which possess two nonsense codons in the same operon supports the idea that reinitiation events are crucial in polarity and that it is the distance to the next available reinitiation site which determines the gradient of polarity. When a double mutant is constructed from nonsense codons located in two successive cistrons of an operon, each nonsense triplet independently reduces by its usual proportion the synthesis of the enzymes coded by the genes distal to it; so that the double mutant is more polar than either of the single mutants. This implies that a reinitiation event takes place at the intercistronic boundary to permit expression of the second mutant.

But different results are obtained when both nonsense codons are located in the same gene. Yanofsky and Ito (1966) found that, in this situation, the polarity exhibited is that typical of strains possessing only the mutation closer to the operator. This can be explained if translation stops at the first nonsense codon and does not start again before the second nonsense mutation; in this case, the second mutation can have no effect on protein synthesis. If the first reinitiation event to take place after termination at a nonsense mutation does not happen until the next inter-cistronic boundary, then further mutations located between the first site and the boundary are ineffectual.

But if the degree of polarity depends on the distance from the nonsense mutation to the initiation site at the start of the next gene, then it should be possible to counteract polarity by creating a further mutation which allows reinitiation to take place between the polar nonsense mutation and the end of the gene. Grodzicker and Zipser (1968) have identified such revertants of a polar mutant in the *z* gene; when a reinitiation site is created by mutation close to the site of nonsense mutation, translation of the distal portion of the *z* gene is resumed so that the activity of the *y* and *a* genes in making permease and transacetylase is restored.

The polar behaviour of double nonsense mutants in the *z* gene of the lactose operon depends upon their precise location, for Newton (1969) found that

---

gradient of polarity, although mutants at the beginning always have large effects on the expression of subsequent genes and mutants at the end of each gene have little effect. Map positions, although determined by recombination frequencies, have been converted in base pairs distant from the start of the gene; because recombination frequencies do not coincide exactly with physical lengths of DNA, the position of these mutants in the gene is subject to some error. When points are located vertically above each other, they represent different mutations which have not been separated by recombination. Data of Yanofsky et al. (1971)

when the sites lie in the region of the gene far from the operator, the double mutants shows the same polarity as a strain containing only the mutation closer to the operator. But if the first mutant site lies within the early or middle part of *z*, the introduction of a more distal mutation causes an increase in polarity. Indeed, the nearer is the first mutant to the operator end of the gene, the greater is the influence of the second mutant. This implies that there is some chance that a reinitiation of polypeptide synthesis can occur after the ribosome has passed the first mutant site but before it reaches the next mutant site; the proability that this reinitiation will occur depends upon the location of the first site. If such a reinitiation does occur, the second mutant is read as a chain terminator and can thus exert a polar effect on synthesis from subsequent genes.

The form of the gradient of polarity can therefore be explained by the presence of natural reinitiation sites. Such sites only act as initiators, of course, when a previous termination event has taken place. This idea is supported by the finding of Zipser et al. (1970) that the reinitiation mutants found to revert polar mutants by themselves have almost no effect on the expression of the *z* gene. When only a few nonsense mutants had been isolated in the *z* gene it appeared that the gradient of polarity was smooth, but the isolation of further mutants has identified the peaks of figure 9.7. These peaks appear to result

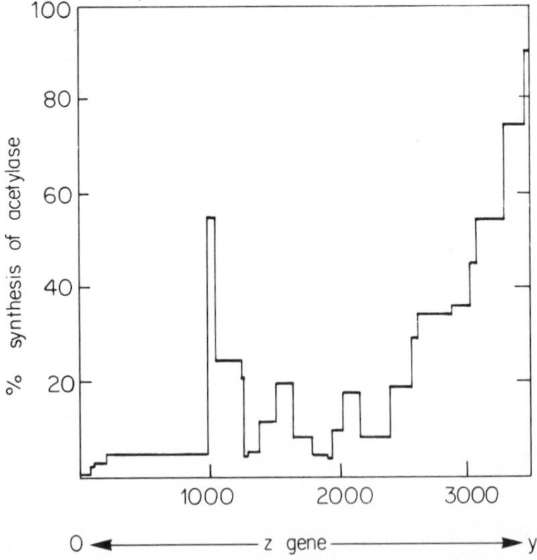

Figure 9.7: polarity in the z gene of the lactose operon. The proportion of wild type synthesis of acetylase is plotted as a function of the position of nonsense mutations in the z gene. Mutants have been located within the regions covered by a series of deletions and each horizontal line represents the average polarity of all those mutations located in this length of the gene. These values are connected by vertical lines for convenience. Data of Zipser et al. (1970)

from the presence of internal natural initiators, which permit translation to restart after termination at a nonsense codon; when a nonsense mutant lies before such a reinitiator, the critical distance in establishing the polarity of the mutant is that to the re-initiation site, not that to the end of the gene.

Reinitiation sites—sometimes called π sites—have been mapped in the *z* gene by examining double nonsense mutants which show an increase in polarity compared with that shown by the single mutant site nearer to the operator. Michels and Zipser (1969) detected two such internal reinitiators by this approach (reviewed by Zipser 1970). That only two efficient re-initiators are found, although β-galactosidase has 24 methionine residues, and therefore 24 AUG codons in phase as well as others out of phase, suggests that the signal for reinitiation may be more complex than just an AUG triplet and may perhaps involve a longer sequence of nucleotides. Another possibility is that the mRNA may need to take up some particular secondary structure around the AUG triplet to allow it to function as an initiator.

The presence of the two natural reinitiators explains why the polarity gradient in the *z* gene falls into three parts; each part reflects the distance to the next available site for starting translation, in the first two instances the internal initiators, in the last the inter-cistronic boundary. According to Zipser et al. (1970), each reinitiation peak has a characteristic height, which implies that the degree of polarity may depend on the kind of π site as well as the distance of a mutation from it.

By identifying new initiator signals as revertants of nonsense mutants, Sarabhai and Brenner (1967) found that an ochre mutant created by a frameshift in the *rIIB* cistron of phage T4 can be reverted by a point mutation which causes reinitiation and synthesis of a polypeptide chain corresponding to the rest of the gene. Sarabhai and Lamfrom (1970) found that the reinitiator does not usually function in the wild type gene but can be activated either by a preceding nonsense mutation or by the deletion of three nucleotides to the left of the reinitiation site. This shows that the presence as such of ribosomes translating the gene does not stop intiation at internal signals, for the deletion does not cause chain termination. The effect of the deletion must therefore be to produce a sequence which mimics the initiation signal.

Although the *z* gene codes for only one polypeptide chain, it can be divided into three regions by inter-alleleic complementation; the α-region is close to the operator, the β region is intermediate, and the ω region is at the far end of the gene. In cells which are diploid for the *z* gene, mutants with deletions in the α region, for example, complement other mutant genes which have intact the α part of the gene. All deletions extending from *o* into *z* complement all mutants located beyond the barrier which marks the end of the α region. This complementation takes place by association of the mutant subunits and does not require any covalent linkage of the different parts of the *z* protein.

The β-galactosidase protein is therfore active provided that all its parts are present, even though they are no longer covalently linked into one polypeptide chain as usual. But the active enzyme produced by complementation with α and ω components has physico-chemical proterties distinct from the wild type, as might be expected from the independence of its different protein chains (reviewed by Ullman and Perrin, 1970). Wild type enzyme seems to contain a certain proportion of free α and ω polypeptides and the amount of these components can be increased by proteolytic treatment. One possible explanation is that some peptide bonds in the completed polypeptide chain may be labile, generating the complementing polypeptides; another is that a certain amount of spontaneous termination and reinitiation may take place when the gene is translated.

That one, or perhaps both, of these events takes place in the cell is supported by the results of Morrison, Zipser and Goldschmidt (1971), who found that when cell extracts containing β-galactosidase activity are autoclaved, an inactive $-NH_2$ terminal fragment is released. This fragment is about 7,400 daltons in weight and has been called auto-α because it has the complementing properties of an α complementation group donor. Auto-α is useful because nonsense mutants of *z* which produce an inactive β-galactosidase protein can nevertheless be tested for their auto-α activity; in effect, this means that inactive polypeptide chains corresponding to β-galactosidase mutants can be identified by their ability to generate the auto-α fragments.

Using this identification, Morrison et al. found that all nonsense mutants contain a polypeptide corresponding to synthesis from the $-NH_2$ terminus to the site of mutation (which provides another demonstration of the colinearity of gene and protein). All those polypeptides larger than a certain size also have a polypeptide, αU1, of about 46,000 daltons, corresponding to the synthesis of protein from the normal $-NH_2$ terminus to a point coinciding with the π site at the beginning of the β region where the second complementation group starts. This fragment might be generated either by a natural weak termination—at about the 10% level—of RNA or protein synthesis, or by cleavage of the RNA or protein. But the coincidence of complementation groups with the gradient of polarity suggests that some termination and reinitiation occurs during translation of the z^+ gene; and when a polar mutant site is present, the natural reinitiators at the beginnings of the β and ω regions may function to restore protein synthesis.

Ribosome Movement on Polycistronic Messengers

Translation of polycistronic messengers demands initiation, elongation and termination of each protein coded by the mRNA. The kinetics of enzyme appearance upon induction or derepression show that the genes are translated sequentially. That polarity depends upon the distance from a nonsense mutation to the next initiation site demonstrates that the critical events which establish the relationship between translation of successive genes take place between termination of one protein and initiation of the next. One important point to resolve is therefore the fate of ribosomes which start to translate a messenger comprising more than one gene. A second question, prompted by the coincidence of transcription and translation, is whether the presence of ribosomes on the messenger influences its synthesis or degradation.

Two of the interactions of protein synthesis suggest that an individual 70S ribosome does not translate the successive genes of a polycistronic messenger. Release from the messenger appears to be a consequence of termination. By translating an mRNA phage in vitro, Webster and Zinder (1969) showed that nonsense mutations cause ribosomes to dissociate from the message at the mutant site; this argues that nonsense triplets alone are sufficient to cause release of ribosomes as well as chain termination, that is, no longer nucleotide sequence is needed for the release reaction. And since initiation of new protein chains demands the participation of 30S subunits, a 70S ribosome translating one gene must dissociate after termination before it can start synthesis of another protein. Release of the 70S ribosomes and association with an f3 dissociation factor to yield a free 30S subunit must therefore take place between termination and initiation if polycistronic messengers are translated in the same way as the templates which have been utilised in vitro.

Early models to explain why termination of protein synthesis in one gene prevents synthesis of proteins coded by subsequent genes in the messenger relied upon events restricted to the level of translation. One model to explain polarity in terms of aberrant translation was proposed by Martin et al.(1966a,b) and Yanofsky and Ito (1966) as an expansion of the original theory of Ames and Hartman (1963). This postulated that ribosomes can attach to a polycistronic messenger only at the first cistron and not at the subsequent initiation sites at the beginning of each cistron.

According to this model, ribosomes commence translation at the first initiation site but remain attached to the messenger after termination and continue to travel along to synthesize each of the proteins subsequently coded by the messenger. This idea is consistent with the conservation of polarity, which shows that the chain termination event influences all subsequent genes to the same extent; and it was thought to explain the observation that many operons appear to produce equimolar quantities of each gene product. Although in some operons the later genes synthesize less protein than the earlier genes—which could be explained by loss of some ribosomes from the messenger—there never seems to be any increase in protein synthesis along the cluster of structural genes.

Polarity was explained by this model by supposing that a ribosome remains attached to its messenger after termination at a nonsense mutation and slips down in it in a disorientated manner with no defined phase. Proper reading of the message starts again when the ribosome is realigned by the next reinitiation region which it encounters. The gradient of polarity was explained by the supposition that there is a high probability that a ribosome may dissociate from the messenger when in a disorientated phaseless state, somewhat proportional to the distance it must travel. The resumption of protein synthesis might also depend upon the efficiency of the next reinitiation site in correctly aligning the phaseless ribosome.

Although this model can be reconciled with the demand for the 30S subunits for initiation by attributing to the 30S subunit some of the properties formerly thought to reside in 70S ribosomes, it is not consistent with the observation that ribosomes are released from the messenger at termination. Whereas it might be possible to postulate that equimolar translation results from failure of the 30S subunit to be released at termination at the end of a gene—perhaps because of the close proximity of the next initiation site which it might utilise to bind a 50S subunit and start synthesis of the next protein—the concept that 70S ribosomes are released at nonsense codons implies that any synthesis of genes subsequent to a polar nonsense mutation must be undertaken by new ribosomes binding afresh to the next initiation site.

But the polarity of nonsense mutations in early genes of the operon demonstrates that, even though different ribosomes are probably implicated in translating successive genes, each cistron is not translated independently. The levels of translation of the genes of wild type messengers and the polarity of nonsense mutations may be a consequence of the closely related kinetics of transcription and translation. That degradation of messengers rapidly succeeds their synthesis suggests that initiation sites for ribosome binding may be available only during a restricted period of time; this may limit the number of times each gene on the messenger can be translated.

Ribosomes attach to the 5′ region of a messenger almost immediately upon its synthesis, before subsequent genes have been transcribed. The attached ribosomes translate the message by following as a group very closely behind the polymerase; degradation seems to follow the ribosomes very rapidly, removing the initiation site and preventing further translation. When the ribosomes reach the end of the first gene they terminate protein synthesis and dissociate from the messenger. By this time the initiation site for the next gene has been transcribed and can promote attachment of ribosomes and translation; but since degradation continues to proceed along the message, this site too may be available for ribosome attachment for only a limited time. Of course, it is not possible to define in more than outline how this sequence of events establishes the level of translation until the mechanism of degradation is characterized.

That the presence of ribosomes on the messenger influences its synthesis or degradation is suggested by comparisons of the messenger sequences present in wild type and polar mutant cells. In strong polar mutants, the accumulation of sequences representing genes beyond the nonsense mutation is much reduced. The extent of the reduction correlates well with the polarity of the mutation; this implies that it is the deficiency in messenger RNA sequences which causes failure of translation of genes subsequent to the mutation, rather than events solely at the level of translation. Lack of messenger RNA sequences has been demonstrated only for strong polar mutants; it is possible that changes in messenger RNA also take place in weaker polar mutants, although are not detected with present hybridization assays, or that polarity may also result from events solely at translation (in which case it again becomes necessary to postulate some relationship at the level of the ribosome between translation of successive genes).

The first observations that changes in the translation of a messenger can influence its transcription were made by Attardi et al. (1963) and Kiho and Rich (1965) who found that nonsense mutations in the *z* gene of the lactose operon may abolish or decrease the size of lactose messengers in induced bacteria. Polar mutants in the lactose *z* gene contain less lactose specific messenger—assayed by hybridization to saturation with ϕ80d*lac*—and the total amount of remaining lactose messenger correlates with the polarity of the mutants judged by their synthesis of transacetylase (reviewed by Contesse, Crepin and Gros, 1970).

The reduction in lactose messenger content results from a lack of sequences for the part of the operon distal to the polar mutation; the sequences prior to the nonsense codon are present in normal concentration. Sedimentation studies show that the lactose sequences present in the polar mutants are shorter than usual; one mutant which produces only 20% of the usual amount of transacetylase has a peak of RNA which sediments at 8–10S in addition to the usual 30S RNA; and a polar mutant early in the *y* gene has a peak of RNA at 22S, which may correspond to a transcript of the *z* gene alone. Hybridization of these short messengers with the DNA of ϕ80*Ez1*, a phage which contains the DNA of only the *z* gene and a short sequence of *y*, have confirmed that they represent the operator proximal regions of the operon.

Pulse label experiments show that polar mutants of the tryptophan operon produce normal numbers of tryptophan messengers (relative to total RNA synthesis) upon derepression. But as Imamoto, Ito and Yanofsky (1966) and Imamoto and Yanofsky (1967a,b) found, the overall amount of messenger RNA hybridizing with the tryptophan operon is reduced in strong polar mutants. Sucrose density gradient profiles show that the reason for the reduction is that most of the tryptophan messengers of strong polar mutants are smaller than the transcripts of the tryptophan operon in wild type cells. The size of the short tryptophan messengers depends upon the location of the polar nonsense codon. Although the precise point where the short messengers terminate cannot be demonstrated by hybridization—the assays with the deletion strains available are not sufficiently precise—the 3′ end of the messenger appears to correspond to a point in the region of the nonsense mutation. The strong polar mutants possess both full length and short tryptophan messengers; and the relative proportions of normal and short molecules seems to correlate reasonably well with the extent of polarity, although the proportions cannot at present be precisely enough defined to say whether deficiency of distal sequences alone is responsible for polarity or whether events at translation itself must also be implicated.

Coupling of Transcription and Translation

Transcription past Nonsense Mutations

Two models have been proposed to explain the lack of messenger RNA representing the region of the operon beyond the site of a polar nonsense mutation. One is to suppose that cessation of translation and dissociation of ribosomes at the nonsense codon prevents transcription; distal messenger is therefore absent because it is not transcribed. An alternative is to postulate that transcription continues to the end of the operon, but that the regions of mRNA beyond the nonsense mutation are degraded rapidly and are therefore unavailable for translation. Figure 9.8 illustrates these two models, the first proposing that translation is ncessary for proper transcription, the second that transcription is independent of translation but that the stability of the messenger depends upon its translation by ribosomes.

One model to couple transcription with a need for simultaneous translation was proposed by Stent (1964); a ribosome might assist the release of nascent

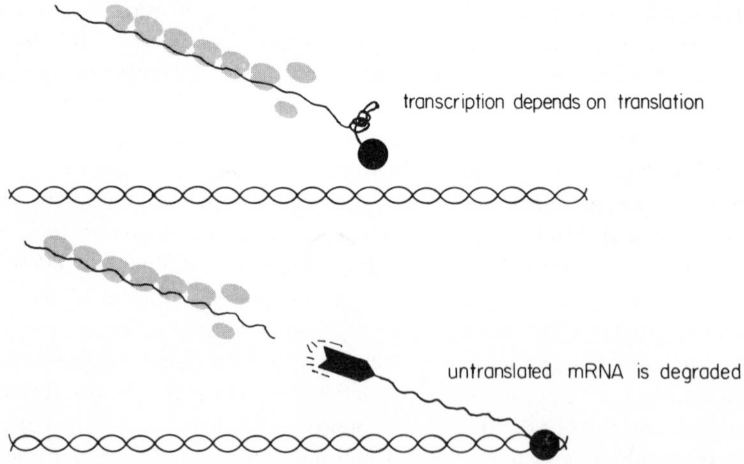

Figure 9.8: models for polarity in bacterial operons. Polarity is caused by the dissociation from the message of ribosomes at sites of nonsense mutation. Upper: this might inhibit transcription, perhaps because the free messenger becomes entangled with the polymerase. Lower: transcription continues, but the free messenger is broken by an endonuclease enzyme (not shown) and then degraded by the exonuclease shown attached to it before it can be translated

messenger RNA from DNA by translating during its transcription to help "pull" it off the genome. Another possibility is that translation might be "forcing" on transcription; by attaching to the messenger immediately behind RNA polymerase, ribosomes might "push" the polymerase along in front of them. In either case, the presence of a nonsense codon would abolish the coupling and the distance from a nonsense mutant to the next site where ribosomes might reinitiate and continue could be important in determining the extent of the inhibition of transcription.

It is unlikely, however, that there is any "mechanical" coupling between transcription and translation, for the process of RNA synthesis is probably the same whether the RNA product is messenger RNA or the stable transfer and ribosomal RNAs, which of course are not translated into protein by ribosomes. If coupling is needed for messenger RNA synthesis but not for the synthesis of stable RNA, then there must be two different sorts of transcription in the bacterial cell. This seems unlikely. It is true, however, that translation and transcription are temporally related in the cell; and the breakdown of this relationship may explain why RNA which is not translated is not stable or why RNA polymerase might be inhibited by untranslated RNA.

That transcription of mRNA in general is not coupled to translation is suggested by the results of Forchhammer and Kjelgaard (1968), who argued that if transcription depends upon simultaneous translation, a decrease in the overall rate of translation must result in a decrease in the overall rate of transcription. The concentration of mRNA in bacteria is determined by the equilibrium between its synthesis and breakdown, so that measurement of the true rate of synthesis requires the use of conditions under which there is no breakdown. This was achieved by using an E.coli strain auxotrophic for uracil; starvation for the pyrimidine causes loss of mRNA, which breaks down rapidly because of its low stability. Messenger synthesis is resumed when uracil is added and the initial rate of synthesis can then be measured independently of mRNA degradation. This rate proved to be the same under conditions of both protein synthesis and amino acid starvation, so that the overall level of mRNA transcription, at least, must be independent of concomitant protein synthesis.

But in experiments using antibiotics to halt translation, Imamoto (1973) observed two different classes of messenger. Some half of the synthesis of messenger RNA was inhibited by chloramphenicol or tetracycline; whereas the remaining messenger fraction was synthesized at the same rate even during inhibition of protein synthesis. Imamoto therefore suggested that synthesis of some messengers (including that of the tryptophan operon) depends upon simultaneous translation, whereas transcription of other messengers is independent of translation. It is not clear why messenger RNA synthesis should be divided in this way by the cell.

If polar nonsense mutants influence the synthesis or degradation of mRNA because ribosomes fail to follow RNA polymerase along the messenger, any treatment which stops ribosome movement or releases ribosomes from the messenger should change its level. Morse (1971) found that addition of chloramphenicol—which halts ribosome movement on the messenger—to cells shortly after derepression of the tryptophan genes prevents accumulation of mRNA representing the distal genes of the operon. This confirms that polarity is a direct consequence of the failure of translation.

That the state of the ribosomes influences the stability of mRNA has been suggested by the results of Varmus, Perlman and Pastan (1971) on transcription of the lactose operon. Addition of chloramphenicol stabilises messengers under translation, presumably because the presence of the ribosomes inhibits access of the enzymes responsible for degradation. And puromycin—which discharges ribosomes from the messenger by prematurely terminating protein synthesis—has the reverse effect and reduces the half life of lactose mRNA from its usual 2.3 minutes to about 0·5 minutes. The stability of messenger RNA thus depends upon whether it is loaded with ribosomes (see also Imamoto, 1973).

The idea that failure of the ribosomes to translate a message beyond a nonsense mutation leads to its degradation predicts that the half life of the distal gene sequences should be reduced. Although the distal parts of the messenger cannot usually be detected in polar mutants, a technique which may allow their assay is to use very short pulse doses of radioactive nucleotide precursors; if the pulse dose is short enough, it may be incorporated into RNA but is not degraded if the RNA is immediately extracted and examined. An increasing proportion of a radioactive label should therefore be recovered from mRNA as the pulse time is reduced.

Estimates of the functional half life of lactose messenger by following the synthesis of enzyme products show that active messengers decay at the same rate in wild type and polar mutant strains. Carter and Newton (1969) measured the rate at which the *z* and *a* genes lose the ability to synthesize β-galactosidase or transacetylase after a short period of induction; when induction ceases, the capacity of the cell to synthesize the enzymes must reflect the activity of the mRNA synthesized previously. Results of such experiments show that the mRNA segments for the *z* and *a* (that is the proximal and distal parts of the operon) have similar half lives in both wild type and polar mutant strains, although the polar mutants produce much less of the transacetylase enzyme.

But although the overall kinetics of synthesis and degradation of tryptophan mRNA—judged by assay of enzyme activities—also appears to be normal in polar mutants, Morse and Yanofsky (1969) observed that many of the messengers suffer degradation of the distal region. What happens in essence is that some messenger RNAs appear to be translated in spite of their nonsense mutation, so that ribosomes must have initiated synthesis further along the message; these messengers retain their usual stability because the ribosomes which they carry protect them against degradation. Most of the messengers of strong polar mutants, however, are degraded beyond the nonsense mutation, but at such a rapid rate that they are never translated into protein, so that their existence is ignored in assays of messenger half life made by following the synthesis of protein product. But experiments using very short pulse doses of radioactive label have detected labile messengers corresponding to the regions beyond nonsense mutations in the lactose as well as the tryptophan operon (reviewed by Contesse, Crepin and Gros, 1970).

One model to explain polarity by degradation is to suppose that an endonuclease cleaves the messenger in a region beyond the nonsense mutation which is left unprotected by the discharge of ribosomes. After this initial attack, an exonuclease might "chase" RNA polymerase along the messenger, degrading the polynucleotide chain as it is synthesized. If ribosomes manage to attach to the initiation site of the next gene when it is transcribed, degradation is inhibited so that the distal regions regain their usual stability; but if degradation follows very closely after the polymerase, the initiation site may be degraded before ribosomes can attach to it, so that all genes located beyond the polar mutation are unable to synthesize protein.

The risk that degradation may take place before reinitiation may depend upon the distance from the nonsense mutation to the next reinitiation site, for the chance that an endonuclease may attack the message is likely to be proportional to length exposed without ribosomes. Another factor contributing to the gradient of polarity might be the ability of an exonuclease to travel along the message faster than polymerase enzyme, so that the chance the nuclease may catch up with the polymerase depends upon the distance from the nonsense mutation to the next reinitiation site. An exonuclease alone, initially attaching to the 5′ end and forced to follow the ribosomes by their slower movement, might therefore account for messenger degradation. That an endonuclease is implicated, however, is suggested by the pseudo-polar effect of chloramphenicol, when the early regions of the messenger may be stabilized but the late segments degraded. But none of the enzymes which may be involved has been identified, in particular any exonuclease able to proceed from 5′ to 3′. Nor do we know whether degradation of messenger in normal circumstances relies upon the same mechanisms which create polarity or whether polar degradation is mediated by enzymes additional to those which usually degrade messenger RNA.

A note of caution in this model has been raised by Hiraga and Yanofsky (1972), who observed that the super-labile mRNA in polar mutants of the tryptophan operon corresponds to the distal part of the gene which is mutated. But they were unable to detect at all any messenger corresponding to the sequences for the subsequent genes. The distal gene sequences might therefore be degraded even more rapidly so that they are undetectable, or their absence may have some other cause, for example a failure in synthesis of the messenger.

Suppression of Polarity

Some tentative support for the combined degradation model is provided by the existence of the *suA* mutation of E.coli, which suppresses the polar effects of nonsense mutations, although it does not suppress the mutation itself to yield an active gene product; *suA* merely restores the amount of synthesis from the subsequent genes. Amber mutations in *suA* have been isolated by Morse and Guertin (1972), who have shown that the suA phenotype is recessive which implies that the wild type allele synthesises some enzyme which is missing from the *suA* mutant. According to Morse and Primakoff (1970), *suA* influences RNA metabolism directly; this idea is supported by the observation that the *suA* allele also reverses the pseudo-polarity caused by chloramphenicol in both the tryptophan and lactose operons. Another effect of *suA* is to alleviate the reduction in the half life of lactose mRNA which is caused by puromycin.

Two possible roles have been suggested for the *suA* gene product. One is to suppose that it is an endonuclease which degrades mRNA unprotected by ribosomes. By inactivating this enzyme, the *suA* mutation would increase the stability of unprotected mRNA and counteract the polar effect of the nonsense mutation. Another possibility is that the *suA* gene codes for some protein factor which links transcription and translation so that the mutation abolishes the need for proper translation if mRNA is to be synthesised. It is more difficult to think of a plausible action for the enzyme in this instance. Extracts from E.-coli cells contain endonuclease activities which have hampered attempts to see whether the *suA* gene might indeed code for an endonuclease; but Kuwano, Schlessinger and Morse (1971) have reported that the degradation of mRNA of phage T4 proceeds more rapidly in the presence of a soluble extract from wild type cells than from *suA* cells.

The idea that the products of more than one gene are involved in determining the degradation of messenger RNA in polar mutants is supported by the isolation of two other suppressors of polarity, *su27* and *su78*. It is possible that *su78* may be a mutation in the same gene as *suA*, for it maps in the same region of the genome; but *su27* maps at a different location. Carter and Newton (1971) isolated these mutants by their ability to suppress polarity in extremely polar mutants of the *z* gene of the lactose operon which carry two nonsense mutations; this ensures that the screening does not yield revertants. Both the suppressors isolated by this procedure lack codon specificity and act on polar mutations created by UGA, UAG, UAA and frameshift mutants. Although the suppressors restore the RNA which is missing in the polar mutants, they do not restore the synthesis of active β-galactosidase, which suggests that they are acting on RNA metabolism and not on translation itself.

The relief of polarity is not restricted to the lactose operon, for the suppressors also increase the expression of *trpA* genes in *trpE* polar mutants. The polar mutant in the lactose operon which Carter and Newton used has about 1% of the normal level of lactose mRNA, but *su78* cells have about 20% whilst *su27* cells have as much as 30–40%. Both suppressors seem to restore the synthesis of the *y* gene mRNA more efficiently than might be expected from the extent of restoration of transacetylase activity, which suggests that polarity may reduce the level of translation directly in addition to its effect on transcription. The suppressors act directly on the stability of messenger RNA, for the half life of lactose mRNA in wild type cells is decreased considerably by the addition of puromycin, but *su78* partially reverses the rapid degradation and *su27* completely reverses it.

An alternative explanation for polarity, however, has been proposed by Imamoto, Kano and Tani (1970), who have reported that the labile mRNA of polar mutants in the tryptophan operon corresponds to the region preceding the mutation; they suggested that the RNA for the following region is not synthesised. By hybridizing the tryptophan mRNA produced after derepression of the tryptophan operon with a series of $\phi80$ phages bearing specific deletions, they found little mRNA corresponding to the region beyond the mutation but were able to detect the region before the mutation by using increasingly short pulse doses. One explanation for the difference in results between this experiment and others which have detected polar-distal mRNA may be that different methods have been used to derepress the tryptophan operon; Morse and Yanofsky have used the tryptophan analogue, 3-indolyl propionic acid, whereas Imamoto et al. have transferred cells from rich medium to one lacking tryptophan.

By blocking protein synthesis through incubation at high temperature of cells with a temperature sensitive mutation in the P10 ribosomal protein needed for initiation, Imamoto and Kano (1971) found that a similar polar result is achieved. Synthesis of tryptophan messenger seems to be inhibited beyond the *trpE* gene by the failure in translation (see also Imamoto, 1973). The model which they suggested to account for these results proposes that the failure of ribosomes to translate mRNA produces a free RNA chain beyond the mutant site; this chain becomes entangled with RNA polymerase and prevents the enzyme from undertaking any further transcription. The *suA* gene must code for some extra enzyme involved in this process.

The different roles ascribed by the two models for polarity to the *suA* and to any other suppressors may prove a means for distinguishing between them. According to either model, the critical event in creating polarity must happen in the interval between the nonsense mutant site and the next reinitiation site where ribosomes may reattach to the messenger. The efficiency of the reinitiation site in binding ribosomes plays some role in setting the level of polarity, which does not depend solely upon the distance between nonsense codon and reinitiation site. Whether degradation of unprotected mRNA or a failure in synthesis is responsible for the lack of messenger segments beyond the mutation, the effect of nonsense mutations upon transcription accounts for polarity in at least the more polar mutants. We do not know whether polarity in weak mutants—in which no change in RNA metabolism is detected—relies upon interference with the transcriptional process at a level too low to be detected, or whether messenger synthesis is normal but translation is defective.

Rho-dependent Termination in Bacterial Operons

The action of rho factor was originally discovered and has since been characterized with templates of phage DNA (see phage 252). Since the protein is present in uninfected bacteria, it is reasonable to suppose that it also acts upon bacterial DNA. Transcription of polycistronic mRNA must terminate at the end of a gene cluster and we may ask whether this response is mediated by rho. Another pertinent question is whether the nucleotide sequences of nonsense codons are implicated in the rho recognition site, for in this case we might expect at least some polar mutations to influence transcription through the mediation of rho factor.

Transcription of λp*gal* DNA in vitro takes place from the galactose promotor, for it is inhibited by mutations at this site, depends upon cyclic AMP and CRP protein and is repressed by the galactose repressor (see page 330). Whether the entire length of the operon is transcribed can be tested by hybridizing the RNA product with the DNA of two phages, λp*gal*8 which contains the sequences of all the three *galKTE* genes and λp*gal* 13 which possesses only the last two genes, *galKT*. The ratio of the RNA hybridizing with the two DNAs is 3:2, the same as the ratio of the number of galactose genes which they carry; and this implies that the complete operon is usually transcribed into RNA.

Addition of rho factor terminates transcription within the bacterial regions of this template. De Crombrugghe et al. (1973) found that the ratio of RNA corresponding to *galKT* compared with *galKTE* drops from 60% to less than 1% when rho is added. This implies that rho halts transcription before the *galKT* genes, which are scarcely transcribed at all under these conditions. Hybridization with two λp*gal* templates derived from deletion strains which contain only the operator-proximal half or third of the *galE* gene shows that rho has no effect on transcription within this region; this shows that its action must take place at a site later in the *galE* gene and most probably at its end.

In the absence of rho, very large molecules of RNA are synthesised which sediment to the bottom of a density gradient. These must be derived by read-through into the genes of λ itself. At low concentrations of rho, the largest RNA molecules sediment at about 22-25S, which corresponds to the size of transcript expected from the complete *galKTE* gene cluster; the smallest RNA molecules in this gradient sediment at about 12-15S, which is about the size of a transcript of *galE* alone. Hybridization experiments have confirmed that the large molecules represent the whole operon and the small molecules only its first gene. This suggests that at low concentrations, rho causes some polymerase molecules to terminate transcription at the end of the *galE* gene whilst others are allowed to read past this site to a second termination sequence located at the end of the operon.

At high concentrations of rho, a symmetrical peak at about 14S is the sole product; all the polymerases must terminate transcription at the end of *galE*. One implication of these results is that polymerases which terminate at the first site do not reinitiate transcription, for there are no RNA molecules corresponding to transcripts of *galKT* alone; this argues against models for rho action which suppose that polymerase can continue transcription after rho has acted (see page 261), although conditions in vivo may differ from those in vitro and so change the response of rho-dependent sites.

Rho factor is active with the lactose operon as well as with the galactose genes. In the absence of rho, the lactose operon is transcribed in vitro into a large range of heterogeneous RNA molecules, presumably entailing read-through past the end of the operon. Low concentrations of rho produce transcripts which appear to correspond to the length of the operon. But high amounts of rho cause an internal termination; the principal RNA product sediments at some 12–14S, which is much smaller than might be expected for a transcript of the *z* gene alone. The length of this RNA corresponds roughly with the distance to the first effective reinitiation site in the *z* gene.

When bacterial cells are induced for the galactose enzymes, they produce messenger RNA molecules corresponding to the complete operon; the polar effects of nonsense mutations in the first gene confirm that these messengers are translated in this polycistronic state. It seems likely that the synthesis of these messengers is terminated at the end of the operon by the action of rho. However, we do not know why this site should be active whereas that after *galE* is presumably inactive. One possible explanation is that the presence of termination sites at the ends of genes within operons may reflect the evolution of gene clusters from independent genes; perhaps these internal termination sequences have diverged from those retained at the ends of operons and are inactive in vivo although they can be activated in vitro. Alternatively, both sites may be used in vivo so that some *galE* and some *galKTE* messengers are produced, their ratio depending upon the relative efficiencies of the two sites. The termination site located within the *z* gene of the lactose operon may reflect the unusual organization of this large gene; if it is utilised in vitro, the *z* gene may in effect be divided into two genes in at least this situation. The rapid degradation of mRNA which takes place in vivo makes it difficult to distinguish models for the action of these sites by characterizing the messengers; the extent of their utilization could be resolved by the isolation of mutants in the rho factor.

Insertions of foreign DNA into the galactose operon cause polarity; the insertion mutants are always extremely polar and their polarity is independent of the location of the insertion within the gene. When one such insertion in the galactose operon, located close to the promotor/operator site, is transcribed in vitro, it produces RNA the same as that transcribed from the normal operon. Addition of rho causes transcription to terminate within the sequence of the insertion; the insertion is very sensitive to rho and reacts at even the lowest concentrations of the factor. In this instance, therefore, polarity is caused by the termination of transcription at a site introduced into the operon by mutation.

Natural polarity is exhibited by some operons, such as the lactose genes which produce decreasing amounts of their successive products; we may speculate that this polarity is caused by rho-dependent termination of some messengers within the operon. Polar ochre mutants in the galactose operon have no effect on transcription, however, in either the absence or presence of rho. This suggests that the polarity caused by nonsense mutations is exercised only in systems undertaking both transcription and translation; it must result from aberrant translation and not from some direct effect of nonsense codon sequences on transcription. Polarity in the expression of bacterial operons can therefore be caused in more than one way.

Reproduction of DNA

CHAPTER 10

Replication of the Bacterial Genome

Movement of the Growing Points

Components of the Replication Apparatus

Definition of the molecular interactions by which DNA is replicated has become possible only recently because of the problems which were encountered in early attempts to characterise the replication apparatus of the bacterial cell. The general mode of DNA synthesis during replication is now reasonably well understood: the production of daughter duplexes is semi-conservative; new chains grow only in the 5′ to 3′ direction; and this growth is discontinuous for new DNA strands are synthesized in segments of short length, each starting with a sequence of RNA which is later removed, the remaining sequences of DNA only subsequently being joined together covalently.

Less is known about the enzyme actions which are responsible for replication; the DNA replicase of E.coli has only recently been been isolated and previously the lack of a purified enzyme system made it difficult to analyse the steps of DNA synthesis in vitro. The DNA replicase itself appears to undertake only the reaction of elongating new segments of DNA; other enzymes are needed to help unwind the DNA, to initiate each new segment of DNA with a sequence of RNA which can be extended by the replicase, to remove the RNA subsequently, and to join together the new lengths of DNA.

Replication within the cell may take place in an enzyme complex containing some or all of these activities. Achieving true replication in vitro may therefore depend upon assembling this complex of enzymes in an aggregate sufficiently like that prevailing in vivo. Within the cell, the complex may be attached to the membrane both at the origin where DNA synthesis is initiated and at the growing points where replication proceeds.

A common feature of replication in all cells is that it starts at an *origin* which is fixed for each replicating unit—or replicon—from which it proceeds in both directions along the duplex. Each bacterial chromosome constitutes one replicon, for there is only one origin, but the much larger chromosomes of eucaryotic cells each comprise many replicons. In bacteria growing rapidly, a second round of replication may commence before the first has been completed; the initiation of new rounds of replication is linked to the growth cycle of the cell (see chapter 13). In eucaryotic cells, synthesis of DNA is restricted to a small part of the cell cycle, but the different replicons of each chromosome are probably each always replicated at a characteristic point during the synthetic phase (see chapter two of volume two).

Direct approaches to elucidate the mechanisms of replication have recently been achieved by the development of systems which can replicate DNA in vitro. Analysis of replication in vitro is complicated by the presence in the bacterial cell of several enzymes which can synthesize DNA; much of this activity, however, is concerned with the repair of DNA damaged by, for example, ultraviolet irradiation and not with replication to produce new copies of the genome. The ability of an enzyme to synthesize DNA in vitro does not imply, therefore, that its function in the cell is concerned with replication; indeed, only a small part of the cellular capacity for DNA synthesis is devoted to replication.

Once the characteristics of replication in vivo have been established (repair is discussed in the next chapter), in vitro systems for synthesizing DNA must be examined for their ability to reproduce DNA in the same manner. But proof that any enzyme is implicated in replication demands the isolation of mutants deficient in its action which cannot undertake replication; and to exclude other enzymes from this role similarly requires characterization of the functions inhibited in cells mutant in their action. Several classes of mutant unable to initiate or elongate DNA chains have been isolated; and several different repair pathways have been defined by mutation. However, some enzymes may be involved in more than one system which acts on DNA.

Semi-Conservative Separation of DNA Strands

The model which is commonly accepted for replication supposes that the two strands comprising the parental duplex unwind and each acts as template for the synthesis of its complementary strand by utilising Watson-Crick hydrogen bonding to assemble the appropriate nucleotides. Each of the two daughter duplexes has an identical content of genetic information, that is the same sequence of base pairs, as the parental molecule. Each daughter duplex consists of one of the original strands of the parent molecule and one strand which has been newly synthesized. Replication is therefore said to be *semi-conservative* since the physico-chemical unit conserved between parent cell and progeny is one of the two single strands comprising the parent duplex.

The prediction that each daughter duplex should consist of one "old" and one "new" strand was first tested by Meselson and Stahl in 1958. When bacteria are grown in in a medium containing the heavy isotope N^{15} their DNA displays an altered buoyant density on CsCl density gradients. After growing bacteria for some time in heavy medium, culture conditions can be changed by substituting the normal (N^{14}) isotope and DNA extracted at intervals during the next few generations. Figure 10.1 shows that after one generation, all the DNA of the progeny bacteria shows a hybrid buoyant density in between that of the heavy N^{15} DNA and the normal N^{14} DNA.

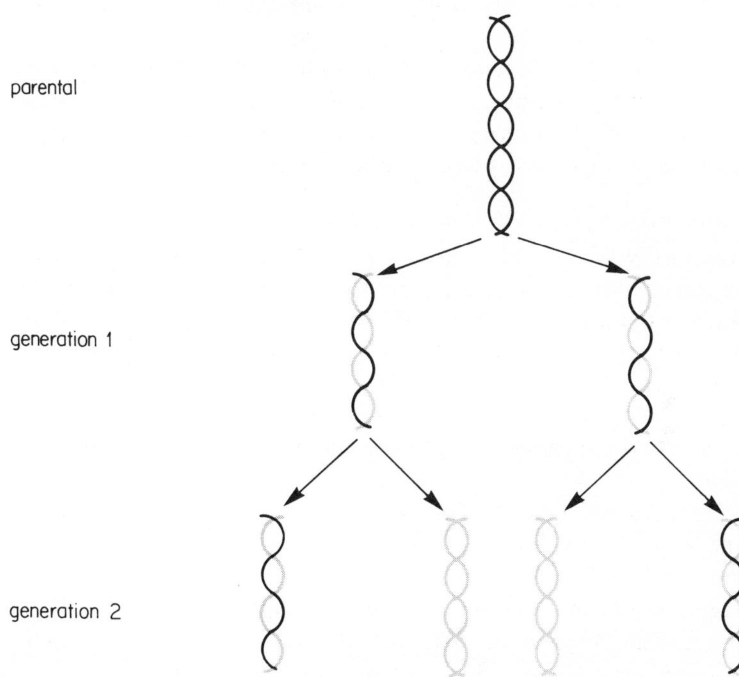

Figure 10.1: semi-conservative replication of DNA. Parental DNA is heavy (dark lines) since bacteria have been grown in medium containing N^{15}. When bacteria are transferred to normal medium, the newly synthesised DNA is of light density (shaded lines). After one generation of growth on normal medium all the DNA has hybrid density; that it consists of one heavy and one light strand has been confirmed by denaturation to give single strands. After the second generation of growth on normal medium, this DNA has replicated again, giving two types of duplex; half are hybrid in density but half are light since both their strands of DNA have been synthesized in normal medium

This intermediate value can best be accounted for by the semi-conservative mode of replication in which each of the progeny DNA duplexes in the first daughter generation has one parental (heavy) strand and one newly synthesised (light) strand. As expected from this pattern of replication, after the next generation, half the DNA duplexes remain hybrid in buoyant density and the other half sediment at the light density corresponding to duplexes with both strands of normal isotope. When replication was followed through several generations, the proportion of hybrid to light duplex DNA always agreed with the predictions of semi-conservative replication.

Denaturation of the hybrid species confirmed that the conserved unit comprises a single polynucleotide strand. Since entirely light chains—duplex molecules consisting of two newly synthesised strands—appear only after all the parental molecules have replicated to the hybrid form, under these growth conditions the second round of replication does not commence until the first has been completed. This suggests that replication is sequential; synthesis commences at a fixed point on the genome—the *origin*—and proceeds until it reaches the *terminus* where replication ceases.

Sequential Movement of the Replicating Fork

If replication proceeds by movement of growing points along the chromosome, and if these growing points always start from the same origin, then any particular site on the genome should always be replicated at the same point in time during the replication cycle. This conclusion has been directly confirmed by Lark, Repko and Hoffman (1963) in an experiment which combined the techniques of radioactive and density labelling. They used a strain of E.coli which is unable to synthesize thymidine and must therefore obtain it from the medium.

After an exponentially growing culture had been labelled with H^3-thymidine for one tenth of a generation, the cells were transferred to a culture containing the analogue of thymine, 5-bromouracil (BUdR). This is incorporated into DNA in place of thymine, but an essential difference is that BUdR is much heavier, so that DNA containing it can be identified by centrifugation through density gradients of CsCl. By removing samples from the culture at various times after its transfer into the BUdR medium and extracting DNA in the form of small fragments, the distribution of radioactivity can be measured in both the normal and the BUdR substituted DNA.

The radioactivity is at first located in a stretch of DNA which has normal weight on both strands—that is light density—but as figure 10.2 shows, when this region is replicated again the strand complementary to the radioactive region becomes labelled with BUdR so that the radioactive label is transferred to fragments of hybrid density. In other words, transfer of the radioactivity from normal to hybrid density demonstrates that the region originally labelled by the radioactive pulse has replicated again. As the figure shows, replication must continue for a complete generation before this transfer occurs—it is sequential. If parts of the chromosome were replicated in a random order, or if a different origin were used for each round of replication, this consistency would not be found.

A similar experiment has shown that the same pattern of sequential replication continues through several generations. Nagata and Meselson (1968) labelled cells with a pulse dose of H^3-thymidine and then grew them on normal medium for several generations before pulse labelling again, this time with BUdR. The radioactive label should be found at hybrid density only if replication of this segment of the genome is proceeding at the moment when the density label is added. As might be expected, the H^3 label associates with the BUdR only when the interval between the addition of the radioactive and density

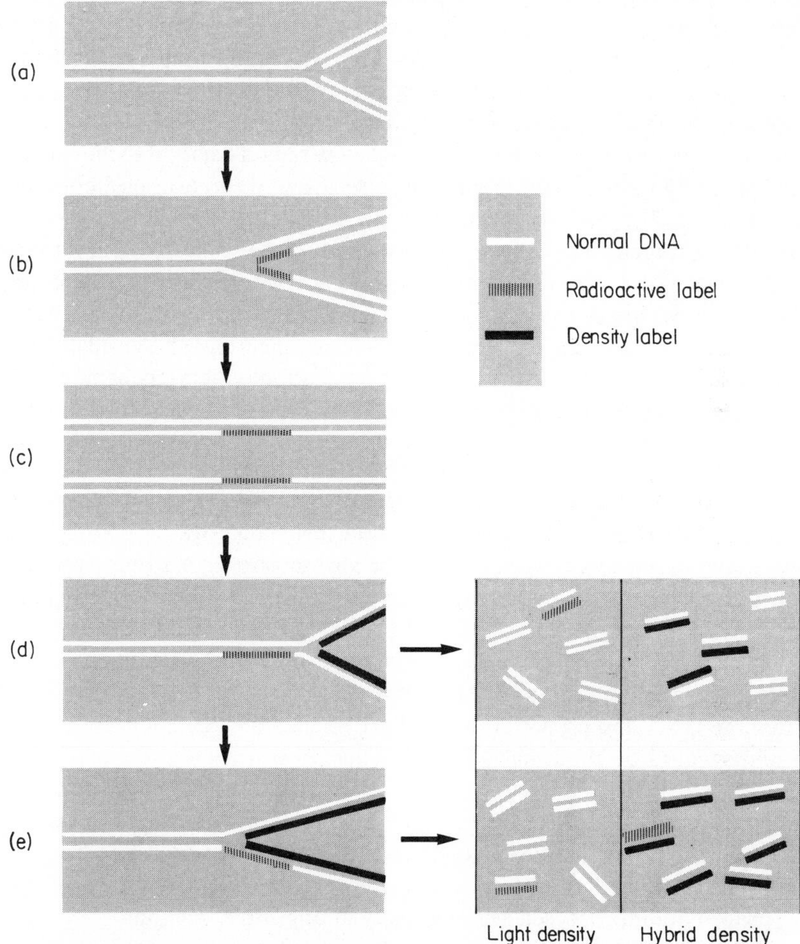

Figure 10.2: sequential replication of DNA. (a) chromosome engaged in replication; (b) incorporation of radioactive label; (c) daughter chromosomes bearing radioactive label separate after completion of synthesis; (d) next round of replication commences in presence of density label; the density label at first partners only unlabelled strands; (e) when replication proceeds as far as the radioactive region, density labels and radioactivity coincide when fragments are analysed on gradients. The radioactive label does not enter the hybrid density peak until one generation time has elapsed between the incorporation of the two labels

labels is an integral multiple of the generation time. That the interval between successive replications of any given segment of the chromosome is always one generation demonstrates that synthesis of DNA must occur sequentially from the same origin in each cycle.

If protein synthesis is inhibited in E.coli, for example by starving an auxotroph for the amino acid which it requires, the current round of replication is completed, but the cell cannot initiate another cycle (see below). When amino acids are restored to the culture, protein synthesis resumes and a new round of replication can be initiated. Lark, Repko and Hoffman (1963) made use of this procedure to obtain synchronous initiation in all the cells of a culture. The segment of the chromosome first replicated—the region adjacent to the origin—was labelled by adding an H^3-thymidine pulse dose at the time when amino acids were restored. The cells were then grown in an unlabelled medium for several generations, after which the starvation procedure was repeated, but with the essential difference that when amino acids were restored BUdR was provided in place of thymidine. The segment of the chromosome previously labelled by the H^3 isotope then appears in the DNA sedimenting at hybrid density, demonstrating directly that the origin for replication remains the same over (at least) several generations.

Bidirectional Replication from a Fixed Origin

Two patterns of replication would be consistent with sequential synthesis from an established origin. One replication fork—or growing point—might start at the origin and continue along the chromosome. In a linear replicon, replication would have to proceed from one end of the DNA to the other; whereas in a circular replicon the origin and terminus would be adjacent.

But there might instead be two growing points, one proceeding in each direction away from the origin. In a linear replicon, replication would start within the DNA and proceed towards the two ends; in a circular replicon, the terminus would be about halfway round the circle from the origin. That the chromosome of E.coli is circular has been established both by genetic mapping and by autoradiography of intact bacterial genomes. To distinguish between undirectional and bidirectional replication we must therefore define the origin and determine whether one replicating fork moves away from it or whether two growing points proceed symmetrically in opposite directions.

Autoradiography of replicating genomes shows that replication is sequential but can be interpreted in terms of either undirectional or bidirectional movement. Cairns (1963) radioactively labelled the DNA of E.coli cells for $1\frac{1}{2}$ generations and then lysed the cells and examined their genomes by autoradiography. This technique displays some one per cent of the chromosomes, as tangled circles engaged in replication. (Circular molecules of DNA are found only comparatively rarely because extraction procedures are likely to break the DNA.) This experiment provided the first demonstration that the chromosome of E.coli can not only be represented by a circular genetic map but also has a circular physical form. Figure 10.3 shows an abbreviated version of the genetic map.

The stage of replication captured by the autoradiography is that of the second round of replication after addition of the radioactive label. The results

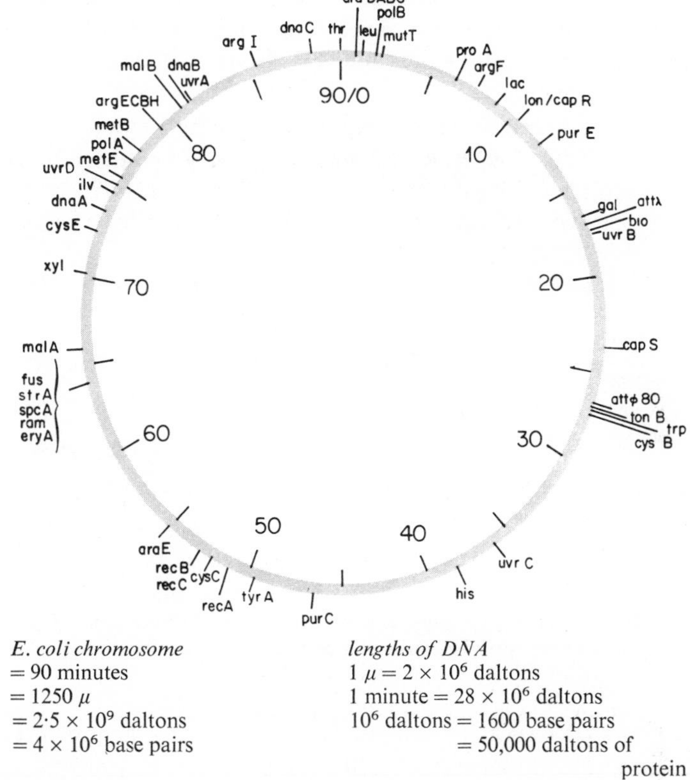

Figure 10.3: abbreviated map of the chromosome of E.coli. About 500 genes have been identified by mutations, but only the positions of some of the genes discussed in this book are marked; a complete map is given by Taylor and Trotter (1972). The numbers give the time in minutes at which each part of the chromosome is transferred by conjugation from an HfrH strain to a recipient; the origin for this transfer is located close to thr and completion of conjugation takes about 90 minutes. The figures given for converting the size of the chromosome measured in minutes of transfer, microns of duplex DNA, daltons or base pairs are approximate. The average size of a gene is often reckoned to be about 10^6 daltons; according to this estimate, the E.coli genome may comprise some 2500 genes

show a single level of radioactivity along the entire length of the chromosome—resulting from the first round of replication—and a double dose along the length replicated during the second round before extraction of the DNA. Figure 10.4 shows the pattern observed; this was originally interpreted in terms of a single replication fork moving in one direction from the end of the labelled segment, but is also consistent with the movement of two replicating forks in opposite directions from the centre of the unlabelled region.

The electron micrographs of replicating polyoma DNA shown in figure 10.5 which Hirt (1969) obtained also clearly show the replication of one circular genome into two by sequential movement of the replicating fork(s). Each strand visible in the photograph represents a duplex of DNA. Once again, however, these results can be interpreted in terms of either unidirectional or bidirectional replication.

Unidirectional and bidirectional replication can be resolved in autoradiographic experiments which utilise the successive use of thymine labelled at low and high specific activity with H^3. Prescott and Kuempel (1972) synchronized cells at the start of a round of replication by starvation for essential amino

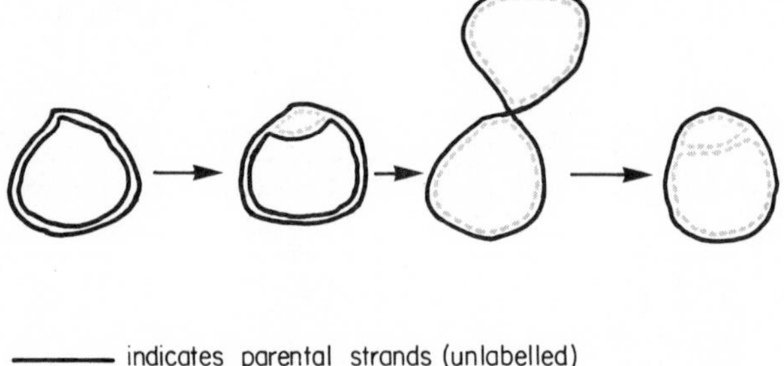

Figure 10.4: sequential replication of a circular chromosome. The last stage, which represents the beginning of the second replication cycle, was identified by the autoradiography of Cairns (1963)

acids and then allowed DNA synthesis to commence in the presence of thymine labelled with tritium at low specific activity. After varying times of incubation, the precursor was replaced by a pulse label of thymine containing high specific activity tritium.

If the cells replicate unidirectionally, they should contain lengths of DNA (corresponding to the time of label with low activity H^3 thymine) poorly labelled, succeeded at one end by a short length of high label (corresponding to the pulse dose of high activity H^3-thymine). Instead, the DNA exhibits poorly labelled regions succeeded at both ends by short lengths of greater label. This corresponds to the symmetrical movement of two growing points away from one origin. Similar experiments have been reported by Rodriguez et al. (1973).

If replication of the chromosome occurs from a fixed origin which is the same in all cells, genes near the origin should be present in a greater frequency than markers near the terminus. In a growing culture, the region near the origin is replicated early in the division cycle so that there are two copies of each gene in this region until the next cell division; regions near the terminus

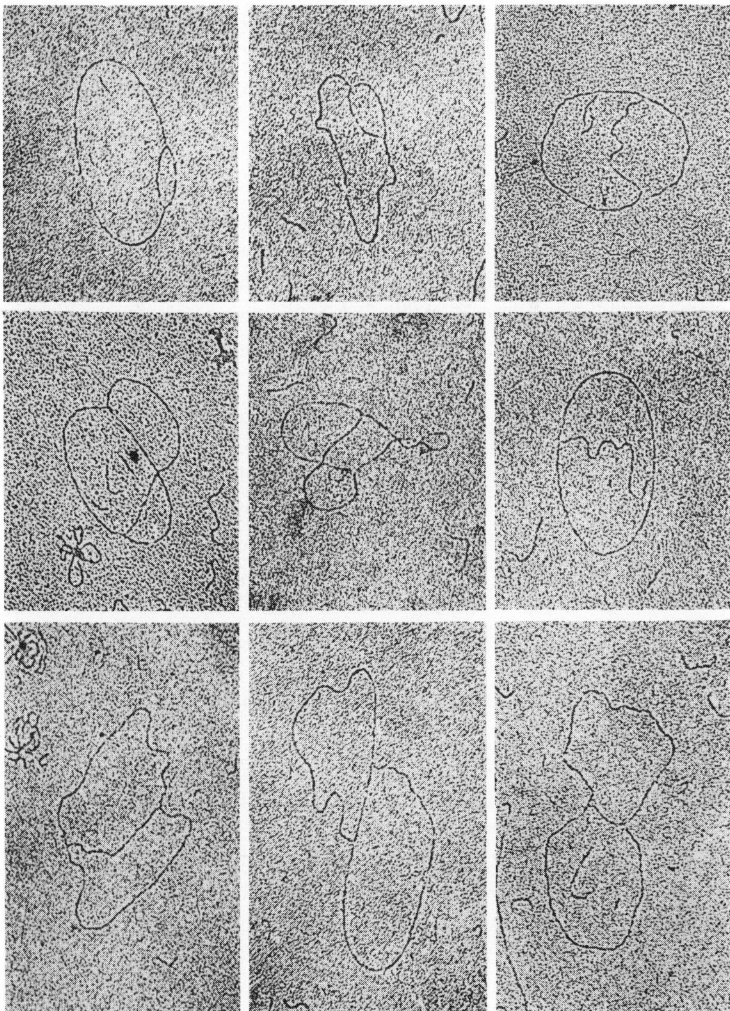

Figure 10.5: electron microscopy of replicating polyoma virus DNA. The growing points move sequentially round the circle in opposite directions until two daughter chromosomes have been synthesized. Each visible strand represents a duplex of DNA. Data of Hirt (1969)

are not replicated until much closer to the time of cell division so that there should usually be only one copy of each gene. In a synchronized culture of bacteria, the frequency of occurrence of each gene should double at a time in the cell cycle related to its distance from the origin. In a culture which is not synchronized, there should be a gradient of frequencies, ranging from two copies of each gene near the origin to one copy of each gene near the terminus.

Three experimental techniques have been used to take advantage of these predictions. Pato and Glaser (1968) subjected synchronized cells to a brief induction period for various enzymes at different times during the cell division cycle. The rate of enzyme synthesis upon induction appears to be related directly to the number of gene copies present in the cell; when a gene is replicated the level of enzyme produced is doubled. The time at which this doubling occurs measures the map position of the gene relative to the origin.

Another approach was used by Cerda-Olmeda et al. (1968), who treated cells with the powerful mutagen nitroso-guanidine; this mutates the replication point of the chromosome with a much higher efficiency than any other region. A large increase in the mutation rate for any particular gene should therefore occur when it is being replicated. Using a synchronized culture, they tested various bacterial functions for a change in activity after mutagenic treatment at various times during the growth cycle.

The most commonly used approach has been to measure gene frequencies themselves, although on a relative rather than absolute basis. Sueoka and Yoshikawa (1963) determined the frequencies of genes located at different sites on the genome of B. subtilis by preparing DNA from growing cultures and testing its ability to transform recipient cells for particular markers. It is not possible to deduce marker frequencies directly from a preparation of transforming DNA because the extent of transformation varies from marker to marker, depending upon differences in their efficiencies of integration. But if a stationary phase culture is used as a control, all its genes are present with equal frequency, so the differences in transforming frequencies observed between a growing and stationary culture can be attributed to differences in the gene frequencies of the replicating cells—the integration efficiency of any particular marker remains the same in both preparations.

The last two approaches have been spanned by the experiments in which Reiter and Ramareddy (1970) starved B. subtilis cells for thymine whilst they are replicating DNA; this causes a loss of DNA for the region immediately behind a growing point, which is degraded. A decrease in the transforming activities of DNA should accompany the breakdown of particular parts of the genome; after starvation, the DNA extracted from cells loses its ability to transfer the gene markers which were being replicated.

Gene frequencies may be determined by transformation of bacterial strains which exhibit this mode of genetic exchange, but can also be determined by transduction with a generalised transducing phage, such as Pl, which has many sites of integration on the E.coli chromosome. Berg and Caro (1967) analysed E.coli cells by this method, again comparing the frequencies of transduction of specific markers between a growing culture and a stationary control. A variation on this theme was used by Caro and Berg (1968) when they labelled a synchronized culture of cells with BUdR at different times during the division cycle. After infecting cells with phage Pl, the resulting lysate was centrifuged to equilibrium on a CsCl density gradient. This isolates the transducing particles possessing hybrid density—that is replicated—DNA; and their relative transducing activities can then be measured for various markers.

The same conclusion was drawn from all these experiments; replication starts at a fixed origin and the growing point moves along the chromosome in one direction only until it reaches a terminus which is adjacent to the origin on a circular chromosome. But in many instances the number of gene functions which could be assayed was quite small; although it is certainly clear from these results that replication proceeds away from a fixed origin, much of the data would also be consistent with a bidirectional replication in which two growing points move away from one origin.

That bidirectional replication is the norm in E.coli was suggested by the results which Masters and Broda (1971) obtained by comparing the frequencies of Pl-mediated transduction for many genes between two cell cultures grown at different rates. Figure 10.6 shows that the ratios of gene frequencies fall into a continuous bidirectional gradient, declining in both directions away from the peak which defines the origin towards a minimum which identifies the terminus. (Since the same results were obtained when only one cell preparation was utilised, without a control, the transduction frequencies may directly reflect the frequencies of occurrence of genes and do not appear to depend to any great extent in differences of the ability of the phage to integrate at particular sites; see also Jonasson, 1973).

Experiments which have made use of phage μ as a marker also suggest that replication is bidirectional. Phage μ has the unusual property that it integrates at random sites on the E.coli chromosome. Host genes may be inactivated when μ is inserted within them; when a gene codes for some metabolic enzyme, this insertion creates an auxotroph which cannot grow unless it is provided with the metabolite product of the enzyme activity. Bird, Louarn and Caro (1972) have isolated a series of auxotrophic strains of E.coli, each of which contains μ inserted at the site coding for the inactivated function.

The number of copies of μ present in each of these strains can be measured by hybridizing their DNA with μ DNA. The amount of λ DNA in the same cells provides a control, for λ is integrated at one site only and should therefore be present in a fixed amount in an exponentially growing culture. But the content of μ DNA should depend on the position at which it is integrated—there should be more copies of μ when the phage has integrated at sites closer to the origin. The results of figure 10.6 show a bidirectional gradient of μ frequencies; the greatest frequency, at *ilv* which is located at 74 minutes, identifies the origin and the smallest frequency, close to *trp*, identifies the terminus diametrically opposite at about 29 minutes. Similar results, but with a more pronounced gradient, were obtained by slowing the rate of DNA synthesis through starving cells for thymine—this increases the number of initiations (and thus gene frequencies at the origin) relative to terminations. The order in which phage μ genomes integrated at different sites are replicated also agrees with these results.

There is a discrepancy in estimates of the location of the origin, for Masters and Broda identified a point between 60 and 65 minutes, whereas Bird, Louarn and Caro found a location at about 74 minutes. The reason for this apparent difference probably reflects the different methods used to measure gene frequencies; as the figure shows, these assays are not precise and a small

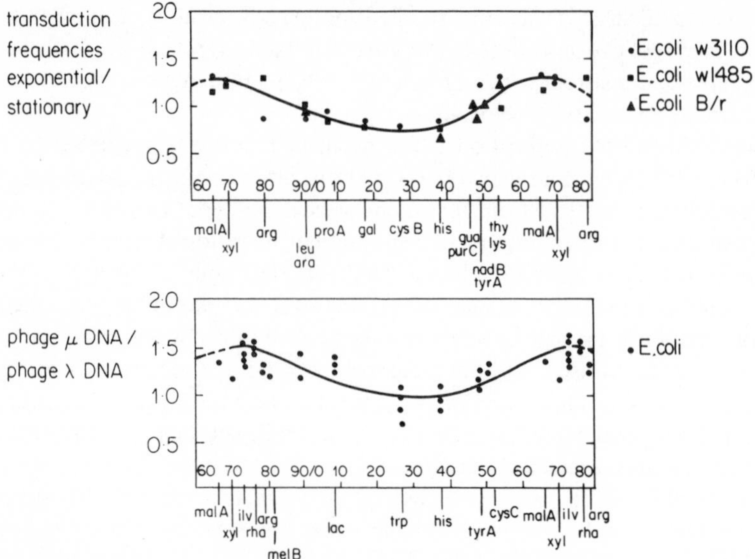

Figure 10.6: bidirectional gradient of replication in E.coli. Gene frequencies are maximal at the origin located around 70 minutes and decline in both directions to a minimum at a terminus located at about 30 minutes. Note that the region of the genetic map from 60–80 minutes is repeated at each end of the figure and is dotted in the repeat of data beyond the origin. The upper curve shows the ratio of transduction frequencies in exponentially growing cells relative to a stationary culture determined by Masters and Broda (1971) for three strains of E.coli. This suggests an origin at about 65 minutes, close to malA and a slightly asymmetrical gradient with a terminus at 28 minutes close to cysB. The lower curve shows the content in an exponential culture of phage μ integrated at different sites, expressed relative to a phage λ control, determined by Bird, Louarn and Caro (1972). This suggests an origin at about 75 minutes, close to ilv, declining symmetrically to a terminus at 29 minutes near trp

error could suggest a displacement of the origin by ten minutes. Both techniques locate the terminus in the same region, at about 29 minutes. If the origin is located at 74 minutes, each growing fork must move at about the same velocity to reach the symmetrically placed terminus. This therefore seems a more probable site than the alternative at about 65 minutes, for in the latter situation one growing point must move more slowly than the other so that they both reach the terminus together. (Because the rate of DNA synthesis remains constant during one round of replication, the two replicating forks must continue moving until they meet each other; if one were to halt at the terminus whilst the other continued to move towards the same point, the rate of DNA synthesis would halve at this point in the growth cycle.) We do not know whether the terminus comprises some special sequence or whether the two growing forks simply continue until they meet each other.

Whether replication is bidirectional in replicons of all species is not known, but the evidence at present available suggests that a common feature of all origins may well be the ability to support initiation by two growing forks which move in opposite directions. Nishioka and Eisenstark (1971) have presented data which may indicate that replication is bidirectional in S. typhimurium, for the origin and terminus appear to be located in different regions of the chromosome. Autoradiography performed by Gyurasits and Wake (1973) suggests that replication is bidirectional in germinating cells of B. subtilis (see also Wake, 1973).

Replication of phage λ, which provides a circular template, and of phage T7, which is a linear DNA molecule, appears to be bidirectional (Kaiser, 1971; Wolfson et al., 1972). When a second round of replication starts before the first has been completed, loops within loops are generated by bidirectionally moving growing points; such structures have been observed by Delius et al. (1971) in electron microscopy of replicating T4 DNA (see figure 13.4). Both bacteria and their phages thus support bidirectional replication.

Less is known about replication in the cells of higher organisms, where each chromosome is divided into many replicons. The lack of genetic markers which can be correlated with biochemical functions means that we cannot tell whether the same origin is always used in each round of replication or, indeed, whether the replicon itself is an invariable structure. But it seems likely that the order of replication of the different replicons remains constant in successive generations (see chapter two of volume two).

Replication of eucaryotic DNA is semi-conservative and in general exhibits the same characteristics as replication of bacterial DNA. The first evidence for the mode of bidirectional replication was in fact derived from a eucaryote and not from a bacterium. Huberman and Riggs (1968) showed by autoradiography of pulse labelled Chinese hamster cells that each chromosome consists of many replication sections, with adjacent sections arranged in opposite orientation. A replication segment has its origin located in the centre of a loop of replicating DNA and replication occurs at fork-like growing points which proceed from the origin of a section in both directions to its two termini. Each replicon thus shares its termini with the termini of the two adjacent replicons. Huberman and Riggs therefore suggested the model for bidirectional replication shown in figure 10.7.

In a subsequent series of experiments using sequential incubation of cells with labelled thymidines of different specific activities, Huberman and Tsai

(1973) confirmed that the symmetrical growing points proceed in opposite directions. They noted that the activities of the two labelled thymidine precursors must be sufficiently distinct since otherwise the two levels of incorporation cannot be clearly distinguished and ambiguous results may suggest unidirectional replication, as in the experiments of Lark et al. (1971).

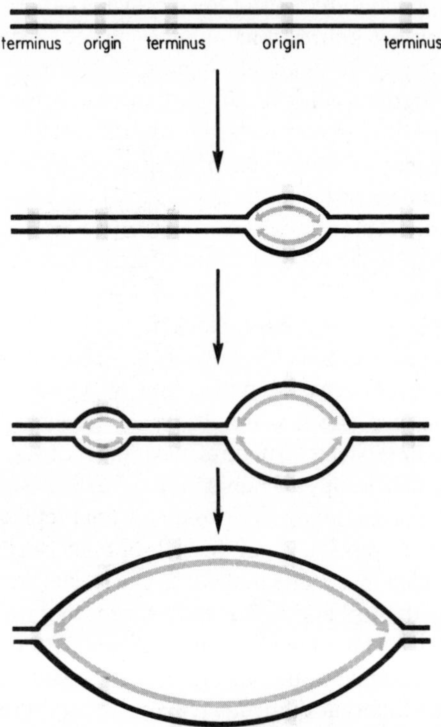

Figure 10.7: replication of mammalian DNA inferred from autoradiography. Replication proceeds bidirectionally from an origin to termini on either side. The two replicons commence their replication at different times. When the rightward moving growing point of the left replicon and the leftward moving growing point of the right replicon meet at their common terminus, they appear as one large loop of replicated DNA. After Huberman and Riggs (1968)

One point of confusion over replication in E.coli has been whether the same origin is used for replication in all cell strains. Most experiments have suggested this conclusion, but Vielmetter, Messer and Schutte (1968) suggested that replication may proceed instead from the point of integration in Hfr strains in which an F factor has been inserted into the bacterial chromosome. Caro and Berg (1968) also found that replication does not start from the usual origin of replication in some Hfr strains of E.coli, although the origin used in these strains does not seem to be the site of integration. These results can probably be reconciled if replication usually starts from the same origin in all strains of E.coli, whether they are F^-, are F^+ with an episome free in the cell, or Hfr with an F factor integrated in the genome. But it is possible that the conditions used in particular experiments with Hfr cells may alter the usual pattern of replication so that some other site—in general one of the possible sites of integration—is used instead.

That such a shift in replication can occur in certain mutants has been shown by Nishimura et al. (1971) by making use of temperature sensitive strains of E.coli which fail to initiate new rounds of replication at high temperature, although they are normal under the usual conditions of growth. The presence of an F factor in these cells does not change the temperature sensitivity of the strain or alter its origin of replication. The F factor is replicated independently of the bacterial chromosome when in its free form, but there is evidently no complementation of replication functions between the episome and the mutated bacterial genome. But the presence of the F factor increases the reversion to temperature resistance of the sensitive mutants.

The revertant bacteria all fall into the Hfr category; the F factor must therefore become integrated into the chromosome before it can overcome the temperature sensitivity of the mutant strain. But the replication system of the F factor is usually repressed in Hfr strains. In these revertants, however, two events occur; one is the insertion of the F factor into the bacterial chromosome and the second is the suppression of the system which usually prevents the replication control of the F factor from functioning. Indeed, insertion of the F factor as such does not confer resistance to temperature. The origin of replication has not yet been determined for these strains, but it seems possible that when the bacterial chromosome is replicated under the control system of the F factor instead of under its usual control, a different origin—probably associated with the F factor itself and therefore with the site of integration—may be used. Events of this nature may explain the occasions on which replication has seemed to take place from origins other than the usual bacterial site.

DNA Replication in the Bacterial Cell

Association of DNA with the Cell Membrane

Although bacteria do not possess a system of the complexity of the mitotic apparatus which ensures the even segregation of chromosomes in higher organisms, some system must be responsible for the distribution of newly replicated chromosomes into the two daughter cells of a division (see page 571). The need for such an apparatus first led to the idea that there may be some association between the bacterial genome and the cell surface, most probably at the membrane. The chromosome of E.coli appears to be attached to the cell membrane both at the origin and at its growing points. DNA synthesising activities are found in the membrane fractions of cells; and mutants which affect the membrane of the cell may inhibit replication. This concentration of activities suggests that a replicating apparatus, perhaps comprising several other enzymes as well as the DNA replicase itself, may be located in the membrane to form an organized structure.

One important point to emphasize is that bacterial DNA is not free in the cell as an extended polynucleotide chain, for simple calculation shows that the DNA of E.coli must be very compact in organisation. The genome of E.coli comprises a DNA duplex of some $2{\cdot}5 \times 10^9$ daltons—about 4×10^6 base pairs. When extended in the double helix form, this DNA would stretch for about 1250μ if strictly linear. A cell of E.coli is some 2μ long and 1 μ wide, so that the extended circular DNA is several hundred times too long to fit into the cell.

Even under the microscope, the bacterial genome appears as a compact body located in one part of the cell—this has sometimes been called the bacterial nuclear or nucleoid body, in analogy with the formal nuclear structure of eucaryotic cells. Although this structure is not confined by membranes as is a eucaryotic nucleus, it has the finely fibrillar appearance also characteristic of eucaryotic nucleoplasm. According to Kellenberger (1960) and Fuhs (1965), these fibrils may be some 20–60Å across and organised into bundles containing up to five hundred parallel fibres—these fibres are strands of duplex DNA, folded back on themselves many times to generate the compact bundle. The shape of the nuclear body is extremely variable, so its structure must be flexible rather than rigid.

That not only DNA is contained in such bodies has been demonstrated by Stonington and Pettijohn (1971), who isolated a complex from E.coli cells which contains the genome; this structure sediments at 3200S, is 80% DNA by weight and contains small amounts of protein and RNA. Most of the protein is RNA polymerase and some of the RNA, at least, comprises nascent molecules of ribosomal RNA still attached to DNA. But treatment with ribonuclease unfolds the structure to a less compact form, as does the protein unfolding agent SDS (sodium dodecyl—or lauryl—sulphate). This suggests that the genome is stabilised in its compact structure by other components of the cell.

This compact body appears to be organized into a definite structure whose size depends upon the stage of the cell cycle. Worcel and Burgi (1972) found that in their preparations the genome sedimented at a range of values from 1300S to 2200S. Cells in which replication is inhibited accumulate smaller bodies of about 1300S, which suggests that the larger structures may represent chromosomes more advanced in the cell cycle, containing more than a single genome of DNA.

When single strands of DNA are obtained by denaturation of the genome, they usually contain about two breaks each (corresponding to about four breaks in the duplex chromosome). However, all the DNA molecules in the preparation of folded genomes behave as though comprising perfect circular duplexes, with no breaks in either strand, when they are treated with ethidium bromide. (This dye intercalates between the base pairs of DNA and creates superhelical turns in circular, but not in linear molecules of DNA.) When breaks are made by treating the folded preparation with DNAase, it is not immediately unwound but sediments more slowly until a minimum unfolding is reached when 6–40 nicks have been introduced. Treatment with ribonuclease unfolds the complex in a different way from DNAase—the effect is all or none; some chromosomes are completely unfolded by ribonuclease whilst others are completely unaltered.

The failure of nicks introduced spontaneously (during isolation) to abolish the response to ethidium bromide and the limited effect of nicks introduced by DNAase suggests that the structures of different parts of the chromosome are independent. Worcel and Burgi therefore suggested a model in which the E.coli chromosome consists of about fifty loops, each of the same superhelical concentration. Rotational events must be unable to propagate from one loop to the next so that nicks can relax the coiled structure only of the loop in which they occur.

Each loop, if comprising an extended DNA duplex, would be much larger (by about twenty times) than the diameter of the nuclear body of the bacterium; the loops themselves must therefore be composed of DNA sequences which are themselves coiled structures. Worcel and Burgi have suggested that the loops are held together by a core which may be an RNA molecule; any distake the form of coiled structures. Worcel and Burgi have suggested that the must unfold the entire chromosome. We do not know whether the construction of the nuclear body is specific and relies upon the interactions of particular sequences of DNA, or whether a more general pattern of coiling along these lines is established as the chromosome is synthesised by replication; the nature of the forces which might establish the compact structure are not known.

The morphological studies of Ryter, Hirota and Jacob (1968) with E.coli and B. subtilis suggested that there is an association between the bacterial nuclear complex and the cell membrane. The isolation of DNA has often produced fractions which are associated with the membrane, but it is usually difficult to exclude the possibility that their association may have taken place during extraction. The demonstration that specific regions of the genome are associated with the membrane supports the idea that association is not merely fortuitous; the use of techniques which isolate membranes and not DNA as such also suggests that the DNA is found in these fractions by virtue of a genuine association with the cell membrane.

When spheroplasts—or cells made fragile by some other method—are mixed with the detergent sarkosyl in the presence of magnesium ions, the cells break in the detergent to form crystals to which the membranes become attached. Tremblay, Daniels and Schaechter (1969) isolated these crystals by sedimentation through a sucrose gradient to give a band which contains virtually all the DNA of the cell, about one third of its RNA—but including two thirds of a pulse label in nascent RNA—proteins and phospholipids. The amount of DNA found in the band can be reduced by shearing before sedimentation; this implies that the genome is attached to the membrane only at specific points. RNA is released from the band together with DNA, which suggests that it comprises nascent molecules under synthesis. Since DNA alone does not interact with the detergent, its presence in this fraction must result from its association with the membrane.

If gently lysed preparations of E.coli or B.subtilis are centrifuged through sucrose density gradients, most of the cellular DNA sediments relatively slowly, but a small fraction sediments much more rapidly to form a pellet. The DNA in this fraction appears to be membrane bound since treatment with ionic detergents converts it to the more slowly sedimenting free form. If H^3-thymidine is added to cultures in exponential growth, it first appears in the DNA of the membrane fractions; the addition of a chase of cold thymidine causes the tritium label to assume the same profile as the bulk of the DNA. This suggests that the membrane fractions include the newly synthesised DNA at the replication fork. Firshein (1972) has obtained similar results with a membrane fraction of Pneumococcus.

This technique was refined by Smith and Hanawalt (1967), who added a "shelf" of concentrated sucrose at the bottom of the density gradient. The membrane bound DNA then sediments onto the shelf and is separated not only from the bulk of the DNA sedimenting on the gradient but also from the cellular debris which pellets at the bottom of the centrifuge tube. The DNA banding at the shelf fraction can then be separated from the other components of the interface fraction and subjected to density centrifugation in CsCl. When a pulse of BUdR is used to label the growing point, the shelf DNA is revealed as a peak banding between the unreplicated parental DNA and the newly replicated hybrid density species. This is the behaviour to be expected of a fork-shaped growing structure in which the forks themselves have been replicated and are therefore of hybrid structure, whereas the "handle" comprises unreplicated parental DNA.

Fairly drastic treatment is needed to free DNA from the shelf fraction, which suggests that it may be complexed with the other components. Digestion with pronase—which is used during the extraction of the fraction from E.coli cells—does not free the DNA from the shelf fraction, but the DNA is sensitive to disaggregation with deoxycholate—which solubilises membrane materials. This suggests that the DNA is associated with lipid. On the other hand, exposure of the B. subtilis complex to pronase releases the DNA from its association with the other shelf components. One explanation for these results is that the shelf complex arises from attachment of the growing point of the DNA, and possibly other points as well, to the cell wall; the different chemical nature of the cell surface in different bacteria therefore means that different treatments are needed to release DNA.

An important disadvantage of both the sarkosyl extraction, which relies on the ionic detergent, and the sucrose sedimentation, which uses pronase in preparation, is that they involve the destruction of enzyme activities. But by releasing the replication fork from E.coli lysates through sonication and addition of a non-ionic detergent, Fuchs and Hanawalt (1970) have been able to prepare a fraction from which it should prove possible to extract enzyme activities. The complex containing pulse labelled DNA sediments at some 100–150S and the amount of pulse DNA in this region varies greatly with the conditions of extraction. The pulse is rapidly chased out of the fraction, for all the H^3 label in a 10 second pulse can be recovered, but when the pulse is extended to 60 seconds only 75% remains in the shelf structure; 25% has already passed into bulk DNA. After a 2 minute chase, all the pulse label shifts to bulk DNA. This suggests that the complex corresponds to the growing point. When cells are labelled with C^{14}-thymine for extended periods before extraction of the complex, some 0·5–1·0% of the label is recovered in the complex; this corresponds to about 20,000 base pairs.

The origin also may be attached to the membrane, for when Sueoka and Quinn (1968) extracted DNA from the membrane fraction of B. subtilis they found it to be enriched in markers close to the origin. The genes conveyed by this DNA upon transformation with recipient cells include markers from all segments of the chromosome, but those on either side of the origin are transferred at the highest frequencies. The origin can be labelled directly by germinating B. subtilis spores in the presence of H^3-thymidine; the first round of replication then takes place synchronously in the cell population so that only the DNA of the origin incorporates a radioactive pulse label. In this situation, the tritium label appears in the shelf fraction and cannot be chased into the bulk of the DNA by allowing the cells to grow in cold thymidine. This suggests that the origin is permanently attached to the membrane, the attachment site extending for some genes on either side. In subsequent experiments, O'Sullivan and Sueoka (1972) demonstrated that in cells undertaking more than one round of replication all the origins are found in the membrane fraction. Ivarie and Pene (1973) suggested that the B. subtilis chromosome may also be attached to the membrane at many other points.

Unwinding the Double Helix

One of the most difficult problems of DNA replication is to explain how the double helix unwinds; this problem is also encountered in transcription of RNA from DNA and in recombination between DNA molecules, although to a lesser extent because only local regions of the molecule need be unwound. But unwinding a duplex molecule of the length of the E.coli genome must require considerable expenditure of energy and may demand a particular topological organisation; such a process may be better undertaken by a replication complex than by a replicase enzyme alone. We might imagine that DNA is unwound by enzyme activities of the replication apparatus in preparation for the DNA replicase to synthesize new strands.

The circular form of the chromosome of E.coli poses an additional question of topology, for a circle cannot simply be unwound as can a linear duplex. The introduction of nicks in the DNA may enable unwinding to proceed, or a "swivel" might exist at some point. An obvious location for the swivel is at the origin of replication, where it might comprise part of the replication apparatus joined to the membrane. A protein which can untwist super-helical coils in lambda DNA has been found in E.coli by Wang (1971); and Champoux and Dulbecco (1972) have shown that nuclei of mouse embryos contain an enzyme activity which can untwist the supercoils of polyoma DNA. Enzymes with this kind of capacity might constitute a swivel to untwist the duplex as it enters the replication apparatus.

One of the proteins which may be involved in a replication complex and appears to have the role of promoting dissociation of the double helix into single strands has been prepared from E.coli cells infected with phage T4. Its isolation depends on a technique developed by Alberts et al. (1968) in which crude cell extracts are passed through a column consisting of DNA—either native or denatured—absorbed onto an inert matrix of cellulose. These DNA-cellulose columns retain only proteins with a high affinity for DNA at concentrations of NaCl above 0·05M. Indeed, under these conditions most of the DNA-binding proteins carry a net negative charge—so their recognition of DNA must be specific and not due to ionic neutralization.

The bound proteins can be eluted from the DNA-cellulose columns by increasing ionic strength, however, which implies that electrostatic interactions are important in their binding to DNA. The cell extracts must be treated with DNAase before chromatography to remove endogenous DNA which otherwise binds to the proteins; and when E.coli divides to give mini-cells—which lack DNA—the binding proteins are not found in the abnormal products of division. This suggests that binding of the proteins to DNA in vitro reflects their functions in vivo. DNA polymerase I and RNA polymerase are amongst the proteins which can be isolated from E.coli cells.

Using cells infected with phage T4 has the advantage that a greater proportion of proteins binds to DNA, as might be predicted from the large number of phage functions which are concerned with DNA itself. Alberts (1970) found that when infected cells are labelled with radioactive amino acids during the early period of infection—between 5–20 minutes at 25°C—the proteins which bind to the columns correspond to the early functions concerned with the metabolism of DNA. The proteins labelled during a later period—35–45 minutes—include the structural proteins of the phage itself. About twenty such proteins in all are found in cells infected with T4. An advantage of using phage infected cells is that the absence of particular proteins can be correlated with mutations in phage functions. Comparison of the elution profiles of the proteins extracted from cells infected with wild type phages and those infected with T4 DNA bearing a mutation in gene 32 has identified the 32-protein.

The product of gene 32 is required for both replication and recombination of the phage DNA; mutants in this locus are defective in both functions. Gene dosage experiments suggest that the 32-protein is required stoichiometrically rather than catalytically; for when Sinha and Snustad (1970) controlled the number of functional copies of gene 32 by infecting cells with unequal proportions of wild type and mutant phages, they found that the burst size—the number of phages produced by each infected cell—decreases rapidly as the proportion of mutant genes is increased. Alberts and Frey (1970) found that about 10,000 molecules of 32-protein are made in an infected cell; the extent of synthesis of this protein seems to control directly the number of progeny phages produced.

The affinity of 32-protein for columns of DNA-cellulose depends upon its concentration. At low levels, it can be eluted by 0.6 M NaCl, but when a greater amount of the protein is present it does not elute until 2·0 M NaCl. A similar change in the elution pattern is caused by the substitution of denatured DNA for native DNA in the column; the protein is eluted from duplex DNA at 0·6 M NaCl, but does not elute from single strands until 2.0 M NaCl. This suggests that individual molecules of 32-protein bind in a cooperative manner to single strands of DNA; when one molecule binds to a site on DNA it enables further molecules to bind to adjacent sites much more readily. The affinity of the protein for sites adjacent to those already occupied is some eighty times greater than its affinity for new sites.

The 32-protein comprises a polypeptide chain of some 35,000 daltons. It binds single strand DNA in large amounts which correspond to one molecule of protein for every 10 nucleotides. Because the protein appears to have an asymmetrical shape with one axis much longer than the other, this implies that adjacent molecules bound to DNA may overlap. The saturated complex of protein and DNA sediments only slightly more rapidly than free DNA itself, in spite of its much greater mass. This suggests that the 32-protein extends the normally folded form of DNA.

Monomers of 32-protein associate readily to give dimers, which may in turn associate to form larger aggregates containing up to at least 10 molecules. Carroll, Neet and Goldthwait (1972) have characterized these aggregates by sedimentation through sucrose gradients. Aggregation proceeds most readily at a pH of about 8 and an ionic strength of about 0·1. High salt allows the dimers to form, but prevents their association into larger aggregates. This suggests that the forces of attraction between the monomeric subunits of each dimer are more stable than those between dimers. One model for the action of 32-protein is shown in figure 10.8, in which dimers interact with single strands of

DNA to maintain the two strands in their unwound state. Binding of the protein at one site on DNA increases the probability that other proteins will bind at this site, adjacent molecules overlapping with each other and maintaining the single strands of DNA in an extended state.

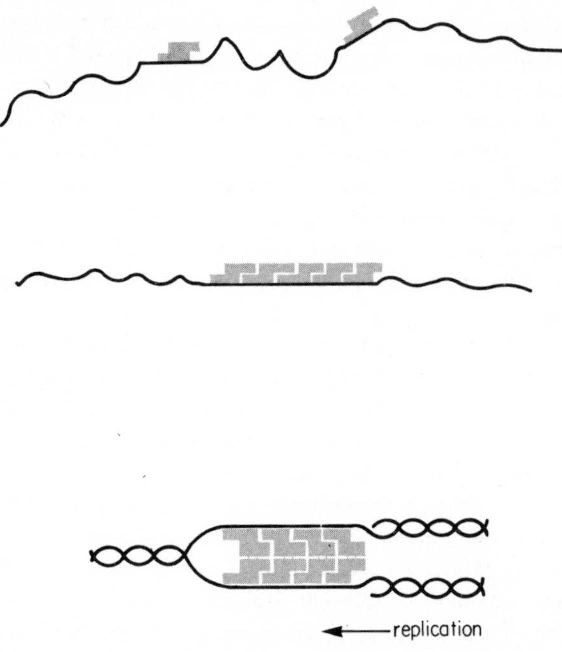

Figure 10.8: interaction of 32-protein with DNA. Upper: at low concentrations monomers of the protein bind to isolated sites on DNA and have little effect upon the structure of the single strands. Centre: at high concentrations protein molecules bind in a cooperative manner at adjacent sites to maintain the DNA as an extended single strand. Adjacent protein molecules overlap each other. Lower: a model for the action of 32-protein to hold the strands of duplex DNA apart at a replication complex. This model allows monomers to interact to form dimers; the postulated details of the interaction are speculative and it might instead take place along the axis of the DNA. In addition, dimers interact with each other to form larger aggregates, the individual protein molecules overlapping each other along the DNA. Data of Alberts (1970) and Carroll, Neet and Goldthwaite (1972)

The biochemical view of the activity of 32-protein has been confirmed by electron microscopy of complexes of the protein bound to the single strands of fd DNA. Delius, Mantell and Alberts (1972) found that the protein forms a linear structure with DNA in which the nucleic acid shows a uniform diameter of 50–70Å. The 10 nucleotides which each monomer covers extend for 46Å, a distance much shorter than the long axis of the protein subunits. The 32-

protein by itself can denature poly-dAT but is unable to separate the strands of poly-GC under physiological conditions.

What is the role of 32-protein in vivo? Its different activities with AT-rich and GC-rich polynucleotides suggests that its action in recombination might be to denature susceptible sequences of T4 DNA and thus to create regions of local unwinding which would be necessary to start recombination. Its failure to denature T4 DNA under physiological conditions in vitro suggests that in replication—when the duplex must be completely unwound—its role may be to maintain the single stranded state of DNA strands which have previously been unwound. This implies that some other protein may be needed to act, either independently or in association with 32-protein, to unwind the duplex, after which the 32-protein acts to stabilise the unwound state.

The DNA polymerase specified by T4 functions efficiently with single stranded templates of DNA to synthesize a complementary strand in a repair like capacity similar to that of DNA polymerase I of E.coli (see page 516). DNA degraded by exonuclease III of E.coli is therefore a suitable template for the T4 polymerase; the nuclease removes nucleotides sequentially from both ends of duplex DNA in the 3′ to 5′ direction to generate a short duplex sequence from which long single strands project in either direction (see figure 11.1). Each DNA chain may then be extended by T4 polymerase, using the other strand as template, and the extent of synthesis is equal to the extent of degradation. Using lambda DNA degraded by exonuclease III as a primer, Huberman, Kornberg and Alberts (1971) found that 32-protein stimulates the action of T4 DNA polymerase. The DNA which is synthesised appears to have the same characteristics both in the absence and presence of 32-protein; this implies that the function of the protein may be to increase the rate of synthesis from each active growing point.

This stimulation probably results from removal of duplex regions in the template DNA ahead of the growing point. The synthetic activity of DNA polymerase itself may therefore be enhanced by a distinct "unwindase" activity which prepares the DNA for replication. The 32-protein must comprise at least part, if not the whole, of this activity. According to this view, we may imagine that the replication apparatus contains a certain number of molecules of 32 protein, organised in a specific structure which assists the dissociation of the two strands of duplex DNA by stabilising them in the single strand state. Some other component of the replication complex may be needed to assist the initial separation of the two strands. Gene 5 of the small phage fd codes for a protein which has properties similar to those of 32-protein of phage T4; 5-protein is smaller, with a molecular weight of about 10,000 daltons, and is synthesized in even larger amounts than 32-protein, at about 120,000–150,000 copies in each infected cell. Oey and Knippers (1972) and Alberts et al. (1972) found that 5-protein binds to single stranded DNA of any source but not to duplex DNA. The stoichiometry of binding suggests that one molecule of protein binds to about 4·6 nucleotides of DNA. The affinity of the protein for DNA increases with concentration, so its binding also is cooperative. The catalytic role of this protein in ensuring production of single strands of fd DNA from the duplex replicating intermediate is not entirely clear, but one possible role may be for it to help remove single strands as they are made on a rolling circle type of structure (see below).

If replication takes place by similar processes for all chromosomes, counterparts to 32-protein and 5-protein should exist in uninfected E.coli cells and, indeed, perhaps in eucaryotic cells also. By characterizing the E.coli proteins binding to DNA-cellulose, Sigal et al. (1972) isolated a protein with properties closely analogous to the 32-protein. The E.coli unwinding protein has a weight of some 22,000 daltons, can form oligomers in solution or upon binding to DNA; and at saturation the mass ratio of protein to DNA is about eight, corresponding to one protein for every 8 bases. Electron microscopy shows that the E.coli protein has a spacing when bound to DNA somewhat different from that of 32-protein, which implies that each unwinding protein may form a complex with DNA of characteristic topology. The E.coli unwinding protein stimulates the activity of E.coli polymerase II but not the other two DNA polymerases or T4 polymerase; the topology of its complex with DNA may therefore be specific for the activity of only certain enzymes. There are about 800 molecules of E.coli unwinding protein in a bacterial cell containing 6 growing points, about the same number of molecules per fork as that of 32-protein in T4-infected cells.

A DNA binding protein analogous to the 32-protein, 5-protein and E.coli proteins has been found in meiotic cells of Lilium, but seems to be implicated only in recombination since it is not found in somatic tissues (see page 549). Definition of the role played by putative unwindase activities must depend upon the isolation of replication complexes in which they can act in their proper structural milieu. And it will of course be necessary to isolate mutants in E.coli which lack or have a defective unwinding protein in order to decide whether it is part of the replication complex, as seems likely at present.

In cells with a circular chromosome, at least, the replication apparatus may demand the participation of two types of activity concerned with ensuring that DNA reaches the replicase in a state suitable for replication. The first activity must be concerned with providing some form of swivel mechanism to allow the DNA to proceed through the complex as a linear duplex. The second activity, comprising the unwindase enzyme(s), must then separate the strands of the duplex and maintain them in their unwound state so that the replicase can utilise them as templates for synthesis of their complements.

A model which provides an alternative to the need for a swivel mechanism has been proposed by Gilbert and Dressler (1968). This demands a circular DNA template and ensures that all its genetic information is preserved by always copying more than one genome of information from the circle. Figure

10.9 shows that the rolling circle model postulates that synthesis begins by opening one strand of the original circular duplex—in phage systems the positive strand—and attaching the newly exposed 5′ phosphate end to the membrane. Chain elongation then begins at the free 3′-hydroxyl end, the closed

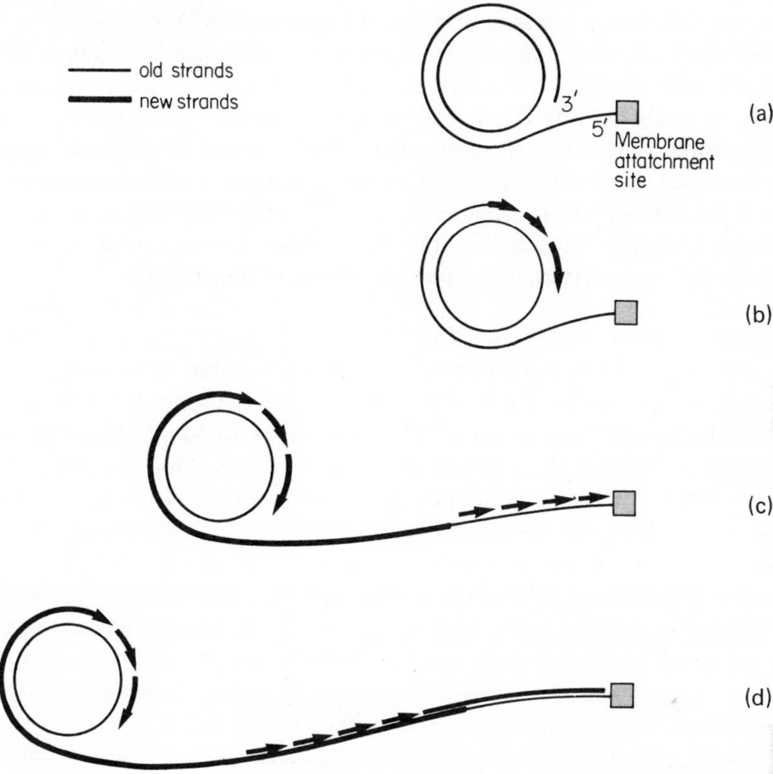

Figure 10.9: the rolling circle model for replication. (a) the resting state with the 5′ end of one strand attached to a membrane site and its 3′ end open; the other strand is a closed circle. (b) Synthesis starts at the free 3′ end using the circular strand as template. Growth of the new strand displaces the old tail strand. (c) Synthesis of the strand around the circle continues and a complement is synthesized on the now exposed tail. (d) Synthesis continues and there are now several genomes' worth of information in the tail. After Gilbert and Dressler (1968)

circular strand acting as template. As synthesis proceeds, it may displace the strand previously complementary to the closed circle, so that a long complement "peels off" ahead of the growing point. The "tail" attached to the membrane may therefore contain several genomes of information in sequence.

Because the rolling circle model demands that one strand remains circular but the other is linear, it predicts that only one of the two origins should be capable of reinitiation when a second round of replication commences before the first has been completed. That bacterial chromosomes instead show symmetrical initiation of replication at all available chromosome origins has been demonstrated in germinating cells of B. subtilis by Quinn and Sueoka (1970) and in E.coli by Caro (1970) and Fritsch and Worcel (1971). Another prediction of the rolling circle model, that one of the new strands is extended from the end of the old strand around the circular template, has been tested by Stein and Hanawalt (1972), who demonstrated that both new strands are unlinked to previously existing DNA. Taken together with the demonstrations of bidirectional replication, these results imply that bacterial chromosomes must unwind via a swivel mechanism which allows growing points to proceed from it in both directions. It remains possible, however, that certain small phages may utilise the rolling circle as a means to produce many copies of one strand for inclusion in mature phage particles (see Dressler, 1970).

Accuracy of Replication of DNA

Although the structure of the double helix is maintained by complementary base pairing between the two strands of DNA, the specificity of hydrogen bond formation does not seem to be sufficient to account for the accuracy of replication. Rather does correct base pairing depend upon the environment provided by the replication apparatus and by the demands of the replicase itself. Of course, correct base recognition must also be maintained when damaged DNA is repaired, so we may expect similar mechanisms for ensuring accurate replication to be characteristic of all enzymes which synthesize DNA. The fidelity of replication is very high, for errors in the selection of bases complementary to the template appear to occur less frequently than 1 in 10^6 times. In other words, the chromosome of E.coli can be replicated virtually without error. Although the template directs base selection, DNA polymerases must play an active role in recognising the correct A-T and G-C pairs and exluding others; hydrogen bonded base pairing, although necessary, is not itself precise enough to account for this degree of accuracy.

The accuracy of replication is under genetic control, for temperature sensitive mutants of the DNA polymerase of phage T4 may change the fidelity of the enzyme very considerably. When Speyer (1965) compared the fidelity of replication of these mutants at both normal and increased temperatures by examining the types of plaque formed—this provides a measure of the extent of faulty replication in the *r* cistrons—he found that the accuracy of replication in the mutants is lowered by an increase in temperature. Hall and Lehman (1968) tested this conclusion in vitro by purifying the polymerase from bacteria infected with either normal T4 or one of the temperature sensitive mutants. When the enzyme is provided with poly-dC as template and both dGTP and dTTP as substrate inaccurate replication can be measured by the ratio of misincorporation of dTTP to proper pairing of dGTP. With the wild type enzyme, the ratio is $2.4 \times 10,^{6}$ at either temperature; but with the mutant the misincorporation is increased about four fold to 8.3×10^{-6} at the higher temperature. Replacing the magnesium ions needed for enzyme action with manganese ions in the incubation mixture causes a five to twenty fold increase in ambiguity with either wild type or mutant enzyme, indicating that factors extrinsic to base pairing itself can influence fidelity.

Although many temperature sensitive mutants in its polymerase increase the rate of mutations in phage T4, some alleles shown an anti-mutator activity and reduce the misincorporation of bases during DNA synthesis. Drake et al. (1969) found that at least eleven of twenty one mutant sites in the polymerase gene cause an increase in the mutation rate; but the test used to detect mutation relied upon an examination of plaques formed by *rII* mutants—that is the ability to revert to wild type—so that it is limited by the number of *rII* mutants available to assay. It seems probable, therefore, that mutator activity is actually shown by rather more sites, possibly up to 80%. At least two of the mutants have the reverse effect and act as anti-mutators.

Both mutator and anti-mutator alleles of the polymerase show specificity in their action. Some mutants exhibit activity in reversion tests but not in forward mutation; this suggests that their action is specific for particular mutational pathways which are not often represented amongst forward mutations. Two anti-mutator alleles strongly suppress mutagenesis by the base analogues 2-aminopurine and BUdR, moderately suppress the mutagenic effect of the alkylating agent ethane methylsulfate, but have no influence upon the action of hydroxylamine. Another test of specificity can be provided by comparing the ability of anti-mutator or mutator alleles to act on some particular phage mutant with the influence of chemical mutagens on the same site. Such assays suggest that the two anti-mutator alleles of the polymerase tend to suppress transitions in DNA which replace an A-T base pair with a G-C pair.

Antimutagenic DNA polymerase may therefore achieve an enhanced ability to discriminate against incorrect base pairing in some instances at least. Indeed, the specificity of these effects implies that the differences observed in response to mutagenic treatments of different organisms may in part represent the different characteristics of their DNA polymerases. But the discovery of anti-mutator alleles of T4 DNA polymerase raises the question: why have such improved polymerases not evolved during evolution? The answer must presumably be that an increase in the accuracy of replication carries a selective disadvantage of some kind, perhaps because it reduces the ability of the organism to induce mutations which may confer a favorable response to environmental changes.

Two contrasting models have been proposed to account for the active role which a DNA polymerase must play in base selection. One suggests that the polymerase itself is responsible for choosing the incoming base; the template would influence the enzyme in a allosteric manner to ensure that only the correct bases are presented to it. But DNA polymerase I of E.coli—which does not replicate the genome for cell division but nonetheless provides an important part of the DNA synthesising capacity of the cell (see page 459)—has only one site at which it can accept triphosphates; if the enzyme must select the incoming base, this site must be able to exist in any one of four specific conformations. In this instance at least, therefore, the enzyme probably fails to preselect base pairs.

An alternative model proposes that the base enters the binding site and the polymerase only then judges its accuracy of fit with the template; Kornberg (1969) has suggested that synthesis of a phosphodiester bond follows only if the base pair is an A-T or G-C combination, perhaps because the enzyme responds to the presence of a correct base pair in the active site by changing its conformation to allow the subsequent catalytic step to proceed. This model demands only that the enzyme can distinguish A-T and G-C pairs from other combinations; both Watson-Crick base pairs contain regions of identical dimensions and geometry which the enzyme could use to establish this discrimination.

Another mechanism may also operate within the enzyme to ensure specificity of DNA synthesis, for DNA polymerase I has a 3′ to 5′ exonuclease activity which can excise mismatched bases from a DNA chain. With a template such as $dT_{260}:dA_{4000}$, the polymerase extends the short primer segment by using the longer strand as template for synthesis of its complement. However, Brutlag and Kornberg (1972) obtained a different result when they prepared mispaired templates for the enzyme by using a terminal transferase to add single nucleotides to the 3′ end of the short primer segment. With polymers such as $dT_{200}H^3\text{-}dC_1:dA_{4000}$, E.coli DNA polymerase I or T4 polymerase removes the mismatched base before extending the chain. All types of mispaired termini—purine-purine, pyrimidine-pyrimidine, purine-pyrimidine—can be removed. Hershfield and Nossal (1972) have come to a similar conclusion on the basis of different assays. Hershfield (1973) observed that one mutator polymerase of T4 is deficient only in base pairing and retains its ability to excise mismatched bases in vitro.

These results suggests that two mechanisms may be utilised by these polymerases to control the accuracy of DNA synthesis (and either might be mutated to change the enzyme activity.) As figure 10.10 shows, first only the correct base is usually allowed to pair; but when a mistake is made the mispaired base is excised by the exonuclease activity of the enzyme. We do not know whether other DNA polymerases, including the replicase of E.coli, utilise similar mechanisms, but this is possible since all the polymerases of E.coli possess 3′ to 5′ exonuclease activities in addition to their abilities in synthesising DNA (see page 469).

The specificity of replication is under genetic control in E.coli—and in

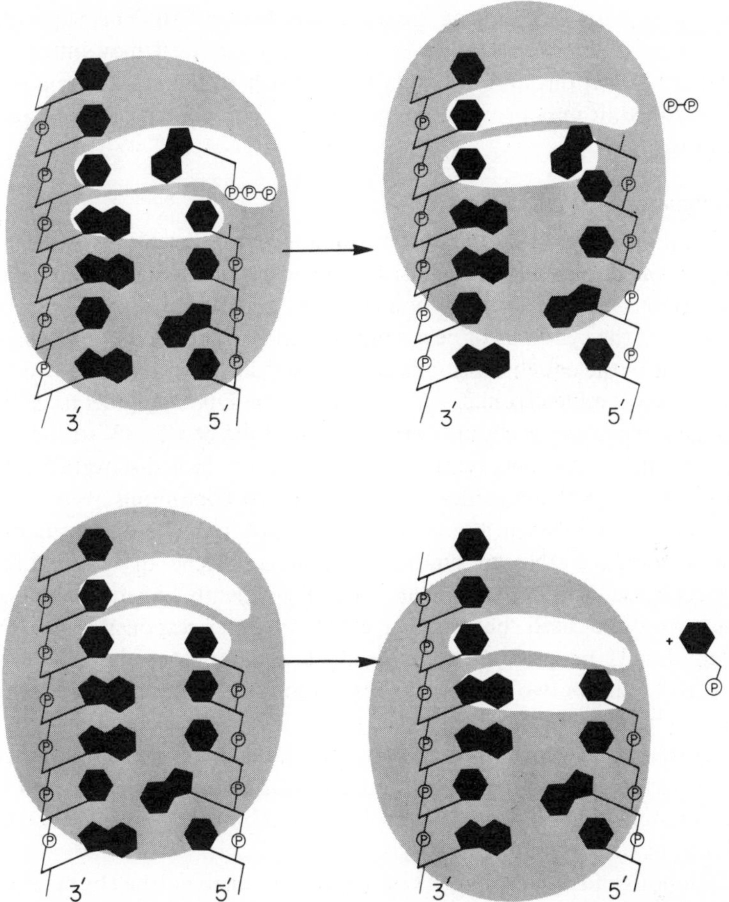

Figure 10.10: model for control of accuracy of DNA synthesis by DNA polymerase I of E.coli. Upper: the triphosphate site usually recognises only A-T and G–C base pairs, presumably by some stereochemical criteria. When a correct fit has been made, a phosphodiester bond is synthesised and the enzyme moves on to accept the next triphosphate. Lower: if a mistake is made, the primer terminus site becomes distorted and in turn prevents the triphosphate binding site from adding further bases to the polynucleotide chain. The $3' \rightarrow 5'$ exonuclease activity of the enzyme then removes the mispaired base as a monophosphate. The chain may then be elongated in the usual way

other organisms also—for loci have been identified at which mutation changes the accuracy of reproduction of DNA. At least three such mutator genes exist in E.coli and Cox (1970) showed that *mutT1*—the Treffers mutator gene—acts only during replication. This allele increases the mutation rate of E.coli at least one hundred fold by stimulating the replacement of A-T base pairs by G-C base pairs (see Gibson, Scheppe and Cox, 1970). This supports the idea that cell components other than the replicase itself may influence the fidelity of replication; and although the product of the *mutT1* gene has not been identified, it may prove to comprise some component of the replication apparatus.

Discontinuous Synthesis of DNA

Replication of the bacterial chromosome occurs sequentially so that both strands of the duplex template must be copied simultaneously from initiation at the origin until the growing points meet at the terminus. At first, it was assumed that the synthesis of new strands must be continuous; this would imply that one of the daughter strands is synthesized in the direction 5′→3′ whilst the other is assembled from 3′→5′. But whilst the DNA polymerase activities identified in vitro can account for the growth of the new 5′→3′ strand, no enzymic action for catalysing synthesis from 3′→5′ has been discovered so that it is impossible to explain synthesis of this strand by continuous assembly.

Replication of DNA in the bacterial cell appears to be discontinuous, so that new strands of DNA are synthesised in short stretches in a 5′→3′ direction by the replicase. The short segments are not covalently linked to each other as they are synthesised, but another enzyme is later responsible for forging covalent bonds between them to produce a continuous daughter strand. Figure 10.11 shows two models for discontinuous synthesis, one in which it applies to both strands and the other which invokes discontinuous synthesis only for the strand which must grow overall in the direction 3′→5′ and allows the 5′→3′ synthesis to proceed continuously. Opinion has been divided about these two models, but it now seems probable that both new strands of DNA are produced by discontinuous synthesis.

Continuous and discontinuous synthesis can be distinguished by determining the structure of the most recently replicated part of the chromosomes—this is the segment selectively labelled by an extremely short pulse dose of radioactivity. If synthesis is discontinuous, the label should be found in unconnected short chains, whereas if it is continuous the label should be present in the bulk of the DNA which must sediment much more rapidly. Okazaki et al. (1968) tested these predictions by using a technique in which cells synthesising either bacterial or phage DNA are exposed to H^3-thymidine for a short period, after which the pulse labelling is terminated by the addition of KCN and ice. DNA is then denatured and characterized by sedimentation on alkaline sucrose gradients (when it remains as single strands).

After a 2 second pulse, the radioactivity is found almost entirely in short fragments which sediment slowly between 7S and 11S; this suggests that the discontinuous unit is some 1000–2000 nucleotides long—about the size of a gene. Both strands of DNA seem to be synthesised discontinuously, since virtually all the label is recovered in the slowly sedimenting fragments—

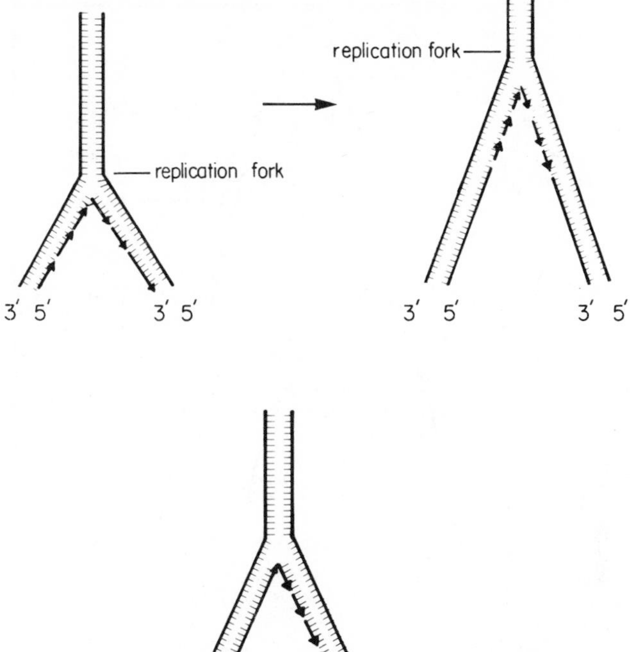

Figure 10.11: discontinuous synthesis of DNA. The upper model shows the discontinuous synthesis of both strands of DNA. Short segments are synthesized in the direction from 5′ to 3′ and the gaps between them are filled in and the segments linked together only later, as the replicating fork moves further along the DNA. The lower model shows an alternative in which one strand is synthesised continuously from 5′ to 3′, with no breaks; whereas the other new strand is synthesised discontinuously in short segments from 5′ to 3′, which are only subsequently linked together to generate the overall growth of the strand in the 3′ to 5′ direction

if one strand were synthesised continuously, half the label should be in the rapidly sedimenting bulk DNA. Figure 10.12 shows that as the length of the pulse dose is increased, a greater proportion of the label enters the rapidly sedimenting DNA of the genome; most of the label in a two minute pulse sediments at 45S, so that the short fragments must have been covalently linked together into a continuous strand. Sugimoto et al. (1969) found that Okazaki fragments of T4-infected cells hybridize with both of the separated single strands of the phage DNA, which supports the idea that both daughter strands are synthesized discontinuously. Replication is probably discontinuous in all organisms, including Chinese hamster cells (Schandl and Taylor, 1969) and Hela cells (Painter and Schaeffer, 1969).

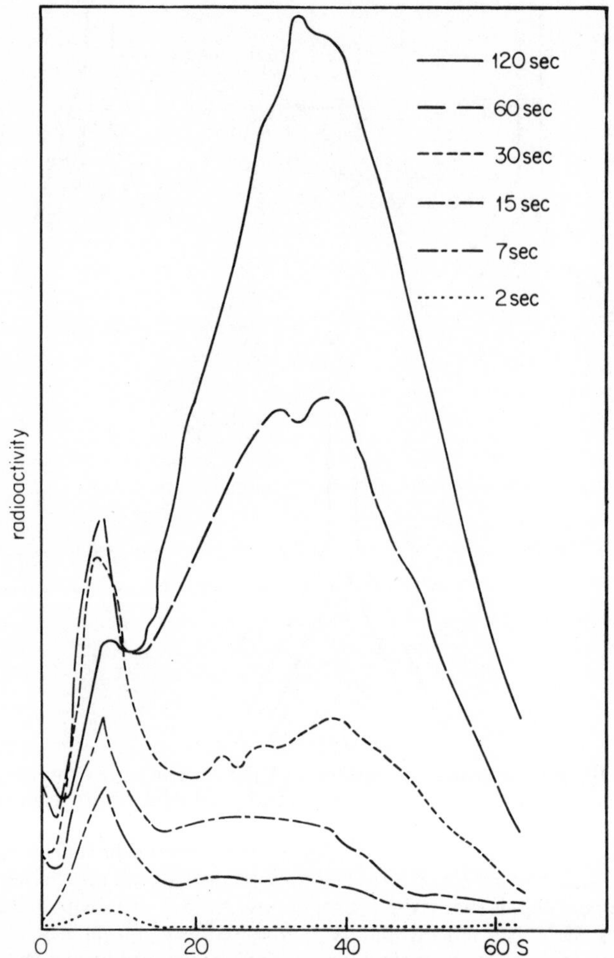

Figure 10.12: sedimentation of single strands of DNA after short radioactive pulse doses. At short pulse times most of the label enters a fraction sedimenting at about 10S (the Okazaki fragments); as the time of incubation with the label is increased a greater proportion sediments in the 40S peak of bulk DNA. Data of Okazaki et al. (1968)

Although the most attractive interpretation of the discovery of short fragments of newly synthesized DNA is that they arise by discontinuous synthesis, some experiments have suggested that they result wholly or in part from the introduction during extraction of nicks into DNA which has been continuously synthesized. (For example, Kozinski and Kozinski, 1969 proposed that Okazaki fragments of T4 might result from nicks made by the phage endonuclease; but Iwatsuki and Okazaki, 1970 showed that although the endonuclease can nick DNA it is not responsible for production of the fragments.) Jacobson and Lark (1973) noted that not all of the small DNA chains identified as Okazaki fragments in E.coli represent intermediates of transcription and suggested that although part of this fraction serves as the precursor to continuous DNA some of it may arise by other processes which act on DNA.

To show that at least one strand must be synthesized discontinuously, Okazaki and Okazaki (1969) confirmed that both new strands must be synthesized in the direction $5' \rightarrow 3'$; this excludes the possibility of continuous synthesis at least for the strand which grows overall from $3' \rightarrow 5'$. Their experimental protocol involved labelling cells infected with T4 during growth at 8°C, when DNA synthesis proceeds normally but at the reduced rate necessary for success of the labelling technique. One culture of infected cells is pulse labelled with H^3-thymidine for 6 seconds and another is labelled for 150 seconds under the same conditions but with C^{14}-thymidine. After the newly synthesized DNA has been extracted, Okazaki fragments can be isolated by centrifugation through alkaline sucrose density gradients; the fragments complementary to each of the parental strands of T4 DNA can be separated by hybridization with the individual phage strands and reisolated so that they can be tested for their susceptibilities to degradation by nucleases. By analysing the two radioactive preparations together, the locations of the two labels can be directly compared.

Different results are produced when the chains are degraded with the enzymes exonuclease I of E.coli (which degrades single strands of DNA stepwise from the 3′ end) and the nuclease of B. subtilis (degrades single strands from the 5′ end). Exonuclease I releases the H^3 label before the C^{14} label; with B. subtilis nuclease the order is reversed. This shows that the most recently incorporated nucleotides—the H^3 label given during the short pulse must be located at the growing ends of the chains—are found at the 3′ ends of the Okazaki fragments. Since the same results are obtained with the fragments which hybridize to either strand of T4 DNA, both of the new strands of DNA must be synthesized in the direction 5′ to 3′. This conclusion has been supported by the further experiments of Sugino and Okazaki (1972), who used other nuclease activities to degrade the short and long pulse labelled fragments.

Some experiments have suggested that synthesis of one strand of DNA may be continuous. Iyer and Lark (1970), for example, found that some of a short pulse label of cells of E.coli $15T^-$ can be extracted as Okazaki fragments and some as bulk DNA. One possible explanation for results such as these is that although both strands are synthesized discontinuously, on one the Okazaki fragments are joined into a covalent strand more rapidly. This contention is lent support by the experiments of Ginsberg and Hurwitz (1970), who hybridized the Okazaki fragments of either phage T4 or λ DNA with the separated strands of the mature phage.

Some of a radioactive label incorporated by cells infected with λ enters Okazaki fragments but some appears to enter bulk DNA. The labelled Okazaki fragments hybridize preferentially with the *l* strand of the phage whereas the newly synthesized bulk DNA hybridizes with the *r* strand. But as the length of the pulse dose is decreased, fragments hybridise increasingly well with the *r* as well as the *l* strand. Extrapolation of the experimental results suggests that when all a radioactive label enters Okazaki fragments and none appears in bulk DNA there would be equal hybridization with both strands. This implies that both strands are synthesized discontinuously but that the newly synthesized *l* strand—that is the fragments which hybridize with the *r* strand—is linked into continuous DNA more rapidly than the newly synthesized *r* strand. But this asymmetry in joining the fragments is not displayed by all replicating phages, for with T4 the Okazaki segments hybridize equally well with both strands under all pulse lengths.

The "knife and fork" model for discontinuous synthesis is illustrated in figure 10.13. This arose out of attempts to explain how the enzyme DNA polymerase I might replicate DNA and takes into account the observation that this enzyme produces frequent hairpins in replicating DNA (see Lewin, 1970). We now know, of course, that DNA polymerase I does not replicate DNA in vivo (see page 459), but it is possible that some other enzyme might use this mechanism. The model postulates that the replicating enzyme moves along one strand in the 5′→3′ direction in which synthesis takes place overall; but after synthesizing a certain length of DNA switches strands and proceeds back along the complementary strand which has been exposed.

The fork produced when the enzyme switches strands must then be nicked, presumably by an endonuclease associated with the replication apparatus. Some experiments have suggested that Okazaki fragments of E.coli or T7 replicating DNA can readily acquire a duplex structure, perhaps as the result of hybridization between complementary lengths which have not yet been nicked (Pauling and Hamm, 1969; Barzilai and Thomas, 1970). But Okazaki and Okazaki (1969) found that the Okazaki fragments of T4 DNA are highly susceptible to exonuclease I, an enzyme which degrades only single stranded DNA and is inhibited by hairpin structures.

The knife and fork model predicts that initiation of Okazaki fragments should take place only on one strand of the DNA; the other strand should be synthesized only as the result of enzyme switching. The observation that each Okazaki fragments starts with a sequence of RNA (see below) may therefore make it possible to test whether the initiation sequences correspond to only one strand or to both. If fragments corresponding to both parental strands contain RNA initiation sequences, then strand switching models may be excluded as an explanation for discontinuous synthesis

Whether or not the enzyme switches strands, a reasonable model is is to suppose that the two new strands of DNA are not synthesised simultaneously. As the two strands of the parental duplex unwind, the daughter strand which grows overall in the direction 5′→3′ may be synthesised immediately behind the replicating fork. The corresponding sequence of the complementary parental strand, which is to direct synthesis of a daughter in the overall direction 3′→5′, may be delayed in replication until some length of it has been exposed. This length can then be used to direct synthesis in the 5′→3′ direction.

As figure 10.13 shows, this model requires that one strand of the parental DNA is replicated immediately upon unwinding, but that the other may remain single stranded for some distance behind the growing point. Results

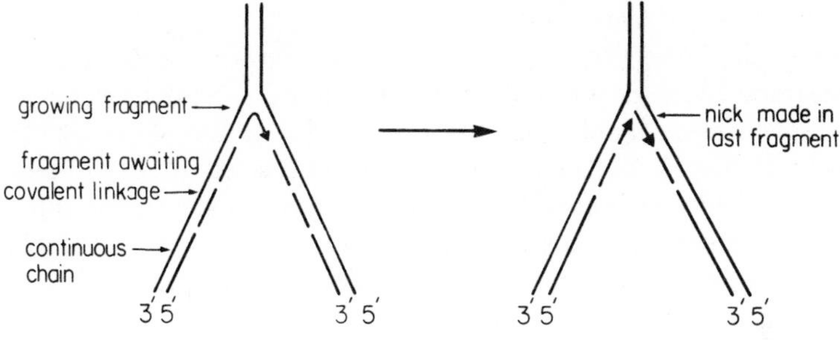

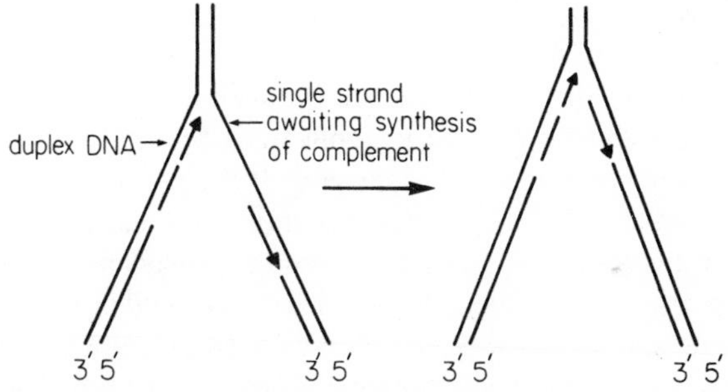

Figure 10.13: models for discontinuous synthesis of DNA. Upper: the knife and fork model. Polymerase synthesizes DNA from 5′ to 3′ on one strand and then switches template strands to synthesize the complementary sequence. The covalent linkage between the newly synthesized complements of both strands is then nicked by an endonuclease, creating two Okazaki fragments. Synthesis by polymerase initates only on the strand growing 5′ to 3′ overall and terminates only on the strand growing 3′ to 5′ overall. Lower: independent synthesis of Okazaki fragments on both template strands; polymerase initiates and terminates on both growing chains. Because synthesis on the strand growing overall from 3′ to 5′ cannot be initiated until the DNA has unwound, the corresponding region on the strand growing 5′ to 3′ may be synthesized first. This means that immediately behind the growing point the strand growing from 5′ to 3′ overall is duplex, but its partner growing overall from 3′ to 5′ is single stranded. The same prediction is made by the knife and fork model and by models in which the 5′ to 3′ strand grows continuously

suggesting such a model have been reported with phage λ by Inman and Schnos (1971) and with phage T7 by Wolfson and Dressler (1972). The replicating T7 chromosome, for example, when visualised in the electron microscope can be seen to contain regions of single stranded DNA (which appear thinner and less rigid than duplex DNA) on one daughter strand of the Y shaped replicating chromosome. The length of the single stranded region is usually less than about 2500 bases, about the length of the Okazaki fragments.

Covalent Linkage of Okazaki Fragments

Since chains can grow only from 5′→3′, synthesis must be discontinuous on the strand which grows overall in the direction 3′→5′; but it is not obvious why a polymerase should be unable to proceed continuously along the other strand from 5′→3′. The knife and fork model predicts that synthesis along both strands must be discontinuous, but demands that some event must cause the enzyme to switch strands; and switching would have to be a regular event in order to account for the production of Okazaki fragments. Models in which synthesis of Okazaki fragments is initiated on both strands demand that each fragment is terminated as well as initiated on both growing strands.

Incorporation of the Okazaki fragments into a continuous length of DNA rapidly succeeds their synthesis. Sealing the gaps between the segments is not a function of the replicase itself, but requires additional enzyme activities. Both E.coli and phage T4 code for an enzyme, polynucleotide ligase, which makes good breaks in the covalent chain between adjacent nucleotides; this enzyme undertakes the final stage in replication by linking together the 3′ end of one fragment and the 5′ end of the next. But polynucleotide ligase is not the only enzyme involved. The gap between adjacent Okazaki fragments may extend for several nucleotides, so that a two stage process may be necessary to join successive fragments. One enzyme may first undertake a repair-like action to add nucleotides to the 3′ end of one Okazaki segment until the chain extends to the 5′ end of the next segment; ligase can then seal this gap.

Polynucleotide ligase was first detected as an enzyme activity needed for the successful replication of phage λ. When E.coli cells are infected with λ DNA, a large fraction of its linear molecules are converted into a circular form, held together by hydrogen bonding between the complementary ends of the linear molecule, so that there is one break in the covalent integrity of each polynucleotide strand. Gellert (1967) and Gefter et al. (1967) found that polynucleotide ligase can seal these single strand nicks and can also repair similar breaks at other sites.

Although their overall reactions are the same, the ligase enzymes of E.coli and phage T4 differ in their demands for cofactors (see Gellert et al., 1968; Olivera et al., 1968a, b; Richardson et al., 1968). The enzyme induced by phage T4 during infection requires ATP whilst the enzyme of the E.coli cell demands NAD. Both enzymes undertake a two step reaction. The first stage is formation of an enzyme—AMP complex; the T4 ligase cleaves pyrophosphate from ATP and the E.coli ligase splits NMN from NAD. Gumport and Lehman (1971) demonstrated that the adenyl group used by E.coli ligase is attached to the ε-amino group of a lysine residue in the enzyme by a P–N bond.

Incubating the ligase–AMP complex from either source with duplex DNA carrying single strand breaks seals the nicks; and one mole of AMP is released for every mole of phosphodiester bonds synthesized. Both enzymes probably display the same reaction in this second step. Preparations of DNA containing single strand breaks with 3′-OH and 5′-P ends cause breakdown of the enzyme complex and the gap is filled but breaks with 3′-P and/or 5′-OH ends do not react.

The enzyme-AMP complex might bind to either the 3′-OH or the 5′-P terminus and figure 10.14 shows the attachment of the AMP to the free phosphate. That this is the reaction sequence is revealed by the action of exonuclease I on DNA which has bound the AMP; this enzyme degrades one strand of DNA sequentially from 3′-OH terminus, releasing monophosphates, but leaves the 5′ terminal dinucleotide intact. If AMP is attached to the 5′-P it must therefore be released in the form of a trinucleotide—that is attached to the terminal dinucleotide—but if it is attached to the 3′-OH terminus it should be released as a mononucleotide by the normal action of the enzyme. Since only the trinucleotide product is isolated when the DNA-AMP complex is treated with exonuclease, the subsequent bond formation must take place by a mechanism similar to that responsible for chain elongation, attack by a free hydroxyl group on the phosphate of the incoming group.

The activity of ligase in the bacterial cell is presumably to repair single strand breaks in duplex DNA, but in vitro it can also undertake other functions. Fareed et al. (1971) observed that T4 ligase can repair single strand breaks in either strand of hybrid polymers comprising one chain of deoxynucleotides and one of ribonucleotides. But its activity is much lower with the hybrid substrates than with duplex DNA. Another intriguing interaction of T4 ligase is the ability reported by Sgaramella et al. (1970) to join together two completely base paired duplex molecules, presumably by bond formation between their ends. Such an activity might be useful in recombination, but we do not know whether it is used in vivo.

Gene 30 of phage T4 has been identified as the structural gene which codes for the ligase; and the behaviour of temperature-sensitive mutants at this locus suggests that the ligase is implicated in joining together the discontinuous Okazaki fragments produced in infected bacteria. The radioactive pulse which first appears to sediment between 7S and 11S is usually transferred very rapidly to larger units sedimenting between 30S and 60S. Newman and Hanawalt (1968) and Okazaki et al. (1968) argued that if the ligase represents an activity which is needed to link together the short segments, the impairment of a temperature sensitive ligase by increase in temperature should inhibit the transition and cause the short segments to accumulate.

Because the synthesis of host DNA is inhibited by infection with T4, only

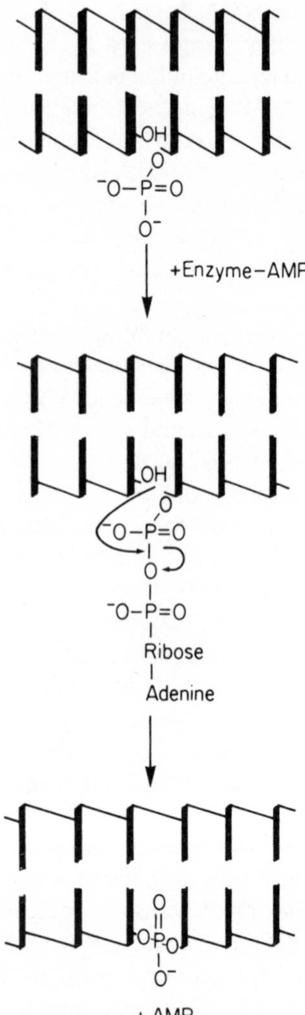

Figure 10.14: action of polynucleotide ligase. The enzyme-AMP complex binds to a gap terminating in 3′-OH and 5′-P; and AMP reacts with the free phosphate group. Attack by the 3′-OH group on this moiety forms a phosphodiester bond which seals the gap. After Olivera, Hall and Lehman (1968)

phage DNA is under synthesis in infected cells which are reproducing the phage. After infection at 20°C for long enough to allow phage replication to commence, the temperature can be raised to 43°C and cells given a pulse label of H^3-thymidine. The rate of incorporation of the label is the same in cells infected with either wild type or mutant phages. But cells infected with wild type phages transfer a pulse label into bulk DNA within about one minute; after infection with the mutant, the radioactivity remains in the 7-11S fraction and is not transferred to more rapidly sedimenting DNA.

When the entire procedure is performed at 20°C, however, there is no difference in behaviour between wild type and mutant phages for both transfer a pulse label into bulk DNA. When cells infected with the mutant phage are labelled at the high temperature and then incubated at low temperature, the label in the short segments gradually disappears and is transferred to bulk DNA. These experiments suggest that T4 ligase is needed to join Okazaki fragments into a continuous polynucleotide chain, but that when the ligase is inactive the fragments which accumulate can later be linked together by restoring the enzyme activity.

Mutants in the ligase of E.coli have proved more difficult to isolate and characterize. Gellert and Bullock (1970) isolated mutants by selecting host cells for their ability to support the reproduction of T4 phages which have a defective ligase. Such mutant hosts synthesize an excessive amount of ligase. Revertants of these bacteria can then be isolated by selecting for cells which can no longer support the growth of *lig*-phages but which can support the growth of lig^{+} T4 DNA. One of these revertants, *lop-8*, *lig-4*, is a double mutant which synthesizes a large amount of a defective bacterial ligase.

The ligase enzyme also plays a role in the repair of radiation damage in the cell (see page 522) for the *lig ts7* mutant isolated by Pauling and Hamm (1969) by its deficiency in a late stage of a repair process also proved to have reduced ligase activity in vitro at high temperature. And at 25°C a pulse label enters Okazaki segments of low molecular weight, but with a longer period of labelling much of the radioactivity is transferred to bulk DNA. At the non-permissive temperature of 40°C, the mutant converts very little of a radioactive label to bulk DNA and accumulates Okazaki fragments.

The two ligase mutations have been placed in standard backgrounds and compared by Gottesman et al. (1973) and Konrad et al. (1973). The mutations in *lop-8*, *lig-4* cells have been separated by recombination; *lop-8* probably identifies a promotor mutation, increasing synthesis of ligase, and the *lig-4* mutation identifies the structural gene for the enzyme. Cells with the *lig-4* mutation alone seal Okazaki fragments normally at 30°C but ten times more slowly than usual at 42°C. In spite of the ligase defect, strains carrying *lig-4* grow well at high temperature and show few defects.

The *lig-ts7* mutation is located in the same gene as *lig*-4 but has a more drastic effect upon the cell; it is a conditional lethal which causes cell death at the non-permissive temperature. These cells are sensitive to ultraviolet irradiation and accumulate Okazaki fragments, joining them together perhaps forty times more slowly than wild type cells.

Both the *lig-4* and *lig-ts7* mutations have much less ligase activity than wild type cells when extracts are assayed in vitro. At the non-permissive temperature of 42°C, the *lop-8*, *lig-4* double mutant enzyme activity is 5% and the *lig-4* single mutant is 1% of wild type activity. Even at the low permissive temperature the *lig-ts7* mutant has only 3% of wild type activity; and Modrich and Lehman (1971) demonstrated a ten fold decrease when the temperature is raised to 40°C.

Because the *lig-4* cells grow well at both low and high temperature and the *lig-ts7* cells grow well at low temperature in spite of their deficiencies of ligase, we may deduce that of the order of 1% of the cellular ligase activity is sufficient to fulfill properly its role in replication in vivo. We do not know why the cell should usually synthesize so much more ligase or what other pathways it may be needed for, but the enzyme is probably implicated in repair as well as replication of DNA.

Conflicting reports have appeared about the ability of the E.coli mutant which lacks DNA polymerase I (see page 460) to replicate its DNA. Kuempel and Veomett (1970) found that newly synthesised DNA of the mutant does not seem to be incorporated into high molecular weight DNA, but Okazaki, Arizawa and Sugino (1971) found that joining does take place, although ten times more slowly than in the parent strain. Since the mutant appears to be able to grow normally in spite of its lack of DNA polymerase I, it must be able to replicate its DNA, so we may conclude that joining of the fragments is slower, but that this is not deleterious to the cell. These results suggest that joining the Okazaki fragments together may be the two step process shown in figure 10.17, with DNA polymerase I first filling the gap between the fragments and polynucleotidase ligase finally sealing the breaks between contiguous nucleotides at the end of one segment and the beginning of the next. The inviability of *lig⁻ polA⁻* double mutants also implies that the catalytic pathways of the two enzymes are related (see page 523).

Initiation of DNA Synthesis with RNA Primers

One of the most pressing problems of replication has been how synthesis of new strands of DNA is initiated. All of the known DNA polymerases of both bacteria and eucaryotic cells can extend a DNA chain from a free 3′-OH terminus but none appears to be able to start synthesis of a chain anew (see page 469). When a single stranded template directs synthesis of its complement, a primer—a complementary oligonucleotide longer than eight or ten residues—must be added before synthesis can start. The same requirement for a 3′-OH group from which to start synthesis presumably exists for replication of duplex DNA.

Two classes of model have been proposed for the start of replication. The first supposes that the two strands of the duplex chromosome are unwound at the origin—by mediation of a nick or swivel—after which synthesis of daughter strands commences, using the parental strands only as templates. The second postulates that the parental strands are nicked so that new strands can extend from the termini which are freed by the breaks; each growing point therefore uses one parental strand to provide a primer activity for the start of synthesis on the other, template strand. Since DNA chains grow only from 5′→3′, only the two parental 3′-OH termini are available for direct extension; this model therefore leaves unanswered the problem of how the discontinuous fragments are initiated for growth in the other direction.

Models which invoke covalent extension of parental chains demand that the new daughter strands must be linked to the old parental strands at the origin. This prediction can be tested by changing the precursors provided for DNA synthesis between two cycles of intiation. By growing bacteria first in BUdR and then initiating the next round of replication with a pulse dose of H^3-thymidine, Stein and Hanawalt (1972) followed the appearance of strands of intermediate density containing the radioactive label. These should correspond to the linkage between old and new strands. Such linkage appears to exist, but proved to result from an artefact of the isolation procedure in which G-C rich Okazaki fragments separate from those of average density. When this aretefact is excluded by examining only DNA produced by covalent linkage of the Okazaki fragments, no transition segments are found at the origin. Pre-existing parental strands are therefore used only as templates and not as primers for synthesis of new strands.

Any primer needed to start synthesis of a segment of DNA must therefore be provided by some other means. One possible primer would be short oligonucleotides of DNA which are complementary to the starting sequence—such primers can initiate replication of circular single strands of ϕX174 DNA in vitro. But an alternative is that the primer might be a sequence of RNA. This sequence might be provided by an oligonucleotide fragment of RNA; or, since RNA polymerase can initiate transcription without a primer, might be synthesised at the starting point. One such model is therefore to suppose that RNA polymerase synthesizes a short sequence of RNA, using the DNA as template, and that the free 3′-OH terminus of the RNA is used as a primer which is covalently extended with deoxynucleotides by the DNA replicase; the RNA must later be removed and replaced with DNA.

Transcription of RNA has been implicated in replication by the inhibition exerted by rifampicin, a drug which prevents bacterial RNA polymerase from synthesizing RNA on DNA templates. Although rifampicin inhibits conversion of the single stranded DNA circles of phage M13 DNA to the duplex form, chloramphenicol does not inhibit the initiation of replication in infected cells; this implies that the effect of rifampicin is exerted upon transcription of RNA as such rather than upon the synthesis of some necessary protein(s). Wickner et al. (1972a) found that conversion in vitro of M13 single strands to the duplex form demands some RNA synthesis; and a radioactive P^{32} label in a deoxynucleotide precursor is transferred to a ribonucleotide after the product of replication in vitro is cleaved with alkali. This suggests that a primer sequence of RNA is covalently linked to DNA.

Although the synthesis of double stranded ϕX174 DNA from the single strand of the mature phage is not sensitive to rifampicin, RNA synthesis seems to be necessary also for replication of this phage. Schekman et al. (1972) found that replication of the single strands of both M13 and ϕX174 DNA is inhibited by actinomycin and that ribonucleotides are needed in an in vitro system. The RNA primer sequence utilised by ϕX174 also seems to be linked to the newly synthesized DNA strand; there is at present no explanation for the different effects of rifampicin on the two phages.

One experiment essential to confirm the need for the presence of an RNA primer in vitro is to demonstrate that ribonuclease H—an enzyme which specifically destroys the RNA moiety of DNA-RNA hybrids—prevents conversion of single strands to the duplex form. Otherwise it remains possible that RNA synthesis is implicated in only a subsidiary and not a direct role. A requirement for the presence of RNA has been shown in this way by Keller (1972), but in a system which relies upon replication of E.coli DNA by a polymerase derived from human tumour KB cells. It is difficult to draw conclusions about replication within the bacterial cell from such heterologous systems.

RNA may also be involved in initiating new cycles of replication in E.coli cells for Lark (1972a) has found that new cycles of replication cannot start when RNA synthesis is inhibited by the addition of rifampicin up to 10 minutes before DNA synthesis is due to start. The addition of chloramphenicol at this time does not inhibit the initiation of DNA synthesis. Although the cells have therefore synthesised the proteins which they need to start a new cycle of replication, they are nonetheless unable to do so if prevented from synthesising RNA. Working with a different strain of E.coli, Messer (1972) has found that a step sensitive to inhibition by rifampicin takes place when replication is initiated. The obvious role for this RNA synthesis is to act as a primer for DNA synthesis, although it may of course play some other function.

But the discontinuous mode of DNA synthesis implies that provision of a primer to initiate a cycle of replication may not be sufficient to ensure that the replicase can continue to synthesise DNA. For how are the Okazaki fragments themselves initiated? Even if a primer is invoked to start the initial movement of the replicating fork, the enzyme does not move continuously along the DNA but must reinitiate synthesis at the beginning of each Okazaki fragment. Does each reinitiation event demand a primer, or can the replicase enzyme in some way continue once it has started to synthesize DNA?

The buoyant density of Okazaki fragments depends upon how soon they are isolated after their synthesis has commenced. By extracting the fragments after giving E.coli cells a very short pulse dose of H^3-thymine, Sugino, Hirose and Okazaki (1972) found that their buoyant density may be greater than that of DNA itself; this suggests that a length of RNA may be associated with the DNA of the fragments. A model which would explain these results is to suppose that each fragment is initiated by synthesis of a short length of RNA, after which the DNA chain is extended from the 3'-OH terminus of the RNA.

A typical experimental procedure is to label cells with H^3-thymine for 15 seconds at 14°C, after which the synthesis of DNA is immediately terminated by adding an ethanol-phenol mixture which also inactivates any nucleases

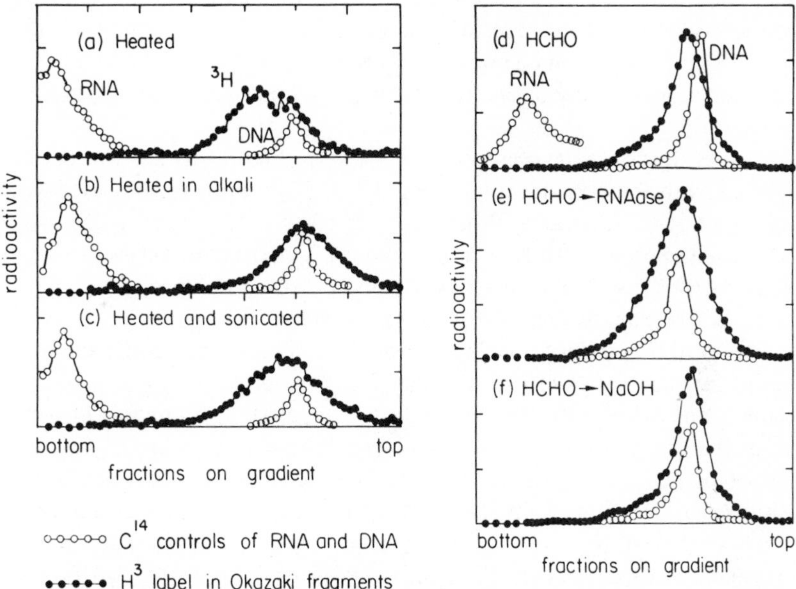

Figure 10.15: isolation of nascent Okazaki fragments containing an RNA primer sequence. A 15 second pulse label of H^3-thymine was given to cells growing at 14°C; the pulse labelled DNA was recovered and banded in a Cs_2SO_4 density gradient. Controls labelled with C^{14} mark the positions of free RNA and DNA on the gradient. The left three experiments show fragments isolated by heat denaturation; the right three experiments utilised formaldehyde. Parts a and d show that the fragments have a buoyant density greater than that of DNA; part c shows that sonication does not change the density. Destruction of RNA, in parts b, e and f, restores the density of the fragments to that of DNA, presumably by removing the RNA part of the polynucleotide chain. Data of Sugino, Hirose and Okazaki (1972)

present. When the DNA extracted from these cells is denatured to yield single stranded Okazaki fragments, the pulse label sediments heterogeneously, but with an average density greater than that of single strands of DNA of E.coli. Figure 10.15 shows two series of experiments, one using denaturation by heating to isolate the fragments (parts *a*, *b* and *c*) and the other relying upon treatment with formaldehyde (parts *d*, *e* and *f*).

When either preparation is treated with alkali (parts *b* and *f*)—which hydrolyses RNA to nucleotides and may also cause some fragmentation of DNA—

the pulse label sediments at the density characteristic of single strands of DNA. Incubation with ribonuclease (part *e*) also removes the RNA part of the Okazaki fragment so that the remainder can band at the position of DNA. These experiments suggest that a length of RNA is covalently linked to the Okazaki fragments, for the denaturation procedures would rupture any structures maintained only by hydrogen bonds. Destruction of the RNA, by alkali or ribonuclease is therefore necessary to generate fragments containing DNA alone. This interpretation is supported by part *c* of the figure, which shows that sonication alone—to break the fragments at random sites—does not cause the density shift.

Direct evidence for the presence of RNA in the Okazaki fragments has been provided by the demonstration that an H^3-label in uridine is associated with a pulse label of C^{14}-thymine. When cells are given a 15 second pulse of C^{14}-thymine in the presence of H^3-uridine, about 1% of the tritium label sediments together with the C^{14} label at a position more dense than that of DNA itself. The H^3 label is rapidly chased out of the fragments bearing the C^{14} label after incubation in the presence of unlabelled thymine; and the density of the C^{14} labelled fragments then returns to that of DNA itself.

This is consistent with the observation that Okazaki fragments of density greater than that of DNA itself can be labelled only by means of very short pulse doses; when the pulse is longer than 30 seconds, the label bands at the position characteristic of single strands of DNA. This implies that the RNA primer is removed from the DNA fragment very soon after its extension with deoxynucleotides has begun. The noticeable effect of the RNA portion of the fragments on their density suggests that the RNA must comprise a relatively large length of the very nascent molecules; and this may be of the order of 50–100 nucleotides. Magnusson et al. (1973) observed that 4-5S replicating intermediates of polyoma DNA may also commence with a short sequence of RNA; this mechanism may be universal in view of the inability of all DNA polymerases to initiate DNA synthesis (see page 469).

A plausible model to account for these results is to suppose that each Okazaki fragment is initiated by the synthesis of a short stretch of RNA by RNA polymerase. This synthesis might be mechanically linked to the movement of the chromosome through the replication apparatus, or might be directed by frequent initiation and termination signals along the chromosome. If the knife and fork model of figure 10.13 accounts for the action of the replicase, the RNA initiator sequences should correspond to only one strand of the DNA. If synthesis of Okazaki fragments is initiated separately on each DNA strand, as seems more likely and is shown in the lower model of the figure, then RNA fragments must be used to start DNA sequences of both strands. The DNA replicase must extend the chain started by RNA polymerase but, of course, using deoxynucleotides to yield a DNA chain.

That the switch from the RNA primer sequence to the DNA chain is specific is suggested by the results of Sugino and Okazaki (1973) and Hirose et al. (1973). When toluene treated cells of E.coli are incubated with labelled precursors to DNA, Okazaki fragments containing the RNA starter sequence can be isolated containing the label. Figure 10.16 shows that when dCTP, labelled with P^{32} in the α position, is provided as a precursor, the label is added to the RNA chain. Alkaline hydrolysis of the Okazaki fragment yields mostly labelled ribo-UMP, with some labelled ribo-GMP, AMP, CMP. Because alkaline hydrolysis results in the transfer of phosphate from a nucleotide to its 5′ neighbour, this implies that the RNA primer sequence terminates in uridine.

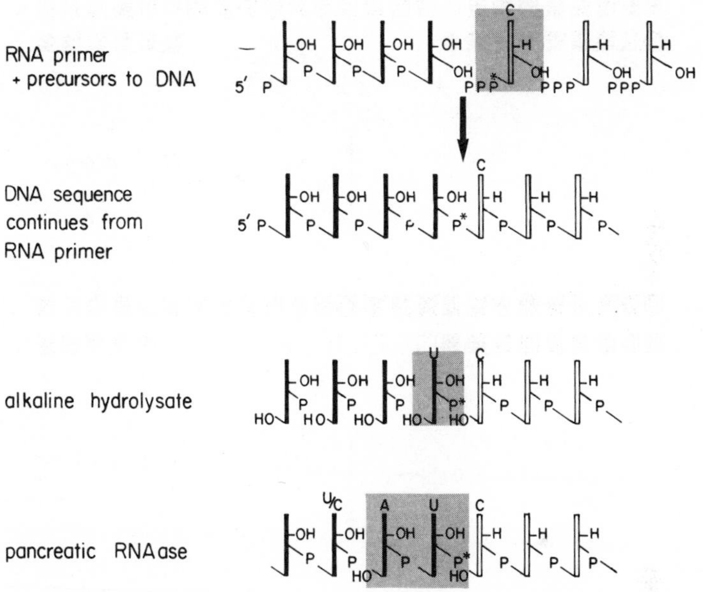

RNA primer + precursors to DNA

DNA sequence continues from RNA primer

alkaline hydrolysate

pancreatic RNAase

Figure 10.16: sequence of junction between RNA and DNA in Okazaki fragments. Toluene treated E.coli cells are given precursors to DNA labelled with P^{32} in the α position. Okazaki fragments containing the RNA primer sequence are extracted and transfer of the P^{32} to RNA is followed by alkaline hydrolysis and treatment with pancreatic ribonuclease. The fraction recovered with the radioactive label is shown shaded. Alkaline hydrolysis splits phosphodiester bonds on the 3′ side of nucleotides containing a 2′-OH (that is RNA); because the phosphodiester bonds are split on their 3′ side, the phosphate in the first base of the DNA sequence is transferred to a ribonucleotide. Most of this transfer occurs when P^{32} dCTP is provided as substrate, and is made to UMP. This shows that the last base of the RNA is a U and the first base of the DNA sequence is a C. Pancreatic RNAse splits bonds in the same way as alkaline hydrolysis, but only when the nucleotide on the 5′ side of the bond attacked is a pyrimidine (U or C). The radioactive label is recovered in a dinucleotide, A-U; this shows that the RNA primer must terminate in A-U and that the base on its 5′ (left) side must be a pyrimidine. Data of Sugino and Okazaki (1973)

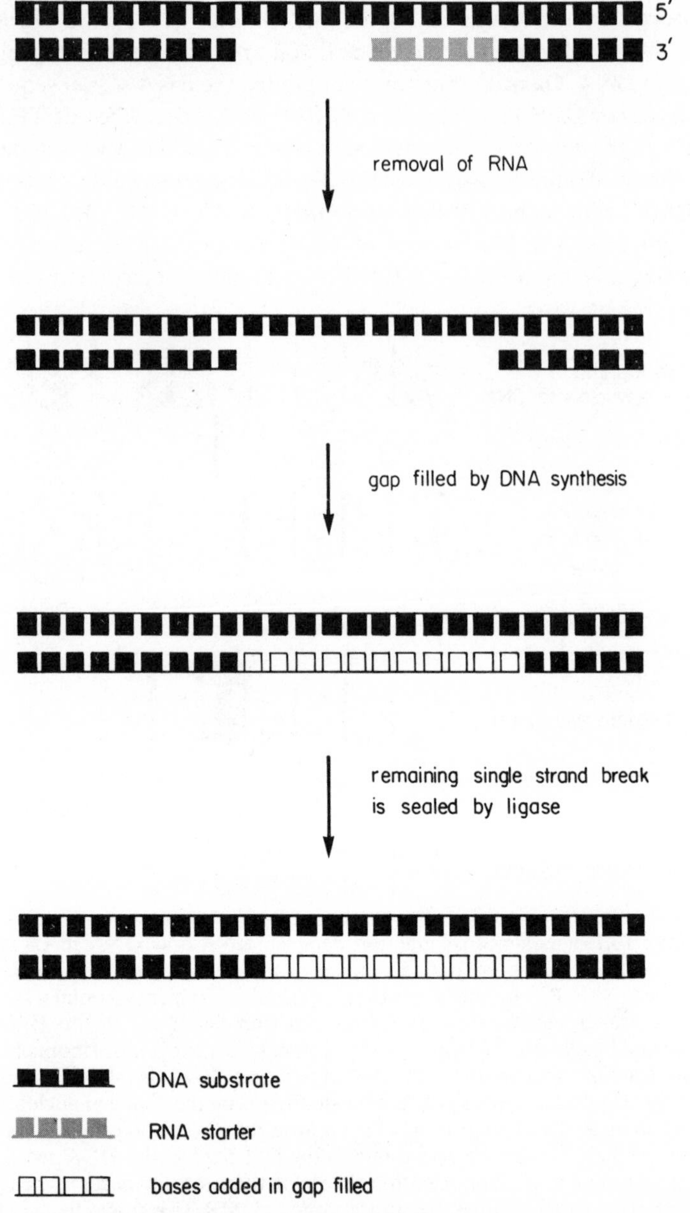

Figure 10.17: model for covalently linking Okazaki fragments. As the upper sequence shows, each fragment commences with a primer sequence of RNA. This RNA is removed by a 5′ to 3′ exonuclease action, after which a repair like DNA synthesis activity extends the fragment on the left until the gap is filled. Both these actions might be undertaken, simultaneously, by DNA

Pancreatic ribonuclease splits phosphodiester bonds only between a pyrimidine ribonucleotide and the base on its 3′ side, again causing transfer to the 5′ phosphate of this neighbour to the released fraction. The label originally present in dCTP is recovered in a dinucleotide, A-U. This implies that the sequence at the end of the RNA primer is: pyrimidine–A–U. Some of the label is recovered as a dinucleotide of sequence A–C; this suggests that the last base in the RNA primer chain is always a pyrimidine, usually U but sometimes (up to one third) C.

No transfer to RNA takes place when the labelled DNA precursor is dATP or dTTP. About 25% of the total counts recovered in ribonucleotides however, is generated by transfer from dGTP to UMP according to the results of alkaline hydrolysis. However, pancreatic ribonuclease does not generate unique fragments from this transfer and the RNA sequence prior to the final U residue appears to be random. This suggests that this transfer may be an artefact of the toluene treated cell system and does not reflect transfer in vivo. It therefore seems likely that the RNA primer sequence always terminates in U/C–A–U/C and the DNA chain always starts with C.

The RNA primer sequences must be very rapidly removed by nuclease action, since they remain associated with DNA for only a short period of time. Even if there is no gap between the end of the DNA sequence of one Okazaki fragment and the RNA starting the next, removal of the RNA generates a gap which must be filled with deoxynucleotides. (And there may, of course, also be a gap between the end of one fragment and the start of the next.) Figure 10.17 illustrates the sequence of events demanded to link the discontinuous fragments of DNA into a continuous covalent chain.

We do not know what enzyme activities remove the RNA primer sequence and replace it with DNA, but one enzyme which could undertake one or both functions is DNA polymerase I. Roychoudry and Kossel (1973) demonstrated that it can degrade the RNA moiety of RNA-DNA hybrids; and Westergaard et al. (1973) showed that in vitro it can excise and replace the RNA primer sequence of phages M13 and ϕX174. DNA polymerase I and polynucleotide ligase together can convert the intermediate containing the RNA primer into a covalently closed duplex of DNA. The role of removing and/or replacing RNA may explain the slower joining of Okazaki fragments in E.coli cells defective for this enzyme. The effect of polynucleotide ligase in causing fragments to accumulate implies that this enzyme is responsible for making

---

polymerase I. Finally, a gap of only one phosphodiester bond remains between the two fragments and this is sealed by the action of polynucleotide ligase. The RNA primer sequence is probably about 50–100 nucleotides long; we do not know what sort of gap may lie between the end of one Okazaki fragment and the RNA sequence commencing the next

the final link of a phosphodiester bond once two adjacent Okazaki fragments of DNA are separated only by a single strand nick.

This model for the role of RNA synthesis in DNA replication predicts that continuation of replication, and not solely its initiation at the origin, should be sensitive to the effect of rifampicin in inhibiting RNA polymerase. One possible explanation for the failure of rifampicin to inhibit replication is that the synthesis of RNA primer sequences takes place by some different mechanism, perhaps involving a new enzyme associated only with the replication apparatus. and is not catalysed by the known RNA polymerase.

Rate of Replication of the Bacterial Chromosome

The rate at which the chromosome of E.coli is replicated is constant at any temperature; replication takes 40 minutes under optimal conditions at 37°C, no matter whether the cells are dividing every 27 minutes or every 60 minutes (see chapter 13). The rate of replication can be reduced, however, by growing cells on poor medium in which concentrations of precursors may become limiting; Bird and Lark (1970) found that cells of E.coli $15T^-$ take 60 minutes to replicate their chromosome when the doubling time is reduced to 70 minutes and take 80 minutes for DNA synthesis when a cell division cycle occupies 120 minutes.

The rate of replication depends directly upon the provision of deoxynucleotide precursors, for Beacham et al. (1971) found that in mutants which cannot synthesize thymidylic acid the rate of replication depends on the level of precursors in the medium; the pattern of dependence varies with the strain of E.-coli, but the maximum replication rate is about 55 minutes, somewhat slower than the rate of wild type cells. Decreases in temperature—which affect all enzymically catalysed metabolic reactions—also reduce the rate of replication.

The chromosome of E.coli contains about 4×10^6 base pairs so that when replication takes 40 minutes, its overall rate must be about 1700 base pairs per second. If replication is bidirectional, each growing point must move at a rate of some 850 base pairs each second. This rate is very rapid indeed and is, for example, very much faster than the rate of transcription, which proceeds at between 35 and 40 nucleotides each second.

The rate of replication within the cell must be the sum of several activities. First, synthesis of an Okazaki fragment must be initiated by the RNA primer sequence; if this polynucleotide is synthesised by RNA polymerase at the usual rate, this will be a slow step in replication. The first deoxynucleotide must be added to the primer, after which the DNA chain must be extended by the replicase. This is probably a very rapid process. After the Okazaki fragment has been completed, it must be linked to the previous fragment; this must involve excision of the length of RNA primer, initiation and synthesis of a length of DNA to replace it, and the mediation of polynucleotide ligase to link the two fragments together by a covalent bond. An Okazaki fragment is of the order of size of a gene, about 1000 nucleotides, so this entire sequence must be repeated each second, on each of the strands of the DNA duplex. Of course, these steps may not take place strictly sequentially—some of them may proceed at the same time so that one fragment is being joined to the last whilst the next is under synthesis.

But the overall rate of movement of the replication fork must be maintained at about 850 nucleotides per second and the limiting factor in its movement is probably initiation or termination of the Okazaki segments rather than their elongation. Manor, Deutscher and Littauer (1971) have determined the thymidylic step time—the time taken to add one thymidylic acid residue to the growing DNA chain—which in E.coli B is 4–7 msecs at 20·5°C. This figure varies with the strain of E.coli, but gives a chain growth rate of about 140–250 nucleotides per second, which is close to the speed at which the replication fork probably moves at this low temperature. This suggests that one new strand of DNA must grow on each arm of the replicating fork at the same time; and it shows that we must expect the DNA replicase to be able to add nucleotides to the chain at about the same rate as the velocity of movement of the replicating fork itself in the cell.

Enzyme Systems for DNA Synthesis

Kornberg DNA Polymerase I

When extracts of E.coli cells are assayed for their ability to synthesize DNA, the predominant enzyme activity is that of DNA polymerase I; indeed, the catalytic activity of this enzyme is greater by an order of magnitude than that of other bacterial DNA polymerases. This enzyme was first isolated by Kornberg and appeared to be the replicase of E.coli. The enzyme can convert single stranded circles of ϕX174 DNA to a duplex form when a primer is added; but attempts to replicate duplex DNA in vitro lead only to the synthesis of an abnormal DNA product (reviewed by Kornberg, 1969).

This DNA is not easily denatured and renatures rapidly; under the electron microscope it appears as branched fibres. A model to explain how this unusual DNA product is generated supposes that the polymerase copies one strand of a duplex in the $5' \rightarrow 3'$ direction, but at some point during synthesis the unreplicated complementary strand competes as template for the polymerase so that the enzyme switches strands. When a second event of this nature occurs, so that the polymerase uses as template the strand which it has just synthesised, a hairpin is generated.

It was this action of the polymerase which led to the formulation of models of the class shown in figure 10.13 for replication by strand switching. But DNA polymerase I switches strands in an irregular manner and can progress along duplex DNA templates only at a rate some hundred times below that of replication in vivo. Another property which argues that the Kornberg polymerase does not replicate DNA in the bacterial cell is its many catalytic activities other than that of synthesizing phosphodiester bonds; the enzyme has both 5'→3' and 3'→5' exonuclease activities and can also participate in vitro in the excision from DNA of the products of ultraviolet irradiation. These activities suggest that the enzyme may be implicated in repair synthesis of DNA rather than in semi-conservative replication, for the repair reaction demands removal of one damaged strand of DNA and synthesis of a substitute by using the remaining single strand as template (see page 502).

The ability of cells to grow normally and to replicate DNA when they lack DNA polymerase I implies that this enzyme is not implicated in bacterial replication in an essential capacity. DNA polymerase I is usually assayed by its ability to incorporate nucleoside triphosphate precursors into DNA in vitro; and it was only by screening the polymerase activity in vitro of extracts of many E.coli colonies treated with mutagens that de Lucia and Cairns (1969) were able to isolate the *polA*$^-$ mutant lacking the enzyme. Cells of this mutant multiply at the same rate as their parental strain and the major defect appears to lie in an increased sensitivity to the damaging effects of ultraviolet irradiation—this supports the idea that DNA polymerase I may be involved in vivo in repair rather than replication (see page 516).

The *polA1* mutation is an amber recessive and Kelley and Whitfield (1971) have isolated the DNA polymerase produced by another mutation, *polA6*, at this locus. The protein found in these cells elutes at a different position from phosphocellulose columns than that of the enzyme of wild type cells; and the mutant polymerase is inhibited by increases in temperature which do not affect the wild type protein. This suggests that the *polA* locus is the structural gene which codes for polymerase I.

Although these experiments suggest that DNA polymerase I is not the replicase, they do not prove the point for it remains possible that the small amount of residual activity found in *polA* mutant cells might be sufficient to replicate DNA. Another caveat is that the enzyme is usually assayed in solution and its activity under these conditions may be different from its behaviour within the cell; for example, the *polA* mutant enzyme might be inactive in solution in vitro, but active in association with some replication complex within the cell. However, complexes which can replicate DNA in vitro in a semi-conservative manner can be isolated from *polA*$^-$ cells of E.coli; and they appear to contain an enzyme activity other than DNA polymerase I.

Membrane Systems for DNA Synthesis

The activity of DNA polymerase I in extracts of E.coli cells is so great that it obscures any other enzyme activities which can incorporate nucleotide precursors into DNA. But the cells of the Cairns *polA*$^-$ mutant do not contain this activity and so can be used to isolate other enzymes which can synthesize DNA. The rationale for many attempts to isolate the replicase has been that a replication complex may be associated with the membrane; and this apparatus may only function efficiently in replication when all its components are together in the appropriate structure (see page 426). The Cairns mutant has therefore been used to isolate membrane systems which can synthesize DNA and from which the replicase may be purified rather than to extract soluble DNA synthesising enzymes as such.

When E.coli cells are embedded in a matrix of agar, the agar can be fragmented and the cells transformed into fragile spheroplasts within the matrix. After the cells have been lysed, soluble components—including DNA polymerase I—can be washed out of the matrix to leave a complex of DNA and membranes. Smith, Schaller and Bonhoeffer (1970) found that when these agar fragments are incubated with deoxynucleotide precursors, DNA is synthesised in a semi-conservative manner. When *polA*$^-$ cells which lack DNA polymerase are used, the agar treatment is unnecessary and replication can be observed in vitro in lysates of spheroplasts. Whether DNA polymerase I is washed out of cells or whether it is absent because of a genetic defect, then, DNA membrane complexes can replicate DNA in vitro in its absence. The rate of chain elongation in this system is of the order of 2000 nucleotides per second, although synthesis continues for only one or two minutes before ceasing abruptly.

The DNA of ϕX174 can be used to test for the replicating activity of infected bacteria and Knippers and Stratling (1970) isolated a membrane complex from cells of the Cairns mutant; this complex incorporates precursors into DNA using an endogenous template but does not utilise added DNA. Most of an added radioactive label is incorporated into E.coli DNA rather than into phage DNA by the complex, but the ϕX174 DNA which is replicated appears to suffer semi-conservative synthesis.

The membrane fraction derived from *polA*$^-$ cells appears to produce Okazaki fragments, although Okazaki et al. (1970) found that these are slightly larger than those synthesised in vivo by the cell. The label in the fragments can be chased into DNA of greater size, which implies that the preparation may have the capacity to join fragments together. Stratling and Knippers (1971b) found that an H^3 label added in vitro becomes covalently joined to a C^{14} label given to the cells before preparation of the membrane system. This implies that the reaction catalysed by the extract may be extension of chains which were under synthesis.

Cells of E.coli treated with toluene retain many of their physiological functions, but become permeable to compounds of low molecular weight, including nucleoside triphosphates. Although the cells are no longer viable, they can replicate and repair DNA. Moses and Richarson (1970a) found that a radioactive precursor is incorporated into DNA. With cells containing DNA polymerase I, treating the toluenised cells with DNAase increases their DNA synthetic activity, which suggests that the enzyme is undertaking repair and is stimulated when nicks are made in DNA. But DNAase treatment of toluenised cells of *polA*$^-$ inhibits their activity. The rate of replication in these cells is about 1500 nucleotides per second at 35°C and is semi-conservative; when the cells used are derived from mutant strains which cannot synthesise DNA at high temperature, the toluene system is also inactivated by increase in temperature.

Toluene treated cells synthesize only the DNA which the growing point was about to synthesize in vivo. Burger (1971) found by density transfer experiments that when cells are grown in a C^{14} thymine label before treatment with toluene, about 6% of the label is found in a region of hybrid density when they are given a BUdR label together with H^3-dCTP in vitro. The H^3 label coincides with the hybrid density and transfer of the original C^{14} label to this region can only take place when the pre-labelling has lasted for one generation or longer—this shows that the transfer results from replication and not repair. When NEM (N-ethyl maleimide) is added, there is no transfer of the H^3 label to hybrid density, although some is found in regions of light density; this reveals a repair process but shows that the major activity of the system—in the absence of NEM—is continuation of replication.

Other treatments have also produced non viable cells of the Cairns mutant which are permeable to nucleotide precursors and can replicate DNA from them. Treatment with ether has been used by Geider and Hoffman-Berling (1971) to reveal a system which can replicate DNA semi-conservatively to give Okazaki fragments which are later joined into a product of higher molecular weight. Wickner and Hurwitz (1972) found that E.coli *polA*$^-$ cells plasmolyzed by treatment with high concentrations of sucrose can then synthesise DNA semi-conservatively. When cells of B. subtilis are treated with sodium azide and made permeable to nucleotides by a non-ionic detergent, synthesis of DNA takes place when ATP is present as an energy source. Ganesan (1971) noted that this synthesis is probably not due to the soluble DNA polymerase of the bacteria—the counterpart to DNA polymerase I of E.coli—because almost all of the enzyme is leached out of the cells during the process of making them permeable.

DNA Synthesis by DNA Polymerase II

The replicase activity of the membrane systems, DNA polymerase II, can be made soluble by non-ionic detergents; Knippers (1970) found that it has a molecular weight in the region of 60,000-90,000 daltons, and uses triphosphates as precursors to synthesize DNA in vitro. That this enzyme is distinct from DNA polymerase I is confirmed by its resistance to antisera which inactivate the Kornberg protein. Another difference is that DNA polymerase II is very sensitive to concentrations of p-chloro-mercuri-benzoate which have little effect on DNA polymerase I; sulfhydryl groups are therefore important in the action of DNA polymerase II.

The same enzyme, DNA polymerase II, has been isolated as the active component of toluenized cells by Moses and Richardson (1970b,c) and can be recovered from either wild type or *polA*$^-$ cells of E.coli. The enzyme has also been isolated by an independent procedure developed by T. Kornberg and Gefter (1970) to purify from *polA*$^-$ E.coli cells a DNA synthesising activity. This enzyme appears to represent the residual activity in DNA synthesis of cells lacking DNA polymerase I; its activity corresponds to about one or two per cent of the total capacity of the cell for DNA synthesis. Kornberg and Gefter (1971) have purified DNA polymerase II to homogeneity and estimated that there are about 100 molecules of the protein present in each bacterium.

When the membrane system is prepared by sedimenting a cell extract through sucrose onto a 60% sucrose shelf, it contains most of the bacterial chromosome, structures of the cell wall and the cytoplasmic membrane. Stratling and Knippers (1971a) found that treating this fraction with ultrasonication—which shears DNA but does not influence the membrane—releases DNA from the fraction. The polymerase II activity remains with the released DNA on top of the gradient and fails to sediment to the shelf. The enzyme must therefore be attached to the chromosome rather than to the membrane.

Is DNA polymerase II the replicase of the cell, or is it another polymerase which is active in DNA synthesis in vitro but has some repair function in vivo? The enzyme demands little species specificity in DNA to use as template, but prefers duplex DNA of a size of about $1\text{-}2 \times 10^6$ daltons to single stranded DNA, with which it is inactive. The rate of DNA synthesis by the purified enzyme is about 20% of the rate of replication in the in vitro systems; it is therefore much slower than the rate of replication within the cell.

DNA polymerase II shows the same demand as DNA polymerase I for a free 3′-OH group to use as a primer from which to extend the growing polynucleotide chain. Both enzymes are unable to initiate DNA synthesis de novo (of course, the initiation of Okazaki fragments with sequences of RNA means that the replicase does not need to be able to initiate DNA synthesis). Accordingly, Gefter et al. (1972) have found that DNA polymerase II is most effective with a template of duplex DNA treated with exonuclease III, which contains a short duplex sequence of overlapping complementary single strands ending in 3′-OH groups (see figure 11.1); these are used as primers for chain extension along the protruding single strands. DNA polymerase II also resembles DNA polymerase I and T4 DNA polymerase in its possession of an intrinsic 3′–5′ exonuclease activity which degrades DNA to give mononucleotides.

By mutagenizing the *polA*$^-$ mutant of E.coli, Campbell, Soll and Richardson (1972) isolated a mutant, *polB*$^-$, which lacks DNA polymerase II activity when extracts are assayed in vitro. These cells appear normal in growth and in replication of DNA, which suggests that the polymerase II enzyme does not provide the replicase activity of the cell. Mutants in polymerase II have also been isolated by Hirota, Gefter and Mindich (1972), by using the rationale that although this enzyme is not needed for replication of the host chromosome it may be implicated in replication of infecting phages. Screening mutants of E.coli unable to replicate the single stranded phage ϕX 174 DNA allowed the isolation of *polB*$^-$ mutants lacking DNA polymerase II activity. The mutation is recessive; the general characteristics of the two *polB*$^-$ mutants are the same and they almost certainly represent mutations in the same gene.

Since DNA polymerase II does not appear to be the DNA replicase of E.coli and since it provides at least the major part of the synthetic capacity of the membrane systems, it is unlikely that the synthesis of DNA in these systems represents the activity of the replication apparatus of the cell. Although DNA polymerase II can synthesise DNA as a soluble enzyme in vitro and is the active component of membrane systems, we do not know what function it plays in the cellular metabolism of DNA.

Temperature Sensitive Mutants in Replication

Mutants which are defective in the replication of DNA fall into two general classes. When the temperature of incubation of cells is raised during replication, some mutants can complete the current round of replication although they cannot start another; these mutations must affect the control of replication so that new cycles cannot be initiated. The second class of mutants suffers an immediate cessation or severe reduction in the rate of DNA synthesis; these must represent mutations in the replication apparatus itself, or in enzymes which produce the precursors necessary for DNA synthesis.

Temperature sensitive mutants in replication map at seven different loci (although some of these loci may comprise more than one gene). Table 10.1 shows the division of these mutants into the initiation-defective and DNA synthesis-defective classes distinguished by Wechsler and Gross (1971). The *dnaA* and *dnaC* classes appear to comprise alterations in the ability of cells to initiate new rounds of DNA synthesis; the remaining classes, *dnaB*, *dnaE*, *dnaF*, *dnaG* all prevent the completion of current rounds of replication. None of these mutations influences the activities of the polymerase I or polymerase II enzymes, which appear to be coded by the loci *polA* and *polB*.

The proteins specified by all of these loci except *dnaC* have now been partially purified. The experiments of Shapiro et al. (1970) and Siccardi et al. (1971) discussed in chapter 13 show that the cells of *dnaA* or *dnaB* mutants have altered membrane proteins and appear to suffer changes in what may be the same membrane protein.

Another assay for the *dna* proteins has been developed by Schekman et al. (1972), who took advantage of the inability of ϕX174 DNA to be replicated by cell free extracts from any of the *dna* (except *dnaF*) mutant cells. When an extract from one of these mutants is supplemented with proteins extracted from wild type cells, replication of the ϕX174 DNA becomes possible; fractionation of the wild type extract thus allows purification of the protein

Table 10.1: properties of mutants in enzymes of DNA replication.

locus	characteristics of mutant cells	molecular basis of mutation
dnaA	residual synthesis of DNA allows completion of current round of replication after shift to 42°C	defective in initiation; membrane altered in at least one protein; extract unable to replicate øX174 DNA
dnaB	DNA synthesis ceases at 42°C and DNA may subsequently be degraded (some mutants show considerable decrease in synthesis instead of halt; more than one gene may be involved)	replication apparatus defective; membrane altered in at least one protein; extract unable to replicate øX 174
dnaC	current round of replication completed at 42°C; no further rounds occur	defective in initiation; extract unable to replicate øX 174
dnaD	current round of replication completed at 42°C; no further rounds occur	mutation in same gene as *dnaC* mutants
dnaF	immediate reduction in rate of DNA synthesis at 42°C	defective in ribonucleoside reductase, causing lack of precursors; renamed *nrd*
dnaG	some mutants halt DNA synthesis immediately at 42°C; others complete round of replication. May be more than one gene	defective in initiation of Okazaki fragments; extract unable to replicate øX 174
polA	defective in repair of radiation damage	structural gene for DNA polymerase I; mutants defective in filling gaps between Okazaki fragments
polB	normal	lack DNA polymerase II
polC	formerly identified as *dnaE*; immediate halt or severe reduction in DNA synthesis at 42°C	defective in DNA polymerase III at 42°C
lig	accumulate Okazaki fragments at 42°C; may show radiation sensitivity	defective in polynucleotide ligase at 42°C

Data of Wechsler and Gros (1971) and: *dnaA*—Shapiro et al. (1970), Schekman et al. (1972); *dnaB*—Siccardi et al. (1971), Schekman et al. (1972); *dnaC*—Wickner et al. (1973); *dnaD*—quoted in Taylor and Trotter (1972); *dnaF*—Fuchs et al. (1971); *dnaG*—Lark (1972), Wickner et al. (1973); *polA*—de Lucia and Cairns (1969), Kelley and Whitfield (1970), *polB*—Campbell et al. (1972), Hirota et al. (1972); *polC*—Gefter et al. (1971); *lig*—Gottesman et al. (1973), Konrad et al. (1973).

making good the deficiency in the mutant extract. Both the *dnaA* and *dnaB* proteins have been partially purified by this assay. Wickner et al (1973) have used this in vitro complementation assay to purify the *dnaG* protein, which appears to have a function related in some way to that of the *dnaA* product. In these experiments, the *dna* proteins behave as soluble factors needed for DNA synthesis; the *dnaA* and *dnaB* proteins identified by the in vitro assay cannot at present be related to those changed in the membranes of mutant cells.

The *dnaF* mutation inactivates the enzyme ribonucleoside diphosphate reductase according to Fuchs et al. (1972). This enzyme catalyses the first step in the pathway which leads to the synthesis of precursors specific for DNA so that its inactivation brings a halt to DNA synthesis.

The locus originally identified by the *dnaE* mutation has since been found to code for the enzyme DNA polymerase III, which appears to be the DNA replicase of E.coli; the locus has been renamed *polC*.

The catalytic functions of the enzymes coded by all loci other than *dnaE* and *dnaF* remain unknown. The functions of all the *dna* loci are presumably required for successful replication of the bacterial chromosome. However, not all these functions are required for the replication of all phage DNAs in vitro according to the assay of Wickner et al. (1972b). Since different loci are implicated in the replication of phages ϕX174, M13 and fd, these assays may reflect the different specialised functions of the cell which are needed by each particular phage. These differences therefore imply that at least some of these loci are concerned not with the synthesis of DNA itself but with other functions, for example the attachment of DNA to the membrane.

Although we know that *polA* and *polB* code for DNA polymerases I and II, the role of these enzymes in the cellular metabolism of DNA is not clear. DNA polymerase I appears to play some role in the repair of E.coli DNA damaged by ultraviolet irradiation (see page 516). Either or both enzymes may be implicated in replication in the capacity of filling gaps between Okazaki fragments, although neither undertakes reproduction per se of the chromosome of the cell. These enzymes may also be implicated to different extents in the replication of infecting phage DNAs, for polymerase I, although dispensable within the cell itself, is required for the reproduction of factor *colE1* (Goebel, 1972).

Replication by DNA Polymerase III

All the membrane extract systems devised for DNA replication in vitro differ from the characteristics of replication within the cell in some respect, whether in the overall kinetics of DNA synthesis, the rate of formation and joining of Okazaki fragments, or their failure to show temperature sensitivity when derived from cells which are mutant in replication at high temperature. Of course, if replication depends upon both a membrane bound complex and soluble components, it is possible that these systems may have lost some of the soluble factors and are therefore defective even though they may contain other parts of the replication apparatus. A system developed by Schaller et al. (1972) offers the advantage that all the macromolecular components of the cell which may be involved in replication are kept together and none is washed away.

A highly concentrated bacterial lysate is placed on a cellophane membrane disc which is layered on a buffer containing all four nucleotide triphosphates. The function of the disc is to prevent the macromolecular components of the lysate from dilution by the incubation mixture. DNA precursors diffuse through the membrane to the lysate and are incorporated into DNA. DNA synthesis starts after the brief lag period taken by the diffusion and is proportional to the number of lysed bacteria covering the disc; synthesis continues for at least 60 minutes at a rate which is about 20% of the expected velocity of movement of a growing point in vivo.

Replication is semi-conservative, for when the newly synthesised DNA is labelled by BUdR and H^3-dCTP the radioactive label is found at hybrid density; and denaturation of the hybrid DNA gives single strands which are either heavy or light in density. A pulse label is incorporated into Okazaki fragments and can be chased to DNA of larger size. Replication in this system appears to be caused by a continuation of movement of those replicating forks which were active within the cell, although at a lower rate, presumably because conditions are less than optimal.

The action of polynucleotide ligase in this system can be blocked by addition of NMN, which prevents activation of the enzyme by NAD. Olivera and Bonhoeffer (1972) found that under these conditions the newly synthesised DNA falls into two equal classes; about half of the DNA accumulates as Okazaki fragments sedimenting at about 9S, whereas the other half sediments in a broad distribution with an average of about 38S. When NAD is provided, the fragments are chased into high molecular weight DNA. That the fragments correspond to synthesis of one strand of DNA and the larger molecules of DNA represent synthesis of the other strand has been shown by Herrman, Huf and Bonhoeffer (1972), who found that they can hybridize with each other.

Incubation in the presence of NMN allows the effect of different conditions upon the size of the Okazaki fragments to be followed. A decrease in the rate of chain elongation, produced by lowering the concentration of nucleotide precursors, considerably reduces the size of the larger (38S) fragments. Freezing the cells before the cellophane disc system is prepared causes an increase in the size of the fragments.

These results suggest an explanation for the differences between systems which appear to replicate both new strands discontinuously and those in which one strand is assembled discontinuously and the other by continuous growth. Olivera and Bonhoeffer suggested that initiation of Okazaki fragments and chain elongation may compete with each other on the strand which grows overall from 5′ to 3′ (this must be the one which is assembled in large fragments in the cellophane disc system). If initiation takes place rapidly, synthesis of one fragment may commence before the last has been completed. If initiation takes place slowly, the enzyme synthesising the last Okazaki fragment may have time to continue movement over the initiation site for the next fragment, thereby assembling an intermediate which is longer than the usual Okazaki fragment.

This must be the situation in the membrane disc system, when initiation does not compete effectively with chain elongation on this strand of DNA. (The other strand must be assembled discontinuously because synthesis takes place in the opposite direction to its unwinding and is therefore dictated by the availability of initiation sites). When chain elongation is reduced, initiation occurs more frequently, so that the size of the larger fragments is reduced. Freezing, which increases their size, must in some way preferentially inactivate the initiation apparatus vis a vis the elongation system.

The behaviour of a lysate of *dnaG* cells on cellophane discs supports this model. Lark (1972b) found that as the temperature of this system is raised, the Okazaki pieces synthesised in vitro become progressively larger. This suggests that the *dnaG* locus may specify the enzyme(s) which initiate synthesis of Okazaki fragments. As initiation is increasingly inhibited by raising the temperature of mutant cells, the fragments synthesised on the strand which grows from 5′ to 3′ become larger in size because the replicase elongates chains past the initiation sites where initiation has failed to take place. The other strand of DNA presumably fails to be synthesised at all.

That this system represents the activity of the replication apparatus of the cell is suggested by the observation of Nusslein et al. (1971) that cellophane disc lysates are inactive when prepared from *dnaE* mutants, which behave as though defective in the synthesis of DNA (see table 10.1). This locus appears to code for a soluble component of the system, for addition of soluble factors prepared from wild type cells restores activity to lysates of *dnaE* cells. The complementing activity appears to be a protein of about 150,000 daltons which has the same properties as the enzyme DNA polymerase III of E.coli.

When extracts of the Cairns *polA*$^-$ mutant are assayed for their ability to synthesize DNA in vitro, most of the remaining activity is found in DNA polymerase II; but another peak activity, DNA polymerase III, is also eluted from phosphocellulose columns. This activity is sensitive to NEM (the sulfhydryl group reagent N-ethyl-maleimide) and resistant to antisera against DNA polymerase I.

By comparing the activities at 30°C and 45°C of DNA polymerases II and III in several of the temperature sensitive *dna* mutants, Gefter et al. (1971) demonstrated that both enzymes are normal at both temperatures in all strains except *dnaE* mutants; DNA polymerase III extracted from these cells shows no activity at higher temperatures. This implies that *dnaE*, since renamed *polC*, codes for the polymerase III enzyme, which must therefore play an essential role in replication.

These experiments do not reveal whether DNA polymerase III represents the replicase itself or plays some other role crucial for replication. That the enzyme is concerned with semi-conservative replication is indicated by the observation of Staudenbauer et al. (1973) that phage M13 DNA can suffer conversion from single strands to duplex form in infected *polC*$^-$ cells at high temperature; but the duplex form cannot undertake semi-conservative replication to generate more copies of the phage.

Polymerase III has catalytic activities in vitro very similar to those of polymerases I and II; it synthesizes DNA from 5′→3′ when a 3′-OH primer is provided. The enzyme also has a 3′→5′ exonuclease activity specific for single strands in common with the other two polymerases. All three E.coli polymerases and T4 polymerase therefore have the same synthetic and exonucleolytic capabilities, although polymerase I has in addition a 5′→3′ exonuclease activity. Kornberg and Gefter (1972) and Otto et al. (1973) found that polymerase III is distinguished from the other DNA polymerases by its lower pH optimum, its sensitivity to salt (it is most active at low ionic strength), its requirement for free –SH groups and its stimulation by ethanol. The enzyme comprises a single polynucleotide chain of 140,000 daltons.

Polymerase III functions efficiently with single strands of DNA as template when provided with a primer; it does not replicate native duplex DNA. If this enzyme is the DNA replicase it must therefore be active only in association with other protein activities which unwind the DNA and provide free primer ends. Both polymerases II and III have now been obtained in soluble form without sonication or detergent treatment, which suggests that they are not maintained in the cell only as part of a membrane complex.

In contrast with the 400 molecules in each cell of polymerase I and the 100 molecules of polymerase II, there are only about 10 molecules of polymerase III in each E.coli cell. But polymerases I and II function much more slowly than polymerase III, which has a rate of DNA synthesis in vitro of 15,000 nucleotides per minute at 30°C. The polymerase III enzyme is therefore some 15 times faster than polymerase I and 300 times faster than polymerase II; the cellular capacity of DNA polymerase III appears to be adequate to undertake replication. These three types of DNA polymerase activity may be common in bacteria, for Ganesan et al. (1973) demonstrated that B. subtilis has three polymerases in about the same relative proportions of these E.coli; we do not yet know whether they are analogous to the E.coli enzymes.

Few DNA polymerases of eucaryotic cells have been isolated in sufficiently purified form to determine their catalytic activities; but what evidence is available suggests that they may share the ability of the bacterial polymerases to extend synthesis from a free 3′–OH primer but not to initiate chains. Weissbach et al. (1971) and Schlachbach et al. (1971) showed that Hela cells contain two DNA polymerase activities; DNA polymerase I is found only in the nucleus but DNA polymerase II is found both in nucleus and cytoplasm. Chang and Bollum (1972) and Smith and Gallo (1972) observed that polymerase I of rabbit bone marrow and human lymphocytes sediments at a rate of about 8S; polymerase II sediments between 3S and 4S and may correspond to a protein size of 30,000–45,000 daltons. Wicha and Stockdale (1972) found that the small polymerase of embryonic muscle cells is most active with templates of native DNA and the larger enzyme functions most efficiently with denatured or nicked DNA; Stavrianopoulos et al. (1972) have purified a DNA polymerase enzyme of chick embryos which is a protein of 27,000 daltons with a preferred template of RNA–DNA hybrids, upon which it extends the shorter strand by using the longer strand as template. Mammalian DNA polymerases therefore take forms different from those of the bacterial enzymes.

CHAPTER 11

Modification and Repair of DNA

Host Modification and Restriction

Enzymes of DNA Metabolism

Bacteria contain several enzyme systems which act upon DNA in addition to those which replicate the chromosome; the responsibility for achieving and maintaining the proper structure of the genome does not lie with the replication apparatus alone. The host modification and restriction system—so called because it was discovered by its effect on phage DNA—adds groups to DNA which form a characteristic pattern in the genome of any particular strain of E.coli; any DNA lacking these groups is degraded in the cell. The modifications usually comprise the addition of methyl groups to the bases of replicated DNA; this introduces a note of specificity into the sequence of the chromosome which depends not upon complementary base pairing as such but upon the interactions of enzymes with the nucleotides of DNA. The modified bases have the same pairing properties as their parent nucleotides, so the sequence of replication to give DNA containing the usual four bases and subsequent modification to introduce further species is repeated during each cell division.

Repair systems exist to rectify damage introduced into DNA by environmental agents such as ultraviolet irradiation; these systems remove the damaged sequences and synthesize replacements for them. DNA polymerases which are not involved in semi-conservative replication of the chromosome may play a role in this repair-replication. The systems which are involved in recombination between DNA received from a donor and that of the recipient bacterium have some functions in common with those of repair; some enzymes participate in both repair and recombination so that the cell has more than one way to remove damaged sequences of DNA.

In addition to restriction and repair enzymes with clearly defined functions, cells may contain nucleases which appear to act on any free DNA; in most instances the particular functions of the nuclease in cellular metabolism are not known. But the exonucleases and endonucleases derived from a number of different cell types have proved useful in deducing the actions of other

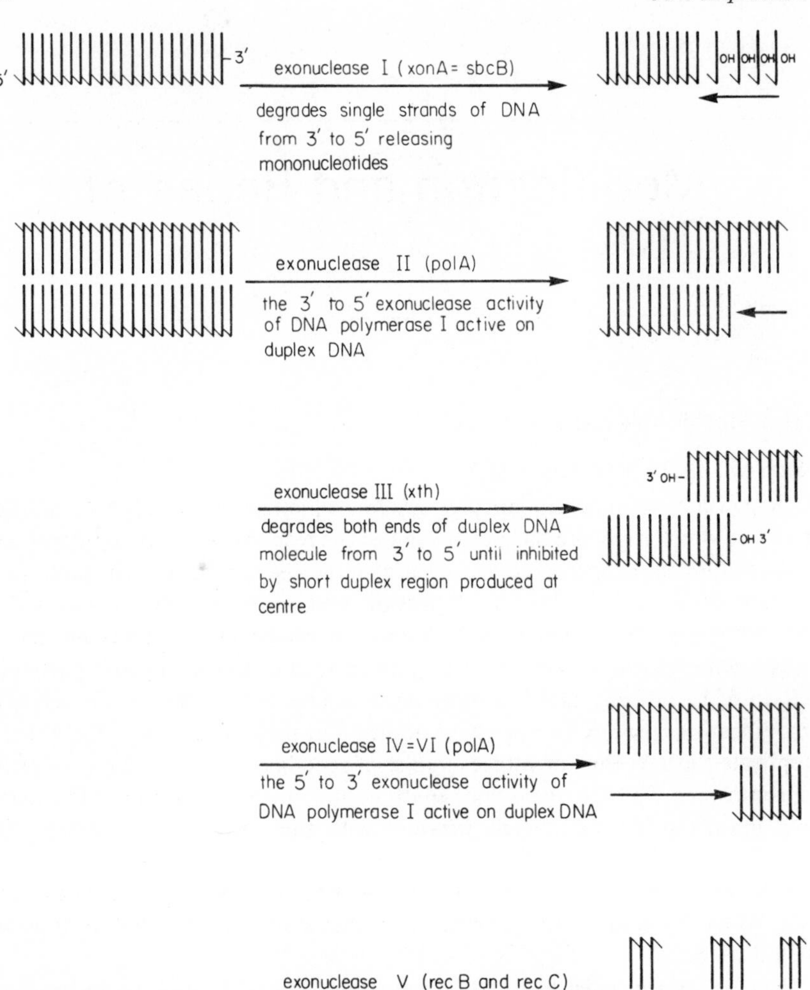
5′
3′
exonuclease I (xonA = sbcB)
degrades single strands of DNA from 3′ to 5′ releasing mononucleotides
OH
OH
OH
OH
exonuclease II (polA)
the 3′ to 5′ exonuclease activity of DNA polymerase I active on duplex DNA
exonuclease III (xth)
degrades both ends of duplex DNA molecule from 3′ to 5′ until inhibited by short duplex region produced at centre
3′ OH–
–OH 3′
exonuclease IV = VI (polA)
the 5′ to 3′ exonuclease activity of DNA polymerase I active on duplex DNA
exonuclease V (rec B and rec C)
ATP–dependent 3′ to 5′ activity which releases short oligonucleotides largely 3—5 bases long

Figure 11.1: exonuclease enzymes of E.coli. Exonuclease I acts on single stranded DNA; the other enzymes all utilise duplex DNA as substrate. The genetic loci which code for the enzymes are noted in brackets. See text for details

enzyme systems on DNA substrates in vitro; the activities of the DNA nuclease enzymes of E.coli are illustrated in figure 11.1

Methylation of the Bacterial Chromosome

The DNA of E.coli contains small amounts of the methylated bases 6-methyl-aminopurine and 5-methyl-cytosine; methylation is accomplished by enzymes which transfer methyl groups from the donor S-adenosyl-methionine to adenine or cytosine bases in the DNA, respectively. Similar DNA methylases can be found in many cell types and may be located in the nucleoplasm of eucaryotic cells. The DNAs of some phages may carry other types of modification; T-even phages, for example, induce an enzyme during infection which glucosylates the phage DNA. Another modification which is used by these phages is the production of 5-hydroxymethyl-cytosine.

When DNA is isolated from any particular cell type, it is usually fully methylated and cannot accept any further methyl groups through enzymes from the same source. DNA from one source is therefore sometimes used as substrate in vitro for the methylating enzymes of some other cell type; although the DNA has been fully methylated by its own enzymes, it can accept further methyl groups through catalysis by heterologous enzymes. Another technique, which has the advantage that it can also be used in vivo, is to starve bacteria which require methionine for the amino acid; their DNA is not methylated during this deprivation but can be modified when methionine is restored. Studies with cell free systems have shown that both the DNA substrate and the methylating enzymes influence the pattern of methylation.

The timing of methylation in the cell cycle has been examined by the procedure shown in figure 11.2. A culture of E.coli cells is pulse labelled for 5 minutes with C^{14}-thymine and then transferred to a medium which contains normal thymine but has H^3-methionine. After ten minutes of incubation in this medium to label the methyl groups of DNA, the culture is transferred to non-radioactive medium which contains BUdR in place of thymine. DNA is extracted after various periods of growth in the density label and the radioactivity is then counted in the hybrid density DNA which contains the newly replicated sequences. Lark (1968a) found that neither the C^{14} nor the H^3 label enters hybrid DNA until almost one generation after the initial pulse dose; and the H^3 label replicates immediately after the C^{14} label. This suggests that the methylated segment of the chromosome is located immediately after the region labelled with C^{14}, which implies that DNA is methylated as or very soon after it is synthesized.

But although replication and methylation usually take place at the same time, the two processes can be divorced. When methionine is removed from the culture medium of cells which cannot synthesise it, the replicating DNA can be labelled with BUdR during the period of starvation but contains no methyl groups. When the cells are subsequently cultured in a normal medium which contains H^3-methionine, the tritium label associates with the density label; this indicates that unmethylated DNA can be converted to its normal state when methionine is restored. Methylation and replication can therefore take place independently.

Only the newly replicated strand of DNA is methylated—the methylating enzyme does not add methyl groups to the parental strand, which was methylated in the previous cycle when it was synthesized. Billen (1968) found that when cells are grown in BUdR and then shifted to a medium containing H^3-

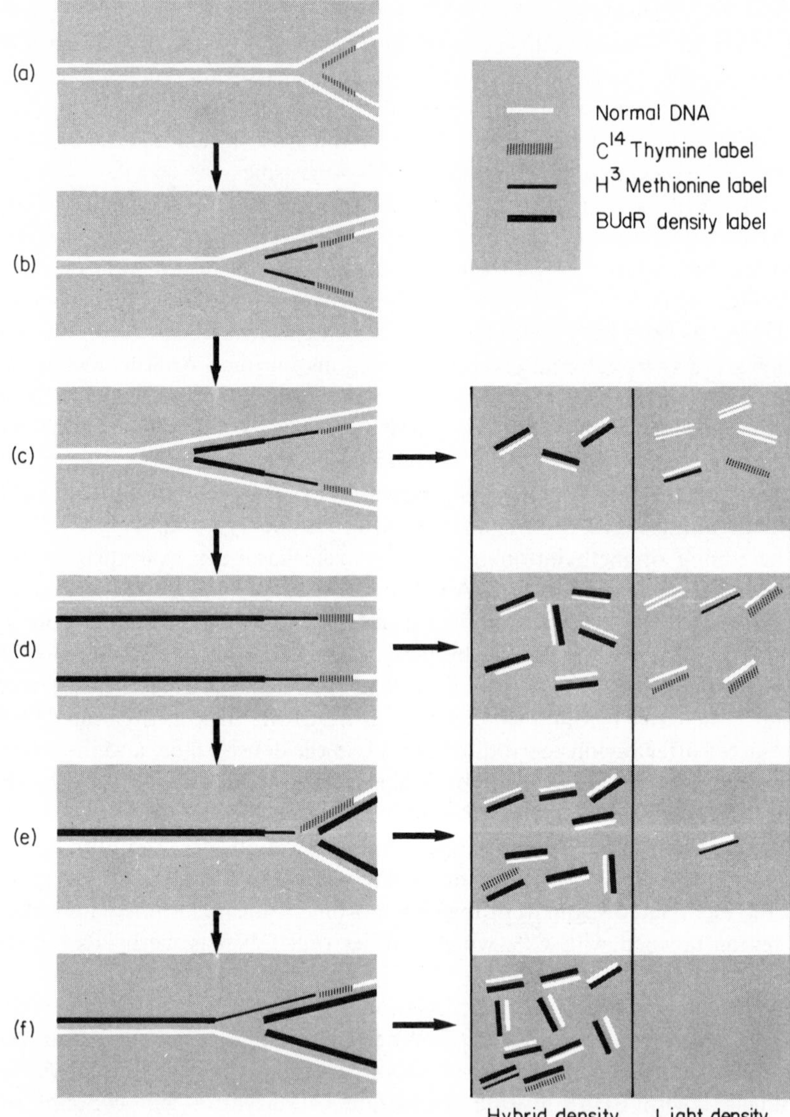

Figure 11.2: coincidence of replication and modification. (a) Five minute pulse incorporation of C^{14}-thymine. (b) Ten minute pulse incorporation of H^3-methionine. (c) Switch to density label of BUdR. (d) Completion of round of synthesis. Because neither radioactive segment has replicated again, both labels are found in light density fragments. (e) Replication of the first radioactive label transfers the C^{14} isotope to hybrid density. (f) The H^3 label is found at hybrid density immediately after transfer of the C^{14} label. This demonstrates that the H^3 must have been incorporated immediately after the C^{14}. Methylation must therefore take place at or immediately after the growing point

methionine, denaturation of the DNA to single strands locates the tritium label only in the heavy fraction corresponding to the newly synthesized strand.

Genes for Host Modification and Restriction in E.coli K and B

One function of this methylation is to provide a mechanism by which the cell may distinguish its own genome from other DNA; the DNA molecules of different strains of E.coli bear characteristic patterns of modification. When phage λ is grown in different strains of E.coli, it is subjected to the modification characteristic of its host; phages grown on strain K are modified in one way, those grown on strain B cells suffer a different modification, whereas E.coli C imposes no detectable modification. That the modification resides in the pattern of methylation of DNA was suggested by Arber's observation (1965) that bacteria fail to impose the usual host specificity on infected phages when the infected culture is starved for methionine. Phage λ, other phages and episomes, and the bacterial chromosome itself are all substrates for the modifying methylation. This early research has been reviewed by Arber (1968) and Glover and Colson (1969).

A host cell can accept the entry only of phage DNA bearing its own characteristic pattern of modification. Whereas $\lambda \cdot$K (phage λ modified by growth on E.coli K) can infect other cells of E.coli K successfully, it cannot infect the cells of different strains of E.coli, such as strain B. When $\lambda \cdot$K enters E.coli B cells, it is degraded, or *restricted*; cells of E.coli B degrade all DNA which enters them except that carrying B-specific modification and E.coli K cells display an analogous demand for K-modified DNA. When DNA is not appropriately modified, therefore, the restriction system recognises that it is "foreign" and degrades it; this means that $\lambda \cdot$K is restricted to growth on cells of E.coli K and this is the origin of the term "restriction".

The restriction system is very efficient, for when phages bearing one pattern of modification infect bacteria of another type, almost all the phage DNAs are destroyed during or shortly after entry into the cell. In a few exceptional bacteria of the culture, however, the restriction system fails to function so that the alien phage DNA is accepted by the cell. This DNA and its progeny are then modified according to the pattern of the new host; the modification pattern of a phage is therefore not inherited from the parent DNA which infected the cell but is imposed by the host bacterium.

Two classes of mutation in the host modification and restriction system arise with equal frequency. Mutants of the type r^-m^+ cannot restrict DNA but continue to modify it; r^-m^- cells lack both activities and can neither modify nor restrict their own or foreign DNA. Mutants lacking only the ability to modify DNA, that is of the class r^+m^- have never been found; one reason for their absence may be that such mutation would be lethal since bacteria would be unable to modify their own DNA and would consequently restrict it, killing the cell.

By making use of a technique which permits isolation of F factors carrying particular regions of the bacterial chromosome, Boyer and Roulland-Dussoix (1969) were able to construct partial permamant diploids of E.coli with different arrangements of mutant alleles for modification and restriction. The r^-m^+ and r^-m^- mutants complement each other, for $\frac{r_1^-m_1^+}{r_2^-m_2^-}$ diploids are wild type and can both modify and restrict their DNA. This shows that each mutation must lie in a different gene.

The equal frequency with which the two classes of mutant arise suggests that both are single gene mutations, mutation in one gene causing loss of restriction activity only and mutation in the other gene causing loss of both restriction and modification activities. As this conclusion predicts, the two genes can recombine in crosses to give wild type bacteria. Because the r^-m^- class itself falls into two complementing groups—so that some $\frac{r_1^-m_1^-}{r_2^-m_2^-}$ mutants are wild type—it seems that mutation in either one of two genes can cause loss of both modification and restriction activities, implicating three genes in all in the system.

These three genes—all of which map close together—might code for polypeptides concerned with recognition, modification and restriction. One model for the action of their three protein products is shown in figure 11.3. Arber and Linn (1969) suggested the nomenclature for the three genes now commonly used. The *hss* cistron specifies the polypeptide responsible for recognising the sites on DNA which are to be either modified or restricted; *hss*$^-$ mutants therefore lack both activities and have the phenotype r^-m^-. The *hsr* locus codes for the polypeptide which has the nuclease activity of degrading unmodified DNA; *hsr*$^-$ mutants fall into the class r^-m^+. The *hsm* gene directs synthesis of the polypeptide with the methylase activity which modifies DNA; *hsm*$^-$ mutants of the class r^+m^- do not exist, but some r^-m^- mutants appear to be caused by mutation to *hsm*$^-$. This implies that mutations in the modification polypeptide may influence restriction as well as modification.

Because only two phenotypic activities—modification and restriction—can be assayed, but three loci appear to be involved, it is more difficult than usual to assign genotypes to particular mutants. But by assuming that three loci are involved, it is possible to build a consistent framework which accounts for the properties of the mutants isolated, the frequencies with which they occur, and the complementation of different mutations to give wild type activity.

Complementation between mutants of the type r^-m^+ and r^-m^- can be explained by supposing that the first class is *hss*$^+$ *hsr*$^-$ *hsm*$^+$ and the second class is *hss*$^-$ *hsr*$^+$ *hsm*$^+$; the diploid has wild type copies of all three genes. Glover (1970) showed that mutants of the type r^-m^- may be isolated in either

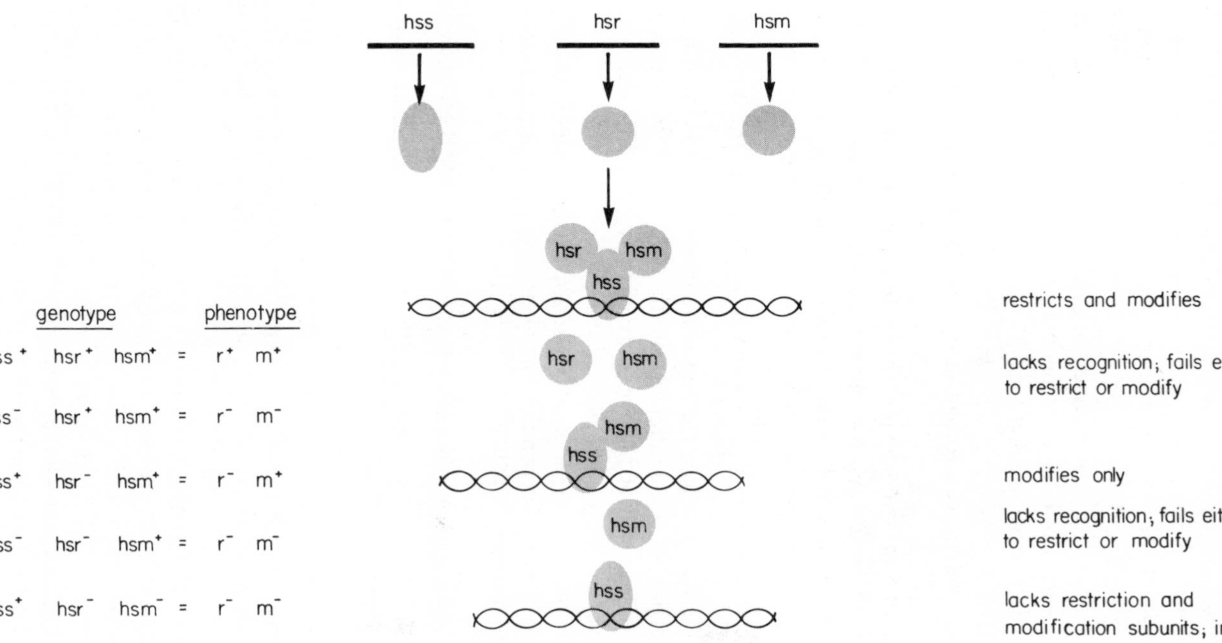

Figure 11.3: genetic character of restriction and modification mutants of E.coli (see text for discussion)

of two ways. The single step mutants which occur with the same frequency as the r^-m^+ type have lost their recognition ability and are *hss*⁼. But r^-m^-mutants may also be isolated by introducing a second mutation into the r^-m^+ mutant class. This mutation may take either of two forms. It may once again be in the *hss* locus, in which the mutation converts the cell from $hss^+\ hsr^-\ hsm^+$ to $hss^-\ hsr^-\ hsm^+$.
The lack of an active *hss* product ensures that neither modification nor restriction occurs. But r^-m^- double mutants should also arise by mutation in the modification activity so that the $hss^+\ hsr^-\ hsm^+$ cell is converted to $hss^+\ hsr^-\ hsm^-$.
Although this mutant can recognise sites on DNA, it can take no action at them.

Both types of r^-m^- two-step mutant are found when r^-m^+ cells are mutated. The activity of the *hsr* and *hsm* gene products can be followed by making diploids between the E.coli B set of modification and restriction alleles and the corresponding genes of E.coli K (which map at the same location). Glover found that cells of the type $\frac{r_K^+\ m_K^+}{r_B^+\ m_B^+}$ can restrict phages of the type λ·K, λ·B or λ·C. The different specificity systems can therefore exist in the same cell and must presumably act upon different sites in the phage DNA. The few phages which survive infection of such cells are of the type λ·KB—they carry both the K-specific and B-specific modifications.

The behaviour of r^-m^- mutants of strain B when made diploid with the r^+m^+ genes of strain K depends upon whether they are one step or two step. The one step mutants, which appear to be $hssB^-$, can only be complemented by an active *hssB* gene; so the diploid $\frac{r_B^-m_B^-}{r_K^+m_K^+}$ fails to show B specificity because the cells are $\frac{hssB^-\ hsr^+\ hsm^+}{hssK^+\ hsr^+\ hsm^+}$. Some of the two step r^-m^- mutants of B show the same behaviour; we expect these to be of the type $hssB^-\ hsr^-\ hsm^+$.

But the remaining two step r^-m^- cells regain their ability to restrict and modify according to B specificity when they are made diploid. These cells must have the genotype $\frac{hssB^+\ hsr^-\ hsm^-}{hssK^+\ hsr^+\ hsm^+}$. This shows that the *hsr* and *hsm* genes are not strain specific; for the hsr^+ and hsm^+ gene products of E.coli K can be used by the $hssB^+$ product to achieve B-specific restriction and modification. A similar conclusion follows from the demonstration that although neither cells of the wild type $r_K^+m_K^+$ nor the mutant type $r_B^-m_B^+$ can degrade DNA of the K type, the diploid between them can do so. The deficiency in restriction of the B system can therefore be made good by the active *hsr* product of the K system.

The restriction and modification activities of either specificity can therefore be directed by the recognition activity of their own or the other cell type. The

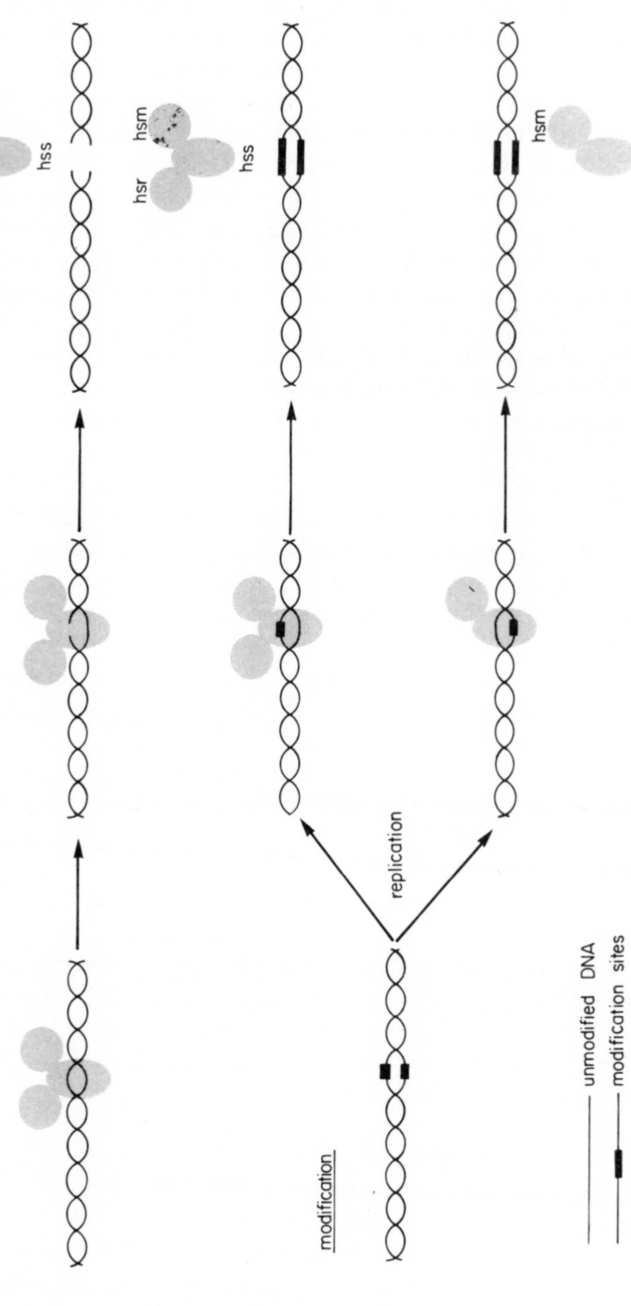

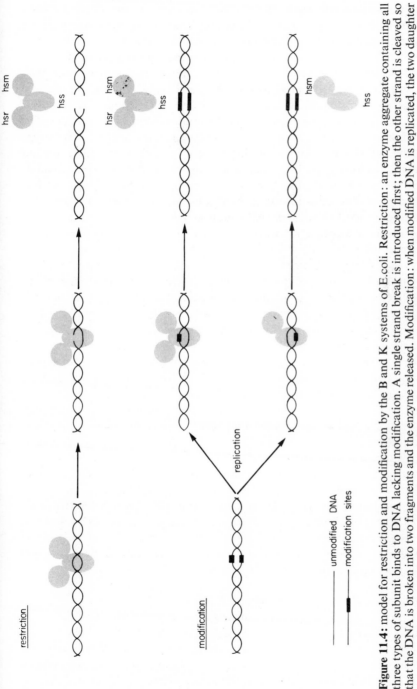

Figure 11.4: model for restriction and modification by the B and K systems of E.coli. Restriction: an enzyme aggregate containing all three types of subunit binds to DNA lacking modification. A single strand break is introduced first; then the other strand is cleaved so that the DNA is broken into two fragments and the enzyme released. Modification: when modified DNA is replicated, the two daughter duplex molecules each have one modified strand and one (newly synthesised) unmodified strand. Either the full enzyme aggregate (upper) or an aggregate containing only hss recognition and hsm modification subunits (lower) may bind to the replicated molecules and modify the new strand

complementation behaviour of the r^-m^+ mutants and the different classes of r^-m^- mutant all support the interpretation that three genes are responsible for achieving restriction and modification, one for each enzyme activity and one to direct these activities according to the specific pattern of the host cell.

By isolating temperature sensitive mutants in the restriction and modification system, Hubacek and Glover (1970) identified a mutation in the *hsm* gene which is temperature sensitive in restriction as well as modification; this means that the *hsm* gene product must be implicated in both activities. The mutation in *hsm* which creates the $r^-m^\pm$ temperature sensitive phenotype partially inhibits the activity of the *hsm* polypeptide in modification but abolishes completely its participation in restriction. In fact, some mutants which are located in the *hsm* gene—as judged by their ability to be complemented by the introduction of an hsm^+ gene from another cell—show the phenotype r^-m^+. This suggests that mutation in *hsm* may on occasion abolish some function essential for restriction, although leaving modification unimpaired.

It is impossible to isolate mutants in *hsm* which retain their ability to restrict foreign DNA; this provides another explanation for the failure of experiments to isolate cells of the phenotype r^+m^-. In teleological terms, it is possible to speculate that this one way interaction between the *hsm* locus and restriction—no mutants in *hsr* affect modification—has evolved because otherwise mutants in *hsm* could generate the lethal ability to restrict but not modify.

The behaviour of mutants in restriction and modification therefore suggests that the recognition polypeptide coded by *hss* must be involved in both restriction and modification; the modification polypeptide coded by *hsm* plays some role in restriction as well as modification, but the restriction polypeptide coded by *hsr* is implicated only in restriction. Figure 11.4 shows that all three polypeptides might aggregate to form one enzyme which undertakes restriction when foreign DNA enters the cell, and which imposes the appropriate modification pattern by methylating new strands of DNA as they are synthesized on a template of the previously modified host cell DNA. Studies of isolated enzymes (see below) support the idea that the cell forms one oligomer of all three polypeptides which can undertake both restriction and modification. In addition, however, the modification and recognition polypeptides appear to be able to form an enzyme lacking the restriction activity which can undertake modification.

Chromosomal and Episomal Systems of E.coli 15

Cells of E.coli 15 carry a host restriction and modification system, type A, which is different from that of either E.coli K or B but which is allelic to them and maps in the same region of the bacterial chromosome. These cells may also carry a second set of genes, type 15, on a plasmid which is closely related to the genome of phage P1. Arber and Wauters-Willems (1970) noted that the two systems are independent and additive; the DNA of these cells carries both A-specific and 15-specific modification and DNA of either type alone is degraded upon entering the cell. This demonstrates that two different systems may co-exist in a stable manner in the same cell and act upon different sites in its DNA.

Inhibiting methylation of the DNA of cells of E.coli $15T^-$—T^- indicates that the cells need to be supplied with thymine in their growth medium—has drastic effects. Lark (1968b) noted that starvation for methionine allows one cycle of replication of the chromosome to be completed, but the next cannot be initiated, presumably because the cell is unable to synthesize the proteins needed for initiation. This means that it is not possible to follow the replication of unmethylated DNA, which does not take place until the cycle of replication after the first round of DNA synthesis in the absence of methyl groups. This difficulty can be overcome, however, by replacing methionine with one of its analogues, ethionine or norleucine; these cannot substitute for methionine in the methylation of nucleic acids but can be incorporated into protein in place of the proper amino acid and so allow enough protein synthesis to permit the next cycle of initiation to be replicated.

When cells are incubated with either of these analogues the cell number and the content of DNA double over the first 60 minutes of growth. During the next 30 minutes the level of DNA remains constant and then about half of it is lost during the next 120 minutes. The cells lose viability in parallel with the degradation of DNA. What seems to happen is that the cells can replicate their chromosomes once in the absence of methionine; this gives progeny genomes which have one (parental) strand methylated and one (newly synthesised) strand lacking methyl groups. When this DNA is replicated in the next cycle of cell division, the unmethylated strand cannot act as a template for DNA synthesis and is degraded. Whilst one strand of the cellular DNA is methylated, it is immune from attack by the restriction system; but as soon as both strands lack proper modification, the chromosome is degraded.

This conclusion has been confirmed by experiments which show directly that methylated DNA can act as a template for further synthesis whereas unmethylated DNA cannot. A pulse label of C^{14} thymine is given to a culture growing in the presence of methionine. The cells are then transferred to a medium containing ethionine instead of methionine and labelled with a pulse of H^3-thymine. After growth in a non-radioactive medium containing thymine and ethionine for ten minutes, the thymine is replaced by a BUdR density label. When DNA is extracted after different periods of incubation, the radioactivity in the hybrid density region can be counted; whereas the C^{14} label appears in this fraction, the H^3 label does not. This shows that the C^{14} DNA, which was synthesised in the presence of methionine, can be replicated; but the adjacent H^3 segment, which was replicated under conditions of methionine deprivation, cannot be used in DNA synthesis. Replication therefore ceases in the unmethylated region of the chromosome, for any newly synthesised DNA is (falsely) recognised as foreign and hence degraded.

Strains of E.coli 15T$^-$ which retain the A-specific system but have a mutant r^-m^+ 15-specific system do not degrade their DNA when grown in ethionine medium. Lark and Arber (1970) have shown that the A, B and K systems all fall into the same category; they do not degrade DNA when cells are grown in ethionine. It is the restriction system of the 15-specific type located on the plasmid which confers this behaviour; when the 15-specific genes are introduced into E.coli K, the response of degradation in response to methionine starvation is found. The DNA of cells which lack the 15-specific system can be labelled either before or after its transfer to medium containing ethionine instead of methionine; DNA labelled before the transfer can be replicated, but DNA labelled after the transfer is less readily replicated, the ease of replication decreasing with the length of time which the cells spent in the ethionine medium.

In cells of E.coli 15T$^-$ there are therefore two responses to deprivation of methionine. DNA synthesised in its absence ceases to be replicated, although this may be either because the DNA is not methylated or because the ethionine does not substitute efficiently enough for methionine in proteins needed to synthesise DNA. A second effect, mediated by the r^+ 15-specific system, degrades any DNA which is synthesised in the absence of methionine.

Two Host Specificity Systems of Hemophilus influenzae

Host modification and restriction is a common attribute of bacteria and the ability of cells of H.influenzae to restrict the phage HP1 suggests that there are at least two different types of specificity system to be found in different cells. Using the Ra type of cell, Piekarowicz and Glover (1972) found that they could not isolate mutants of the type r^-m^+ and r^-m^-; all the mutants which they isolated can be described as $r^{\pm}m^+$ or $r^{\pm}m^{\pm}$ for they retain some activity in modification and restriction. When phage grown on each of the $r^{\pm}m^{\pm}$ mutants is then grown on one of the other mutants, however, DNA of the type HP1–Ra3 is destroyed by mutant cells Ra5 and DNA of the type HP1–Ra5 is destroyed by mutant cells Ra3.

This suggests that the Ra specificity system includes two sets of genes; Ra3 mutant cells are $r^+_{A1}m^+_{A1}$ $r^-_{A2}m^-_{A2}$ whereas Ra5 mutant cells are $r^-_{A1}m^-_{A1}$ $r^+_{A2}m^+_{A2}$. This explains why each class of mutant retains some activity in host modification and restriction. Hemophilus influenzae therefore contains two different chromosomal specificity systems and only when both are mutated do the cells fail completely to restrict or modify DNA. The two systems presumably act upon different sites in the DNA substrate. Whether it is an accident of evolution or whether some cells have a selective advantage by gaining a second system can only be a matter for speculation, but restriction of alien DNA is more efficient in bacteria which contain two specificity systems and fewer phages survive such infections.

Restriction and Modification Enzymes of E.coli

Degradation of foreign DNA takes place very soon after its entry into the cell; it is possible that the restriction activity may be located at the cell surface. The specific action of the restriction system is confined to making double stranded breaks in the alien duplex; some fragments of DNA may survive this cleavage for a short time before they are further degraded, for Dussoix and Arber (1962) were able to rescue donor genes from a foreign phage. Marker rescue seems to compete with the second phase of degradation, which does not appear to be specific but comprises attack by the nucleases of the cell on the fragments of DNA released by the restriction system. Some of the breakdown products diffuse into the culture medium and others are reutilized by the cell.

Restriction endonuclease activities have been isolated from several cells. That the activity of introducing a break in duplex DNA represents the host specificity system is suggested by the isolation from E.coli of an enzyme fraction which cleaves duplex circles of the phage fd. Linn and Arber (1968) showed that the presence of this activity in vitro corresponds to the ability of cells to restrict fd DNA in vivo. Strain C and K cells—which do not restrict fd DNA—have little activity, but the enzyme fraction of strain K degrades the DNA. Mutants of the r^-m^+ and r^-m^- classes are inactive, but a mixture of their extracts is active so that complementation can take place in vitro.

The DNA of phage λ is another good substrate to test for restriction activity and Meselson and Yuan (1968) first characterized the activity of E.coli K cells which makes double stranded breaks in this template. The attack takes place in two stages, for when twisted circles of λ DNA are used as substrate for the enzyme the first product is a form nicked in one strand which sediments more slowly; a second break is then introduced to cleave the duplex. Several breaks altogether are made across the duplex, for degradation proceeds to give a limit product consisting of duplex fragments of DNA. Figure 11.5 shows that the enzyme can attack the DNA of phage $\lambda \cdot 0$ (phage λ which has not been modified) but does not attack the DNA of $\lambda \cdot K$. The general action of the enzyme, then is to recognise appropriate sites on DNA, make a break in one strand of the DNA at this site and then complete scission of the duplex by breaking the other strand.

Recognition of foreign DNA requires both strands to lack modification, for hybrid λ DNA derived by renaturing the separated single strands of $\lambda \cdot K$ and $\lambda \cdot 0$ is not attacked by the enzyme. This agrees with the behaviour in vivo of the restriction systems of E.coli which seem to act on their own DNA only when it has two unmodified strands. By measuring the ability of the restriction endonuclease to bind DNA to nitrocellulose membrane filters, Yuan and Meselson (1970) showed that ATP, Mg^{2+} ions and S-adenosylmethionine are all required for activity; the same cofactors are demanded by the fraction which Arber and Linn (1968) used to cleave fd duplex DNA circles. The demand for the methyl group donor is consistent with the genetic demonstra-

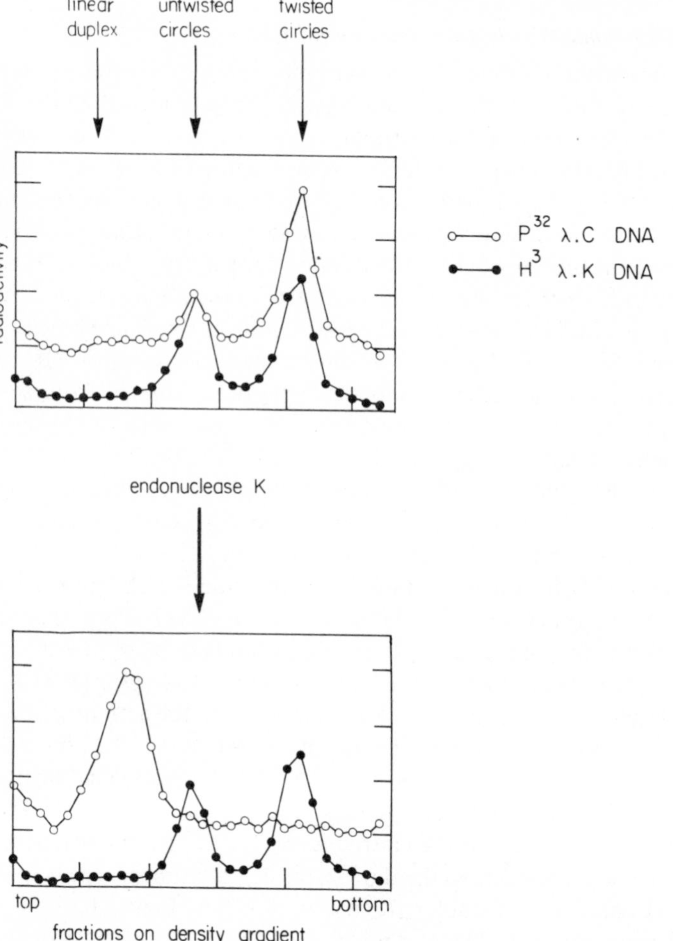

Figure 11.5: action of endonuclease K of E.coli. The upper figure shows the substrate used for the enzyme, consisting of a mixture of P^{32}-labelled λ.C (non-modified) DNA and H^3-labelled λ.K (K-specific modified DNA), both in the form of twisted and non-twisted circles. The lower figure shows the effect of endonuclease K; the H^3-λ.K is not degraded by the enzyme, but the P^{32}-λ.C is restricted to give linear duplex molecules. This demands breakage of both strands of the DNA. Data of Meselson and Yuan (1968)

tion that the *hsm* gene product is required for restriction, although we do not know what role it plays.

Cells of E.coli K which are lysogenic for phage P1 make use of the P1 specificity system as well as the K system. When $\lambda \cdot$K/P1—which carries both K and P1 modifications—is grown on cells of E.coli K, most of the progeny phages are of the $\lambda \cdot$K variety but a few retain both specificity modifications and are $\lambda \cdot K/P1$. When bacteria of the K strain are infected with an average of only one copy of the K/P1 phage each, an average of 100–250 copies of $\lambda \cdot K$ may be produced, but only one $\lambda \cdot K/P1$ DNA results on average. This suggests that it is the material of the parental phage which retains the joint K/P1 characteristics.

By labelling $\lambda \cdot K/P1$ phages with P^{32} before infection of cells of E.coli K, Arber and Dussoix (1962) found that the progeny $\lambda \cdot K$ phages are resistant to inactivation by radioactive decay but the $\lambda \cdot K/P1$ progeny are destroyed. The sensitivity of the progeny $\lambda \cdot K/P1$ to the breakdown of P^{32} is about half of that of the parental phages which were originally labelled; this suggests that the $\lambda \cdot K/P1$ parent phages have replicated so that they contain one parental strand, which bears both K and P1 modifications, and one newly synthesised strand, which bears only the K specificity. The successive rounds of replication of the new strands bearing only K specificity generates the large number of progeny phages which are $\lambda \cdot K$; the much smaller number of phages which retain one parental strand continue to be $\lambda \cdot K/P1$.

Another experiment which points to the conclusion that modification of one strand of DNA is adequate to ensure protection from restriction is association of a density label with the K/P1 characteristic. Arber and Dussoix grew $\lambda \cdot K/P1$ phages on E.coli K/P1 living on D_20 medium and then used the labelled phages to infect bacteria of the K variety growing on normal H_2O medium. The progeny phages can be isolated by their density on a CsCl gradient; assay of their specificities reveals that the $\lambda \cdot K/P1$ phages have a heavier density than usual whereas the $\lambda \cdot K$ phages have normal density. The phages carrying both specificities appear to comprise one strand of parental DNA and one strand of newly synthesised DNA. Since it is not necessary for the phage genome to retain all the parental DNA sequences to have immunity to P1 restriction, only certain sites must be involved.

The purified restriction endonuclease of E.coli K dissociates in SDS to give three polypeptide chains of 135,000, 62,000 and 52,000 daltons. Each enzyme aggregate seems to have two molecules of the largest chain and one each of the smaller. The enzyme binds rapidly and tightly to unmodified but not to modified DNA. In agreement with the concept that only DNA unmodified on both strands is a substrate for restriction, the enzyme does not bind to heteroduplex molecules of λ which have one modified and one unmodified strand (for review see Meselson, Yuan and Heywood, 1972).

The restriction endonuclease of the B-system is very similar. Eskin and Linn (1972) found that it consists of three types of subunit, of 135,000, 60,000 and 55,000 daltons. The enzyme shows the same demand as the K-specific endonuclease for SAM, ATP and Mg^{++} ions. The modification activity of E.coli B has been isolated by Lautenberger and Linn (1972) and has two subunits, of 60,000 and 55,000 daltons. These can aggregate in various forms, with from one to three molecules of the larger subunit associated with one molecule of the 55,000 subunit. All these forms have the same catalytic activity and use SAM as a methyl donor to produce 6-methyl-adenine in DNA; ATP and Mg^{++} ions stimulate the reaction.

These biochemical results support the model of figure 11.4 suggested by genetic analysis; the similarity between the K-specific and B-specific enzymes

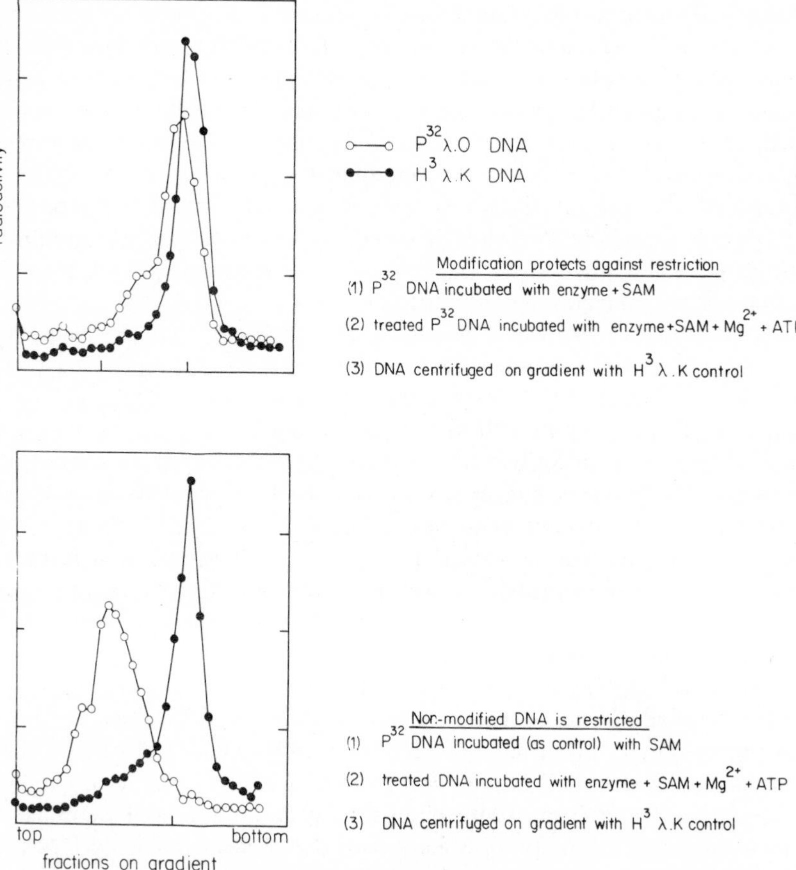

Figure 11.16: modification and restriction are catalysed by same enzyme at specific sites on λ DNA. The upper reaction shows that λ DNA is not degraded under conditions of restriction (SAM + Mg^{++} + ATP) if it has previously been incubated with the enzyme under conditions of modification (SAM cofactor only). The lower reaction shows that if the enzyme is omitted from the modification stage, the λ DNA is not protected against restriction and can be degraded to smaller fragments. This shows that the same enzyme catalyses both modification and restriction and implies that both reactions take place at the same site on the λ DNA. A control of λ.K DNA is not modified in the reaction. Data of Haberman, Heywood and Meselson (1972)

also supports the conclusion that either recognition subunit can direct the activities of the modification and restriction subunits. The large subunit of 135,000 daltons must possess the restriction activity specified by *hsr*. The proteins of 62–60,000 and 52–55,000 daltons must be the products of *hsm* and *hss* (although we do not know which protein has the modification and which the recognition activity).

The aggregate of all three subunits can undertake both restriction and modification. Haberman, Heywood and Meselson (1972) found that when unmodified λ DNA is incubated with the K-enzyme, the provision of SAM, ATP and Mg^{++} ions allows restriction to proceed. However, because ATP and Mg^{++} ions are essential for restriction but only stimulate modification, the provision of SAM alone allows the enzyme to undertake modification; as figure 11.6 shows, the λ DNA therefore becomes resistant to attack when Mg^{++} and ATP are later added to promote the restriction activity. Although the triple enzyme aggregate can undertake modification, the *hss* and *hsm* subunits must be able to exhibit the methylase activity. We do not know whether there is any difference in the modification activities of the triple and double aggregates.

The other restriction and modification systems of E.coli do not have the same composition as the K- and B-specific enzymes. The endonuclease specified by phage P1 has a much lower molecular weight and must therefore have a different structure (for review see Meselson, Yuan and Heywood, 1972). The RI and RII restriction activities specified by the drug resistance transfer factors RTF-1 and RTF-2 have molecular weights of 80,000 and 100,000 and require only Mg^{++} ions for activity. Yoshimori et al. (1972) found that all restriction mutants of these plasmids are r^-m^+; there are none of the class r^-m^-. This implies that the restriction enzyme may be controlled by one gene and the modification enzyme by another, with no common recognition polypeptide.

Mutation in Recognition Sites of Phage DNA

The proportion of the methylated bases of bacterial DNA which is implicated in the modification and restriction system appears to be quite small, which explains the failure of early attempts to demonstrate a correlation between methylation and modification. Cells of E.coli K and B differ in their patterns of methylated bases, for example; 6-methyl-adenine (more correctly described as 6-methyl-aminopurine) is found in both strains whereas 5-methyl-cytosine is found only in K cells. But Mamelak and Boyer (1970) found that there is no change in these patterns when $r_K^+m_K^+$ alleles are introduced into E.coli B. The cytosine methylase gene of E.coli K12 can be introduced into B cells, in which case they produce 5-methyl-cytosine in DNA; it is also possible to stop methylation of cytosine in E.coli K12. But these activities bear no relation to the host specificity restriction and modification system. The number of sites at which the host specificity systems act must be much fewer than the number of methylated bases; different enzyme systems must be involved in host modification and in other methylations and the purpose of the additional methylation is unknown.

The excess of methylation makes it difficult to observe specific events on the bacterial chromosome concerned with the host specificity systems. Phage DNAs provide better substrates because they have much lower backgrounds of methylation. Two approaches have been used to identify recognition sites. One is to select phage mutants which are not restricted when they enter cells of some other host specificity; these mutations can be mapped and locate the sites at which the specificity system of the new host acts. An alternative is to pin down the sites of modification biochemically by growing infected cultures in radioactive methionine or by using non-modified phage DNA as a substrate for cell extracts in vitro. Another experiment possible in vitro is to cleave the non-modified DNA with a restriction enzyme and to characterize the sites of cleavage.

Comparison of the restriction of phages λ and $\phi 80$ by E.coli K12 shows that there are more recognition sites in λ DNA than in $\phi 80$ DNA. Franklin and Dove (1969) found that the targets for restriction appear to be located in small regions of the genome which can be exchanged by genetic recombination; by using recombinants between λ and $\phi 80$, Murray et al. (1973) mapped three of the K-targets in λ DNA. By obtaining mutants of λ which are resistant to the A-restriction system, Arber et al. (1972) mapped the recognition site for this system at a locus between genes *cII* and *0*, probably lying within a gene. Whereas $\lambda \cdot 0$ (non-modified DNA) is labelled with methionine when exposed to the modification enzyme of the A system, $\lambda \cdot A$ and the resistant mutant $\lambda \cdot SA^{\circ}$ are not labelled. This confirms that modification comprises a methylation event and that mutation may prevent recognition of the specificity site. Murray and Brammar (1973) observed that a recognition site for the K-system lies within the *trpE* gene of E.coli; recognition sites must therefore include sequences which code for protein.

The number of recognition sites in toto in a phage DNA appears to depend roughly on its length, for by using different DNAs as substrates for the restriction enzyme activity of E.coli B, Roulland-Dussoix and Boyer (1969) found that one break is made in fd DNA, 3–4 breaks in λ DNA and about 100 in the bacterial chromosome. The DNA of λ has a different number of specificity sites for each host system; there is one site for the A system, 3–4 for the B system, there may be about 6 sites for the K system; and Arber and Linn (1969) have suggested that there may be between 2 and 10 sites for each of the 15, P1 and RTF-2 systems. There appears to be of the order of one recognition site of any particular specificity for every 10^4 base pairs—about 10 genes—which suggests that the minimum sequence needed to constitute a specific host recognition site is about 6–8 base pairs.

Phage fd DNA is restricted by B strain bacteria of E.coli but not by strains C or K. Smith et al. (1972) and Kuhnlein and Arber (1972) observed that when fd is grown on bacteria in medium containing C^{14} methyl groups, 6-methyladenine is the only modified base in the phage. Phage fd DNA grown on E.coli O—"O" denotes any strain of E.coli which lacks host modification and restriction, in this instance r^-m^- mutants—contains on average 1·8 methyl groups on the single DNA strand found in mature phage particles; fd DNA grown on cells of E.coli K itself is labelled to precisely the same extent. But fd modified by E.coli B has 3·8 methyl groups on each single strand of DNA. This suggests that four methylation events take place on the single stranded DNA genome of phage fd grown on B type bacteria; two are the product of the host specific modification and restriction system and two are concerned not with this system but represent a methylation which occurs in all bacteria and whose function is unknown. This latter addition of methyl groups may be characteristic of the "excess" methylation which occurs in bacterial DNA itself.

Mutants of fd which are resistant to degradation by the B specific system can be isolated by growing the phage for alternate generations on E.coli B and E.coli O. Each successive infection of strain B then favors mutants which are resistant to B restriction; such mutants become enriched and may eventually comprise the majority of the phage population. Resistance is conferred by mutation at two sites; the single mutants are SB-1° and SB-2° so that the completely resistant double mutant is SB-1°, SB-2°. When the double mutant phages are grown on cells of E.coli B, their level of methylation is that found in unmodified phage; the phage therefore has two sites at which either a modification may take place to methylate an adenine or restriction may take place to degrade the DNA.

Phage DNA can be methylated in vitro by an enzyme purified from E.coli B cells which require S-adenosyl-methionine; DNA treated in this way is no longer a substrate for B-specific restriction. Only duplex DNA of the replicating form of fd DNA can react with the enzyme; DNA of phage fd grown on E.coli is not methylated but unmodified fd duplex is methylated to yield 6-methyladenine. An average of 4·5 methyl groups are accepted by each wild type fd duplex; only 2·2 methylations take place on the single mutant SB-1° and there is virtually no methylation on the double mutant SB-1°, SB-2°.

That twice as many methyl group are placed on each form of duplex fd DNA as are found on isolated single strands suggests that each site of modification demands methylation on both strands of DNA. This implies that fd DNA has two B-specific recognition sites; at each site one 6-methyl-adenine is created on each strand of DNA, so that each single strand of fd DNA suffers two B-specific methylations and duplex DNA suffers four. In addition, there are two other sites—on the single stranded DNA at least—where another type of methylation event unrelated to the B system also yields 6-methyl-adenine. Modification of the two strands at each B-specific recognition site cannot take place in only one base pair, for 6-methyl–adenine is the product on both strands; one possible model for the recognition site is to suppose that it comprises a sequence of DNA which is symmetrical about a central point, one methylation taking place on each side of this point in complementary DNA strands.

Symmetrical Sequences of Recognition Sites

The sites recognised by three restriction endonucleases have been sequenced and have in common the characteristic of rotational symmetry about a central point. Two of the sites are palindromic, that is the nucleotide sequence is identical on both strands of DNA (although of course of reversed order in the complementary strands). The symmetry of the recognition site means that each strand of DNA must present the same appearance to the enzyme and this may be needed to allow the enzyme to recognise and then to modify or restrict both strands of DNA at analogous sites. The enzymes make different uses of the symmetry, for one cleaves the centre of the palindromic sequence on both strands of DNA and the other two cleave the two strands on either side of the centre of the sequence.

The endonuclease of H. influenzae is inactive with homologous DNA as substrate but produces a limited number of 5′-P, 3′-OH ends by cleaving foreign molecules of duplex DNA. The limit product generally has a chain length of the order of 1000 base pairs and contains no single strand breaks; this is very similar to the product of the K and B restriction endonucleases of E.coli. The length of the product implies that the sequence recognised by the enzyme may occur by chance about once in every thousand base pairs; since any particular sequence of 6 bases must occur by chance about 1/1024 times, we may expect the recognition site to be of about this length.

When T7 DNA is used as substrate for the enzyme, about forty breaks are made in the duplex. Kelly and Smith (1970) used T7 DNA labelled uniformly with P^{33}; after the DNA has been restricted by the endonuclease, the free 5′-P end groups can be replaced by P^{32} by the protocol shown in figure 11.7. Degradation with different DNAases can then be used to release mononucleotides, dinucleotides, trinucleotides from the treated DNA. These fragments can be separated and identified by chromatography using the P^{33} control label; the P^{32} label identifies the nucleotides derived from the sites cleaved by the enzyme.

Degradation with pancreatic DNAase followed by snake venom phosphodiesterase releases the P^{32} label in the form of 63% dAMP and 37% dGMP. Treatment with pancreatic DNAase followed instead by exonuclease I of E.coli degrades away from 3′–OH ends to give mononucleotides, but leaves intact the 5′ terminal dinucleotide; 62% of the P^{32} label is located in pApA and 38% in pGpA. This suggests that the restriction endonuclease cleaves

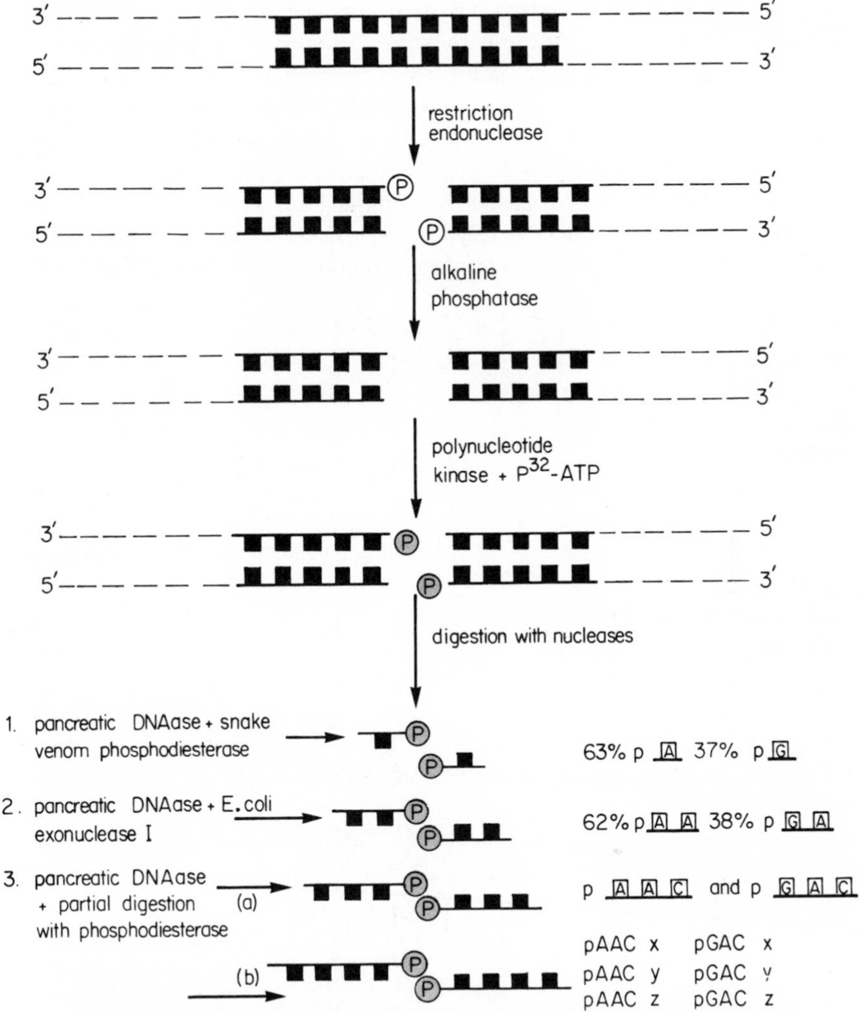

Figure 11.7: analysis of sequence cleaved by restriction endonuclease of H. influenzae. A template of T7 DNA is degraded with the enzyme, which makes about 40 breaks. The 5′-phosphate generated by the cleavage is removed with alkaline phosphatase and is then replaced with P^{32} by the use of polynucleotide kinase and labelled ATP. The labelled DNA is then degraded with appropriate nucleases to generate mono, di-, tri-, and tetranucleotides; those carrying the radioactive label are identified by chromatography. Data of Kelly and Smith (1970)

the duplex on the 5′ side of sequences ending in $\text{p}^{\text{A}}_{\text{G}}\text{pA}$, so that either purine is acceptable at the cleavage point.

When the limit product is digested with pancreatic DNAase alone, oligonucleotides with an average chain length of about 5 residues are generated.

After an average of 1·4 residues has been removed from the 3′ end of each fragment by partial digestion with snake venom phosphodiesterase, the mixture can be fractionated according to its chain length. The P^{32} label found in trinucleotides is located only in pApApC and pGpApC. When tetranucleotides are examined, the P^{32} label is found in six different tetramers, compared with the eight which should be present if the fourth base is irrelevant. This implies that the fourth base is not unique and, within the limits of the experimental technique, is probably random. These results therefore suggest that the restriction enzyme cleaves each strand of DNA on the 5′ side of a

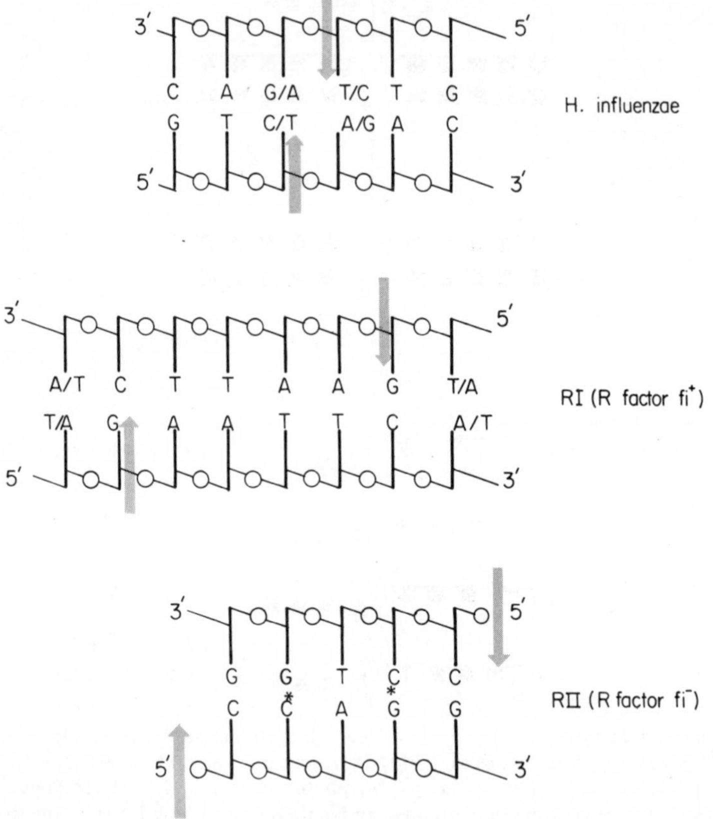

Figure 11.8: symmetrical sequences recognised by restriction endonucleases. Arrows identify points of attack; circles represent phosphate groups. The H. influenzae cleaves the centre of a palindromic siz base pair sequence. The RI endonuclease cleaves points four base pairs apart in an eight nucleotide long palindromic sequence. The RII enzyme cleaves at points five bases apart, symmetrical about a central A-T base pair; the same sequence is recognised in modification, when the two cytosines marked with asterisks are methylated. Data of Kelly and Smith (1970), Hedgpeth et al. (1972), Bigger et al. (1973), Boyer et al. (1973)

sequence ending in $p^{A}_{G}pApC$. The total length of the recognition site is therefore six base pairs.

These experiments do not prove that both strands are cleaved at precisely the same point, for the two cleavage sequences might be located on complementary strands a few bases apart. But if only one site is used, the 3′ ends of the cleaved DNA should be complementary to the 5′ sequence. Digestion of DNA with the enzyme Micrococcal nuclease yields mononucleotides and dinucleotides of the form pX and pXpY in about equal proportions; but the ends of DNA give a unique structure, a dinucleoside monophosphate of the form XpY. Two such species are found in the digest of T7 limit products, TpC and TpT. This suggests an even break in the duplex sequence shown in figure 11.8. The recognition sequence is palindromic, with either an A or G purine on the 5′ side of the break and, correspondingly, a T or C pyrimidine on the 3′ side. By recognising one single sequence of DNA, the enzyme is therefore able to cleave both strands.

Duplex DNA of the eucaryotic virus SV40 has been used as a substrate for the restriction endonuclease specified by the RI system of E.coli. Morrow and Berg (1972) and Mulder and Delius (1972) found that the enzyme makes a break at only site in SV40 DNA, converting the circular molecules to linear duplex molecules. A denaturation map of the linear SV40 DNA can be made by treatment with gene 32 protein or with low concentrations of alkali; the susceptible regions which become denatured can be identified by their strand separation under the electron microscope; such regions are always found in fixed locations relative to the ends of the molecule. Another technique which has shown that the SV40 DNA is cleaved at only one point is heteroduplex mapping of hybrids formed using the virus Ad2^{+}ND1, which contains a small amount of SV40 DNA in an adenovirus genome. This also shows that the homologous sequence is fixed at one particular position of the linear molecule.

The cleaved linear molecules can be converted to covalently closed circles by incubation with E.coli ligase. Mertz and Davis (1972) found when the linear duplex products are denatured, they are also able to reanneal with each other, not only "head to tail" but also "head to head" and "tail to tail". This implies that the RI enzyme cleaves the two strands of DNA at different sites to produce symmetrical cohesive ends by the action

——— a–b–b–a ↓ ——— → ———a–b–b–a + ———
——— ↑ A–B–B–A——— ——— A–B–B–A———

where a,A and b,B are complementary bases.

By using protocols analogous to those shown in figure 11.7, Hedgpeth, Goodman and Boyer (1972) have sequenced the site cleaved in λ DNA by the RI endonuclease, coded by the *fi*$^{+}$ R factor of E.coli. About five breaks are made in each λ molecule, all at sites of identical sequence. Pancreatic DNAase and snake venom phosphodiesterase were used to digest the λ

fragments labelled with P^{32} at the 5′ ends into oligonucleotides of 3–5 residues in length. The fractions of each length can be separated by two dimensional electrophoresis and their base compositions determined. This gives the sequences:

pA
pApA
pApApT
pApApTpT
pApApTpTpC
pApApTpTpCp$^{T}_{A}$.

The cohesive ends which are generated by the action of the RI endonuclease can be "repaired" by the DNA polymerase of Rous sarcoma virus, which extends a primer along a complementary strand of DNA in the action:

```
————a–b–b–a      ————a–b–b–a
                →
————              ————A–B–B–A
```

where A,B are the new bases inserted by pairing with a,b. If a P^{32} label is provided in the α position of the triphosphate precursors, the newly synthesised sequence can be degraded by enzyme action so that the label is released associated with the neighbour of the base which originally contained it. A label provided in ATP is transferred to either A or G; a label provided in dTTP is transferred to either T or A. Labels in dGTP or dCTP are not utilised atall.

This suggests the recognition sequence shown in figure 11.8, in which each strand is cleaved between G and A, two bases from the centre of the symmetrical sequence. When the two molecules produced by the break separate, each therefore has a protruding single stranded sequence of

```
pApApTpT————3′
        ————5′
```

Two of these cohesive ends can base pair under appropriate conditions. Repair of this end demands incorporation of only A and T bases, the first A having for its neighbour the G at the 3′ side of the break, the second having the A incorporated before it. In the same way, the T incorporated has as its neighbours A and then T.

The RII restriction endonuclease specified by the fi^- R factor of E.coli introduces about 20 breaks into λ DNA. The recognition sequence has been determined by Bigger et al. (1973) by replacing the 5′-phosphate groups freed by the breakage with labelled (P^{32}) groups. Analysis of the labelled oligonucleotides released by pancreatic DNAase suggests the pentanucleotide recognition sequence of figure 11.8. Breakage probably occurs on the two strands at the ends of this sequence, that is at sites five base pairs apart. Although not palindromic, the recognition sequence is rotationally symmetrical about the central A-T base pair.

By modifying DNA in vitro with RII enzyme, Boyer et al. (1973) demonstrated directly that the site subject to modification comprises the same pentanucleotide sequence which suffers restriction. Labelled methyl groups enter cytosine alone; digestion of the methylated DNA and characterization of the fragments possessing 5-methylcytosine suggests that the two cytosine residues immediately adjacent to the centre of symmetry—marked by asterisks in figure 11.8—provide the points of modification.

Rotational symmetry about a central axis is therefore a feature common to the restriction and modification sites of H. influenzae, RI and RII enzymes. Although the length of the recognition site, its possession of ambiguous bases, and the location of the cleavage points within it all may differ, all restriction enzymes probably recognise their substrate by the same general class of mechanism, presumably involving a symmetry in the enzyme to match the symmetry in DNA. Although recognition sites for other proteins which recognise DNA have not been determined, speculations have been made that sequences where recombination is initiated may be palindromic (see page 539) and that symmetrical sequences may provide the basis for recognition of operators by repressor proteins (see page 289). The biological significance of symmetrical sequences has been reviewed by Sobell (1973).

Excision and Repair of Mutant DNA Sequences

Influence of Ultraviolet Irradiation

Ultraviolet irradiation of bacteria is mutagenic and, depending upon the dose and the bacterium, may be lethal. Irradiation damages DNA by covalently linking adjacent pyrimidine residues on the same strand to form the cyclobutane ring shown in figure 11.9; the principal product of the reaction is thymine-thymine dimers, but some cytosine-cytosine and cytosine-thymine dimers are also produced. Damage to DNA is the only cellular event caused by ultraviolet irradiation. Formation of the dimers distorts the structure of the DNA duplex since the two bases are no longer connected only through the phosphodiester backbone. The dimers inhibit both replication and transcription; the replicase appears to be held up when it must pass a dimer and continuation of DNA synthesis often results in the introduction of breaks and mutations. RNA polymerase ceases transcription at ultraviolet damaged sites and is released from the template, resulting in the production of short RNA molecules.

All cells, both bacterial and eucaryotic, appear to have enzyme systems which can recognise distortions in the DNA double helix and respond by excising them and synthesising a replacement stretch of DNA. The degree of

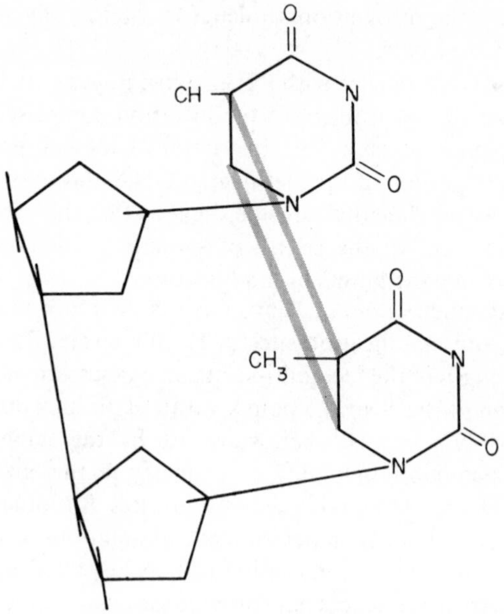

Figure 11.9: Formation of the cyclobutane ring structure of thymine dimers induced by ultraviolet irradiation

resistance to ultraviolet irradiation varies with the organism, depending upon the efficiency of its particular repair systems. Only the enzymes of E.coli and Micrococcus bacteria have been characterized in detail. Micrococcus is especially efficient in removing thymine dimers from DNA and can tolerate large doses of ultraviolet irradiation; E.coli offers the advantage that a large number of mutant bacteria have been isolated which lack the ability to overcome irradiation damage so that their properties can be compared with relatively more resistant strains.

Correction Systems of E.coli

Many genetic loci are concerned with determining the degree of sensitivity of E.coli to radiation and mutations have been mapped and identified in both the K12 and B strains (early research leading to the definition of the different systems has been reviewed by Witkin, 1967 and Hanawalt, 1968). Mutants exhibit a range of influences upon the characteristics of sensitivity to ultraviolet irradiation, sensitivity to X-rays, genetic recombination and the ability to undertake cell division after irradiation.

Phages which do not specify their own repair systems may make use of some of those of the bacterial host and the ability of bacteria to repair damaged phage DNA can be used as an assay for the host correction systems. Bacteria which lose the ability to repair damage either to the DNA of their own genomes or to the DNA of infecting phages are described as hcr$^-$; this stands for *host cell reactivation*, the ability of the bacteria to act upon phage DNA. Mutants of the hcr$^-$ phenotype show a ten or twenty fold increase in sensitivity to both the lethal and mutagenic effects of ultraviolet irradiation; and there is a high correlation between the survival of colony forming ability in ultraviolet irradiated cells and ability in plaque formation when the cells are used as hosts for irradiated phages. A second mutant phenotype is sensitive to ultraviolet because cells cannot repair damage in their own genomes but retain the ability to support host cell reactivation; these are described as uvr$^-$.

The mutations identified in E.coli fall into several genetic classes corresponding to the various systems which can repair the damage caused by ultraviolet and other radiations. In addition to the loci described here which clearly constitute each system, other mutants with related properties have been isolated but the molecular nature of their defects is usually unknown; some of these loci have been tentatively assigned to known systems but which systems other loci belong to are unknown. Whether or not all these loci code for proteins which participate in known repair systems, it is likely that there is only a small number of repair systems, each locus coding for a protein which belongs to such a system rather than functions independently. Cells bearing combinations of mutant loci have been constructed to test the relationships of different mutations.

The *phr* system comprises one gene which specifies an enzyme which can split thymine dimers when cells are growing in visible light; this involves action on irradiation damage in situ.

The *uvr* system is responsible for removal of thymine dimers from DNA by a mechanism which involves excision of a stretch of the single strand of DNA containing a dimer and its replacement by synthesis of a new sequence of DNA. Howard-Flanders et al. (1966) reported that mutation in any one of three unlinked loci, *uvrA,B,C*, causes inability to remove thymine dimers from DNA by this mechanism. Another locus which may belong to this system has been reported by Ogawa et al. (1968) and Ogawa (1970). The nomenclature for describing radiation sensitive cells of E.coli is at times confused, for the *uvr* mutants do not fall into the uvr$^-$ phenotypic class but show the behaviour of hcr$^-$ cells; in other words, the *uvr*$^+$ system acts on phage as well as bacterial DNA.

Mutations in the *rec* system, which comprises three loci—*recB* and *recC* which are closely linked and *recA* which maps at a different location—cause an increased sensitivity to both ultraviolet and X-irradiation (Howard-Flanders and Theriot, 1966). Other mutations which may be part of the *rec* system have been reported by Storm et al. (1971) and Storm and Zaunbrecher (1972). Rec$^-$ cells are deficient in genetic recombination and the primary effect of the *rec* system appears to be concerned with exchanging sequences of duplex DNA between genomes.

Cells bearing the *polA*$^-$ mutation which lack DNA polymerase I are more sensitive to ultraviolet light and their behaviour is similar in some respects to certain *rec*$^-$ mutants. *PolA* and the *rec* genes appear to identify two different systems which to some extent can undertake the same function; we do not know to which system DNA polymerase I belongs or what other enzymes may be implicated in the pathway in which it participates.

Increased sensitivity to X-rays is conferred by mutation at the *exr* locus; *exr*$^-$ mutants are also sensitive to ultraviolet irradiation. This system may be concerned with filling gaps in DNA.

Another class of mutant affects the ability of cells to divide after ultraviolet irradiation or X-ray treatment; these have been termed *lon*$^-$ in E.coli K12 or *fil*$^-$ in E.coli B. The cell division mechanisms of *lon*$^+$ and *fil*$^+$ strains are vulnerable to irradiation and their cells cannot produce septa and divide if their DNA contains unrepaired radiation damage; but the mutants can do so. The mutation changes only the cell division ability and not the capacity to excise dimers or overcome their effects.

Photoreactivation of Thymine Dimers

When E.coli cells are grown in the light, thymine dimer formation may be reversed in situ. Thymine dimers are stable to acid and enzymic hydrolysis, so it is possible to follow their fate after a cell has been irradiated. Both E.coli and yeast cells are *photoreactivable*; after exposure to a source of visible light above 330 mμ—most effectively at about 500 mμ—the damage to DNA is reversed.

Ability in photoreactivation can be lost by mutation in a single gene; *phr*$^+$ cells possess an enzyme which monomerizes the dimers by splitting the bonds which join them and this activity is absent from *phr*$^-$ bacteria. Setlow and Setlow (1963) found that this repair can be catalysed by a cell extract in vitro and restores biological activity to DNA by splitting dimers present in a transforming preparation from bacteria lacking the enzyme. The extract from *phr*$^-$ cells lacks this activity. The biochemical specificities of the enzymes of yeast and E.coli are identical (reviewed by Hanawalt, 1968); the enzyme binds specifically to ultraviolet irradiated but not to unirradiated DNA, in the dark, to form a stable complex. After exposure to light, the complex dissociates to release the active enzyme and a repaired DNA which lacks the dimers.

Excision and Repair of Damaged DNA in E.coli

The *uvr*$^+$ strains of E.coli are relatively resistant to ultraviolet irradiation and the block which it imposes on replication is only temporary; after a period in the dark, cells recover their ability to synthesize DNA and form colonies. This dark repair system does not work by splitting dimers, for when Setlow, Swenson and Carrier (1963) measured the total number of dimers in cells at various times after irradiation they found that the level remains constant; during photoreactivation, of course, the level decreases. This implies that the *uvr* system renders the dimers unable to hamper DNA replication, but without changing the covalent linkage between the thymine bases.

The role of the *uvr* system is to excise the dimers from DNA in the form of small oligonucleotides which enter the soluble fraction of the cell. When Setlow and Carrier (1964) and Boyce and Howard-Flanders (1964a) measured the content of thymine dimers in the acid-insoluble cell fraction (containing macromolecular DNA) and the acid soluble fraction (containing small oligonucleotides), they found that the fate of the dimers depends upon the genotype of the cells. In radiation sensitive *uvr*$^-$ strains of E.coli, the dimers remain in the acid-insoluble fraction; but in *uvr*$^+$ resistant strains they leave DNA and appear in the acid soluble fraction.

The dimers are found in oligonucleotides of three or four residues in length, but this does not correspond to the size of the fragment removed from DNA because further degradation may have taken place after excision. However, the loss of thymine monomers from DNA relative to the loss of dimers suggests that between twenty and fifty nucleotides are removed on average for each dimer. But this figures hides a wide range of lengths of DNA excised for individual dimers (see later).

Although several resistant strains of E.coli remove dimers from DNA at the same rate, they differ in the efficiencies of the subsequent steps of repair, in which more than one enzyme may be involved. Pettijohn and Hanawalt (1964) examined the steps after excision by replacing the thymine of the culture medium with BUdR after the cells had been irradiated. When the cells used cannot produce their own thymine but must depend upon the supply in the medium, any DNA synthesised during the recovery from radiation damage must be density labelled. The arrangement of the BUdR on the chromosome can be followed by denaturing the DNA into single strands which are fragmented and then sedimented through a density gradient.

Semi-conservative replication of DNA takes place by the mechanism of figure 10.1 in which each growing point moves sequentially along the chromosome; a switch in medium therefore generates DNA of hybrid density. But repair-replication takes place by the *non-conservative* processes of figure 11.10 in which the newly synthesised DNA appears to comprise short sequences distributed randomly throughout the chromosome. Because new sequences are fairly short, incorporation of BUdR does not change the density of DNA. Some DNA molecules are repaired in only one strand and others in both. This suggests that after the region around a thymine dimer has been removed, complementary base pairing is used to replace it to restore the original structure of the damaged strand. Although both strands of DNA can undertake repair, it seems likely that only one strand is repaired at any particular site, since otherwise the deleted regions on the complementary strands might overlap so that the chromosome breaks and genetic information is lost.

The DNA repaired by non-conservative replication can subsequently undertake cycles of semi-conservative replication. Repair replication and semi-conservative replication can be separated by ensuring that cells cannot perform normal replication whilst the repair activities are proceeding. Hanawalt (1967) and Hanawalt et al. (1968) achieved this situation by starving cultures of E.coli TAU-bar for amino acids for 90 minutes, which allows the entire population of cells to complete their present replication cycle but prevents them from starting another. The cells can be irradiated in this state and then allowed to

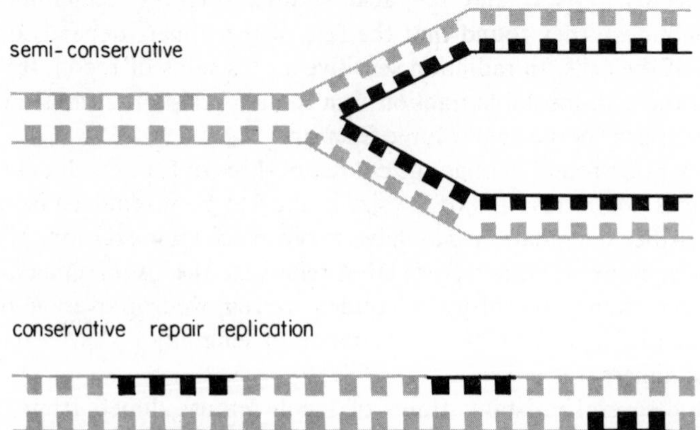

Figure 11.10: semi-conservative sequential replication and conservative repair replication. The usual replication of the chromosome in the division cycle (upper) produces two new chromosomes each of which has one newly synthesised and one parental strand (see also figure 10.1). If new synthesis takes place in heavy medium (indicated by dark lines) the daughter chromosomes are of hybrid density. Conservative repair replication after thymine dimers have been removed insertes only comparatively short lengths of new DNA into the pre-existing chromosome; each strand therefore remains largely composed of its original, light material (shaded lines) and does not suffer a density change

make good the damage by non-conservative repair replication. When the cells are then transferred to a medium containing all their metabolic requirements, a normal round of semi-conservative replication of the chromosome takes place. The DNA produced by repair replication is therefore biologically normal.

Repair systems which act upon dimers created by ultraviolet irradiation can also repair other types of damage to DNA. Hanawalt and Haynes (1965) observed that the relative sensitivities of the ultraviolet resistant E.coli B/r and the sensitive B_{s-1} are similar for both ultraviolet inactivation and nitrogen mustard treatment. After treatment with nitrogen mustard—which introduces cross links between guanines on opposite strands—labelling with BUdR shows the same pattern of non-conservative repair replication which results from ultraviolet treatment. Boyce and Howard-Flanders (1964b) showed that mitomycin C, which also introduces cross links between DNA strands, causes damage which is repaired by ultraviolet-resistant but not by ultraviolet sensitive strains of E.coli.

This suggests that repair systems do not recognise the exact chemical nature of the damage, but rather provide general error-correcting mechanisms which recognise some feature of the distortion of the secondary structure of the double helix. Hanawalt (1968) has suggested that the step in excision-repair which involves recognition of the damage may be formally equivalent to "threading" DNA through a close fitting "sleeve" which gauges closeness of fit to the Watson-Crick structure. Of course, repairing cross-links in DNA may require enzyme actions different from those which excise damage to one strand alone, although some of the same enzymes may be involved (see page 510).

Enzymes for Excision-Repair in M. luteus

Several enzyme actions are implicated in recovery from ultraviolet damage by excision of dimers and Hanawalt and Haynes (1967) proposed the two models shown in figure 11.11 for the order in which they may act. The "cut and patch" model supposes that an endonuclease excises a short oligonucleotide fragment containing the dimer. The resulting short gap may then be enlarged by attack of an exonuclease on the exposed 3′-OH terminus; a polymerase undertakes repair replication by inserting complementary nucleotides opposite the single stranded region which is revealed and finally the ligase connects the last nucleotide added to the exposed 5′-phosphate terminus of the original strand.

The "patch and cut" scheme supposes that repair is initiated by a single strand break introduced near the dimer. Repair replication commences immediately and the defective strand "peels back" simultaneously with the insertion of complementary nucleotides. Repair is terminated by a second nick, which releases the segment containing the damage. This model has the advantage that it would be possible for a single enzyme complex to catalyse all the repair processes as it moves along the DNA. Of course, "cut and patch" and "patch and cut" are extreme models and modes of repair between them are possible.

The enzymes of the *uvr* excision-repair system of E.coli have not been isolated. But some of the enzymes involved in repair replication in M.luteus have been identified and characterized with DNA substrates in vitro. This bacterium has been used as a source for repair nucleases because it has only low levels of other nucleases and this assists identification of the enzymes involved in repair. Takagi et al. (1968) and Grossman et al. (1968) found that a crude extract from this bacterium degrades ultraviolet-irradiated DNA much more rapidly than unirradiated DNA.

The extract contains two enzymes, an endonuclease and an exonuclease, which act in sequence to remove dimers from DNA. The first step in repair

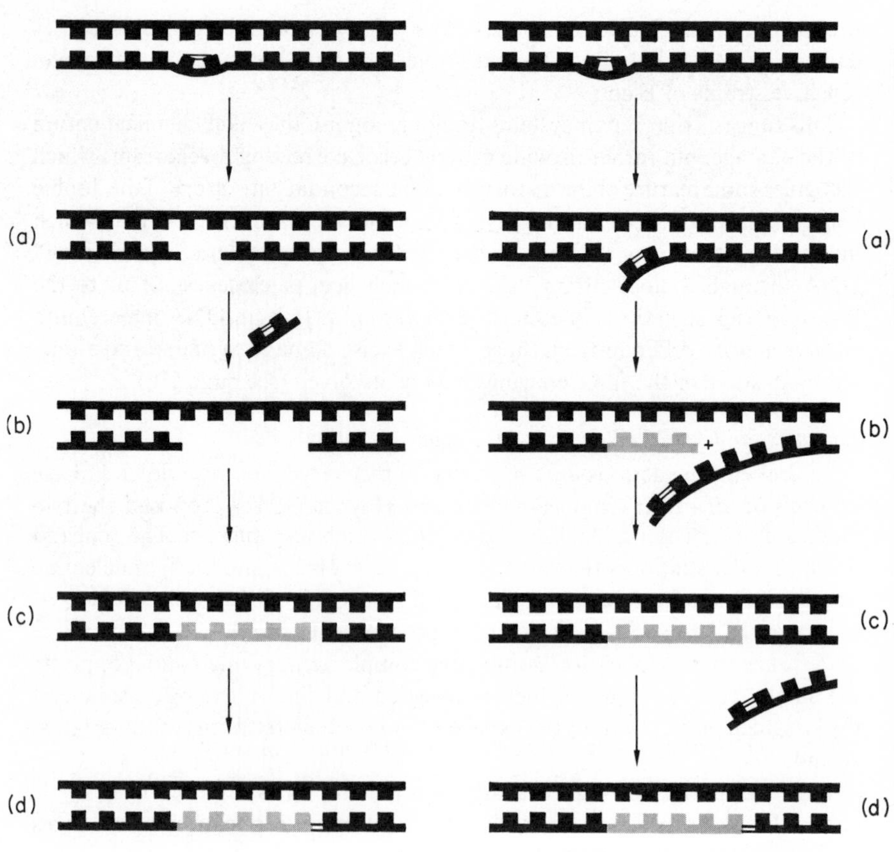

Figure 11.11: models for repair of ultraviolet irradiation damage to DNA. Cut and Patch: (a) endonuclease excises thymine dimer from DNA, (b) exonuclease attack extends region of single strand degradation, (c) repair replication fills the gap, (d) newly synthesized segment is joined covalently to old strand by ligase. Patch and Cut: (a) nick is made adjacent to dimer, (b) repair replication commences, displacing the damaged strand, (c) second nick releases the segment bearing the thymine dimer, terminating replication, (d) ligase action joins new to old segments. Note: the polynucleotide length is usually much larger than illustrated

is the introduction of a single strand break in DNA adjacent to the dimer; Kaplan, Kushner and Grossman (1969) found that the endonuclease is completely specific for ultraviolet irradiated DNA and inactive when normal duplex DNA is provided as substrate; and there is a linear relationship between the dose of irradiation and the number of phosphodiester bonds which it breaks. When DNA which has been irradiated and treated with this enzyme is sedimented through neutral gradients—when it remains in duplex form—

there is no change in its properties; but sedimentation through alkaline sucrose gradients—when the DNA is denatured to single strands—reveals a new peak in irradiated DNA attacked by the enzyme which corresponds to molecules of about one tenth of the length of the unirradiated DNA. The endonuclease therefore makes only single strand breaks in DNA, so that the duplex remains intact.

The endonuclease cleaves the damaged DNA strand on the 5′ side of the dimer, for irradiated DNA containing single strand breaks cannot be used as substrate in vitro for the phosphodiesterase of calf spleen. This enzyme digests DNA from a 5′-OH terminus, but is inhibited by dimers. If the endonuclease makes a break on the 3′ side of the dimer, the phosphodiesterase should therefore be able to degrade DNA; but Kushner et al. (1971a) found that the enzyme is inactive on this substrate, which implies that the break is 5′ to the dimer and fairly close to it. Polynucleotide kinase can place a phosphate group on irradiated DNA which has been broken by the endonuclease, which implies that the 5′ terminus released by scission bears a hydroxyl group; the kinase is inhibited with this substrate, however, which supports the idea that the dimers are close to the break point.

The exonuclease can only act upon DNA which has suffered single strand breaks, for it is inactive with unirradiated DNA, or even irradiated DNA per se, but is active on irradiated DNA which has been treated with the endonuclease. The exonuclease is also active with denatured DNA, either irradiated or non-irradiated. Kaplan, Kushner and Grossman (1971) found that the products of hydrolysis by exonuclease are 5′ mononucleotides, but with a small content of di- or tri-nucleotides of the form pXpX and pXpXpX, which presumably contain the dimers. The exonuclease appears first to release fragments from DNA incised by the endonuclease and then to degrade them, so its overall mode of action is not that of an exonuclease alone but also has features of endonucleolytic action. Overall, about six moles of phosphate are released by the exonuclease for every phosphodiester bond broken by the endonuclease.

This order of events corresponds closely to the cut and patch scheme of figure 11.11. Mahler, Kushner and Grossman (1971) confirmed that the endonuclease provides the first step in repair within the cell by isolating mutants of M. luteus which have lowered levels of the endonuclease. When DNA is isolated from these cells after irradiation, its separated strands retain their usual length, whereas the strands of DNA extracted from cells which have an active nuclease is much shorter. There is a direct correlation between the level of endonuclease in the cell and the extent of subsequent degradation of DNA, which implies that excision of dimers can take place only after the initial incision has been made by the endonuclease.

One difference between the action of the enzymes in vitro and in vivo is that considerably more than six bases are excised for each dimer within the cell. This suggests that the excess degradation may perhaps be due to other nucleases in the cell and is not necessarily essential for the repair of ultraviolet damage. The enzymes which repair the excised DNA have not been isolated, but the DNA polymerase of M. luteus—which is analogous to DNA polymerase I of E.coli—is one candidate for this role.

Phage Coded Excision Enzymes

Although the smaller phages do not code for their own correction systems, the larger phages may specify enzymes concerned with recombination and repair. At least two genes of phage T4 control its sensitivity to ultraviolet light, for mutants of the class v^- or x^- are readily damaged by irradiation and x^- phages are also inhibited in genetic recombination. The two step excision of thymine dimers from DNA may be a common mechanism, for extracts of cells infected with T4 appear to contain two enzyme activities which may be similar to the endonuclease and exonuclease of M. luteus. Sekiguchi et al. (1970) showed that extracts of phage infected cells can remove thymine dimers from DNA, but that this activity is missing from uninfected cells.

The v^+ gene appears to control the first function in the excision of dimers. Friedberg and King (1969) found that extracts of E.coli cells infected with v^+ T4 contain a DNAase activity which acts on ultraviolet irradiated DNA in vitro; this activity is missing from cells infected with v^- phages. Yasuda and Sekiguchi (1970) showed that the active extracts induce breaks in DNA which reduce the length of single strands; this is comparable to the action of the endonuclease of M. luteus.

When irradiated DNA which has been treated with the v^+ enzyme activity is then used as substrate for an extract derived from T4 v^- cells, dimers can be released from the DNA. This enzyme can only act on DNA which has been incised first and releases about ten nucleotides for every dimer. No locus for the control of this enzyme has been identified yet, but its activity may be analogous to the M. luteus exonuclease.

Excision-Repair in Eucaryotic Cells

The ability to repair the damage caused by ultraviolet irradiation or other treatments with similar effects appears to be common to all cell types, although we do not know, of course, whether the same types of enzyme system are involved in all cases. Microorganisms tend to display both photoreactivation and dark repair, for Davies (1965, 1967) has shown that both mechanisms may be used to remove thymine dimers from the DNA of the green alga Chlamydomonas reinhardii; and mutants with an enhanced sensitivity to radiation which fall into phenotypic categories similar to those of bacteria have been isolated in yeast by Snow (1967) and by Nakai and Matsumoto (1967) and in the fungus Ustilago maydis by Holliday (1967).

Repair systems exist also in mammalian cells, for non-conservative repair replication may take place in response to ultraviolet irradiation or treatment with nitrogen mustard; Cleaver and Painter (1968) and Roberts, Crathorn and Brent (1968), for example, have shown that HeLa cells have this capacity and Cleaver (1970) has shown that Chinese hamster cells also do so. Repair replication of this nature appears to occur in almost all mammalian cells, the one exception being rodent cells which Painter and Cleaver (1970) found to have a much lower activity of this type.

Examination of the DNA of unirradiated mammalian cells has suggested that they may have a continuing repair action in addition to their response to irradiation; this is sometimes detected as "unscheduled" replication, when DNA appears to be synthesised during the G1 and G2 phases of the cell cycle (semi-conservative replication takes place only during S phase). Such replication has been attributed with the role of maintaining the integrity of the genetic material. However, Gautschi, Young and Painter (1972) have shown that the apparent non-conservative incorporation of label into DNA is an artefact resulting from the unusual sedimentation behaviour of newly synthesised strands of DNA.

Xeroderma pigmentosum is a human skin disease which follows an autosomal recessive pattern of inheritance, in each known case as though due to a single gene mutation; exposure to sunlight causes a variety of phenotypic changes in skin cells, depending in severity upon the particular form of the disease. We do not know whether the mutations found in different families represent different alterations to one gene or mutations in different genes, but the molecular defect of mutant cells appears to be located in all the instances tested in an excision-repair system which removes thymine dimers from DNA.

Fibroblasts derived from normal skin can excise thymine dimers from DNA, but Setlow et al. (1969) found that cells from the skin of patients with the disease do so at very much less than the usual rate. Normal cells respond to ultraviolet light by showing first an increase and then a decrease in the molecular size of single strands of DNA as breaks are made, dimers excised, and the DNA repaired; xeroderma pigmentosum cells fail to show this series of changes, which suggests that they may be defective in the endonuclease which makes the initial incision close to the dimer.

The ability of the mutant cells to perform repair replication seems to vary with the extent of the disease; Bootsma et al. (1970) showed that there is a correlation between the overall ability of the cells to perform repair replication of damaged DNA and the degree of expression of the mutant. The system which is responsible for repair of ultraviolet irradiation damage also has a capacity for host cell reactivation, for Aaronson and Lyttle (1970) found that functions of SV40 virus which has been irradiated can be expressed in infection of normal cells but not in Xeroderma pigmentosum cells. That the mutant cells are defective only in an early step of the excision-repair system is suggested by the finding of Kleijer et al. (1970) that the single strand breaks introduced by X-rays can be repaired. And other systems must be present in the Xeroderma pigmentosum cell in addition to the ultraviolet excision-repair system which is defective, for Cleaver (1971) found that alkylation damage can be repaired as efficiently as in normal cells.

Few enzymes which may participate in excision-repair have been isolated from mammalian cells and it is not usually possible to correlate any activities which are found with mutations in specific systems. But Burt and Brent (1971) found a DNAase in Hela cells which specifically degrades the DNA of ultraviolet-irradiated E.coli and Lindahl (1971) detected an enzyme activity in mammalian cell nuclei which can liberate dimers in oligonucleotide form from irradiated B. subtilis DNA.

Most of the eucaryotic repair activities which have been noted appear to correspond—albeit only approximately—to the *uvr* excision-repair system of E.coli. The patches inserted, however, appear to be different from those of bacterial systems, for Edenberg and Hanawalt (1972) found that they are confined to some 30 nucleotides in length. Rodent cells seem to contain an activity which may be a counterpart to the recombination-repair systems of bacteria, for Lehmann (1972a) found that gaps are introduced into DNA when irradiated DNA which has not been repaired is replicated; these gaps presumably correspond to the sites where the dimers were located on the complementary parent strand. They are later filled by another enzyme in an action which seems to involve de novo synthesis of a replacement stretch of DNA; this is different from the system which appears to fill gaps in bacterial DNA (see below). In mouse cells and in Chinese hamster cells, the gaps are filled so rapidly that their presence is only transient (Lehmann, 1972b).

Recombination-Repair Systems of E.coli

Post Replication Repair of DNA Containing Thymine Dimers

The presence of unexcised thymine dimers in DNA inhibits its ability to act as template for DNA or RNA synthesis in vivo or in vitro; and the presence of dimers has a mutagenic effect on the expression of genes in which they are located. But the inability of *uvr*$^-$ cells to excise and repair ultraviolet irradiation damage is not necessarily lethal, for DNA containing the dimers can be copied—although more slowly and less efficiently than usual—and sequences lacking the dimers are reconstructed from the replicated chromosomes.

When *uvr*$^-$ cells are allowed to replicate their DNA after ultraviolet irradiation, replication seems to proceed along the chromosome at the usual rate until a dimer is reached, when there is a delay of some 10 seconds. The DNA which is synthesised after irradiation contains many single strand breaks, for Rupp and Howard-Flanders (1968) and Howard-Flanders et al. (1968)

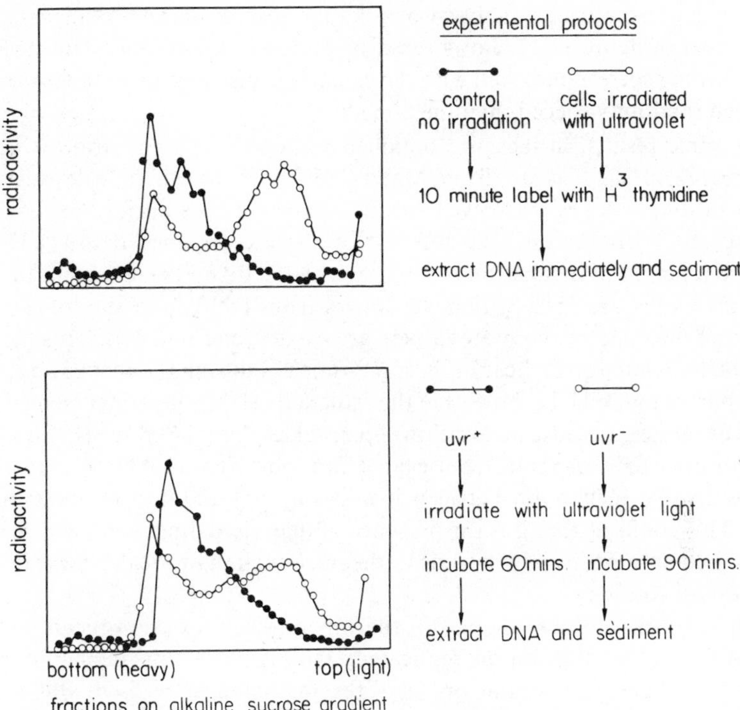

Figure 11.12: effect of unexcised thymine dimers in DNA upon replication. Sedimentation through gradients of alkaline sucrose is used to analyse the fate of an H^3-thymidine label in the newly synthesised single strands of DNA. Upper: if a pulse label is given to unirradiated control cells it enters DNA of high molecular weight. If the cells have just previously been irradiated, they contain thymine dimers which cause breaks in the newly synthesised DNA which therefore is shorter and sediments closer to the top of the centrifuge tube. The same results are obtained with uvr^+ and uvr^- cells since the uvr system does not have time to act during a 10 minute incubation after irradiation. Lower: the same pulse label is given to irradiated cells but they are incubated for an appreciable time in the dark. Under this condition, the photoreactivation system is inactive but the uvr system, if present, can excise thymine dimers. After incubation, DNA is extracted and sedimented through the gradient. The DNA of uvr^+ cells is restored to high molecular weight; the thymine dimers must have been removed during the incubation so that they no longer are present to cause breaks when DNA is replicated. In uvr^- cells, the DNA continues to show a large proportion sedimenting at low molecular weight because the dimers have not been removed and therefore remain an impediment to proper replication. Data of Smith and Meun (1970)

found that a 10 minute pulse of H^3-thymidine enters DNA which is much smaller than usual when analysed on alkaline sucrose density gradients. The upper part of figure 11.12 shows the distribution of these short sequences of DNA, which corresponds well with the calculated distribution of the distances between randomly spaced thymine dimers.

The same results, in fact, are obtained whatever bacterial strain is used, whether *uvr*$^-$ or *uvr*$^+$, as Smith and Meun (1970) have shown, for little excision-repair of dimers takes place even within competent cells during the first ten minutes after irradiation. But different results are obtained if the cells are allowed a period of incubation before the tritiated thymidine is added. In this case, *uvr*$^+$ cells are able to remove dimers from DNA, and the longer the period of incubation, the more dimers are excised and repaired. This means that the template for replication becomes normal during the incubation; the lower part of figure 11.12 shows that the radioactive label then enters sequences of the usual size. Because no excision-repair takes place in *uvr*$^-$ cells, however, they continue to incorporate a pulse label into short lengths of DNA even after a considerable incubation between irradiation and addition of the tritium label. This confirms that it is the presence of unexcised dimers in the parental DNA used as template which leads to the introduction of breaks in the newly synthesised strands.

If these breaks are located opposite the dimers, the *uvr* excision-repair system should be unable to act on the replicated DNA because of its demand for an intact complementary strand opposite the region to be excised. But if the dimers themselves are replicated and the gaps are located further along the new strand of DNA, then a complementary sequence opposite the dimer should be available to the excision-repair system. Howard-Flanders et al. (1968) found that when the DNA of an episome carrying the lactose genes is allowed to replicate in *uvr*$^-$ cells after irradiation with ultraviolet light and is then transferred to recipient cells which are *uvr*$^+$, the damaged DNA cannot be repaired as judged by the ability of the lactose genes to be expressed. But if the cells are exposed to visible light, the photoreactivation system removes the dimers in the donor DNA. This shows that the replicated DNA which has been transferred contains thymine dimers and that the breaks in the newly synthesised strand appear to be opposite the dimers.

The breaks in DNA are not short sequences restricted to the immediate vicinity of the dimers, but appear to comprise long gaps. Iyer and Rupp (1971) made use of columns of benzoylated naphthoylated DEAE-cellulose, which reacts preferentially with single stranded regions of nucleic acids. DNA containing the gaps induced by replication of thymine dimers is retained by the column. Increasing concentrations of caffeine elute increasingly long stretches of single stranded DNA from the column; and the DNA with the gaps elutes at a position which suggests that the breaks in the duplex extend for some $2{\cdot}7 \times 10^5$ daltons—this is about 800 nucleotides.

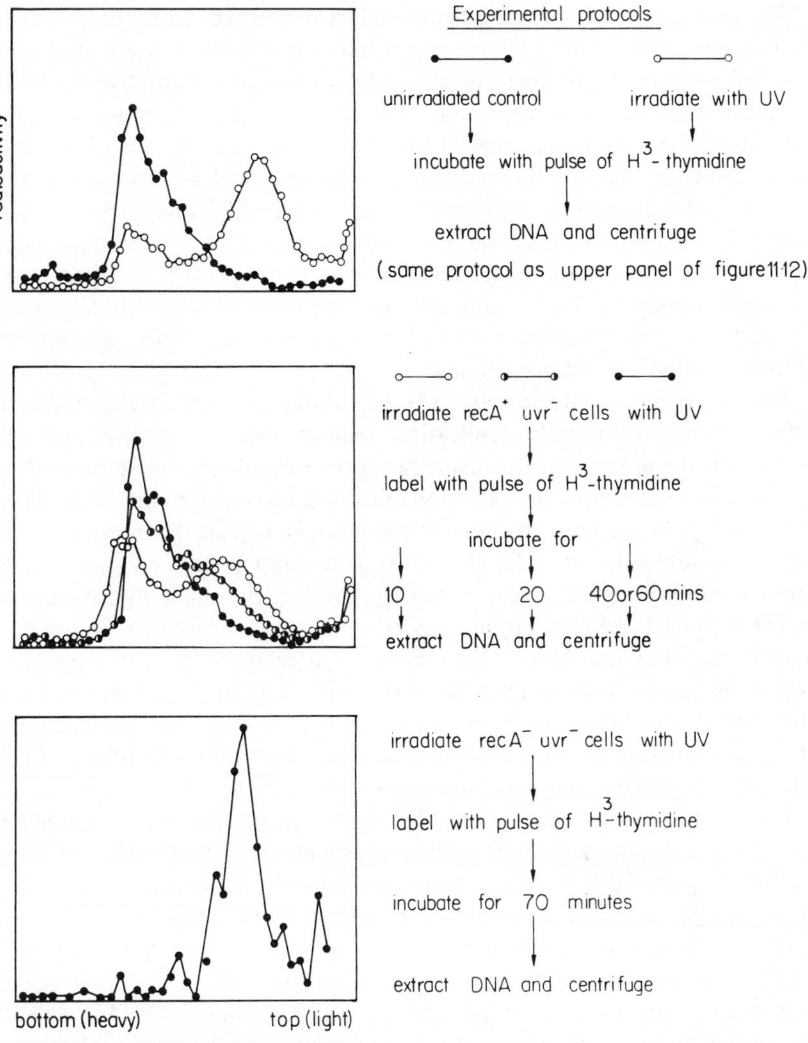

Figure 11.13: role of the rec system in recombination-repair of gaps induced in DNA by replication of thymine dimers. The incorporation of a pulse label of H^3-thymidine into newly synthesised DNA is followed by sedimenting single strands through alkaline sucrose gradients. Upper: control sedimentations showing incorporation of label into unirradiated DNA or into DNA immediately after irradiation. Centre: if cells are incubated after the label has been incorporated, the short strands are progressively joined into longer strands. These cells are uvr$^-$ so the repair is not caused by the uvr system. Lower: cells carrying the recA$^-$ mutation cannot join the short strands into long strands even after extensive incubation. Data of Smith and Meun (1970)

The gaps do not persist in replicated DNA; for as the central part of figure 11.13 shows, when *uvr*$^-$ bacteria which have been irradiated and then pulse labelled with H^3 thymidine are allowed to grow, the short fragments are progressively joined into DNA strands of normal length. The cells must therefore be able to undertake replication during this period; this observation prompted Rupp et al. (1971) to suggest that the process of recovery may involve exchange of sequences of newly synthesised DNA which are normal with the parental sequences containing the thymine dimers. This exchange could reconstitute a normal duplex. That a recombination process of this nature is involved is suggested by the unusually high extent of incorporation of donor DNA into a recipient genome which is experienced when DNA is transferred from irradiated *uvr*$^-$ bacteria to *uvr*$^-$ recipients.

Several models could account for gap filling by exchange between the chromosomes produced by replication. Duplex regions might be exchanged so that the dimers accumulate in one chromosome and the other is free of them. Given the random direction of action of the enzymes which correct mispaired bases in DNA (see next chapter), it seems unlikely that the enzymes would act in one direction only. An alternative is to suppose that regions of single strands are exchanged; the gap opposite a dimer is thus filled by the sequence of DNA which lies on the parental strand which was not mutated. Figure 11.14 shows that after the strands have been exchanged the gaps lie opposite a normal length of DNA which can act as template to fill them. One reason for the operation of such a mechanism may be that the gaps induce the action of the recombination system; and the exchange of strands and filling of gaps may all take place in the same operation.

If strand exchange takes place, the newly synthesised strands should be covalently linked to the old parental strands with which they exchange. Rupp et al. (1971) tested this prediction by growing *uvr*$^-$ bacteria on a heavy medium containing C^{14}-thymidine, irradiating the cells with ultraviolet light, and transferring them to light medium containing H^3-thymidine. Some of the C^{14} label and some of the H^3 label enters single strands which are of intermediate density and must therefore consist of segments of parental DNA covalently linked to newly synthesised regions. The mean distance between the segments containing heavy isotopes in the light chains is about 18×10^6 daltons; this compares with an average spacing between dimers—under these conditions of irradiation—of about 11×10^6 daltons, which suggests that one exchange event takes place for every one or two dimers. Only single strand exchanges seem to be involved, so that the dimers must remain in the original parental strand; but the exchange process ensures that the new strands lack any damage and are therefore themselves good templates for the next round of replication.

The recombination-repair system is also implicated in the repair of the damage caused when psoralen and light treatment is used to introduce cross

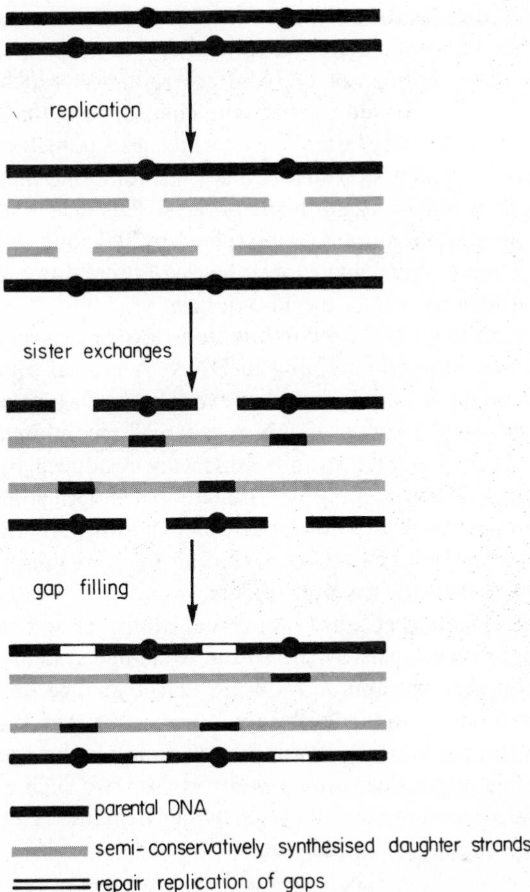

Figure 11.14: gap filling by exchange of sister strands. When DNA containing thymine dimers is replicated, gaps are left opposite the dimers in the new strands of DNA. These gaps are filled by the supply of the homologous length of the non mutated single strand of the parent. After strand exchange, the gaps are opposite a normal length of DNA, which acts as template for their filling by repair replication

links in DNA. Cole (1973) found that repair of this damage involves cleavage of the DNA into discrete pieces which are later joined together by retrieval of information from a homologous duplex.

Enzymes of the rec *System for Recombination and Repair*

Mutants of E.coli which are defective in recombination are also sensitive to ultraviolet light and fall into two classes. The "reckless" mutants in *recA* suffer a drastic reduction in recombination and are very sensitive to ultraviolet irradiation, which causes them to degrade an abnormally large amount of their genomes. The "cautious" mutants in *recB* or *recC* have reduced but detectable recombination, increased sensitivity to ultraviolet light, and a smaller extent of breakdown of DNA after irradiation which is comparable in amount to that of the wild type strain. This suggests that the product of the *recB* and *recC* genes may be a nuclease which is responsible for much of the breakdown of DNA which succeeds ultraviolet irradiation, whereas the product of the *recA* gene in some way limits the process. This idea is supported by the behaviour of the double mutants constructed by Barbour and Clark (1970); $recA^- \ recB^-$ or $recA^- \ recC^-$ have lower levels of breakdown after irradiation than those exhibited by $recA^-$ or wild type cells.

The *rec* mutants identify the recombination-repair pathway of E.coli, which acts on unexcised dimers remaining in DNA. A crucial difference between the *uvr* and *rec* systems is therefore that excision-repair acts on dimers themselves in the parental strands of DNA, whereas recombination-repair acts on the gaps left in daughter strands which are produced by replication of unexcised dimers. That the *rec* genes are concerned with some process other than excision-repair has been confirmed by the experiments of Shlaes, Anderson and Barbour (1972), who have shown that all the *rec* mutant classes display normal activity in excising thymine dimers.

Mutants in *recB* or *recC* show the usual ability to overcome unexcised thymine dimers in uvr^- cells by the strand exchange system which operates after replication. But mutants in *recA* are defective. The importance of the *recA* locus for repair is indicated by the observation that E.coli cells which are uvr^- can survive as many as fifty unexcised dimers in their genomes; but the uvr^- *recA* double mutant can survive only one or two such events. That the *recA* locus is implicated in a repair system which demands replication of DNA is indicated by two types of experiment.

When irradiated cells of E.coli are incubated in a buffer medium in the dark for some time after irradiation, they recover the ability to replicate their DNA normally. This is the result of the action of the *uvr* system in excising dimers shown in figure 11.12; and the process has been termed "liquid holding recovery" when performed in buffer solution or "minimal medium recovery" when performed on poor growth medium. The essential condition for this recovery is that cells are unable to replicate their DNA; for if they attempt to synthesize DNA whilst unexcised dimers remain in the template, they produce the gaps shown in figure 11.14.

A smaller number of cells therefore survive irradiation when they are plated on complex medium (which allows DNA synthesis) rather than on poor medium (when DNA synthesis is prevented). Ganesan and Smith (1968) found that with $recA^-$ mutants the ratio is reduced even further, for it becomes vital to allow excision of dimers from these cells before replication. The effect of DNA synthesis is therefore extremely deleterious to $recA^-$ cells which have not yet had time to excise thymine dimers from their DNA. This implies that the *recA*$^+$ product is responsible for helping to overcome the damage which is caused when cells with unexcised dimers attempt to replicate their DNA. In *recA*$^-$ cells which lack this protein, DNA replication must therefore introduce damage which cannot be repaired at all and is therefore lethal.

Direct experiments with the DNA of mutant cells provide a second line of evidence which confirms that the *recA* product acts after replication and involves overcoming the gaps which are made in DNA synthesis from templates containing dimers. Smith and Meun (1970) found that *recA*$^-$ mutants cannot join together the short strands of DNA which are produced by the replication of damaged DNA. As the lowest panel of figure 11.13 shows, *uvr*$^-$ *recA*$^-$ cells produce short lengths of DNA which remain in this form. This contrasts with the results of the central panel in which the *rec* system causes the short strands to be joined together as the gaps between them are filled. This explains why *recA*$^-$ cells are susceptible to the presence of only one or two dimers in their genomes.

We do not know what enzyme activity resides in the product of the *recA*$^+$ gene; but its absence in *recA*$^-$ cells permits some of the steps of recombination to take place, leading to degradation of the genome instead of strand exchange. The *recA*$^+$ enzyme may be concerned with linking fragments together, thus preventing their degradation. That the product of the *recB* and *recC* genes is involved both in this degradation and in the extensive degradation which follows ultraviolet irradiation of *ras*$^-$, *polA*$^-$, *exrA*$^-$ or *uvrD*$^-$ mutants is indicated by the observation of Youngs and Bernstein (1973) that degradation in double mutants of *recB* or *recC* with any of these loci as well as with *recA*$^-$ mutants is greatly reduced compared with the single mutant which is *recB*$^+$ *recC*$^+$.

The product of the *recB*$^+$ and *recC*$^+$ genes is an ATP-dependent DNAase; cells which are either *recB*$^-$ or *recC*$^-$ lack this enzyme activity in vitro; Barbour et al. (1970) showed that *rec*$^+$ revertants recover the activity. The enzyme consists of two subunits, of 140,000 and 128,000 daltons, one of which is presumably specified by each gene. The enzyme possesses several catalytic activities. Goldmark and Linn (1972) found that it has ATPase activity (see Karu and Linn, 1972), can act as an endonuclease which attacks single stranded DNA at random sites, and acts as an exonuclease to degrade duplex DNA sequentially, releasing short oligonucleotides mostly of about 3–5 residues. The endonuclease action is relatively independent of ATP, but the exonuclease activity demands its presence. When DNA is degraded exonucleolytically, ATP is cleaved; Wright, Buttin and Hurwitz (1971) found that although more ATP is cleaved than phosphodiester bonds of DNA broken, hydrolysis is proportional to the extent of degradation of DNA. They have suggested the name exonuclease V for the enzyme.

Mutants of E.coli in *recB* or *recC* may revert to wild type either by reversal of the original mutation or by gaining a mutation in either of two loci, *sbcA* or *sbcB*. The double mutants have normal activity in recombination, compared with the hundred fold decrease seen in *recB*$^-$ or *recC*$^-$ cells. Kushner et al. (1971) showed that *sbcA* mutants possess an ATP-independent exonuclease, which may act as an analogue of the *recB/C* product to restore the function of the *rec* pathway.

Mutants in *sbcB* appear to lack the enzyme activity of exonuclease I; this enzyme must therefore prevent recombination in *recB*$^-$ or *recC*$^-$ cells which lack exonuclease V. Mutants which suppress the ultraviolet sensitivity of *recB/C* cells have been isolated by Kushner, Nagaishi and Clark (1972); these *xonA*$^-$ mutants also lack exonuclease I and may be located in the same gene as the *sbcB* mutants. The *xonA* mutation does not overcome the deficiency of the *recB/C*$^-$ cells in recombination. One possible explanation is that the *sbc* and *xonA* mutations lie in the same structural gene, the *sbcB* mutants lacking exonuclease I activity in toto but the *xonA* class retaining sufficient activity to repair ultraviolet irradiation damage but insufficient to act in recombination. An alternative model is to suppose that the mutations lie in different genes and that both *xonA* and *sbcB* activities are required for activity of exonuclease I, but the *sbcB* product is required in addition for some other activity. A formally equivalent model in terms of its prediction of the activities of mutant cells is that *sbcB* identifies a control element which is needed for the synthesis of both exonuclease I and some other enzyme which is essential in recombination.

One conclusion to which these results point is that although the recombination and repair pathways share some activities, they are not identical. Exonuclease I removes free 3′-OH groups from DNA, which implies that such termini may be needed for survival of cells after ultraviolet irradiation. When the enzyme removes these groups, the repair system which operates in the absence of the *recB/C* product is inhibited; when the enzyme is missing, the repair pathway can function even in the absence of the *recB/C* exonuclease. The failure of the *recB/C*$^-$ mutants which also carry a *xonA* mutation to carry out recombination as well as repair implies that the presence of 3′-OH groups, although it may be necessary for recombination, is not sufficient.

Induction of Mutations by Ultraviolet Light

When the *uvr* excision-repair system is active, thymine dimers are excised and replaced by a normal sequence of bases. There is no reason to suppose that the operation of this system should lead to the introduction of mutations in DNA, which led to the early suggestions that the lethal and mutagenic effects of ultraviolet light upon the cell might be due to different reactions. But rather does it seem that mutations are induced when DNA containing unexcised thymine dimers is replicated. Bridges and Munson (1968) found that an E.coli strain which lacks the ability to excise dimers shows a low level of mutation in each replication cycle after irradiation.

If mutation results from the presence of unexcised dimers, it should be prevented by subjecting the cells to photoreactivation. When the cells are exposed to light after growth in the dark, mutation ceases as the dimers are removed; after four generations of growth in the dark, the induction of mutations ceases and photoreactivation loses its effect. This implies that unexcised dimers may persist in the cell for some four cycles of growth and have a low probability of causing a mutation in each cycle. Some step concerned with the replication of DNA sequences containing thymine dimers may therefore be error-prone and induces mutations.

The mutagenic effect cannot be replication of the dimers themselves, for the replicase leaves a large gap when it progresses past an unexcised dimer. But an obvious step at which mutation might occur is the filling of this gap. One stage of this process is the single strand exchange catalysed by the *recA* gene; exchange of DNA sequences should not in itself induce mutations, for it restores the original duplex structure of two parent strands over the region of exchange. But the new and old strands must be joined together after exchange, and the gap which has been created on the duplex which has lost its sequence of newly synthesised DNA must then be filled. Mutations might be induced at any of these stages. One indication of the reaction which may be involved is the finding of Meistrich and Drake (1972) that thymine dimers in phage T4 DNA induce both frameshift mutations and GC → AT transitions. This implies that the mutations are not produced at the site of the dimer itself, for misreading of a T-T sequence should induce AT → GC transitions.

The *exr* system has been implicated by genetic means in the induction of mutations by ultraviolet, although we do not know what biochemical role its product may play. Mutants which are exr^- are defective in their response to X-ray irradiation. Two types of damage seem to result from this irradiation. Single strand breaks are made in DNA and these are repaired by a fast process, probably just a simple ligase action. In support of this contention, Dean and Pauling (1970) and Konrad et al. (1973) observed that the *lig-ts7* mutant of E.coli is more sensitive to X-irradiation than its parent strain.

In addition to the creation and repair of breaks, E.coli cells respond to X-irradiation by degrading a considerable amount of DNA; this degradation is not related to the enzyme action of the excision-repair system for Emmerson and Howard-Flanders (1964) and Billen et al. (1967) found that mutants unable to excise thymine dimers continue to degrade their genomes in response to X-irradiation. Some 6000 nucleotides appear to be released from the genome by each irradiation event. Mutants which are exr^- display increased degradation of DNA and a depression of incorporation of nucleotides into DNA after irradiation with either ultraviolet or X-rays. The degradation may in part at least be mediated by exonuclease V, for Youngs and Bernstein (1973) showed that it is greatly reduced in $recB^-$ or $recC^-$ mutants.

A connection between the *exr* system and the systems which repair ultraviolet damage is indicated by the observation that although mutants which are sensitive to ultraviolet light are not usually more sensitive to X-rays, many X-ray sensitive mutants are also sensitive to ultraviolet. And in addition to causing an increased lethality rate in response to ultraviolet irradiation, the *exr*$^-$ mutation has a profound effect upon the induction of mutations by ultraviolet light. If *exr*$^-$ strains are subjected to doses of ultraviolet which are mutagenic in *exr*$^+$ strains, fewer bacteria survive; but there are no ultraviolet-induced mutations amongst the survivors.

This indicates that the *exr* system provides a repair mechanism which is error-prone and so tends to introduce mutations into DNA; Witkin (1967, 1969) has proposed that because *exr*$^-$ mutants lack this system, there are fewer survivors after irradiation, but the bacteria that do survive must have overcome their damage by some other, more accurate system which does not induce mutations. One obvious role for the *exr* product is to help fill the gaps which are left when DNA containing thymine dimers is replicated; the system is probably implicated in a late stage of repair because it seems to be implicated with the healing of short breaks in DNA rather than extensive gaps.

If the *exr* system is concerned with repair of this nature, mutation from *exr*$^+$ to *exr*$^-$ should interfere with the mutagenic effect of any agents which act by causing single strand gaps. In accordance with this prediction, Bridges, Law and Munson (1969) found that both gamma irradiation and thymine deprivation have a severely curtailed effect in *exr*$^-$ bacteria compared with the *exr*$^+$ parent strains.

Catalytic Activities of DNA Polymerase I

Cells of the *polA*$^-$ mutant of E.coli which lack DNA polymerase I are about five times more sensitive to ultraviolet light than cells of their parent strain. Although apparently normal in growth and reproduction, the cells appear to join together the Okazaki fragments of replication more slowly than wild type cells (see page 450). Another deficiency of the mutant seems to lie in a slowed activity in repairing single strand breaks, for Kanner and Hanawalt (1970) showed that single strands of DNA—analysed on gradients of alkaline sucrose—are always of lower molecular weight than those of the parent strain.

The absence of DNA polymerase I is unlikely to cause failure to seal simple breaks, the action catalysed by polynucleotide ligase, but is probably concerned rather with the synthesis of short stretches of DNA to fill single strand gaps. It is not apparent what function the enzyme plays in repair of thymine dimers within the cell and whether it has similar functions in repair and in replication; but in vitro the enzyme possesses many of the abilities which must be needed by enzymes of the excision-repair system.

In addition to its ability to synthesize DNA on a single strand template from a primer sequence, the enzyme can degrade the strands of a duplex in either the $3' \rightarrow 5'$ direction (an activity shared with all other bacterial DNA polymerases) or from 5′ → 3′ (a unique activity). In reviewing the structure and functions of DNA polymerase I, Kornberg (1969) observed that it comprises a single polypeptide chain of 109,000 daltons which possesses all its catalytic activities. The protein appears to have (at least) the five active sites illustrated in figure 11.15; the template and primer sites bind the two strands of a duplex DNA substrate and the triphosphate site binds incoming nucleotides.

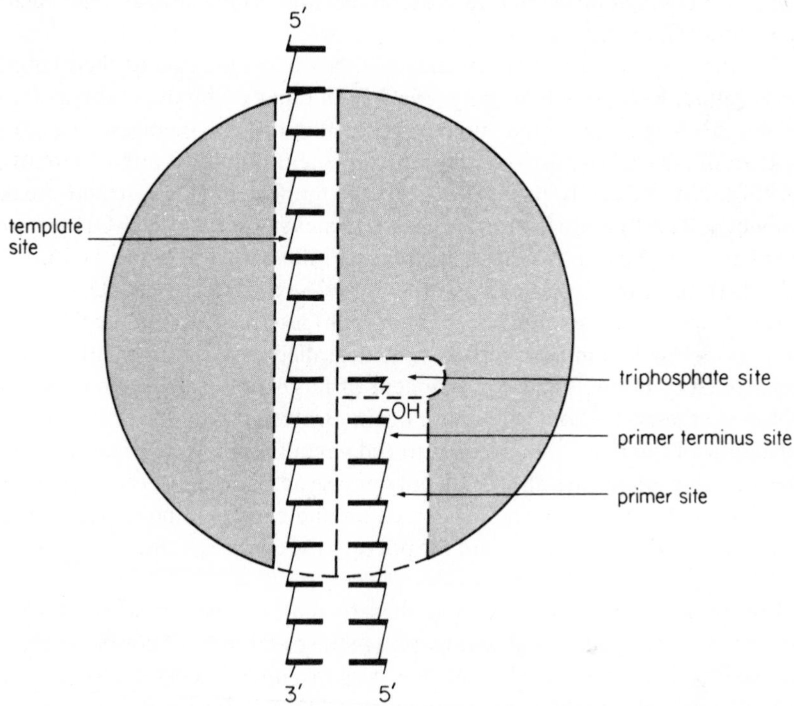

Figure 11.15: catalytic sites in the active centre of DNA polymerase. In addition to the four sites shown, a further independent site must be involved in 5′→3′ exonuclease activity. After Kornberg (1969)

One site appears to be implicated in both 5′ → 3′ DNA synthesis and 3′ → 5′ exonuclease degradation, for Deutscher and Kornberg (1969) found that the presence of dideoxy-TTP (a dTTP in which the 3′-OH has been replaced by hydrogen) at the 3′ end of the chain in the primer terminus makes the primer inert to both further extension and 3′→5′ hydrolysis; 5′→3′ degradation is not inhibited. The ability of monophosphates to bind to the primer terminus site demonstrates that it is distinct from the triphosphate binding site and that it demands a 3′-OH group at the end of the primer chain. Figure 10.10 illustrates a model for how these sites may be involved in ensuring that replication is accurate.

This interpretation of the location of catalytic activities in the polymerase is supported by the properties of the two fragments which are produced when a proteolytic enzyme is used to split the protein chain. Setlow, Brutlag and Kornberg (1972) found that the large fragment of 76,000 daltons retains the polymerase and $3' \rightarrow 5'$ exonuclease activities; whereas the small fragment possesses only the $5' \rightarrow 3'$ exonuclease activity. Since the large fragment can synthesize DNA, it must contain the template and primer sites and the nucleoside triphosphate binding site.

The enzyme binds to single strands of DNA in proportion to their length; but Englund, Kelly and Kornberg (1969) found that no binding takes place to duplex DNA lacking strand interruptions. But the polymerase can bind at nicks in duplex DNA, for the amount of protein bound to duplex circles of ϕX174 DNA is directly proportional to the number of single strand breaks. Kelly et al. (1970) reported that when these nicks possess free 3'-OH termini, the enzyme engages in the "nick translation" illustrated in figure 11.16.

Nick translation demands both the $5' \rightarrow 3'$ exonuclease and DNA polymerase activities; the exonuclease action degrades one strand of the DNA and the polymerase action then replaces the degraded material by synthesizing a new stretch of DNA. The large fragment of DNA polymerase can undertake DNA synthesis without concomitant hydrolysis; but the characteristic behaviour of the intact enzyme is restored when the small fragment is added. The $5' \rightarrow 3'$ exonuclease and DNA polymerase activities must therefore reside in independent regions of the protein chain and can combine to degrade and resynthesize DNA whether or not the polymerase comprises one or two protein units.

The ability to bind to nicks and then to remove one strand of DNA by exonuclease action and replace it by polymerase action is, of course, precisely the ability demanded of the enzyme systems undertaking excision-repair. Indeed, Kelly et al. (1969) have shown that DNA polymerase I can catalyse both the excision and repair steps when acting on irradiated DNA in vitro. With a template consisting of one strand of poly-A annealed to one strand which is poly-T culminating in four cytosine residues, an incision is made in the base paired region on the 3' side of the C_4 sequence and the mismatched bases are removed in the form of an oligonucleotide, usually of the form C_4T or C_4T_2.

With a template which consists of irradiated poly-T, the $3' \rightarrow 5'$ exonuclease action can proceed along the single strands until it reaches a dimer, when hydrolysis is inhibited. But if poly-A is annealed to the poly-T to form a duplex, thymine dimers can be removed; the demand for a duplex substrate suggests that this action is the prerogative of the $5' \rightarrow 3'$ exonuclease activity, which can function only with double stranded DNA substrates. Dimers are released in the form of oligonucleotide fragments, generally of 5–8 residues.

This suggests a model for the action of the enzyme in which it proceeds in a

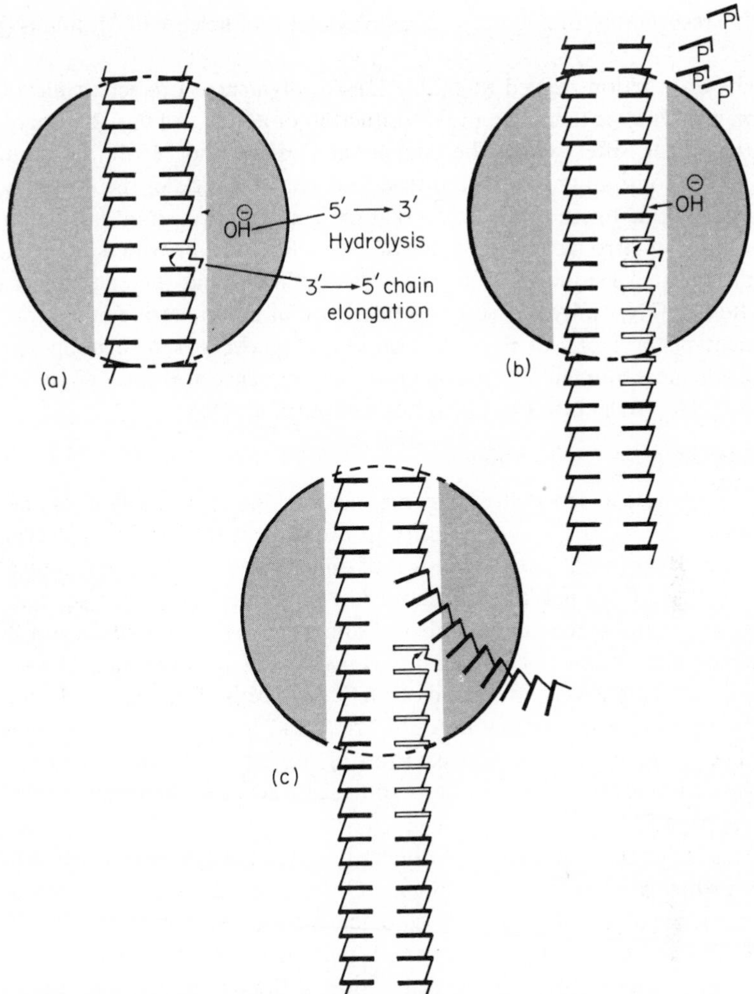

Figure 11.16: chain elongation by DNA polymerase I on duplex template. (a) as the primer chain is extended by addition of nucleotides to its free 3′-OH end, the 5′-P terminating chain ahead of it is degraded from 5′ to 3′ by exonucleolytic attack. (b) chain elongation continues at the expense of degradation of the 5′-P terminating chain ahead. (c) the 5′-P terminus is displaced from the template strand and chain elongation continues without concomitant degradation

5′→3′ direction along DNA up to one ortwo bases before the dimer; when the distorted structure is recognised by its lack of complementary base pairing, the enzyme makes an incision on its far (3′) side. This action is analogous to its recognition of mismatched C-A segments and implies that the excision capability of the polymerase is neither wholly exonucleolytic nor endonucleolytic;

in this respect it is analogous to the ultraviolet exonuclease of M. luteus (page 503).

The only action needed to enable DNA polymerase I to repair ultraviolet damaged DNA is therefore the introduction of a nick in DNA on the 5′ side of the dimer, after which the enzyme can excise and repair the damaged sequence, continuing to a nick introduced on the 3′ side of the dimer, where polynucleotide ligase is needed to seal the gap. Confirmation that repair can take place in vitro by this sequence of events has been provided by an experiment in which Heijneker et al. (1971) have shown that the endonuclease of M. luteus, DNA polymerase I and the ligase of E.coli, can repair ultraviolet damage in the DNA of B. subtilis so that it recovers transforming activity. This scheme is equivalent to the patch and cut sequence of events of figure 11.11, with DNA polymerase I fulfilling both steps (b) and (c).

Role of the polA *Locus in Repair*

The increase in sensitivity to ultraviolet irradiation of $polA^-$ cells does not appear to be caused by an inability to excise thymine dimers, however, for although their initial rate of removal of dimers after irradiation is only about half of that of the parent strain, excision continues for longer so that after 40 minutes almost the same number of dimers have been removed. But Boyle, Paterson and Setlow (1970) found that $polA^-$ cells degrade their DNA more rapidly and extensively after irradiation than do cells of the parent strain. This response is of course analogous to the "reckless" behaviour of $recA^-$ cells and implies in the same way that excision of dimers starts but does not stop. Degradation seems to be due to an exonuclease activity, probably exonuclease V (see page 513).

Excision-repair replication of dimers involves the synthesis of stretches of DNA which are too small to sediment at hybrid density when they are made in the presence of BUdR; the average patch is about 30 nucleotides long (page 499). But the distribution of single strands of DNA in an alkaline density gradient is skewed towards the heavy side of the parental strand, indicating that some patches may be large enough for the heavy sequences to change their density. By shearing DNA strands before centrifugation, Cooper and Halawalt (1972a) have been able to isolate segments which have long patches of repaired DNA.

When E.coli cells are exposed to ultraviolet light and then irradiated with BUdR, unsheared DNA—which consists of fragments of about 20×10^6 daltons—sediments at light density. But when the fragments are sheared to a duplex weight of 0·5–0·8 $\times 10^6$ daltons before denaturation, many fragments sediment on an alkaline gradient between the light parental density and a heavy density. This implies that some of these fragments, which are about 1000 nucleotides long, must have extensive regions containing BUdR inserted by repair replication. When the patches are obtained from repaired DNA without the shearing procedure, some appear to have molecular weights in excess of 3×10^6 daltons.

The origin of these patches cannot be the activity of the recombination-repair system, because the same results are obtained when cells which are temperature sensitive in replication are allowed to repair DNA at high temperature. Formation of the large patches must therefore be due to the operation of a system which does not need replication of DNA. Comparing the extent of large patch formation with the average length of repaired DNA suggests that most dimers are excised and replaced by a very short patch of about 5 nucleotides; but a few dimers are replaced by synthesis of an extensive patch which may be as long as 3000 bases.

One possible explanation is that E.coli cells contain two excision-repair systems, one which responds to most dimers with a short patch and one which responds to a few dimers with an extensive patch. It is difficult to follow the repair of irradiation damage in *polA*⁻ cells because the nucleotides released by degradation of DNA enter precursor pools and are reutilised in repair synthesis; they therefore compete with any labelled nucleotides added to the cells. When DNA is labelled with P^{32}, cells irradiated, and then H^3-BUdR is added during repair, *polA*⁻ cells show the same ratio of H^3/P^{32} in DNA. This suggests that the mutant can undertake the same extent of repair as wild type cells, but by taking into account the degradation and reutilisation of nucleotides, Cooper and Hanawalt (1972b) were able to calculate the *polA*⁻ cells in fact undertake four times as much repair synthesis in response to ultraviolet light. Because the same number of dimers appears to be repaired eventually, this implies that the length of the patches must on average be four times greater.

When *recA*⁻ *recB*⁻ cells are irradiated—it is necessary to use the double mutant, which has the cautious phenotype, for otherwise the reckless degradation obscures repair activities—dimers can be excised but the extent of non-conservative repair synthesis is as low as in *uvr*⁻ cells; and no large patches can be found. This suggests that the *polA* system and the *rec* system may compete for the same sites on irradiated DNA. After an incision has been made near a dimer, the *polA* system may excise and repair a small length of DNA; alternately, the *rec* system may undertake excision-repair, but removes and replaces a much greater length of DNA. Most dimers are repaired by the more efficient *polA* system and only a few by the *rec* system. This also accords with the observation of Town, Smith and Kaplan (1971) that E.coli cells appear to have two systems which can rejoin the single strand breaks introduced by X-irradiation; the *polA* system which acts rapidly and the *rec* system which acts more slowly.

The *polA* gene may belong to the *uvr* system, for Monk, Peacey and Gross (1971) noted that although the *polA*⁻ mutant is 4–5 times more sensitive to ultraviolet than wild type cells and *uvrA*⁻ mutants are about 12 times more sensitive, the double mutant is only 13–16 times more sensitive. This implies that the *uvrA* mutation prevents expression of the *polA* mutation; in other words, *polA* might act at a later stage of the same pathway in which *uvrA* participates. This agrees with the idea that the role of DNA polymerase I may be to help fill gaps after excision. The report of Shizuya and Dykhuizen (1972) that the double mutant combination *polA*⁻ *uvrB*⁻ is inviable implies that some essential role in DNA metabolism must be played either by *polA*⁺ or by *uvrB*⁺.

That *polA* and the *rec* genes may, to some extent at least, fulfill the same functions in E.coli is suggested by the failure of attempts to construct double mutants of *polA*⁻ *recA*⁻. When Gross, Grunstein and Witkin (1971) attempted to introduce a *recA*⁻ gene into *polA*⁻ recipients, they found that the double mutant progeny seem to be inviable; this cannot be due to a sum of the effects of the two individual mutations but suggests rather that some essential function of E.coli cells must be supplied by either the *polA* or *recA* genes. Whatever function the *recA* gene plays in introducing long patches of synthesis in response to thymine dimers is, of course, different from although presumably related to its role in the post-replication recombination-repair which takes place in *uvr*⁻ cells. This implies that the *recA*⁺ gene product can act at two stages in repair: it helps in the repairs which follows the initial excision of a dimer by making long patches; and it catalyses the strand exchange which acts on progeny DNA produces by replication in cells which have not excised thymine dimers.

The double mutant *polA*⁻ *recB*⁻ also appears to be inviable (Monk and Kinross, 1972); but Strike and Emmerson (1972) found that they could isolate this combination when the *sbcA*⁻ mutation is also present. Cells of the type *polA*⁻ *recB*⁻ *sbcA*⁻ are therefore viable. This implies that it is essential for cells of E.coli to possess one of the nucleases specified by these genes; figure 11.1 shows that *polA* codes for exonucleases II and IV and *recB/C* codes for exonuclease V. The cellular roles of exonucleases I and III are not known; mutants of the *sbcB* locus coding exonuclease I relieve the effects of *recB/C*⁻ mutations and Milcarek and Weiss (1972) isolated mutants in exonuclease III which have no apparent defects. But the *sbcB*⁻ and *xth*⁻ mutations have not yet been tested in combination with *polA*⁻ and other repair mutations.

The ligase mutation *lig-ts7* which greatly reduces the amount of ligase is itself a conditional lethal and so cannot be tested in combination with other conditional lethal mutations. But the less damaging *lig-4* mutation can be used to construct double mutants; Gottesman et al. (1973) reported that *lig-4 polA*⁻ cells also are inviable. Mutation in *polA* therefore appears to be lethal in combination with any one of the *uvrB*⁻, *recA*⁻, *recB*⁻, *lig*⁻ mutations. One explanation is to suppose that the polymerase I enzyme may be able to substitute for some of the activities exercised by the repair genes; however, it is unlikely that it can undertake the sealing action catalysed by ligase. We do not know whether the DNA polymerase plays one catalytic function in the cell (whose absence is lethal in combination with any one of the repair mutations) or whether it has several different functions which substitute for the different defects in repair (in which case the various double mutants are lethal for different reasons).

CHAPTER 12

Recombination between DNA Duplexes

Reciprocal Exchange of Genetic Information by Crossing Over

Physical Exchange of Genetic Material

Genetic recombination was first discovered as a process involving reassortment of different mutant genes in Drosophila. Following the discovery of linkage—the tendency of some characters to fail to assort independently in the manner observed by Mendel but instead to stay together in their parental array—it was noticed that any particular two loci exchange their parental arrangement for a new one with a characteristic frequency. The additive relationship of the frequencies of recombination when three mutants are followed—the sums of the frequencies of recombination between *a-b* and *b-c* are close to that between *a-c*—led to the idea that the recombination frequency of two loci reflects their physical distance apart on the chromosome.

Recombination appears to be a symmetrical event, for reciprocal recombinants always occur in equal frequencies in populations; this suggested that both recombinants might be generated by a single exchange. The production of recombinant classes appears to be correlated with the crossing over which occurs at the chiasmata formed between homologous chromosomes at meiosis. If the probability of chiasma formation between two loci on a chromosome depends upon their distance apart, genes located close to each other should tend to stay together; and as the distance between the loci increases so must the frequency of formation of recombinants.

Evidence that crossing over corresponds to genetic recombination was obtained by correlating cytological observations with genetic inheritance. Using translocations in which the parent chromosomes bear distinguishing structural features well as as genetic mutations shows that the formation of genetic recombinants corresponds to a physical exchange which forms new chromosomes with a different arrangement of structural features.

A correlation between the behaviour of chiasmata and the formation of crossovers which affords further support for the concept that they are the sites of genetic exchange is the phenomenon of positive interference. The occurrence of one genetic exchange diminishes the probability that another will occur nearby; this has the effect that the frequency of recombination between two outside markers, *a-c*, is less than the sum of the frequencies between them and an intermediate marker, *a-b* + *b-c*. A similar pattern is found in chiasma formation; there is only a low probability that two chiasmata will occur near each other.

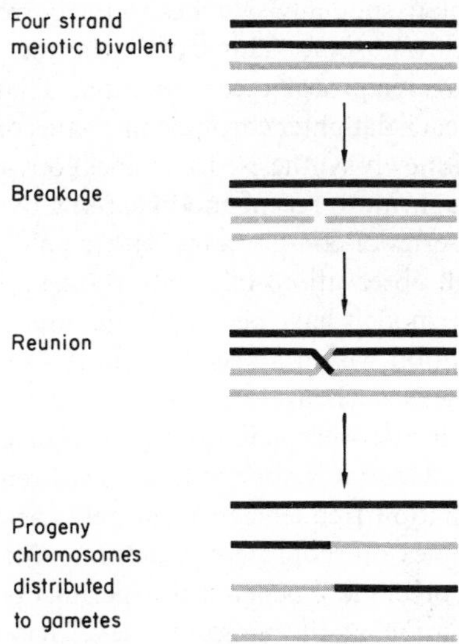

Figure 12.1: recombination by breakage and reunion. Each of the two copies of each homologue (one homologue is dark, one is shaded) consists of a duplex molecule of DNA. Breakage and crosswise reunion between two of these duplex molecules generates reciprocal recombinants, each of which has part of its genetic information derived from one parent and part from the other

Because recombination occurs on only two of the four strands at a chiasma point—each strand is a duplex of DNA—two strands carry reciprocal recombinants and two carry the parental arrangements; this means that recombination frequencies cannot rise above the 50% characteristic of independent assortment of genes on different chromosomes. Cytological observations of the nature of the pairing (*synapsis*) between homologues which precedes exchange of genetic material at meiosis are discussed in chapter two of volume two.

The observation of physical exchange of corresponding parts of chromosomes at meiosis suggests that recombination might occur by a *breakage and reunion* in which the two parental duplex molecules of DNA are broken at corresponding positions and then joined crosswise. As figure 12.1 shows, such an event yields chromosomes which are reciprocal recombinants, each with its total genetic material derived some from each parent. In its original form, this theory encounters difficulties in explaining events at the molecular level—see below—and alternative theories in vogue during the fifties and early sixties suggested that recombination might be linked to the replication process.

These *copy-choice* theories imply a conservative mode of replication in which progeny chromosomes are synthesised de novo from the parental species; during the replication of paired chromosomes, synthesis of the daughter genome on one template suddenly switches to the corresponding position on the other parental chromosome. This causes the reciprocal replica also to switch strands, giving reciprocal recombination. That replication is semi-conservative, so that each daughter chromosome gains one parental strand and only its complement is newly synthesised, excludes copy-choice theories in their original form, although more recent modifications overcome this difficulty. But these demand a series of complex, and rather unlikely, assumptions; and in view of more recent observations of events at the molecular level in recombination, copy-choice models have become more improbable.

Because recombination within a specified short stretch of DNA is a rare event, higher organisms such as Drosophila, which proved so useful for analysing recombination between genes, are more difficult to use for observing events at closely related loci; it is often impossible to generate enough progeny to estimate recombination frequencies accurately. And eucaryotic chromosomes comprise complex nucleoprotein structures which are not well understood and therefore preclude biochemical experiments. Most of the genetic analysis of recombination at the molecular level has been performed with fungal systems; and the isolation of intermediates of DNA and of enzymes which may be involved has taken place very largely with bacteria and bacteriophages.

Recombination takes place at the four strand stage of meiosis in eucaryotic cells and the four progeny gametes usually separate, so that it is impossible to distinguish the events of one meiosis and a statistical analysis of many progeny must be employed. But in fungi, the four products of one meiosis segregate into a linear order of spores in a single ascus. This permits analysis of all the DNA molecules involved in a single recombination event.

Recombination in bacteria and phages takes many specialised forms. Acceptance of transforming DNA by a bacterium, for example, involves the integration into the host genome of a single strand of DNA and not recombination between duplex molecules. Phages possess recombination systems which enable them to integrate into the bacterial chromosome at their attachment sites and to be released from this state. In addition, phages may possess generalised systems which support recombination during lytic infection; these systems may be more typical of those functional in bacterial conjugation and eucaryotic recombination.

Bacteriophages undertake generalized recombination during infection of host bacteria in a pool which contains replicating phages; so the milieu of the recombining DNA may differ from that of a meiotic chromosome and many rounds of reproduction may take place before progeny phages are released. But intermediates in phage recombination have been isolated and comprise molecules of DNA consisting partly of one parental duplex and partly of the other; this accords with the predictions of models of breakage and reunion and excludes copy-choice models in which recombinant genomes are synthesized from new material and inherit only genetic information from their parents.

The results obtained in phage systems are best analysed in terms of the hybrid DNA models discussed in the next section, although other interpretations are of course possible. Genetic analysis of fungi accords with the conclusion that recombination depends upon a two stage breakage and reunion event; and studies with bacterial mutants defective in recombination suggest that the enzymes responsible for recombination may have features in common with and may in part be the same as those responsible for the correction of damage to DNA.

Formation of Hybrid DNA

Since recombinants rarely contain additions or deletions in the genetic material, the recombination event must take place between precisely corresponding nucleotides in each parental DNA molecule. Homologous pairing requires identical parts of each genome to be brought into apposition at synapsis, although the exterior of any duplex segment appears to be structurally equivalent to any other. In the cells of higher organisms, where DNA is associated with protein, synapsis is a function of chromosomes rather than DNA molecules as such. Recognition of attachment sites on bacterial DNA by a phage DNA which is to undertake integration appears to be mediated by a protein which recognises the appropriate sequences of DNA on each molecule. This leaves only recombination between bacterial DNAs themselves or between phage DNAs during reproduction to depend upon pairing of homologous DNA duplex sequences as such; and we do not know how this may be achieved.

Although synapsis mediated by proteins may account for the initial pairing between homologous chromosomes of eucaryotic cells, the precision of the breakage and reunion event itself cannot be explained by recognition between identical sequences of duplex DNA and simple breakage and reunion. But there is, of course, a precise mechanism for the association of single polynucleotide chains by base pairing between complementary sequences. More recent models for recombination have proposed that a breakage and reunion event occurs in two stages which utilise complementary base pairing to ensure accuracy. Only one strand is at first broken in each duplex; and reunion takes place not by an end to end association of broken duplex molecules but by crosswise pairing between the single strands released by the initial breakage and the unbroken strands. This creates recombinant duplexes which have one chain from each parent duplex for the region around the crossover event—this has been termed *hybrid DNA*.

Two models have been proposed for the formation of hybrid DNA and they differ in the pattern of initial breakages which they stipulate. Holliday (1964) suggested that each duplex is probably broken in a strand of the same polarity.

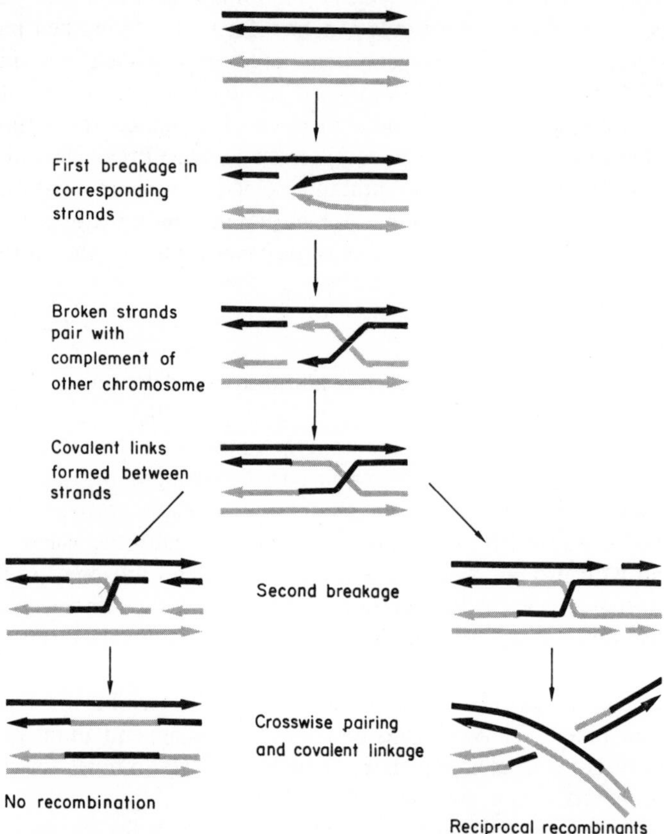

Figure 12.2: formation of hybrid DNA by breakage and reunion in strands of the same polarity. The figure shows only the two duplex molecules of DNA which are involved in the breakage and reunion; the other two parental molecules are not included. After breakage, the strands pair with their complement in the other parent molecule and become covalently linked to the other strand of the same polarity. A second breakage is required to release the two DNA duplexes from their crosswise linkage. If the second breakage occurs in the same two strands implicated in the first exchange as shown on the left, there is no recombination but each chromosome nevertheless bears a region of hybrid DNA. If the second breakage takes place in the other two strands, as shown on the right, reciprocal recombinants are produced, each possessing a sequence of hybrid DNA in the region of the crossover event. After Holliday (1964)

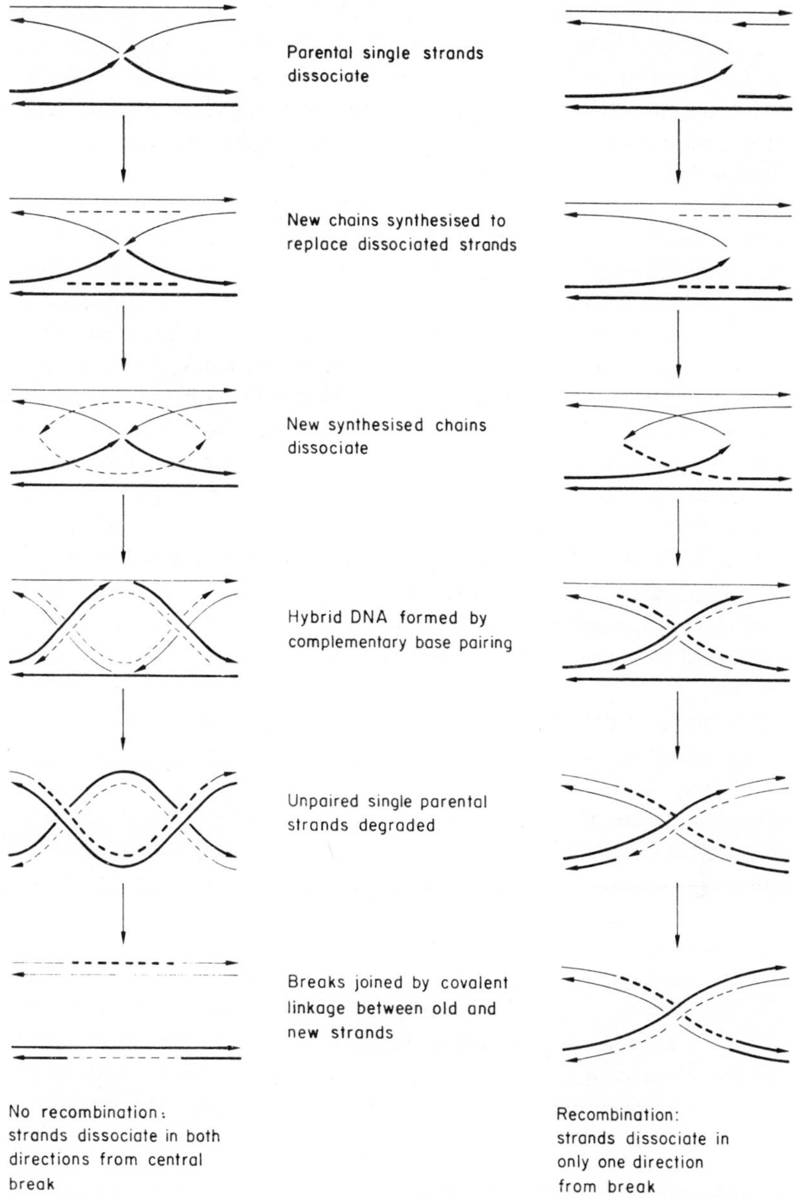
Parental single strands
dissociate
New chains synthesised to
replace dissociated strands
New synthesised chains
dissociate
Hybrid DNA formed by
complementary base pairing
Unpaired single parental
strands degraded
Breaks joined by covalent
linkage between old and
new strands
No recombination:
strands dissociate in both
directions from central
break
Recombination:
strands dissociate in
only one direction
from break

Figure 12.3: recombination by formation of hybrid DNA after breakage in strands of opposite polarity. After Whitehouse (1965)

After breakage, the broken strands separate from their partners in the same direction along each chromosome; figure 12.2 shows that crosswise annealing forms the hybrid DNA. If recombinants are to result, a second break must be introduced, this time in the hitherto unbroken strands; if breakage takes place again in the strands which have been broken previously, the original chromosomes can separate without recombination, although each possesses a region of hybrid DNA.

This model is less complex than the previous model of Whitehouse (1963) which proposes that although the initial breakages occur at the same site on each of the parental DNA duplexes they take place in strands of opposite polarity. Figure 12.3 shows that the broken ends then fall away in the same direction from their partner strands and new strand segments are synthesised to replace them using the unbroken strands as templates. These newly synthesised strands then fall away in turn to base pair with the corresponding parental strands which have broken away on the other DNA duplex. This forms a sequence of hybrid DNA in which each of the two strands of the duplex again carries genetic information from one of its parents. The final step is to delete segments of the hitherto unbroken chains so that there has been no net synthesis of DNA and to join the ends of the hybrid DNA segments covalently to the new parental ends exposed by this degradation. The single parental strands can fall away from the initial break in either direction, but if they fall away in both there is no recombination although a stretch of hybrid DNA is generated.

The topology of the formation of hybrid DNA has not been defined. Sigal and Alberts (1972) have pointed out that geometric restraints prohibit pairing with all four strands intact. When only two strands—one of each duplex—are implicated, models can be built so that a single strand is transferred from one duplex to the other within one residue of the sugar phosphate backbone. The exchange of strands can therefore take place without requiring the loss of any base pairing. The length of exchanged strands can then diffuse along the two cross linked duplex molecules by a zipper like reaction in which the strands exchange partners. Meselson (1972) proposed that this might be achieved by a rotary diffusion in which both duplex molecules are rotated about their helical axes in the same sense; the length of hybrid DNA would increase with time until the cross connection is broken. According to models such as these, therefore, the formation of lengths of hybrid DNA takes place gradually.

Recombination without Crossing Over

Gene Conversion in Fungi

When recombination between alleles was discovered, it was assumed that it takes place by the same mechanism of reciprocal exchange which takes place

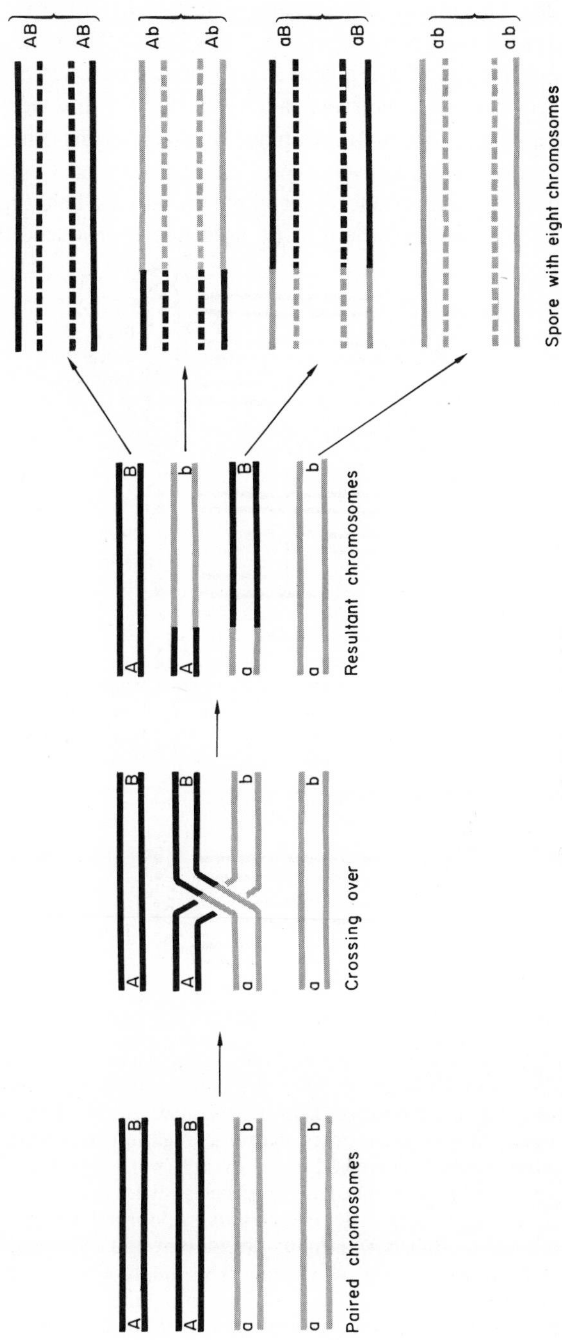

Figure 12.4: spore formation in Ascomycetes. The eight chromosomes in the ascus are arranged in linear sequence so that each represents the genetic characteristics of one of the eight single DNA strands of the four chromosomes produced by meiosis. The spores are therefore arranged in pairs. When recombination takes place as shown, the A/a markers lie in the parental order AAAAaaaa and the B/b markers in the order BBbbBBbb

between different genes. This pictures the chromosome as a length of DNA in which recombination can occur at any point, whether within or between genes. That mutations can be mapped within a gene in a definite order, just as genes can be ordered on the chromosome, was taken to support this view. But recombination between alleles usually takes place by mechanisms other than the conventional reciprocal crossing over.

In the Ascomycetes fungi, the products of a single meiosis are held together in a single large cell, the ascus, and the four haploid nuclei produced by the

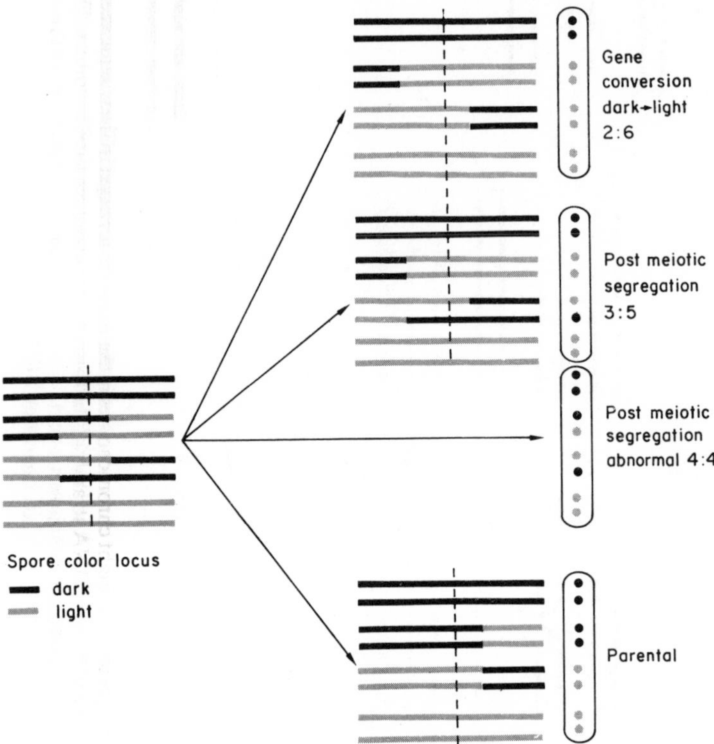

Figure 12.5: segregation of spore color mutants in Sordaria. Gene conversion: both of the regions of hybrid DNA are converted to the mutant (light) characteristic to give a 2:6 ratio. If both are converted to wild type, a 6:2 ratio is produced (not shown). Post-meiotic segregation: if only one of the hybrid chromosomes is corrected a 3:5 or 5:3 (not shown) ratio is produced. If no correction takes place, there is the usual 4:4 ratio of wild type to mutant characters, but their order in the ascus is abnormal. Parental: one duplex is converted so that both strands are mutant and the other is converted so that both strands are wild type. The diagram shows the creation of hybrid DNA when genetic recombination takes place, but hybrid DNA may also be generated and corrected without concomitant genetic recombination

meiosis are arranged in linear order. A mitosis occurs after the meiosis has been completed so that eight haploid nuclei are arranged in linear order. Figure 12.4 shows that each of these final products effectively represents the genetic character of one of the eight single strands of the four chromosomes produced by the meiosis. If the chromosomes resulting from the meiosis are genetically homogeneous—that is both strands of the chromosome carry the same genetic information—the characters should be arranged in pairs in the final ascus.

A system that has been widely used for such analysis is that of spore color in Sordaria fimocola; this has the advantage that the alternate parental characters can be observed directly. If meiosis is normal, each ascus should have four haploid nuclei with the color of one parent and four with the color of the other. But the abnormal patterns of figure 12.5 are sometimes found in which the ratio of spore colors is unbalanced, although mutations located in outside genes continue to segregate in the usual 4:4 ratio and arrangement.

A situation in which there are two copies of one spore color allele and six of the other was observed by Lindegren (1953) in Saccharomyces and he made the first modern proposal for a mechanism of genetic change involving the gene itself without reciprocal recombination. This *gene conversion* was also observed by Olive (1959) with spore color mutants of Sordaria, when a further type of abnormality was noticed; some asci have five spores of one color and three spores of the other color. The abnormal ratios may occur in either direction; there may be 6 mutant:2 wild type or 2 mutant:6 wild type, or there may be either 5 mutant:3 wild type or 3 mutant: 5 wild type.

The difference in color between two partner spores which is found in the 5:3 and 3:5 ratios implies that the DNA resulting from the preceding meiosis must be genetically heterozygous—the two strands of the duplex must carry different genetic information, one for each spore color. This has been termed *post-meiotic segregation* since the characters must then segregate at the mitosis succeeding the meiosis. Another abnormal segregation of this kind is found in asci which contain the usual 4:4 proportions of the two segregating characters but have an abnormal sequence because the spores are not arranged in pairs.

Gene conversion and post-meiotic segregation do not depend upon crossing over but appear to be correlated with it. In an analysis of spore colour mutants of Sordaria, Kitani, Olive and El-Ani (1962) found that in 10,000 asci there were:

		black		grey mutant
post meiotic segregation	8 abnormal	4	:	4
	52	5	:	3
	10	3	:	5
gene conversion	47	6	:	2
	6	2	:	6

When the spores are germinated after the asci have been counted, the characteristics of genetic markers outside the spore color gene can be followed. Using two mutations which bracket the spore color gene and which are 4% apart in recombination frequency, about 36% of the aberrant asci show recombination. This frequency is far greater than would be expected from random recombination between genes 4 units apart; but it is much less than the 100% recombination which should be found if gene conversion and post meiotic segregation result only from crossing over. In almost all cases, when a recombination event does take place, it involves the same strands of DNA which are involved in forming the aberrant spores. This shows a clear link between formation of aberrant asci and recombination.

This correlation suggests that the unusual ratios found in aberrant asci reflect the processes involved in recombination (extensively reviewed by Whitehouse, 1969). When hybrid DNA is formed as the first step of recombination, the two chromosomes produced each have one strand from each parent over the hybrid region. If one parent is mutant for a site located within this sequence of DNA, the chromosome is heterozygous for it and must have mispaired bases. Correcting enzymes, similar to those which recognise and repair ultraviolet damage may act upon the mispaired bases to restore normal Watson-Crick hydrogen bonding between the two complementary DNA strands; after excision of one of the strand segments, repair replication inserts bases correctly hydrogen bonded to the bases on the unexcised strand.

If the repair process acts symmetrically, so that one hybrid chromosome is corrected to wild type and the other is corrected to mutant, no record is left of the hybrid DNA. But if the process of excision and repair fails to take place, the abnormal situation of the 4:4 ratio with unpaired sequences in the ascus is achieved when the mispaired strands separate at the replication which succeeds the meiosis. Figure 12.5 shows that if only one of the two mispaired chromosomes is corrected, a 5:3 or 3:5 ratio results, depending upon whether correction is to wild type by excision of a mutant strand or vice-versa. If both chromosomes are corrected in the same direction, so that both become wild type or both become mutant, a 6:2 or 2:6 ratio is generated.

The correlation of gene conversion and post-meiotic segregation with crossing over suggests that the formation of hybrid DNA sets in train the correction process but does not always lead to recombination; as figures 12.2 and 12.3 show, it may lead to restoration of the parental arrangement of genes outside the hybrid DNA region, so that no crossing over takes place; although a region of hybrid DNA is created and offers the opportunity for correction processes to act. Aberrant asci may therefore be produced even in the absence of concomitant recombination.

Recombination between Alleles

When two different mutants in the same gene recombine, reciprocal crossing over is rarely responsible; recombination within short stretches of DNA appears to take place by the same mechanisms which are responsible for the formation of aberrant asci in gene conversion and post meiotic segregation. The lack of reciprocal recombination is revealed when the arrangement of outside markers is observed in the progeny which have recombined mutants in a gene. Mitchell (1955) followed the recombination of two mutants in the pyridoxine locus (*pdx*) of Neurospora by selecting for wild type progeny of the cross. The parents carried two outside markers, pyrimidine (*pyr*) and colonial (*co*), so that the only $+^{pdx}$ wild type progeny of the cross:

$$\frac{pyr \quad pdx_1 \quad + \quad co}{+ \quad + \quad pdx_2 \quad +}\ \times$$

should be of the form: $+ \quad + \quad + \quad co$

if a single crossover takes place between the two central mutants in the *pdx* gene. (This assumes that pdx_1 is to the left of pdx_2; if the order is reversed, then the ++ recombinants in the *pdx* gene should be $pyr +^{co}$ for the outside markers). But the $+^{pdx_1} +^{pdx_2}$ wild type recombinants contained *all* the parental classes, in proportions

5	+	*co*
7	*pyr*	*co*
7	+	+
13	*pyr*	+

The event which recombines the two mutants in the *pdx* gene therefore does not take place by a single exchange between the two mutant sites—which would generate only the + *co* class of outside markers—and is not reciprocal for no double mutants of the type $pdx_1\ pdx_2$ could be found.

Non reciprocal recombination can result from the formation of hybrid DNA at only one of two mutant sites within a gene. When two such mutants are crossed, selection for wild type copies of the gene—that is recombinants which are wild type at both loci—yields the products of both gene conversion and post meiotic segregation as well as those in which reciprocal recombination has apparently taken place. Figure 12.6 shows that if the hybrid DNA of the chromosome which contains a copy of the wild type sequence at the first site is not corrected at the second site, then one of the eight spores is wild type; if it is corrected at the second site to yield a wild type sequence on both strands, two of the eight spores are wild type, giving a 2:6 ratio in the ascus.

Correction of the other hybrid chromosome does not influence the production of wild type recombinants, for it bears the mutation at the left end of the gene. But reciprocal recombination between the two genes *appears* to take place if this hybrid chromosome is corrected to mutant sequence whilst the other is corrected to wild type. The parental arrangement of mutants is regenerated if the hybrid containing the left mutant site is converted to mutant whilst the hybrid which is wild type at the left side is converted to wild type at the right.

Correction usually shows a bias in one direction, however, and wild type recombinants may be produced between two mutants in a gene without the reverse event occurring. The formation of hybrid DNA within a gene is correlated with the liklihood that there will be a genuine crossover event, and examination of outside markers reveals whether this has taken place. But as a general rule, recombination between alleles takes place by a non-reciprocal recombination which involvesthe formationof hybrid DNA and the occurrence of a crossover between the two mutant sites is rare.

Gradients of Gene Conversion

Recombination between seven pairs of cysteine requiring mutants of Neurospora was found to follow this pattern by Stadler and Towe (1963); recombination within this gene is always non-reciprocal and outside markers are found in all four classes in appreciable frequencies, excluding the possibility that a single crossover event takes place. The *cys* mutants fall into two groups, one at the left end of the gene and one at the right end. When a cross is made between one mutant of each class, the wild type $+^{cys}$ recombinants which contain the outside marker beyond the *cys* mutant at the right end of the gene predominate. When the cross is:

$$\begin{gathered} a + \text{cys}_r + \\ \times \\ +\text{cys}_1 + b \end{gathered}$$

where *a* and *b* represent outside genes and the two inner loci are both located within the *cys* gene, the $+^{cys_1} + {}^{cys_r}$ wild type recombinants in the *cys* gene contain all four parental classes; however, the two classes *a*+ and ++ are found in greater proportion than the classes *ab* and +*b*. In other words, the formation of $+^{cys}$ DNA takes place more frequently on the chromosome which has the outside marker $+^{b}$. If no recombination event takes place (apart that is from the hybrid DNA formation and correction within the gene) the outside marker class associated with $+^{cys}$ is the parental *a*+; if a recombination event occurs between the outside markers (as is often associated with the formation of hybrid DNA), the outside marker arrangement is ++.

Similar results were obtained by Murray (1963) with crosses in the methionine requiring locus of Neurospora. The cross

$$\begin{gathered} a + {}^{me_1}\text{me}_2 + \\ \times \\ + \text{me}_1 + {}^{me_2} b \end{gathered}$$

gives wild type $+^{me_1}+^{me_2}$ recombinants which can be isolated and then examined for their outside marker patterns. All four classes were found, in the ratios:

$$\text{parental}\ \begin{cases} a+ & 11 \\ +b & 44 \end{cases}$$

$$\text{recombinant}\ \begin{cases} ab & 9 \\ ++ & 36 \end{cases}$$

When the cross was performed with the outside markers in the reverse orientation as:

$$+\ +^{me_1}\ \mathrm{me}_2\ b$$
$$\times$$
$$a\ \mathrm{me}_1\ +^{me_2}\ +$$

the outside marker combinations of the $+^{me}$ class were reversed in the ratios:

a+ 43
+*b* 7
ab 44
++ 7

The progeny which are wild type in the *me* gene therefore most frequently have the outside marker located to the left of the me_1 mutation in the parental chromosomes. This suggests that the most frequent way for recombination to take place between the two methionine mutants is for the one on the left to be converted to a wild type sequence.

This asymmetry suggests that hybrid DNA is more frequently formed at one end of a gene; in the cysteine locus the right end of the gene usually suffers conversion, in the methionine locus the left end more frequently is converted. A polarity of conversion frequencies seems to apply in many genes; Fincham (1967) found that the likelihood of gene conversion appears to increase from one end of the *am* gene of Neurospora to the other. Lissouba and Rizet (1960) crossed pairs of mutants in a gene for spore colour in Ascobolus immersus and looked at the arrangement of the two mutant loci in the aberrant asci containing two wild type (black) recombinants. Of the six pale spores (mutants), four always have the mutation on the left and two have the mutation on the right. The 6:2 asci therefore seem to result from conversion of the mutation on the right to wild type; the mutation on the left remains unchanged. This agrees with the model of figure 12.6, and implies that hybrid DNA is formed from the right end of the gene, progressing towards the left end. This polarity gradient is a common phenomenon but is not always observed, for Krusweska and Gajewski (1967) observed that conversion frequencies do not seem to be correlated with map position in the white spore locus, *Y*, of Ascobolus.

By crossing mutants in the *b2* locus of Ascobolus immersus, in which each single mutant, the double mutant and wild type recombinants all have different spore colours, Leblon and Rossignol (1973) directly observed conversion events at each mutant site. The correction process appears to be induced by the

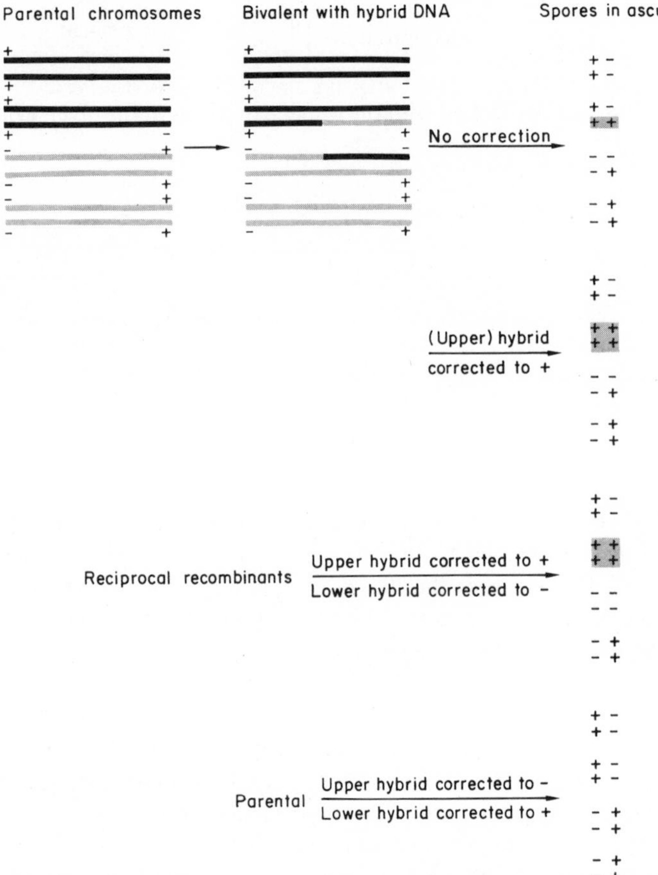

Figure 12.6: recombination by correction of hybrid DNA. Each line represents a single strand of the DNA of one gene. The upper parent (dark) is wild type at the left but mutant at a site on the right; the lower parent (light) is mutant at the site on the left and wild type at the right. Hybrid DNA forms the right end of the gene only, covering the right mutant. If no correction takes place, one of the eight spores is wild type at both sites so that the gene as a whole is functional. A 2:6 ratio of wild type to mutant genes is produced if the upper hybrid is corrected to wild type (correction of the lower hybrid does not change this ratio because the left site is mutant). If the upper hybrid is corrected to wild type and the lower hybrid is corrected to mutant, reciprocal recombination appears to have taken place between the two mutant sites. The reverse order of corrections produces chromosomes of the parental genotype

mutant site, with a characteristic frequency and direction of conversion, and often extends to a second site. In double mutants, excision-repair usually starts only at one of the sites. Since single site conversions can be found at both mutant loci, it must be possible to form hybrid DNA starting from either end of the gene. The results of these crosses imply that hybrid DNA is always formed over two chromatids, in contrast to the observation of Stadler and Towe (1971) that 5:3 or 3:5 post-meiotic segregation ratios may be achieved by the formation of hybrid DNA in only one chromatid.

Models for hybrid DNA formation can explain polarized conversion frequencies by requiring that recombination is initiated only from fixed points rather than from random sites on the chromosome. Whitehouse (1966) suggested that the end of a gene might be a suitable place for this interaction to start; this would conform with the observation that the probability of conversion falls from one end of the gene to the other. Adjacent genes need not necessarily possess the same orientations of polarity in recombination; and the formation of hybrid DNA may extend from one gene into the next, so that the end of a gene may have a probability of gene conversion which relies both upon initiation at its own end and upon initiation of recombination in the adjacent gene.

A model in which palindromic sequences of DNA provide recognition sites for initiating recombination has been proposed by Sobell (1972). A lengthy palindromic sequence may exist either in the usual form of an extended double helix or may be able to take up a clover leaf structure in which each of the strands of the double helix pairs with itself. Sobell proposed that such a sequence might be converted from its extended form to a clover leaf by a recombination protein. By nicking each clover leaf, perhaps in a loop extended of single strand DNA, homologous clover leafs can come together by base pairing. The intermediate generated in this way can readily be converted to the structures proposed in the Holliday model of hybrid DNA or to the branched intermediates observed in phage systems (see below). This model therefore implies that recombination can start only at specific sites palindromic in sequence, from where it spreads with a zipper like action along the parental strands.

Recombination by gene conversion was discovered by crossing two different mutations located in the same gene and selecting the progeny for wild type; outside markers show all four possible arrangements instead of the one which would result from reciprocal recombination. By crossing mutants in the two adjacent genes *me-7* and *me-9* of Neurospora, Murray (1970) found that wild type recombinants show all four possible combinations of flanking outside markers. This suggests that gene conversion may be responsible for recombination between closely sited genes as well as within a gene. Both the *me* genes appear to have a gradient of conversion frequency which runs from left to right, with a discontinuity between them; but events initiated in the right gene, *me-9*, seems to be able to extend into the left gene, *me-7*.

Amongst the aberrant tetrads which are found when two mutants in one gene are crossed are those in which no recombination has occurred, but there are six spores of one mutant type and two spores of the other mutant. Figure 12.7 shows that this situation arises by *co-conversion*, when the length of hybrid DNA covers both mutant sites so that both are corrected in the same

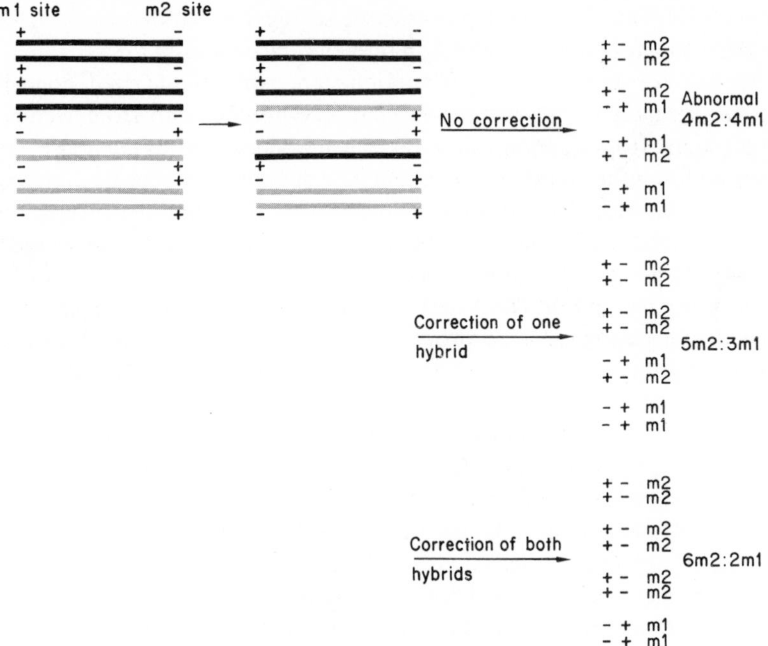

Figure 12.7: co-conversion of closely linked mutant sites. Chromosomes with the mutant (–) allele at the left site show the ml mutant genotype; those with the mutant allele at the right site are m2 mutants. If hybrid DNA covers both mutant sites on two of the four meiotic chromosomes, the lack of correction generates the abnormal arrangement of 4:4 characteristics of post-meiotic segregation. A 5:3 ratio is generated by the correction of one hybrid; a 6:2 ratio (the most frequent occurrence) results from the correction of both hybrid DNAs in the same direction

direction. The 6:2 ratio is the most common, but conversion of only one of the hybrid DNAs, instead of both, can generate 5:3 ratios (reviewed by Fogel and Mortimer, 1971).

Hybrid DNA may therefore extend for some distance, for Hurst, Fogel and Mortimer (1972) found that co-conversion can cover two, three or even four sites within a gene. Fogel and Mortimer (1969) have shown that the closer together two mutant sites lie, the greater is the probability that both will be converted; when they lie far apart they must suffer independent conversion. This explains why maps based upon recombination within a gene give a consistent order, even though they do not depend upon the same events—that is reciprocal recombination—which give linear maps of genes within chromosomes. As the distance between two mutant sites in a gene or in adjacent genes increases, the probability of co-conversion—that is no recombination—decreases and the probability of independent conversion, with associated recombination between the sites, increases (see also Leblon and Rossignol, 1973).

Some of the peculiarities of maps based upon recombination by gene conversion have been discussed by Fincham and Holliday (1970). One characteristic of these maps is the *map expansion* which results from negative interference; although positive interference inhibits one reciprocal exchange event from occurring near another, precisely the opposite effect is found over small distances. This negative interference stimulates recombination events to occur close to each other, with the result that recombination frequencies are enhanced so that the map of a small region is expanded.

Recombination by gene conversion differs in several ways from classical genetic recombination by exchange events; the postulates upon which genetic mapping is based do not seem to hold at the molecular level. Negative instead of positive interference between closely sited recombination events is one difference. Another is the failure of the principle used in constructing genetic maps of chromosomes that the particular mutants used do not themselves influence recombination frequencies; two loci recombine only in proportion to their distance apart, no matter which particular mutants are present at those loci.

But this principle does not hold for recombination at the molecular level, when pronounced marker effects may be encountered. The recombination frequency between very close mutants depends upon the particular base substitutions involved; and Norkin (1970) found that the introduction of a third mutation between two others in E.coli may change the recombination frequency between them. Gutz (1971) also found a site specific effect in which a particular mutation at the *ade-6* locus in yeast shows a much higher frequency of gene conversion and recombination than neighbouring sites. We may attribute these difference to the different influences which particular mispaired base sequences exercise on the correction enzymes.

The inequality in directions of gene conversion at any particular site shows that the particular base pairs involved may influence the correction as well as the formation of hybrid DNA. The idea that conversion is achieved by correction of hybrid DNA sequences is supported by experiments in which Kitani and Olive (1970) added different DNA bases to supplement the medium on which Sordaria mutants were crossed; this treatment may change the frequencies and directions of gene conversion, presumably by distorting the nucleotide precursor pools on which the correction enzymes may draw. Conversion is in general an accurate process, however, for Fogel and Mortimer

(1970) found that suppression patterns of mutants of yeast are not changed by gene conversion; this implies that correction to the mutant genotype copies the mutant strand of DNA in mutant-wild type hybrids accurately.

The segregation of the products of meiosis in most eucaryotic cells makes tetrad analysis impossible, but Chovnick et al. (1970, 1971) have conducted a half tetrad analysis of Drosophila by using a mutant chromosome which has a compound structure comprising the attachment of homologous regions to one centromere. This means that the products of a meiosis must remain in pairs in the progeny.

Rosy mutants of D. melanogaster lack the enzyme xanthine dehydrogenase, specified by a locus on the third chromosome, and therefore have an abnormal eye color. The metabolic defect of these flies allows progeny which are wild type to be selected by growing eggs on appropriate medium. When two different *rosy* mutants were crossed, of the 18 flies with wild type eye color which survived, only 6 could have arisen by classical exchange as judged by the arrangement of outside markers. In another 10, one of thr *rosy* mutations has been converted to wild type; and the other mutation suffers conversion in the two remaining flies. In a similar analysis of the *maroon-like* locus on the X-chromosome—using attached X chromosomes—Smith, Finnerty and Chovnick (1970) found that all 29 of the wild type recombinants arose by gene conversion and not by classical exchange.

The mechanism of recombination is probably similar, therefore, in all organisms, occurring by formation of hybrid DNA so that closely linked mutant sites usually recombine by a correction event and not an exchange. It is worth noting at this point that alternative models for recombination which do not involve creation of hybrid DNA have been proposed; the disadvantages of these models have been discussed by Holliday and Whitehouse (1970).

One important feature of recombination systems is that recombination frequencies are not absolute but may themselves depend upon genetic factors; strains of Drosophila, for example, may be derived by artificial selection procedures with increased or reduced frequencies of recombination. And one difficulty encountered by Stadler, Towe and Rissignol (1970) in analysing an apparent conversion frequency gradient in Ascobolus was that the strains of fungi appear to segregate for genes which influence recombination and conversion. These effects presumably depend upon changing the efficiencies of the enzyme systems which create and correct hybrid DNA.

Recombination Intermediates of Phage DNA

Heterozygosis in Phage T4

Wild type (r^+) phages are characterized by small plaques which have a turbid halo, whereas r^- mutants exhibit larger and clearer plaques. Hershey and Chase (1952a) discovered that some 2% of the plaques arising from the individual particles produced by a cross of $r^+ \times r^-$ have the mottled appearance which is usually obtained after mixed infection of a bacterium by both species. When the particles from these mottled plaques are used to infect host cells, they give about equal numbers of pure r^+ and r^- plaques and the same 2% of mottled plaques. Since the mottled plaques arise from infection by only a single phage particle, this result implies that the parent genome must have been heterozygous, carrying both the r^+ and r^- sequences.

Such heterozygotes have since been found to arise with equal frequency for all genetic markers; and the same characteristic 2% frequency is generated under a range of conditions of infection. The average length of the heterozygous region is quite short and less than the size of the *rII* cistrons. The most obvious interpretation of this discovery is that the heterozygous phages are intermediates in recombination which possess a length of hybrid DNA. The replication and recombination of phages takes place in a single pool from which genomes are randomly withdrawn for insertion into mature phage particles; the constant frequency with which the heterozygotes arise suggests that they represent random withdrawal from the pool of an intermediate which is continually generated by recombination events.

Phage heterozygosis arises from two causes, only one of which depends upon the creation of hybrid DNA. Recombinant molecules of phage DNA contain a length of hybrid DNA joining the sequences of two different parents; this is an intermediate in phage reproduction and disappears by segregation at the next round of replication if it is not withdrawn from the pool. There is therefore an equilibrium in the pool between creation of heterozygotes by recombination events and their removal when the strands of the duplex are separated by replication. At any given time, 2% of the pool must be in the heterozygous state. The second class of heterozygotes arises from terminal redundancy; these phage molecules may carry duplicate copies of some genes at the ends of the genome and these may represent different alleles if the ends are derived from different parents.

Experiments in which host cells suffer mixed infection by two phages have demonstrated that recombination must occur by a breakage and reunion process rather than by copy-choice. Meselson and Weigle (1961) showed that recombination between molecules of phage λ can occur independently of the replication which would be demanded by copy-choice. Phages can be labelled by growth on bacteria cultured in a heavy medium containing C^{13} and N^{15} isotopes; after two different mutants have been density labelled in this way, they can be crossed by mixed infection of hosts growing on normal medium. The cross generates three classes of phage DNA, which can be separated on density gradients into heavy, hybrid and light positions corresponding to 0, 1 or 2 rounds of replication. Genetic recombinants are present in all three classes, including the unreplicated heavy phages; breakage and reunion must therefore be responsible. This conclusion has been confirmed by crossing two phage λ species each of which bears a different radioactive label; only physical incorporation of the DNA of both parents can produce recombinants containing both types of isotope. Meselson (1964) also found that a small amount of DNA synthesis seems to take place during recombination; this may correspond to the excision-repair of hybrid DNA sequences.

When DNA synthesis is inhibited after bacterial cells have been infected with phage T4, the incidence of r^+/r^- heterozygotes is increased several fold; this suggests that DNA synthesis is required to remove the heterozygous sequences. Tomizawa and Anraku (1965) found that in these conditions mixed infection with two phage DNAs, one labelled with P^{32} and the other with BUdR, generates linear phages which have genomes comprising the two types of parental material joined end to end but not covalently linked. They suggested that these *joint molecules* represent the primary product of the recombination process and are later converted to covalently linked phage recombinants by a process which requires DNA synthesis.

Recombination Activities in T4 Infected Cells

When protein synthesis is inhibited during phage infection, recombination is prevented. Kozinski et al. (1967) and Kozinski (1968) used a protocol in which P^{32} labelled phage particles are allowed to infect E.coli cells growing in a medium containing heavy isotopes. Parental phage DNA can then be identified by its radioactive label and newly synthesised progeny phage DNA by its density label. Chloramphenicol can be added during the phage infection to inhibit protein synthesis and the DNA later extracted and analysed both on CsCl density gradients, when it remains in duplex form, and on alkaline sucrose density gradients, when it is denatured to single strands.

In the first 3–10 minutes after infection there is extensive replication of the injected parental phage DNA, so that the radioactive label sediments at hybrid density. At about 20 minutes after infection, the replicated parental DNA at hybrid density and the entirely heavy progeny strands recombine so that the radioactive label is displaced towards a heavier location on the density gradient. If chloramphenicol is added during the first five minutes after infection, the injected DNA cannot replicate. Addition of the inhibitor after 5 minutes allows replication of phage DNA, but it does not subsequently undertake recombination. If protein synthesis is not inhibited until 7 minutes after infection, molecular exchanges take place to form joint molecules, but the integrity of the polynucleotide chain is not restored—when the recombinant molecules are denatured they release pure parental fragments. The addition of chloramphenicol between 7 and 9 minutes is without effect, for it allows the parental fragment to become covalently bonded to the adjacent progeny strand so that it can no longer be released by denaturation.

The reduced effects of adding chloramphenicol at successively later times during infection suggest that a series of enzymes coded by the phage are sequentially synthesised during infection to undertake the various steps of recombination. Both degradative and synthetic enzyme activities seem to be involved, for concomitant with the inhibitory effect of chloramphenicol on recombination there is an inhibition of endonucleolytic activity directed against the phage DNA. The addition of the inhibitor at various times after infection allows degradation to proceed to different extents, which suggests that more than one nuclease is involved. About 5 minutes after infection, DNA suffers single strand breaks and further breaks are introduced about 2 minutes later opposite to the primary nicks to cause double strand breaks. At about 15 minutes after infection, single strand breaks are introduced into these fragments and it is at this time that recombination between parents and progeny begins to be expressed.

This suggests a sequence of events in which breaks must be introduced into the phage DNA by phage-coded nucleases, after which recombination to form joint molecules is catalysed, also by enzymes specified by the phage. The final step in recombination is the covalent linking of the joint molecules. The obvious candidate for the enzyme to catalyse the repair of gaps remaining in polynucleotide strands is the T4 polynucleotide ligase. But by using mutants which are defective in ligase activity, Kozinski and Kozinski (1969) showed that covalent recombinants can be formed apparently without mediation of this enzyme. It seems unlikely that the joining reaction can be catalysed by the host ligase because the addition of chloramphenicol before 7 minutes after infection inhibits the conversion of joint molecules to recombinants; this implies that some phage coded enzyme other than ligase must be needed to seal the nicks.

When the phages used to infect bacteria lack the T4 DNA polymerase, they are hampered in replication but can manage to form both joint and recombinant molecules. But Anraku and Lehman (1969) found that phages which are mutant in both the T4 DNA polymerase and the ligase genes produce joint molecules almost exclusively. When two wild type phages are crossed and one carries a P^{32} label and the other a BUdR density label, the radioactivity is found at an intermediate density when the phages recombine to form molecules containing both parental types of sequence. When the DNA of the intermediate density molecules is denatured to single strands and centrifuged through an alkaline gradient of CsCl, it remains at intermediate density, showing that the two types of molecule have been covalently linked together. The same result is found for phages which are mutant in DNA polymerase. But when phages lack both polymerase and ligase, although joint molecules of intermediate density are formed, the radioactive and density labels are separated by the alkaline centrifugation. This shows that the lack of both enzymes prevents the conversion of joint molecules to recombinants.

These experiments imply that both DNA polymerase and ligase are implicated in recombination of phage T4 genomes, although the lack of either activity alone can be substituted by other enzymes. Joint molecules can be converted to recombinants in vitro by the action of both enzymes; neither is sufficient alone. Anraku, Anraku and Lehman (1969) found that when joint molecules are isolated from cells infected with phages which are mutant in both polymerase and ligase, there are about 24 gaps for the length of every T4 genome. By allowing T4 DNA polymerase to repair the gaps in vitro, their length can be estimated from the incorpoation of radioactive nucleotides; the gaps have an average length of 300–400 nucleotides. This explains why both polymerase and ligase activities are needed to seal the interruptions.

Creation of these lengthy gaps depends upon the products of genes 46 and 47: Prashad and Hosoda (1972) found that the single strand breaks made during the earlier stages of infection remain as nicks if the phage has been mutated in either of these genes, but are converted to gaps when they are active. That this activity is necessary for recombination is suggested by the reduction in recombination frequencies observed in phages which are mutant in these functions. Direct assay of the production of joint molecules shows that although phages mutant in both polymerase and ligase generate joint molecules, those with a further mutation in gene 46 are unable to do so. This argues that genes 46 and 47 code for products which play an essential role in recombination by generating single strand regions where one strand is in part degraded.

Following the formation of the breaks, strand exchange must take place and branched molecules which may be the immediate products of this reaction have been observed in preparations of phage T4 DNA by Broker and Lehman (1971). That the branches are an intermediate in recombination, linking together the two parental molecules, is confirmed by experiments using parental phages one of heavy and one of light DNA; the hybrid fraction produced by joint infection, which contains the recombinant molecules, is enriched in its content of branches. Mutations in gene 32 or in the polymerase of T4 greatly reduce the number of branches; however, phages lacking the ligase function have an increased number of branches. This suggests that unrepaired nick—maintained in this state by the absence of ligase—is a necessary step preceding formation of the branched molecules. The role of the gene 32 protein, which is needed to help unwind DNA for both replication and recombination, may be to maintain the single stranded regions of the recombining molecules in an appropriate conformation.

At least four types of enzyme activity must therefore be required for the recombination of T4 genomes. First, breaks must be made in DNA at the sites which will promote recombination; more than one nuclease activity appears to be necessary, since these nicks must then be enlarged to gaps of appreciable size. Single strands of the two recombining molecules must pair crosswise to form a branched molecule and this may be the stage at which gene 32 protein is involved. The branched molecules must then be reduced to linear structures; these are the joint molecules. The interruptions in the integrity of the recombinant molecules must then be filled in by a repair like activity exercised by T4 polymerase. The nicks remaining must then be filled by ligase activity.

Redundant Joint Molecules of Phage λ

Generalised recombination between molecules of λ DNA takes place when the phage reproduces during lytic infection (reviewed by Signer, 1971). The phage can suffer recombination in *rec*$^-$ bacteria of E.coli, which suggests that it codes for its own recombination enzymes. Mutants classified as *red*$^-$ λ, however, cannot undertake this recombination. The *red* system appears to specify two gene products, one an exonuclease and the other the β-protein, whose function is unknown.

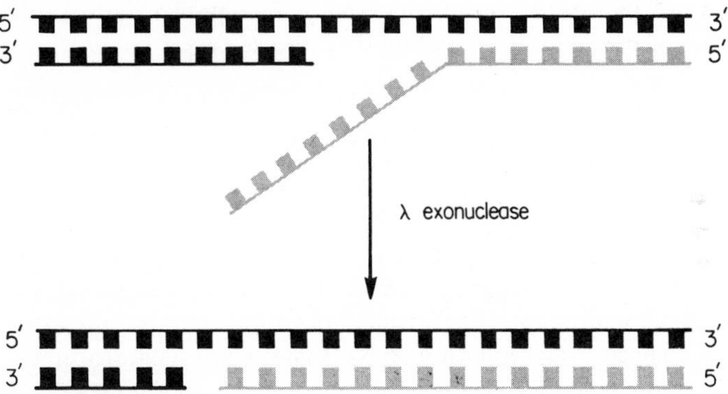

Figure 12.8: action of the λ exonuclease. The upper molecule contains a redundant joint, which might be the immediate product of a recombination event. The λ exonuclease degrades one partner strand of the duplex just far enough to allow the redundant single strand to participate in complementary base pairing

The exonuclease is inactive on duplex DNA molecules which have internal nicks, but is active on the structures shown in figure 12.8 which have a redundant single strand protruding from the nick. Cassuto and Radding (1971) showed that the action of the exonuclease is to degrade one strand of the duplex DNA from 5′ to 3′ to allow the redundant single strand to base pair with its complement. A circular substrate can be prepared by annealing a short length of P^{32} labelled DNA to one strand of phage λ DNA and then adding the other strand. When the λ molecule forms a circle, there are no free ends to the duplex but a redundant joint is formed where the short P^{32} sequence prevents the second strand from base pairing all the way round the circle. The release of P^{32} from this structure by the exonuclease corresponds to the loss of the redundant joint, which allows the molecule to sediment more rapidly as an intact circle.

This action could play a role in trimming the loose ends of a recombinant produced by exchange of strands between λ DNA molecules. Cassuto et al.

(1971) shown have that single strand assimilation by the exonuclease is perfect, for the product of enzyme action can be converted into a covalently linked polynucleotide chain by the action in vitro of ligase alone. This suggests that the exonuclease might recognise the nick which is produced by loss of the redundant strand as a signal to stop degradation.

The successor to the redundant joint molecule must therefore be a joint molecule, although it contains only a single strand interruption and not the lengthy break found in T4. This may mean that although the general process of recombination is similar in phages T4 and λ, the immediate product of strand exchange is different, containing lengthy breaks in T4 and overlaps in λ. The ability of the λ exonuclease to assimilate single strands at redundant joints in vitro does not prove, of course, that this is its action in vitro; and before defining the mechanism of recombination of λ we must know what role the β-protein plays.

Correlation between Recombination and Repair Activities

The similarities between the enzyme activities required to create and correct hybrid DNA sequences and those which are concerned with the repair of damage to DNA suggests that some enzymes may play a role in both processes. Both require an endonuclease to nick DNA as an initial step: in repair this is triggered by recognition of a distortion in the duplex; in recombination it may take place at specific regions although we do not know what features they may present to the enzyme. Repair demands the excision of damaged regions and synthesis of a replacement stretch of DNA; recombination may demand degradation of strands as a preliminary step in exchange and perhaps also to ensure a smooth fit between the two recombining duplexes. The removal and replacement of mispaired bases in hybrid DNA sequences is very similar to the excision-repair of thymine dimers. Both recombination and repair are terminated by gap filling, which may involve some synthesis of DNA and then a final sealing of single strand breaks.

The effects of the mutations in the *rec* and *sbc* loci in E.coli (page 514) implies that there must be at least some overlap of recombination and repair systems in that the recombination activities also undertake repair (although the *uvr* repair activities do not undertake recombination). The apparent role of the *rec* system in promoting exchange of single strands as a repair mechanism also accords with our ideas of the enzyme activities which must be involved in recombination. The need for certain exonuclease activities if cells are to be viable and the implication of these activities in both recombination and repair shows that features of the degradation of DNA may be the same in each process.

That the same enzymes may be involved in both processes is indicated also by the obervation that repair and recombination compete for the same activities in B. subtilis. When DNA is introduced into recipient cells by recombination, one strand is integrated into the chromosome, but mixed clones do not form as a result, even when we may expect a hybrid DNA sequence with different information on each strand to have been prepared. This implies that correction of such mispairing takes place before the chromosome can replicate to perpetuate its heterozygosity. But ultraviolet irradiation abolishes the correction, and this implies that repair of thymine dimers may compete with recombination for the same enzymes.

Little is known about the enzymes which undertake recombination in eucaryotic cells, but some activities appear to reside specifically in meiotic cells. The lilly is a good organism for such experiments, for the micro-sporocytes which comprise its meiotic cells develop synchronously and can be separated from the surrounding somatic tissues. Howell and Stern (1971) found that an endonuclease activity appears at a specific time during meiosis—it is first detected at leptotene, reaches a peak at zygotene and pachytene, and disappears by diplotene—and Hotta and Stern (1971a) showed that a protein equivalent to 32-protein of phage T4 is synthesised by meiotic cells but not by somatic cells. Repair synthesis of DNA seems to take place during meiosis and Hotta and Stern (1971b) suggested that this activity may be implicated in recombination. Although incomplete, these data would be consistent with the idea that the same sort of general processes, involving nicking, unwinding and strand exchange, are involved in recombination in all organisms.

CHAPTER 13

The Cell Division Cycle

Initiation of New Cycles of Replication

Control of the Replicon

Bacterial cells replicate their chromsomes once for each time they divide and cycles of replication are initiated at the origin only at the appropriate time in the cell cycle. The rate of DNA synthesis in the cell is constant, so that it is the number of initiation events which determines how often the chromosome is reproduced; because replication takes place sequentially from a fixed origin, a single initiation event is sufficient to trigger each complete cycle of replication. In eucaryotic cells, of course, many separate initiation events must take place to replicate each individual chromosome.

When donor fragments of DNA are introduced into a recipient bacterium by conjugation, they cannot be reproduced in this form but must first be inserted by recombination into the host genome. This led Jacob, Brenner and Cuzin (1963) to suggest that each genetic element—such as a bacterial chromosome, phage genome or episome—constitutes a unit of replication, the *replicon*, which is characterized by its possession of the control elements needed for replication. Any segments of DNA lacking these control functions—such as the parts of a chromosome transferred during conjugation—must be unable to replicate. Episomes or phages constitute separate replicons during infection of bacteria so that their reproduction is independent of that of the bacterial chromosome.

When an episome or phage becomes integrated into a bacterial chromsome, however, its own replication control is switched off and it is reproduced only as part of the bacterial replicon. Bacterial mutants which do not replicate their chromosomes do not replicate integrated episomes or prophages, although the mutation does not usually influence the replication of free extra-chromosomal DNA. One way for the cell to overcome such mutations is the integrative suppression discovered by Nishimura et al. (1970), in which the replication system of the F factor becomes active so that the entire bacterial chromosome is replicated as part of the episomal replicon (see page 425). Lindahl, Hirota and Jacob (1971) have shown that the phage P2 replicon can also replicate the chromosome of *dnaA* mutants of E.coli which have lost the ability to initiate new rounds of synthesis.

Each chromosome of a bacterial cell therefore constitutes one replicon, as does each independent episome or phage genome. Each chromosome of a eucaryotic cell, by contrast, comprises many replicons, whose initiation may start at different times. Each replicon is characterized by an origin from which DNA synthesis commences in both directions and a terminus at which it ceases. As figure 10.4 shows, in circular replicons, such as those of bacteria, the two growing forks of one replicon continue until they meet at its terminus. In eucaryotic chromosomes, as figure 10.7 shows, each replicon has two termini and the growing forks of adjacent replicons continue until they meet at their common terminus. Possession of an origin which initiates bidirectional replication may therefore be characteristic of all replicons.

Synthesis of Initiator Proteins

Two strains of E.coli in particular have been used to study the replication cycle, one consisting of cell lines derived from the $15T^-$ (thymine-requiring) mutant, which has two significant properties. First, DNA synthesis ceases if thymine is removed from the incubation medium. Second, the absence of thymine causes the cells to lose viability in a characteristic manner, revealed by the loss of ability to form colonies. This arises from either of two causes. If the bacteria carry lysogenic prophage, thymine starvation induces the prophage, thus destroying the cells. However, thymineless death still occurs in strains "cured" of prophage and there has been a long debate about the reason for this effect, for which many explanations have been proposed. Whatever the cause, the susceptibility of cells to thymineless death has proved a useful tool for analysing the start of cycles of replication.

Using certain derivatives of the $15T^-$ strain, in particular the $T^-A^-U^-$ mutant—often termed TAU-bar—which requires as growth factors thymine, arginine and uracil, it is possible to block selectively any or all of DNA, RNA or protein synthesis by withholding the appropriate precursor. Maaløe and Hanawalt (1961) found that when an exponentially growing culture of these cells is transferred from a complete medium into one lacking thymine but containing arginine and uracil (–T, +AU), there is a lag period followed by an exponential loss of cell viability which leaves no survivors. But if the culture of cells is instead transferred to a medium which lacks all three growth factors (–T, –AU), the survival curve is almost identical but tails off to a plateau corresponding to a level of about 3% survivors. Although most cells therefore die when DNA synthesis is blocked, irrespective of whether RNA and protein synthesis can proceed, a small fraction survive when these last two processes are prevented. Maaløe and Hanawalt suggested that this immune fraction might represent cells at some particular stage of the growth cycle.

This theory can be tested by removing arginine and uracil from exponentially growing cultures to block RNA and protein synthesis and then allowing varying periods of time before thymine is removed to block DNA synthesis. Figure 13.1 shows that the proportion of cells in the immune fraction depends upon

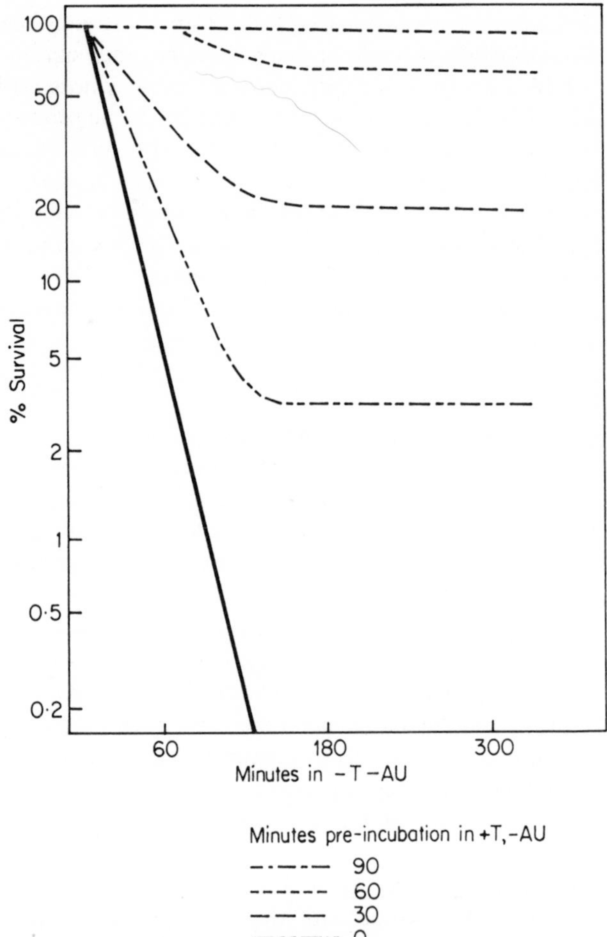

Figure 13.1: thymineless death of TAU-bar cells in (−T, −AU) medium after various periods of incubation in (+T, −AU) medium. As the time of preincubation is increased, a greater proportion of cells complete their current replication cycle but cannot start another; in this state they are immune to thymineless death and so the proportion of survivors increases. The heavy curve represent survival upon transfer to (−T, +AU) medium. Data of Maaløe and Hanawalt (1961)

the length of time spent in (+T, −AU) medium before transfer to (−T, −AU) medium. The longer the preincubation with thymine to allow DNA but *not* RNA and protein synthesis before the transfer to block DNA synthesis, the greater the fraction of cells which is immune to thymineless death. After 90 minutes of preincubation, the entire cell population enters the immune state.

These results suggest that cells may be immune to thymineless death when they are not engaged in DNA synthesis at the time of transfer into the medium lacking thymine. If thymine starvation kills only cells which are synthesising DNA when the nucleotide is withdrawn, those cells which have just completed one replication cycle but have not yet commenced the next will be immune. Synthesis of RNA and/or proteins must be needed before the new round of replication can be initiated; this explains why there are no survivors when cells are transferred from (+T, +AU) normal medium to (−T, +AU) medium—all the cells can undertake RNA and protein synthesis, attempt to start another replication cycle and suffer thymineless death.

But when cells are transferred from their (+T, +AU) medium to (−T, −AU) medium, the necessary RNA and/or protein synthesis cannot take place; although all those cells which are replicating DNA are killed by the transfer, the 3% between replication cycles are unable to commence a new cycle and therefore remain immune. If cells are preincubated in (+T, −AU) medium before their transfer to the (−T, −AU) medium, they can complete their current replication cycles successfully but cannot start another. As the preincubation period is lengthened, more cells complete chromosome replication and are therefore trapped between cycles; in this state they are immune to thymineless death. When immune cells are transferred into a culture in which RNA and protein synthesis is permitted, they lose their immunity as this model predicts by initiating another cycle of DNA synthesis.

These conclusions have been examined experimentally by autoradiographic analyses, when Hanawalt et al. (1961) found that individual cells continue DNA synthesis for differing periods of time after RNA and protein synthesis is halted. Whatever the fraction of immune cells, a similar fraction lacks the ability to synthesize DNA under conditions of RNA and protein synthesis. Lark, Repko and Hoffman (1963), using the technique of introducing a radioactive label into a portion of the bacterial chromosome and following its next replication by density labelling (figure 10.2), showed that after amino acid starvation a labelled segment is not transferred to hybrid density. This is consistent with the idea that it cannot replicate again because cells cannot initiate the next replication cycle unless they are allowed to synthesise RNA and/or proteins.

Initiator proteins must be synthesised by particular times during the replication cycle if a new cycle is to take place. In E.coli $15T^-$, high concentrations (150 μg/ml) of chloramphenicol inhibit initiation as effectively as starvation for amino acids, but low concentrations (25 μg/ml) are somewhat less effective. Lark and Renger (1969) have shown that two separate processes requiring protein synthesis are needed for initiation; one is inhibited by even the low concentration of chloramphenicol—termed the CM-sensitive step—but the other is only inhibited by the higher level—the CM-resistant step. These two events are separated temporally in the replication cycle.

Initiation of replication can be followed by subjecting cells to amino acid starvation so that they complete their current rounds of replication but do not start the next; when amino acids are restored in the presence of a pulse of radioactive thymine, the cells start replication synchronously and the label is incorporated into the region of the origin of the chromosome. After the pulse label, cells are grown in normal medium for a short time and then transferred to a medium containing the heavy isotopes C^{13} and N^{15}. Appearance of the radioactivity in the hybrid density fraction of DNA marks the replication of the origin, that is the occurrence of the next initiation event.

By adding inhibitor to cells before transfer to heavy medium, it is possible to define the time when initiator proteins are synthesized. If the inhibitor is added before synthesis of the protein, a new cycle of replication cannot start; but if it is added after the protein has been made, it is without effect. Addition of chloramphenicol to a concentration of 150 μg/ml prevents the initiation of new cycles of replication if added more than 15 minutes before initiation would take place; addition after this time does not prevent transfer of the radioactive label to hybrid density. When the experiment is performed with only 25 μg/ml chloramphenicol, the inhibitor loses its effect about 30 minutes before the transfer of radioactive label.

This suggests that two steps are demanded for initiation, a CM-sensitive step taking place 30 minutes before initiation and a CM-resistant step taking place 15 minutes later. Once initiation has occurred, cells lose their immunity to inhibitor and again require protein synthesis before another cycle of replication can be initiated.

Phenethyl alcohol also inhibits the initiation of DNA synthesis in bacteria. During treatment with this inhibitor, the CM-sensitive protein is synthesised and sufficient accumulates to allow several rounds of replication to take place even in the presence of chloramphenicol. This suggests that phenethyl alcohol acts upon the CM-resistant protein and this conclusion is supported by the discovery that it can exert its inhibitory effect only if added during the period of the replication cycle preceding synthesis of the CM-resistant species. Since phenethyl alcohol influences the cell membrane—mutants of E.coli resistant to it exhibit abnormal membrane functions and fail to divide normally—it seems likely that this initiation step may involve the cell membrane.

Cells of E.coli B/r—another strain often used for experiments in replication—are much more sensitive to chloramphenicol than cells of E.coli $15T^-$. The CM-sensitive step can be blocked by 2 μg/ml and the CM-resistant step can be blocked by concentrations greater than 20 μg/ml. Different times have been reported for these two steps. Ward and Glaser (1969) observed that the CM-sensitive step takes place about 20 minutes before intiation and the CM-resistant step coincides with initiation. Messer (1971) found the same time for the CM-sensitive step but noted that the CM-resistant step occurs at 5-10 minutes before initiation.

Because there is a lag of 15 minutes between synthesis of the second initiator protein and the time of initiation in E.coli $15T^-$, Lark and Renger (1969) suggested that a third step—independent of chloramphenicol—may take place to trigger the start of DNA synthesis. Lark (1972a) observed that rifampicin can inhibit initiation at any time prior to start of the round of replication in the $15T^-$ cells; and Messer (1971) reported a similar observation for B/r cells. One model is to suppose that this event depends upon the synthesis of an RNA which is not translated into protein (see page 452).

Initiation Mutants in Membrane Proteins

Bacterial DNA is clearly attached to the membrane of the cell at the origin where replication of the chromosome is initiated, and possibly at other sites as well (page 425). We do not know what role this association may play in replication itself, but it may be implicated both in initiating new cycles of replication and in ensuring even segregation of replicated chromosomes to daughter cells. That the association is important is suggested by the isolation of mutants which fail to initiate new cycles of replication at high temperature and which have altered membrane proteins under these conditions. Mutants of the *dnaA* class complete their current rounds of replication when raised to high temperature, but then cease to synthesize DNA. Cell division continues at a decreasing rate after the arrest of DNA synthesis, producing bacteria which lack DNA. Bacteria in which replication has ceased can reproduce phage λ, which supports the idea that it is only the initiation of new cycles which is prevented.

The membranes of mutant cells are abnormal at high temperature according to several criteria. Shapiro et al. (1970) found that some *dnaA* mutants are unusually sensitive to deoxycholate at 41°C and are lysed by a concentration of 0·1 %, although at 30°C they show the usual resistance to a concentration of 0·5 %. These mutants also have an alteration in the reactivity of their surface with a fluorescent agent, for when grown at high temperature they bind more anilino-naphthalene sulphonic acid and fluoresce, whereas there is little reaction at low temperature. By labelling one set of cells with H^3 leucine and another with C^{14} leucine, the membrane proteins of mutant and wild type cells can be compared. At high temperature, two *dnaA* mutants tested suffer a decrease in a peak which migrates on gel electrophoresis at a position corresponding to a molecular weight of about 60,000. At the same time, the amount of another peak of about 34,000 daltons is increased. Hirota, Mordoh and Jacob (1970) noted that the change in the membrane takes much longer to occur after a temperature increase than does the cessation of replication.

The electrophoretic separation resolves polypeptide chains only by weight, so the extent of the change in the mutant and the number of proteins affected are not revealed. But the same change is also found in *dnaB* strains grown at high temperature; the decrease in the peak of 60,000 daltons is fairly constant, but the increase in the 30,000 peak is more variable. Siccardi et al. (1971) found that one of the *dnaB* mutants has its defect in initiating replication overcome by growth in 2% salt; but the pattern of membrane proteins remains mutant. This suggests that the alteration in the membrane proteins is the primary effect of the *dnaB* mutation although whatever defect they cause can in some instances be remedied by growth in high salt concentrations.

Any one of several defects might cause the reduction in the 60,000 peak of membrane proteins of the mutant. The simplest is that synthesis of some protein has been prevented or reduced; an alternative is that its synthesis is normal, but its ability to integrate into the membrane is defective. Another possibility is that the membrane becomes disordered at high temperature and releases or degrades this component. When cultures of a *dnaB* mutant are labelled with C^{14} leucine at 30°C and then shifted to 40°C, many changes are seen in the pattern of the label; these changes are too complex to suggest a simple explanation, but that protein synthesis is implicated is suggested by their abolition by chloramphenicol.

When the proteins of the membranes of the mutants are disaggregated under much harsher conditions, using 100°C instead of 40°C in a solution of 1% SDA and mercaptoethanol a different distribution of proteins is found, with components of lower molecular weight. Mutants of the *dnaB* class show little change from usual. But mutants of the *dnaA* type have a deficiency at 35,000 daltons and a corresponding increase at 25,000 daltons. This suggests that the *dnaA* strains lack some protein which is usually found at the 60,000 dalton peak; *dnaB* strains have the component but it fails to aggregate to this molecular size. We do not know what role this membrane component may play in replication, but the mutant character of *dnaA* and *dnaB* confirms that the membrane is implicated in some way in allowing new cycles of initiation to take place.

Link Between Replication Cycles and Cell Division

DNA Synthesis During the Cell Cycle

Replication of the chromosome must be coordinated with division of the cell so that cycles of replication occur with the same frequency as cycles of cell division. Bacteria grow at widely differing rates depending upon their conditions of culture—E.coli cells may vary from doubling times of 20-180 minutes or more—and this raises the question of whether the rate of replication is proportional to the growth rate or whether replication takes place at a constant rate which is independent of the doubling time. We might expect that in the first instance cells are always in a state of replication, so that one round of replication occupies one cell division cycle. On the other hand, if the rate of replication remains constant, at slower growth rates there should be gaps in the cell division cycle where no replication occurs.

Using a synchronous culture of E.coli cells growing upon glucose medium with doubling times from 37–55 minutes, Clark and Maaløe (1967) found that cycles of replication are initiated about half way through the division cycle and not at the beginning. But in spite of the lack of coincidence between initiation and cell division, the periodicity between successive initiations of replication is the same as the division time.

The rate of DNA synthesis can be measured during the division cycle of E.coli cells by pulse labelling an exponentially growing culture with H^3-thymidine and measuring the incorporation of label into cells of different ages by counting the radioactivity in their progeny. Helmstetter and Cooper (1968) developed a technique of this nature in which the pulse labelled culture is bound to the surface of a millipore membrane filter; irrigating the surface of the membrane with medium allows the bound cells to grow and to release progeny cells into the effluent after a cell division. The new born cells are collected and their radioactivity measured as a function of the time at which they are eluted; this gives the rate of DNA synthesis in their ancestors as a function of the time in the division cycle when they were labelled—the sooner a new born cell is eluted from the filter, the older its parent must have been when bound to the membrane.

The start of a round of replication is revealed by a sudden increase in incorporation of label into DNA in cells of that age; completion of a round is signalled by a fall in the level of synthesis. The time taken for one round of replication is fairly constant at faster growth rates; for doubling times between 22 and 60 minutes, one round of DNA synthesis takes about 41 minutes. This agrees with the finding of Clark and Maaløe (1967) that the rate of DNA synthesis was more or less constant per growing point within their range of generation times. At slower growth rates, the rate of replication is lower, probably because the cells suffer limitations on precursor or energy supplies (see page 458).

On the basis of the results shown in figure 13.2, Cooper and Helmstetter (1968) proposed a model to link the replication cycle with the cell division cycle. The replication cycle can be defined in terms of two constants. The rate of replication is the time which it takes for a growing point to proceed from the origin to the terminus of the chromosome; since we now know that replication is bidirectional, we may say more precisely that the rate of replication is the time taken for both growing points to proceed round the chromosome to their common terminus. Both growing points must set out at the same time and reach the terminus at the same time, for there is only one doubling and halving of the amount of DNA synthesis during each replication cycle. The rate of replication, termed C, occupies about 40 minutes in E.coli. The second constant, D, is the fixed time which elapses after completion of a round of replication before the cells division which results from it. This is about 20 minutes (more precisely 22 minutes) for doublings taking from 22–60 minutes.

The left side of figure 13.2 shows the rate of DNA synthesis for a period of

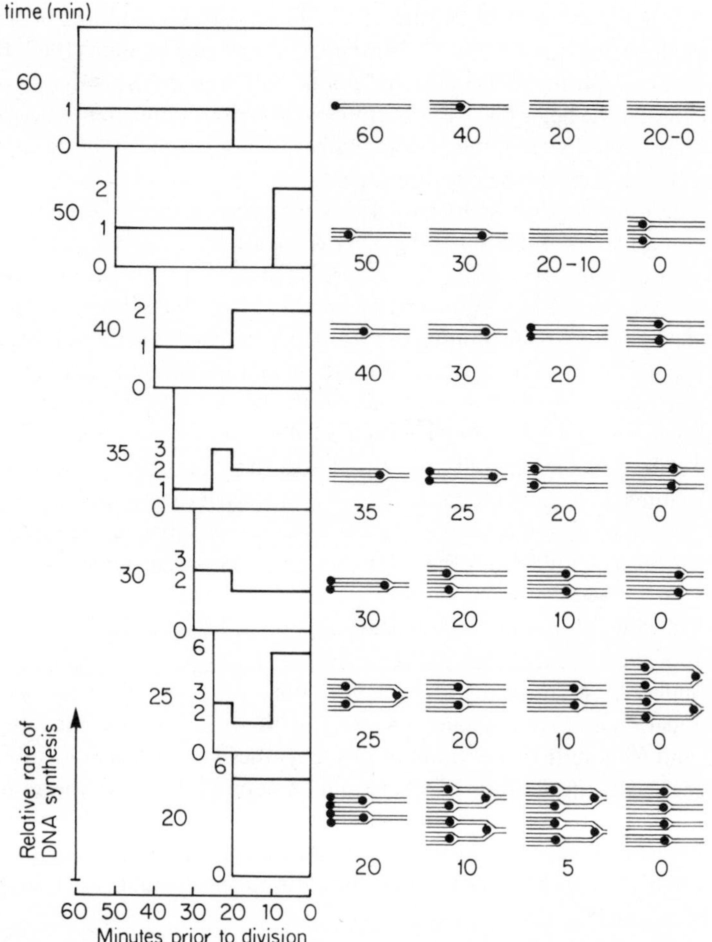

Figure 13.2: the rate of DNA synthesis in E.coli at various doubling times. Left: the rate of DNA synthesis during the division cycle in terms of the number of growing points. Increases represent the starts of rounds of replication and decreases mark their termination. At 40 and 20 minutes per doubling these occur simultaneously and obscure each other. At doubling times of more than 40 minutes there are gaps in the replication cycle when there is no DNA synthesis. Rounds of replication always take 40 minutes and terminate 20 minutes before a cell division. Right: state of the chromosomes during the division cycle. Growing points are represented by the black circles. At doubling times faster than 40 minutes chromosomes have multiple replication forks because a second round of replication starts before the first has finished. The figure depicts only half of the chromosome, that is the growing points moving in only one direction, as a linear structure; chromosomes in fact are circular and thus have a matching set of growing points moving in the opposite direction. Data of Cooper and Helmstetter (1968)

one division cycle in terms of the time before the next cell division. Initiations and completions of rounds of replication are observed as increases or decreases in the amount of DNA synthesis, except for doubling times of 20 and 40 minutes when an initiation and termination event occur simultaneously and obscure each other. The constant value of D is revealed by the decrease in DNA synthesis which always takes place 20 minutes before a cell division. The right side of the figure illustrates the state of the chromosomes according to the model, although for simplicity it shows only one half of the chromosome, as a linear molecule, with the origin at the left and one growing point moving away from it to the right.

A round of chromosome replication must be initiated a fixed time, C + D = 60 minutes, before a cell division. Initiations cannot coincide with division, therefore, but as the growth rate increases—doubling time shortens—initiation takes place earlier in the division cycle. If cells are dividing more frequently than every 60 minutes, the next round of replication must be initiated before the present division cycle is completed. For example, in cells with a doubling time of 50 minutes, initiation of replication for the next but one cell division must occur 10 minutes before the immediate cell division—10 minutes of the present division cycle + 50 minutes of the next give the required 60 minutes for replication and the gap before division.

In cells dividing less frequently than every 40 minutes, there is a period devoid of DNA synthesis during the division cycle because the cycle of replication is completed in less time than it takes to complete the division cycle. In cells growing at 40 minutes per doubling the time taken to replicate the chromosome is precisely the same as the time between cell divisions so that the starts and ends of cycles of replication coincide in the middle of the division cycle, 20 minutes before division.

When cells divide more frequently than every 40 minutes, they must initiate new rounds of replication before there has been sufficient time to complete the previous one. New growing points must therefore set out in each direction from the origin before the present ones have met at the terminus; this means that the chromosome must bear multiple forks. For example, in cells growing at 35 minutes per doubling, the current replication cycle must end D = 20 minutes before the next cell division. But the replication cycle for the division after next must be initiated C + D = 60 minutes prior to the division, that is 25 minutes before the immediate division (25 minutes of this division cycle + 35 minutes of the next gives the required 60 minutes). This is 5 minutes before the present round of replication is completed. Similarly, doubling times of 30, 25 and 20 minutes require successive rounds of replication to be initiated 10, 15 and 20 minutes respectively before the end of the present cycle.

One consequence of the constant time taken to replicate the chromosome is that as the growing rate of the cells increases—doubling time decreases—replication is initiated earlier and earlier in terms of generations prior to the cell

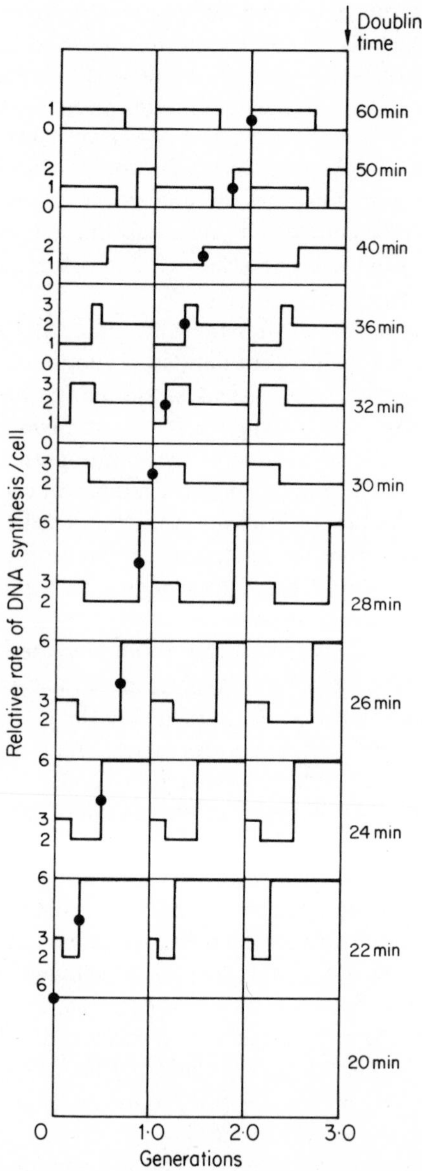

Figure 13.3: rate of DNA synthesis in terms of growing points during three successive division cycles. Cell age is plotted in generation times instead of real time. The set number is the time from the filled circles to the final division. Data of Cooper and Helmstetter (1968)

division D minutes after its end. If the doubling time is greater than 60 minutes, initiation of a replication cycle takes place in the same division cycle as that in which the replication is completed. Between doubling times of 60 and 30 minutes, replication is initiated in the division cycle one round before that in which it is completed. Between 30 and 20 minutes, replication cycles are initiated two division cycles before their completion. This conclusion is illustrated

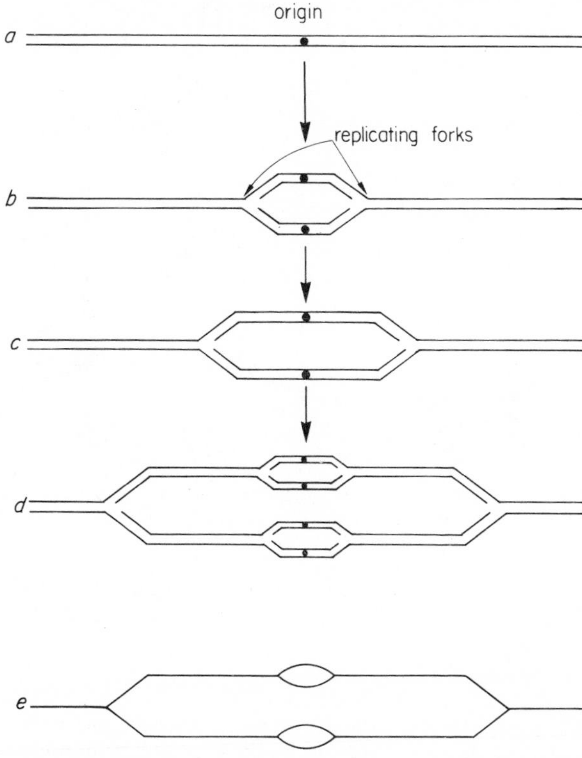

Figure 13.4: bidirectional replication of linear chromosome. The movement of the two replicating forks away from the origin create the loop of structures *b* and *c*, which increases in size as the forks move farther away from each other. Loops within this loop are created in structure *d* when a second round of replication is initiated before the first has been completed. In the electron microscope, this appears in the form of structure *e*

in figure 13.3, which plots the rate of DNA synthesis against cell age in terms of fractions of a generation rather than real time.

Cultures of cells growing at any particular rate cannot be characterized as possessing one or two chromosomes per cell, for as the right side of figure 13.2 shows, at growth rates faster than 40 minutes per doubling some cells have one replicating chromosome and some have two, depending upon their position in the cell cycle. The chromosome configurations for any particular

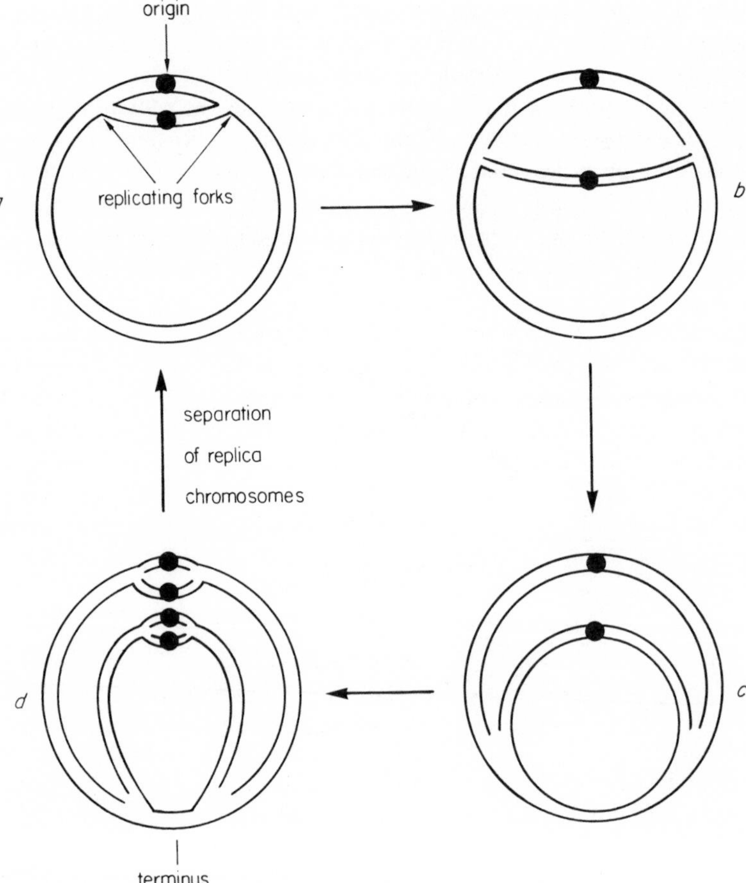

Figure 13.5: structures of circular chromosomes with multifork bidirectional replication. A loop is created by the two replicating forks moving away from the origin; this increases in size from structures *a* to *c* as replication proceeds around the chromosome. Loops within the loop are generated in structure *d* when a second initiation even takes place before the previous two replicating points have met each other. When replication is completed at the terminus, the two chromosomes separate to give two structures of the form *a* which already have replicating forks. These types of structures would be shown in an E.coli cell growing at 30 minutes per doubling

doubling time can only be characterized by the chromosome "set", the series of states of the chromosome shown in the figure. Any chromosome set can be defined by the "set number", which is the time in fractions of a generation between the start of a new round of replication and the cell division following completion of that round. Numerically this is (C + D)/T, where T is the doubling time. This describes the configurations of the whole culture; the distance between the filled circles of figure 13.3 and the division indicated by the arrow gives the set numbers for various doubling times.

Although the replication system of the F factor (and other episomes) is independent of that of the bacterial chromosome, its synthesis also appears to lie under a control related to the cell cycle. Cooper (1972) and Davis and Helmstetter (1973) observed that replication of F′lac factors—measured by an increase in their synthesis of β-galactosidase—is initiated when cells reach a mass which is a constant multiple of the number of episomes per cell at each growth rate.

The chromosome structures illustrated in figure 13.2 do not show the true physical state of the chromosome since the genome is circular and is replicated bidirectionally. As figure 13.4 shows, chromosomes which are replicated in this manner possess loops within loops when a second round of replication is initiated before the first has been completed. Structures of this kind can be generated only by bidirectional replication and not by unidirectional movement of one growing point; Delius, Howe and Kozinski (1971) have observed loops within loops in electron micrographs of replicating phage T4 DNA. The structures generated by bidirectional replication of a circular chromosome are illustrated in figure 13.5, which predicts the class of structure for the chromosome of E.coli cells doubling every 30 minutes; at this growth rate, a second cycle of replication is initiated before the first has begun, so that the chromosomes which separate after completion of the first cycle each already possess two growing points.

The model for linking the replication cycle to the division cycle does not appear to hold for slower growth rates. For doubling times greater than 60 minutes there should be one gap at the beginning of the division cycle, before initiation of DNA synthesis, and another gap of 20 minutes after the completion of replication, before cell division occurs. But for generation times longer than 60 minutes per doubling, C and D are no longer constant in time but instead occupy fixed proportions of the division cycle. Replication is initiated at the beginning of a division cycle and the chromosome is synthesised during the first two thirds of the generation, with a gap devoid of DNA synthesis occupying the one third remaining to cell division. Under these very slow conditions, the provision of necessary precursors may become limiting so that the usual timing of events is altered.

Titration of Cell Mass

Chromosome replication and cell division show a reciprocal relationship which depends upon the ratio of DNA content to cell mass. Cell division can take place only after the cell has completed a cycle of replication and has an appropriate content of DNA for its mass. New cycles of replication can be initiated only when the cell achieves an appropriate ratio of cell mass to the number of origins. These interactions ensure that replication and division take place in a coordinate manner in which the frequency of cycles of replication equals that of cell divisions.

By comparing the time of initiation with the cell mass, Donachie (1968) was able to show that initiation occurs at a constant ratio of cell mass to the number of chromosome origins according to the relationship:

$$\log_e M_d = \log_e M_i/N_i + (C + D)/T$$

where M_d and M_i are the cell masses at division and initiation of replication and N_i is the number of chromosome origins at initiation. The mass of cells at division can be measured experimentally, and the mass at intiation must then be their mass $C + D = 60$ minutes earlier. The equation holds if the ratio M_iN_i is constant; this implies that the cell can in some way titrate its mass to judge when it is ready for initiation.

Cells could titrate their mass if a critical amount of some protein controls initiation of new cycles of replication. Two classes of model can provide this ability. One suggests that an initiator of replication is synthesised during the cell cycle and that accumulation of a critical amount triggers initiation. An alternative is to suppose that an inhibitor protein which represses initiation is synthesised at some point in the cell cycle and is then diluted out by cell growth; initiation takes place when its value falls below some critical value. The responses of cells to treatments which change their DNA/mass ratios can reveal whether general models of this nature—involving accumulation or dilution—apply, but do not distinguish between them.

If an initiator protein is synthesised continually and one unit of initiator per origin is required to start a new replication cycle under any growth conditions, the generation time of cells growing exponentially in any particular medium reflects the time required to synthesise sufficient units of initiator at that growth rate. According to the model proposed by Helmstetter et al. (1968) and Pierucci (1969), one unit of initiator per origin is synthesised during each replication cycle. The constancy of the cell mass per origin at the time of initiation demands that the amount of initiator must be related to the mass or volume of the cell; and this could be achieved if a gene located near the origin of the chromosome is expressed at a rate which is inversely proportional to the doubling time. The total amount of this protein should therefore depend upon the number of copies of its gene—that is the number of origins—and the growth rate of the cell.

Changes in cell division following shift up experiments—when cells are transferred from a medium on which they grow slowly to one which stimulates more rapid growth—are consistent with this interpretation. After an increase in growth rate, the time of the next initiation should be advanced to the moment when sufficient initiator protein has accumulated. The time of initiation can be determined by measuring the time of cell division and substracting $C + D = 60$ minutes. For example, suppose that cells growing at 40 minutes per doubling are shifted up to a growth rate of 20 minutes per doubling. Under the slow conditions of growth they must synthesise one unit of initiator per origin in 40 minutes, but under the fast conditions this same amount of synthesis will take only 20 minutes.

A cell which has initiated a cycle of replication 10 minutes before the shift up must therefore have synthesised $10/40 = \frac{1}{4}$ of a unit of initiator per chromosome origin at the time when its growth conditions are changed. It should synthesise the remaining 3/4 of a unit in the next $3/4 \times 20 = 15$ minutes, whereas in the absence of the shift up this event would not have taken place for another 30 minutes. Different cells are at different stages of their replication cycles when an unsynchronised culture is shifted up, but the distribution of the ensuing cell divisions with time can be calculated from this model; theoretical predictions are in good accord with the experimental results.

This model predicts that if DNA synthesis is inhibited selectively in a culture in which RNA and protein synthesis remain unaffected, initiator protein should continue to accumulate; when DNA synthesis is permitted again, new initiations take place if the cell mass has increased sufficiently and the consequent cell division can be followed subsequently. After DNA synthesis has been stopped, cell division should continue normally for D minutes as cells which have completed their current rounds of replication undertake the consequent division. Division then stops but initiator protein continues to accumulate. After the restoration of DNA synthesis, the rate of division should be the same as the previous rate for the next $C + D$ minutes as the replication forks which were halted by the inhibition now restart and continue to the terminus. After this time, there should be a wave of cell divisions resulting from the multiple initiations which were caused by the accumulation of initiator during the block to DNA synthesis.

That the initiation of new rounds of replication does not depend upon DNA synthesis but reflects the growth of the cell is shown by the ability of cells to commence new cycles of initiation after treatment with nalidixic acid, starvation for thymine, or incubation of a temperature sensitive mutant at high temperature. When DNA replication is inhibited by one of these means for some length of time, cells immediately re-initiate new cycles of replication as soon as replication is permitted again. Comparable results are obtained when the rate of movement of the replicating fork is slowed by the substitution of BUdR for thymine.

In all these experiments the block to replication was imposed for a long enough time for sufficient initiator to accumulate to trigger initiation as soon as DNA synthesis is allowed again. A more critical test of the model has been provided by Ward and Glaser (1970) in experiments in which a block to replication was imposed by adding nalidixic acid for only a short time. A premature initiation relative to the present initiation cycle should still occur after restoration of DNA synthesis (the initiation is premature, because the growing points which are halted by the block to DNA synthesis have not moved as far as usual when the next round is initiated); but the initiation should not take place immediately upon resumption of replication, as it does after long blocks, because insufficient initiator is synthesised during the short block. The critical amount of initiator is not achieved until after some time of resumed DNA synthesis; the longer the block to replication, the sooner initiation takes place when DNA synthesis resumes.

According to the model, cells should initiate new rounds of replication at the age at which they would have done so anyway had DNA synthesis not been blocked. But initiation is not completely independent of DNA synthesis, for there is a delay in the time of initiation when a block to replication is imposed and then removed. This delay seems to correspond to the time taken to complete the current cycle of replication, that is to let the replication forks which were halted by the block move to the terminus. One way to explain this result is to suppose that the cell controls the number of growing forks in any given medium and there may be a limit to the number which can be active under particular growth conditions. When the new premature forks are initiated, the original replicating forks appear to be inactivated. The inability of the cell to contain more than a certain number of active replicating forks may therefore delay the initiation of new ones.

By measuring the incorporation of labelled thymidine into DNA, Donachie (1969) showed that the increased rate of DNA synthesis in reinitiating cells persists until the total content of DNA is restored to the level usually characteristic of the culture. Figure 13.6 shows that the cells then continue to synthesise DNA at the usual rate. The longer the period of thymine starvation, the higher the initial rate of DNA synthesis; and the correlation suggests that when the cell mass reaches a critical amount relative to the number of chromosome origins, the cell reinitiates DNA synthesis at all available origins.

When thymine is removed from this medium, the division of these cells ceases as well as DNA synthesis; when thymine is restored division recommences after a delay. This delay is not constant but depends upon the length for which the cells have been deprived of thymine; the longer the deprivation, the greater the delay. After this delay, every cell in the population divides during an interval which is shorter than the usual generation time—as figure 13.6 shows, this division is therefore partly synchronised. This suggests a model in which the initiation of the extra rounds of replication (as a result of thymine starvation) is correlated with a delay in division until D minutes after the termination of the extra round. Completion of the extra round restores the DNA/mass ratio of the cell culture to its usual value. Donachie et al. (1968) suggested that this form of control should be displayed by cells which are unable to divide when the DNA/mass ratio is too low and are therefore delayed in division until it has been restored; this may therefore represent a mechanism

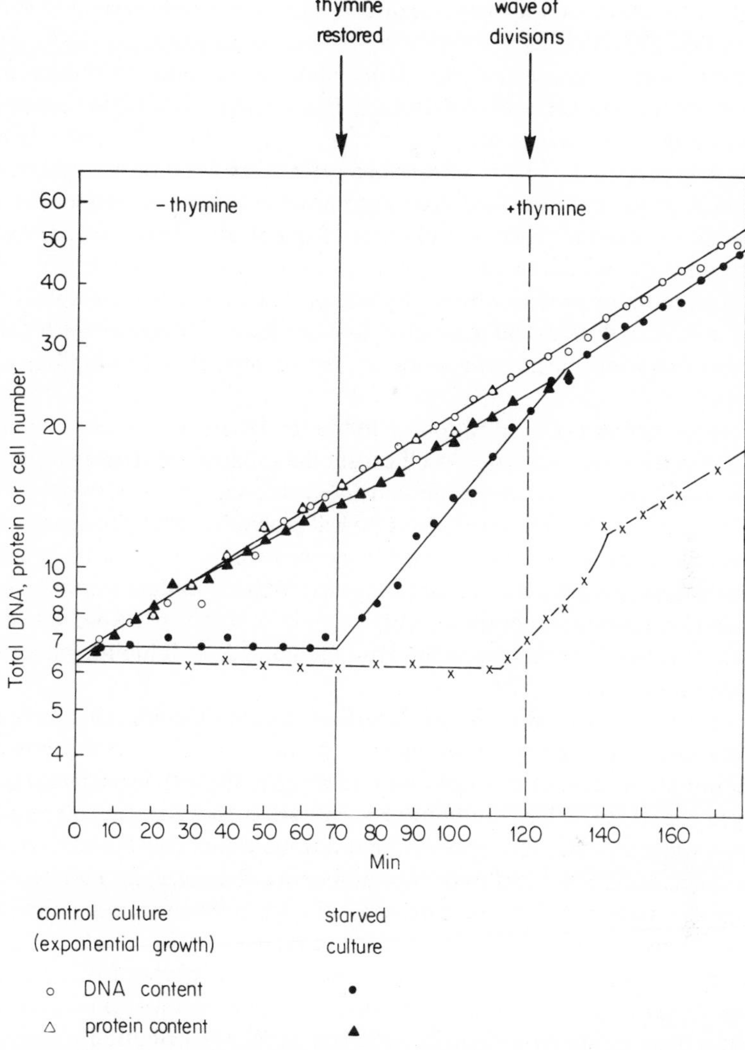

Figure 13.6: control of initiation of DNA synthesis and cell division in E.coli. Cells which have been starved for thymine show an increased rate of DNA synthesis, due to new initiation events, when the nucleotide is restored; protein—that is mass—content of the cells increases normally during the period of starvation and the increased rate of DNA synthesis continues until the normal DNA/mass ratio is restored. Division is delayed after restoration of thymine until the DNA/mass ratio is returning to its usual value. The ratio of DNA to protein is therefore critical both in the initiation of new rounds of replication (extra initiations take place when the DNA/mass ratio is too low) and in setting conditions for cell division (division is prevented if the DNA/mass ratio is too low). Data of Donachie et al. (1968)

for helping to maintain the proper growth conditions of cells in spite of variations in DNA/mass ratios.

Another way to titrate cell mass is to dilute an inhibitor of initiation in concentration as the cell grows. Pritchard, Barth and Collins (1969) proposed that an inhibitor protein is coded by a gene located adjacent to the origin—or which is part of it—and is transcribed only at the time of its replication. Each such gene, and therefore each chromosome origin, is responsible for synthesising a fixed number of molecules of inhibitor protein irrespective of growth rate. The inhibitor interacts either with the chromosome origin itself, or with an initiator protein which is synthesised at a constant rate during the cell cycle—formally this is the same model—in such a way that a two fold change in its concentration effects a transition between complete and no inhibition of initiation.

When the cellular concentration of inihibitor falls below a critical value, initiation is triggered. Initiation itself causes the production of more inhibitor protein as the relevant gene is replicated and transcribed, thus raising the level of inhibitor above the critical value. As the cell grows, its volume increases, and since no more inhibitor is synthesised its concentration falls due to dilution. When it passes below the critical value, the next round of initiation commences. In an equilibrium situation, the concentration of inhibitor must oscillate over a two-fold range, irrespective of the absolute number of inhibitor molecules produced at each pulse.

The frequency of initiation is therefore determined by the dilution rate of the inhibitor, that is the reciprocal of the growth rate. This means that successive acts of initiation occur after successive doublings of the cell. Since initiation is achieved at a critical value of inhibitor concentration, that is at a constant value of a/V_i—where a is the number of the inhibitor molecules and V_i is the volume of the cell at initiation—and since the number of inhibitor molecules depends upon the number of chromosome origins, V_i/a, the cell volume per number of origins is constant at initiation for all growth rates.

An important feature of this model is that it is self regulatory; it requires that the *mean* time interval between successive acts of initiation equals the doubling time, although individual cells can show a distribution about this mean. If initiation takes place before the inhibitor value has fallen to its critical value, the daughter cells are smaller than usual so that the pulse of inhibitor synthesised at the initiation event produces a higher concentration than usual. This means that it takes longer before the next initiation occurs, restoring the average between initiation events.

Growth of the Unit Cell

When E.coli cells are plated on agar their growth can be followed relative to fixed markers. Using this technique, Donachie and Begg (1970) observed that on minimal medium the cells appear to grow at one end only until they divide after some 60-70 minutes of growth. The cells produced by a division are 1·7 μ long and they double in length to 3·4 μ before the next division. A different pattern is seen in rich medium, when cells may grow beyond the size of 3·4 μ and then appear to elongate in both directions relative to the agar surface.

The minimum size of an E.coli cell appears to demand a length of 1·7 μ and Donachie and Begg suggested that this can be regarded as *unit cell*; cells which are more than 3·4 μ long can therefore be regarded as comprising two unit cells. The unit cell provides the basic growth unit of E.coli, each unit cell growing autonomously at a single growth site. Synthesis of new membrane takes place asymmetrically from this site which is located at one end of the unit cell.

The model initially proposed for the growth of the unit cell suggested that when a single unit cell grows from one end to reach a size of 3·4 μ and then divides, each of the daughter cells grows at the end newly created by division. The same pattern is repeated when these cells later divide so that cells of a length less than 3·4 μ always grow only from one end. When E.coli cells are grown under conditions in which division does not take place at the 3·4 μ length of two unit cells, duplication of the growth site takes place as though the daughter unit cells had divided; cells longer than 3·4 μ therefore consist of two unit cells each of which contains one growth site. The extension of these longer cells in both directions appeared to be achieved by growth at two sites in the centre of the cell (where it would have divided into two cells under conditions suitable for division).

More recent results, however, suggest an alternate pattern for the growth of E.coli cells, also consistent with the earlier data. By using a strain of E.coli which is temperature sensitive for the synthesis of the receptor for phage T6, Begg and Donachie (1973) located the positions of old and new membrane sections by switching cells from one temperature to the other during growth; after addition of T6, the membrane sites to which it binds can be visualized by electron microscopy. Use of this technique confirmed the earlier conclusion that cells of E.coli consist of independently growing unit cells each of which lays down new membrane from only one growth site. However, as the model of figure 13.7 shows, each unit appears to grow first at one end and then, when a daughter unit cell has been formed, at the other end.

As the figure shows, growth of the unit cell outer membrane alternates between the two ends. When the initial unit cell "0" has doubled in length, its growth continues (between doublings 1 and 2) from the other end; the new unit cell produced by its growth, "1", is extended from its newly formed end. When the second doubling has been completed, each of the new unit cells, "2", is extended from the newly completed end; and the reversal of growth sites in the earlier unit cells "1" and "2" causes growth at the central point of the cell. An analogous pattern of reversal is followed in subsequent generations.

Replication can be linked to cell growth if new cycles of DNA synthesis are initiated whenever the number of unit cells doubles. At this time, a single round of replication is initiated in each unit cell. This means that the ratio of the number of unit cells—or number of growth sites—to the number of chromosome origins is always unity. The unit cell therefore constitutes the initiation

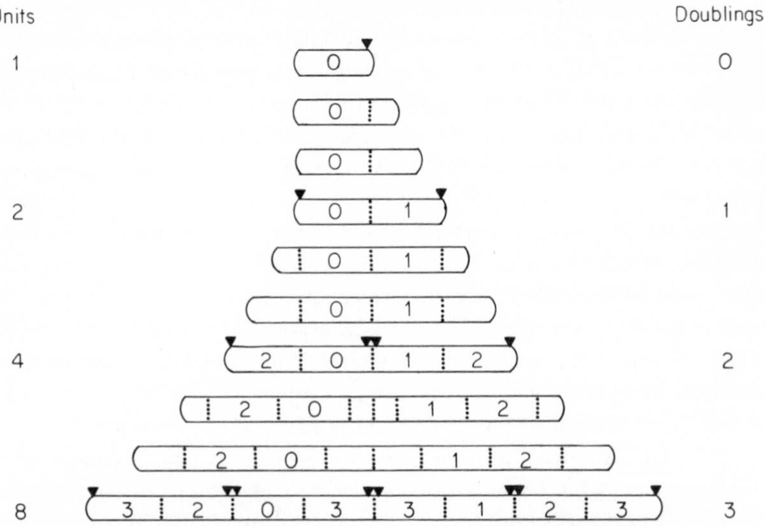

Figure 13.7: unit cell model for growth of E.coli. An initial unit cell "0" grows through three doublings whilst division is inhibited; daughter unit cell "1" is produced by the first doubling, unit cells "2" and "3" result from the second and third doublings. The dotted lines indicate the boundaries between unit cells and the triangles represent the locations of growth sites. The direction of growth of each unit cell reverses after each division. Cells shorter than 3.4 μ therefore grow in one direction only; cells of greater length show autonomous growth of each constituent unit cell, the result being symmetrical growth about the central axis. If cell division had been permitted, divisions would have taken place at the dotted lines marking the boundaries between unit cells. Data of Begg and Donachie (1973)

mass, M_i, of the equation discussed in the last section. This provides a topological model to show how the number of chromosome origins relates to the mass of the cell as a whole; the cell contains one unit cell for each origin. Chromosomes could become spatially separated into daughter cells if they are physically attached to the growth sites. Such attachment may play some role in allowing division to succeed separation of daughter chromosomes after termination events; and if division is prevented when a site is occupied by a prematurely initiating chromosome, this initiation might delay the next cell division.

If the rate of growth of each unit cell is constant, the volume of an E.coli cell which is shorter than 3·4 μ should increase in a linear way until the size of two unit cells is reached; the growth rate should then double as each growth site contributes to the enlargement of the cell. If the signal to initiate new rounds of replication is the duplication of the growth site on the membrane, the point in the life cycle at which DNA synthesis is initiated should correspond to the time when the cell growth rate doubles.

E.coli B/r cells growing in glucose medium show an increase in growth rate of 1·7 times after 28 minutes and then another increase of 1·9 times after 80 minutes of growth. Ward and Glaser (1971) found that DNA synthesis doubles at 30 and 82 $\pm$ 5 minutes. These figures suggest that cycles of replication are usually initiated when the growth site is duplicated. But if nalidixic acid is added to block replication, cells can still undertake a doubling in their growth rate; this means that duplication of the growth site does not depend on replication. If nalidixic acid is added for only 15 minutes, the next round of replication is delayed for 18 minutes and the next doubling in the growth rate is postponed for 35 minutes. Replication can therefore be initiated before the growth site has been duplicated. The ability of the two events to occur independently demonstrates that although duplication of the growth site and initiation of replication may usually take place simultaneously, neither is in fact the trigger for the other. In other words, we cannot regard the duplication of the growth site as equivalent to the synthesis of initiator protein; rather is it an event which is usually temporally linked with initiation of cyles of replication, but which is not mechanically coordinated with them.

Trigger for Cell Division

Replication must be linked to the apparatus responsible for division if the cell is to divide only when replication of its chromosome has been completed. And the cell membrane may be implicated in the segregation of replicated chromosomes to daughter cells. Jacob et al. (1966) suggested that if each of the two daughter chromosomes resulting from a round of replication is attached to the membrane, the growth of a septum between their attachment sites could automatically segregate them into different progeny cells as illustrated in figure 13.8. Ryter et al. (1968) observed that such a mechanism might influence the pattern of segregation after cell division. If DNA strands are distributed at random, the separation of labelled strands should follow the pattern in figure 13.9. If there is some definite mode of segregation, only one of the progeny shown in the figure should be generated.

These predictions can be tested by labelling E.coli or B. subtilis cells with H^3-thymidine, growing the culture for several generations in unlabelled medium, and then determining the position of the label in the progeny by autoradiography. Results obtained with both bacterial species suggest that segregation is random, so that at each cell division each of the old strands has the same probablity of distribution into either of the progeny. This excludes models which require some attachment at initiation in which the previously existing or newly synthesized strands are always attached to the old or newly synthesized growth sites. (But we should note that Lark, 1966 obtained results suggesting a definite pattern of segregation in E.coli $15T^-$ cells under certain growth conditions.)

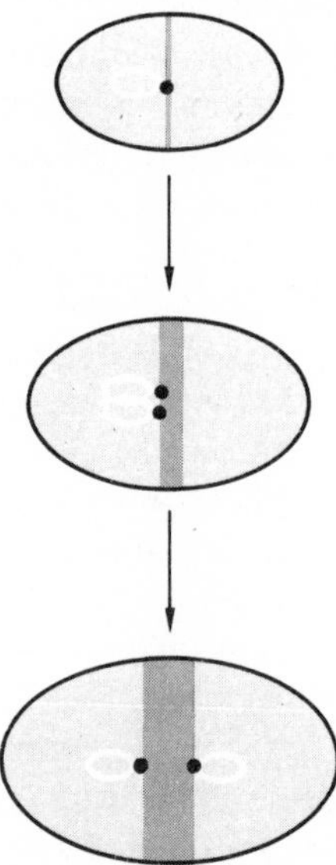

Figure 13.8: segregation of daughter chromosomes into two progeny cells by growth of a septum between their membrane attachment sites. After Jacob, Brenner and Cuzin (1966)

The large number of mutants which are defective in cell division confirms that it must involve many processes. Clark (1968b) used two approaches to delineate the different stages of division. The susceptibility of cells to killing when they are infected with mutants of phage T4 which cannot multiply in the host measures septum formation. If a host cell retains its integrity, infection is lethal; but if a septum has formed, the distal end of a cell infected with a single phage particle is immune from the effects of infection. The time when a rigid cross wall is formed can be estimated by subjecting cells to sonication. When an osmotically rigid wall has developed, there is a sudden increase in the number of survivors of sonication because one end of the cell can survive event when the other is damaged by the treatment.

With a culture of E.coli doubling every 45 minutes, a round of replication is completed at 20 minutes, septum formation is achieved at 30 minutes and a rigid cross wall forms at 37 minutes. The *physiological division* resulting from septum formation, in which the cytoplasmic events at opposite ends of the cell become independent of each other, therefore occurs well after the completion of replication and well before the physical separation of a parent into two progeny cells. Regulation of cell division probably depends only upon events taking place prior to this compartmentation.

Cell division depends upon completion of a round of replication. Helmstetter and Pierucci (1968) and Clark (1968a,b) showed that once a round of replication has been terminated, further DNA synthesis is not necessary for

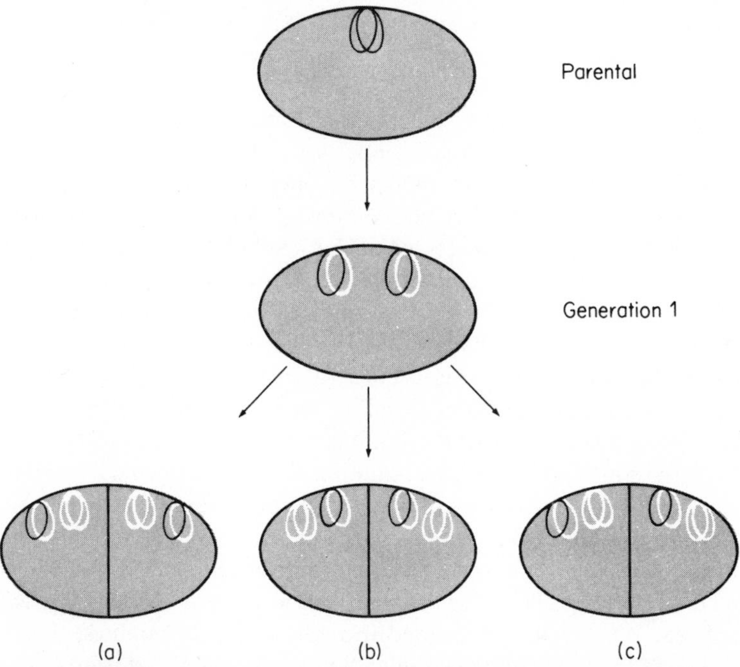

Figure 13.9: possible modes of chromosomes segregation at division. (a) parental strands segregate to far ends of cell, (b) parental strands remain near septum, (c) both parental strands segregate to left (or, not illustrated, to right). If segregation of chromosomes is random, *a*, *b c* should be found with frequencies of 25%, 25% and 50%; this agrees with experimental observations. After Ryter, Hirota and Jacob (1968)

cell division. When DNA synthesis is inhibited by ultraviolet irradiation, treatment with mitomycin C, or addition of nalidixic acid, cell divisions continue for 20 minutes after DNA synthesis has ceased. This represents the ensuing divisions of cells which have completed rounds of replication. Gross et al. (1968) obtained analogous results by incubating temperature sensitive mutants in replication at their non-permissive temperatures.

Results of shift up experiments also are consistent with a sequence of events in which termination of replication provides a trigger essential for cell division. Lark (1966) and Cooper (1969) observed that after a shift, cells complete their current round of replication in the manner characteristic of the old medium and display a response to their new, richer conditions only in the next cycle. When culture conditions are changed from a slow to a faster growth rate, an initiation of replication takes place shortly after the shift up. But the next cell division takes place after the interval characteristic of the slower medium. This suggests that it depends upon completion of replication by the current growing points to trigger the D minute wait for division. (Under these conditions there was only one growing point moving in each direction at the time of transfer.) The second cell division follows more rapidly after an interval characteristic of the new medium; and takes place D minutes after the enzyme which has initiated replication upon the shift up reaches the chromosome terminus.

Completion of replication is essential for cell division because it supports the synthesis of necessary and protein(s). Jones and Donachie (1973) synchronized cells by a two stage procedure: inhibition of protein synthesis first causes all cells of the population to accumulate between replication cycles; and allowing protein synthesis to resume whilst DNA synthesis is blocked then allows the proteins needed for initiation to accumulate. When thymine is restored to the medium to allow DNA synthesis, all cells start replication synchronously; and division follows very shortly after the completion of replication, that is at about 45 minutes after initiation, without the usual D minute wait.

Both RNA and protein synthesis are needed during the ensuing replication cycle if division is to take place. Addition of chloramphenicol or rifampicin at times up to about 35-40 minutes after initiation prevents division. Addition and subsequent removal of rifampicin before this time does not delay division. Synthesis of RNA and protein essential for division must therefore occur at about 5-10 minutes before the division, that is at the time when replication is completed. When cells are allowed to replicate for 20 minutes, thymine removed for 45 minutes to block replication whilst RNA and protein synthesis continue, and thymine then restored, division takes place at the time when the interrupted cycles of replication should be completed. Addition of rifampicin prevents division.

Synthesis of the "termination protein(s)" therefore takes place only when replication is completed, perhaps because they are coded by gene(s) located at or very close to the terminus. This link between replication and division thus ensures that cells divide only when they possess completed chromosomes ready for segregation into the daughter cells.

In the normal cell cycle, a 20 minute delay separates the completion of replication from the subsequent cell division. The absence of this delay in the cells synchronised by sequential deprivation of protein synthesis and DNA synthesis suggest that the events which usually occur during the 20 minute period can take place earlier in the cycle when cells have been allowed to synthesize proteins for some time before initiation takes place. Synthesis of these division proteins must be independent of replication.

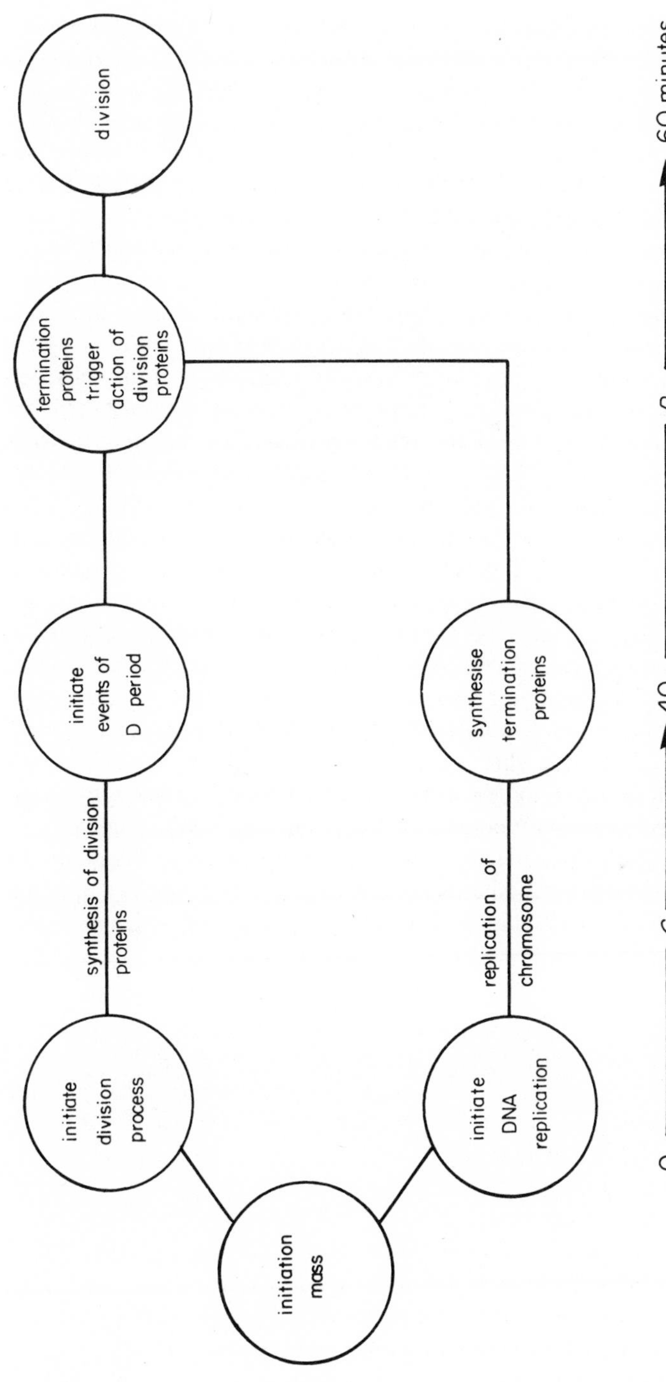

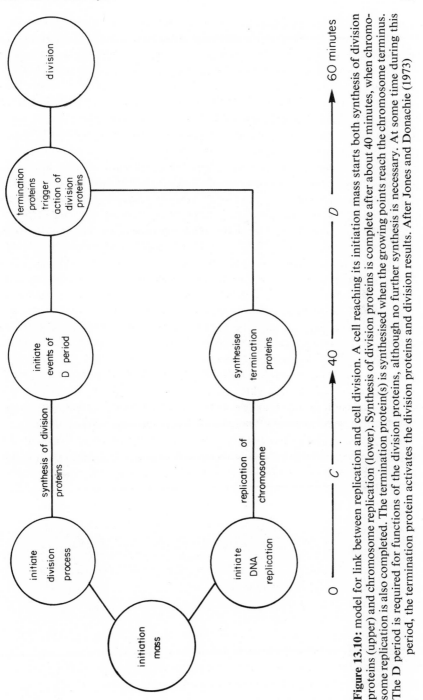

Figure 13.10: model for link between replication and cell division. A cell reaching its initiation mass starts both synthesis of division proteins (upper) and chromosome replication (lower). Synthesis of division proteins is complete after about 40 minutes, when chromosome replication is also completed. The termination protein(s) is synthesised when the growing points reach the chromosome terminus. The **D** period is required for functions of the division proteins, although no further synthesis is necessary. At some time during this period, the termination protein activates the division proteins and division results. After Jones and Donachie (1973)

Two parallel series of events may therefore be needed for cell division and the model proposed by Jones and Donachie is illustrated in figure 13.10. Initiation of replication sets in train the synthesis of specific proteins which are needed for division and which accumulate during the cycle. Because synthesis of DNA, RNA and protein is not necessary during the 20 minute D period, these proteins must be synthesized by the end of the 40 minute replication period. The 20 minute delay before division must therefore be needed to allow these proteins to exercise their functions and not for their production. The second class of event also starts with initiation of replication, for when the growing points reach the terminus 40 minutes later they trigger synthesis of some protein(s) needed for division. That these proteins can act immediately is shown by the rapid succession of division after completion of replication in the cells synchronized by the two step procedure; the termination protein(s) must therefore be involved in some late stage of the actions mediated by the other division proteins.

That protein(s) essential for cell division accumulate during the cycle has been suggested by Smith and Pardee (1970), who observed that shifting cells to 45°C for 16 minutes synchronises division in an unsynchronized population. Since cells of the temperature shocked population are in different stages of their cell cycles, this implies that cells of different ages are influenced to different extents by the temperature shock. A 6 minute temperature shock does not synchronize the cell population; but a 10 minute exposure to para-fluorophenylalanine followed by 6 minutes at high temperature does so. The treatment appears to have little effect on cells early in the cycle but delays cells closer to division for much longer, so that all cells divide together. One model to explain these results is to suppose that cell division requires the accumulation during the cell cycle of a protein which is destroyed by heat shock; when the protein has been inactivated, all cells take the same time to replace it and thus to divide. One possible candidate for this role is the protein(s) which makes up the septum.

References

Aaronson S.A. and Lyttle C.D. (1970). Decreased host cell reactivation of irradiated SV40 virus in Xeroderma pigmentosum. *Nature*, **228**, 359–361.

Abelson J.N., Gefter M.L., Barnett L., Russell R.L. and Smith J.D. (1970). Mutant tyrosine tRNAs. *J. Mol. Biol.*, **47**, 15–28.

Adams A., Lindahl T. and Fresco J.R. (1967). Conformational differences between the biologically active and inactive forms of a tRNA. *Proc. Nat. Acad. Sci.* **57**, 1684–1691.

Adams J.M. (1968). On the release of the formyl group from nascent protein. *J. Mol. Biol.*, **33**, 571–590.

Adams J.M. and Capecchi M.R. (1966). N-formylmethionyl-tRNA as the initiator of protein biosynthesis. *Proc. Nat. Acad. Sci.*, **56**, 147–155.

Adams J.M., Jeppesen P.G.V., Sanger F. and Barell B.G. (1969). Nucleotide sequence from the coat protein cistron of R17 phage RNA. *Nature*, **223**, 1009–1015.

Adesnik M. and Levinthal C. (1969). Synthesis and maturation of rRNA in E.coli. *J. Mol. Biol.*, **46**, 281–304.

Adhya S.L. and Shapiro J.A. (1969). The galactose operon of E.coli K12. I. Structural and pleiotropic mutations of the operon. *Genetics*, **62**, 231–247.

Adler K., Beyreuther K., Fanning E., Geisler N., Gronenborn B., Klemm A., Muller-Hill B., Pfahl M. and Schmitz A. (1972). How *lac* repressor binds to DNA. *Nature*, **237**, 322–327.

Ahmed A. (1968). Organization of the histidine-3 region of Neurospora. *Mol. Gen. Genetics*, **103**, 185–193.

Ahmed A., Case M.E. and Giles N.H. (1964). The nature of complementation among mutants in the histidine-3 region of N. crassa. *Brookhaven Symp. Biol.*, **17**, 53–65.

Alberts B.M. (1970). Function of the gene-32 protein, a new protein essential for the genetic recombination and replication of T4 phage DNA. *Fed. Proc.*, **29**, 1154–1163.

Alberts B.M., Amodio F.J., Jenkins M., Gutman E.D. and Ferris R.L. (1968). Studies with DNA-cellulose chromatography. I. DNA binding proteins from E.coli. *Cold Spring Harbor Symp. Quant. Biol.*, **33**, 289–306.

Alberts B.M. and Frey L. (1970). T4 phage gene-32; a structural protein in the replication and recombination of DNA. *Nature*, **227**, 1313–1317.

Alberts B.M., Frey L. and Delius H. (1972). Isolation and characterization of gene-5 protein of filamentous bacterial viruses. *J. Mol. Biol.*, **68**, 139–152.

Altman S., Brenner S. and Smith J.D. (1971). Identification of an ochre suppressing anticodon. *J. Mol. Biol.*, **56**, 195–198.

Altman S. and Smith J.D. (1971). Tyrosine tRNA precursor molecule polynucleotide sequence. *Nature New Biol.*, **233**, 35–39.

Ames B.N. and Hartman P.E. (1963). The histidine operon. *Cold Spring Harbor Symp. Quant. Biol.*, **28**, 349–356.

Anderson K.W. and Smith J.D. (1972). Still more mutant tyrosine tRNAs. *J. Mol. Biol.*, **69**, 349–356.

Anderson W.B., Schneider A.B., Emmer M., Perlman R.L. and Pastan I. (1971). Purification and properties of the cyclic AMP receptor protein which mediates cyclic AMP dependent gene transcription in E.coli. *J. Biol. Chem.*, **246**, 5929–5937.

Anderson W.F. (1969) Evolutionary conservation of the synthetase recognition site of alanine tRNA. *Biochem.*, **8**, 3687–3691.

Anraku N., Anraku Y. and Lehman I.R. (1969). Enzymic joining of polynucleotides. VIII. Structure of hybrids of parental T4 DNA molecules. *J. Mol. Biol.*, **46**, 481–492.

Anraku N. and Lehman I.R. (1969). Enzymic joining of polynucleotides. VII. Role of the T4-induced ligase in the formation of recombinant molecules. *J. Mol. Biol.*, **46**, 467–479.

Anton D.N. (1968). Histidine regulatory mutants in S. typhimurium. V. Two new classes of histidine regulator mutants. *J. Mol. Biol.*, **33**, 553–546.

Arber W. (1965). Host specificity of DNA produced by E.coli. V. The role of methionine in the production of host specificity. *J. Mol. Biol.*, **11**, 247–262.

Arber W. (1968). Host Controlled restriction and modification of bacteriophage. *Symp. Soc. Gen. Mic.*, **18**, 295–314.

Arber W. and Linn S. (1969) DNA modification and restriction. *Ann. Rev. Biochem.*, I. Host controlled modification of phage λ. *J. Mol. Biol.*, **5**, 18–36.

Arber W. and Linn S. (1969). DNA modification and restriction. *Ann. Rev. Biochem.*, **38**, 467–498.

Arber W., Rifat A., Wauters-Willems D. and Kuhnlein V. (1972). Host specificity of DNA produced by E.coli. XVI. Phage λ carries a single site of affinity for A-specific restriction and modification. *Mol. Gen. Gen.*, **115**, 105–207.

Arber W. and Wauters-Willems D. (1970). Host specificity of DNA produced by E.coli. XII. The two restriction and modification systems of strain $15T^-$. *Mol. Gen. Gen.*, **118**, 203–217.

Arditti R.R., Scaife J.G. and Beckwith J.R. (1968). The nature of mutants in the *lac* promotor region. *J. Mol. Biol.*, **38**, 421–426.

Atkins J.F. and Loper J.C. (1970). Transcription initiation in the histidine operon of S. typhimurium. *Proc. Nat. Acad. Sci.*, **65**, 925–932.

Attardi G., Naono S., Rouviere J., Jacob F. and Gros F. (1963). Production of messenger RNA and regulation of protein synthesis. *Cold Spring Harbor Symp. Quant Biol.*, **28**, 363–374.

Averner M.J. and Pace N.R. (1972). The nucleotide sequence of a marsupial 5S rRNA. *J. Biol. Chem.*, **247**, 4491–4493.

Avery O.T., Macleod C.M. and McCarty M. (1944). Studies on the chemical nature of the substance inducing transformation of Pneumococcal types. I. Induction of transformation by a DNA fraction isolated from Pneumococcus type III. *J. Expt. Med.*, **79**, 137–158.

Bagdasarian M., Ciesla Z. and Sendecki W. (1970). Polyribosomes of the histidine operon in S. typhimurium. *J. Mol. Biol.*, **48**, 53–65.

Baguley B.C. and Staehelin M. (1968). The specificity of tRNA methylases from rat liver. *Biochem.*, **7**, 45–50.

Baguley B.C., Wehrli W. and Staehelin M. (1970). In vitro methylation of yeast serine tRNA. *Biochem.* **9**, 1645–1650.

Baich A. and Johnson M. (1968). Evolutionary advantage of control of a biosynthetic pathway. *Nature*, **218**, 464–465.

Baker R. and Yanofsky C. (1970). Transcription initation frequency for the tryptophan operon of E.coli. *Cold Spring Harbor Symp. Quant. Biol.*, **35**, 467–470.

Baker R. and Yanofsky C. (1972). Transcription initiation frequency and translational yield for the tryptophan operon of E.coli. *J. Mol. Biol.*, **69**, 89–102.

Baldwin A.N. and Berg P. (1966). Transfer RNA induced hydrolysis of valyl-adenylate bound to isoleucyl-tRNA synthetase. *J. Biol. Chem.*, **241**, 839–845.
Baliga B.S., Schechtman M.G. and Munro H.N. (1973). Competitive binding of EF1 and EF2 by mammalian ribosomes; role of GTP hydrolysis in overcoming inhibition by EF2 of aminoacyl-tRNA binding. *Biochem. Biophys. Res. Commun.*, **51**, 406–413.
Barbour S.D. and Clark A.J. (1970). Biochemical and genetic studies of recombination proficiency in E.coli. I. Enzymatic activity association with *recB*$^+$ and *recC*$^+$ genes. *Proc. Nat. Acad. Sci.*, **65**, 955–961.
Barbour S.D., Nagaishi H., Templin A. and Clark A.J. (1970). Biochemical and genetic studies of recombination proficiency in E.coli. II. *rec*$^+$ revertants caused by indirect suppression of *rec*$^-$ mutations. *Proc. Nat. Acad. Sci.*, **67**, 128–135.
Barnett L., Brenner S., Crock F.H.C., Schulman R.G. and Watts-Tobin R.J. (1967). Phase shift and other mutants in the first part of the *rII* cistron of phage T4. *Phil. Trans. Roy. Soc. B*, **252**, 487–560.
Barzilai R. and Thomas C.A.jr. (1970). Spontaneous renaturation of newly synthesized phage T7 DNA. *J. Mol. Biol.*, **51**, 145–156.
Bauerle R.H. and Margolin P. (1967). Evidence for two sites for initiation of gene expression in the tryptophan operon of S. typhimurium. *J. Mol. Biol.*, **26**, 423–436.
Baumberg S., Bacon D.F. and Vogel H.J. (1965), Individually repressible enzymes specified by clustered genes of the arginine system. *Proc. Nat. Acad. Sci.*, **53**, 1029–1032.
Bautz E.K.F. and Bautz F.A. (1970). Studies on the function of the RNA polymerase factor in promotor selection. *Cold Spring Harbor Symp. Quant. Biol.*, **35**, 227–232.
Bautz E.K.F., Bautz F.A. and Dunn J.J. (1969). E.coli σ factor: a positive control element in phage T4 development. *Nature*, **223**, 1022–1024.
Beacham I.R., Beachan K., Zaritsky A. and Pritchard R.H. (1971). Intracellular thymidine triphosphate concentrations in wild type and in thymine requiring mutants of E.coli 15 and K12. *J. Mol. Biol.*, **60**, 75–86.
Beadle G.W. and Tatum E.L. (1941). Genetic control of biochemical reactions in Neurospora. *Proc. Nat. Acad. Sci.*, **27**, 499–506.
Beardsley K., Tao T. and Cantor C.R. (1970). Studies on the conformation of the anticodon loop of phenylalanine tRNA. Effect of environment on the fluorescence of the Y base. *Biochem.*, **9**, 3524–3533.
Beaudet A.L. and Caskey C.T. (1970). Release factor translation of RNA phage terminator codons. *Nature*, **227**, 38–41.
Beaudet A.L. and Caskey C.T. (1971). Mammalian peptide chain termination. II. Codon specificity and GTPase activity of release factors. *Proc. Nat. Acad. Sci.*, **68**, 619–624.
Beaudet A.L. and Caskey C.T. (1972). Polypeptide chain termination. In L. Bosch (Ed.), *The Mechanism of protein synthesis and its regulation.* North Holland, Amsterdam. pp. 133–172.
Beckwith J.R. (1964). A deletion analysis of the *lac* operator region in E.coli. *J. Mol. Biol.*, **8**, 427–430.
Beckwith J.R. (1970). *Lac*: the genetic system. In Beckwith J.R. and Zipser D. (Eds.), *The Lactose Operon*, Cold Spring Harbor Laboratory, New York. pp 5–26.
Beckwith J.R., Grodzicker T. and Arditti R, (1972). Evidence for two sites in the lac promotor region. *J. Mol. Biol.*, **69**, 155–160.
Beller R.J. and Davis B.D. (1971). Selective dissociation of free ribosomes of E.coli by sodium ions. *J. Mol. Biol.*, **55**, 477–486.
Begg K.J. and Donachie W.D. (1973). Personal communication.

Benzer S. and Champe S.P. (1961). Ambivalent *rII* mutants of phage T4. *Proc. Nat. Acad. Sci.*, **47**, 1025–1038.

Benzer S. and Champe S.P. (1962). A change from nonsense to sense in the genetic code. *Proc. Nat. Acad. Sci.*, **48**, 1114–1121.

Berg C.M. and Caro L.G. (1967). Chromosome replication in E.coli. I. Lack of influence of the integrated F factor. *J. Mol. Biol.*, **29**, 419–431.

Berger H., Brammar W.J. and Yanofsky C. (1968). Analysis of amino acid replacements resulting from frameshift and missense mutations in the tryptophan synthetase A gene of E.coli. *J. Mol. Biol.*, **34**, 219–238.

Bergman F.H., Berg P., and Dieckmann M. (1961). The enzymic synthesis of amino-acyl derivatives of RNA. II. The preparation of leucyl-, valyl-, isoleucyl-, and methionyl-tRNA synthetases from E.coli. *J. Biol. Chem.*, **236**, 1735–1740.

Bernardi A. and Spahr P.F. (1972). Nucleotide sequence at the binding site for coat protein on RNA of phage R17. *Proc. Nat. Acad. Sci.*, **69**, 3033–3037.

Bick M.H., Lee C.S. and Thomas C.A.jr. (1972). Local destabilization of DNA during transcription. *J. Mol. Biol.*, **71**, 1–10.

Bickle T.A., Hershey J.W.B. and Traut R.R. (1972). Spatial arrangement of ribosomal proteins: reaction of the E.coli 30S subunit with bis-imido esters. *Proc. Nat. Acad. Sci.*, **69**, 1237–1331.

Bigger C.H., Murray K. and Murray N.E. (1973). Recognition sequence of a restriction enzyme. *Nature New Biol.*, **244**, 7–10.

Billen D. (1968). Methylation of the bacterial chromosome: an event at the 'replication point'? *J. Mol. Biol.*, **31**, 477–486.

Billen D., Hewitt R.R., Laphthisophon T. and Achey M. (1967). DNA repair replication after ultraviolet light of X-ray exposure of bacteria. *J. Bacteriol.*, **94** 1538–1545.

Bird R.E. and Lark K.G. (1970). Chromosome replication in E.coli $15T^-$ at different growth rates: rate of replication of the chromosome and the rate of formation of small pieces. *J. Mol. Biol.*, **49**, 343–366.

Bird R.E., Louarn J., Martuscelli J. and Caro L. (1972). Origin and sequence of chromosome replication in E.coli. *J. Mol. Biol.*, **70**, 549–566.

Birge E.A., Craven G.R., Hardy S.J.S., Kurland C.G. and Voynow A. (1969). Structural determinant of a ribosomal protein. *Science*, **164**, 1285–1286.

Birnboim H.C. and Coakley B.V. (1971). Adenylate rich oligonucleotides of ribosomal and ribosomal precursor RNA from Hela cells. *Biochem. Biophys. Res. Commun.*, **42**, 1169–1176.

Birnsteil M.L., Sells B.H. and Purdom I.F. (1972). Kinetic complexity of RNA molecules. *J. Mol. Biol.*, **63**, 21–39.

Birnsteil M., Spiers J., Purdom I., Jones K. and Loening U.E. (1968). Properties and composition of the isolated ribosomal DNA satellite of Xenopus laevis. *Nature*, **219**, 454–463.

Biswas D.K. and Gorini L. (1972). Restriction, de-restriction and mistranslation in a missense suppression. Ribosomal discrimination of tRNAs. *J. Mol. Biol.*, **64**, 119–134.

Bjare U. and Gorini L. (1971). Drug dependence reversed by a ribosomal ambiguity mutation, *ram*, in E.coli. *J. Mol. Biol.*, **57**, 423–436.

Blank H.V. and Soll D. (1971a). Purification of five leucine tRNA species from E.coli and their acylation by heterologous leucyl-tRNA synthetase. *J. Biol. Chem.*, **246**, 4947–4950.

Blank H.V. and Soll D. (1971b). The nucleotide sequence of two leucine tRNA species from E.coli K12. *Biochem. Biophys. Res. Commun.*, **43**, 1192–1197.

Blattner F.R. and Dahlberg J.E. (1972). RNA synthesis startpoints in phage lambda: are the promotor and operator transcribed? *Nature New Biol.*, **237**, 227–232.
Blattner F.R., Dahlberg J.E., Boettiger J.K., Fiandt M. and Szybalski W. (1972). Distance from a promotor mutation to an RNA synthesis startpoint on phage lambda DNA. *Nature New Biol.*, **237**, 232–237.
Bleyman M., Kondo M., Hecht N. and Woese C. (1969). Transcriptional mapping: functional organization of the ribosomal and transfer RNA cistrons in the B. subtilis genome. *J. Bacteriol.*, **99**, 535–543.
Blobel G. (1971). Isolation of a 5S-protein complex from mammalian ribosomes. *Proc. Nat. Acad. Sci.*, **68**, 1881–1885.
Blobstein S., Grunberger D., Weinstein I.N. and Nakanishi K. (1973). Isolation and structure determination of the fluorescent base from bovine liver phenylalanine tRNA. *Biochem.* **12**, 188–193.
Blumenthal T., Landers T.A. and Weber K. (1972). Bacteriophage Qβ replicase contains the protein biosynthetic elongation factors Tu and Ts. *Proc. Nat. Acad. Sci.*, **69**, 1313–1317.
Bodley J.W. and Lin L. (1972). Studies on the nature of the G binding site on the 50S ribosomal subunit. *Biochem.*, **11**, 782–785.
Bodley J.W., Zieve F.J. and Lin L. (1970). Studies on translocation. IV. The hydrolysis of a single round of GTP in the presence of fusidic acid. *J. Biol. Chem.*, **245**, 5662–5667.
Bodley J.W., Zieve F.J., Lin L. and Zieve S.T. (1970). Studies on translocation. III. Conditions necessary for the formation and detection of a stable ribosome–G-factor–GDP complex in the presence of fusidic acid. *J. Biol. Chem.*, **245**, 5656–5661.
Bollen A., Davies J., Ozaki M. and Mizushima S. (1969). Ribosomal protein conferring sensitivity to the antibiotic spectinomycin in E.coli. *Science*, **165**, 85–86.
Bollen A., Faelen M., Lecocq J.P., Herzog A., Zengel J. and Hahan L. (1973). The structural gene for the ribosomal protein S18 in E.coli. I. Genetic studies on a mutant having an alteration in the protein S18. *J. Mol. Biol.*, **76**, 463–472.
Bootsma D., Mullder M.P., Pot F. and Cohen J.A. (1970). Different inherited levels of DNA repair replication in Xeroderma pigmentosum cell strains after exposure to ultraviolet irradiation. *Mut. Res.*, **9**, 507–516.
Bothwell A.L.M. and Apirion D. (1971). Is RNAase V a manifestation of RNAase II? *Biochem. Biophys. Res. Commun.*, **44**, 844–851.
Bourgeois S. (1971). The *lac* repressor. *Current Topics Cell Regulat.*, **4**, 39–76.
Bourgeois S., Cohn M. and Orgel L. (1965). Suppression of and complementation among mutants of the regulatory gene of the lactose operon of E.coli. *J. Mol. Biol.*, **14**, 300–302.
Boyce R.P. and Howard-Flanders P. (1964a). Release of ultraviolet light induced thymine dimers from DNA in E.coli K12. *Proc. Nat. Acad. Sci.*, **51**, 293–300.
Boyce R.P. and Howard-Flanders P. (1964b). Genetic control of DNA breakdown and repair in E.coli treated with mitomycin C or ultraviolet. *Z. Vererbungsl.*, **95**, 345–350.
Boyer H.W., Chow L.T., Dugaiczyk A., Hedgpeth J. and Goodman H.M. (1973). DNA substrate site for the Eco_{RII} restriction endonuclease and modification methylase. *Nature New Biol.*, **244**, 40–43.
Boyer H.W. and Roulland-Dussoix D.A. (1969). A complementation analysis of the restriction and modification of DNA in E.coli. *J. Mol. Biol.*, **41**, 459–472.
Boyle J.M., Paterson M.C. and Setlow R.B. (1970). Excision repair properties of an E.coli mutant deficient in DNA polymerase. *Nature*, **226**, 708–711.

Boyle J.M. and Setlow R.B. (1970). Correlations between host cell reactivation, ultraviolet reactivation and pyrimidine dimer excision in the DNA of phage lambda. *J. Mol. Biol.*, **51**, 131–142.

Bram S. (1971). Secondary structure of DNA depends on base composition. *Nature New Biol.*, **232**, 174–176.

Bremer H., Berry L. and Dennis P.P. (1973). Fractions of RNA polymerase engaged in the synthesis of stable RNA at different steady state growth rates. *J. Mol. Biol.* **75**, 161–180.

Bremer H. and Konrad M.W. (1964). A complex of enzymatically synthesized RNA and template DNA. *Proc. Nat. Acad. Sci.*, **51**, 801–810.

Brenner M. and Ames B.N. (1972). Histidine regulation in S. typhimurium. IX. Histidine tRNA of the regulatory mutants *J. Biol. Chem.*, **247**, 1080–1088.

Brenner M., Lewis J.A., Strauss D.A., de Lorenzo F. and Ames B.N. (1972). Interaction of the his-tRNA synthetase with his-tRNA. *J. Biol. Chem.*, **247**, 4333–4340.

Brenner S. and Beckwith J.R. (1965). Ochre mutants, a new class of suppressible nonsense mutants. *J. Mol. Biol.*, **13**, 629–637.

Brenner S., Jacob F. and Meselson M. (1961). An unstable intermediate carrying information from genes to ribosomes for protein synthesis. *Nature*, **190**, 576–580.

Brenner S., Stretton A.O.W. and Kaplan S. (1965). Genetic Code: the nonsense triplets for chain termination and their suppression. *Nature*, **206**, 994–998.

Bretscher M.S. (1966). Polypeptide chain initiation and the characterization of ribosomal binding in E.coli. *Cold Spring Harbor Symp. Quant. Biol.*, **31**, 289–296.

Bretscher M.S. (1968a). Translocation in protein synthesis in a hybrid structure model. *Nature*, **218**, 675–677.

Bretscher M.S. (1968b). Polypeptide chain termination: an active process. *J. Mol. Biol.*, **34**, 131–136.

Bretscher M.S. (1968c). Direct translation of a circular messenger DNA. *Nature*, **220**, 1088–1091.

Bretscher M.S. (1969). Direct translation of phage fd DNA in the absence of neomycin B. *J. Mol. Biol.*, **42**, 595–598.

Bretscher M.S. and Marcker K.A. (1966). Polypeptidyl-tRNA and aminoacyl-tRNA binding sites on ribosomes. *Nature*, **211**, 380–384.

Bridges B.A., Law J. and Munson R.J. (1969). Mutagenesis in E.coli. II. Evidence for a common pathway for mutagenesis by ultraviolet light, ionizing radiation, and thymine deprivation. *Mol. Gen. Genet.*, **103**, 266–273.

Bridges B.A. and Munson R.J. (1968). The persistence through several replication cycles of mutation producing pyrimidine dimers in a strain of E.coli. *Biochem. Biophys. Res Commun.*, **30**, 620–624.

Brimacombe R., Trupin J., Nirenberg M., Leder P., Bernfield M. and Jaouni T. (1965). *Proc. Nat. Acad. Sci.*, **54**, 954–960.

Brody E.R., Sederoff R., Bolle A. and Epstein R.H. (1970). Early transcription in T4 infected cells. *Cold Spring Harbor Symp. Quant. Biol.*, **35**, 203–212.

Broker T.R. and Lehman I.R. (1971). Branched DNA molecules: intermediates in T4 recombination. *J. Mol. Biol.*, **60**, 131–150.

Brostoff S.W. and Ingram V.M. (1968). Chemical modification of yeast alanine tRNA with a radioactive carbodiimide. *Science*, **158**, 666–669.

Brot N., Redfield B. and Weissbach H. (1970). Studies on the reaction of the aminoacyl–Tu–GTP complex with ribosomal subunits. *Biochem. Biophys. Res. Commun.*, **41**, 1388–1395.

Brot N., Spears C. and Weissbach H. (1971). The interaction of transfer factor G, ribosomes and guanosine nucleotides in the presence of fusidic acid. *Arch. Biochem. Biophys.*, **143**, 286–296.

Brot N., Yamasaki E., Redfield B. and Weissbach H. (1971). The properties of an E.coli ribosomal protein required for the function of factor G. *Arch. Biochem. Biophys.*, **148**, 148–155.

Brown C.E. and Hogg R.W. (1972). A second transport system for L-arabinose in E.coli B/r controlled by the *araC* gene. *J. Bacteriol.*, **111**, 606–613.

Brown D.D. and Weber C.S. (1968). Gene linkage by RNA-DNA hybridization. II. Arrangement of the redundant gene sequences for 21S and 18S ribosomal RNA. *J. Mol. Biol.*, **34**, 681–698.

Brown G.M. and Attardi G. (1965). Methylation of nucleic acids in Hela cells. *Biochem. Biophys. Res. Commun.*, **20**, 298–302.

Brown J.C. and Smith A.E. (1970). Initiator codons in eucaryotes. *Nature*, **226**, 610–612.

Brownlee G.G. and Cartwright E. (1971). Sequence studies on precursor 16S ribosomal RNA of E.coli. *Nature New Biol.*, **232**, 5–52.

Brownlee G.G., Sanger F. and Barrell B.G. (1967). Nucleotide sequence of 5S ribosomal RNA from E.coli. *Nature*, **215**, 735–736.

Brownlee G.G., Sanger F. and Barrell B.G. (1968). The sequence of 5S ribosomal RNA. *J. Mol. Biol.*, **34**, 379–412.

Bruner R. and Cape R.E. (1970). The expression of two classes of late genes of phage T4. *J. Mol. Biol.*, **53**, 69–90.

Brutlag D. and Kornberg A. (1972). Enzymatic synthesis of DNA. XXXVI. A proof reading function for the 3′ to 5′ activity of DNA polymerase. *J. Biol. Chem.*, **247**, 241–248.

Buchanan C.E., Hua S.S., Avni H. and Markovitz A. (1973). Transcriptional control of the galactose operon by the *capR* (*lon*) and *capT* genes. *J. Bacteriol.*, **114**, 891–894.

Buanocore V. and Schlesinger S. (1972). Interactions of tyrosyl-tRNA synthetase from E.coli with its substrates. *J. Biol. Chem.*, **247**, 1343–1348.

Burger R. M. (1971). Toluene treated E.coli replicate only that DNA which was about to be replicated in vivo. *Proc. Nat. Acad. Sci.*, **68**, 2124–2126.

Burgess R.P., Travers A.A., Dunn J.J. and Bautz E.K.F. (1969). Factor stimulating transcription by RNA polymerase. *Nature*, **221**, 43–47.

Burt D.H. and Brent T.P. (1971). A DNAase activity of Hela cells specific for UV-irradiated DNA. *Biochem. Biophys. Res. Commun.*, **43**, 1382–1387.

Cabrer B., Vazquez D. and Modollel J. (1972). Inhibition by elongation factor EF G of aminoacyl-tRNA binding to ribosomes. *Proc. Nat. Acad. Sci.*, **69**, 733–736.

Caffier H., Raskas H.J., Parsons J.T. and Green M. (1971). Initiation of mammalian viral protein synthesis. *Nature New Biol.*, **229**, 239–241.

Cairns J. (1963). The chromosome of E.coli. *Cold Spring Harbor Symp. Quant. Biol.*, **28**, 43–46.

Callahan R., Blume A.J. and Balbinder E. (1970). Evidence for the order promotor-operator-first structural gene in the tryptophan operon of Salmonella. *J. Mol. Biol.*, **52**, 709–716.

Cameron V. and Uhlenbeck O.C. (1973). Removal of Y-37 from $tRNA^{phe}_{yeast}$ alters oligomer binding to two loops. *Biochem. Biophys. Res Commun.*, **50**, 635–640.

Campbell J.L., Soll L. and Richardson C.C. (1972). Isolation and partial characterization of a mutant of E.coli deficient in DNA polymerase II. *Proc. Nat. Acad. Sci.*, **69**, 2090–2094.

Capecchi M.R. and Gussin G.N. (1965). Suppression in vitro: identification of a serine tRNA as a nonsense suppressor. *Science*, **149**, 416–422.

Capecchi M.R. and Klein H.A. (1969). Characterization of three proteins involved in polypeptide chain termination. *Cold Spring Harbor Symp. Quant. Biol.*, **34**, 469–478.

Capecchi M.R. and Klein H.A. (1970). Release factor mediating termination of complete protein. *Nature*, **226**, 1029–1033.

Carbon J., Squires C. and Hill C.W. (1970). Glycine tRNA of E.coli. II. Impaired GGA recognition in strains containing a genetically altered tRNA: reversal by secondary suppressor mutation. *J. Mol. Biol.*, **52**, 571–584.

Caro L.G. (1970). Chromosome replication in E.coli. III. Segregation of chromosomal strands in multiforked replication. *J. Mol. Biol.*, **48**, 329–338.

Caro L.G. and Berg C.M. (1968). Chromosome replication in some strains of E.coli K12. *Cold Spring Harbor Symp. Quant. Biol.*, **33**, 559–576.

Carroll R.B., Neet K.E. and Holdthwait D.A. (1972). Self association of gene-32 protein of phage T4. *Proc. Nat. Acad. Sci.*, **69**, 2741–2744.

Carter T. and Newton A. (1969). Messenger RNA stability and polarity in the *lac* operon of E.coli. *Nature*, **223**, 707–710.

Carter T. and Newton A. (1971). New polarity suppressors in E.coli: suppression and messenger RNA stability. *Proc. Nat. Acad. Sci.*, **68**, 2962–2966.

Case M.E. and Giles N.H. (1968). Evidence for nonsense mutations in the *arom* gene cluster of N. crassa. *Genetics*, **60**, 49–58.

Case M.E. and Giles N.H. (1971). Partial enzyme aggregates formed by pleiotropic mutants in the *arom* gene cluster of N. crassa. *Proc. Nat. Acad. Sci.*, **68**, 58–62.

Cashel M. and Gallant J. (1968). Control of RNA synthesis in E.coli. I. Amino acid dependence of the synthesis of the substrates of RNA polymerase. *J. Mol. Biol.*, **34**, 317–330.

Cashel M. and Gallant J. (1969). Two compounds implicated in the function of the *RC* gene of E.coli. *Nature*, **221**, 838–841.

Cashmore T. (1971). Interaction between loops I and III in the tyrosine suppressor tRNA. *Nature New Biol.*, **230**, 236–239.

Caskey C.T., Beaudet A. and Nirenberg M. (1968). Codons and protein synthesis. XV. Dissimilar responses of mammalian and bacterial tRNA fractions to mRNA codons. *J. Mol. Biol.*, **37**, 99–118.

Cassuto E., Lash T., Sriprakash K.S. and Radding C.M. (1971). Role of exonuclease and β protein of phage lambda in genetic recombination. V. Recombination of lambda DNA in vitro. *Proc. Nat. Acad. Sci.*, **68**, 1639–1643.

Cassuto E. and Radding C. M. (1971). Mechanism for the action of lambda exonuclease in genetic recombination. *Nature New Biology*, **239**, 13–16.

Celis J.E., Smith J.D. and Brenner S. (1973). Correlation between genetic and translational maps of gene 23 in phage T4. *Nature New Biol.*, **241**, 130–132.

Celma M.L., Vazquez D. and Modollel J. (1972). Failure of fusidic acid and siomycin to block ribosomes in the pretranslocated state. *Biochem. Biophys. Res. Commun.*, **48**, 1240–1246.

Cerda-Olmeda E., Hanawalt P.C. and Guerola N. (1968). Mutagenesis of the replication point by nitrosoguanidine: map and pattern of replication of the E.coli chromosome. *J. Mol. Biol.*, **33**, 705–720.

Chadwick P., Pirrotta V., Steinberg R., Hopkins N. and Ptashne M. (1970). The lambda and 434 phage repressors. *Cold Spring Harbor Symp. Quant. Biol.*, **35**, 283–294.

Chae Y.B., Mazumder R. and Ochoa S. (1969a). Polypeptide chain initiation in E.coli: isolation of homogeneous initiation factor f2 and its relation to ribosomal protein. *Proc. Nat. Acad. Sci.*, **62**, 1181–1188.

Chae Y.B., Mazumder R. and Ochoa S. (1969b). Polypeptide chain initiation in E.coli: studies on the function of initiation factor f1. *Proc. Nat. Acad. Sci.*, **63**, 828–833.

Chamberlin M. and Berg P. (1962). DNA directed synthesis of RNA by an enzyme from E.coli. *Proc. Nat. Acad. Sci.*, **48**, 81–93.

Chamberlin M.J. and Ring J. (1972). Studies of the binding of E.coli RNApoly merase to DNA. V. T7 chain initiation of enzyme DNA complexes. *J. Mol. Biol.*, **70**, 221–238.

Champoux J.J. and Dulbecco R. (1972). An activity from mammalian cells that untwists superhelical DNA—a possible swivel for DNA replication. *Proc. Nat. Acad. Sci.*, **69**, 143–156.

Chan T.S. and Garen A. (1970). Amino acid substitutions resulting from suppression of nonsense mutations. V. Tryptophan insertion by the $su9^+$ gene, a suppressor of the UGA nonsense triplet. *J. Mol. Biol.*, **49**, 231–234.

Chang B. and Irr J. (1973). Maturation of ribosomal RNA in stringent and relaxed bacteria. *Nature New Biol.*, **243**, 35–37.

Chang F.N. and Flaks J.G. (1970). Topography of the E.coli 30S ribosomal subunit and streptomycin binding. *Proc. Nat. Acad. Sci.*, **67**, 1321–1328.

Chang F.N. and Flaks J.G. (1971). Topography of the E.coli ribosome. II. Preliminary sequence of 50S subunit protein attack by trypsin and its correlation with functional activities. *J. Mol. Biol.*, **61**, 387–400.

Chang L.M.S. and Bollum F.J. (1972). Low molecular weight DNA polymerase from rabbit bone marrow. *Biochem.*, **11**, 1264–1272.

Chang S.E. (1973). Selective modification of cytidine and uridine residues in E.coli fmet-tRNA. *J. Mol. Biol.*, **75**, 533–548.

Chapeville F., Lipmann F., Von Ehrenstein G., Weisblum B., Ray W.J. and Benzer S. (1962). On the role of soluble RNA in coding for amino acids. *Proc. Nat. Acad. Sci.*, **48**, 1086–1092.

Chapeville F. and Rouget P. (1972). Aminoacyl tRNA synthetases. In Bosch L. (Ed.), *The mechanism of protein synthesis and its regulation*, North Holland. pp 5–32.

Chargaff E. (1971). Preface to a grammar of biology. *Science*, **171**, 637–642.

Charlier J. (1972). Isoleucyl-tRNA synthetase from B. stearothermophilus. II. Dynamics of the enzyme substrate interactions. *Europ. J. Biochem.*, **25**, 175–180.

Chen B., De Crombrugghe B., Anderson W.B., Gottesman M.E., Pastan I. and Perlman R.L. (1971). On the mechanism of action of *lac* repressor. *Nature New Biol.*, **233**, 67–70.

Chen C.M. and Ofengand J. (1970). Inactivation of the Tu–GTP recognition site in aminoacyl-tRNA by chemical modification of the tRNA. *Biochem. Biophys. Res. Commun.*, **41**, 190–198.

Choi Y.C. and Busch H. (1970). Studies on the 5′ terminal and alkali resistant dinucleotides of nucleolar high molecular weight RNA. *J. Biol. Chem.*, **245**, 1954–1961.

Chovnick A., Ballantyne G.H., Baillie D.L. and Holm D.G. (1970). Gene conversion in higher organisms: half tetrad analysis of recombination within the rosy cistron of D. melanogaster. *Genetics*, **66**, 315–329.

Chovnick A., Ballantyne G.H. and Holm D.G. (1971). Studies on gene conversion and its relationship to linked exchange in D. melanogaster. *Genetics*, **69**, 179–209.

Clark B.F.C. and Marcker K.A. (1966). The role of N-formyl-methionyl-tRNA in protein biosynthesis. *J. Mol. Biol.*, **17**, 394–406.

Clark D.J. (1968a). Regulation of DNA replication and cell division in E.coli B/r. *J. Bacteriol.*, **96**, 1214–1225.

Clark D.J. (1968b). The regulation of DNA replication and cell division in E.coli B/r. *Cold Spring Harbor Symp. Quant. Biol.*, **33**, 823–838.

Clark D.J. and Maaløe O. (1962). DNA replication and the division cycle in E.coli *J. Mol. Biol.*, **23**, 99–112.

Cleary P.P. and Campbell A. (1972). Deletion and complementation analysis of the biotin gene cluster of E.coli. *J. Bacteriol.*, **112**, 830–839.

Cleary P.P., Campbell A. and Chang R. (1972). Location of promotor and operator sites in the biotin gene cluster of E.coli. *Proc. Nat. Acad. Sci.*, **69**, 2219–2223.

Cleaver J.E. (1970). Repair replication in Chinese hamster cells after damage from ultraviolet light. *Photochem. Photobiol.*, **12**, 12–28.

Cleaver J.E. (1971). Repair of alkylation damage in UV-sensitive (Xeroderma pigmentosum) human cells. *Mutat. Res.*, **12**, 455–462.

Cleaver J.E. and Painter R.B. (1968). Evidence for repair replication of Hela cell DNA damaged by UV light. *Biochim. Biophys. Acta*, **161**, 552–554.

Cohen P.T., Yaniv M. and Yanofsky C. (1973). Nucleotide sequences from mRNA transcribed from the operator proximal portion of the tryptophan operon of E.coli. *J. Mol. Biol.*, **74**, 163–178.

Cole P.E. and Crothers D.M. (1972). Conformational changes of tRNA. Relaxation kinetics of the early melting transition of methionine tRNA (E.coli). *Biochem.*, **11**, 4368–4374.

Cole R.S. (1973). Repair of DNA containing interstrand cross links in E.coli. *Proc. Nat. Acad. Sci.*, **70**, 1064–1068.

Colli W., Smith I. and Oishi M. (1971). Physical linkage between 5S, 16S and 23S rRNA genes in B. subtilis. *J. Mol. Biol.*, **56**, 117–128.

Comb D.G. and Sarkar N. (1967). The binding of 5S ribosomal RNA to ribosomal subunits. *J. Mol. Biol.*, **25**, 317–330.

Contesse G., Crepin M. and Gros F. (1970). Transcription of the lactose operon E.coli. In Beckwith J.R. and Zipser D., *The Lactose Operon*, Cold Spring Harbor Laboratory, New York. pp 111–142.

Contreras R., Ysebaert M., Jou W.M. and Fiers W. (1973). Phage MS2 RNA: nucleotide sequence of the end of the A protein gene and the intercistronic region. *Nature New Biol.*, **241**, 99–101.

Cooper P.K. and Hanawalt P.C. (1972a). Heterogeneity of patch size in repair replicated DNA in E.coli. *J. Mol. Biol.*, **67**, 1–10.

Cooper P.K. and Hanawalt P.C. (1972b). Role of DNA polymerase I and the *rec* system in excision repair in E.coli. *Proc. Nat. Acad. Sci.*, **69**, 1156–1160.

Cooper S. (1969). Cell division and DNA replication following a shift to a richer medium. *J. Mol. Biol.*, **43**, 1–12.

Cooper S. (1972). Relationship of *Flac* replication and chromosome replication. *Proc. Nat. Acad. Sci.*, **69**, 2706–2710.

Cooper S. and Helmstetter C.E. (1968). Chromosome replication and the division cycle of E.coli B/r. *J. Mol. Biol.*, **31**, 519–540.

Cory S., Adams J.M., Spahr P.F. and Rensing U. (1972). Sequence of 51 nucleotides at the 3′ end of R17 phage RNA. *J. Mol. Biol.*, **63**, 41–56.

Cory S. and Marcker K.A. (1970). The nucleotide sequence of methionine $tRNA_m$. *Europ. J. Biochem.*, **12**, 177–194.

Cotter R.I. and Gratzer W.B. (1971). Accessibility of RNA and proteins in the ribosome. Investigation by hydrogen exchange and solvent perturbation. *Europ. J. Biochem.*, **23**, 468–496.

Cotter R.I., McPhie P. and Gratzer W.B. (1967). Internal organization of the ribosome. *Nature*, **216**, 864–868.

Cox E.C. (1970). Mutator gene action and the replication of bacteriophage lambda DNA. *J. Mol. Biol.*, **50**, 129–136.
Cox E.C., White J.R. and Flaks J.G. (1964). Streptomycin action and the ribosome. *Proc. Nat. Acad. Sci.*, **51**, 703–709.
Cramer F., Doepner H., Haar F.V.D., Schimme E. and Seidel H. (1968). On the conformation of tRNA. *Proc. Nat. Acad. Sci.*, **61**, 1384–1391.
Cramer F. and Erdmann V.A. (1968). Amount of adenine and uracil base pairs in E.coli 23S, 16S and 5S ribosomal RNA. *Nature*, **218**, 92–93.
Cramer F. and Gauss D.H. (1972). The dimensional structure of tRNA. In Bosch L., *The Mechanism of Protein synthesis and its regulation*, North Holland, Amsterdam. pp 219–242.
Craven G.R. and Gupta V. (1970). Three dimensional organization of the 30S ribosomal proteins from E.coli. I. Preliminary classification of the proteins. *Proc. Nat. Acad. Sci.*, **67**, 1329–1336.
Craven G.R., Voynow P., Hardy S.J.S. and Kurland C.G. (1969). The ribosomal proteins of E.coli. II. Chemical and physical characterization of the 30S ribosomal proteins. *Biochem.*, **8**, 2906–2915.
Crick F.H.C. (1958). On protein synthesis. *Symp. Soc. Exp. Biol.*, **12**, 138–163.
Crick F.H.C. (1966). Codon-anticodon pairing: the wobble hypothesis. *J. Mol. Biol.*, **19**, 548–555.
Crick F.H.C. (1968). The origin of the genetic code. *J. Mol. Biol.*, **38**, 367–380.
Crick F.H.C. (1970). Central dogma of molecular biology. *Nature*, **227**, 561–562.
Crick F.H.C., Barnett L., Brenner S. and Watts-Tobin R.J. (1961). General nature of the genetic code for proteins. *Nature*, **192**, 1227–1232.
Crick F.H.C., Griffith J. and Orgel L. (1957). Codes without commas. *Proc. Nat. Acad. Sci.*, **43**, 416–427.
Cunin R., Elseviers D., Sand., Freundlich G. and Glansdorff N. (1969). On the functional organization of the *argECBH* cluster of genes in E.coli K12. *Mol. Gen. Genet.*, **106**, 32–47.
Dalgarno L. and Gros F. (1968a). Completion of ribosome particles in E.coli during inhibition of protein synthesis. *Biochim. Biophys. Acta*, **157**, 52–63.
Dalgarno L. and Gros F. (1968b). RNA synthesis in relaxed and stringent E.coli. Breakdown of preformed ribonucleoprotein particles and subsequent RNA synthesis. *Biochim. Biophys. Acta*, **157**, 64–75.
Dausse J.P., Sentenac A. and Fromageot P. (1972a). Interaction of RNA polymerase from E.coli with DNA. Selection of initiation sites on T7 DNA. *Europ. J. Biochem.*, **26**, 43–49.
Dausse J.P., Sentenac A. and Fromageot P. (1972b). Interaction of RNA polymerase from E.coli with DNA. Selection of initiation sites on T7 DNA. *Europ J. Biochem*, **26**, 43-49.
Dausse J.P., Sentenac A. and Fromageot P. (1972b). Interaction of RNA polymerase from E.coli with DNA. Influence of DNA scissions on RNA polymerase binding and chain initiation. *Europ. J. Biochem.*, **31**, 394–404.
Davies D.F. (1965). Repair mechanisms of variation in UV sensitivity within the cell cycle. *Mutat. Res.*, **2**, 477–486.
Davies D.R. (1967). UV sensitive mutants in C. reinhardii. *Mutat. Res.*, **4**, 765–770.
Davies J. (1964). Studies on the ribosomes of streptomycin sensitive and resistant strains of E.coli. *Proc. Nat. Acad. Sci.*, **51**, 659–663.
Davies J. (1966). Streptomycin and the genetic code. *Cold Spring Harbor Symp. Quant. Biol.*, **31**, 665–670.

Davies J., Jones D.S. and Khorana H.G. (1965). A further study of misreading of codons induced by streptomycin and neomycin using ribonucleotides containing nucleotides in alternating sequence as templates. *J. Mol. Biol.*, **18**, 48–57.

Davis B.D. (1971). Role of subunits in the ribosome cycle. *Nature*, **231**, 153–157.

Davies D.B. and Helmstetter C.E. (1973). Control of *F′lac* replication in E.coli B/r. *J. Bacteriol.*, **114**, 294–299.

Davis R.W. and Hyman R.W. (1970). Physical locations of the in vitro RNA initiation site and termination sites of T7M DNA. *Cold Spring Harbor Symp. Quant. Biol.*, **35**, 269–282.

Davies R.W. and Hyman R.W. (1971). A study in evolution: the DNA base sequence homology between coliphages T7 and T3. *J. Mol. Biol.*, **62**, 287–302.

Davies R.W., Simon M. and Davidson N. (1971). Electron microscope duplex method for mapping regions of base sequence homology in nucleic acids. *Methods Enzymol.*, **21D**, 413–428.

Dean C. and Pauling C. (1970). Properties of a DNA ligase mutant of E.coli: X-ray sensitivity. *J. Bacteriol.*, **102**, 588–589.

De Crombrugghe B., Adya S., Gottesman M. and Pastan I. (1973). Effect of rho on transcription of bacterial operons. *Nature New Biol.*, **241**, 260–264.

De Crombrugghe B., Chen B., Anderson W.B., Gottesman M.E. and Perlman R.L. (1973c). Role of cyclic AMP and the cyclic AMP receptor protein in the initiation of *lac* transcription. *J. Biol. Chem.*, **246**, 7343–7348.

De Crombrugghe B., Chen B., Anderson W.B., Nissley P., Gottesman M., Pastan I. and Perlman R.L. (1971b). *Lac* DNA, RNA polymerase and cyclic AMP receptor protein, cyclic AMP, *lac* repressor and inducer are the essential elements for controlled *lac* transcription. *Nature New Biol.*, **231**, 139–142.

De Crombrugghe B., Chen B., Gottesman M., Pastan I., Varmus H.E., Emmer M. and Perlman R.L. (1971a). Regulation of lac mRNA synthesis in a soluble cell free system. *Nature New Biol.*, **230**, 37–40.

Dekio S. and Takata R. (1969). Genetic studies of the ribosomal proteins in E.coli. II. Altered 30S ribosomal protein component specific to spectinomycin resistant mutants. *Molec. Gen. Genet.*, **105**, 219–224.

Dekio S., Takata R. and Osawa S. (1970). Genetic studies of the ribosomal proteins in E.coli. VI. Determination of the chromosomal loci for several ribosomal components using a hybrid strain between E.coli and S. typhimurium. *Molec. Gen. Genet.*, **109**, 131–141.

Delius H., Howe C. and Kozinski A.W. (1971). Structure of the replicating DNA from phage T4. *Proc. Nat. Acad. Sci.*, **68**, 3049–3053.

Delius H., Mantell N.J. and Alberts B. (1972). Characterization by electron microscopy of the complex formed between T4 phage gene-32 protein and DNA. *J. Mol. Biol.*, **67**, 341–350.

De Lorenzo F. and Ames B.N. (1970). Histidine regulation in S. typhimurium. VII. Purification and general properties of the histidyl-tRNA synthetase. *J. Biol. Chem.*, **245**, 1710–1716.

De Lucia P. and Cairns J. (1969). Isolation of an E.coli strain with a mutation affecting DNA polymerase. *Nature*, **224**, 1164–1166.

Dennis P. (1972). Regulation of ribosomal and tRNA synthesis in E.coli. *J. Biol. Chem.*, **247**, 2842–2845.

Deusser E., Stoffler G., Wittman H.G. and Apirion D. (1970). Ribosomal proteins. XVI. Altered S4 proteins in E.coli revertants from streptomycin dependence to independence. *Molec. Gen. Genet.*, **109**, 298–302.

Deutscher M.P. and Kornberg A. (1969). Enzymatic synthesis of DNA. XXIX. Hydrolysis of DNA from the 5′ terminus by an exonuclease function of DNA polymerase. *J. Biol. Chem.*, **244**, 3029–3027.

Dickerman H.W. and Smith B.C. (1971). Formylmethionyl-tRNA transformylase: the specific interaction of the enzyme with its tRNA substrate. *J. Mol. Biol.*, **59**, 425–446.

Dickerman H.W., Steers E.jr., Redfield B.G. and Weissbach H. (1966). Formylation of E.coli methionyl-tRNA. *Cold Spring Harbor Symp. Quant. Biol.*, **31**, 287–288.

Di Mauro E., Snyder L., Marino P., Lamberti A., Coppo A. and Tocchini-Valentini G.P. (1969). *Nature*, **222**, 533–537.

Dintzis H.M. (1961). Assembly of the peptide chains of haemoglobin. *Proc. Nat. Acad. Sci.*, **47**, 247–260.

Donachie W.D. (1968). Relationship between cell size and time of initiation of DNA replication. *Nature*, **219**, 1077–1079.

Donachie W.D. (1969). Control of cell division in E.coli: experiments with thymine starvation. *J. Bacteriol.*, **100**, 260–268.

Donachie W.D. and Begg K.J. (1970). Growth of the bacterial cell. *Nature*, **277**, 1220–1224.

Donachie W.D., Hobbs D.G. and Masters M. (1968). Chromosome replication and cell division in E.coli 15T$^-$ after growth in the absence of DNA synthesis. *Nature*, **219**, 1079–1080.

Donini P. (1972). Turnover of ribosomal RNA during the stringent response in E.coli. *J. Mol. Biol.*, **72**, 553–570.

Donner D. and Kurland C.G. (1972). Changes in the primary structure of a mutationally altered ribosomal protein S4 of E.coli. *Molec. Gen. Genet.*, **115**, 49–53.

Doolittle W.F. and Pace N.R. (1971). Transcriptional organization of the ribosomal RNA cistrons in E.coli. *Proc. Nat. Acad. Sci.*, **68**, 1786–1790.

Dottin R.P. and Pearson M.L. (1973). Regulation by *N* gene protein of phage lambda of anthranilate synthesis in vitro. *Proc. Nat. Acad. Sci.*, **70**, 1078–1082.

Drake J.W., Allen E.F., Forsberg S.A., Preparata R.M. and Greening E.O. (1969). Genetic control of mutation rates in phage T4. *Nature*, **221**, 1128–1131.

Dressler D. (1970). The rolling circle for øX replication. II. Synthesis of single stranded circles. *Proc. Nat. Acad. Sci.*, **67**, 1934–1942.

Drews J., Grasmuk H. and Weil R. (1972). Utilization of methionine accepting tRNA species from E.coli, ascites tumour cells, and yeast in homologous and heterologous cell free systems. *Europ. J. Biochem.*, **29**, 119–127.

Dube S.K. and Rudland P.S. (1970). Control of translation by T4 phage: altered binding of disfavoured messengers. *Nature*, **226**, 820–824.

Dubnoff J.S., Lockwood A.H. and Maitra U. (1972). Studies on the role of GTP in polypeptide chain initiation in E.coli. *J. Biol. Chem.*, **247**, 2884–2894.

Dudock B., DiPeri C., Scileppi K. and Reszelbach (1971). The yeast phenylalanyl-tRNA synthetase recognition site: the region adjacent to the dihydrouridine loop. *Proc. Nat. Acad. Sci.*, **68**, 681–684.

Dunn J.J. and Studier F.W. (1973). T7 early RNAs are generated by site specific cleavages. *Proc. Nat. Acad. Sci.*, **70**, 1559–1563.

Dussoix D. and Arber W. (1962). Host specificity of DNA produced by E.coli. II. Control over acceptance of DNA from infecting phage lambda. *J. Mol. Biol.*, **5**, 37–49.

Echols H. (1971). Regulation of lytic development. In Hershey A.D. (Ed.), *The bacteriophage lambda*, Cold Spring Harbor Laboratory, New York, pp 247–270.

Echols H. and Green L. (1971). Establishment and maintenance of repression by phage lambda: the role of the *cI*, *cII* and *cIII* proteins. *Proc. Nat. Acad. Sci.*, **68**, 2190–2194.

Edenberg H. and Hanawalt P.C. (1972). Size of repair patches in the DNA of ultraviolet irradiated Hela cells. *Biochim. Biophys. Acta.*, **272**, 361–372.

Edlin G. and Broda P. Physiology and genetics of the RNA control locus in E.coli. *Bacteriol. Rev.*, **32**, 206–226.

Eisenstadt J.M. and Brawerman G. (1967). The role of the native subribosomal particles of E.coli in polypeptide chain initiation. *Proc. Nat. Acad. Sci.*, **58**, 1560–1565.

Eleuterio M., Griffin B. and Sheppard D.E. (1972). Characterization of strong polar mutations in a region immediately adjacent to the L-arabinose operon in E.coli B/r *J. Bacteriol.*, **111**, 383–391.

Elseviers D., Cunin R., Glansdorff N., Baumberg S. and Ashcroft E. (1972). Control regions within the *argECBH* gene cluster of E.coli. *Molec. Gen. Genet.*, **117**, 349–365.

Emmer M., De Crombrugghe B., Pastan I. and Perlman R. (1970). Cyclic AMP receptor protein of E.coli: its role in the synthesis of inducible enzymes. *Proc. Nat. Acad. Sci.*, **66**, 480–487.

Emmerson P.T. and Howard-Flanders P. (1964). Post irradiation degradation of DNA following exposure of UV-sensitive and resistant bacteria to X-rays. *Biochem. Biophys. Res. Commun.*, **18**, 24–29.

Englesberg E., Irr J., Power J. and Lee N. (1965). Positive control of enzyme synthesis by gene *C* in the L-arabinose system. *J. Bacteriol.*, **90**, 946–957.

Englesberg E., Sheppard D., Squires C. and Meronk F. (1969a). An analysis of revertants of a deletion mutant in the *C* gene of the L-arabinose gene complex in E.coli B/r: isolation of initiator constitutive mutants. *J. Mol. Biol.*, **43**, 281–298.

Englesberg E., Squires C. and Meronk F.jr. (1969b). The L-arabinose operon in E.coli B/r: a genetic demonstration of two functional states of the product of a regulator gene. *Proc. Nat. Acad. Sci.*, **62**, 1100–1107.

Englund P.T., Kelly R.B. and Kornberg A. (1969). Enzymatic synthesis of DNA. XXXI. Binding of DNA to DNA polymerase. *J. Biol. Chem.*, **244**, 3045–3052.

Epstein C.J. (1966). Role of the amino acid code and of selection for conformation in the evolution of proteins. *Nature*, **210**, 25–28.

Erbe R.W., Nau M.N. and Leder P. (1969). Translation and translocation of defined RNA messengers. *J. Mol. Biol.*, **39**, 441–460.

Erdmann V.A., Fahnestock S., Higo K. and Nomura M. (1971). Role of 5S RNA in the functions of 50S ribosomal subunits. *Proc. Nat. Acad. Sci.*, **68**, 2932–2936.

Eron L., Beckwith J.R. and Jacob F. (1970). Deletion of translation start signals in the *lac* operon of E.coli. In Beckwith J.R. and Zipser D. (eds.), *The Lactose Operon*, Cold Spring Harbor Laboratory, New York. pp 353–358.

Eron L. and Block R. (1971). Mechanism of initiation and repression of in vitro transcription of the *lac* operon of E.coli. *Proc. Nat. Acad. Sci.* **68**, 1828–1832.

Eron L., Morse D., Reznikoff W. and Beckwith J. (1971). Fusions of the *lac* and *trp* regions of E.coli: covalently fused mRNA. *J. Mol. Biol.*, **60**, 203–210.

Ertel R., Redfield B., Brot N. and Weissbach H. (1968). Role of GTP in protein synthesis: interaction of GTP with soluble transfer factors from E.coli. *Arch. Biochem. Biophys.*, **128**, 331–338.

Eskin B. and Linn S. (1972). The DNA modification and restriction enzymes of E.coli B. II. Purification, subunit structure and catalytic properties of the modification methylase. *J. Biol. Chem.*, **247**, 6183–6191.

Fahnestock S., Erdmann V. and Nomura M. (1973). Reconstitution of 50S ribosomal subunits from protein free RNA. *Biochem.*, **12**, 220–223.

Fahnestock S., Weissbach H. and Rich A. (1972). Formation of a ternary complex of phenyllactyl-tRNA with the transfer factor Tu and GTP. *Biochim. Biophys. Acta*, **269**, 62–66.
Fiarfield S.A. and Barnett W.E. (1971). On the similarities between the tRNAs of organelles and procaryotes. *Proc. Nat. Acad. Sci.*, **68**, 2972–2976.
Fakunding J.L. and Hershey J.W.B. (1973). The interaction of radioactive initiation factor IF2 with ribosomes during initiation of protein synthesis. *J. Biol. Chem.*, **248**, 4206–4212.
Falvey A.K. and Staehelin T. (1970). Structure and function of mammalian ribosomes. II. Exchange of ribosomal subunits at various stages of in vitro polypeptide synthesis. *J. Mol. Biol.*, **43**, 21–34.
Fareed G.C., Will E. and Richardson C.C. (1971). Hybrids of RNA and DNA homopolymers as substrates for polynucleotide ligase of phage T4. *J. Biol. Chem.*, **246**, 925–932.
Favre A., Michelson A.M. and Yaniv M. (1971). Photochemistry of 4-thiouridine in E.coli $tRNA_1^{val}$. *J. Mol. Biol.*, **58**, 367–380.
Felicetti L., Tocchini-Valentini G.P. and Di Matteo G.F. (1969). The role of G factor in protein synthesis. Studies on a temperature sensitive E.coli mutant with an altered G factor. *Biochem.*, **8**, 3428–3432.
Fellner P. (1969). Nucleotide sequences from specific areas of the 16S and 23S rRNAs of E.coli. *Europ. J. Biochem.*, **11**, 12–27.
Fellner P., Ehresmann C. and Ebel J.P. (1970). Nucleotide sequences in a protected area of the 16S RNA within 30S ribosomal subunits from E.coli. *Europ. J. Biochem.*, **13**, 583–588.
Fellner P., Ehresmann C., Stiegler P. and Ebel J.P. (1972). Partial nucleotide sequence of 16S rRNA from E.coli. *Nature New Biol.*, **239**, 1–5.
Fellner P. and Sanger F. (1968). Sequence analysis of specific areas of the 16S and 23S ribosomal RNAs. *Nature*, **219**, 236–238.
Ferretti J.R. (1971). Low level reading of the UGA triplet in S. typhimurium. *J. Bacteriol.*, **106**, 691–693.
Feunteun J., Jordan B.R. and Monier R. (1972). Study of the maturation of 5S RNA precursor in E.coli. *J. Mol. Biol.*, **70**, 465–474.
Fiandt M., Hradneca Z., Lozeron H.A. and Szybalski W. (1971). Electron micrographic mapping of deletions, insertions, inversions and homologies in the DNAs of coliphages lambda and phi 80. In Hershey A.D. (Ed.), *The Bacteriophage Lambda*, Cold Spring Harbor Laboratory, New York. pp 329–354.
Fiil N.P., Meyenburg V.K. and Friesen J.D. (1972). Accumulation and turnover of guanosine tetraphosphate in E.coli. *J. Mol. Biol.*, **71**, 769–784.
Fincham J.R.S. (1967). Recombination within the *am* gene of N. crassa. *Genet. Res.*, **9**, 49–62.
Fincham J.R.S. and Holliday R. (1970). An explanation of fine structure map extension in terms of excision-repair. *Molec. Gen. Genet.*, **109**, 309–322.
Fink G.R. (1966). A cluster of genes controlling three enzymes in histidine biosynthesis in S. cerevisiae. *Genetics*, **53**, 445–459.
Fink G.R. and Martin R.G. (1967). Translation and polarity in the histidine operon. II. Polarity in the histidine operon. *J. Mol. Biol.*, **30**, 97–108.
Fink G.R. and Roth J.R. (1968). Histidine regulatory mutants in S. typhimurium. VI. Dominance studies. *J. Mol. Biol.*, **33**, 547–558.
Firshein W. (1972). The DNA membrane fraction of Pneumococcus contains a DNA replication complex. *J. Mol. Biol.*, **70**, 383–398.
Flaks J.G., Leboy P.S., Birge E.A. and Kurland C.G. (1966). Mutations and genetics concerned with the ribosome. *Cold Spring Harbor Symp. Quant. Biol.*, **31**, 623–631.

Fogel S. and Mortimer R.K. (1969). Informational transfer in meiotic gene conversion. *Proc. Nat. Acad. Svi.*, **62**, 96–103.
Fogel S. and Mortimer R.K. (1970). Fidelity of gene conversion in yeast. *Molec. Gen. Genet.*, **109**, 177–185.
Fogel S. and Mortimer R.K. (1971). Recombination in yeast. *Ann. Rev. Genet.*, **5**, 219–235.
Fogel S. and Sypherd P.S. (1968). Chemical basis for heterogeneity of ribosomal proteins, *Proc. Nat. Acad. Sci.*, **59**, 1329–1336.
Folk W.R. and Yaniv M. (1972). Coding properties and nucleotide sequences of E.coli glutamine tRNAs. *Nature New Biol.*, **237**, 165–166.
Forchhammer J., Jackson E.N. and Yanofsky C. (1972). Different half lives of mRNA corresponding to different segments of the tryptophan operon of E.coli. *J. Mol. Biol.*, **71**, 687–700.
Forchhammer J. and Kjelgaard N.O. (1968). Regulation of messenger RNA synthesis in E.coli. *J. Mol. Biol.*, **37**, 245–256.
Ford P.J. and Southern E.M. (1973). Different sequences for 5S RNA in kidney cells and ovaries of X. laevis. *Nature New Biol.*, **241**, 7–12.
Forget B.G. and Varricchio (1970). A naturally occurring 43S ribosomal precursor particle in E.coli: nature and RNA composition. *J. Mol. Biol.*, **48**, 409–419.
Franklin N.C. and Dove W.F. (1969). Genetic evidence for restriction targets in the DNA of phages λ and ø80. *Genet. Res.*, **14**, 151–158.
Fresco J.R., Adams A., Ascione R., Henley D. and Lindahl T. (1966). Tertiary structure in transfer RNAs. *Cold Spring Harbor Symp. Quant. Biol.*, **31**, 527–538.
Friedberg E.C. and King J.J. (1969). Endonucleolytic cleavage of UV-irradiated DNA controlled by the v^+ gene in phage T4. *Biochem. Biophys. Res. Commun.*, **37**, 646–651.
Friedman S.M., Berezney R. and Weinstein I.B. (1968). Fidelity in protein synthesis. The role of the ribosome. *J. Biol. Chem.*, **243**, 5044–5055.
Fritsch A. and Worcel A. (1971). Symmetric multifork chromosome replication in fast growing E.coli. *J. Mol. Biol.*, **59**, 207–212.
Fuchs E. and Hanawalt P. (1970). Isolation and characterization of the DNA replication complex from E.coli. *J. Mol. Biol.*, **52**, 301–322.
Fuchs J.A., Karlstrom H.O., Warner H.R. and Reichard P. (1972). Defective gene product in *dnaF* mutant of E.coli. *Nature New Biol.*, **238**, 69–71.
Fuhs G.W. (1965). Fine structure of replication and of bacterial nucleoids. *Bacteriol. Rev.*, **29**, 277–293.
Fuller W. and Hodgson A. (1967). Conformation of the anticodon loop in tRNA. *Nature*, **215**, 817–821.
Fuller W.D., Sanchez R. and Orgel L.E. (1972). Studies in prebiotic synthesis. VI. Synthesis of purine nucleosides. *J. Mol. Biol.*, **67**, 25–34.
Funatsu G., Nierhaus K. and Wittman-Liebold B. (1972a). Ribosomal proteins. XXII. Studies on the altered protein S5 from a spectinomycin resistant mutant of E.coli. *J. Mol. Biol.*, **64**, 201–210.
Funatsu G., Puls W., Schiltz E., Reinbolt J. and Wittman H.G. (1972b). Ribosomal proteins. XXXI. Comparative studies on altered proteins S4 of six E.coli revertants from streptomycin dependence. *Molec. Gen. Genet.*, **115**, 131–139.
Funatsu G., Schiltz E. and Wittman H.G. (1972c). Ribosomal proteins. XXVII. Localization of the amino acid exchanges in protein S5 from two E.coli mutants resistant to spectinomycin. *Molec. Gen. Genet.*, **114**, 106–111.
Furano A.V., Bradley D.F. and Childers L.G. (1966). The conformation of the RNA in ribosomes. Dye stacking studies. *Biochem.*, **5**, 3044–3056.

Gallant J. and Cashel M. (1967). On the mechanism of amino acid control of RNA biosynthesis. *J. Mol. Biol.*, **25**, 545–553.

Gallant J., Irr J. and Cashel M. (1971). The mechanism of amino acid control of guanylate and adenylate biosynthesis. *J. Biol. Chem.*, **246**, 5812–5816.

Gallant J., Margason G. and Finch B. (1972). On the turnover of ppGpp in E.coli. *J. Biol. Chem.*, **247**, 6055–6058.

Galper J.B. and Darnell J.E. (1971). Mitochondrial protein synthesis in Hela cells. *J. Mol. Biol.*, **57**, 363–368.

Ganesan A.K. and Smith K.C. (1968). Recovery of recombination-deficient mutants of E.coli K12 from ultraviolet irradiation. *Cold Spring Harbor Symp. Quant. Biol.*, **33**, 235–242.

Ganesan A.T. (1971). ATP-dependent synthesis of biologically active DNA by azide-poisoned bacteria. *Proc. Nat. Acad. Sci.*, **68**, 1296–1300.

Ganesan A.T., Yehle C.O. and Yu C.C. (1973). DNA replication in a polymerase I deficient mutant and the identification of DNA polymerases II and III in B. subtilis. *Biochem. Biophys. Res. Commun.*, **50**, 155–163.

Gangloff J., Keith G., Ebel J.P. and Dirheimer G. (1971). Structure of aspartate-t-RNA from Brewer's yeast. *Nature New Biol.*, **230**, 125–126.

Garen A. (1968). Sense and nonsense in the genetic code. *Science*, **160**, 149–159.

Garen A. and Siddiqui O. (1962). Suppression of mutations in the alkaline phosphatase structural cistron of E.coli. *Proc. Nat. Acad. Sci.*, **48**, 1121–1126.

Garrett R.A., Rak K.H., Daya L. and Stoffler G. (1972). Ribosomal proteins. XXIX. Specific protein binding sites on 16S RNA of E.coli. *Molec. Gen. Genet.*, **114**, 112–124.

Gautschi J.R., Young B.R. and Painter R.B. (1972). Evidence for DNA repair replication in unirradiated mammalian cells—is it an artefact? *Biochim. Biophys. Acta*, **281**, 324–328.

Gavrilova L.P., Ivanov D.A. and Spirin A.S. (1966). Studies on the structure of ribosomes. III. Stepwise unfolding of the 50S particles without loss of ribosomal protein. *J. Mol. Biol.*, **16**, 473–489.

Geider K. and Hoffman-Berling H. (1971). DNA synthesis in nucleotide permeable E.coli cells. Chain elongation in specific regions of the bacterial chromosome. *Europ. J. Biochem.*, **21**, 374–384.

Gellert M. (1967). Formation of covalent circles of lambda DNA by E.coli extracts. *Proc. Acad. Nat. Sci.*, **57**, 148–155.

Gellert M. and Bullock M.L. (1970). DNA ligase mutants of E.coli. *Proc. Nat. Acad. Sci.*, **67**, 1580–1587.

Gellert M.W., Little J.W., Oshinsky C.K. and Zimmerman S.B. (1968). Joining of DNA strands by DNA ligase of E.coli. *Cold Spring Harbor Symp. Quant. Biol.*, **33**, 21–26.

Georgopoulos C.P. (1971). Bacterial mutants in which the *N* gene function of phage lambda is blocked have an altered RNA polymerase. *Proc. Nat. Acad. Sci.*, **68**, 2977–2981.

Gerard G.F., Johnson J.C. and Boezi J.A. (1972). Release of the sigma subunit of Pseudomonas putida DNA dependent RNA RNA polymerase. *Biochem.*, **11**, 989–997.

Ghosh H.P. and Ghosh K. (1972). Specificity of the initiator methionine tRNA for terminal and internal recognition. *Biochem. Biophys. Res. Commun.*, **49**, 550–557.

Ghosh, H.P. and Khorana H.G. (1967). Studies on polynucleotides. LXXXIV. On the role of ribosome subunits in protein synthesis. *Proc. Nat. Acad. Sci.*, **58**, 2455–2467.

Ghosh H.P., Soll D. and Khorana H.G. (1967). Studies on polynucleotides. LXVII. Initiation of protein synthesis in vitro as studied by using polyribonucleotides with repeating nucleotide sequences as messengers. *J. Mol. Biol.*, **25**, 275–298.

Ghosh K. and Ghosh H.P. (1970). Role of modified nucleoside adjacent to 3′ end of anticodon in codon anticodon interaction. *Biochem. Biophys. Res. Commun.*, **40**, 135–143.

Ghosh K. and Ghosh H.P. (1972). Effect of removal of the modified base adjacent to 3′ end of the anticodon in codon-anticodon interaction. *J. Biol. Chem.*, **247**, 3369–3375.

Gibson T.C., Scheppe M.L. and Cox E.C. (1970). Fitness of an E.coli mutator gene. *Science*, **169**, 686–688.

Gielow L., Largen M. and Englesberg E. (1971). Initiator constitutive mutants of the L-arabinose operon (*OIBAD*) of E.coli B/r. *Genetics*, **69**, 289–302.

Gilbert W. (1963). Polypeptide synthesis in E.coli. I. Ribosomes and the active complex. *J. Mol. Biol.*, **6**, 374–388.

Gilbert W. and Dressler D. (1968). DNA replication: the rolling circle model. *Cold Spring Harbor Symp. Quant. Biol.*, **33**, 473–484.

Gilbert W. and Muller-Hill B. (1967). The lac operator is DNA. *Proc. Nat. Acad. Sci.*, **58**, 2415–2421.

Gilbert W. and Muller-Hill B. (1970). The lactose repressor. In Beckwith J.R. and Zipser D. (Eds.), *The Lactose Operon*, Cold Spring Harbor Laboratory, New York. pp 93–110.

Gill M. and Dinius L.L. (1973). The elongation factor 2 content of mammalian cells. Assay method and relation to ribosome number. *J. Biol. Chem.*, **248**, 654–658.

Gilmore R.A., Stewart J.W. and Sharman F. (1971). Amino acid replacement resulting from super-suppression of nonsense mutants of iso-l-cytochrome *c* from yeast. *J. Mol. Biol.*, **61**, 157–174.

Ginsberg B. and Hurwitz J. (1970). Unbiased synthesis of pulse labelled DNA fragments of phage and T4. *J. Mol. Biol.*, **52**, 265–280.

Ginsberg T., Rogg H. and Staehelin M. (1971). Nucleotide sequences of rat liver serine tRNA. III. The partial enzymatic digestion of serine tRNA and the derivation of total primary structure. *Europ. J. Biochem.*, **21**, 249–257.

Glover S.W. (1970). Functional analysis of host specificity mutants in E.coli. *Genet. Res.*, **15**, 237–250.

Glover S.W. and Colson C. (1969). Genetics of host-controlled restriction and modification in E.coli. *Genet. Res.*, **13**, 227–240.

Goebel W. (1972). Replication of the DNA of the colicin factor E1 at the restrictive temperature in a DNA replication mutant thermosensitive for DNA polymerase III. *Nature New Biol.*, **237**, 67–71.

Goff C.G. and Weber K. (1970). A T4 induced RNA polymerase α subunit modification. *Cold Spring Harbor Symp. Quant. Biol.*, **35**, 101–108.

Goldberg A.R. and Hurwitz J. (1972). Studies on termination of in vitro RNA synthesis by rho factor. *J. Biol. Chem.* **247**, 5637–5645.

Goldmark P.J. and Linn S. (1972). Purification and properties of the *recB* DNAase of E.coli K12. *J. Biol. Chem.*, **247**, 1849–1860.

Goldstein J.L. and Caskey C.T. (1970). Peptide chain termination: effect of protein S on ribosomal binding of release factors. *Proc. Nat. Acad. Sci.*, **67**, 537–543.

Goodman H.M., Abelson, J., Landy A., Brenner S. and Smith J.D. (1968). Amber suppression: a nucleotide change in the anticodon of a tyrosine tRNA. *Nature*, **217**, 1019–1024.

Goodman H.M., Abelson J.N., Landy A., Zadrazil S. and Smith J.D. (1970b). The nucleotide sequences of tyrosine tRNAs of E.coli. Sequences of the amber suppressor $su3^+$ tRNA, the wild type $su3^-$ tRNA and the tyrosine tRNA species I and II. *Europ. J. Biochem.*, **13**, 461–483.

Goodman H.M., Billeter M.A., Hindley J. and Weissman C. (1970a). The nucleotide sequence at the 5′ terminus of the Qβ RNA minus strand. *Proc. Nat. Acad. Sci.*, **67**, 921–928.

Gopinathan K.P. and Garen A. (1970). A leucyl-tRNA specified by the amber suppressor gene $su6^+$. *J. Mol. Biol.*, **47**, 393–401.

Gordon J. (1968). A stepwise reaction yielding a complex between a supernatant fraction from E.coli, GTP, and aminoacyl-tRNA. *Proc. Nat. Acad. Sci.*, **59**, 179–183.

Gordon J. and Weissbach H. (1970). Immunochemical distinction between E.coli polypeptide chain elongation factor Tu and Ts. *Biochem.*, **9**, 4231–4236.

Gorini L. and Kataja E. (1964). Streptomycin induced over-suppression in E.coli. *Proc. Nat. Acad. Sci.*, **51**, 995–1001.

Gorini L. and Kataja E. (1965). Suppression activated by streptomycin and related antibiotics in E.coli strains. *Biochem. Biophys. Res. Commun.*, **18**, 656–663.

Gottesman M.M., Hicks M.L. and Gellert M. (1973). Genetics and function of DNA ligase in E.coli. *J. Mol. Biol.*, **77**, 531–549.

Gray P.N., Garrett R.A., Stoffler G. and Monier R. (1972). An attempt at the identification of the proteins involved in the incorporation of 5S RNA during 50S ribosomal subunit assembly. *Europ. J. Biochem.*, **28**, 412–428.

Greenberg H. and Penman S. (1966). Methylation and processing of ribosomal RNA in Hela cells. *J. Mol. Biol.*, **21**, 527–535.

Greenblatt J. (1973). Regulation of the expression of the *N* gene of phage λ. *Proc. Nat. Acad. Sci.*, **70**, 421–424.

Greenblatt J. and Schleif R. (1971). Arabinose *C* protein: regulation of the arabinose operon in vitro. *Nature New Biol.*, **233**, 166–170.

Greenleaf A.L., Linn T.G. and Losick R. (1973). Isolation of a new RNA polymerase binding protein from sporulating B. subtilis. *Proc. Nat. Acad. Sci.*, **70**, 490–494.

Griffith F. (1928). Significance of Pneumococcal types. *J. Hyg. Camb.*, **27**, 113–159.

Grodzicker T. and Zipser D. (1968). A mutation which creates a new site for the reinitiation of polypeptide synthesis in the *z* gene of the lac operon of E.coli. *J. Mol. Biol.*, **38**, 305–314.

Gros F., Hiatt H., Gilbert W., Kurland C.G., Risebrough R.W. and Watson J.D. (1961). Unstable RNA revealed by pulse labelling of E.coli. *Nature*, **190**, 581–585.

Grosjean D.L., Garrett R.A., Pongs O., Stoffler G. and Wittman H.G. (1972). Properties of the interaction of ribosomal protein S4 and 16S RNA in E.coli. Revertants from streptomycin dependence to independence. *Molec. Gen. Genet.*, **119**, 277–286.

Gross J., Grunstein J. and Witkin E.M. (1971). Inviability of $recA^-$ derivatives of the DNA polymerase mutant of de Lucia and Cairns. *J. mol. Biol.*, **58**, 631–634.

Gross J.D., Karamata D. and Hempstead P.G. (1968). Temperature sensitive mutants of B. subtilis defective in DNA synthesis. *Cold Spring Harbor Symp. Quant. Biol.*, **33**, 307–312.

Grossman L., Kaplan J., Kushner S. and Mahler I. (1968). Enzymes involved in the early stages of repair of ultraviolet irradiated DNA. *Cold Spring Harbor Symp. Quant. Biol.*, **33**, 229–234.

Guha A., Saturen Y. and Szybalski W. (1971b). Divergent orientation of transcription from the biotin locus of E.coli. *J. Mol. Biol.*, **56**, 53–62.

Guha A., Szybalski W., Salser W., Bolle A. and Geiduschek E.P. (1971a). Controls and polarity of transcription during phage T4 development. *J. Mol. Biol.*, **59**, 329–350.
Gumport R.I. and Lehman I.R. (1971). Structure of the DNA ligase adenylate intermediate: lysine (ε-amino) linked AMP. *Proc. Nat. Acad. Sci.*, **68**, 2559–2563.
Gupta N.K., Chatterjee N.K., Woodley C.L. and Bose K.K. (1971b). Protein synthesis in rabbit reticulocytes. Factors controlling internal and terminal methionine codon recognition by the tRNA species. *J. Biol. Chem.*, **246**, 7460–7469.
Gupta S.L., Waterson J., Sopori M.L., Weissman S. and Lengyel P. (1971a). Movement of the ribosome along the mRNA during protein synthesis. *Biochem.*, **10**, 4410–4412.
Guthrie C., Nashimoto H. and Nomura M. (1969a). Studies on the assembly of ribosomes in vivo. *Cold Spring Harbor Symp. Quant. Biol.*, **34**, 69–76.
Guthrie C., Nashimoto H. and Nomura M. (1969b). Structure and function of E.coli ribosomes. VIII. Cold sensitive mutants defective in ribosome assembly. *Proc. Nat. Acad. Sci.*, **63**, 384–391.
Guthrie C. and Nomura M. (1968). Initiation of protein synthesis: a critical test of the 30S subunit model. *Nature*, **219**, 232–236.
Guttman B. and Novick A. (1963). A messenger RNA for β-galactosidase in E.coli. *Cold Spring Harbor Symp. Quant. Biol.*, **28**, 373–374.
Gutz H. (1971). Gene conversion: remarks on the quantitative implications of hybrid DNA models. *Genet. Res.*, **17**, 45–52.
Gyurasits E.B. and Wake R.G. (1973). Bidirectional chromosome replication in B. subtilis. *J. Mol. Biol.*, **73**, 55–63.
Haberman A., Heywood J. and Meselson M. (1972). DNA modification methylase activity of E.coli restriction endonucleases K and P. *Proc. Nat. Acad. Sci.*, **69**, 3138–3141.
Haenni A. and Lucas-Lenard J. (1968). Stepwise synthesis of a tripeptide. *Proc. Nat. Acad. Sci.*, **61**, 1363–1369.
Hagen D.C. and Magasanik B. (1973). Isolation of the self regulated repressor protein *hut* operons of S. typhimurium. *Proc. Nat. Acad. Sci.*, **70**, 808–812.
Hall B.D. and Spiegelman S. (1961). Sequence complementarity of T2 DNA and T2 specific RNA. *Proc. Nat. Acad. Sci.*, **47**, 137–161.
Hall B. and Gallant J. (1972). Defective translation in RC^- cells. *Nature New Biol.*, **237**, 131–135.
Hall Z.W. and Lehman I.R. (1968). An in vitro transversion by a mutationally altered T4 induced DNA polymerase. *J. Mol. Biol.*, **36**, 321–334.
Hamlin J. and Zabin I. (1972). β-galactosidase: immunological activity of ribosome bound growing polypeptide chains. *Proc. Nat. Acad. Sci.*, **69**, 412–416.
Hanawalt P.C. (1967). Normal replication of DNA after repair replication in bacteria. *Nature*, **214**, 269–270.
Hanawalt P.C. (1968). Cellular recovery from photochemical damage. *Photophysiol.*, **4**, 203–251.
Hanawalt P.C. and Haynes R.H. (1965). Repair replication of DNA in bacteria: irrelevance of the chemical nature of base defect. *Biochem. Biophys. Res. Commun.*, **19**, 462–467.
Hanwalt P.C. and Haynes R.H. (1967). The repair of DNA. *Scient. Amer.*, **216**, **2**, 36–43.
Hanawalt P.C., Maaløe O., Cummings D.J. and Schaechter M. (1961). The normal DNA replication cycle. *J. Mol. Biol.*, **3**, 156–165.
Hanawalt P.C., Pettijohn D.E., Pauling E.C., Brunk C.F., Smith D.W., Kanner L.C. and Couch J.L. (1968). Repair replication of DNA in vivo. *Cold Spring Harbor Symp. Quant. Biol.*, **33**, 187–194.

Hangii U.J. and Zachau H.G. (1971). Partial nuclease digestion of tRNAs and aminoacylated tRNAs. *Europ. J. Biochem.*, **18**, 496–502.
Harada F. and Nishimura S. (1972). Possible anticodon sequences of $tRNA^{his}$, $tRNA^{asn}$, and $tRNA^{asp}$ from E.coli B. Universal presence of nucleoside Q in the first position of the anticodons of these tRNAs. *Biochem.*, **11**, 301–308.
Hardesty B., Culp W. and McKeehan W. (1969). The sequence of reactions leading to the synthesis of a peptide bond on reticulocyte ribosomes. *Cold Spring Harbor Symp. Quant. Biol.*, **34**, 331–346.
Haseltine W.A. (1972). In vitro transcription of E.coli rRNA genes. *Nature*, **235**, 329–333.
Hseltine W.A. and Block R. (1973). Synthesis of guanosine tetra and pentaphosphate requires the presence of a codon specific uncharged tRNA in the acceptor site of ribosomes. *Proc. Nat. Acad. Sci.*, **70**, 1564–1568.
Haseltine W.A., Block R., Gilbert W. and Weber K. (1972). MSI and MSII made on ribosome in idling step of protein synthesis. *Nature*, **238**, 381–384.
Hatfield D. (1972). Recognition of nonsense codons in mammalian cells. *Proc. Nat. Acad. Sci.*, **69**, 3014–3018.
Hayashi H. and Soll D. (1971). Purification of E.coli leucine suppressor tRNA and its aminoacylation by the homologous leu-tRNA synthetase. *J. Biol. Chem.*, **246**, 4951–4954.
Hayashi M.N. and Hayashi M. (1968). The stability of native DNA-RNA complexes during in vivo ϕX174 transcription. *Proc. Nat. Acad. Sci.*, **61**, 1107–1115.
Hayes F., Hayes D., Fellner P. and Ehresmann C. (1971). Additional nucleotide sequences in precursor 16S ribosomal RNA from E.coli. *Nature New Biol.*, **232**, 54–55.
Hedgpeth J., Goodman H.M. and Boyer H.W. (1972). DNA nucleotide sequence restricted by the RI endonuclease. *Proc. Nat. Acad. Sci.*, **69**, 3448–3452.
Heijneker H.L., Pannekoek H., Oosterbaan R.A., Pouwels P.H., Bron S., Arwert F. and Venema G. (1971). In vitro excision repair of ultraviolet irradiated transforming DNA from B subtilis. *Proc. Nat. Acad. Sci.*, **68**, 2967–2971.
Helene C., Brun and Yaniv M. (1971). Fluorescence studies of interactions between E.coli valyl-tRNA synthetase and its substrate. *J. Mol. Biol.*, **58**, 349–366.
Helmstetter C.E. and Cooper S. (1968). DNA synthesis during the division cycle of rapidly growing E.coli B/r. *J. Mol. Biol.*, **31**, 507–518.
Helmstetter C.E., Cooper S., Pierucci O. and Revelas E. (1968). On the bacterial life sequence. *Cold Spring Harbor Symp. Quant. Biol.*, **33**, 809–822.
Helmstetter C.E. and Pierucci O. (1968). Cell division during inhibition of DNA synthesis in E.coli. *J. Bacteriol.*, **95**, 1627–1633.
Helser T.L., Davies J.E. and Dahlberg J.E. (1971). Change in methylation of 16S ribosomal RNA associated with mutation to kasugomycin resistance in E.coli. *Nature New Biol.*, **233**, 12–14.
Helser T.L., Davies J.E. and Dahlberg J.E. (1972). Mechanism of kasugomycin resistance in E.coli. *Nature New Biol.*, **235**, 6–9.
Herrmann R., Huf J. and Bonhoeffer F. (1972). Cross hybridization and rate of chain elongation of the two classes of DNA intermediates. *Nature New Biol.*, **240**, 235–237.
Hershey A.D. and Chase M. (1952a). Genetic recombination and heterozygosis in bacteriophage. *Cold Spring Harbor Symp. Quant. Biol.*, **16**, 471–479.
Hershey A.D. and Chase M. (1952b). Independent functions of viral protein and nucleic acid in growth of bacteriophage. *J. Gen. Physiol.*, **26**, 36–56.
Hershey J.W.B., Dewey K.F. and Thach R.E. (1969). Purification and properties of initiation factor f1. *Nature*, **222**, 944–947.

Hershey J.W.B. and Thach R.E. (1967). Role of GTP in the initiation of peptide synthesis. I. Synthesis of formyl-methionine puromycin. *Proc. Nat. Acad. Sci.*, **57**, 759–766.

Hershfield M.S. (1973). On the role of DNA polymerase in determining mutation rates. Characterization of the defect in the T4 DNA polymerase caused by the ts L88 mutation. *J. Biol. Chem.*, **248**, 1417–1423.

Hershfield M.S. and Nossal N.G. (1972). Hydrolysis of template and newly synthesized DNA by the 3′ to 5′ exonuclease activity of the T4 DNA polymerase. *J. Biol. Chem.*, **247**, 3393–3404.

Herskowitz I. and Signer E. (1970). Control of transcription from the *r* strand of phage lambda. *Cold Spring Harbor Symp. Quant. Biol.*, **35**, 355–368.

Highland J.H., Bodley J.W., Gordon J., Hasenvank R. and Stoffler G. (1973). Identity of the ribosomal proteins involved in the interaction with elongation factor G. *Proc. Nat. Acad. Sci.*, **70**, 147–150.

Hill C.W., Squires C. and Carbon J. (1970). Glycine tRNA of E.coli. I. Structural genes for two glycine tRNA species. *J. Mol. Biol.*, **52**, 557–570.

Hinckle D.C. and Chamberlin M. (1970). The role of sigma subunit in template site selection by E.coli RNA polymerase. *Cold Spring Harbor Symp. Quant. Biol.*, **35**, 65–72.

Hinckle D.C. and Chamberlin M.J. (1972a). Studies of the binding of E.coli RNA polymerase to DNA. I. The role of sigma subunit in site selection. *J. Mol. Biol.*, **70**, 157–186.

Hinckle D.C. and Chamberlin M.J. (1972b). Studies of the binding of RNA polymerase to DNA. II. The kinetics of the binding reaction. *J. Mol. Biol.*, **70**, 187–196.

Hinckle D.C., Mangel W.F. and Chamberlin M.J. (1972). Studies of the binding of E.coli RNA polymerase to DNA. IV. The effect of rifampicin on binding and on RNA chain initiation. *J. Mol. Biol.*, **70**, 209–220.

Hinckle D.C., Ring J. and Chamberlin M.J. (1972). Studies of the binding of E.coli RNA polymerase to DNA. III. Tight binding of the RNA polymerase holoenzyme to single strand breaks in T7 DNA. *J. Mol. Biol.*, **70**, 197–208.

Hindley J. and Staples D.H. (1969). Sequence of a ribosome binding site in phage Qβ RNA. *Nature*, **224**, 964–967.

Hiraga S. and Yanofsky C. (1972). Hyperlabile messenger RNA in polar mutants of the tryptophan operon of E.coli. *J. Mol. Biol.*, **72**, 103–110.

Hirose S., Okazaki R. and Tamanoi F. (1973). Mechanism of DNA chain growth. XI. Structure of RNA linked DNA fragments of E.coli. *J. Mol. Biol.*, **77**,

Hirota Y., Gefter M. and Mindich L. (1972). A mutant of E.coli defective in DNA polymerase II activity. *Proc. Nat. Acad. Sci.*, **69**, 3238–3242.

Hirota Y. Mordoh J. and Jacob F. (1970). On the process of cellular division in E.coli III. Thermosensitive mutants of E.coli altered in the process of DNA initiation. *J. Mol. Biol.*, **53**, 369–388.

Hirsh D. (1971). Tryptophan tRNA as the UGA suppressor. *J. Mol. Biol.*, **58**, 439–458.

Hirsh D. and Gold L. (1971). Translation of the UGA triplet in vitro by tryptophan tRNAs. *J. Mol. Biol.*, **58**, 459–468.

Hirt B. (1969). Replicating molecules of polyoma virus DNA. *J. Mol. Biol.*, **40**, 141–144.

Hoagland M.B., Keller E.B. and Zemcnik P.C. (1956). Enzymatic carboxyl activation of amino acids. *J. Biol. Chem.*, **218**, 345–358.

Hoagland M.B., Stephenson M.L., Scott J.E., Hecht L.I. and Zamecnik P.C. (1958). A soluble RNA intermediate in protein synthesis. *J. Biol. Chem.*, **231**, 241–257.

Hoagland M.B., Zamecnik P.C., Sharon N., Lipmann F., Stulberg M.P. and Boyer P.D. (1957). Oxygen transfer to AMP in the enzymatic synthesis of the hydroxamate of tryptophan. *Biochim. Biophys. Acta*, **26**, 215–217.

Holley R.W., Apgar J., Everett G.A., Madison J.T., Marquise H., Merrill S.H., Penswick J.R. and Zamir A. (1965). Structure of an RNA. *Science*, **147**, 1462–1465.

Holliday R. (1964). A mechanism for gene conversion in fungi. *Genet. Res.*, **5**, 282–304.

Holliday R. (1967). Altered recombination frequencies in radiation sensitive strains of Ustilago. *Mutat. Res.*, **4**, 275–288.

Holliday R. and Whitehouse H.L.K. (1970). The wrong way to think about gene conversion. *Molec. Gen. Genet.*, **107**, 85–93.

Holmes R.K. and Singer M.F. (1971). Inability to detect RNAase V in E.coli and comparison of other ribonucleases before and after infection. *Biochem. Biophys. Res. Commun.*, **44**, 837–843.

Homann H.E. and Nierhaus K.H. (1971). Ribosomal proteins. Protein compositions of biosynthetic precursors and artificial subparticles from ribosomal subunits in E.coli K12. *Europ. J. Biochem.*, **20**, 249–257.

Hopkins N. and Ptashne M. (1971). Genetics of virulence. In Hershey A.D. (Ed.), *The bacteriophage Lambda*, Cold Spring Harbor Laboratory, New York. pp 571–574.

Hosokawa K. (1970). Binding of 5S rRNA to the unfolded 50S ribosome of E.coli. *J. Biol. Chem.*, **245**, 5880–5887.

Hotta Y. and Stern H. (1971a). A DNA binding protein in meiotic cells of Lilium. *Devel. Biol.*, **26**, 87–99.

Hotta Y. and Stern H. (1971b). Analysis of DNA synthesis during meiotic prophase in Lilium. *J. Mol. Biol.*, **55**, 337–356.

Howell S.H. and Stern H. (1971). The appearance of DNA breakage and repair activities in the synchronous meiotic cycle of Lilium. *J. Mol. Biol.*, **55**, 357–378.

Howard-Flanders P., Boyce R.P. and Theriot L. (1966). Three loci in E.coli K12 that control the excision of pyrimidine dimers and certain other mutagen products from DNA. *Genetics*, **53**, 1119–1136.

Howard-Flanders P., Rupp W.D., Wilkins B.M. and Cole R.S. (1968). DNA replication and recombination after UV-irradiation. *Cold Spring Harbor Symp. Quant. Biol.*, **33**, 195–208.

Howard-Flanders P. and Theriot L. (1966). Mutants of E.coli defective in DNA repair and in genetic recombination. *Genetics*, **53**, 1137–1150.

Hsu T. and Weiss B. (1969). Selective translation of T4 template RNA from T4 infected E.coli. *Proc. Nat. Acad. Sci.*, **63**, 345–351.

Hua S.S. and Markovitz A. (1972). Multiple regulator gene control of the galactose operon in E.coli K12. *J. Bacteriol.*, **110**, 1089–1099.

Hubacek J. and Glover S.W. (1970). Complementation analysis of temperature sensitive host specificity mutants in E.coli. *J. Mol. Biol.*, **50**, 111–127.

Huberman J.A., Kornberg A. and Alberts B.M. (1971). Stimulation of T4 phage DNA polymerase by the protein product of T4 gene 32. *J. Mol. Biol.*, **62**, 39–52.

Huberman J.A. and Riggs D.A. (1968). On the mechanism of DNA replication in mammalian chromsomes. *J. Mol. Biol.*, **32**, 327–341.

Huberman J.A. and Tsai A. (1973). Direction of DNA replication in mammalian cells. *J. Mol. Biol.*, **75**, 5–12.

Hunter A.R. and Jackson R.J. (1970). Miscoding by E.coli tRNAS for methionine, cysteine and valine in the synthesis of rabbit globin. *Europ. J. Biochem.*, **15**, 381–390.

Hurst P.D., Fogel F. and Mortimer R.K. (1972). Conversion associated recombination in yeast. *Proc. Nat. Acad. Sci.*, **69**, 101–105.

Iaccarino M. and Berg P. (1969). Requirement of sulfhydryl groups for the catalytic and tRNA recognition functions of isoleucyl-tRNA synthetase. *J. Mol. Biol.*, **42**, 151–169.

Imamoto F. (1968a). On the initiation of transcription of the tryptophan operon in E.coli. *Proc. Nat. Acad. Sci.*, **60**, 305–312.

Imamoto F. (1968b). Immediate cessation of transcription of the operator region of the tryptophan operon of E.coli. *Nature*, **220**, 31–35.

Imamoto F. (1969). Intragenic initiations of transcription of the tryptophan operon following dinitrophenol treatment with tryptophan. *J. Mol. Biol.*, **43**, 51–70.

Imamoto F. (1970). Immediate cessation of an operator proximal segment of the tryptophan operon in E.coli following repression of the operon. *Molec. Gen. Genet.*, **106**, 123–138.

Imamoto F. (1973). Diversity of genetic transcription. I. Effect of antibiotics which inhibit the process of translation of RNA metabolism in E.coli. *J. Mol. Biol.*, **74**, 113–136.

Imamoto F. and Ito J. (1968). Simultaneous initiation of transcription and translation at internal sites in the tryptophan operon of E.coli. *Nature*, **220**, 27–31.

Imamoto F., Ito J. and Yanofsky C. (1966). Polarity in the tryptophan operon of E.coli. *Cold Spring Harbor Symp. Quant. Biol.*, **31**, 235–250.

Imamoto F. and Kano Y. (1971). Inhibition of transcription of the tryptophan operon in E.coli by a block in initiation of translation. *Nature New Biol.*, **232**, 169–173.

Imamoto F., Kano Y. and Tani S. (1970). Transcription of the tryptophan operon in nonsense mutants of E.coli. *Cold Spring Harbor Symp. Quant. Biol.*, **35**, 471–490.

Imamoto F., Morikawa N. and Sato K. (1965b). On the transcription of the tryptophan operon. III. multicistronic messenger RNA and polarity for transcription. *J. Mol. Biol.*, **13**, 169–182.

Imamoto F., Morikawa N., Sato K., Mishima S., Nishimura S., Nishimura T. and Matsushiro A. (1965a). On the transcription of the tryptophan operon. II. Production of the specific messenger RNA. *J. Mol. Biol.*, **13**, 157–168.

Imamoto F. and Yanofsky C. (1967a). Transcription of the tryptophan operon in polarity mutants of E.coli. I. Characterization of the tryptophan messenger RNA of polar mutants. *J. Mol. Biol.*, **28**, 1–23.

Imamoto F. and Yanofsky C. (1967b). Transcription of the tryptophan operon in polarity mutants of E.coli. II. Evidence for normal production of *trp*-mRNA molecules and for premature termination of transcription. *J. Mol. Biol.*, **28**, 25–35.

Infante A.A. and Baierlein R. (1971). Pressure induced dissociation of sedimenting ribosomes: effect on sedimentation patterns. *Proc. Nat. Acad. Sci.*, **68**, 1780–1785.

Ingram V.M. (1957). Gene mutations in human haemoglobin: the chemical difference between normal and sickle cell haemoglobin. *Nature*, **180**, 326–328.

Inman R.B. and Schnos M. (1971). Structure of branch points in replicating DNA: presence of single stranded connections in λ DNA branch points. *J. Mol. Biol.*, **56**, 319–326.

Ippen K., Miller J.H., Scaife J. and Beckwith J. (1968). New controlling element in the *lac* operon of E.coli. *Nature*, **217**, 825–826.

Irr J. and Englesberg E. (1971). Control of expression of the L-arabinose operon in temperature sensitive mutants of gene *araC* in E.coli B/r. *J. Bacteriol.*, **105**, 136–141.

Ishikura H. and Nishimura S. (1968). Fractionation of serine tRNAs from E.coli and their coding properties. *Biochim. Biophys. Acta*, **155**, 72–81.

Ishitsuka H. and Kaji A. (1970). Release of tRNA from ribosomes by a factor other than G factor. *Proc. Nat. Acad. Sci.*, **66**, 168– 173.

Itoh T., Otaka E. and Osawa S. (1968). Release of ribosomal proteins from E.coli ribosomes with high concentrations of LiCl. *J. Mol. Biol.*, **33**, 109–122.

Ivarie R.D. and Pene J.J. (1973). Association of many regions of the B. subtilis chromosome with the cell membrane. *J. Bacteriol.*, **114**, 571–576.

Iwakura Y., Ishihama A. and Yura T. (1973). RNA polymerase mutants of E.coli. II. Streptolydigin resistance and its relation to rifampicin resistance. *Molec. Gen. Genet.*, **121**, 181–196.

Iwatsuki N. and Okazaki R. (1970). Mechanism of DNA chain growth. V. Effect of chloramphenicol on the formation of T4 nascent short chains. *J. Mol. Biol.*, **52**, 37–44.

Iyer V.N. and Lark K.G. (1970). DNA replication in E.coli: location of recently incorporated thymidine within molecules of high molecular weight DNA. *Proc. Nat. Acad. Sci.*, **67**, 629–636.

Iyer V.N. and Rupp W.D. (1971). Usefulness of benzoylated naphthoylated DEAE-cellulose to distinguish and fractionate double stranded DNA bearing different extents of single stranded regions. *Biochim. Biophys. Acta*, **228**, 117–126.

Jackson E.N. and Yanofsky C. (1972). Internal promotor of the tryptophan operon of E.coli is located in a structural gene. *J. Mol. Biol.*, **69**, 307–314.

Jackson E.N. and Yanofsky C. (1973). The regions between the operator and first structural gene of the tryptophan operon of E.coli may have a regulatory function. *J. Mol. Biol.*, **76**, 89–102.

Jackson R. and Hunter T. (1970). Role of methionine in the initiation of haemoglobin synthesis. *Nature*, **227**, 672–676.

Jacob F., Brenner S. and Cuzin F. (1963). On the regulation of DNA replication in bacteria. *Cold Spring Harbor Symp. Quant. Biol.*, **28**, 329–348.

Jacob F. and Monod J. (1961). Genetic regulatory mechanisms in the synthesis of proteins. *J. Mol. Biol.*, **3**, 318–356.

Jacob F., Ryter A. and Cuzin F. (1966). On the association between DNA and the membrane in bacteria. *Proc. Roy. Soc. B.*, **164**, 267–278.

Jacob F., Ullman A. and Monod F. (1964). Le promoteur, element genetique necessaire a l'expression d'un operon. *Comptes Rendue Acad. Sci.* (*Paris*), **258**, 3125–3128.

Jacobson M.K. and Lark K.G. (1973). DNA replication in E.coli; evidence for two classes of small DNA chains. *J. Mol. Biol.*, **73**, 371–396.

Jacoby A. (1971). Mapping the gene determining ornithine transcarbamylase and its operator in E.coli B. *J. Bacteriol.*, **108**, 645–651.

Jacoby G.A. (1972). Control of the *argECBH* cluster in E.coli. *Molec. Gen. Genet.*, **117**, 337–349.

Jacoby G.A. and Gorini L. (1969). A unitary account of the repression mechanism of arginine biosynthesis in E.coli. I. The genetic evidence. *J. Mol. Biol.*, **39**, 73–87.

Jayamaran R. (1972). Transcription of phage T4 DNA by E.coli RNA polymerase in vitro: identification of some immediate early and delayed early genes. *J. Mol. Biol.*, **70**, 253–264.

Jeanteur P., Amaldi F. and Attardi G. (1968). Partial sequence analysis of ribosomal RNA from Hela cells. II. Evidence for sequences of non-ribosomal type in 45S and 32S ribosomal RNA precursors. *J. Mol. Biol.*, **33**, 757–776.

Jeanteur P. and Attardi G. (1969). Relationship between Hela cell rRNA and its precursors studied by high resolution RNA-DNA hybridization. *J. Mol. Biol.*, **45**, 305–324.

Jeppesen P.G.N., Steitz J.A., Gesteland R.R. and Spahr P.F. (1970). Gene order in the phage R17 RNA. *Nature*, **226**, 230–237.

Jobe A. and Bourgeois S. (1972). *Lac* repressor-operator interaction. VI. The natural inducer of the *lac* operon. *J. Mol. Biol.*, **69**, 397–408.

Jobe A. and Bourgeois S. (1973). *Lac* repressor-operator interaction. VIII. Lactose is an anti-inducer of the *lac* operon. *J. Mol. Biol.*, **75**, 303–314.

Jobe A., Riggs A.D. and Bourgeois S. (1972). *Lac* repressor-operator interaction. VI. Characterization of super and pseudo-wild repressors. *J. Mol. Biol.*, **64**, 181–200.

Jonasson J. (1973). Evidence for bidirectional chromosome replication in E.coli C based on marker frequency analysis by DNA/DNA hybridization with P2 and λ prophages. *Molec. Gen. Genet.*, **120**, 69–90.

Jones D.A., Nishimura S. and Khorana H.G. (1966). In vitro synthesis of co-polypeptides. *J. Mol. Biol.*, **16**, 454–472.

Jones N.C. and Donachie W.D. (1973). Chromosome replication, transcription and control of cell division in E.coli. *Nature New Biol.*, **243**, 100–103.

Jordan B.R. (1971). Studies on 5S RNA conformation by partial ribonuclease analysis. *J. Mol. Biol.*, **55**, 423–440.

Jou W.M., Haegeman G., Ysebaert M. and Fiers W. (1972). Nucleotide sequence of the gene coding for the phage MS2 coat protein. *Nature*, **237**, 82–88.

Kadner R.J. and Maas W.K. (1971). Regulatory gene mutations affecting arginine biosynthesis in E.coli. *Molec. Gen. Genet.*, **111**, 1–14.

Kaempfer R. (1970). Dissociation of ribosomes on polypeptide chain termination and origin of single ribosomes. *Nature*, **228**, 534–537.

Kaempfer R. (1971). Control of single ribosome formation by an initiation factor for protein synthesis. *Proc. Nat Acad. Sci.*, **68**, 2458–2462.

Kaempfer R. (1972). Initiation factor IF-3: a specific inhibitor of ribosomal subunit association. *J. Mol. Biol.*, **71**, 583–598.

Kaempfer R. and Meselson M. (1968). Permanent association of 5S RNA molecules with 50S ribosomal subunits in growing bacteria. *J. Mol. Biol.*, **34**, 703–708.

Kaempfer R. and Meselson M. (1969). Studies of ribosomal subunit exchange. *Cold Spring Harbor Symp. Quant. Biol.*, **34**, 209–220.

Kaempfer R., Meselson M. and Raskas H.J. (1968). Cyclic dissociation into stable subunits and reformation of ribosomes during bacterial growth. *J. Mol. Biol.*, **31**, 277–289.

Kahan L., Zengel J., Nomura M., Bollen A. and Herzog A. (1973). Structural gene for the ribosomal protein S18 in E.coli. II. Chemical studies on the protein S18 having an altered electrophoretic mobility. *J. Mol. Biol.*, **76**, 473–484.

Kaiser D. (1971). Lambda DNA replication. In Hershey A.D. (Ed.), *The bacteriophage lambda*, Cold Spring Harbor Laboratory, New York. pp 195–210.

Kaji H. and Kaji A. (1965). Specific binding of tRNA to ribosomes: effect of streptomycin. *Proc. Nat. Acad. Sci.*, **54**, 213–218.

Kaltschmidt E. and Wittman H.G. (1970). Ribosomal proteins. XII. Number of proteins in small and large ribosomal subunits of E.coli as determined by two dimensional gel electrophoresis. *Proc. Nat. Acad. Sci.*, **67**, 1276–1282.

Kanner L. and Hanawalt P. (1970). Repair deficiency in a bacterial mutant defective in DNA polymerase. *Biochem. Biophys. Res. Commun.*, **39**, 149–155.

Kaplan J.C., Kushner S.R. and Grossman L. (1969). Enzymatic repair of DNA. I. Purification of two enzymes involved in the excision of thymine dimers from UV-irradiated DNA. *Proc. Nat. Acad. Sci.*, **63**, 144–151.

Kaplan J.C., Kushner S.R. and Grossman L. (1971). Enzymatic repair of DNA. III. Properties of the UV endonuclease and UV exonuclease. *Biochem.*, **10**, 3315–3325.

Kaplan S. (1971). Lysine suppressor in E.coli. *J. Bacteriol.*, **105**, 984–987.

Karlstrom O. and Gorini L. (1969). A unitary account of the repression mechanism of arginine biosynthesis in E.coli. II. Application to the physiological evidence. *J. Mol. Biol.*, **39**, 89–94.

Karu A.E. and Linn S. (1972). Uncoupling of recBC ATPase from DNAase by DNA cross linked with psoralen. *Proc. Nat. Acad. Sci.*, **69**, 2855–2859.

Kellenberger E. (1960). The physical state of the bacterial nucleus. *Symp. Soc. Gen. Mic.*, **10**, 39–66.

Keller W. (1972). RNA primed synthesis in vitro. *Proc. Nat. Acad. Sci.*, **69**, 1560–1564.

Kelley W.S. and Whitfield H.J. jr. (1971). Purification of an altered DNA polymerase from an E.coli strain with a *pol* mutation. *Nature*, **230**, 33–36.

Kelly R.B., Atkinson M.R., Huberman J.A. and Kornberg A. (1969). Excision of thymine dimers and other mismatched sequences by DNA polymerase of E.coli *Nature*, **224**, 495–501.

Kelly R.B., Cozarelli N.R., Deutscher M.P., Lehman I.R. and Kornberg A. (1970). Enzymatic synthesis of DNA. XXXII. Replication of duplex DNA by polymerase at a single strand break. *J. Biol. Chem.*, **245**, 39–45.

Kelly T.J. and Smith H.O. (1970). A restriction enzyme from Hemophilus influenzae. II. Base sequence of the recognition site. *J. Mol. Biol.*, **51**, 393–410.

Kennedy E.P. (1970). The lactose permease system of E.coli. In Beckwith J.R. and Zipser D. (Eds.), *The lactose operon*, Cold Spring Harbor Laboratory, New York. pp 49–92.

Kerwar S.S., Spears C. and Weissbach H. (1970). Studies on the initiation of protein synthesis in animal tissues. *Biochem. Biophys. Res. Commun.*, **41**, 78–84.

Khorana H.G., Bucchi H., Ghosh H., Gupta N., Jacob T.M., Kossel H., Morgan R., Narang S.A., Ohtsuka E. and Wells R.D. (1966). *Cold Spring Harbor Symp. Quant. Biol.*, **31**, 39–49.

Kiho Y. and Rich A. (1966). A polycistronic messenger RNA associated with *β*-galactosidase induction. *Proc. Nat. Acad. Sci.*, **54**, 1751–1758.

Kim S.H. and Rich A. (1969). Crystalline tRNA: the three dimensional Patterson function at 12Å resolution. *Science*, **166**, 1621–1624.

Kimura F., Harada F. and Nishimura S. (1971). Primary sequence of $tRNA_1^{val}$ from E.coli B. II. Isolation of large fragments by limited digestion with RNAases and overlapping of the fragments to deduce the primary sequence. *Biochem.*, **10**, 3277–3284.

King J.L. and Jukes T.H. (1969). Non Darwinian evolution. *Science*, **164**, 788–797.

Kingdon H.S., Webster L. and Davie E. (1958). Enzymatic formation of adenyl isolation and identification. *Proc. Nat. Acad. Sci.*, **44**, 757–765.

Kitani Y. and Olive L.S. (1970). Alteration of gene conversion patterns in Sordaria fimicola by supplementation with DNA bases, *Proc. Nat. Acad. Sci.*, **66**, 1290–1297.

Kitani Y., Olive L.S. and El-Ani A.S. (1962). Genetics of Sordaria fimicola. V. Aberrant segregation at the *g* locus. *Amer. J. Bot.*, **49**, 697–706.

Kleijer W.J., Lohman P.H.M., Mullder M.P. and Bootsma D. (1970). Repair of X-ray damage in DNA of activated cells from patients having Xeroderma pigmentosum. *Mutat. Res.*, **9**, 517–523.

Klein H.A. and Capecchi M.R. (1971). Polypeptide chain termination. Purification of the release factors R1 and R2 from E.coli. *J. Biol. Chem.*, **246**, 1055–1061.

Klem E.B., Hsu W.T. and Weiss S.B. (1970). The selective inhibition of protein initiation by T4 phage-induced factors. *Proc. Nat. Acad. Sci.*, **67**, 696–701.

Knippers R. (1970). DNA polymerase II. *Nature*, **228**, 1050–1055.

Knippers R. and Stratling W. (1970). The DNA replicating capacity of isolated E.coli cell wall membrane complexes. *Nature*, **226**, 713–717.

Kohler R.E., Ron E.Z. and Davis B.D. (1968). Significance of the free 70S ribosomes in E.coli extracts. *J. Mol. Biol.*, **36**, 71–82.

Konrad E.B., Modrich P. and Lehman I.R. (1973). Genetics and enzymatic characterization of conditional lethal mutant of E.coli with a tempetature sensitive DNA ligase. *J. Mol. Biol.*, **77**, 519–530.

Kornberg A. (1969). Active center of DNA polymerase. *Science*, **163**, 1410–1418.

Kornberg T. and Gefter M.L. (1970). DNA synthesis in cell free extracts of a DNA polymerase defective mutant. *Biochem. Biophys. Res. Commun.*, **40**, 1348–1355.

Kornberg T. and Gefter M.L. (1971). Purification and DNA synthesis in cell free extracts: properties of DNA polymerase II. *Proc. Nat. Acad. Sci.*, **68**, 761–764.

Kornberg T. and Gefter M.L. (1972). DNA synthesis in cell free extracts. IV. Purification and catalytic properties of DNA polymerase III. *J. Biol. Chem.*, **247**, 5369–5375.

Kosakowski H.M. and Bock A. (1971). Substrate complexes of phe-tRNA synthetase from E.coli. *Europ. J. Biochem.*, **24**, 190–200.

Kossel H., Morgan A.R. and Khorana H.G. (1967). Studies on polynucleotides. LXXIII. Synthesis in vitro of polypeptides containing repeating tetrapeptide sequences dependent upon DNA like polymers containing repeating tetranucleotide sequences; direction of reading of messenger RNA. *J. Mol. Biol.*, **26**, 449–475.

Kourilsky P., Bourguignon M.F., Bouquet M. and Gros F. (1970). Early transcription controls after induction of prophage λ. *Cold Spring Harbor Symp. Quant. Biol.*, **35**, 305–315.

Kourilsky P., Bourguignon M.F. and Gross F. (1971). Kinetics of viral transcription after induction of prophage. In Hershey A.D. (Ed.), *The bacteriophage lambda*, Cold Spring Harbor Laboratory, New York, pp 647–666.

Kovach J.S., Phang J.M., Blasi F., Barton R.W., Ballesteros-Olmo A. and Goldberger R.F. (1970). Interaction between his tRNA and the first enzyme for histidine biosynthesis. *J. Bacteriol.*, **104**, 787–792.

Kovach J.S., Phang J.M., Ference M. and Goldberger R.F. (1969). Studies on repression of the histidine operon. II. The role of the first enzyme in control of the histidine system. *Proc. Nat. Acad. Sci.*, **63**, 481–488.

Kozak M. and Nathans D. (1972). Translation of the genome of an RNA phage. *Bacteriol. Rev.*, **36**, 109–134.

Kozinski A.W. (1968). Molecular recombination in the ligase negative T4 amber mutant. *Cold Spring Harbor Symp. Quant. Biol.*, **33**, 375–392.

Kozinski A.W. and Kozinski P.B. (1969). Covalent repair of molecular recombinants in the ligase negative amber mutant of T4 phage. *J. Virol.*, **3**, 85–88.

Kozinski A.W., Kozinski P.B. and James R. (1967). Molecular recombination in T4 phage DNA. I. Tertiary structure of an early replicative and recombining DNA. *J. Virol.*, **1**, 758–770.

Krakow J.S. and Von der Helm K. (1970). Azobacter RNA polymerase transitions and the release of sigma. *Cold Spring Harbor Symp. Quant. Biol.*, **35**, 73–84.

Krell K., Gottesman M.E., Parks J.S. and Eisenberg M.A. (1972). Escape synthesis of the biotin operon in induced λb2 lysogens. *J. Mol. Biol.*, **68**, 69–82.

Kroon A.M., Agsteribbe E. and De Vries H. (1972). Protein synthesis in mitochondria and chloroplasts. In Bosch L. (Ed.), *The Mechanism of protein synthesis and its regulation*, North Holland, Amsterdam, pp 539–582.

Kruszweska A. and Gajewski W. (1967). Recombination within the Y locus of A, immersus. *Genet. Res.*, **9**, 159–177.

Kryzek R. and Rogers P. (1972). Arginine control of transcription of *argECBH* mRNA in E.coli. *J. Bacteriol.*, **110**, 945–954.

Kuempel P.L. and Voemett H.G.E. (1970). A possible function of DNA polymerase in chromosome replication. *Biochem. Biophys. Res. Commun.*, **41**, 973–980.

Kuhnlein V. and Arber W. (1972). Host specificity of DNA produced by E.coli. XV. The role of nucleotide methylation in in vitro B-specific modification. *J. Mol. Biol.*, **63**, 9–20.

Kumar S., Calef E. and Szybalski W. (1970). Regulation of the transcription of E.coli phage λ by its early genes *N* and *tof*. *Cold Spring Harbor Symp., Quant. Biol.*, **35**, 331–340.

Kumar S. and Szybalski W. (1969). Orientation of transcription of the *lac* operon and its repressor gene in E.coli. *J. Mol. Biol.*, **40** 145–151.

Kurland C.G., Nomura M. and Watson J.D. (1962). The physical properties of the chloromycetin particles. *J. Mol. Biol.*, **4**, 388–394.

Kurland C.G., Voynow P., Hardy S.J.S., Randall L. and Lutter L. (1969). Physical and functional heterogeneity of E.coli polyribosomes. *Cold Spring Harbor Symp. Quant. Biol.*, **34**, 17–24.

Kushner S.R., Kaplan J.C., Ono H. and Grossman L. (1971a). Enzymatic repair of DNA. IV. Mechanism of photoproduct excision. *Biochem.*, **10**, 3325–3335.

Kushner S.R., Nagaishi H. and Clark A.J. (1972). Indirect suppression of *recB* or *recC* mutations by exonuclease deficiency. *Proc. Nat. Acad. Sci.*, **69**, 1366–1370.

Kushner S.R., Nagaishi H., Templin A. and Clark A.J. (1971b). Genetic recombination in E.coli: the role of exonuclease I. *Proc. Nat. Acad. Sci.*, **68**, 824–827.

Kuwano M., Apirion D. and Schlessinger D. (1970). RNAase V of E.coli. V. Requirement for ribosome translocation but not for polypeptide formation. *J. Mol. Biol.*, **51**, 453–458.

Kuwano M., Schlessinger D. and Apirion D. (1970a). Ribonuclease V of E.coli requires ribosomes and is inhibited by drugs. *Nature*, **226**, 514–517.

Kuwano M., Schlessinger D. and Apirion D. (1970b). RNAase V of E.coli. IV. Exonucleolytic cleavage in the 5′ to 3′ direction with production of 5′ nucleotides. *J. Mol. Biol.*, **51**, 75–82.

Kuwano M., Schlessinger D. and Morse D.E. (1971). Loss of dispensable endonuclease activity in relief of polarity by *suA*. *Nature New Biol.*, **231**, 214–217.

Lacey J.C.jr. and Pruitt K.M. (1969). Origin of the genetic code. *Nature*, **223**, 799–804.

Lagerkvist U., Rymo L. and Waldenstrom J. (1966). Structure and function of tRNA. II. Enzyme-substrate complexes with val-tRNA synthetases from yeast. *J. Biol. Chem.*, **241**, 5391–5400.

Lai C.J., Wiesblum B., Fahnestock S.R. and Nomura M. (1973). Alteration of 23S rRNA and erythromycin-induced resistance to lincomycin and spiramycin in S. aureus. *J. Mol. Biol.*, **74**, 67–72.

Laiken S.L., Gross C.A. and Von Hippel P.H. (1972). Equilibrium and kinetic studies of E.coli *lac* repressor-inducer interactions. *J. Mol. Biol.*, **66**, 143–156.

Lake J.A. and Beeman W.W. (1968). On the conformation of yeast tRNA. *J. Mol. Biol.*, **31**, 115–125.

Lark C. (1968a). Studies on the in vivo methylation of DNA in E.coli 15T$^-$. *J. Mol. Biol.*, **31**, 389–400.

Lark C. (1968b). Effect of methionine analogues, ethionine and norleucine on DNA synthesis in E.coli 15T$^-$. *J. Mol. Biol.*, **31**, 401–414.

Lark C. and Arber W. (1970). Host specificity of DNA produced by E.coli. XIII. Breakdown of cellular DNA upon growth in ethionine of strains with r^+_{15}, r^+_{P1} or r^+_{N3} restriction phenotypes. *J. Mol. Biol.*, **52**, 337–348.

Lark K.G. (1966). Regulation of chromosome replication and segregation in bacteria. *Bacteriol. Rev.*, **30**, 1–32.

Lark K.G. (1972a). Evidence for the direct involvement of RNA in the initiation of DNA replication in E.coli 15T$^-$. *J. Mol. Biol.*, **64**, 47–60.
Lark K.G. (1972b). Genetic control over the initiation of the synthesis of the short deoxynucleotide chains in E.coli. *Nature New Biol.*, **240**, 237–240.
Lark K.G., Consigli R. and Toliver A. (1971). DNA replication in Chinese hamster cells: evidence for a single replication fork per replicon. *J. Mol. Biol.*, **58**, 873–876.
Lark K.G. and Renger H. (1969). Initiation of DNA replication in E.coli 15T$^-$: chronological dissection of three physiological processes required for initiation. *J. Mol. Biol.*, **42**, 221–236.
Lark K.G., Repko T. and Hoffman E.J. (1963). The effect of amino acid deprivation on subsequent DNA replication. *Biochim. Biophys. Acta*, **76**, 9–24.
Lautenberger J.H. and Linn S. (1972). The DNA modification and restriction enzymes of E.coli B. I. Purification, subunit structure and catalytic properties of the modification methylase. *J. Biol. Chem.*, **247**, 6176–6183.
Lazzarini R.A., Cashel M. and Gallant J. (1971). On the regulation of GTP levels in stringent and relaxed strains of E.coli. *J. Biol. Chem.*, **246**, 4381–4385.
Lazzarini R.A. and Dahlberg A.E. (1971). The control of RNA synthesis during amino acid deprivation in E.coli. *J. Biol. Chem.*, **246**, 420–429.
Leader D.P. and Wool I.G. (1972). Partial purification and characterization of an initiation factor from rat liver which promotes the binding of phe-tRNA to 40S ribosomal subunits. *Biochim. Biophys. Acta*, **262**, 360–370.
Leblon G. and Rossignol J.L. (1973). Mechanism of gene conversion in A. immersus. III. The interaction of heteroalleles in the conversion process. *Molec. Gen. Genet.*, **122**, 165–182.
Leboy P.S., Cox E.C. and Flaks J.G. (1964). The chromosomal site specifying a ribosomal protein in E.coli. *Proc. Nat. Acad. Sci.*, **52**, 1367–1374.
Leder P. and Nau M.N. (1967). Initiation of protein synthesis. III. Factor-GTP-codon dependent binding of fmet-tRNA to ribosomes. *Proc. Nat. Acad. Sci.*, **58**, 774–781.
Leder P., Skogerson L.E. and Roufa D.J. (1969). Translocation of mRNA codons. II. Properties of an anti-translocase antibody. *Proc. Nat. Acad. Sci.*, **62**, 928–933.
Lee-Huang S. and Ochoa S. (1971). Messenger discriminating species of initiation factor f3. *Nature New Biol.*, **234**, 236–239.
Lee-Huang S. and Ochoa S. (1972). Specific inhibitors of MS2 and T4 RNA translation in E.coli. *Biochem. Biophys. Res. Commun.*, **48**, 371–376.
Legrain C., Halleux P., Stalon V. and Glasndorff N. (1972). The dual genetic control of ornthinine carbamyl transferase in E.coli. A case of bacterial hybrid enzymes. *Europ. J. Biochem.*, **27**, 93–102.
Lehman A.R. (1972a). Post replication repair of DNA in ultraviolet irradiated mammalian cells. *J. Mol. Biol.*, **66**, 319–338.
Lehman A.R. (1972b). Post replication repair of DNA in ultraviolet irradiated mammalian cells. No gaps in DNA synthesized late after ultraviolet irradiation. *Europ. J. Biochem.*, **31**, 438–445.
Leive L. and Kollin V. (1967). Synthesis, utilization and degradation of lactose operon mRNA in E.coli. *J. Mol. Biol.*, **24**, 247–259.
Lelong J.C., Cousin M.A., Gros D., Grunberg-Manago M. and Gros F. (1971). Streptomycin induced release of fmet-tRNA from the ribosomal initiation complex. *Biochem. Biophys. Res. Commun.*, **42**, 530–537.
Lelong J.C., Grunberg-Manago M., Dondon J., Gros D. and Gros J. (1970). Interaction between guanosine derivatives and factors involved in the initiation of protein synthesis. *Nature*, **226**, 505–511.
Lengyel P. and Soll D. (1969). Mechanism of protein synthesis. *Bacteriol. Rev.*, **33**, 264–301.

Levin D.H., Kyner D. and Acs G. (1972). Formation of a mammalian initiation complex with reovirus mRNA, methionyl-tRNA$_f$ and ribosomal subunits. *Proc. Nat. Acad Sci.*, **69**, 1234–1238.
Levin D.H., Kyner D. and Acs G. (1973). Protein initiation in eucaryotes: formation and function of a ternary complex composed of a partially purified ribosomal factor, met-tRNA$_f$ and GTP. *Proc. Nat. Acad. Sci.*, **79**, 41–45.
Levitt M. (1969). Detailed molecular model for tRNA. *Nature*, **224**, 759–763.
Levy J. and Biltonen R. (1972). Thermodynamic model for the thermal unfolding of yeast specific tRNA. *Biochem.*, **11**, 4146–4152.
Lewin B. (1970). *The Molecular Basis of Gene Expression*, John Wiley and Sons, London.
Lewin S. (1967). Some aspects of hydration and stability of the native state of DNA. *J. Theor. Biol.*, **17**, 181–212.
Lewin S. (1973). *Water displacement and its control of biochemical reactions.* Academic Press, London.
Lewis E.B. (1945). The relation of repeats to position effects in D. melanogaster. *Genetics*, **30**, 137–166.
Lewis J.A. and Ames B.N. (1972). Histidine regulation in S. typhimurium. XI. The percentage of tRNA^{his} charged in vivo and its relation to the repression of the histidine operon. *J. Mol. Biol.*, **66**, 131–142.
Lewis J.B. and Doty P. (1970). Derivation of the secondary structure of 5S RNA from its binding of complementary oligonucleotides *Nature*, **225**, 510–512.
Lieb M. (1969). Allosteric properties of the λ repressor. *J. Mol. Biol.*, **39**, 379–382.
Lin S.Y. and Riggs A.D. (1971). *Lac* repressor binding to operator analogues: comparison of polydAT, polydBRU and polydAU. *Biochem. Biophys. Res. Commun.*, **45**, 1542–1547.
Lin S.Y. and Riggs A.D. (1972a). *Lac* operator analogues: bromodeoxyuridine substitution in the *lac* operator affects the rate of dissociation of the *lac* repressor. *Proc. Nat. Acad. Sci.*, **69**, 2574–2576.
Lin S.Y. and Riggs A.D. (1972b). *Lac* repressor binding to non operator DNA: detailed studies and a comparison of equilibrium and rate competition methods. *J. Mol. Biol.*, **72**, 671–690.
Lindahl G., Hirota Y. and Jacob F. (1971). On the process of cellular division in E.coli: replication of the bacterial chromosome under control of prophage P2. *Proc. Nat. Acad. Sci.*, **69**, 2407–2411.
Lindahl T. (1971). Excision of pryimidine dimers from UV irradiated DNA by exonuclease from mammalian cells. *Europ. J. Biochem.*, **18**, 407–414.
Lindegren C.C. (1953). Gene Conversion in Saccharomyces. *J. Genet.*, **51**, 625–637.
Ling V. (1972). Pyrimidine sequences from the DNA of phages fd, fl and øX 174. *Proc. Nat. Acad. Sci.*, **69**, 742–746.
Linn S. and Arber W. (1968). Host specificity of DNA produced by E.coli. X. In vitro restriction of phage fd replicative form. *Proc. Nat. Acad. Sci.*, **59**, 1300–1306.
Linn T.G., Greenleaf A.L., Shorenstein R.G. and Losick R. (1973). Loss of the sigma activity of RNA polymerase of B. subtilis during sporulation. *Proc. Nat. Acad. Sci.*, **70**, 1865–1869.
Lipmann F. (1963). Messenger RNA. *Progress Nucleic Acid Res.*, **1**, 135–163.
Lipmann F. (1969). Polypeptide chain elongation in protein biosynthesis. *Science*, **164**, 1024–1031.
Lissouba P. and Rizet G. (1960). Sur l'existence d'une genetique polarisee ne subissant que des echanges non reciproques. *Comptes Rendus Acad. Sci.* (*Paris*), **250**, 3408–3410.

Littlefield J.W., Keller E.B., Gross J. and Zamecnik P.C. (1955). Studies on cytoplasmic ribonucleoprotein particles from the liver of the rat. *J. Biol. Chem.*, **217**, 111–123.
Lockwood A.H., Chakraborty P.R. and Maitra U. (1971). A complex between initiation factor IF-2, GTP and fmet-tRNA$_f$: an intermediate in initiation complex formation. *Proc. Nat. Acad. Sci.*, **68**, 3122–3126.
Lockwood A.H., Hattman S. and Maitra U. (1969). The nature of T factor-guanine nucleotide complexes. *Cold Spring Harbor Symp. Quant. Biol.*, **34**, 433–436.
Lockwood A.H., Sarkar P. and Maitra U. (1972). Release of polypeptide chain initiation factor IF2 during initiation complex formation. *Proc. Nat. Acad. Sci.*, **69**, 3602–3605.
Lodish H.F. (1968). Phage f2 RNA: control of translation and gene order. *Nature*, **220**, 345–350.
Lodish H.F. (1970). Specificity in bacterial protein synthesis: role of initiation factors and ribosomal subunits. *Nature*, **226**, 705–707.
Lodish H.F. and Robertson H.D. (1969). Cell free synthesis of phage f2 maturation protein. *J. Mol. Biol.*, **45**, 9–22.
Loftfield R.B. (1963). The frequency of errors in protein biosynthesis. *Biochem. J.*, **89**, 82–92.
Loftfield R.B. (1972). The mechanism of aminoacylation of tRNA. *Progress Nucleic Acid Res. Mol. Biol.*, **12**, 87–128.
Loftfield R.B. and Vanderjagt D. (1972). The frequency of errors in protein biosynthesis. *Biochem. J.*, **128**, 1353–1356.
Lowry C.V. and Dahlberg J.E. (1971). Structural differences between the 16S ribosomal RNA of E.coli and its precursor. *Nature New Biol.*, **232**, 52–54.
Lucas-Lenard J. and Haenni A.L. (1968). Requirement of GTP for ribosomal binding of aminoacyl-tRNA. *Proc. Nat. Acad. Sci.*, **59**, 554–561.
Lucas-Lenard J. and Haenni A.L. (1969). Release of transfer RNA during peptide chain elongation. *Proc. Nat. Acad. Sci.*, **63**, 93–97.
Lucas-Lenard J. and Lipmann F. (1967). Initiation of polyphenylalanine synthesis by N-acetylphenylalanine-tRNA. *Proc. Nat. Acad. Sci.*, **57**, 1050–1057.
Lucas-Lenard J., Tao P. and Haenni A.L. (1969). Further studies on bacterial polypeptide elongation. *Cold Spring Harbor Symp. Quant. Biol.*, **34**, 455–462.
Lund E. and Kjelgaard N.O. (1972). Metabolism of ppGpp in E.coli. *Europ. J. Biochem.*, **28**, 316–326.
Lutter L.C., Zeichardt H., Kurland C.G. and Stoffler G. (1972). Ribosomal protein neighbourhoods. I. S18 and S21 as well as S5 and S8 are neighbours. *Molec. Gen. Genet.*, **119**, 357–366.
Luzzati D. (1970). Regulation of λ exonuclease synthesis: role of the *N* gene product and λ repressor. *J. Mol. Biol.*, **49**, 515–519.
Luzzato L., Apirion D. and Schlessinger D. (1968b). Mechanism of action of streptomycin in E.coli: interruption of the ribosome cycle at the initiation of protein synthesis. *Proc. Nat. Acad. Sci.*, **60**, 873–881.
Luzzato L., Apirion D. and Schlessinger D. (1969). Polyribosome depletion and blockage of the ribosome cycle by streptomycin in E.coli. *J. Mol. Biol.*, **43**, 315–336.
Luzzato L., Schlessinger D. and Apirion D. (1968a). E.coli: high resistance or dependence on streptomycin produced by the same allele. *Science*, **161**, 478–479.
Maaløe O. and Hanawalt P.C. (1961). Thymine deficiency and the normal DNA replication cycle. *J. Mol. Biol.*, **3**, 144–155.
Maas W.K. and Clark A.J. (1964). Studies on the mechanism of repression of arginine biosynthesis in E.coli. II. Dominance of repressibility in diploids. *J. Mol. Biol.*, **8**, 365–370.

Maas W.K., Maas R., Wiame J.R. and Glansdorff N. (1964). Studies on the mechanism of repression of arginine biosynthesis in E.coli. I. Dominance of repressibility in zygotes. *J. Mol. Biol.*, **8**, 359–364.

Mackie G. and Wilson D.B. (1972). Regulation of the *gal* operon of E.coli by the *capR* gene. *J. Biol. Chem.*, **247**, 2973–2978.

Maden B.E.H., Salim M. and Summers D.F. (1972). Maturation pathway for ribosomal RNA in the Hela cell nucleus. *Nature New Biol.*, **237**, 5–9.

Madison J.T., Everett G.A. and Kung H.K. (1966). The nucleotide sequence of yeast tyrosine transfer RNA. *Cold Spring Harbor Symp. Quant, Biol.*, **31**, 409–418.

Magnusson G., Pigiet V., Winnacker E.L., Abrams R. and Reichard P. (1973). RNA-linked short DNA fragments during polyoma replication. *Proc. Nat. Acad. Sci.*, **70**, 412–415.

Mahler I., Kushner S.R. and Grossman L. (1971). In vivo role of the UV endonuclease from M. luteus in the repair of DNA. *Nature New Biol.*, **234**, 47–50.

Maitra U., Lockwood A.H., Dubnoff J.S. and Guha A. (1970). Termination, release and reinitiation of RNA chains from DNA templates by E.coli RNA polymerase. *Cold Spring Harbor Symp. Quant. Biol.*, **35**, 143–156.

Maitra U., Nakata Y. and Hurwitz J. (1967). The role of DNA in RNA synthesis. XIV. A study of the initiation of RNA synthesis. *J. Biol. Chem.*, **242**, 4908–4918.

Malkin L.I. and Rich A. (1967). Partial resistance of nascent polypeptide chains to proteolytic digestion due to ribosome shielding. *J. Mol. Biol.*, **26**, 329–346.

Mamelak L. and Boyer H.W. (1970). Genetic control of the secondary modification of DNA in E.coli. *J. Bacteriol.*, **104**, 57–62.

Mangiarotti G., Apirion D., Schlessinger D. and Silengo L. (1968). Biosynthetic precursors of 30S and 50S ribosomal particles in E.coli. *Biochem.*, **7**, 456–471.

Mangiarotti G. and Schlessinger D. (1966). Polyribosome metabolism in E.coli. I. Extraction of polyribosomes and ribosomal subunits from fragile growing E.coli. *J. Mol. Biol.*, **20**, 123–143.

Mangiarotti G. and Schlessinger D. (1967). Polyribosome metabolism in E.coli. II. Formation and lifetime of mRNA molecules, ribosome subunit couples and polyribosomes. *J. Mol. Biol.*, **29**, 355–418.

Maniatis T. and Ptashne M. (1973). Multiple repressor binding at operators in phage λ. *Proc. Nat. Acad. Sci.*, **70**, 1531–1535.

Marcker K. A., Clark B.F.C. and Anderson J.S. (1966). N-formyl-methionyl tRNA and its relation to protein synthesis. *Cold Spring Harbor Symp. Quant. Biol.*, **31**, 279–286.

Marcus A., Weeks D.P., Leis J.P. and Keller E.B. (1970). Protein chain initiation by met-tRNA in wheat embryos. *Proc. Nat. Acad. Sci.*, **67**, 1681–1687.

Marmur J., Rownd R. and Schildkraut C.L. (1963). Denaturation and renaturation of DNA. *Progress Nucleic Acid Res.*, **1**, 232–300.

Marsh R.C. and Parmeggiani A. (1973). Requirement of protein S5 and S9 from 30S subunits for the ribosome dependent GTPase activity of elongation factor G. *Proc. Nat. Acad. Sci.*, **70**, 151–155.

Marshall R. and Nirenberg M. (1969). RNA codons recognised by tRNA from amphibian embryos and adults. *Develop. Biol.*, **19**, 1–11.

Martin R.G., Whitfield H.J.jr., Berkowitz D.B. and Voll M.J. (1966a). A molecular model of the phenomenon of polarity. *Cold Spring Harbor Symp. Quant. Biol.*, **31**, 215–220.

Martin R.G., Silbert D.F., Smith D.W.E. and Whitfield H.J.jr. (1966b). Polarity in the histidine operon. *J. Mol. Biol.*, **21** 357–369.

Maruta H., Tsuchiya Y. and Mizuno D. (1971). In vitro reassembly of functionally active 50S ribosomal particles from ribosomal proteins and RNA of E.coli. *J. Mol. Biol.*, **61**, 123–134.

Masters M. and Broda P. (1971). Evidence for the bidirectional replication of the E.coli chromosome. *Nature New Biol.*, **232**, 137–140.

Matsubara M., Takata R. and Osawa S. (1972). Chromosomal loci for 16S rRNA in E.coli. *Molec. Gen. Genet.*, **117**, 311–318.

Matsuura S., Tashiro Y., Osawa S. and Otaka E. (1970). Electron microscopic studies on the biosynthesis of the 50S ribosomal subunit in E.coli. *J. Mol. Biol.*, **47**, 383–391.

Mazumder R., Chae Y.B. and Ochoa S. (1969). Polypeptide chain initiation in E.coli: sulfhydryl groups and the function of the initiation factor f2. *Proc. Nat. Acad. Sci.*, **63**, 98–102.

McClain W.H. and Champe S.P. (1970). Genetic alterations of the *rIIB* cistron polypeptides of phage T4. *Genetics*, **66**, 11–21.

McKeehan W. (1972). The ribosomal subunit requirements for GTP hydrolysis by reticulocyte elongation factors EF-1 and EF-2. *Biochem. Biophys. Res. Commun.*, **48**, 1117–1123.

Meistrich M.L. and Drake J.W. (1972). Mutagenic effects of thymine dimers in phage T4. *J. Mol. Biol.*, **66**, 107–114.

Mertz J.E. and Davis R.W. (1972). Cleavage of DNA by restriction endonuclease generates cohesive ends. *Proc. Nat. Acad. Sci.*, **69**, 3370–3374.

Meselson M. (1964). On the mechanism of genetic recombination between DNA molecules. *J. Mol. Biol.*, **9**, 734–745.

Meselson M. (1972). Formation of hybrid DNA by rotary diffusion during genetic recombination. *J. Mol. Biol.*, **71**, 795–798.

Meselson M., Nomura M., Brenner S., Davern C. and Schlessinger D. (1964). Conservation of ribosomes during bacterial growth. *J. Mol. Biol.*, **9**, 696–711.

Meselson M. and Stahl F.W. (1958) The replication of DNA in E.coli. *Proc. Nat. Acad. Sci.* **44**, 671–682.

Meselson M. and Weigle J.J. (1961). Chromosome breakage accompanying genetic recombination in bacteriophage. *Proc. Nat. Acad. Sci.*, **47**, 857–868.

Meselson M. and Yuan R. (1968). DNA restriction enzyme from E.coli. *Nature*. **217**, 1110–1114.

Meselson M., Yuan R., Heywood J. (1972). Restriction and modification of DNA. *Ann. Rev. Biochem.*, **41**, 447–466.

Messer W. (1972). Initiation of DNA replication in E.coli B/r: chronology of events and transcriptional control of initiation. *J. Bacteriol.*, **112**, 7–12.

Michels C.A. and Reznikoff W.S. (1971). The gradient of polarity of *z* gene nonsense mutations in *trp-lac* fusion strain of E.coli. *J. Mol. Biol.*, **55**, 119–122.

Michels C.A. and Zipser D. (1969). Mapping of polypeptide reinitiation sites within the β-galactosidase structural gene. *J. Mol. Biol.*, **41**, 341–348.

Midgley J.E.M. (1962). The nucleotide base composition of RNA from several microbial species. *Biochim. Biophys. Acta*, **61**, 513–525.

Midgley J.E.M. and McIlreavy D.J. (1967a). The chemical structure of bacterial ribosomal RNA. II. Growth conditions and polynucleotide distribution in E.coli rRNA. *Biochim. Biophys. Acta.*, **142**, 345–354.

Midgley J.E.M. and McIlreavy D.J. (1967b). The isinicotinyl hydrazones of E.coli rRNAs. *Biochim. Biophys. Acta.*, **145**, 512–514.

Migita L.K. and Doi R.H. (1970). The aminoterminal residues of B. subtilis proteins made in vitro. *J. Biol. Chem.*, **245**, 2005–2010.

Milanesi G., Brody E.N., Grau O. and Geiduschek E.P. (1970). Transcription of the phage T4 template in vitro: separation of delayed early from immediate early transcription. *Proc. Nat. Acad. Sci.*, **66**, 181–188.

Milcarek C. and Weiss B. (1972). Mutants of E.coli with altered DNAases. I. Isolation and characterization of mutants for exonuclease III. *J. Mol. Biol.*, **68**, 303–318.
Miller C.G. and Roth J.R. (1971). Recessive lethal nonsense suppressors in S. typhimurium. *J. Mol. Biol.*, **59**, 63–76.
Miller D.L. (1972). Elongation factors EF Tu and EF G interact at related sites on ribosomes. *Proc. Nat. Acad. Sci.*, **69**, 752–755.
Miller D.L., Hachman J. and Weissbach H. (1971). The reactions of the sulfhydryl groups on the elongation factors Tu and Ts. *Arch. Biochem. Biophys.*, **144**, 115–121.
Miller D.L. and Weissbach H. (1970). Interactions between the elongation factors: the displacement of GDP from the Tu-GDP complex by factor Ts. *Biochem. Biophys. Res. Commun.*, **38**, 1016–1022.
Miller J.H. (1970). Transcription starts and stops in the lactose operon. In Beckwith J.R. and Zipser D.(Eds.), *The Lactose Operon*, Cold Spring Harbor Laboratory, New York. pp 173–188.
Miller O.L., Hamkalo B.A. and Thomas C.A.jr. (1970). Visualization of bacterial genes in action. *Science*, **169**, 392–395.
Miller Z., Varmus H.E., Parks J.S., Perlman R.L. and Pastan I. (1971). Regulation of *gal* messenger RNA synthesis in E.coli by cyclic AMP. *J. Biol. Chem.*, **246**, 2898–2903.
Millette R.L., Trotter C.D., Herrlich P. and Schweiger M. (1970). In vitro synthesis, termination and release of active messenger RNA. *Cold Spring Harbor Symp. Quant. Biol.*, **35**, 135–142.
Mirzabekov A.D. and Griffin B.E. (1972). 5S RNA conformation. Studies of its partial T1-RNAase digestion by gel electrophoresis and two dimensional thin layer chromatography. *J. Mol. Biol.*, **72**, 633–644.
Mitchell M.B. (1955). Aberrant recombination of pyridoxine mutants of Neurospora. *Proc. Nat. Acad. Sci.*, **41**, 215–220.
Miura K. (1962). The nucleotide composition of RNAs of soluble and particulate fractions in several species of bacteria. *Biochim. Biophys. Acta*, **55**, 62–70.
Mizushima S. and Nomura M. (1970). Assembly mapping of 30S ribosomal proteins from E.coli. *Nature*, **226**, 1241–1219.
Model P., Webster R.E. and Zinder N.D. (1969). The UGA codon in vitro: chain termination and suppression. *J. Mol. Biol.*, **43**, 177–190.
Modolell J., Cabrer B., Parmeggiani A. and Vazquez D. (1971). Inhibition by siomycin and thiostrepton of both aminoacyl-tRNA and factor G binding to ribosomes. *Proc. Nat. Acad. Sci.*, **68**, 1796–1800.
Modolell J. and Davis B.D. (1970). Breakdown by streptomycin of initiation complexes formed on ribosomes of E.coli. *Proc. Nat. Acad. Sci.*, **67**, 1148–1155.
Modolell J. and Vazquez D. (1973). Inhibition by aminoacyl-tRNA of elongation factor G-dependent binding of guanosine nucleotide to ribosomes. *J. Biol. Chem.*, **248**, 488–493.
Modrich P. and Lehman I.R. (1971). Enzymatic characterization of a mutant of E.coli with an altered DNA ligase. *Proc. Nat. Acad. Sci.*, **68**, 1002–1005.
Moldave K. (1972). Protein synthesis in the cytoplasm of eucaryotic cells. In Bosch L. (Ed.), *The mechanism of protein synthesis and its regulation*, North Holland, Amsterdam. pp 465–486.
Moldave K., Galasinki W., Rao P. and Siller J. (1969). Studies on the peptidyl-tRNA translocase from rat liver. *Cold Spring Harbor Symp. Quant. Biol.*, **34**, 347–356.
Momose H. and Gorini L. (1971). Genetic analysis of streptomycin dependence in E.coli. *Genetics*, **67**, 19–38.

Monk M. and Kinross J. (1972). Conditional lethality of *recA* and *recB* derivatives of a strain of E.coli K12 with a temperature sensitive DNA polymerase I. *J. Bacteriol.*, **109**, 971–978.

Monk M., Peacey M. and Gross J. D. (1971). Repair of damage induced by UV light in DNA polymerase defective E.coli cells. *J. Mol. Biol.*, **58**, 623–630.

Monod J. (1972). *Chance and necessity: an essay on the natural philosophy of molecular biology*. Translated from the French by Austryn Wainhouse. Knopf, New York.

Monod J. and Cohen-Bazire G. (1953). L'effect d'inhibition specifique dans la biosynthese de la tryptophane-desmase chez Aerobacter aerogenes. *Comptes Rendus Acad. Sci. (Paris).*, **236**, 530–532.

Monro R.E., Celma M.L., and Vazquez D. (1969). The peptidyl transferase activity of ribosomes. *Cold Spring Harbor Symp. Quant. Biol.*, **34**, 357–368.

Moore C.H., Farron F., Bohnert D. and Weissman C. (1971). Posisble origin of a minor virus specific protein (A1) in Qβ particles. *Nature New Biol.*, **234**, 204–206.

Moore R.L. and McCarthy B.J. (1967). Comparative study of ribosomal RNA cistrons in Enterobacteria and Myxobacteria. *J. Bacteriol.*, **94**, 1066–1074.

Morgan J. and Brimacombe R. (1972). A series of specific ribonucleoprotein fragments from the 30S subparticle of E.coli ribosomes. *Europ. J. Biochem.*, **29**, 542–552.

Morikawa N. and Imamoto F. (1969). On the degradation of mRNA for the tryptophan operon in E.coli. *Nature*, **223**, 37–40.

Morris D.W. and Kjelgaard N.O. (1968). Evidence for the *non*-coordinate regulation of RNA synthesis in stringent strains of E.coli. *J. Mol. Biol.*, **31**, 145–148.

Morrison S.L., Zipser D. and Goldschmidt R. (1971). Polypeptide products of nonsense mutations. II. Minor fragments produced by nonsense mutations in the *z* gene of the lactose operon of E.coli. *J. Mol. Biol.*, **60**, 485–498.

Morrow J.F. and Berg P. (1972). Cleavage of SV40 DNA at a unique site by a bacterial restriction enzyme. *Proc. Nat. Acad. Sci.*, **69**, 3365–3369.

Morse D.E. (1971). Polarity induced by chloramphenicol and relief by suA. *J. mol. Biol.*, **55**, 113–118.

Morse D.E., Baker R.F. and Yanofsky C. (1968). Translation of the tryptophan mRNA of E.coli. *Proc. Nat. Acad. Sci.*, **60**, 1428–1435.

Morse D.E. and Guertin M. (1972). Amber *suA* mutations which relieve polarity. *J. Mol. Biol.*, **63**, 605–608.

Morse D.E., Mosteller R., Baker R.F., and Yanofsky C. (1969). Direction of in vivo degradation of tyrptophan mRNA—a correction. *Nature*, **223**, 40–43.

Morse D.E. and Primakoff P. (1970). Relief of polarity by *suA*. *Nature*, **226**, 28–31.

Morse D.E. and Yanofsky C. (1968). The internal low efficiency promotor of the tryptophan operon of E.coli. *J. Mol. Biol.*, **38**, 447–452.

Morse D.E. and Yanofsky C. (1969). A transcription inhibiting mutation within a structural gene of the tryptophan operon. *J. Mol. Biol.*, **41**, 317–328.

Moses R.E. and Richardson C.C. (1970a). Replication and repair of DNA in cells of E.coli treated with toluene. *Proc. Nat. Acad. Sci.*, **67**, 674–681.

Moses R.E. and Richardson C.C. (1970b). A new DNA polymerase activity of E.coli. I. Purification and properties of the activity present in E.coli *polAl. Biochem. Biophys. Res. Commun.*, **41**, 1557–1564.

Moses R.E. and Richardson C.C. (1970c). A new DNA polymerase activity of E.coli. II. Properties of the enzyme purified from wild type E.coli and DNA ts mutants. *Biochem. Biophys. Res. Commun.*, **41**, 1565–1571.

Mosteller R.D., Rose J.K. and Yanofsky C. (1970). Transcription initiation and degradation of *trp* mRNA. *Cold Spring Harbor Symp. Quant. Biol.*, **35**, 461–467.

Mosteller R.D. and Yanofsky C. (1970). Transcription of the tryptophan operon in E.coli: rifampicin as an inhibitor of initiation. *J. Mol. Biol.*, **48**, 525–531.

Mowshowitz D.B. (1970). Transfer RNA synthesis in Hela cells. II. Formation of tRNA from a precursor in vitro and formation of pseudovirions. *J. Mol. Biol.*, **50**, 143–151.

Muench K.H. and Safille P.A. (1968). Transfer RNAs in E.coli: multiplicity and variation. *Biochem.*, **7**, 2799–2808.

Mulder C. and Delius H. (1972). Specificity of the break produced by restricting endonuclease RI in SV40 DNA, as revealed by partial denaturation mapping. *Proc. Nat. Acad. Sci.*, **69**, 3215–3219.

Murao K., Tanabe T., Ishii F., Namiki M. and Nishimura S. (1972). Primary sequence of arginine tRNA from E.coli. *Biochem. Biophys. Res. Commun.*, **47**, 1322–1337.

Murray N.E. (1963). Polarized recombination and fine structure within the *me-2* gene of N. crassa. *Genetics*, **48**, 1163–1183.

Murray N.E. (1970). Recombination events that span sites within neighbouring gene loci of Neurospora. *Genet. Res.*, **15**, 109–122.

Murray N.E. and Brammar W.J. (1973). The *trpE* gene of E.coli K contains recognition sequence for the K restriction system. *J. Mol. Biol.*, **77**, 615–624.

Murray N.E., Ritis P.M. and Foster L.A. (1973). DNA targets for the E.coli restriction enzyme analysed genetically in recombinants between phages $\phi 80$ and λ. *Molec. Gen. Genet.*, **120**, 261–282.

Myers G.L. and Salder J.R. (1971). Mutational inversion of control of the lactose operon of E.coli. *J. mol. Biol.*, **58**, 1–28.

Nagata T. and Meselson M. (1968). Periodic replication of DNA in steadily growing E.coli: the localized origin of replication. *Cold Spring Harbor Symp. Quant. Biol.*, **33**, 553–558.

Nakada D. and Magasanik B. (1964). The roles of inducer and catabolic repressor in the synthesis of β-galactosidase by E.coli. *J. Mol. Biol.*, **8**, 105–127.

Nakai S., Matsumoto S. (1967). Two types of radiation sensitive mutants in yeast. *Mutat. Res.*, **4**, 129–136.

Nakanishi S., Adya S., Gottesman M.E. and Pastan I. (1973). In vitro repression of transcription of the *gal* operon by purified *gal* repressor. *Proc. Nat. Acad. Sci.*, **70**, 334–338.

Nanninga N., Garrett R.A., Stoffler G. and Klotz G. (1972). Ribosomal proteins. XXXVIII Electron microscopy of ribosomal protein S4–16S RNA complexes of E.coli. *Molec. Gen. Genet.*, **119**, 175–184.

Naono S. and Tokuyama K. (1970). On the mechanism of λ DNA transcription in vitro. *Cold Spring Harbor Symp. Quant. Biol.*, **35**, 375–382.

Nashimoto H., Held W., Kaltschmidt E. and Nomura M. (1971). Structure and function of bacterial ribosomes. XII. Accumulation of 21S particles by some cold sensitive mutants of E.coli. *J. Mol. Biol.*, **62**, 121–138.

Nashimoto H. and Nomura M. (1970). Structure and function of bacterial ribosomes. XI. Dependence of 50S ribosomal assembly on simultaneous assembly of 30S subunits. *Proc. Nat. Acad. Sci.*, **67**, 1440–1447.

Naughton M.A. and Dintzis H.M. (1962). Sequential biosynthesis of the peptide chains of haemoglobin. *Proc. Nat. Acad. Sci.*, **48**, 1822–1830.

Neidhardt F.C. (1966). Roles of amino acid activating enzymes in cellular physiology. *Bacteriol. Rev.*, **30**, 701–719.

Newman J. and Hanawalt P. (1968). Intermediates in T4 DNA replication in a T4 ligase deficient strain. *Cold Spring Harbor Symp. Quant. Biol.*, **33**, 145–150.

Newton A. (1969). Reinitiation of polypeptide synthesis and polarity in the *lac* operon of E.coli. *J. Mol. Biol.*, **41**, 329–340.

Newton A., Beckwith J.R., Zipser D. and Brenner S. (1965). Nonsense mutations and polarity in the *lac* operon of E.coli. *J. Mol. Biol.*, **14**, 290–295.

Nichols J.L. (1970). Nucleotide sequence from the polypeptide chain termination region of the coat protein cistron in phage R17 RNA. *Nature*, **225**, 147–152.

Nierhaus K.H., Bordasch K. and Homann E. (1973). In vivo assembly of E.coli ribosomal proteins. *J. Mol. Biol.*, **74**, 587–598.

Nierlich D.P. (1972a). Regulation of RNA synthesis in growing bacterial cells. I. Control over the total rate of RNA synthesis. *J. Mol. Biol.*, **72**, 751–764.

Nierlich D.P. (1972b). Regulation of RNA synthesis in growing bacterial cells. II. Control over the composition of the newly made RNA. *J. Mol. Biol.*, **72**, 765–778.

Ninio J., Favre A. and Yaniv M. (1969). Molecular model for tRNA. *Nature*, **223**, 1333–1335.

Nirenberg M., Caskey C.T., Marshall R., Brimacombe R., Kelley D., Doctor B., Hatfield D., Levin J., Rottman F., Pestka S., Wilcox F. and Anderson W.F. (1966). The RNA code and protein synthesis. *Cold Spring Harbor Symp. Quant. Biol.*, **31**, 11–24.

Nirenberg M. and Leder P. (1964). The effect of trinucleotides upon the binding of tRNA to ribosomes. *Science*, **145**, 1399–1400.

Nirenberg M.W. and Mattaei J.H. (1961). The dependence of cell-free protein synthesis in E.coli upon naturally occuring or synthetic polyribonucleotides. *Proc. Nat. Acad. Sci.*, **47**, 1588–1602.

Nishimura S., Weinstein I.B. (1969). Fractionation of rat liver tRNA. Isolation of tyrosine, valine, serine and phenylalanine tRNAs and their coding properties. *Biochem.*, **8**, 832–842.

Nishimura Y., Caro L., Berg C.M. and Hirota Y. (1971). Chromosome replication in E.coli. IV. Control of chromosome replication and cell division by an integrated episome. *J. Mol. Biol.*, **55**, 441–456.

Nishioka Y. and Eisenstark A. (1970). Sequence of genes replicated in S. typhimurium as examined by transduction techniques. *J. Bacteriol.*, **102**, 320–333.

Nishizuka Y. and Lipmann F. (1966). The interrelationship between GTP and amino acid polymerization. *Arch. Biochem. Biophys.*, **116**, 344–351.

Nissley P., Anderson W.B., Gallo M., Perlman R.L. and Pastan I. (1972). The binding of cyclic AMP receptor to DNA. *J. Biol. Chem.*, **247**, 4264–4269.

Nissley S.P., Anderson W.B., Gottesman M., Perlman R.L. and Patsan I. (1971). In vitro transcription of the gal operon requires cyclic AMP and cyclic AMP receptor protein. *J. Biol. Chem.*, **246**, 4671–4678.

Noller H.F. and Chaires J.B. (1972). Functional modification of 16S rRNA by kethoxal. *Proc. Nat. Acad. Sci.*, **69**, 3115–3118.

Nomura M. and Engbaek F. (1972). Expression of ribosomal protein genes as analysed by phage mu induced mutations. *Proc. Nat. Acad. Sci.*, **69**, 1526–1530.

Nomura M. and Erdmann U.A. (1970). Reconstitution of 50S ribosomal subunits from dissociated molecular components. *Nature*, **228**, 744–748.

Nomura M., Lowry C.V. and Guthrie C. (1967). The initiation of protein synthesis: joining of the 50S ribosomal subunit to the initiation complex. *Proc. Nat. Acad. Sci.*, **58**, 1487–1493.

Nomura M., Mizushima S., Ozaki M., Traub P. and Lowry C.V. (1969). Structure and function of ribosomes and their molecular components. *Cold Spring Harbor Symp. Quant. Biol.*, **34**, 49–62.

Nomura M. and Traub P. (1968). Structure and function of E.coli ribosomes. III. Stoichiometry and rate of reconstitution of ribosomes from subribosomal particles and split proteins. *J. Mol. Biol.*, **34**, 609–619.

Nomura M., Traub P. and Bechmann H. (1968). Hybrid 30S ribosomal particles reconstituted from components of different bacterial origins. *Nature*, **219**, 793–799.

Nonomura Y., Blobel. G. and Sabatini G. (1971). Structure of liver ribosomes studied by negative staining. *J. Mol. Biol.*, **60**, 303–324.
Norkin L.C. (1970). Marker specific effects in genetic recombination. *J. Mol. Biol.*, **51**, 633–656.
Norris T.E. and Koch A.L. (1972). Effect of growth rate on the relative rates of synthesis of messenger, ribosomal and tRNA in E.coli. *J. Mol. Biol.*, **64**, 635–650.
Novelli G.D. (1967). Amino acid activation for protein synthesis. *Ann. Rev. Biochem.*, **36**, 449–484.
Nusslein V., Otto B., Bonhoeffer F. and Schaller H. (1971). Function of DNA polymerase III in DNA replication. *Nature New Biol.*, **234**, 285–286.
Ocada Y., Amagaee S. and Tsugita A. (1970). Frameshift mutation in the lysozyme gene of phage T4: demonstration of the insertion of five bases and a summary of in vivo codons and lysozyme activities. *J. Mol. Biol.*, **54**, 219–246.
Oey J.L. and Knippers R. (1972). Properties of the isolated gene 5 protein of phage fd. *J. Mol. Biol.*, **68**, 125–138.
Ogawa H. (1970). Genetic locations of *uvrD* and *pol* genes of E.coli. *Molec. Gen. Genet.*, **108**, 378–381.
Ogawa H., Shimada K. and Tomizawa J. (1968). Studies on repair sensitive mutants of E.coli. I. Mutants defective in the repair synthesis. *Molec. Gen. Genet.*, **101**, 227–244.
Ohashi Z., Saneyoshi M., Harada H., Hara H. and Nishimura S. (1970). Presumed anticodon structure of glutamic acid tRNA from E.coli: a possible location of a 2-thiouridine derivative in the first position of the anticodon. *Biochem. Biophys. Res. Commun.*, **40**, 866–872.
Ohlsson B.M., Strigini P.F. and Beckwith J.R. (1968). Allelic amber and ochre suppressors. *J. Mol. Biol.*, **36**, 209–218.
Ohta T. and Thach R.E. (1968). Binding of formyl-methionyl-tRNA and aminoacyl-tRNA to ribosomes. *Nature*, **219**, 238–242.
Okazaki R., Arisawa M. and Sugino A. (1971). Slow joining of newly replicated DNA chains in DNA polymerase I deficient E.coli mutants. *Proc. Nat. Acad. Sci.*, **68**, 2954–2957.
Okazaki R., Okazaki T., Sakabe K., Sugimoto K., Kainuma R., Sugino R. and Zwatsuki N. (1968). In vivo mechanism of DNA chain growth. *Cold Spring Harbor Symp. Quant. Biol.*, **33**, 129–144.
Okazaki R., Sugimoto K., Okazaki T., Imae Y. and Sugino A. (1970). DNA chain growth: in vivo and in vitro synthesis in a DNA polymerase negative mutant of E.coli. *Nature*, **228**, 223–227.
Okazaki T. and Okazaki R. (1969). Mechanism of DNA chain growth. II. Direction of synthesis of T4 short DNA chains revealed by exonucleolytic degradation. *Proc. Nat. Acad. Sci.*, **64**, 1242–1248.
Olive L.S. (1959). Aberrant tetrads in S. fimicola. *Proc. Nat. Acad. Sci.*, **45**, 727–732.
Olivera B.M. and Bonhoeffer F. (1972). Discontinuous DNA replication in vitro. I. Two distinct size classes of intermediates. *Nature New Biol.*, **240**, 233–235.
Olivera B.M., Hall Z.W., Anraku Y., Chien J.R. and Lehman I.R. (1968a). On the mechanism of the polynucleotide joining reaction. *Cold Spring Harbor Symp. Quant. Biol.*, **33**, 37–36.
Olivera B.M., Hall Z.W. and Lehman I.R. (1968b). Enzymatic joining of polynucleotides. V. A DNA adenylate intermediate in the polynucleotide joining reaction. *Proc. Nat. Acad. Sci.*, **61**, 237–244.
Ono Y. (1968). Peptide chain elongation: discrimination against the initiator tRNA by microbial amino acid polymerization factors. *Nature*, **220**, 1304–1307.

Ono Y., Skoultchi A., Waterson J. and Lengyel P. (1969a). Peptide chain elongation: GTP cleavage catalysed by factors binding aminoacyl-tRNA to the ribosome. *Nature*, **222**, 645–648.

Ono Y., Skoultchi A., Waterson J. and Lengyel P. (1969b). Stoichiometry of aminoacyl-tRNA binding and GTP cleavage during chain elongation and translocation. *Nature*, **223**, 697–701.

Ordal G. W. (1971). Super-virulent mutants and the structure of operator and promotor. In Hershey A.D. (Ed.), *The Bacteriophage Lambda*, Cold Spring Harbor Laboratory, New York, pp. 565–570.

Orgel L.E. (1968). Evolution of the genetic apparatus. *J. Mol. Biol.*, **38**, 381–394.

Osawa S. (1968). Ribosome formation and structure. *Ann. Rev. Biochem.*, **37**, 109–130.

Osawa S., Otaka E., Itoh T. and Fukui T. (1969). Biosynthesis of 50S ribosomal subunit in E.coli. *J. Mol. Biol.*, **40**, 321–352.

Osawa S., Takata R. and Dekio S. (1970). Genetic studies of the ribosomal proteins in E.coli. III. Composition of ribosomal proteins on various strains of E.coli. *Molec. Gen. Genet.*, **107**, 32–38.

O'Sullivan M.A. and Sueoka N. (1972). Membrane attachment of the replication, origins of a multifork (dichotomous) chromosome in B. subtilis. *J. Mol. Biol.*, **69,** 237–248.

Ozaki M., Mizushima S. and Nomura M. (1969). Identification and functional characterization of the protein controlled by the streptomycin resistant locus in E.coli. *Nature*, **222**, 333–339.

Painter R.B. and Cleaver J.E. (1970). Repair replication, unscheduled DNA synthesis and the repair of mammalian DNA. *Radiation Research*, **37**, 451–466.

Painter R.B. and Schaffer A.W. (1971). Variation in the rate of DNA chain growth through the S phase in Hela cells. *J. Mol. Biol.*, **58**, 289–296.

Papas T.S. and Peterkofsky A. (1972). A random sequential mechanism for arg-tRNA synthetase of E.coli. *Biochem.*, **11**, 4602–4608.

Parfait R. and Grosjean H. (1972). Arg-tRNA synthetase from B. stearothermophilus. Purification, properties and mechanism of action. *Europ. J. Biochem.*, **30**, 242–249.

Parks J.S., Gottesman M., Perlman R.L. and Pastan I. (1971a). Regulation of galactokinase synthesis by cyclic AMP in cell free extracts of E.coli. *J. Biol. Chem.*, **246**, 2419–2424.

Parks J.S., Gottesman M., Shimada K., Weisberg R.A., Perlman R.L. and Pastan I. (1971b). Isolation of the *gal* repressor. *Proc. Nat. Acad. Sci.*, **68**, 1981–1895.

Pastan I. and Perlman R.L. (1968). The role of the *lac* promotor locus in the regulation of β-galactosidase synthesis by cyclic AMP. *Proc. Nat. Acad. Sci.*, **61**, 1336–1342.

Pato M.L. and Glaser D.A. (1968). The origin and direction of replication of the chromosome of E.coli B/r. *Proc. Nat. Acad. Sci.*, **60**, 1268–1274.

Pauling C. and Hamm L. (1969). Properties of a temperature sensitive mutant of E.coli. II. DNA replication. *Proc. Nat. Acad. Sci.*, **64**, 1195–1202.

Pearson P., Delius H. and Traut R.R. (1972). Purification and characterization of 50S ribosomal proteins of E.coli. *Europ. J. Biochem.*, **27**, 482–290.

Pedersen F.S., Lund E. and Kjelgaard N.O. (1973). Codon specific, tRNA dependent in vitro synthesis of ppGpp and pppGpp. *Nature New Biol.*, **234**, 13–15.

Pero J. (1971). Deletion mapping of the site of action of the *tof* gene product. In Hershey A.D. (Ed.), *The bacteriophage lambda*, Cold Spring Harbor Laboratory, New York. pp. 599–608.

Person S. and Obsorn M. (1968). The conversion of amber suppressors to ochre suppressors. *Proc. Nat. Acad. Sci.*, **60**, 1030–1038.

Pestka S. (1968). Studies on the formation of tRNA-ribosome complexes. V. On the function of a soluble transfer factor in protein synthesis. *Proc. Nat. Acad. Sci.*, **61**, 726–733.

Pestka S. (1969). Studies on the formation of rRNA-ribosome complexes. VI. Oligopeptide synthesis and translocation on ribosomes in the presence and absence of soluble transfer factors. *J. Biol. Chem.*, **244**, 1533–1539.
Pestka S. (1970). Studies on the formation of tRNA-ribosome complexes. VII. The role of the 3′ hydroxyl terminal end of tRNA for interaction with ribosomes and ribosomal subunits. *J. Biol. Chem.*, **245**, 1497–1503.
Pestka S., Hishizawa T. and Lessard J.L. (1970). Aminoacyl-oligonucleotide binding to ribosomes: characteristics and requirements. *J. Biol. Chem.*, **245**, 6208–6219.
Pestka S., Marshall R. and Nirenberg M. (1965). RNA codewords and protein synthesis. V. Effect of streptomycin on the formation of ribosome-tRNA complexes. *Proc. Nat. Acad. Sci.*, **53**, 639–646.
Petrissant G. (1973). Evidence for the absence of the G–T–ψ–C sequence from two mammalian initiation tRNAs. *Proc. Nat. Acad. Sci.*, **70**, 1046–1049.
Pettijohn D.E., Clarkson K., Kossman C.R. and Stonington O.G. (1970). Synthesis of ribosomal RNA on a protein DNA complex isolated from bacteria: a comparison of ribosomal RNA synthesis in vitro and in vivo. *J. Mol. Biol.*, **52**, 281–300.
Pettijohn D.E. and Hanawalt P.C. (1964). Evidence for repair replication of ultraviolet damaged DNA in bacteria. *J. Mol. Biol.*, **9**, 395–410.
Phillips L.A. and Franklin R.M. (1969). The in vivo distribution of bacterial polysomes, ribosomes and ribosomal subunits. *Cold Spring Harbor Symp. Quant. Biol.*, **34**, 243–254.
Peikarowicz A. and Glover S.W. (1972). Host specificity of DNA in H. influenzae: the two restriction and modification systems in strain *Ra. Molec. Gen. Genet.*, **116**, 11–25.
Pierucci O. (1969). Regulation of cell division in E.coli. *Biophys. J.*, **9**, 90–112.
Pinder J.C. and Gratzer W.B. (1972). Nuclease degradation and the structure of ribosomes. *Europ. J. Biochem.*, **26**, 73–80.
Pirrotta V., Chadwick P. and Ptashne M. (1970). Active forms of the two coliphage repressors. *Nature*, **227**, 41–44.
Pirrotta V. and Ptashne M. (1969). Isolation of the 434 repressor. *Nature*, **222**, 541–544.
Platt T., Weber K., Ganem D. and Miller J.H. (1972). Translation restarts. AUG reinitiation of a *lac* repressor fragment. *Proc. Nat. Acad. Sci.*, **69**, 897–901.
Pongs O. and Reinwald E. (1973). Function of Y in codon-interaction of $tRNA^{phe}$. *Biochem. Biophys. Res. Commun.*, **50**, 357–363.
Portier M.M., Marcaud L., Cohen A. and Gros F. (1972). Mechanism of transcription in the *N* operon of phage λ. *Molec. Gen. Genet.*, **117**, 72–81.
Prashad N. and Hosoda J. (1972). Role of genes 46 and 47 in phage T4 reproduction. II. Formation of gaps in parental DNA of polynucleotide ligase defective mutants. *J. Mol. Biol.*, **70**, 617–636.
Preiss J., Berg P., Ofengand E.J., Bergmann F.H. and Dieckmann M. (1959). The chemical nature of the RNA-amino acid compound formed by amino acid activating enzymes. *Proc. Nat. Acad. Sci.*, **45**, 319–328.
Prescott D.M. and Kuempel P.L. (1972). Bidirectional replication of the chromosome in E.coli. *Proc. Nat. Acad. Sci.*, **69**, 2842–2845.
Pritchard R.H., Barth P.T. and Collins J. (1969). Control of DNA synthesis in bacteria. *Symp. Soc. Gen. Microbiol.*, **19**, 293–298.
Ptashne M. (1967a). Isolation of the λ phage repressor. *Proc. Nat. Acad. Sci.*, **57**, 306–312.
Ptashne M. (1967b). Specific binding of the λ phage repressor to λ DNA. *Nature*, **214**, 232–234.

Ptashne M. (1971). Repressor and its action. In Hershey A.D. (Ed.), *The bacteriophage lambda*, Cold Spring Harbor Laboratory, New York. pp 221–238.

Ptashne M. and Hopkins N. (1968). The operators controlled by the λ phage repressor. *Proc. Nat. Acad. Sci.*, **60**, 1282–1287.

Quinn W.G. and Sueoka N. (1970). Symmetric replication of the B. subtilis chromosome. *Proc. Nat. Acad. Sci.*, **67**, 717–723.

Rabbani E. and Srinivasan P.R. (1973). Role of the translocation factor G in the regulation of RNA synthesis. *J. Bacteriol.*, **113**, 1177–1183.

RajBhandary U.L., Chang S.H., Stuart A., Faulkner R.D., Hoskinson R.M. and Khorana H.G. (1969). Studies on polynucleotides. LXVIII. The primary structure of yeast phenylalanine tRNA. *Proc. Nat. Acad. Sci.*, **57**, 751–758.

Randall-Hazelbauer L.L. and Kurland C.G. (1972). Identification of three 30S proteins contributing to the ribosomal A site. *Molec. Gen. Genet.*, **115**, 234–242.

Reeves R.H., Cantor C.R. and Chambers R.W. (1970). Effect of magnesium ions on the conformation of two highly purified yeast adenine tRNAs. *Biochem.*, **9**, 3993–4002.

Reichardt L. and Kaiser A.D. (1971). Control of λ repressor synthesis. *Proc. Nat. Acad. Sci.*, **68**, 2185–2189.

Reiter H. and Ramareddy G. (1970). Loss of DNA behind the growing point of thymine starved B. subtilis 168. *J. Mol. Biol.*, **50**, 538–548.

Remaut E. and Fiers W. (1972). Studies on phage MS2 RNA. XVI. The termination signal of the A protein cistron. *J. Mol. Biol.*, **71**, 243–262.

Revel M. (1972). Polypeptide chain termination. In Bosch L. (Ed.), *The mechanism of protein synthesis and its regulation*, North Holland, Amsterdam. pp. 87–172.

Reznikoff W.S. and Beckwith J.R. (1969). Genetic evidence that the operator locus is distinct from the *z* gene in the *lac* operon of E.coli. *J. Mol. Biol.*, **43**, 215–218.

Reznikoff W.S., Miller J.H., Scaife J.G. and Beckwith J.R. (1969). A mechanism for repressor action. *J. Mol. Biol.*, **43**, 201–214.

Richardson C.C., Masamune Y., Live T.R., Jacquemin-Sablon A. and Weiss B. (1968). Studies on the joining of DNA by polynucleotide ligase of phage T4. *Cold Spring Harbor Symp. Quant. Biol.*, **33**, 151–164.

Richardson J.P. (1969). RNA polymerase and the control of RNA synthesis. *Prog. Nuc. Acid Res.*, **9**, 75–116.

Richardson J.P. (1970). Reinitiation of RNA chain synthesis in vitro. *Nature*, **225**, 1109–1112.

Richman N. and Bodley J.W. (1972). Ribosomes cannot interact simultaneously with elongation factors EF Tu and EF G. *Proc. Nat. Acad. Sci.*, **69**, 686–689.

Richter D. (1972). Inability of E.coli ribosomes to interact simultaneously with the bacterial elongation factors Tu and G. *Biochem. Biophys. Res. Commun.*, **46**, 1850–1856.

Riddle D.L. and Carbon J. (1973). Frameshift suppression: a nucleotide addition in the anticodon of a glycine tRNA. *Nature New. Biol.*, **242**, 230–234.

Riddle D.L. and Roth J.R. (1970). Suppressors of frameshift mutations in S. typhimurium. *J. Mol. Biol.*, **54**, 131–144.

Riddle D.L. and Roth J.R. (1972a). Frameshift suppressors. II. Genetic mapping and dominance studies. *J. Mol. Biol.*, **66**, 483–495.

Riddle D.L. and Roth J.R. (1972b). Frameshift suppressors. III. Effects of suppressor mutation on tRNAs. *J. Mol. Biol.*, **66**, 495–506.

Riggs A.D., Newby R.F. and Bourgeois S. (1970). *Lac*-repressor-operator interaction. II. Effect of galactosides and other ligands. *J. Mol. Biol.*, **51**, 303–314.

Riggs A.D., Reiness G. and Zubay G. (1971). Purification and DNA-binding properties of the catabiolte gene activator protein. *Proc. Nat. Acad. Sci.*, **68**, 1222–1225.

Ritossa F.M. and Spiegelman S. (1965). Localization of DNA complementary to ribosomal RNA in the nucleolus organiser region of D. melanogaster. *Proc. Nat. Acad. Sci.*, **53**, 737–745.
Riva S., Cascino A. and Geiduschek E.P. (1970a). Coupling of late transcription to viral replication in phage T4 development. *J. Mol. Biol.*, **54**, 85–99.
Riva S., Cascino A. and Geiduschek E.P. (1970b). Uncoupling of late transcription from DNA replication in phage T4 development. *J. Mol. Biol.*, **54**, 103–120.
Roberts J.J., Crathorn A.R. and Brent T.P. (1968). Repair of alkylated DNA in mammalian cells. *Nature*, **218**, 970–972.
Roberts J.W. (1969). Termination factor for RNA synthesis. *Nature*, **224**, 1168–1175.
Roberts J.W. (1970). The rho factor: termination and anti-termination in λ. *Cold Spring Harbor Symp. Quant. Biol.*, **35**, 121–126.
Robertson H.D., Barrell B.G., Weith H.L. and Donelson J.E. (1973). Isolation and sequence analysis of a ribosome protected fragment from phage ϕX174 DNA. *Nature New Biol.*, **241**, 38–40.
Rodriguez R.L., Dalbey M.S. and Davern C.I. (1973). Autoradiographic evidence for bidirectional DNA replication in E.coli. *J. Mol. Biol.*, **74**, 599–604.
Roe B. and Dudock B. (1972). The role of the fourth nucleotide from the 3′ end in the yeast phe-tRNA synthestae recognition site: requirement for adenosine. *Biochem. Biophys. Res. Commun.*, **49**, 399–406.
Rogers P., Krzyzek R., Kaden T.M. and Arfman E. (1971). Effect of arginine anp canavanine on arginine mRNA synthesis. *Biochem. Biophys. Res. Commun.*, **44**, 1220–1226.
Rose J.K., Mosteller R.D. and Yanofsky C. (1970). Tryptophan mRNA elongation rates and steady state levels of tryptophan operon enzymes under various growth conditions. *J. Mol. Biol.*, **51**, 541–550.
Rose J.K. and Yanofsky C. (1972). Metabolic regulation of the tryptophan operon of E.coli: repressor-independent regulation of transcription initiation frequency. *J. Mol. Biol.*, **69**, 103–118.
Rosset R. and Gorini L. (1969). A ribosomal ambiguity mutation. *J. Mol. Biol.*, **39**, 95–112.
Rosset R. and Moner R. (1963). A propos de la presence d'acide ribonucleique de faible poids moleculaire dans les ribosomes d'E.coli. *Biochem. Biophys Acta*, **68**, 653–655.
Roth J.R. (1970). UGA nonsense mutations in S. typhimurium. *J. Bacteriol.*, **102**, 467–475.
Roth J.R. and Ames B.N. (1966). Histidine regulatory mutants in S. typhimurium. II. Histidine regulatory mutants have altered his-tRNA synthetase. *J. Mol. Biol.*, **22**, 325–334.
Roth J.R., Anton D.R. and Hartman P.E. (1966). Histidine regulatory mutants in S. typhimurium. I. Isolation and general properties. *J. Mol. Biol.*, **22**, 305–323.
Roufa D.J., Skogerson L.E. and Leder P. (1970). Translation of phage Qβ mRNA: a test of the two site model for ribosomal function. *Nature*, **227**, 567–570.
Rouget P. and Chapeville F. (1971a). Leucyl-tRNA synthetase. Mechanism of leu-tRNA formation. *Europ. J. Biochem.*, **23**, 443–451.
Rouget P. and Chapeville F. (1971b). Leucyl-tRNA synthetase. Two forms of the enzyme: role of sulfhydryl groups. *Europ. J. Biochem.*, **23**, 452–458.
Roulland-Dussoix D.A. and Boyer H.B. (1969). The E.coli B restriction endonuclease. *Biochim. Biophys. Acta*, **195**, 219–229.
Roy K.L. and Soll D. (1970). Purification of five seryl tRNA species from E.coli and their acylation by homologous and heterologous seryl-tRNA synthetases. *J. Biol. Chem.*, **245**, 1394–1400.

Roychoudhury R. and Kossel H. (1973). Transcriptional role in DNA replication: degradation of RNA primer during DNA synthesis. *Biochem. Biophys. Res. Commun.*, **50**, 259–265.

Rudland P.S. and Clark B.F.C. (1972). Polypeptide chain initiation and the role of a methionine tRNA. In Bosch L. (Ed.), *The Mechanism of protein synthesis and its regulation*, North Holland, Amsterdam. pp 56–86.

Rudland P.S. and Klemperer H.G. (1971). A factor promoting the ejection of deacylated initiator tRNA from ribosomes. *J. Mol. Biol.*, **61**, 377–386.

Rudland P.S., Whybrow W.A., Marcker K.A. and Clark B.F.C. (1969). Recognition of bacterial initiator tRNA by initiation factors. *Nature*, **222**, 750–753.

Rudloff E. and Hilse K. (1971). Properties of isoaccepting species of lysine tRNA from rabbit reticulocytes in codon recognition and in haemoglobin biosynthesis in vitro. *Europ. J. Biochem.*, **24**, 313–320.

Rupp W.D. and Howard-Flanders P. (1968). Discontinuities in the DNA synthesized in an excision defective strain of E.coli following ultraviolet irradiation. *J. Mol. Biol.*, **31**, 291–304.

Rupp W.D., Wilde D.E., Reno D.L. and Howard-Flanders P. (1971). Exchanges between DNA strands in UV-irradiated E.coli. *J. Mol. Biol.*, **61**, 25–44.

Russell R.L., Abelson J.N., Landy A., Gefter M.L., Brenner S. and Smith J.D. (1970). Duplicate genes for tyrosine tRNA in E.coli. *J. Mol. Biol.*, **47**, 1–13.

Ryter A., Hirota Y. and Jacob F. (1968). DNA-membrane complex and nuclear segregation in bacteria. *Cold Spring Harbor Symp. Quant. Biol.*, **33**, 669–676.

Sabol S. and Ochoa S. (1971). Ribosomal binding of labelled initiation factor f3. *Nature New Biol.*, **234**, 233–236.

Sabol S., Sillero M.A.G., Iwasaki K. and Ochoa S. (1970). Purification and properties of initiation factor f3. *Nature*, **228**, 1269–1273.

Sadler J.R. and Smith T.F. (1971). Mapping of the lactose operator. *J. Mol. Biol.*, **62**, 139–170.

Saedler H., Gullon A., Fiethen L. and Starlinger P. (1968). Negative control of the galactose operon in E.coli. *Molec. Gen. Genet.*, **102**, 79–88.

Salas M., Miller J.H., Wahba A.J. and Ochoa S. (1967). Translation of the genetic message. II. Effect of initiation factors on the binding of formyl-methionyl-tRNA to ribosomes. *Proc. Nat. Acad. Sci.*, **57**, 387–394.

Salser W., Bolle A. and Epstein R. (1970). Transcription during bacteriophage T4 development: a demonstration that distinct sub classes of the early RNA appear at different times and that some are turned off at late times. *J. Mol. Biol.*, **49**, 271–296.

Salser W., Fluck M. and Epstein R. (1969). The influence of the reading context upon the suppression of nonsense codons. *Cold Spring Harbor Symp. Quant. Biol.*, **34**, 513–520.

Sambrook J.F., Fan D.P. and Brenner S. (1967). A strong suppressor specific for UGA. *Nature*, **214**, 452–453.

Sanger F. and Brownlee G.G. (1967). Fractionation of radioactive nucleotides. In Shugar D. (Ed.), *Genetic Elements*, Academic Press, London. pp. 303–314.

Sanger F., Brownlee G.G. and Barrell B.G. (1965). A two dimensional fractionation procedure for radioactive nucleotides. *J. Mol. Biol.*, **13**, 373–398.

Sarabhai A., Stretton A.O.W. and Brenner S. (1964). Colinearity of the gene with the polypeptide chain. *Nature*, **201**, 13–17.

Sarabhai A. and Brenner S. (1967). A mutant which reinitiates the polypeptide chain after chain termination. *J. Mol. Biol.*, **27**, 145–162.

Sarabhai A. and Lamfrom H. (1970). Initiation of the polypeptide chain in the *rIIB* cistron of phage T4. *J. Mol. Biol.*, **52**, 131–136.

Sarkar S. and Moldave K. (1968). Characterization of the RNA synthesized during amino acid degradation of a stringent auxotroph of E.coli. *J. Mol. Biol.*, **33**, 213–224.
Scaife J. and Beckwith J.R. (1966). Mutational alteration of the maximal level of *lac* operon expression. *Cold Spring Harbor Symp. Quant. Biol.*, **31**, 403–408.
Schaller H., Otto B., Nusslein V., Huf J., Herrmann R. and Bonhoeffer F. (1972). DNA replication in vitro. *J. Mol. Biol.*, **63**, 183–200.
Schandl E.K. and Taylor J.H. (1969). Early events in the replication and integration of DNA into mammalian chromosomes. *Biochem. Biophys. Res. Commun.*, **34**, 291–300.
Schaup H.W., Green M. and Kurland C.G. (1970). Molecular interactions of ribosomal components. I. Identification of RNA binding sites for individual 30S ribosomal proteins. *Molec. Gen. Genet.*, **109**, 193–205.
Schaup H.W., Green M. and Kurland C.G. (1971). Molecular interactions of ribosomal components. II. Site specific complex formation between 30S proteins and ribosomal RNA. *Molec. Gen. Genet.*, **112**, 1–8.
Schaup H.W. and Kurland C.G. (1972). Molecular interactions of ribosomal components. III. Isolation of the RNA binding site for a ribosomal protein. *Molec. Gen. Genet.*, **114**, 350–357.
Schekman R., Wickner W., Westergaard O., Brutlag D., Geider K., Bertsch L.L. and Kornberg A. (1972). Initiation of DNA synthesis: synthesis of a ϕX 174 replicative form requires RNA synthesis resistant to rifampicin. *Proc. Nat. Acad. Sci.*, **69**, 2691–2695.
Schendl P., Maeba P. and Craven G.R. (1972). Identification of the proteins associated with subparticles produced by mild ribonuclease digestion of 30S ribosomal particles from E.coli. *Proc. Nat. Acad. Sci.*, **69**, 544–548.
Schlachbach A., Fridlander B., Bolden A. and Weissbach A. (1971). DNA dependent DNA polymerase from Hela cell nuclei. II. Template and substrate utilization. *Biochem. Biophys. Res. Commun.*, **44**, 879–885.
Schleif R. (1971). L-arabinose operon messenger of E.coli. Its inducibility and translation efficiency relative to lactose operon messenger. *J. Mol. Biol.*, **61**, 275–280.
Schleif R. (1972). Fine structure deletion map of the E.coli arabinose operon. *Proc. Nat. Acad. Sci.*, **69**, 3479–3484.
Schleif R., Greenblatt J. and Davis R.W. (1971). Dual control of arabinose genes on transducing phage λ*dara*. *J. Mol. Biol.*, **59**, 127–150.
Schlessinger D., Mangiarotti G. and Apirion D. (1967). The formation and stabilization of 30S and 50S ribosome couples in E.coli. *Proc. Nat. Acad. Sci.*, **58**, 1782–1789.
Schlessinger S. and Magasanik B. (1964). Effect of α-methyl histidine on the control of histidine synthesis. *J. Mol. Biol.*, **9**, 670–682.
Schmidt D.A., Mazaitis A.J., Kasai T. and Bautz E.K.F. (1970). Involvement of a phage T4 factor and an anti-terminator protein in the transcription of early T4 genes in vivo. *Nature*, **225**, 1012–1016.
Schrier P.I., Maassen J.A. and Moller W. (1973). Involvement of 50S ribosomal proteins L6 and L10 in the ribosome dependent GTPase activity of elongation factor G. *Biochem. Biophys. Res. Commun.*, **53**, 90–98.
Schulte C. and Garrett R.A. (1972). Optimal conditions for the interaction of ribosomal protein S8 and 16S RNA and studies on the reaction mechanism. *Molec. Gen. Genet.*, **119**, 345–356.
Schwartz T., Craig E. and Kennell D. (1970). Inactivation and degradation of mRNA from the lactose operon of E.coli. *J. Mol. Biol.*, **54**, 299–313.
Scolnick E.H. and Caskey C.T. (1969). Peptide chain termination. V. The role of release factors in mRNA terminator codon recognition. *Proc. Nat. Acad. Sci.*, **64**, 1235–1241.

Scolnick E., Tompkins R., Caskey C.T. and Nirenberg M. (1968). Release factors differing in specificity for terminator codons. *Proc. Nat. Acad. Sci.*, **61**, 768–777.

Sekiguchi M., Yasufa S., Okubo S., Nakayama H., Shimada K. and Takagi Y. (1970). Mechanism of repair of DNA in phage. I. Excision of pyrimidine dimers from UV-irradiated DNA by an extract of T4-infected cells. *J. Mol. Biol.*, **47**, 231–242.

Setlow J.K. and Setlow R.B. (1963). Nature of the photoreactivable ultraviolet lesion in DNA. *Nature*, **197**, 560–562.

Setlow P., Brutlag D. and Kornberg A. (1972). DNA polymerase: two distinct enzymes in one polypeptide. I. A proteolytic fragment containing the polymerase and a 3′-5′ exonuclease function. *J. Biol. Chem.*, **247**, 224–231.

Setlow R.B. and Carrier W.L. (1964). The disappearance of thymine dimers from DNA: an error correcting mechanism. *Proc. Nat. Acad. Sci.*, **51**, 226–231.

Setlow R.B., Regan J.D., German J. and Carrier L. (1969). Evidence that Xeroderma pigmentosum cells do not perform the first step in the repair of ultraviolet damage to their DNA. *Proc. Nat. Acad. Sci.*, **69**, 1035–1041.

Setlow R.B., Swenson P.A. and Carrier W.L. (1963). Thymine dimers and inhibition of DNA synthesis by UV irradiation of cells. *Science*, **142**, 1464–1465.

Sgaramella V., Van de Sande J.H. and Khorana H.G. (1970). Studies on polynucleotides. C. A novel joining reaction catalysed by the T4 polynucleotide ligase. *Proc. Nat. Acad. Sci.*, **67**, 1468–1475.

Shapiro B.M., Siccardi A.G., Hirota Y. and Jacob F. (1970). Membrane protein alterations associated with mutations affecting the initiation of DNA synthesis. *J. Mol. Biol.*, **52**, 75–89.

Shapiro J., MacHattie L., Eron L., Ihler G., Ippen K. and Beckwith J. (1969). Isolation of pure *lac* operon DNA. *Nature*, **224**, 768–774.

Shepherd J. and Maden B.E.H. (1972). Ribosome assembly in Hela cells. *Nature*, **236**, 211–214.

Sherman F., Liebman S.W., Stewart J.W. and Jackson M. (1973). Tyrosine substitutions resulting from suppression of amber mutants of iso-1-cytochrome *c* in yeast. *J. Mol. Biol.*, **78**, 155–166.

Shih A.Y., Eisenstadt J. and Lengyel P. (1966). On the relationship between RNA synthesis and peptide chain initiation in E.coli. *Proc. Nat. Acad. Sci.*, **56**, 1599–1605.

Shizuya I.H. and Dykhuizen D. (1972). Conditional lethality of deletions which include *uvrB* in strains of E.coli lacking DNA polymerase. *J. Bacteriol.*, **112**, 676–681.

Shlaes D.M., Anderson J.A. and Barbour S.D. (1972). Excision-repair properties of isogenic *rec*⁻ mutants of E.coli K12. *J. Bacteriol.*, **111**, 723–730.

Shorey R.L., Ravel J.M., Garner C.W. and Shive W. (1969). Formation and properties of the aminoacyl-tRNA-GTP protein complex. An intermediate in the binding of aminoacyl-tRNA to ribosomes. *J. Biol. Chem.*, **244**, 4555–4565.

Siccardi A.G., Shapiro B.M., Hirota Y. and Jacob F. (1971). On the process of cellular division in E.coli. IV. Altered protein composition and turnover of the membranes of the thermosensitive mutants defective in chromosomal replication. *J. Mol. Biol.*, **56**, 475–490.

Sigal N. and Alberts B. (1972). Genetic recombination: the nature of a crossed strand exchange between two homologous DNA molecules. *J. Mol. Biol.*, **71**, 789–794.

Sigal N., Delius H., Kornberg T., Gefter M.L. and Alberts B. (1972). A DNA unwinding protein isolated from E.coli: its interaction with DNA and with DNA polymerases. *Proc. Nat. Acad. Sci.*, **69**, 3537–3541.

Signer E. (1971). General recombination. In Hershey A.D. (Ed.), *The bacteriophage lambda*, Cold Spring Harbor Laboratory, New York. pp. 139–174.

Silverstone A.E., Arditti R.R. and Magasanik B. (1970). Catabolite insensitive revertants of *lac* promotor mutants. *Proc. Nat. Acad. Sci.*, **66**, 773–779.

Silverstone A.E. and Magasanik B. (1972). Polycistronic effects of catabolite repression on the lac operon. *J. Bacteriol.*, **112**, 1184–1191.
Silverstone A.E., Magasanik B., Renzikoff W.S., Miller J.H. and Beckwith J.R. (1969). Catabolite sensitive site on the *lac* operon. *Nature*, **221**, 1012–1014.
Simon M.N., Davis R.W. and Davidson N. (1971). Heteroduplexes of DNA molecules of lambdoid phages: physical mapping of their base sequence relationships by electron microscopy. In Hershey A.D. (Ed.), *The bacteriophage lambda*, Cold Spring Harbor Laboratory, New York. pp 313–328.
Simsek M. and RajBhandary U.L. (1972). The primary structure of yeast initiator tRNA. *Biochem. Biophys. Res. Commun.*, **49**, 508–515.
Simsek M., Ziegenmeyer J., Heckman J. and RajBhandary U.L. (1973). Absence of the sequence G-T-ψ-C-G-(A) in several eucaryotic cytoplasmic initiator tRNAs. *Proc. Nat. Acad. Sci.*, **70**, 1041–1045.
Singer C.E. and Smith G.R. (1972). Nucleotide sequence of histidine tRNA. *J. Biol. Chem.*, **247**, 2989–3000.
Sinha N.K. and Snustad D.P. (1971). DNA synthesis in phage T4 infected E.coli: evidence supporting a stoichiometric role for gene 32 product. *J. Mol. Biol.*, **62**, 267–271.
Skoultchi A., Ono Y., Waterson J. and Lengyel P. (1969). Peptide chain elongation. *Cold Spring Harbor Symp. Quant. Biol.*, **34**, 437–454.
Slayter H.S., Warner J.R., Rich A. and Hall C.E. (1963). The visualization of polyribosome structure. *J. Mol. Biol.*, **7**, 652–657.
Sly W.S., Rabideau K. and Kolber A. (1971). The mechanism of λ virulence. II. Regulatory mutations in classical virulence. In Hershey A D. (Ed.), *The bacteriophage lambda*, Cold Spring Harbor Laboratory, New York. pp 575–588.
Smith A.E. (1973). The initiation of protein synthesis directed by the RNA from EMC virus. *Europ. J. Biochem.*, **33**, 301–313.
Smith A.E. and Marcker K.A. (1970). Cytoplasmic methionine tRNAs from eucaryotes. *Nature*, **226**, 607–610.
Smith D.W. and Hanawalt P.C. (1967). Properties of the growing point region in the bacterial chromsome. *Biochim. Biophys. Acta*, **149**, 519–531.
Smith D.W., Schaller H.E. and Bonhoeffer F.J. (1970). DNA synthesis in vitro. *Nature*, **226**, 711–713.
Smith G.R. and Magasanik B. (1971). The two operons of the histidine utilization system in S. typhimurium. *J. Biol. Chem.*, **246**, 3330–3341.
Smith K.C. and Meun D.H.C. (1970). Repair of radiation induced damage in E.coli. I. Effect of *rec* mutations on post-replication repair of damage due to UV irradiation. *J. Mol. Biol.*, **51**, 459–472.
Smith H.S. and Pardee A.B. (1970). Accumulation of a protein required for division during the cell cycle of E.coli. *J. Bacteriol.*, **101**, 901–909.
Smith J.D., Abelson J.N., Clark B.F.C., Goodman H.M. and Brenner S. (1966). Studies on amber suppressor tRNA. *Cold Spring Harbor Symp. Quant. Biol.*, **31**, 479–486.
Smith J.D., Arber W. and Kuhnlein V. (1972). Host specificity of DNA produced by E.coli. XIV. The role of nucleotide methylation in in vivo B-specific modification. *J. Mol. Biol.*, **63**, 1–8.
Smith J.D., Barnett L., Brenner S. and Russell R.L. (1970). More mutant tyrosine tRNAs. *J. Mol. Biol.*, **54**, 1–14.
Smith J.D. and Celis J.E. (1973). Mutant tyrosine transfer RNA that can be charged with glutamine. *Nature New Biol.*, **243**, 66–71.
Smith P.D., Finnerty V.G. and Chovnick A. (1970). Gene conversion in Drosophila: non reciprocal events at the maroon like cistron. *Nature*, **228**, 442–445.

Smith R.G. and Gallo R.C. (1972). DNA dependent DNA polymerases I and II from normal human blood lymphocytes. *Proc. Nat. Acad. Sci.*, **69**, 2879–2884.

Snow R. (1967). Mutants of yeast sensitive to ultraviolet light. *J. Bacteriol.*, **94**, 571–575.

Sobell H.M. (1972). Molecular mechanism for genetic recombination. *Proc. Nat. Acad. Sci.*, **69**, 2843–2487.

Sobell H.M. (1973). Symmetry in protein-nucleic acid interactions and its genetic implications. *Adv. Gen.*, **17**, 411–490.

Sogin M., Pace B., Pace N.R. and Woese C.R. (1971). Primary structural relationship of p16 to m16 rRNA. *Nature New Biol.*, **232**, 48–49.

Soll D. (1968). Studies on polynucleotides. LXXXV. Partial purification of an amber suppressor tRNA and studies on in vitro suppression. *J. Mol. Biol.*, **34**, 175–188.

Soll D., Cherayil J.D. and Bock R.M. (1967). Studies on polynucleotides. LXXV. Specificity of tRNA for codon recognition as studied by ribosomal binding techniques. *J. Mol. Biol.*, **29**, 97–112.

Soll D. and RajBhandary V.L. (1967). Studies on polynucleotides. LXXVI. Specificity of tRNA for codon recognition as studied by amino acid incorporation. *J. Mol. Biol.*, **29**, 113–124.

Soll L. and Berg P. (1969a). Recessive lethals: a new class of nonsense suppressors in E.coli. *Proc. Nat. Acad. Sci.*, **63**, 392–399.

Soll L. and Berg P. (1969b). Recessive lethal nonsense suppressor in E.coli which inserts glutamine. *Nature*, **223**, 1340–1342.

Sonneborn T.M. (1965). Degeneracy of the genetic code: extent, nature, and genetic information. In Bryson V. and Vogel H.J. (Eds.), *Evolving genes and proteins*, Academic Press, New York, pp 377–397.

Spadari S. and Ritossa F. (1970). Clustered genes for ribosomal RNAs in E.coli. *J. Mol. Biol.*, **53**, 357–368.

Speyer J.F. (1963). Synthetic polynucleotides and the amino acid code. *Cold Spring Harbor Symp. Quant. Biol.*, **28**, 559–567.

Speyer J.F. (1965). Mutagenic DNA polymerase. *Biochem. Biophys. Res. Commun.*, **21**, 6–8.

Spiegelmann W.G., Heinemann S.F., Brachet P., Da Silva L.P. and Eisen H. (1970). Regulation of the synthesis of phage lambda repressor. *Cold Spring Harbor Symp. Quant. Biol.*, **35**, 325–330.

Spiegelman W.G., Rechardt L.F., Yaniv M., Heinemann S.F., Kaiser A.D. and Eisen H. (1972). Bidirectional transcription and the regulation of phage λ repressor synthesis. *Proc. Nat. Acad. Sci.*, **69**, 3156–3160.

Springer M. and Grunberg-Manago M. (1972). Characteristics of N-acetyl-phe-tRNA binding and its correlation with internal aminoacyl-tRNA recognition. *Biochem. Biophys. Res. Commun.*, **47**, 477–484.

Squires C. and Carbon J. (1971). Normal and mutant glycine tRNAs. *Nature New Biol.*, **233**, 274–277.

Stadler D.R. and Towe A.M. (1971). Evidence for mitotic recombination in Ascobolus involving only one member of a tetrad. *Genetics*, **68**, 401–413.

Stadler D.R., Towe A.M. and Rossignol J.L. (1970). Intragenic recombination of ascospore colour mutants in Ascobolus and its relationship to the segregation of outside markers. *Genetics*, **66**, 429–447.

Stadler L.J. and Towe A.M. (1963). Recombination of allelic cysteine mutants in Neurospora. *Genetics*, **48**, 1323–1344.

Staehelin T., Maglott D. and Monro R.E. (1969). On the catalytic centre of peptidyl transfer: a part of the 50S ribosome structure. *Cold Spring Harbor Symp. Quant. Biol.*, **34**, 39–48.

Staehelin T. and Meselson M. (1966). In vitro recovery of ribosomes and of synthetic activity from synthetically inactive ribosomal subunits. *J. Mol. Biol.*, **16**, 245–249.

Stamato T.D. and Pettijohn D.E. (1971). Regulation of rRNA synthesis in stringent bacteria. *Nature New Biol.*, **234**, 99–102.

Stanley W.M., Salas M., Wahba A.J. and Ochoa S. (1966). Translation of the genetic message: factors involved in the initiation of protein synthesis. *Proc. Nat. Acad. Sci.*, **56**, 290–295.

Staples D.H. and Hindley J. (1971). Ribosome binding site of Qβ RNA polymerase cistron. *Nature New Biol.*, **234**, 211–212.

Staples D. H., Hindley J., Billeter M.A. and Weissman C. (1971). Localization of Qβ maturation cistron ribosome binding site. *Nature New Biol.*, **234**, 202–204.

Stavrianopoulos J.G., Karkas J.D. and Chargaff E. (1972). Mechanism of DNA replication by highly purified DNA polymerase of chicken embryo. *Proc. Nat. Acad. Sci.*, **69**, 2609–2613.

Stein G.H. and Hanawalt P.C. (1972). Initiation of the DNA replication cycle in E.coli: linkage of origin daughter DNA to parental DNA? *J. Mol. Biol.*, **64**, 393–408.

Steinberg R.A. and Ptashne M. (1971). In vitro repression of RNA synthesis by purified λ phage repressor. *Nature New Biol.*, **230**, 76–80.

Steitz J.A. (1969). Polypeptide chain initiation: nucleotide sequences of the three ribosomal binding sites in phage R17 RNA. *Nature*, **224**, 957–963.

Steitz J.A. (1973). Specific recognition of non-initiator regions in RNA phage messengers by ribosomes of B. stearothermophilus, *J. Mol. Biol.*, **73**, 1–16.

Steitz J.A., Dube S.K. and Rudland P.S. (1970). Control of translation by T4 phage: altered ribosome binding at R17 sites. *Nature*, **226**, 824–827.

Stent G.S. (1958). Mating in the reproduction of bacterial viruses. *Adv. Virus Res.*, **5**, 95–149.

Stern C. (1970). The continuity of genetics. *Daedalus*, **99**, 882–908.

Stern R., Zutra L.A. and Littauer U.Z. (1969). Fractionation of tRNA on a methylated albumin silicic acid column. II. Changes in elution profiles following modification of tRNA. *Biochem.*, **8**, 313–321.

Stevens A. (1972). New small polypeptides associated with DNA dependent RNA polymerase of E.coli after infection with phage T4. *Proc. Nat. Acad. Sci.*, **69**, 603–607.

Stewart J.A. and Sherman F. (1972). Demonstration of UAG as a nonsense codon in bakers' yeast by amino acid replacements in iso-1-cytochrome *c*. *J. Mol. Biol.*, **68**. 429–444.

Stewart J.W. and Sherman F. (1973). Confirmation of UAG as a nonsense codon in bakers' yeast by amino acid replacements of glutamic acid 71 in iso-1-cytochrome *c*. *J. Mol. Biol.*, **78**, 167–182.

Stewart J.W., Sherman F., Jackson M., Thomas F.L.X. and Shipman N. (1972). Demonstration of the UAA ochre codon in bakers' yeast by amino acid replacements in iso-1-cytochrome *c*. *J. Mol. Biol.*, **68**, 83–97.

Stewart J.W., Sherman F., Shipman N.A. and Jackson M. (1971). Identification and mutational relocation of the AUG codon initiating translation of iso-1-cytochrome *c* in yeast. *J. Biol. Chem.*, **246**, 7429–7445.

Stoffler G., Daya L., Rak K.H. and Garrett R.A. (1971a). Ribosomal proteins. XXVI. The number of specific protein binding sites on 16S and 23S RNA of E.coli. *J. Mol. Biol.*, **62**, 411–414.

Stoffler G., Daya L., Rak K.H. and Garrett R.A. (1972). Ribosomal proteins. XXX. Specific protein binding sites on 23S RNA of E.coli. *Molec. Gen. Genet.*, **114**, 125–133.

Stoffler G., Deusser E., Wittman H.G. and Apirion D. (1971b). Ribosomal proteins. XIX. Altered S5 ribosomal protein in an E.coli revertant from streptomycin dependence to independence. *Molec. Gen. Genet.*, **111**, 334–341.

Stonington O.G. and Pettijohn D.E. (1971). The folded genome of E.coli isolated in a protein-DNA-RNA complex. *Proc. Nat. Acad. Sci.*, **68**, 6–9.

Storm P.K., Hoekstra P.M., De Haan P.G. and Verhoet C. (1971). Genetic recombination in E.coli. IV. Isolation and characterization of recombination deficient mutants of E.coli. *Mutat. Res.*, **13**, 9–17.

Storm P.K. and Zaunbrecher W.M. (1972). A new radiation sensitive mutant of E.coli K12. *Molec. Gen. Genet.*, **115**, 89–92.

Stratling W. and Knippers R. (1971a). Properties of the DNA synthesizing activity in DNA-membrane complexes from bacterial cell extracts. *Europ. J. Biochem.*, **20**, 330–339.

Stratling W. and Knippers R. (1971b). DNA synthesis in isolated DNA-membrane complexes. *J. Mol. Biol.*, **61**, 471–488.

Streisinger G., Okada Y., Emrich J., Newton J., Tsugita A., Terzhagi E. and Inouye M. (1966). Frameshift mutations and the genetic code. *Cold Spring Harbor Symp. Quant. Biol.*, **31**, 77–84.

Stretton A.O.W.(1965). The genetic code. *Brit. Med. Bull.*, **21**, **3**, 229–235.

Stretton A.O.W. and Brenner S. (1965). Molecular consequences of the amber mutation and its suppression. *J. Mol. Biol.*, **12**, 456–465.

Strigini P. and Gorini L. (1970). Ribosomal mutations affecting efficiency of amber suppression. *J. Mol. Biol.*, **47**, 517–530.

Strike P. and Emmerson P.T. (1972). Coexistence of *polA* and *recB* mutations of E.coli in the presence of *sbc*, a mutation which indirectly suppresses *recB*. *Molec. Gen. Genet.*, **116**, 177–180.

Stubbs J.D. and Hall B.D. (1968a). Level of tryptophan messenger RNA in E.coli. *J. Mol. Biol.*, **37**, 289–302.

Stubbs J.D. and Hall B.D. (1968b). Effects of amino acid starvation upon constitutive tryptophan messenger RNA synthesis. *J. Mol. Biol.*, **37**, 303–312.

Subramanian A.R. and Davis B.D. (1971). Rapid exchange of subunits between free ribosomes in extracts of E.coli. *Proc. Nat. Acad. Sci.*, **68**, 2453–2457.

Subramanian A.R. and Davis B.D. (1973). Release of 70S ribosomes from polysomes in E.coli. *J. Mol. Biol.*, **74**, 45–56.

Subramanian A.R., Davis B.D. and Beller R.J. (1969). The ribosome dissociation, factor and the ribosome polysome cycle. *Cold Spring Harbor Symp. Quant. Biol.*, **34,** 223–230.

Subramanian A.R. and Davis B.D. (1970). Activity of initiation factor f3 in dissociating E.coli ribosomes. *Nature*, **228**, 1273–1276.

Sueoka N. and Quinn W.G. (1968). Membrane attachment of the chromosome replication origin in B. subtilis. *Cold Spring Harbor Symp. Quant. Biol.*, **33**, 695–706.

Sueoka N. and Yoshikawa H. (1963). Regulation of chromosome replication in B. subtilis. *Cold Spring Harbor Symp. Quant. Biol.*, **28,** 43–54.

Sugimoto K., Okazaki T., Imae Y. and Okazaki R. (1969). Mechanism of DNA chain growth. III. Equal annealing of T4 nascent short DNA chains with the separated complementary strands of the phage DNA. *Proc. Nat. Acad. Sci.*, **63,** 1343–1350.

Sugino A., Hirose S. and Okazaki R. (1972). RNA linked nascent DNA fragments in E.coli. *Proc. Nat. Acad. Sci.*, **69,** 1863–1867.

Sugino A. and Okazaki R. (1972). Mechanism of DNA chain growth. VII. Direction and rate of growth of T4 nascent short DNA chains. *J. Mol. Biol.*, **64,** 61–86.

Sugino A. and Okazaki R. (1973). RNA linked DNA fragments in vitro. *Proc. Nat. Acad. Sci.*, **70**, 88–92.

Sugiura M., Okamoto T. and Takanami M. (1970). RNA polymerase σ factor and the selection of initiation sites. *Nature*, **225**, 598–600.

Summers W.C. and Siegel R. (1970). Regulation of phage T7 RNA metabolism in vivo and in vitro. *Cold Spring Harbor Symp. Quant. Biol.*, **35**, 253–258.

Sundarajan T.A. and Thach R.E. (1966). Role of the formyl-methionine codon AUG in phasing transition of synthetic mRNA. *J. Mol. Biol.*, **19**, 74–90.

Sypherd P.S., O'Neill D.M. and Taylor M.M. (1969). The chemical and genetic structure of bacterial ribosomes. *Cold Spring Harbor Symp. Quant. Biol.*, **34**, 77–84.

Szybalski W., Bøvre K., Fiandt M., Hayes S., Hradneca Z., Kumar S., Lozeron H.A., Nijkamp H.J.J. and Stevens W.F. (1970). Transcriptional units and their controls in E.coli phage λ: operons and scriptons. *Cold Spring Harbor Symp. Quant. Biol.*, **35**, 341–355.

Szybalski W., Kubinski H. and Sheldrick P. (1966). Pyrimidine clusters on the transcribing strand of DNA and their possible role in the initiation of RNA synthesis. *Cold Spring Harbor Symp. Quant. Biol.*, **31**, 123–127.

Tai P.C. and Davis B.D. (1972). Transfer RNA content of runoff and complexed ribosomes of E.coli. *J. Mol. Biol.*, **67**, 219–230.

Takagi Y., Sekiguchi M., Okubo S., Nakayama H., Shimada K., Yasuda S., Mishimoto T. and Yoshihara H. (1968). Nucleases specific for ultraviolet light irradiated DNA and their possible role in dark repair. *Cold Spring Harbor Symp. Quant. Biol.*, **33**, 219–228.

Takanami M., Okamoto T. and Sugiura M. (1970). The starting nucleotide sequence and size of RNA transcribed in vitro on phage DNA templates. *Cold Spring Harbor Symp. Quant. Biol.*, **35**, 179–188.

Takanami M., Okamoto T. and Sugiura M. (1971). Termination of RNA transcription on the replicative form DNA of phage fd. *J. Mol. Biol.*, **62**, 81–88.

Takata R. (1972). Mapping of ribosomal protein components by intergenic mating experiments between Serratia marsecens and E. coli. *Molec. Gen. Genet.*, **118**, 363–372.

Takeda M. and Webster R.E. (1968). Protein chain initiation and deformylation in B. subtilis homogenates. *Proc. Nat. Acad. Sci.*, **60**, 1487–1494.

Takeishi K., Nomoto A. and Ukita T. (1972). Histidine tRNA from bakers' yeast: number of isoaccepting species and their coding specificity. *Biochem. Biophys. Acta*, **272**, 262–274.

Taylor A.L. and Trotter C.D. (1972). Linkage map of E.coli strain K12. *Bacteriol. Rev.*, **36**, 504–524.

Thach S.S. and Thach R.E. (1971). Translocation of messenger RNA and "accomodation" of fmet-tRNA. *Proc. Nat. Acad. Sci.*, **68**, 1791–1795.

Thach R.E., Sundarajan T.A., Dewey K.F., Brown J.C. and Doty P. (1966). Translation of synthetic messenger RNA. *Cold Spring Harbor Symp. Quant. Biol.*, **31**, 85–98.

Thiebe R., Harbers K. and Zachau H.G. (1972). Aminoacylation of fragment combinations from yeast $tRNA^{phe}$. *Europ. J. Biochem.*, **26**, 144–152.

Thiebe R. and Zachau H.G. (1968a). A special modification next to the anticodon of phenylalanine tRNA. *Europ. J. Biochem.*, **5**, 546–555.

Thiebe R. and Zachau H.G. (1968b). The role of the anticodon region in homologous and heterologous charging of $tRNA^{phe}$. *Biochem. Biophys. Res. Commun.*, **33**, 260–265.

Thiebe R. and Zachau H.G. (1969). Acceptor activity in homologous and heterologous combinations of half molecules form $tRNA^{phe}_{yeast}$ and $tRNA^{phe}_{wheat}$. *Biochem. Biophys. Res. Commun.*, **36**, 1024–1031.

Thiebe R., Zachau H.G., Baczynskj L., Biemann K. and Sonnenbichler J. (1971). Studies on the properties and structure of the modified base Y* of yeast $tRNA^{phe}$. *Biochim. Biophys. Acta*, **240**, 163–169.

Tomizawa J.I. and Anraku N. (1956). Molecular mechanisms of genetic recombination of bacteriophage. IV. Absence of polynucleotide interruption in DNA of T4 and λ phage particles with special reference to heterozygosis. *J. Mol. Biol.*, **11**, 509–527.

Tompkins R.K., Scolnick E.M., Caskey C.T. (1970). Peptide chain termination. VII. The ribosomal and release factor requirements for peptide release. *Proc. Nat. Acad. Sci.*, **65**, 702–708.

Tongur V.S., Wladytchenskaya N.S. and Kotchkina W.M. (1968). A natural RNA-DNA complex in bacterial cells. *J. Mol. Biol.*, **33**, 451–464.

Town C.D., Smith K.C. and Kaplan H.S. (1971). DNA polymerase required for rapid repair of X-ray induced strand breaks in vivo. *Science* **172**, 851–853.

Traub P. and Nomura M. (1968a). Structure and function of E.coli ribosomes. I. Partial fractionation of the functionally active ribosomal proteins and reconstitution of the artificial subribosomal particles. *J. Mol. Biol.*, **34**, 575–594.

Traub P. and Nomura M. (1968b). Structure and function of E.coli ribosomes. V. Reconstitution of functionally active 30S ribosomal particles from RNA and proteins. *Proc. Nat. Acad. Sci.*, **59**, 777–784.

Traub P. and Nomura M. (1968c). Streptomycin resistance mutation in E.coli: altered ribosomal protein. *Science*, **160**, 198–199.

Traub P. and Nomura M. (1969a). Structure and function of E.coli ribosomes. VI. Mechanism of assembly of 30S ribosomes studied in vitro. *J. Mol. Biol.*, **40**, 391–414.

Traub P. and Nomura M. (1969b). Studies on the assembly of ribosomes in vitro. *Cold Spring Harbor Symp. Quant. Biol.*, **34**, 63–68.

Traub P., Soll D. and Nomura M. (1968). Structure and function of E.coli ribosomes. II. Translational fidelity and efficiency in protein synthesis of a protein deficient subribosomal particle. *J. Mol. Biol.*, **34**, 595–608.

Traut R.R., Delius H., Ahmed-Zadeh C., Bickle T.A., Pearson P. and Tissieres A. (1969). Ribosomal proteins of E.coli: stoichiometry and implications for ribosome structure. *Cold Spring Harbor Symp. Quant. Biol.*, **34**, 25–38.

Traut R.R., Moore P.B., Delius H., Noller H. and Tissieres A. (1967). Ribosomal proteins of E.coli. I. Demonstration of different primary structures. *Proc. Nat. Acad. Sci.*, **57**, 1294–1301.

Travers A.A. (1969). Phage sigma factor for RNA polymerase. *Nature*, **223**, 1107–1111.

Travers A. (1970a). RNA polymerase and T4 development. *Cold Spring Harbor Symp. Quant. Biol.*, **35**, 241–252.

Travers A. (1970b). Positive control of transcription by a phage sigma factor. *Nature*, **225**, 1009–1012.

Travers A. (1973). Control of ribosomal RNA synthesis in vitro. *Nature*, **244**, 15–17.

Travers A., Baillie D.L. and Pedersen S. (1973). Effect of DNA conformation on ribosomal RNA synthesis in vitro. *Nature New Biol.*, **243**, 161–163.

Travers A. and Buckland R. (1973). Heterogeneity of E.coli RNA polymerase. *Nature New Biol.*, **243**, 357–260.

Travers A.A. and Burgess R.R. (1969). Cyclic reuse of the RNA polymerase sigma factor. *Nature*, **222**, 537–540.

Travers A.A., Kamen R.I. and Schleif R.F. (1970). Factor necessary for rRNA synthesis. *Nature*, **228**, 748–752.

Tremblay G.Y., Daniels M.J. and Schaechter M. (1969). Isolation of a cell membrane-DNA-nascent RNA complex from bacteria. *J. Mol. Biol.*, **40**, 65–76.

Uhlenbeck O.C. (1972). Complementary oligonucleotide binding to tRNA. *J. Mol. Biol.*, **65**, 25–42.

Uhlenbeck O.C., Baller J. and Doty P. (1970). Complementary oligonucleotide binding to the anticodon loop of fmet-tRNA. *Nature*, **225**, 508–510.

Ullman A. and Perrin D. (1970). Complementation in β-galactosidase. In Beckwith J.R. and Zipser D. (Eds.), *The Lactose Operon*, Cold Spring Harbor Laboratory, New York. pp 143–172.

Urm E., Yanf H., Zubay G., Kelker N. and Maas W. (1973). In vitro repression of N-α-acetyl ornithinase synthesis in E.coli. *Molec. Gen. Genet.*, **121**, 1–8.

Van Duin J. and Kurland C.G. (1970). Functional heterogeneity of the 30S ribosomal subunit of E.coli. *Molec. Gen. Genet.*, **109**, 169–176.

Van Duin J., Van Knippenberg P.H., Dieben M. and Kurland C.G. (1972). Functional heterogeneity of the 30S ribosomal subunit of E.coli. II. Effect of S21 on initiation. *Molec. Gen. Genet.*, **116**, 181–191.

Varmus H.E., Perlman R.L. and Pastan I. (1970a). Regulation of *lac* mRNA synthesis by cyclic AMP and glucose. *J. Biol. Chem.*, **245**, 2259–2267.

Varmus H.E., Perlman R.L. and Pastan I. (1970b). Regulation of *lac* transcription in E.coli by cyclic AMP. Studies with DNA-RNA hybridization and hybridization competition. *J. Biol. Chem.*, **245**, 6366–6372.

Varmus H.E., Perlman R.L. and Pastan I. (1971). Regulation of *lac* transcription in antibiotic treated E.coli. *Nature New Biol.*, **230**, 41–44.

Vazquez D. (1966). Mode of action of chloramphenicol and related antibiotics. *Symp. Soc. Gen. Microbiol.*, **16**, 169–189.

Vazquez D., Battaner E., Neth R., Heller G. and Monro R.E. (1969). The function of 80S ribosomal subunits and effects of some antibiotics. *Cold Spring Harbor Symp. Quant. Biol.*, **34**, 369–376.

Vesco C. and Colombo B. (1970). Effect of sodium fluoride on protein synthesis in Hela cells: inhibition of ribosome dissociation. *J. Mol. Biol.*, **47**, 335–352.

Vielmetter W., Messer W. and Schutte A. (1968). Growth, direction and segregation of the E.coli chromosome. *Cold Spring Harbor Symp. Quant. Biol.*, **33**, 585–598.

Vogel T., Meyers M., Kovach J.S. and Goldberger R.F. (1972). Specificity of interaction between the first enzyme for histidine biosynthesis and aminoacylated his-tRNA. *J. Bacteriol.*, **112**, 126–132.

Volkin E. and Astrachan L. (1957). RNA metabolism in T2 infected E.coli. In McElroy W.D. and Glass B. (Eds.), *The chemical basis of heredity*, Johns Hopkins Press, Baltimore. pp 686–695.

Voynow P. and Kurland C.G. (1971). Stoichiometry of the 30S ribosomal proteins of E.coli. *Biochem.*, **10**, 517–523.

Wake R.G. (1973). Circularity of the B. Subtilis chromosome and further studies on its bidirectional replication. *J. Mol. Biol.*, **77**,

Wallace B.J. and Davis B.D. (1973). Cyclic blockage of initiation sites by streptomycin damaged ribosomes in E.coli: an explanation for dominance of sensitivity. *J. Mol. Biol.*, **75**, 377–390.

Wallace B.J., Tai P.C. and Davis B.D. (1973). Effect of streptomycin on the response of E.coli ribosomes to the dissociation factor. *J. Mol. Biol.*, **75**,391–400.

Wallace H. and Birnsteil M.L. (1966). Ribosomal cistrons and the nucleolar organiser. *Biochim. Biophys. Acta*, **114**, 296–310.

Waller J.P. (1963). The NH_2-terminal residues of the proteins from cell free extracts of E.coli. *J. Mol. Biol.*, **7**, 483–496.

Waller J.P. (1964). Fractionation of the ribosomal protein from E.coli. *J. Mol. Biol.*, **10**, 319–336.

Wang J.C. (1971). Interaction between DNA and an E.coli protein. *J. Mol. Biol.*, **55**, 523–533.

Ward C.B. and Glaser D.A. (1969). Analysis of the chloramphenicol-sensitive and chloramphenicol-resistant steps in the initiation of DNA synthesis in E.coli B/r. *Proc. Nat. Acad. Sci.*, **64**, 905–912.

Ward C.B. and Glaser D.A. (1970). Control of initiation of DNA synthesis in E.coli B/r. *Proc. Nat. Acad. Sci.*, **67**, 255–262.

Ward C.B. and Glaser D.A. (1971). Correlation between rate of cell growth and rate of DNA synthesis in E.coli B/r. *Proc. Nat. Acad. Sci.*, **68**, 1061–1064.

Ward C.B., Hane M.W. and Glaser D.A. (1970). Synchronous reinitiation of chromosome replication in E.coli B/r after nalidixic acid treatment. *Proc. Nat. Acad. Sci.*, **66**, 365–369.

Warner J., Knopf P.M. and Rich A. (1963). A multiple ribosome structure in protein synthesis. *Proc. Nat. Acad. Sci.*, **49**, 122–129.

Watanabe K. and Imahori K. (1971). The conformational difference between $tRNA_f^{met}$ and formyl-methionyl-$tRNA_f^{met}$ from E.coli. *Biochem. Biophys. Res. Commun.*, **45**, 488–494.

Waterson J., Beaud G. and Lengyel P. (1970). The S1 factor in peptide chain elongation. *Nature*, **227**, 34–38.

Watson J.D. and Crick F.H.C. (1953a). A structure for DNA. *Nature*, **171**, 736–738.

Watson J.D. and Crick F.H.C. (1953b). Genetic implications of the structure of DNA. *Nature*, **171**, 964–967.

Weber K., Platt T., Ganem D. and Miller J.H. (1972). Altered sequences changing the operator binding properties of the *lac* repressor: colinearity of the repressor protein with the *i* gene map. *Proc. Nat. Acad. Sci.*, **69**, 3624–3628.

Webster R.E., Engelhardt D.L. and Zinder N.D. (1966). In vitro protein synthesis: chain initiation. *Proc. Nat. Acad. Sci.*, **55**, 155–161.

Webster R.E. and Zinder N.D. (1969). Fate of the message-ribosome complex upon translation of termination signals. *J. Mol. Biol.*, **42**, 425–440.

Wechsler J.A. and Gross J.D. (1971). E.coli mutants temperature sensitive for DNA synthesis. *Molec. Gen. Genet.*, **113**, 273–284.

Weigert M.G. and Garen A. (1965a). Amino acid substitutions resulting from suppression of nonsense mutations. I. Serine insertion by the *su-1* suppressor gene. *J. Mol. Biol.*, **12**, 448–455.

Weigert M.G. and Garen A. (1965b). Base composition of nonsense codons in E.coli. *Nature*, **206**, 992–994.

Weigert M.G., Lanka E. and Garen A. (1965). Amino acid substitutions resulting from suppression of nonsense mutations. II. Glutamine insertion by *su2* suppressor gene. *J. Mol. Biol.*, **14**, 522–527.

Weinberg R.A. and Penman S. (1970). Processing of 45S nucleolar RNA. *J. Mol. Biol.*, **47**, 169–178.

Weiner A.M. and Weber K. (1971). Natural readthrough at the UGA termination signal of Qβ coat protein cistron. *Nature New Biol.*, **234**, 206–209.

Weinstein I.B., Friedman S.M. and Ochoa M.jr. (1966). Fidelity during translation of the genetic code. *Cold Spring Harbor Symp. Quant. Biol.*, **31**, 671–682.

Weintraub S.B. and Frankel F.R. (1972). Identification of the T4 *rIIB* gene product as a membrane protein. *J. Mol. Biol.*, **70**, 589–616.

Weiss R.L., Kimes B.W. and Morris D.R. (1973). Cations and ribosome structure. III. Effects on the 30S and 50S subunits of replacing bound Mg^{++} by inorganic cations. *Biochem.*, **12**, 450–456.

Weissbach H., Brot N., Miller D., Rosman M. & Ertel R. (1969). Interaction of GTP with E.coli soluble transfer factors. *Cold Spring Harbor Symp. Quant. Biol.*, **34**, 419–432.

Weissbach A., Schlachbach A., Fridlender B. and Bolden A. (1971). DNA polymerases from human cells. *Nature New Biol.*, **231**, 167–170.
Weissbach H., Redfield B. and Brot N. (1971a). Aminoacyl-tRNA-Tu-GTP interaction with ribosomes. *Arch. Biochem. Biophys.*, **145**, 676–684.
Weissbach H., Redfield B. and Brot N. (1971b). Further studies on the role of factors Ts and Tu in protein synthesis. *Arch. Biochem. Biophys.*, **144**, 224–229.
Weissbach H., Redfield B., Yamasaki E., Davis R.C.jr., Pestka S. and Brot N. (1972). Studies on the ribosomal sites involved in factors Tu and G dependent reactions. *Arch. Biochem. Biophys.*, **149**, 110–117.
Westergaard O., Brutlag D. and Kornberg A. (1973). Initiation of DNA synthesis. IV. Incorporation of the RNA primer into the phage replicative form. *J. Biol. Chem.*, **248**, 1361–1364.
Wetekam W., Staack K. and Ehring R. (1971). DNA dependent in vitro synthesis of enzymes of the galactose operon of E.coli. *Molec. Gen. Genet.*, **112**, 14–27.
Whitehouse H.L.K. (1963). A theory of crossing over by means of hybrid DNA. *Nature*, **199**, 1034–1040.
Whitehouse H.L.K. (1966). An operator model of crossing over. *Nature*, **211**, 708–713.
Whitehouse H.L.K. (1969). *The Mechanism of Heredity*, Edward Arnold, London. (Second Edition).
Whitfield H. (1972). Suppression of nonsense, frameshift and missense mutations. In Bosch L. (Ed.), *The mechanism of protein synthesis and its regulation*, North Holland Amsterdam. pp 243–284.
Whitfield H.J.jr., Gutnick D.L., Margolies M.N., Martin R.G., Rechler M.M and Voll M.J. (1970). Relative translation frequencies of the cistrons of the histidine operon. *J. Mol. Biol.*, **49**, 245–250.
Wicha M. and Stockdale F.E. (1972). DNA dependent polymerases in differentiating embryonic muscle cells. *Biochem. Biophys. Res. Common.*, **48**, 1079–1087.
Wickner R. B. and Hurwitz J. (1972). DNA replication in E.coli made permeable by treatment with high sucrose. *Biochem. Biophys. Res. Commun.*, **47**, 202–211.
Wickner R.B., Wright M., Wickner S. and Hurwitz J. (1972b). Conversion of ϕX174 and fd single stranded DNA to replicative form in extracts of E.coli. *Proc. Nat. Acad. Sci.*, **69**, 3233–3237.
Wickner S., Wright M. and Hurwitz J. (1973). Studies on in vitro DNA synthesis. Purification of *dnaG* product from E.coli. *Proc. Nat. Acad. Sci.*, **70**, 1613–1618.
Wickner W., Brutlag D., Schekman R. and Kornberg A. (1972a). RNA synthesis initiates in vitro conversion of M13 DNA to its replicative form. *Proc. Nat. Acad. Sci*, **69**, 965–969.
Wigle D.T. and Dixon G.H. (1970). Transient incorporation of methionine at the N-terminus of protamine newly synthesised in trout testis cells. *Nature*, **227**, 676–680.
Wilcox G., Singer J. and Hefferman L. (1971). DNA-RNA hybridization studies on the L-arabinose operon of E.coli B/r. *J. Bacteriol.*, **108**, 1–4.
Wilcox G., Clemetson K.J., Santi D.V., Englesberg E. (1971). Purification of the *araC* protein. *Proc. Nat. Acad. Sci.*, **68**, 2145–2148.
Wilkins M.F.H., Stokes A.R. and Wilson H.R. (1953). Molecular structure of DNA. *Nature*, **171**, 738–740.
Williamson R. and Brownlee G.G. (1969). The sequence of 5S rRNA from two mouse cell lines. *FEBS Lett.*, **3**, 306–308.
Willson C., Perrin D., Cohn. M., Jacob F. and Monod J. (1964). Non inducible mutants of the regulator gene in the lactose system of E.coli. *J. Mol. Biol.*, **8**, 582–592.
Wilson D.B. and Dintzis H.M. (1970). Protein chain initiation in rabbit reticulocytes. *Proc. Nat. Acad. Sci.*, **66**, 1282–1289.

Wilson D.W. and Hogness D.S. (1969). The enzymes of the galactose operon in E.coli. IV. The frequencies of translation of the terminal cistrons in the operon. *J. Biol. Chem.*, **244**, 2143–2148.

Witkin E.M. (1967). Mutation proof and mutation prone modes of survival in derivatives of E.coli B differing in sensitivity for ultraviolet light. *Brookhaven Symp. Biol.*, **20**, 17–55.

Witkin E.M. (1969). The mutability toward ultraviolet light of recombination-deficient strains of E.coli. *Mutat. Res.*, **8**, 9–14.

Witmer H.J. (1971a). Effect of ionic strength and temperature on the in vitro transcription of T4 DNA. *Biochim. Biophys. Acta*, **246**, 29–43.

Witmer H.J. (1971b). In vitro transcription of T4 DNA by E.coli RNA polymerase. Sequential transcription of immediate and delayed early cistrons in the absence the release factor, rho. *J. Biol. Chem.*, **246**, 5220–5227.

Wittman H.G. and Stoffler G. (1972). Structure and function of bacterial ribosomal proteins. In Bosch L. (Ed.), *The mechanism of protein synthesis and its regulation*, North Holland, Amsterdam. pp 285–352.

Wittman H.G., Stoffler G., Hindenach I., Kurland C.G., Randall-Haschbauer L., Birge E.A., Nomura M., Kaltschmidt E., Mizushima S., Traut R.R. and Bickle T.A. (1971). Correlation of 30S ribosomal proteins of E.coli isolated in different laboratories. *Molec. Gen. Genet.*, **111**, 327–333.

Woese C.R. (1965). On the evolution of the genetic code. *Proc. Nat. Acad. Sci.*, **54**, 1546–1552.

Woese C.R. (1967). The present status of the genetic code. *Prog. Nucleic Acid Res.*, **7**, 107–172.

Woese C.R. (1968). The fundamental nature of the genetic code: the prebiotic interactions between polynucleotides and polyamino acids of their derivatives. *Proc. Nat. Acad. Sci.*, **59**, 110–117.

Woese C. (1969). Models for the evolution of codon assignments. *J. Mol. Biol.*, **43**, 235–240.

Wolfson J. and Dressler D. (1972). Regions of single stranded DNA in the growing points of replicating phage T7 chromosomes. *Proc. Nat. Acad. Sci.*, **69**, 2682–2686.

Wolfson J., Dressler D. and Magazin M. (1972). Phage T7 DNA replication in a linear replicating intermediate. *Proc. Nat. Acad. Sci.*, **69**, 499–504.

Worcel A. and Burgi E. (1972). On the structure of the folded chromosome of E.coli. *J. Mol. Biol.*, **71**, 127–148.

Wright M., Buttin G. and Hurwitz J. (1971). The isolation and characterization from E.coli of an ATP dependent DNAase directed by *recB,C* genes. *J. Biol. Chem.*, **246**, 6543–6555.

Wuesthoff G. and Bauerle R.H. (1970). Mutations creating internal promotor elements in the tryptophan operon of S. typhimurium. *J. Mol. Biol.*, **49**, 171–196.

Yang S.K. and Crothers D.M. (1972). Conformational changes of tRNA. Comparison of the early melting transitions of two tyrosine specific tRNAs. *Biochem.*, **11**, 4375–4381.

Yaniv M. and Barrell B.G. (1969). Nucleotide sequence of E.coli B $tRNA_1^{val}$. *Nature*, **222**, 278–279.

Yaniv M., Chestier A., Gros F. and Favre A. (1971). Biological activity of irradiated $tRNA^{val}$ containing a 40-thiouridine cytosine dimer. *J. Mol. Biol.*, **58**, 381–388.

Yaniv M. and Gros F. (1969). Studies on valyl-tRNA synthetase and $tRNA^{val}$ from E.coli. II. Interaction between valyl-tRNA synthetase and valine acceptor tRNA. *J. Mol. Biol.*, **44**, 17–30.

Yanofsky C., Carlton B.C., Guest J.R., Helsinki D.R. and Henning U. (1964). On the colinearity of gene structure and protein structure. *Proc. Nat. Acad. Sci.*, **51**, 266–272.

Yanofsky C., Drapeau G.R., Guest J.R. and Carlton B.C. (1967). The complete amino acid sequence of the tryptophan synthetase A protein (α-subunit) and its colinear relationship with the genetic map of the A gene. *Proc. Nat. Acad. Sci.*, **57** 296–298.

Yanofsky C. and Horn V. (1972). Tryptophan synthetase alpha chain positions affected by mutations near the ends of the genetic map of *trpA* of E.coli. *J. Biol. Chem.*, **247**, 4494–4499.

Yanofsky C., Horn V., Bonner M. and Stasiowski S. (1971). Polarity and enzyme functions in mutants of the first three genes of the tryptophan operon of E.coli. *Genetics*, **69**, 409–433.

Yanofsky C. and Ito J. (1969). Nonsense codons and polarity in the tryptophan operon. *J. Mol. Biol.*, **21**, 313–334.

Yanofsky S.A. and Spiegelman S. (1962). The identification of the rRNA cistron by sequence complementarity. II. Saturation of and competitive inhibition at the RNA cistron. *Proc. Nat. Acad. Sci.*, **48**, 1466–1472.

Yarus M. (1972). Phe-tRNA synthetase and ile-$tRNA^{phe}$: a possible verification mechanism for aminoacyl-tRNA. *Proc. Nat. Acad. Sci.*, **59**, 1915–1919.

Yarus M. and Berg P. (1969). Recognition of tRNA by ile-tRNA synthetase. Effect of substrates on the dynamics of tRNA-enzyme function. *J. Mol. Biol.*, **42**, 171–190.

Yarus M. and Rashbaum S. (1972). Divalent cations in tRNA and aminoacyl-tRNA synthetase function and structure. *Biochem.*, **11**, 2043–2049.

Yasuda S. and Sekiguchi M. (1970). T4 endonuclease involved in repair of DNA. *Proc. Nat. Acad. Sci.*, **67**, 1839–1854.

Ycas M. (1969). *The biological code*. North Holland, Amsterdam.

Yoshida K. and Osawa S. (1968). Origin of the protein component of the chloramphenicol particles in E.coli. *J. Mol. Biol.*, **33**, 559–570.

Yoshida M. (1973). The nucleotide sequence of $tRNA^{gly}$ from yeast. *Biochem. Biophys. Res. Commun.*, **50**, 779–784.

Yoshida M. and Rudland P.S. (1972). Ribosomal binding of phage RNA with different components of initiation factor f3. *J. Mol. Biol.*, **68**, 465–482.

Yoshida M., Takeishi K. and Ukita T. (1970). Anticodon structure of GAA specific glutamic acid tRNA from yeast. *Biochem. Biophys. Res. Commun.*, **39**, 852–857.

Yoshimori R., Roulland-Dussoix D. and Boyer H.W. (1972). R Factor controlled restriction and modification of DNA: restriction mutants. *J. Bacteriol.*, **112**, 1275–1279.

Youngs D.A. and Bernstein I.A. (1973). Involvement of the recB-recC nuclease (exonuclease V) in the process of X-ray induced DNA degradation in radiosensitive strains of E.coli K12. *J. Bacteriol.*, **113**, 901–906.

Yuan R. and Meselson M. (1970). A specific complex between a restriction endonuclease and its DNA substrate. *Proc. Nat. Acad. Sci.*, **65**, 357–362.

Zabin I. and Fowler A.V. (1970). β-galactosidase and thiogalactoside transacetylase. In Beckwith J.R. and Zipser D. (Eds.), *The Lactose Operon*, Cold Spring Harbor Laboratory, New York. pp 27–48.

Zachau H.G. (1972). Transfer RNAs. In Bosch L. (Ed.), *The mechanism of protein synthesis and its regulation*. North Holland, Amsterdam. pp. 173–218.

Zachau H.G., Acs G. and Lipmann F. (1958). Islation of adenosine amino acid esters from a ribonuclease digest of soluble liver RNA. *Proc. Nat. Acad. Sci.*, **44**, 885–888.

Zachau H.G., Dutting D., Feldmann H., Melchers F. and Karau W. (1966). Serine specific tRNAs. XIV. Comparison of nucleotide sequences and secondary structure models. *Cold Spring Harbor Symp. Quant. Biol.*, **31**, 417–424.

Zamecnik P.C. (1969). An historical account of protein synthesis, with current overtones—a personalized view. *Cold Spring Harbor Symp. Quant. Biol.*, **34**, 1–16.

Zamenhof S. and Eichoran H.H. (1967). Study of microbial evolution through loss of biosynthetic functions: establishement of "defective" mutants. *Nature*, **216**, 456–458.

Ziff E.B., Sedat J.W. and Galibert F. (1973). Determination of the nucleotide sequence of a fragment of phage ϕX174 DNA. *Nature New Biol.*, **241**, 34–37.

Zillig W., Zechel K., Rabussay D., Schachner M., Sethi V.S., Palm P., Heil A. and Seifert W. (1970). On the role of different subunits of DNA-dependent RNA polymerase from E.coli in the transcription process. *Cold Spring Harbor Symp. Quant. Biol.*, **35**, 47–58.

Zimmermann R.A., Garvin R.T. and Gorini L. (1971). Alteration of a 30S ribosomal protein accompanying the *ram* mutation in E.coli. *Proc. Nat. Acad. Sci.*, **68**, 2263–2267.

Zimmermann R.A., Muto A., Fellner P., Ehresmann C. and Branlant C. (1972). Location of ribosomal protein binding sites on 16S rRNA. *Proc. Nat. Acad. Sci.*, **69**, 1282–1286.

Zipser D. (1967). UGA: a third class of suppressible polar mutants. *J. Mol. Biol.*, **29**, 41–45.

Zipser D. (1969). Polar mutations and operon function. *Nature*, **221**, 21–25.

Zipser D. (1970). Polarity and translational punctuation. In Beckwith J.R. and Zipser D. (Eds.), *The Lactose Operon*, Cold Spring Harbor Laboraroty, New York, pp 221–232.

Zipser D. and Newton A. (1967). The influence of deletions on polarity. *J. Mol. Biol.*, **25**, 567–569.

Zipser D., Zabell S., Rothman J., Grodzicker T. and Wenk M. (1970). Fine structure of the gradient of polarity in the *z* gene of the *lac* operon of E.coli. *J. Mol. Biol.*, **49**, 251–254.

Zubay G. (1962). A theory on the mechanism of RNA synthesis. *Proc. Nat. Acad. Sci.*, **48**, 456–461.

Zubay G., Gielow L. and Englesberg E. (1971). Cell free studies on the regulation of the arabinose operon. *Nature New Biol.*, **233**, 164–165.

Zubay G., Morse D.E., Schrenk W.J. and Miller J.H.M. (1972). Detection and isolation of the repressor protein for the tryptophan operon of E.coli. *Proc. Nat. Acad. Sci.*, **69**, 1100–1103.

Zubay G., Schwartz D. and Beckwith J. (1970). Mechanism of activation of catabolite sensitive genes: a positive control system. *Proc. Nat. Acad. Sci.*, **66**, 104–110.

Index